Elementary Differential Equations with Applications

Elementary Differential Equations with Applications

C. H. EDWARDS, JR.

University of Georgia

DAVID E. PENNEY

University of Georgia

Prentice-Hall, Inc., *Englewood Cliffs, New Jersey* 07632

Library of Congress Cataloging in Publication Data

Edwards, C. H. (Charles Henry), (date)
  Elementary differential equations with applications.

  Bibliography: p.
  Includes index.
  1. Differential equations.   I. Penney, David E.
II . Title.
QA371 .E3   1985   515.3'5   84-11588
ISBN 0-13-254129-7

Editorial/production supervision: Maria McKinnon
Designed by: Walter A. Behnke, Judy Winthrop, Jayne Conte, Chris Wolf
Art direction: Linda Conway
Cover design: Susan Behnke
Cover photo: Fred Burrell
Manufacturing buyer: John Hall

Printed in the United States of America

10 9 8 7 6 5 4 3

ISBN 0-13-254129-7 01

Prentice-Hall International, Inc., *London*
Prentice-Hall of Australia Pty. Limited, *Sydney*
Editora Prentice-Hall do Brasil, Ltda., *Rio de Janeiro*
Prentice-Hall Canada Inc., *Toronto*
Prentice-Hall of India Private Limited, *New Delhi*
Prentice-Hall of Japan, Inc., *Tokyo*
Prentice-Hall of Southeast Asia Pte. Ltd., *Singapore*
Whitehall Books Limited, *Wellington, New Zealand*

To Alice and Carol

Summary
Contents

Contents

Preface

We wrote this book to provide a concrete and readable text for the traditional course in elementary differential equations that science, engineering, and mathematics students take following calculus. It includes enough material appropriately arranged for different courses varying in length from one quarter to two semesters. Our treatment is shaped throughout by the goal of an exposition that students will find accessible, attractive, and interesting. We hope that we have anticipated and addressed most of the questions and difficulties that students typically encounter when they study differential equations for the first time.

The book begins (in Section 1.1) and ends (in Section 9.4) with discussions of the mathematical modeling of real-world situations. The fact that differential equations have diverse and important applications is too familiar for extensive comment here. But these applications have played a singular role in the historical development of this subject. Whole areas of the subject exist mainly because of their applications. So in teaching it, we want our students to learn first to solve those differential equations that enjoy the most frequent application.

We therefore make consistent use of appealing applications for both motivation and illustration of the standard elementary techniques of solution of differential equations. A number of the more substantial applications are placed in optional sections, each marked with an asterisk (in the table of contents and in the text). These sections can be omitted without loss of continuity, but their availability can provide instructors with flexibility for variations in emphasis.

While according real-world applications their due, we also think the first course in differential equations should be a window on the world of mathematics. Matters of definition, classification, and logical structure deserve (and receive here) careful attention—for the first time in the mathematical experience of many of the students (and perhaps for the last time in some cases). While it is neither feasible nor desirable to include proofs of the fundamental existence and uniqueness theorems along the way in an elementary course, students need to see precise and clearcut statements of these theorems, and to understand their role in the subject. We do include some existence and uniqueness proofs in Section 9.4, and occasionally refer to them in the main body of the text.

The list of introductory topics in differential equations is quite standard, and a glance at our chapter titles will reveal no major surprises, though in the fine structure we have attempted to add a bit of zest here and there. A number of different permutations in the order of topics are possible, and the table that follows this preface exhibits the logical dependence between chapters. In most chapters the principal ideas of the topic are introduced in the first few sections of the chapter, and the remaining sections are devoted to extensions and applications. Hence the instructor has a wide range of choice regarding breadth and depth of coverage.

Chapter 1 naturally treats first order equations, with separable equations (Section 1.4), linear equations (Section 1.5), substitution methods (Section 1.6), and exact equations (Section 1.7) comprising the core of the chapter. Chapter 2 is devoted to linear equations of higher order. In order to make the concepts of linear independence and general solutions more concrete and tangible, we discuss only second order equations in Section 2.1, and follow with the nth order case in Section 2.2.

Chapter 3 begins with a review of the basic facts about power series that will be needed. The first three sections of the chapter treat the standard power series techniques for the solution of linear equations with variable coefficients. We devote more attention than usual to certain matters—such as shifting indices of summation—that are mathematically routine but nevertheless troublesome for many students. In Section 3.4 (optional) we include for reference more detail on the method of Frobenius than ordinarily will be covered in the classroom. Similarly, we go slightly further than is customary in Section 3.6 (optional) with applications of Bessel functions. Chapter 4 on Laplace transforms is rather standard, though our discussion in Section 4.6 (optional) of impulses and Dirac delta functions may have some merit.

There is much variation in the treatment of linear systems in introductory courses, depending on the background in linear algebra that is assumed. The first two sections of Chapter 5 can stand alone as an introduction to linear systems without the use of linear algebra and matrices. The last four sections of Chapter 5 employ the notation and terminology (though not so much theory) of elementary linear algebra. For ready reference, we have included in Section 5.3 a complete and self-contained account of the needed notation and terminology of determinants, matrices, and vectors.

Chapter 6 on numerical methods requires some special comment. Personal computers are now with us and here to stay. Pocket computers are relatively inexpensive and already in the hands of some students. The great difference (in the perception of students) between personal computing and mainframe computing may not yet be universally appreciated. Students can now envision the numerical approximation of solutions as a routine and commonplace matter. Our viewpoint in Chapter 6 is that understanding and appreciation of numerical algorithms is enhanced (and rendered more concrete to students) by discussion of their computer implementations. We decided to include illustrative programs because no flowchart has the convincing tangibility of a program that actually runs. Our choice of programming language was motivated by the recent adoption of BASIC as the *lingua franca* of personal computers. Moreover, only in BASIC could we include programs that without extensive discussion would be intelligible and informative to students with little or no programming experience. In another vein, it is pointed out in the Chapter 1 summary that the first four sections of Chapter 6 can be covered at any point in the course subsequent to Chapter 1. The increasingly widespread use of computers may provide a motive for covering numerical methods earlier than has been the custom in the past.

Chapters 1 through 6 are devoted to ordinary differential equations. Chapters 7 and 8 treat the applications of Fourier series, separation of variables, and Sturm-Liouville theory to partial differential equations and boundary value problems. After the introduction of Fourier series, the three classical equations—the wave and heat equations and Laplace's equation—are discussed in the last three sections of Chapter 7. The Sturm-Liouville methods of Chapter 8 are developed sufficiently to include some rather significant and realistic applications.

Apart from its final section on existence and uniqueness, Chapter 9 is a brief introduction to qualitative properties and stability of solutions, with numerous applications to competition, survival, and extinction of species.

Probably in no other mathematics course beyond calculus are the exercises and problem sets so crucial to student learning as in the introductory differential equations course. We therefore devoted great effort to the development and selection of the approximately 1750 problems in this book. Each section contains more computational problems ("solve the following equations," and so on) than any class will ordinarily use, plus an ample number of applied problems. We were, however, very sparing in our inclusion of purely theoretical problems. The answer section includes the answers to all odd-numbered problems and to some of the even-numbered ones.

All experienced textbook authors know the value of critical reviewing during the preparation and revision of a manuscript. In writing this book we profited greatly from the advice of the following exceptionally able reviewers:

W. Dan Curtis, Kansas State University; Bruce Conrad, Temple University; James W. Cushing, University of Arizona; James L. Heitsch, University of Illinois at Chicago; Erich Zauderer, Polytechnic Institute of New York; Anthony Peressini, University of Illinois; and William Rundell, Texas A & M University.

We owe special thanks to Professor George Feissner, State University of New York at Cortland, whose detailed and perceptive analysis of our manuscript affected every section of every chapter of the book. Finally, we cannot adequately thank Alice F. Edwards and Carol W. Penney for their continued assistance, encouragement, support, and patience.

C. H. E., Jr.
D. E. P.

Dependence of Chapters

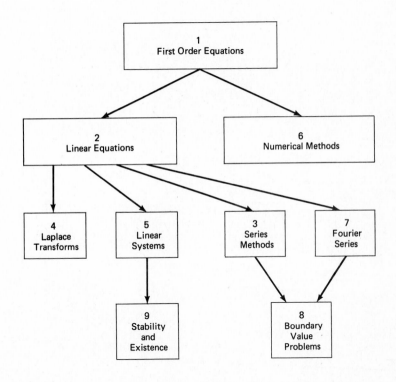

Elementary Differential Equations with Applications

Introduction and First Order Differential Equations

1

The laws of the universe are written largely in the language of mathematics. Algebra is sufficient to solve many static problems, but the most interesting natural phenomena involve change and are best described by equations that relate changing quantities.

Because the derivative $dy/dt = f'(t)$ of the function f may be regarded as the rate at which the quantity $y = f(t)$ changes with respect to the independent variable t, it is natural that equations involving derivatives are those that describe the changing universe. An equation involving an unknown function and one or more of its derivatives is called a **differential equation**, and the study of differential equations has two principal goals:

1. To discover the differential equation that describes a physical situation;
2. To find the appropriate solution of that equation.

Unlike algebra, in which we seek the unknown numbers that satisfy an equation such as $x^3 + 7x^2 - 11x + 41 = 0$, in solving a differential equation we are challenged to find the unknown functions $y = g(x)$ for which an identity such as $g'(x) - 2xg(x) = 0$—in Leibniz notation,

$$\frac{dy}{dx} - 2xy = 0$$

—holds on some interval of real numbers. Ordinarily we will want to find *all* solutions of the differential equation if possible.

The following three examples illustrate the process of translating scientific laws and principles into differential equations, by interpreting rates of change as derivatives. In each of these examples the independent variable is time t, but we will see numerous applications in which some quantity other than time is the independent variable.

EXAMPLE 1 Newton's law of cooling may be stated in the following form: The *time rate of change* (the rate of change with respect to time t) of the temperature $T(t)$ of a body is proportional to the difference between T and the temperature A of the surrounding medium. That is,

$$\frac{dT}{dt} = k(A - T) \tag{1}$$

where k is a positive constant.

Thus the physical law is translated into a differential equation. If we are given the values of k and A, we hope to find an explicit formula for $T(t)$, and then—with the aid of this formula—we can predict the future temperature of the body.

EXAMPLE 2 The *time rate of change* of a population $P(t)$ with constant birth and death rates is, in many simple cases, proportional to the size of the population. That is,

$$\frac{dP}{dt} = kP \tag{2}$$

where k is the constant of proportionality.

EXAMPLE 3 Torricelli's law implies that the *time rate of change* of the volume V of water in a draining tank is proportional to the square root of the depth y of the water in the tank:

$$\frac{dV}{dt} = -ky^{1/2} \tag{3}$$

where k is constant. If the tank is a cylinder with cross-sectional area A, then $V = Ay$, and so $dV/dt = A(dy/dt)$. In this case Eq. (3) takes the form

$$\frac{dy}{dt} = -hy^{1/2} \tag{4}$$

where $h = k/A$.

Let us discuss Example 2 further. Note first that each function of the form

$$P(t) = Ce^{kt} \tag{5}$$

is a solution of the differential equation, Eq. (2),

$$\frac{dP}{dt} = kP.$$

We verify this assertion as follows:

$$P'(t) = Cke^{kt} = k(Ce^{kt}) = kP(t)$$

for all real numbers t. Because substitution of each function of the form given in (5) into Eq. (2) produces an identity, all these functions are solutions of Eq. (2).

Thus, even if the value of the constant k is known, the differential equation $dP/dt = kP$ has *infinitely* many different solutions of the form $P(t) = Ce^{kt}$—one for each choice of the "arbitrary" constant C. This is typical of differential equations in general. It is also fortunate, because it allows us to use additional information to select from all the solutions a particular one that fits the situation under study.

EXAMPLE 4 Suppose that $P(t)$ is the population of a bacterial colony at time t, that the population at time $t = 0$ (hours, h) was 1000, and that the population doubled after 1 h. This additional information about the function $P(t)$ yields the following equations:

$$1000 = P(0) = Ce^0 = C,$$

$$2000 = P(1) = Ce^k.$$

It follows that $C = 1000$ and that $k = \ln 2$. Thus the function $P(t)$ describing the population of this particular bacterial colony is known exactly:

$$P(t) = 1000e^{t \ln 2}.$$

Therefore, we can predict the population at any future time; for example, the population at time $t = 90$ minutes (min) (1.5 h) will be $P(1.5) = 1000e^{(1.5)\ln 2}$, or about 2828 bacteria.

The condition $P(0) = 1000$ is called an **initial condition** because we normally write differential equations for which $t = 0$ is the starting time. Figure 1.1 shows a number of graphs of the form $P(t) = Ce^{kt}$ for which $k = \ln 2$. The graphs of all the solutions of $dP/dt = (\ln 2)P$ in fact fill up the entire two-dimensional plane, and no two intersect. Moreover, the selection of any point on the P-axis amounts to a determination of the value $P(0)$. Because exactly one solution passes through each such point, we see in this case that an initial condition $P(0) = P_0$ may determine a unique solution agreeing with known data.

It is possible that none of these solutions fits the known information. In such a case we must suspect that the differential equation—a mathematical model of the physical phenomenon in question—may not adequately describe the real world. The solutions of Eq. (2) are of the form $P(t) = Ce^{kt}$ where C is a positive constant, but for *no* choice of the constants k and C does $P(t)$ accurately describe the actual growth of the human population of the world over the past hundred years. We must therefore write a more complicated differential equation, one that takes into account the effects of population pressure on the birth rate, the declining food supply, and other factors. This should not be regarded as a failure of the model of Example 2, but as an insight into what additional factors must be considered in studying the growth of populations. Indeed, Eq. (2) is quite accurate under certain circumstances—for

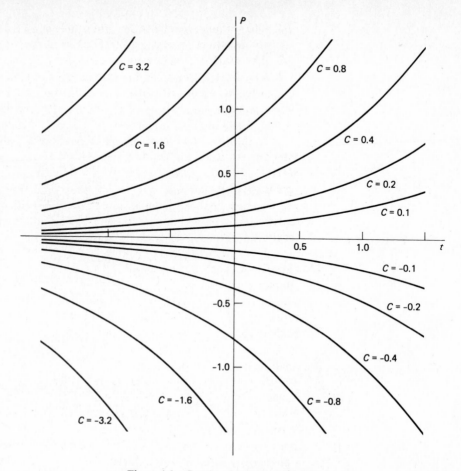

Figure 1.1 Graphs of $P(t) = C \exp (t \ln 2)$.

example, the growth of a bacterial population under conditions of unlimited food and space.

This little discussion of population growth illustrates the crucial process of *mathematical modeling* (see Fig. 1.2), which involves:

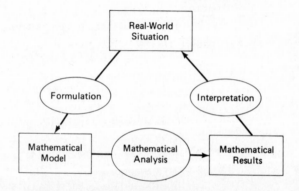

Figure 1.2 The process of mathematical modeling.

1. The formulation of a real-world problem in mathematical terms—that is, the construction of a mathematical model;
2. The analysis or solution of the resulting mathematical problem;
3. The interpretation of the mathematical results in the context of the original real-world situation—for example, answering the question originally posed.

In the population example, the real-world problem is that of determining the population at some future time. A **mathematical model** consists of a list of variables (P and t) that describe the given situation, together with one or more equations relating these variables ($dP/dt = kP$, $P(0) = P_0$) that are known or are assumed to hold. The mathematical analysis consists of solving these equations (here, for P as a function of t). Finally, we apply these mathematical results to answer the original real-world question.

But in our population example we ignored the effects of such factors as varying birth and death rates. This made the mathematical analysis quite simple, perhaps unrealistically so. A satisfactory mathematical model is subject to two contradictory requirements: It must be sufficiently detailed to represent the real-world situation with relative accuracy, and yet it must be sufficiently simple to make the mathematical analysis practical. If the model is so detailed that it fully represents the physical situation, the mathematical analysis may be too difficult to carry out. If the model is too simple, the results may be so inaccurate as to be useless. Thus there is an inevitable trade-off between what is physically realistic and what is mathematically possible. The construction of a model that adequately bridges this gap between realism and feasibility is therefore the most crucial and delicate step in the process. Ways must be found to simplify the model mathematically without sacrificing essential features of the real-world situation.

Mathematical models are discussed throughout this book. The remainder of this introductory section is devoted to simple examples and to standard terminology used in discussing differential equations and their solutions.

EXAMPLE 5 Given the differential equation $y' = y^2$, it is easy to verify that the function $y = 1/(1 - x)$ is a solution on the interval $x < 1$. For

$$\frac{dy}{dx} = \frac{1}{(1 - x)^2} = y^2$$

if $x < 1$. The same computation is valid for the interval $x > 1$. Therefore $y = 1/(1 - x)$ is a solution of the given equation on $(-\infty, 1)$ and on $(1, +\infty)$. There is no (continuous) solution of the equation $y' = y^2$ on the entire real line except for the constant function $y = 0$ (though this is not obvious).

EXAMPLE 6 Verify that the function $y = 2x^{1/2} - x^{1/2} \ln x$ satisfies the differential equation

$$4x^2 y'' + y = 0 \tag{6}$$

for all $x > 0$.

Solution First we compute the derivatives

$$y' = -\tfrac{1}{2} x^{-1/2} \ln x$$

and

$$y'' = \tfrac{1}{4} x^{-3/2} \ln x - \tfrac{1}{2} x^{-3/2}.$$

Then substitution into Eq. (6) yields

$$4x^2 y'' + y = 4x^2(\tfrac{1}{4} x^{-3/2} \ln x - \tfrac{1}{2} x^{-3/2}) + 2x^{1/2} - x^{1/2} \ln x = 0$$

if x is positive, and so the differential equation is satisfied for all $x > 0$.

The fact that we can write a differential equation is not enough to guarantee that it has a solution. For example, it is clear that the differential equation

$$(y')^2 + y^2 = -1 \tag{7}$$

has *no* (real-valued) solution because the sum of nonnegative numbers cannot be negative. For another variation on this theme, note that the equation

$$(y')^2 + y^2 = 0 \tag{8}$$

obviously has only the (real-valued) solution $y(x) \equiv 0$. In our previous examples any differential equation having at least one solution indeed had infinitely many.

The **order** of a differential equation is the order of the highest derivative that appears in it. The differential equation of Example 6 is of second order, our previous examples are first order equations, and $y^{(4)} + x^2 y^{(3)} + x^5 y = \sin x$ is a fourth order equation. The most general form of an **nth order** differential equation with independent variable x and unknown function, or dependent variable, $y = y(x)$ is

$$F(x, y, y', y'', \ldots, y^{(n)}) = 0 \tag{9}$$

where F is a specific real-valued function of $n + 2$ variables.

Our usage of the word *solution* has until now been somewhat informal. More precisely, we say that the function $y = u(x)$ is a **solution** of the differential equation in (9) **on the interval I** provided that the derivatives $u', u'', \ldots,$ $u^{(n)}$ exist and that

$$F(x, u, u', u'', \ldots, u^{(n)}) = 0$$

for all x in I. When brevity is needed, we say that $y = u(x)$ **satisfies** the differential equation in (9) on I.

In both Eqs. (7) and (8), the appearance of y' as an implicitly defined function causes complications. For this reason, we will ordinarily assume that any differential equation under study can be solved explicitly for the highest derivative that appears; that is, that the equation may be written in the form

$$y^{(n)} = G(x, y, y', y'', \ldots, y^{(n-1)}) \tag{10}$$

where G is a real-valued function of $n + 1$ variables. In addition, we will always seek only real-valued solutions unless we warn the reader to the contrary.

All the differential equations we have mentioned so far are **ordinary** differential equations, meaning that the unknown function (dependent variable) depends upon only a *single* independent variable. For this reason only ordinary derivatives appear in the equation. If the dependent variable is a function of two or more independent variables, then partial derivatives are likely to be involved; if so, the equation is called a **partial** differential equation. For example, the temperature $u = u(x, t)$ of a long thin rod at the point x at time t satisfies (under appropriate simple conditions) the partial differential equation

$$\frac{\partial u}{\partial t} = k \frac{\partial^2 u}{\partial x^2},$$

where k is a constant (called the *thermal diffusivity* of the rod). Until Chapter 7 we will be concerned only with *ordinary* differential equations and will refer to them simply as differential equations.

In this chapter we concentrate our attention on first order differential equations of the general form

$$\frac{dy}{dx} = f(x, y), \tag{11}$$

and a few higher order equations that readily can be reduced to first order equations. We will also sample the wide range of applications of such equations. A typical mathematical model of an applied situation will be an **initial value problem**, consisting of a differential equation of the above form together with an **initial condition** $y(x_0) = y_0$. Note that we call $y(x_0) = y_0$ an initial condition whether or not $x_0 = 0$. To solve the initial value problem

$$\frac{dy}{dx} = f(x, y), \qquad y(x_0) = y_0 \tag{12}$$

means to find a differentiable function $y(x)$ that satisfies both conditions in Eq. (12).

The central question of greatest interest to us is this: If we are given a differential equation known to have a solution satisfying a given initial condition, how do we actually *find* or *compute* that solution? And, once found, what can we *do* with it? We will see that a relatively few simple techniques—separation of variables (Section 1.4), solution of linear equations (Section 1.5), substitution methods (Section 1.6), multiplication by integrating factors (Section 1.7)—are enough to enable us to solve a diversity of first order equations having impressive applications.

1.1 Problems

In each of Problems 1–12, verify by substitution that each given function is a solution of the given differential equation.

    1. $y' = 3x^2$; $y = x^3 + 7$.

    2. $y' + 2y = 0$; $y = 3e^{-2x}$.

    3. $y'' + 4y = 0$; $y_1 = \cos 2x, y_2 = \sin 2x$.

    4. $y'' = 9y$; $y_1 = e^{3x}, y_2 = e^{-3x}$.

    5. $y' = y + 2e^{-x}$; $y = e^x - e^{-x}$.

    6. $y'' + 4y' + 4y = 0$; $y_1 = e^{-2x}, y_2 = xe^{-2x}$.

7. $y'' - 2y' + 2y = 0$; $y_1 = e^x \cos x$, $y_2 = e^x \sin x$.

8. $y'' + y = 3 \cos 2x$; $y_1 = \cos x - \cos 2x$, $y_2 = \sin x - \cos 2x$.

9. $y' + 2xy^2 = 0$; $y = \dfrac{1}{1 + x^2}$.

10. $x^2 y'' + xy' - y = \ln x$; $y_1 = x - \ln x$, $y_2 = \dfrac{1}{x} - \ln x$.

11. $x^2 y'' + 5xy' + 4y = 0$; $y_1 = \dfrac{1}{x^2}$, $y_2 = \dfrac{\ln x}{x^2}$.

12. $x^2 y'' - xy' + 2y = 0$; $y_1 = x \cos (\ln x)$, $y_2 = x \sin (\ln x)$.

In each of Problems 13–16, substitute $y = e^{rx}$ into the given differential equation to determine all values of r for which $y = e^{rx}$ is a solution of the equation.

13. $3y' = 2y$.

14. $4y'' = y$.

(15) $y'' + y' - 2y = 0$.

16. $3y'' + 3y' - 4y = 0$.

In each of Problems 17–20, show that $y(x)$ satisfies the given differential equation for all values of the constants A and B. Then find values of A and B so that $y(0) = y'(0) = 1$.

(17) $y'' + 3y' = 0$; $y(x) = A + Be^{-3x}$.

18. $y'' - 2y' + y = 0$; $y(x) = Ae^x + Bxe^x$.

19. $y'' = 4y$; $y(x) = Ae^{2x} + Be^{-2x}$.

20. $y'' - 4y' + 5y = 0$; $y(x) = e^{2x}(A \cos x + B \sin x)$.

In each of Problems 21–25, a function $y = g(x)$ is described by some geometric property of its graph. Write a differential equation of the form $y' = f(x, y)$ having the function $g(x)$ as its solution (or as one of its solutions).

21. The slope of the graph of g at the point (x, y) is the sum of x and y.

22. The tangent line to the graph of g at the point (x, y) intersects the x-axis at the point $(x/2, 0)$.

23. Every straight line normal to the graph of g passes through the point $(0, 1)$.

24. The graph of g is normal to every curve of the form $y = k + x^2$ (k is a constant) where they meet.

25. The line tangent to the graph of g at (x, y) passes through the point $(-y, x)$.

In each of Problems 26–30, write—in the manner of Eqs. (1)–(4) of this section—a differential equation that is a mathematical model of the situation described.

26. The time rate of change of a population P is proportional to the square root of P.

27. The time rate of change of the velocity v of a coasting motorboat is proportional to the square of v.

28. The acceleration dv/dt of a certain sports car is proportional to the difference between 250 kilometers per hour (km/h) and the velocity of the car.

29. In a city having a fixed population of P persons, the time rate of change of the number N of those persons who have heard a certain rumor is proportional to the number of those who have not yet heard the rumor.

30. In a city with a fixed population of P persons, the time rate of change of the number N of persons infected with a certain disease is proportional to the product of the number who have the disease and the number who do not.

In Problems 31–36, determine at least one solution of the given differential equation by inspection.

31. $y'' = 0$.
32. $y' = y$.
33. $xy' + y = 3x^2$.
34. $(y')^2 + y^2 = 1$.
35. $y' + y = e^x$.
36. $y'' + y = 0$.

1.2
Solution by Direct Integration

The first order differential equation $y' = f(x, y)$ takes an especially simple form if the function f is independent of the dependent variable y:

$$\frac{dy}{dx} = f(x). \tag{1}$$

In this special case we need only integrate both sides of Eq. (1) to obtain

$$y = \int f(x)\, dx + C. \tag{2}$$

This is a **general solution** of Eq. (1), meaning that it involves an arbitrary constant C, and for every choice of C it is a solution of the differential equation. If $G(x)$ is a particular antiderivative of $f(x)$—that is, if $G'(x) \equiv f(x)$—then

$$y = G(x) + C. \tag{3}$$

To satisfy an initial condition $y(x_0) = y_0$, we need only substitute $x = x_0$ and $y = y_0$ into Eq. (3) to obtain $y_0 = G(x_0) + C$, so that $C = y_0 - G(x_0)$. With this choice of C, we obtain the **particular solution** of (1) satisfying the initial value problem

$$y' = f(x), \qquad y(x_0) = y_0.$$

We will see that this is the typical pattern for solutions of first order differential equations. Ordinarily we will first find a *general solution* involving an arbitrary constant C. We then can attempt to obtain, by appropriately choosing C, a *particular solution* satisfying a given initial condition $y(x_0) = y_0$.

EXAMPLE 1 Solve the initial value problem

$$\frac{dy}{dx} = \frac{x}{(x^2 + 9)^{1/2}}, \qquad y(4) = 2.$$

Solution Integration immediately yields the general solution

$$y = \int \frac{x}{(x^2 + 9)^{1/2}}\, dx = (x^2 + 9)^{1/2} + C.$$

The substitution $x = 4$, $y = 2$ gives $C = -3$, so the desired particular solution is $y = (x^2 + 9)^{1/2} - 3$.

The observation that the special first order equation $dy/dx = f(x)$ is readily solvable (provided the function $f(x)$ can be integrated) extends to second order differential equations of the special form

$$\frac{d^2 y}{dx^2} = g(x), \tag{4}$$

in which the given function on the right-hand side involves neither the independent variable y nor its derivative y'. We simply integrate once to obtain

$$\frac{dy}{dx} = G(x) + C_1,$$

where $G(x)$ is an antiderivative of $g(x)$ and C_1 is an arbitrary constant. Then another integration yields

$$y = \int G(x)\, dx + C_1 x + C_2$$

where C_2 is a second arbitrary constant. In effect, the second order differential equation in (4) is one that can be solved by solving successively the *first order* differential equations

$$\frac{dv}{dx} = g(x) \quad \text{and} \quad \frac{dy}{dx} = v(x).$$

VELOCITY AND ACCELERATION

Direct integration is sufficient to allow us to solve a number of important problems concerning the motion of a particle (or *mass point*) in terms of the forces acting upon it. The motion of a particle along a straight line (the x-axis) is described by its **position function**

$$x = f(t) \tag{5}$$

giving its x-coordinate at time t. The **velocity** $v(t)$ of the particle is defined to be

$$v(t) = f'(t); \quad \text{that is,} \quad v = \frac{dx}{dt}. \tag{6}$$

Its **acceleration** $a(t)$ is $a(t) = v'(t) = x''(t)$; in alternative notation,

$$a = \frac{dv}{dt} = \frac{d^2x}{dt^2}. \tag{7}$$

Newton's *second law of motion* implies that if a force $F(t)$ acts on the particle and is directed along its line of motion, then

$$ma(t) = F(t); \quad \text{that is,} \quad F = ma, \tag{8}$$

where m is the mass of the particle. If the force $F(t)$ is known, then the equation $x''(t) = F(t)/m$ can be integrated twice to obtain the position function $x(t)$ in terms of two constants of integration. These two arbitrary constants are frequently determined by the **initial position** $x(0) = x_0$ and the **initial velocity** $v(0) = v_0$ of the particle.

For instance, suppose that the force F, and therefore the acceleration $a = F/m$, are constant. Then we begin with the equation

$$\frac{dv}{dt} = a \qquad (a \text{ is a constant}) \tag{9}$$

and integrate both sides to obtain $v = at + C_1$. We know that $v = v_0$ when $t = 0$, and substitution of this information into the last equation yields the information that $C_1 = v_0$. So

$$v = \frac{dx}{dt} = at + v_0. \tag{10}$$

A second integration gives $x = \frac{1}{2}at^2 + v_0 t + C_2$, and the substitution $t = 0$, $x = x_0$ gives $C_2 = x_0$; therefore

$$x = \tfrac{1}{2}at^2 + v_0 t + x_0. \tag{11}$$

Thus with Eq. (10) we can find the velocity, and with (11) the position, of the particle at any time t in terms of its *constant* acceleration a, its initial velocity v_0, and its initial position x_0.

EXAMPLE 2   A lunar lander is falling freely toward the surface of the moon at a speed of 1000 miles per hour (mi/h). Its retrorockets, when fired, provide a deceleration of 20,000 miles per hour per hour (mi/h²) (the gravitational acceleration produced by the moon is assumed to be included in the given deceleration). At what height above the lunar surface should the retrorockets be activated to insure a "soft touchdown" ($v = 0$ at impact)?

Solution   We denote by $x(t)$ the height of the lunar lander above the surface, as indicated in Fig. 1.3. Then $v_0 = -1000$ (mi/h—negative because the height is decreasing), and $a = +20,000$ because an upward thrust increases the velocity v (although it decreases the *speed* $|v|$). Then Eq. (10) and (11) become

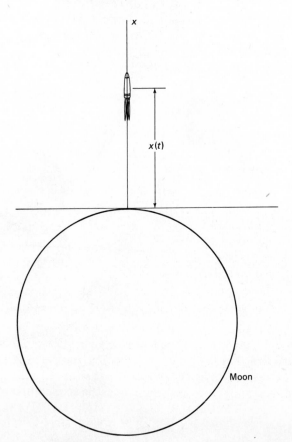

Moon

Figure 1.3   The lunar lander.

$$v = 20,000t - 1000 \tag{12}$$

and

$$x = 10,000t^2 - 1000t + x_0, \tag{13}$$

where x_0 is the height of the lander above the lunar surface at the time $t = 0$ at which the retrorockets should be activated.

From Eq. (12) we see that $v = 0$ (soft touchdown) occurs when $t = \frac{1}{20}$ (h; thus 3 min); then substitution of $t = \frac{1}{20}$, $x = 0$ into (13) yields

$$x_0 = -10,000(\tfrac{1}{20})^2 + 1000(\tfrac{1}{20}) = 25$$

(mi). Thus the retrorockets should be activated when the lunar lander is 25 mi above the surface, and it will reach the surface 3 min later.

A common application of Eqs. (10) and (11) involves vertical motion near the surface of the earth. Recall that the **weight** W of a body is the force exerted by gravity upon the body at the surface of the earth. Hence Newton's second law implies that

$$W = mg \tag{8'}$$

where m is the mass of the body and g is the gravitational acceleration at the surface of the earth. (The value of g depends upon the precise location of its measurement, but it is nearly constant; we assume it constant in the remainder of this section.) Although it was convenient to use units of miles and hours in Example 2, we will ordinarily employ one of the three systems of units summarized in the following table.

	fps System	cgs System	mks System
Force	pound (lb)	dyne (dyn)	newton (N)
Mass	slug	gram (g)	kilogram (kg)
Distance	foot (ft)	centimeter (cm)	meter (m)
Time	second (s)	second (s)	second (s)
g	32 ft/s²	980 cm/s²	9.8 m/s²

The values listed for g are approximations; a more accurate value for g at most locations near sea level is 32.2 ft/sec². All three unit systems are compatible with Newton's law (8'). For example, the weight of a mass of 1 slug is $W = (1 \text{ slug})(32 \text{ ft/s}^2) = 32$ lb, and a force of 1 dyn imparts to a mass of 1 g an acceleration of 1 cm/s². Similarly, a mass of 1 g has a weight of 980 dyn.

Because we intend to deal here with vertical motion, it is natural to choose the y-axis as the coordinate system for position. If we choose the upward direction as the positive direction, then the effect of gravity on the body is to *decrease* its height and also to *decrease* its velocity $v = dy/dt$, so we see that if air resistance is ignored, then the acceleration of the body is

$$a = \frac{dv}{dt} = -g = -32 \quad \text{(ft/s}^2\text{)}.$$

Equations (10) and (11) then take the forms

$$v = -32t + v_0 \qquad (10')$$

and

$$y = -16t^2 + v_0 t + y_0. \qquad (11')$$

Here y_0 is the initial height of the body in feet and v_0 its initial velocity in feet per second.

If the initial $(t = 0)$ values v_0 and y_0 are given, then the subsequent velocity and height of the body are given by Eqs. (10') and (11'). For instance, suppose that a ball is thrown straight upward from the ground $(y_0 = 0)$ with initial velocity $v_0 = 96$ ft/s. Then it reaches its maximum height when its velocity is zero, $v = -32t + 96 = 0$, and thus when $t = 3$ s. Hence the maximum height the ball attains is $y(3) = -(16)(3)^2 + (96)(3) = 144$ ft.

*THE DEFLECTION OF A UNIFORM BEAM

We include now an example of the use of a relatively simple differential equation to explain a complicated physical phenomenon—the shape of a horizontal beam upon which a vertical force is acting.

Consider the horizontal beam shown in Fig. 1.4, uniform both in cross section and in material. If it is supported only at its ends, then the force of its own weight distorts its longitudinal axis of symmetry into the curve shown as a dashed line in the figure. We want to investigate the shape $y = y(x)$ of this curve, the **deflection curve** of the beam. We will use the coordinate system indicated in Fig. 1.5, with the positive y-axis directed downward.

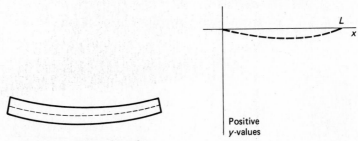

Figure 1.4 Distortion of a horizontal beam.

Figure 1.5 The deflection curve.

A consequence of the theory of elasticity is that, for relatively small deflections of such a beam (so small that $(y')^2$ is negligible in comparison with unity), an adequate mathematical model of the deflection curve is the fourth order differential equation

$$EIy^{(4)} = F(x), \qquad (14)$$

where

*Asterisks denote material that can be omitted without loss of continuity.

E denotes the Young's modulus of the material of the beam,

I denotes the moment of inertia of the cross section of the beam about a horizontal line through the centroid of the cross section, and

$F(x)$ denotes the density of *downward* force acting vertically on the beam at the point x.

Density of force? Yes; this means that the force acting downward on a very short segment $[x, x + \Delta x]$ of the beam is approximately $F(x) \Delta x$. The units of $F(x)$ are units of force per unit length, such as pounds per foot. We will consider here the case in which the only force distributed along the beam is its own weight, w pounds per foot, so that $F(x) \equiv w$. Then Eq. (14) takes the form

$$EIy^{(4)} = w \tag{15}$$

where E, I, and w are all constant.

Note: We assume no previous familiarity with elasticity or with Equations (14) and (15) here. It is important to be able to begin with a differential equation that arises in a specific applied discipline and then analyze its implications; thus we develop an understanding of the equation by examining its solutions. Observe that, in essence, Eq. (15) implies that the fourth derivative $y^{(4)}$ is proportional to the weight density w. This proportionality involves, however, *two* constants: E, which depends only upon the material in the beam, and I, which depends only upon the shape of the cross section of the beam. Values of the Young's modulus E of various materials can be found in handbooks of physical constants; $I = \frac{1}{4}\pi a^4$ for a circular cross section of radius a.

Though (15) is a fourth order differential equation, its solution involves only the solution of simple first order equations by successive simple integrations. One integration of (15) yields

$$EIy''' = wx + C_1;$$

a second yields

$$EIy'' = \tfrac{1}{2}wx^2 + C_1x + C_2;$$

another yields

$$EIy' = \tfrac{1}{6}wx^3 + \tfrac{1}{2}C_1x^2 + C_2x + C_3;$$

a final integration gives

$$EIy = \tfrac{1}{24}wx^4 + \tfrac{1}{6}C_1x^3 + \tfrac{1}{2}C_2x^2 + C_3x + C_4,$$

where C_1, C_2, C_3, and C_4 are arbitrary constants. Thus we obtain a solution of Eq. (15) of the form

$$y(x) = \frac{w}{24EI}x^4 + Ax^3 + Bx^2 + Cx + D, \tag{19}$$

where A, B, C, and D are constants resulting from the four integrations.

These last four constants are determined by the way in which the beam is supported at its ends, where $x = 0$ and $x = L$. Figure 1.6 shows two common types of support. A beam might also be supported one way at one end

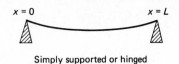

Simply supported or hinged

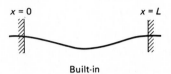

Built-in

Figure 1.6 Two ways of supporting a beam.

but another way at the other end. For instance, Fig. 1.7 shows a **cantilever**—a beam firmly fastened at $x = 0$ but *free* (no support whatsoever) at $x = L$. The following table shows the **boundary**, or **endpoint**, **conditions** corresponding to the three most common cases. We will see that these conditions are applied readily in beam problems, though a discussion here of their origin would take us too far afield.

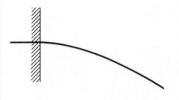

Figure 1.7 The cantilever.

Support	*Endpoint Conditions*
Simply supported	$y = y'' = 0$
Built-in or fixed end	$y = y' = 0$
Free end	$y'' = y^{(3)} = 0$

For example, the deflection curve of the cantilever in Fig. 1.7 would be given by Eq. (16), with the coefficients A, B, C, and D determined by the conditions

$$y(0) = y'(0) = 0 \quad \text{and} \quad y''(L) = y^{(3)}(L) = 0 \tag{17}$$

corresponding to the fixed end at $x = 0$ and the free end at $x = L$. The conditions in (17) together with the differential equation in (16) constitute an **endpoint value problem** (or boundary value problem). An endpoint value problem calls for a solution of a differential equation on a closed interval, satisfying conditions imposed at *both* endpoints of the interval. By contrast, in an initial value problem conditions on the solution are imposed only at a single point.

EXAMPLE 3 Determine the shape of the deflection curve of a uniform horizontal beam of length L and weight w per unit length and simply supported at each end.

Solution We have the endpoint conditions $y(0) = y''(0) = 0 = y(L) = y''(L)$. Rather than imposing them directly on Eq. (16), let us begin with the differential equation $EIy^{(4)} = w$ and determine the constants as we proceed with the four successive integrations. The first two integrations yield

$$EIy^{(3)} = wx + A;$$
$$EIy'' = \tfrac{1}{2}wx^2 + Ax + B.$$

Hence $y''(0) = 0$ implies that $B = 0$, and then $y''(L) = 0$ gives
$$0 = \tfrac{1}{2}wL^2 + AL.$$
It follows that $A = -wL/2$ and thus that
$$EIy'' = \tfrac{1}{2}wx^2 - \tfrac{1}{2}wLx.$$
Then two more integrations give
$$EIy' = \tfrac{1}{6}wx^3 - \tfrac{1}{4}wLx^2 + C,$$
and finally
$$EIy(x) = \tfrac{1}{24}wx^4 - \tfrac{1}{12}wLx^3 + Cx + D. \tag{18}$$
Now $y(0) = 0$ implies that $D = 0$; then, because $y(L) = 0$,
$$0 = \tfrac{1}{24}wL^4 - \tfrac{1}{12}wL^4 + CL.$$
It follows that $C = wL^3/24$. Hence from Eq. (18) we obtain
$$y(x) = \frac{w}{24EI}(x^4 - 2Lx^3 + L^3x) \tag{19}$$

as the shape of the supported beam. It is apparent from symmetry—see also Problem 33—that the *maximum deflection* $y_{\max}$ of the beam occurs at its midpoint $x = L/2$, and thus has the value
$$y_{\max} = y\left(\frac{L}{2}\right) = \frac{w}{24EI}\left(\frac{1}{16}L^4 - \frac{2}{8}L^4 + \frac{1}{2}L^4\right);$$
that is,
$$y_{\max} = \frac{5wL^4}{384EI}. \tag{20}$$

For instance, suppose that we want to calculate the maximum deflection of a simply supported steel rod 20 ft long with a circular cross section 1 inch (in.) in diameter. From a handbook we find that typical steel has density $\delta = 7.75$ g/cm^2, and that its Young's modulus is $E = 2 \times 10^{12}$ g/cm·s^2, so it will be more convenient to work in cgs units. Thus our rod has

$$\text{Length:} \quad L = (20 \text{ ft})\left(30.48 \frac{\text{cm}}{\text{ft}}\right) = 609.60 \text{ cm}$$

and

$$\text{Radius:} \quad a = \left(\frac{1}{2} \text{ in.}\right)\left(2.54 \frac{\text{cm}}{\text{in.}}\right) = 1.27 \text{ cm}.$$

Its linear mass density is

$$\rho = \pi a^2 \delta = \pi (1.27)^2 (7.75) \approx 39.27 \frac{\text{g}}{\text{cm}},$$

so

$$w = \rho g = \left(39.27 \frac{\text{g}}{\text{cm}}\right)\left(980 \frac{\text{cm}}{\text{s}^2}\right)$$

$$\approx 38{,}484.6 \frac{\text{dyn}}{\text{cm}}.$$

The area moment of inertia of a circular disk of radius a about a diameter is $I = \pi a^4/4$, so

$$I = \tfrac{1}{4}\pi (1.27)^4 \approx 2.04 \text{ cm}^4.$$

Therefore Eq. (20) yields

$$y_{max} \approx \frac{(5)(38484.6)(609.60)^4}{(384)(2 \times 10^{12})(2.04)} \approx 16.96 \text{ cm},$$

or about 6.68 in. as the maximum deflection of the rod at its midpoint. It is interesting to note that y_{max} is proportional to L^4, so if our rod were 10 ft long, its maximum deflection would be only one-sixteenth as much—only about 0.42 in. Because $I = \pi a^4/4$, we see from Eq. (20) that the same reduction in maximum deflection could be achieved by doubling the radius a of the rod.

1.2 Problems

In each of Problems 1–10, find a function $y = f(x)$ satisfying the given differential equation and the prescribed initial conditions.

1. $\dfrac{dy}{dx} = 2x + 1$; $y(0) = 3$.

2. $\dfrac{dy}{dx} = (x - 2)^3$; $y(2) = 1$.

3. $\dfrac{dy}{dx} = x^{1/2}$; $y(4) = 0$.

4. $\dfrac{dy}{dx} = \dfrac{1}{x^2}$; $y(1) = 5$.

5. $\dfrac{dy}{dx} = (x + 2)^{-1/2}$; $y(2) = -1$.

6. $\dfrac{dy}{dx} = x(x^2 + 9)^{1/2}$; $y(-4) = 0$.

7. $\dfrac{dy}{dx} = \dfrac{10}{x^2 + 1}$; $y(0) = 0$.

8. $\dfrac{dy}{dx} = \cos 2x$; $y(0) = 1$.

9. $\dfrac{dy}{dx} = (1 - x^2)^{-1/2}$; $y(0) = 0$.

10. $\dfrac{dy}{dx} = xe^{-x}$; $y(0) = 1$.

In Problems 11–17, find the position function $x(t)$ of a moving particle with the given acceleration $a(t)$, initial position $x_0 = x(0)$, and initial velocity $v_0 = v(0)$.

11. $a(t) = 50$, $v_0 = 10$, $x_0 = 20$.

12. $a(t) = -20$, $v_0 = -15$, $x_0 = 5$.

13. $a(t) = 3t$, $v_0 = 5$, $x_0 = 0$.

14. $a(t) = 2t + 1$, $v_0 = -7$, $x_0 = 4$.

15. $a(t) = 4(t + 3)^2$, $v_0 = -1$, $x_0 = 1$.

16. $a(t) = \dfrac{3}{(t + 4)^{1/2}}$, $v_0 = -1$, $x_0 = 1$.

17. $a(t) = \dfrac{1}{(t + 1)^3}$, $v_0 = 0$, $x_0 = 0$.

18. A ball is dropped from the top of a building that is 400 ft high. How long does it take to reach the ground? With what speed does the ball strike the ground?

19. The brakes of a car are applied when it is moving at 100 km/h and provide a constant deceleration of 10 meters per second per second (m/s²). How far does the car travel before coming to a stop?

20. A ball is thrown straight upward from ground level with an initial speed of 160 ft/s. What is the maximum height that the ball attains? How long does it remain aloft?

21. A ball is thrown straight downward from the top of a tall building. The initial speed of the ball is 10 m/s. It strikes the ground with a speed of 60 m/s. How tall is the building?

22. A baseball is thrown straight downward with an initial speed of 40 ft/s from the top of the Washington Monument (555 ft high). How long does it take to reach the ground, and with what speed does the baseball strike the ground?

23. A bomb is dropped from a balloon hovering at an altitude of 800 ft. A gun emplacement is located on the ground directly below the balloon. The gun fires a projectile straight upward toward the bomb exactly 2 s after the bomb is released. With what initial speed should the projectile be fired in order to intercept the bomb at an altitude of exactly 400 ft?

24. A car traveling at 60 mi/h skids 176 ft after its brakes are suddenly applied. Under the assumption that the braking system provides constant deceleration, what is that deceleration? How many seconds does the skid continue?

25. The skid marks made by an automobile indicate that its brakes were fully applied for a distance of 225 ft before it came to a stop. The car in question is known to have a constant deceleration of 50 feet per second per second (ft/s²) under the conditions we describe. How fast was the car traveling when the brakes were first applied?

26. Suppose that a car skids 15 m if it is moving at 50 km/h when the brakes are applied. Assuming that the car has the same constant deceleration, how far will it skid if it is moving at 100 km/h when the brakes are applied?

27. On the planet Gzyx, a ball dropped from a height of 20 ft hits the ground in 2 s. If a ball is dropped from the top of a 200-ft-tall building on Gzyx, how long will it take to hit the ground? With what speed will it hit?

28. A person can thrown a ball straight upward from the surface of the earth to a maximum height of 144 ft. How high could the person throw the ball on the planet Gzyx of Problem 27?

29. A stone is dropped from rest from initial height h above the surface of the earth. Show that the speed at which it strikes the ground is $v = (2gh)^{1/2}$.

30. If a woman has enough "spring" in her legs to jump vertically to a height of 2.25 ft on the earth, how high could she jump on the moon, where the surface gravitational acceleration is (approximately) 5.3 ft/s²?

31. (a) A cantilever beam is fixed at $x = 0$ and free at $x = L$ (its other end). Show that its shape is given by

$$y = \frac{w}{24EI}(x^4 - 4Lx^3 + 6L^2x^2).$$

(b) Show that $y'(x) = 0$ only at $x = 0$, and thus that it follows (why?) that the maximum deflection of the cantilever is $y_{max} = y(L) = wL^4/8EI$.

32. (a) Suppose that a beam is fixed at its ends $x = 0$ and $x = L$. Show that its shape is given by

$$y = \frac{w}{24EI}(x^4 - 2Lx^3 + L^2x^2).$$

(b) Show that the roots of $y'(x) = 0$ are $x = 0$, $x = L$, and $x = L/2$, and so it follows (why?) that the maximum deflection of the beam is $y_{max} = y(L/2) = wL^4/(384EI)$, one-fifth that of a beam with simply supported ends.

33. For the simply supported beam whose deflection curve is given by Eq. (19), show that the only root of $y'(x) = 0$ in $[0, L]$ is $x = L/2$, so it follows (why?) that the maximum deflection is indeed given by Eq. (20).

34. (a) A beam is fixed at its left end, $x = 0$, but is simply supported at the other end, $x = L$. Show that its deflection curve is

$$y = \frac{w}{48EI}(2x^4 - 5Lx^3 + 3L^2x^2).$$

(b) Show that its maximum deflection occurs where $x = (15 - \sqrt{33})L/16$, and is about 41.6% of the maximum deflection that would occur if it were simply supported at each end.

In the case of a general first order differential equation of the form

$$y' = f(x, y), \tag{1}$$

we cannot simply integrate each side as in Section 1.2, because now the right-hand side involves the unknown function $y(x)$. Before one spends much time trying to solve a differential equation, it is best to know that solutions actually *exist*. We may also want to know whether there is only one solution of the equation satisfying a given initial condition—that is, whether solutions are *unique*. For instance, such a simple-looking initial value problem as

$$\frac{dy}{dx} = 2\sqrt{y}, \qquad y(0) = 0 \tag{2}$$

has the two different solutions $y_1(x) = x^2$ and $y_2(x) \equiv 0$.

The questions of existence and uniqueness also bear on the process of mathematical modeling. Suppose we are studying a physical system in which its behavior is completely determined by certain initial conditions, but that our proposed mathematical model involves a differential equation *not* having unique solutions. Then this raises an immediate question as to whether the mathematical model adequately represents the physical system.

In order to investigate the possible behavior of solutions of a differential equation of the form $y' = f(x, y)$, we may think of it in a very geometric way: At various points (x, y) of the two-dimensional plane, the value of $f(x, y)$ determines a slope y'. A *solution* of the above differential equation is a differentiable function with graph having slope $y' = f(x, y)$ at each point (x, y).

EXAMPLE 1 Consider the differential equation

$$y' = y. \tag{3}$$

In geometric language, solving this differential equation means finding curves $y = y(x)$ each with the following property: The slope of the graph at (x, y) is y. Figure 1.8 shows a **slope field** for the equation $y' = y$. Through various points (x, y), we have sketched short line segments with slope y. It is easy to believe that various curves can be drawn for which the short line segments are tangent segments. The slope field also gives qualitative information about *all* the solutions of $y' = y$. For instance, as x approaches $-\infty$, it seems that every solution approaches 0. It also seems that $y \equiv 0$ is the only solution that is constant on any interval. In this particular example, we can see by inspection that $y(x) = Ce^x$ is a general solution of (3), and Fig. 1.8 shows a typical solution curve threading its way through the slope field.

EXAMPLE 2 Consider the differential equation

$$y' = y^2 - x^2. \tag{4}$$

*This section may be deferred to any point later in the chapter.

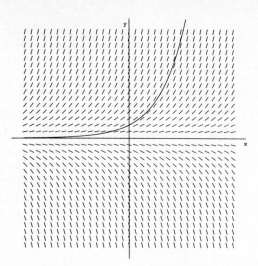

Figure 1.8 The slope field for $y' = y$ and the graph of one solution of the equation.

To construct its slope field (shown in Fig. 1.9), we note by inspection of Eq. (4) that $y' = 0$ at each point of the two lines $y = \pm x$, but that $y' = m$ at each point of the hyperbola $y^2 - x^2 = m$. It appears that the curve shown in the figure, having local extrema at its intersections with the lines $y = \pm x$, is a typical solution curve.

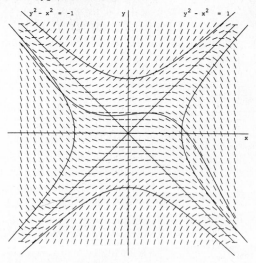

Figure 1.9 The slope field for $y' = y^2 - x^2$.

EXAMPLE 3 Consider the differential equation

$$y^2 + x^2 y' = 0. \tag{5}$$

We show a slope field for this equation in Fig. 1.10. Indeed, because $y' = -(y/x)^2$, we see that each differentiable solution must be decreasing except on the coordinate axes (where x or y is zero). Much additional qualitative information can be read from the slope field of Eq. (5).

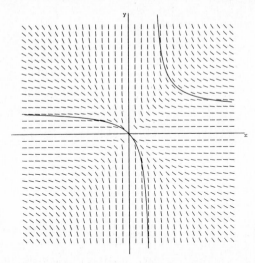

Figure 1.10 Slope field for $y^2 + x^2 y' = 0$ and the graph of one solution of the equation.

In Section 1.4 we will solve equations like this one—known as **separable** differential equations—by separating the variables; we first write

$$\frac{dy}{dx} = -\frac{y^2}{x^2},$$

then

$$-\frac{dy}{y^2} = \frac{dx}{x^2};$$

next, we antidifferentiate to obtain

$$\frac{1}{y} = -\frac{1}{x} + C.$$

Finally, we solve for y to obtain

$$y = y(x) = \frac{x}{Cx - 1}. \tag{6}$$

We have drawn a number of solution curves for various values of C; these are shown in Fig. 1.11. Each solution curve of the above form has asymptotes $y = 1/C$ and $x = 1/C$ provided that $C \neq 0$, and in fact each such solution curve consists of the two branches of a rectangular hyperbola, one branch of which passes through the origin. For $C = 0$ we obtain the "different" solution $y = -x$. Moreover, the above solution process excludes the function $y \equiv 0$, which (by substitution in Eq. (5)) is yet another solution.

The remarkable feature of the differential equation in (5) and its solutions is this: First, there are *infinitely many* different solutions satisfying the

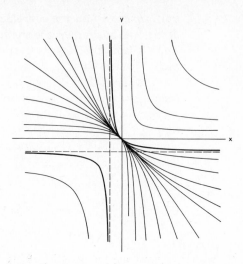

Figure 1.11 Graphs of some solutions of Equation (5).

initial condition $y(0) = 0$. But it follows from Eq. (6) that $y(0) = 0$, so if $b \neq 0$, then there is *no* solution satisfying the initial condition $y(0) = b$. Finally, if $a \neq 0$ and b is arbitrary, there is exactly one solution with $y(a) = b$. All these observations are evident from inspection of Fig. 1.11.

Thus an initial value problem may have either no solution, a unique solution, or many—even infinitely many—solutions. The following theorem provides sufficient conditions to insure existence and uniqueness of solutions, so that neither extreme case (no solution or nonunique solutions) can occur. Methods of proving existence and uniqueness theorems are discussed in Section 9.4.

THEOREM: EXISTENCE AND UNIQUENESS OF SOLUTIONS

Suppose that the real-valued function $f(x, y)$ is continuous on some rectangle in the plane containing the point (a, b) in its interior. Then the initial value problem

$$\frac{dy}{dx} = f(x, y); \qquad y(a) = b \qquad (7)$$

has at least one solution on some open interval J containing the point $x = a$. If, in addition, the partial derivative $\partial f/\partial y$ is continuous on that rectangle, then the solution is unique on some (perhaps smaller) open interval J_0 containing the point $x = a$.

In the case of the differential equation $y' = y$ of Example 1, both the function $f(x, y) = y$ and the partial derivative $\partial f/\partial y = 1$ are continuous everywhere, so the theorem implies the existence of a unique solution for any initial data (a, b). Although the theorem insures existence only on some open interval about $x = a$, each solution $y(x) = Ce^x$ actually is defined for all x.

In the case of the differential equation $y' = 2\sqrt{y}$ in Eq. (2), the function $f(x, y) = 2\sqrt{y}$ is continuous everywhere, but the partial derivative $\partial f/\partial y = 1/\sqrt{y}$ is discontinuous when $y = 0$, and hence at the point $(0, 0)$. This explains the existence of two different solutions $y_1(x) = x^2$ and $y_2(x) \equiv 0$, each of which satisfies the initial condition $y(0) = 0$.

In Example 3 we analyzed the differential equation $y^2 + x^2 y' = 0$ and found that there was no solution passing through $(0, 1)$. If we take $f(x, y) = -(y/x)^2$, we see that the above theorem cannot guarantee existence of a solution through $(0, 1)$ because f is not continuous there. (Note that f is also not continuous at $(0, 0)$, but some solutions *do* pass through this point. Thus continuity of f is a sufficient condition, but not a necessary condition, for the existence of solutions.)

EXAMPLE 4   Consider the initial value problem

$$y' = (1 - y^2)^{1/2}; \qquad y(0) = 0. \tag{8}$$

The theorem implies that this problem has a unique solution on some interval J about $x = 0$. That solution is, in fact, part of the graph of $y = \sin x$. But the initial value problem above does not have a unique solution. One of its many solutions is $y = \sin x$. Another is

$$y = \begin{cases} \sin x & \text{if } x < \dfrac{\pi}{2}; \\ 1 & \text{if } x \geqq \dfrac{\pi}{2}. \end{cases}$$

Indeed, this initial value problem has infinitely many solutions passing through $(0, 0)$; you may construct as many examples as you please by piecing together parts of the graphs of $y = \sin x$, $y \equiv 1$, and $y \equiv -1$ to form a differentiable function.

The point is that the theorem guarantees uniqueness *near* the point (a, b), but the solution curve may branch elsewhere, and uniqueness will be lost. Similarly, the theorem can guarantee existence *near* the point (a, b), but the differential equation may have no solution for some other values of x.

Finally, in Example 5 of Section 1.1, we examined the especially simple differential equation $y' = y^2$. Here we have $f(x, y) = y^2$ and $\partial f/\partial y = 2y$. Each of these functions is continuous everywhere in the plane, and in particular on the rectangle $-2 < x < 2$, $0 < y < 2$. Because the point $(0, 1)$ lies in the interior of this rectangle, the existence and uniqueness theorem guarantees a unique solution—necessarily a continuous function—of the initial value problem

$$y' = y^2, \qquad y(0) = 1$$

on *some* open interval containing $x_0 = 0$. This is the solution

$$y = \frac{1}{1 - x}$$

that we discussed in the above-mentioned example. But $1/(1 - x)$ is discontinuous at $x = 1$, so we do *not* have *existence* of a solution on the entire interval $-2 < x < 2$. This means that the interval J of the theorem may not be as wide as the rectangle; even though the hypotheses of the theorem are satisfied for all x in an interval I (and for all y in an appropriate interval), the solution may, as in this example, exist only on a smaller interval J. Similarly, the interval J_0 of uniqueness may be even smaller than J.

Nevertheless, there is one important case in which *global* existence and uniqueness are assured. The *linear* first order differential equation

$$\frac{dy}{dx} = a(x)y + b(x) \tag{9}$$

is particularly important in applications. It is most common for $a(x)$ and $b(x)$ to be continuous functions on some open interval, and we will see in Section 1.5 that this is enough to guarantee existence and uniqueness of the solution of any initial value problem involving Eq. (9) on that (entire) interval. The equation in (9) is called **linear** because only the first powers of the dependent variable and its derivative y' appear (we do not require that $a(x)$ or $b(x)$ be linear functions of x). The differential equation $y' = 2x^3y + \cos x$ is linear, with $a(x) = 2x^3$ and $b(x) = \cos x$ each continuous on the entire real line, whereas the equations in Examples 2, 3, and 4 above are nonlinear.

1.3 Problems

In each of Problems 1–16, we have provided the slope field of a differential equation (also given). Sketch three or four typical solutions for each. (One method: Photocopy the slope field, and draw your solutions in a second color. Another method: Use tracing paper.)

1. $y' = x^2 + y^2$.

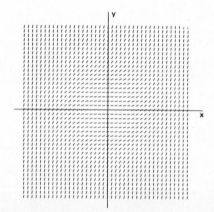

Figure 1.12 Slope field for Problem 1.

2. $y' = \dfrac{-y}{2}.$

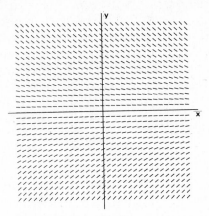

Figure 1.13 Slope field for Problem 2.

3. $y' = y^2 + 1.$

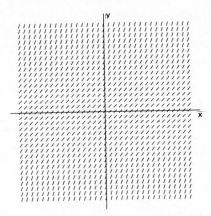

Figure 1.14 Slope field for Problem 3.

4. $y' = x^2 - y^2.$

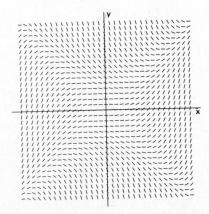

Figure 1.15 Slope field for Problem 4.

5. $y' = xy + 1.$

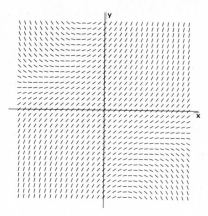

Figure 1.16 Slope field for Problem 5.

6. $y' = x + y.$

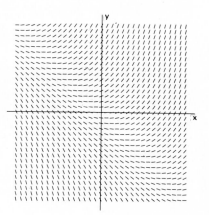

Figure 1.17 Slope field for Problem 6.

7. $y' = x - y.$

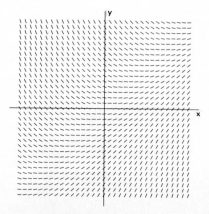

Figure 1.18 Slope field for Problem 7.

8. $y' = xy$.

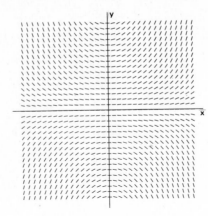

Figure 1.19 Slope field for Problem 8.

9. $y' = \dfrac{1}{y}$.

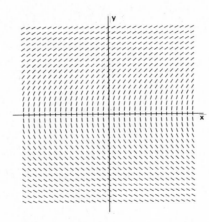

Figure 1.20 Slope field for Problem 9.

10. $y' = \dfrac{x}{y}$.

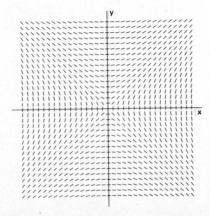

Figure 1.21 Slope field for Problem 10.

11. $y' = y^3$.

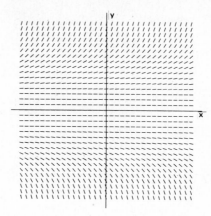

Figure 1.22 Slope field for Problem 11.

12. $y' = y^{2/3}$.

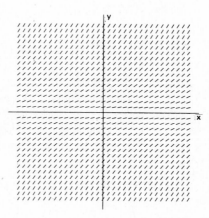

Figure 1.23 Slope field for Problem 12.

13. $y' = \sin y$.

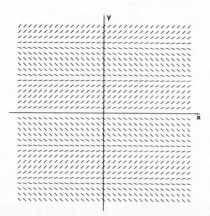

Figure 1.24 Slope field for Problem 13.

14. $y' = \sin xy$.

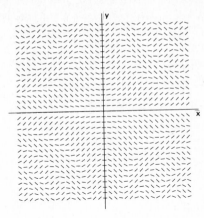

Figure 1.25 Slope field for Problem 14.

15. $y' = xe^{-y}$.

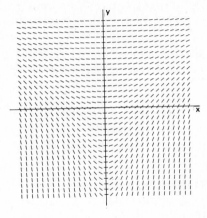

Figure 1.26 Slope field for Problem 15.

16. $y' = \ln(1 + y^2)$.

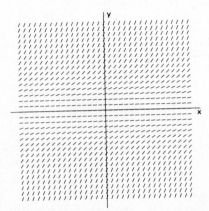

Figure 1.27 Slope field for Problem 16.

In each of Problems 17–26, determine whether the existence and uniqueness theorem of this section does or does not guarantee existence of a solution of the given initial value problem. If existence is guaranteed, then determine whether the theorem does or does not guarantee uniqueness of that solution.

17. $y' = 2x^2y^2$; $y(1) = -1$.

18. $y' = x \ln y$; $y(1) = 0$.

19. $y' = y^{1/3}$; $y(0) = 1$.

20. $y' = y^{1/3}$; $y(0) = 0$.

21. $y' = (x - y)^{1/2}$; $y(2) = 2$.

22. $y' = (x - y)^{1/2}$; $y(2) = 1$.

23. $y' = \dfrac{x}{y}$; $y(0) = 1$.

24. $y' = \dfrac{x}{y}$; $y(1) = 1$.

25. $y' = \ln(1 + y^2)$; $y(0) = 0$.

26. $y' = x^2 - y^2$; $y(0) = 1$.

In each of Problems 27–35, construct a slope field for the given differential equation.

27. $y' = 2x^2y^2$.

28. $y' = y^{1/3}$.

29. $y' = (x - y)^{1/2}$.

30. $y' = \dfrac{y}{x}$.

31. $y' = \ln(1 + x^2y^2)$.

32. $y' = y^4$.

33. $y' = xy^2$.

34. $y' = ye^{-x}$.

35. $y' = x^3 - y^3$.

The next three problems illustrate that if the hypotheses of the existence and uniqueness theorem fail at a point (a, b), then there may be no solutions, finitely many solutions, or infinitely many solution curves passing through (a, b).

36. Draw a slope field for the differential equation $xy' = y$. Apply the existence and uniqueness theorem of this section to find those points (a, b) such that a solution of the differential equation *must* exist on an open interval J containing a. Find those points (a, b) such that a unique solution exists on an open interval J_0 containing a. Then find how many solutions pass through $(0, 0)$.

37. Show that, on the interval $0 \leqq x \leqq \pi$, the functions $y_1(x) \equiv 1$ and $y_2(x) = \cos x$ each satisfy the initial value problem

$$y' + (1 - y^2)^{1/2} = 0, \qquad y(0) = 1.$$

Why does this fact not contradict the existence and uniqueness theorem stated in this section? Explain your answer carefully.

38. Repeat the instructions of Problem 36 for the differential equation $y' = 4xy^{1/2}$ $(y \geq 0)$. (*Suggestion:* Note that $y_1(x) \equiv 0$ is a solution, as is $y_2(x) = (x^2 + C)^2$ if $x^2 + C \geqq 0$.)

1.4
Separable Equations and Applications

The first order differential equation

$$\frac{dy}{dx} = H(x, y) \tag{1}$$

is called *separable* provided that $H(x, y)$ can be written as the product of a function of x and a function of y or, equivalently, as a quotient $H(x, y) = g(x)/f(y)$. In this case the variables x and y can be *separated*—isolated on opposite sides of an equation—by writing informally the equation $f(y) \, dy = g(x) \, dx$, which we understand to be compact notation for the differential equation

$$f(y)\frac{dy}{dx} = g(x), \tag{2}$$

It is easy to solve this special type of differential equation simply by integrating both sides with respect to x:

$$\int f(y(x))\frac{dy}{dx}\,dx = \int g(x)\,dx + C;$$

more concisely,

$$\int f(y)\,dy = \int g(x)\,dx + C. \tag{3}$$

All that is required is that the antiderivatives $F(y) = D_y^{-1}f(y)$ and $G(x) = D_x^{-1}g(x)$ can be found. To see that (2) and (3) are equivalent, note the following consequence of the chain rule:

$$D_x F(y(x)) = F'(y(x))y'(x) = f(y)\frac{dy}{dx} = g(x) = D_x G(x),$$

which in turn is equivalent to

$$F(y(x)) = G(x) + C, \tag{4}$$

because two functions have the same derivative on an interval if and only if they differ by a constant on that interval.

EXAMPLE 1 Solve the differential equation

$$x^2\frac{dy}{dx} = \frac{x^2 + 1}{3y^2 + 1}.$$

Solution We multiply each side by the formal expression $(3y^2 + 1)\,dx/x^2$ to obtain

$$(3y^2 + 1)\,dy = \left(1 + \frac{1}{x^2}\right)dx.$$

Integration of both sides then gives

$$y^3 + y = x - \frac{1}{x} + C.$$

As Example 1 illustrates, it may not be possible or practical to solve Eq. (4) explicitly for y as a function of x. If not, we call (4) an *implicit solution* of the differential equation in (2). Because Eq. (4) contains the arbitrary constant C, we also call it a **general solution** of (2). Given an initial condition $y(x_0) = y_0$, the choice $C_0 = F(y_0) - G(x_0)$ of C yields the equation

$$F(y) = G(x) + C_0$$

that implicitly defines a **particular solution** (if any) of the initial value problem

$$f(y)\frac{dy}{dx} = g(x), \qquad y(x_0) = y_0.$$

For instance, we see from Example 1 that a solution of the initial value problem

$$x^2\frac{dy}{dx} = \frac{x^2 + 1}{3y^2 + 1}, \qquad y(1) = 2$$

is defined implicitly by the equation

$$y^3 + y = x - \frac{1}{x} + 10.$$

Thus the equation $K(x, y) = 0$ is called an **implicit solution** of a differential equation if it is satisfied (on some interval) by some solution $y = y(x)$ of the differential equation. But note that a particular solution $y = y(x)$ of $K(x, y) = 0$ may or may not satisfy a given initial condition. For example, differentiation of $x^2 + y^2 = 4$ yields

$$x + y\frac{dy}{dx} = 0,$$

so $x^2 + y^2 = 4$ is an implicit solution of the differential equation $x + yy' = 0$. But only the first of the two explicit solutions $y = +(4 - x^2)^{1/2}$ and $y = -(4 - x^2)^{1/2}$ satisfies the initial condition $y(0) = 2$.

The argument preceding Example 1 shows that every particular solution of (2) satisfies (4) for some choice of C; *this* is why it is appropriate to call (4) a general solution of (2).

Warning: Suppose, however, that we begin with the differential equation

$$\frac{dy}{dx} = g(x)h(y) \tag{5}$$

and divide by $h(y)$ to obtain the separated equation

$$\frac{1}{h(y)}\frac{dy}{dx} = g(x). \tag{6}$$

If y_0 is a root of the equation $h(y) = 0$—that is, if $h(y_0) = 0$—then the constant function $y(x) \equiv y_0$ is clearly a solution of (5), but may *not* be contained in the general solution of (6). Thus solutions of a differential equation may be lost upon division by a vanishing factor. (Indeed, false solutions may be gained upon multiplication by a vanishing factor. This phenomenon is similar to the introduction of extraneous roots in solving algebraic equations.)

In Section 1.5 we shall see that every particular solution of a *linear* first order differential equation is contained in its general solution. By contrast, it is common for a nonlinear first order differential equation to have both a general solution involving an arbitrary constant C and one or several particular solutions that cannot be obtained by selecting a value for C. These exceptional solutions are frequently called **singular solutions**. In Example 3 of Section 1.3, we found that the solution $y(x) \equiv 0$ was a singular solution of the equation $y^2 + x^2y' = 0$; this solution cannot be obtained from the general solution $y = x/(Cx - 1)$ by any choice of the constant C.

EXAMPLE 2 Find all solutions of the differential equation

$$\frac{dy}{dx} = 2x\sqrt{y - 1}.$$

Solution We note first the constant solution $y(x) \equiv 1$. If $y \neq 1$, we can divide each side by $\sqrt{y - 1}$ to obtain

$$(y - 1)^{-1/2}\frac{dy}{dx} = 2x.$$

Integration gives $2\sqrt{y - 1} = x^2 + C$; upon solving for y we get the general solution

$$y = 1 + \tfrac{1}{4}(x^2 + C)^2.$$

Note that no value of C gives the particular solution $y \equiv 1$. It was lost when we divided by $\sqrt{y - 1}$.

EXAMPLE 3 Solve the initial value problem

$$\frac{dy}{dx} = xy + x - 2y - 2, \qquad y(0) = 2.$$

Solution Sometimes a factorization is not obvious at first glance; here we have

$$\frac{dy}{dx} = (x - 2)(y + 1).$$

The constant function $y \equiv -1$ satisfies the differential equation but does not satisfy the initial condition, so we cannot lose the solution of the initial value problem by dividing by $y + 1$. We therefore perform that division and integrate:

$$\int \frac{dy}{y + 1} = \int (x - 2)\,dx;$$

$$\ln|y + 1| = \frac{1}{2}x^2 - 2x + C;$$

$$\ln(y + 1) = \frac{1}{2}x^2 - 2x + C.$$

In the final step we use the fact that $y + 1 > 0$ near the initial value $y = 2$. Now apply the exponential function to each side of the last equation. With the observation that $e^{\ln z} = z$, we get

$$y + 1 = \exp\left(\tfrac{1}{2}x^2 - 2x + C\right),$$

so that

$$y = A \exp\left(\tfrac{1}{2}x^2 - 2x\right) - 1$$

where $A = e^C$. Note in passing that we did not lose the constant solution $y \equiv -1$; it corresponds to $A = 0$. But to conclude the example, the initial condition $y(0) = 2$ implies that $A = 3$, so the desired solution is

$$y = 3 \exp\left(\tfrac{1}{2}x^2 - 2x\right) - 1.$$

NATURAL GROWTH AND DECAY

The differential equation

$$\frac{dx}{dt} = kx \qquad (k \text{ a constant}) \tag{7}$$

serves as a mathematical model for a remarkably wide range of natural phenomena—any involving a quantity whose time rate of change is proportional to its current value. Here are some examples.

Population Growth

Suppose that $P(t)$ is the number of individuals in a population (of humans, or insects, or bacteria) having *constant* birth and death rates β and δ (in births or deaths per individual per unit of time). Then, during a short time interval Δt, approximately $\beta P(t)\, \Delta t$ births and $\delta P(t)\, \Delta t$ deaths occur, so the change in $P(t)$ is given approximately by

$$\Delta P = (\beta - \delta)P(t)\, \Delta t,$$

and therefore

$$\frac{dP}{dt} = \lim_{\Delta t \to 0} \frac{\Delta P}{\Delta t} = kP \tag{8}$$

where $k = \beta - \delta$.

Compound Interest

Let $A(t)$ be the number of dollars in a savings account at time t, and suppose that the interest is *compounded continuously* at an annual interest rate r. (Note that 10% annual interest means that $r = 0.10$.) Continuous compounding means that, during a short time interval Δt, the amount of interest added to the account is approximately $\Delta A = rA(t)\, \Delta t$, so that

$$\frac{dA}{dt} = \lim_{\Delta t \to 0} \frac{\Delta A}{\Delta t} = rA. \tag{9}$$

Radioactive Decay

Consider a sample of material that contains $N(t)$ atoms of a certain radioactive isotope at time t. It has been observed that a constant fraction of these radioactive atoms will spontaneously decay (into atoms of another element or into another isotope of the same element) during each unit of time. Consequently, the sample behaves exactly like a population with a constant death rate, but with no births occurring. To write a model for $N(t)$, we use Eq. (8) with N in place of P, with $k > 0$ in place of δ, and with $\beta = 0$. We thus get the differential equation

$$\frac{dN}{dt} = -kN. \tag{10}$$

The value of k depends upon the particular radioactive isotope.

The key to the method of *radiocarbon dating* is that a constant proportion of the carbon atoms in any living creature is made up of the radioactive isotope C^{14} of carbon. This proportion remains constant because the fraction of C^{14} in the atmosphere remains almost constant, and living matter is continuously taking up carbon from the air or is consuming other living matter containing the same constant ratio of C^{14} atoms to ordinary carbon atoms. The same ratio permeates all life, because organic processes seem to make no distinction between the two isotopes.

The ratio of C^{14} to normal carbon remains constant in the atmosphere because, though C^{14} is radioactive and slowly decays, the amount is continuously replenished through the conversion of nitrogen to C^{14} by cosmic rays in the upper atmosphere. Over the long history of the planet, this decay and replenishment process has come into nearly steady state.

Of course, when a living organism dies, it ceases its metabolism of carbon, and the process of radioactive decay begins to deplete its C^{14} content. There is no replenishment of C^{14}, and consequently the ratio of C^{14} to normal carbon begins to drop. By measuring this ratio, the amount of time elapsed since the death of the organism can be estimated. For such purposes, it is necessary to measure the decay constant; for C^{14}, it's known that k is approximately 0.0001216.

(Matters are not so simple as we have made them appear. In applying the technique of radiocarbon dating, extreme care must be taken to avoid contaminating the sample with organic matter, or even with ordinary fresh air. In addition, the cosmic ray levels apparently have not been constant, so the ratio of radioactive carbon in the atmosphere has varied over the past centuries. By using independent methods of dating samples, researchers in this area have compiled tables of correction factors to enhance the accuracy of the process.)

Drug Elimination

In many cases the amount $A(t)$ of a certain drug in the bloodstream, measured by the excess over the natural level of the drug, will decline at a rate proportional to the current excess amount. That is,

$$\frac{dA}{dt} = -\lambda A \tag{11}$$

where $\lambda > 0$. The parameter λ is called the **elimination constant** of the drug, and $T = 1/\lambda$ is called the **elimination time**.

The prototype differential equation $x' = kx$ with $x(t) > 0$ and k a constant (either positive or negative) is readily solved by separating the variables and integrating:

$$\int \frac{1}{x}\, dx = \int k\, dt;$$

$$\ln x = kt + C.$$

Then we solve for x:

$$e^{\ln x} = e^{kt+C};$$

$$x = e^C e^{kt} = A e^{kt}.$$

Because C is a constant, so is $A = e^C$. It is also clear that $A = x(0) = x_0$, so the particular solution of Eq. (7) with the initial condition $x(0) = x_0$ is simply

$$x(t) = x_0 e^{kt}. \tag{12}$$

Because of the presence of the natural exponential function in its solution, the differential equation $x' = kx$ is often called the **exponential**, or

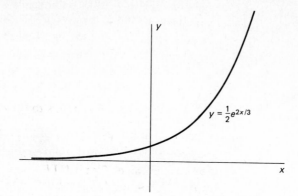

Figure 1.28 Natural growth.

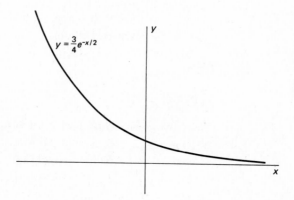

Figure 1.29 Natural decay.

natural growth, equation. Figure 1.28 shows a typical graph of $x(t)$ in the case $k > 0$, and Fig. 1.29 shows a typical graph in the case $k < 0$.

EXAMPLE 4 In mid-1982, the world population was 4.5 billion and it was then increasing at the rate of a quarter million persons each day. Assuming constant birth and death rates, when should a world population of 10 billion be expected?

Solution From Eq. (12) with P in place of x, we know that $P(t) = P_0 e^{kt}$. We measure the world population $P(t)$ in billions and measure time t in years. We shall take $t = 0$ to correspond to 1982, so that $P_0 = 4.5$. The fact that P is increasing by a quarter million, or 0.00025 billion, persons per day at time $t = 0$ means that

$$P'(0) = (0.00025)(365.25) \approx 0.0913$$

billion per year. From Eq. (8) we now obtain

$$k = \frac{P'(0)}{P(0)} \approx \frac{0.0913}{4.5} \approx 0.0203.$$

Thus the population is growing at the rate of about 2.03 % per year.

To find when the population will be 10 billion, we need only solve the equation

$$10 = P(T) = (4.5)e^{(0.0203)T}$$

for

$$T = \frac{\ln(10/4.5)}{0.0203} \approx 39$$

years, which corresponds to the year 2021.

The decay constant of a radioactive isotope is often specified in terms of another empirical constant, the *half-life* of the isotope, because this parameter is more convenient. The **half-life** τ of a radioactive isotope is the time required for *half* of it to decay. To find the relationship between k and τ, we set $t = \tau$ and $N = \frac{1}{2}N_0$ in the equation $N(t) = N_0 e^{-kt}$, so that $\frac{1}{2}N_0 = N_0 e^{-k\tau}$. When we solve for τ, we find that

$$\tau = \frac{\ln 2}{k}.$$

For example, the half-life of C^{14} is $\tau = (\ln 2)/(0.0001216)$, approximately 5700 years.

EXAMPLE 5   A specimen of charcoal found at Stonehenge turns out to contain 63% as much C^{14} as a sample of present-day charcoal of equal mass. What is the age of the sample?

Solution   We take $t = 0$ as the time of the death of the tree from which the Stonehenge charcoal was made and N_0 as the number of C^{14} atoms it contained then. We are given that $N = (0.63)N_0$ now, so we solve the equation $(0.63)N_0 = N_0 e^{-kt}$ with the value $k = 0.0001216$. Thus we find that

$$t = -\frac{\ln(0.63)}{0.0001216} \approx 3800$$

years. The sample is about 3800 years old, and if it has any connection with the builders of Stonehenge, our computations suggest that this observatory, monument, or temple—whichever it may be—dates from 1800 B.C. or earlier.

COOLING AND HEATING

According to Newton's law of cooling (Eq. (1) in Section 1.1), the time rate of change of the temperature $T(t)$ of a body immersed in a medium of constant temperature A is proportional to the difference $A - T$. That is,

$$\frac{dT}{dt} = k(A - T) \tag{14}$$

where k is a positive constant. This is an instance of the linear first order differential equation with constant coefficients:

$$\frac{dx}{dt} = ax + b. \tag{15}$$

It includes the exponential equation as a special case ($b = 0$) and is also easy to solve by separation of variables.

EXAMPLE 6   A 5-lb roast, initially at 50°F, is put into a 375°F oven at 5:00 P.M.; it is found that the temperature $T(t)$ of the roast is 125°F after 75 min. When will the roast be 150°F (medium rare)?

Solution   We take time t in minutes, with $t = 0$ corresponding to 5:00 P.M. We also assume (not altogether realistically) that at any instant the temperature $T(t)$ of the roast is uniform throughout. We have $T(t) < A = 375$, $T(0) = 50$, and $T(75) = 125$. Hence

$$\frac{dT}{dt} = k(375 - T);$$

$$\int \frac{1}{375 - T} dT = \int k \, dt;$$

$$-\ln(375 - T) = kt + C;$$

$$375 - T = Be^{-kt}.$$

Now $T(0) = 50$ implies that $B = 325$, so $T(t) = 375 - 325e^{-kt}$. We also know that $T = 125$ when $t = 75$. Substitution of these values in the last equation yields

$$k = -\tfrac{1}{75} \ln \left(\tfrac{250}{325}\right) \approx 0.0035.$$

Hence we finally solve the equation

$$150 = 375 - 325e^{(-0.0035)t}$$

for $t = -[\ln(225/325)]/(0.0035) \approx 105$ min, the total cooking time required. Because the roast was put in the oven at 5:00 P.M., it should be removed at about 6:45 P.M.

TORRICELLI'S LAW

Suppose that a water tank has a hole with area a at its bottom, from which water is leaking. Denote by $y(t)$ the depth of the water in the tank at time t, and by $V(t)$ the volume of water in the tank then. It is plausible—as well as true under ideal conditions—that the velocity of water exiting through the hole is

$$v = \sqrt{2gy}, \tag{16}$$

which is the velocity a drop of water would acquire in falling freely from the surface of the water to the hole (see Problem 25 of Section 1.2). Under real conditions, taking into account the constriction of a water jet from an orifice, $v = c\sqrt{2gy}$ where c is an empirical constant between 0 and 1 (usually about 0.6 for a small continuous stream of water). For simplicity we take $c = 1$ in the following discussion.

As a consequence of Eq. (16) we have

$$\frac{dV}{dt} = -av = -a\sqrt{2gy}; \tag{17}$$

this is a statement of Torricelli's law for a draining tank. If $A(y)$ denotes the horizontal cross-sectional area of the tank at height y, then the method of volume by cross sections gives

$$V = \int_0^y A(y)\, dy,$$

so the fundamental theorem of calculus implies that $dV/dy = A(y)$ and therefore that

$$\frac{dV}{dt} = \frac{dV}{dy}\frac{dy}{dt} = A(y)\frac{dy}{dt}. \tag{18}$$

From Eqs. (17) and (18) we finally obtain

$$A(y)\frac{dy}{dt} = -a\sqrt{2gy}, \tag{19}$$

the most useful form of Torricelli's law.

EXAMPLE 7   A hemispherical tank has top radius 4 ft and at time $t = 0$ is full of water. At that moment a circular hole of diameter 1 in. is opened in the bottom of the tank. How long will it take for all the water to drain from the tank?

Solution   From the right triangle in Fig. 1.30, we see that

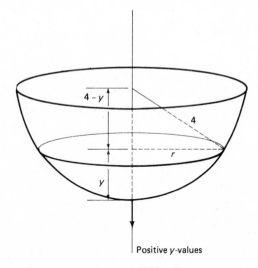

Positive y-values

Figure 1.30   Draining a hemispherical tank.

$$A(y) = \pi r^2 = \pi[16 - (4 - y)^2] = \pi(8y - y^2).$$

With $g = 32$ ft/s², Eq. (19) becomes

$$(8y - y^2)\frac{dy}{dt} = -\left(\frac{1}{24}\right)^2\sqrt{64y};$$

$$\int(8y^{1/2} - y^{3/2})\, dy = -\int\frac{dt}{72};$$

$$\frac{16}{3}y^{3/2} - \frac{2}{5}y^{5/2} = -\frac{1}{72}t + C.$$

Now $y(0) = 4$, so $C = \frac{16}{3}(4)^{3/2} - \frac{2}{5}(4)^{5/2} = \frac{448}{15}$. The tank is empty when $y = 0$, thus when

$$t = (72)(\frac{448}{15}) \approx 2150 \text{ s};$$

that is, about 35 min 50 s. So it takes slightly less than 36 min for the tank to drain.

1.4 Problems

Find general solutions (implicit if necessary, explicit if convenient) of the differential equations in Problems 1–8.

1. $\dfrac{dy}{dx} = xy^3$.

2. $y\dfrac{dy}{dx} = x(y^2 + 1)$.

3. $y^3 \dfrac{dy}{dx} = (y^4 + 1)\cos x$.

4. $\dfrac{dy}{dx} = \dfrac{1 + \sqrt{x}}{1 + \sqrt{y}}$.

5. $\dfrac{dy}{dx} = \dfrac{(x-1)y^5}{x^2(2y^3 - y)}$.

6. $(x^2 + 1)\dfrac{dy}{dx}\tan y = x$.

7. $\dfrac{dy}{dx} = 1 + x + y + xy$.

8. $x^2 y' = 1 - x^2 + y^2 - x^2y^2$.

Find explicit particular solutions of the initial value problems in Problems 9–16.

9. $\dfrac{dy}{dx} = ye^x$; $\quad y(0) = 2e$.

10. $\dfrac{dy}{dx} = 3x^2(y^2 + 1)$; $\quad y(0) = 1$.

11. $2y\dfrac{dy}{dx} = x(x^2 - 16)^{-1/2}$; $\quad y(5) = 2$.

12. $\dfrac{dy}{dx} = 4x^3y - y$; $\quad y(1) = -3$.

13. $\dfrac{dy}{dx} + 1 = 2y$; $\quad y(1) = 1$.

14. $y'\tan x = y$; $\quad y\left(\dfrac{\pi}{2}\right) = \dfrac{\pi}{2}$.

15. $x\dfrac{dy}{dx} - y = 2x^2y$; $\quad y(1) = 1$.

16. $\dfrac{dy}{dx} = 2xy^2 + 3x^2y^2$; $\quad y(1) = -1$.

17. (Population growth)   A certain city had a population of 25,000 in 1960 and a population of 30,000 in 1970. Assume that its population will continue to grow exponentially at a constant rate. What population can its city planners expect in the year 2000?

18. (Population growth)   In a certain culture of bacteria, the number of bacteria increased sixfold in 10 h. How long did it take for the population to double its initial number?

19. (Radiocarbon dating)   Carbon extracted from an ancient skull contained only one-sixth as much radioactive C^{14} as carbon extracted from present-day bone. How old is the skull?

20. (Radiocarbon dating)   Carbon taken from a purported relic of the time of Christ contained 4.6×10^{10} atoms of C^{14} per gram. Carbon extracted from a present-day specimen of the same substance contained 5.0×10^{10} atoms of C^{14} per gram. Compute the approximate age of the relic. What is your opinion as to its authenticity?

21. (Continuously compounded interest)   Upon the birth of their first child, a couple deposited $5000 in a savings account that draws 8% interest compounded continuously. The interest payments are allowed to accumulate. How much will the account contain on the child's eighteenth birthday?

22. (Continuously compounded interest) Suppose that you discover in your attic an overdue library book on which your great-great-grandfather owed a fine of 30¢ 100 years ago. If an overdue fine grows exponentially at a 5% annual rate compounded continuously, how much would you have to pay if you returned the book today?

23. (Drug elimination) Suppose that sodium pentobarbitol is used to anesthetize a dog: The dog is anesthetized when its bloodstream concentration contains at least 45 milligrams (mg) of sodium pentobarbitol per kilogram of the dog's body weight. Suppose also that sodium pentobarbitol is eliminated exponentially from the dog's bloodstream, with a half-life of 5 h. What single dose should be administered in order to anesthetize a 50-kg dog for 1 h?

24. The half-life of radioactive cobalt is 5.27 years. Suppose that a nuclear accident has left the level of cobalt radiation in a certain region at 100 times the level acceptable for human habitation. How long will it be before the region is again habitable? (Ignore the likely presence of other radioactive elements.)

25. Suppose that a mineral body, formed in an ancient cataclysm—perhaps the formation of the earth itself—originally contained the uranium isotope U^{238} (which has a half-life of 4.51×10^9 years) but no lead, the end product of the radioactive decay of U^{238}. If today the ratio of U^{238} atoms to lead atoms in the mineral body is 0.9, when did the cataclysm occur?

26. A certain moon rock was found to contain equal numbers of potassium and argon atoms. Assume that all the argon is the result of radioactive decay of potassium (its half-life is about 1.28×10^9 years) and that one of every nine potassium atom disintegrations yields an argon atom. What is the age of the rock, measured from the time it contained only potassium?

27. A pitcher of buttermilk initially at 25°C is to be cooled by setting it on the front porch, where the temperature is 0°C. Suppose that the temperature of the buttermilk has dropped to 15°C after 20 min. When will it be at 5°C?

28. When sugar is dissolved in water, the amount A that remains undissolved after t minutes satisfies the differential equation $dA/dt = -kA$ $(k > 0)$. If 25% of the sugar dissolves after 1 min, how long does it take for half the sugar to dissolve?

29. The intensity I of light at a depth x meters below the surface of a lake satisfies the differential equation $dI/dx = (-1.4)I$. (a) At what depth is the intensity half the intensity I_0 at the surface (where $x = 0$)? (b) What is the intensity at a depth of 10 m (as a fraction of I_0)? (c) At what depth will the intensity be $\frac{1}{100}$ of that at the surface?

30. The barometric pressure p (in inches of mercury) at an altitude x miles above sea level satisfies the initial value problem $dp/dx = (-0.2)p$; $p(0) = 29.92$. (a) Calculate the barometric pressure at 10,000 ft and again at 30,000 ft. (b) Without prior conditioning, few people can survive when the pressure drops to less than 15 in. of mercury. How high is that?

31. Consider a savings account that contains A_0 dollars initially and earns interest at the annual rate r compounded continuously. Suppose that deposits are added to this account at the rate of Q dollars per year. To simplify the mathematical model, assume that these deposits are made continuously rather than (for instance) monthly. (a) Derive the differential equation for the amount $A(t)$ in the account at time t years. (b) Suppose that you wish to arrange, at the time of her birth, for your daughter to have $40,000 available for college expenses at her eighteenth birthday. You plan to do so by making frequent small—essentially continuous—deposits in a savings account, at the rate of Q thousand dollars each year. This account accumulates interest at 11% annual interest compounded continuously. What should Q be so that you may achieve your goal?

32. According to one cosmological theory, there were equal amounts of the two uranium isotopes U^{235} and U^{238} at the creation of the universe in the "big bang." At present there are 137.7 U^{238} atoms for each atom of U^{235}. Using the half-lives 4.51 billion years for U^{238} and 0.71 billion years for U^{235}, calculate the age of the universe.

33. A cake is removed from an oven at 210°F and left to cool at room temperature, which is 70°F. After 30 min, the temperature of the cake is 140°F. When will it be 100°F?

34. (a) Payments are made on a mortgage (original loan) of P_0 dollars continuously at the constant rate of c dollars per month. Let $P(t)$ denote the principal (amount still owed) after t months, and let r denote the monthly interest rate paid by the borrower (for instance, $r = 0.12/12 = 0.01$ if the annual interest rate is 12%). Derive the differential equation

$$\frac{dP}{dt} = rP - c, \qquad P(0) = P_0.$$

(b) An automobile loan of $10,800 is to be paid off continuously over a period of 60 months. Determine the monthly payment required if the annual interest rate is (i) 12%; (ii) 18%.

35. A certain piece of dubious information about phenylthiourea in the drinking water began to spread one day in a city with a population of 100,000. Within a week, 10,000 people had heard this rumor. Assume that the rate of increase of the number who have heard the rumor is proportional to the number who have not yet heard it. How long will it be before half the population of the city has heard this piece of information?

36. A tank is shaped like a vertical cylinder; it initially contains water to a depth of 9 ft, and a bottom plug is pulled at time $t = 0$ (hours). After 1 h the depth has dropped to 4 ft. How long does it take for all the water to run out of this tank?

37. Suppose that the tank of Problem 36 has a radius of 3 ft and that its bottom hole is circular with radius 1 in. How long will it take the water (initially 9 ft deep) to drain completely?

38. A cylindrical tank with length 5 ft and radius 3 ft is situated with its axis horizontal. If a circular bottom hole with a radius of 1 in. is opened and the tank is initially half full of xylene, how long will it take for the liquid to drain completely?

39. A spherical tank with radius 4 ft is full of gasoline when a circular bottom hole with radius 1 in. is opened. How long will be required for all the gasoline to drain from the tank?

40. (The Clepsydra, or water clock) A 12-h water clock is to be designed with the dimensions shown in Fig. 1.31, shaped like the surface obtained by revolving the curve $y = f(x)$ around the y-axis. What should be this curve, *and* what should be the radius of the circular bottom hole, in order that the water level will fall at the *constant* rate of 4 inches per hour (in./h)?

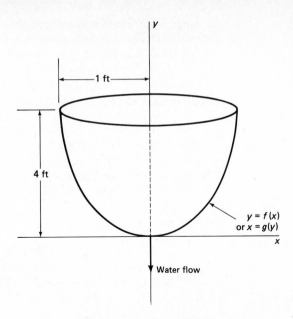

Figure 1.31 The clepsydra.

41. Suppose that a cylindrical tank initially containing V_0 gallons of water drains (through a bottom hole) in T minutes. Use Torricelli's law to show that the volume of water in the tank after $t \leq T$ minutes is $V = V_0[1 - (t/T)]^2$.

42. Early one morning it began to snow at a constant rate. At 7:00 A.M. a snowplow set off to clear a road. By 8:00 A.M. it had traveled 4 mi, and by 9:00 A.M. it had gone an additional 3 mi. What time did it start snowing? (*Note:* Assume that the snowplow clears snow from the road at a constant rate (in cubic feet per hour, say).) (*Answer:* 4:27 A.M.)

43. At time $t = 0$ the bottom plug (at the vertex) of a full conical water tank 16 ft high is removed. After 1 h the water in the tank is 9 ft deep. When will the tank be empty?

1.5
Linear First Order Equations

In Section 1.4 we saw how to solve a separable differential equation by integrating *after* multiplying each side by an appropriate factor. For instance, to solve the equation

$$\frac{dy}{dx} = 2xy \qquad (y > 0), \tag{1}$$

we multiply each side by the factor $1/y$ to get

$$\frac{1}{y}\frac{dy}{dx} = 2x; \quad \text{that is,} \quad D_x[\ln y] = D_x[x^2]. \tag{2}$$

Because each side of the equation in (2) is recognizable as a *derivative* (with respect to the independent variable x), all that remains is a simple integration, which yields $\ln y = x^2 + C$. For this reason, the function $\rho = 1/y$ is called an *integrating factor* for the original equation in (1). An **integrating factor** for a differential equation is a function $\rho(x, y)$ such that multiplication of each side of the differential equation by $\rho(x, y)$ yields an equation in which each side is recognizable as a derivative.

With the aid of the appropriate integrating factor, there is a standard technique for solving the **linear first order equation**

$$\frac{dy}{dx} + P(x)y = Q(x) \tag{3}$$

on an interval where the coefficient functions $P(x)$ and $Q(x)$ are continuous. We multiply each side in Eq. (3) by the integrating factor

$$\rho = \rho(x) = e^{\int P(x)\, dx}. \tag{4}$$

The result is

$$e^{\int P(x)\, dx} \frac{dy}{dx} + P(x)e^{\int P(x)\, dx}y = Q(x)e^{\int P(x)\, dx}. \tag{5}$$

Because $D_x \left[\int P(x)\, dx \right] = P(x)$, the left-hand side is the derivative of the *product* $y \cdot e^{\int P(x)\, dx}$, so (5) is equivalent to

$$D_x(y(x)e^{\int P(x)\, dx}) = Q(x)e^{\int P(x)\, dx}.$$

Integration of both sides of this equation gives

$$y(x)e^{\int P(x)\, dx} = \int \left(Q(x)e^{\int P(x)\, dx} \right) dx + C.$$

Finally solving for y, we obtain the general solution of the linear first order equation in (3):

$$y = y(x) = e^{-\int P(x)\, dx} \left[\int \left(Q(x)e^{\int P(x)\, dx} \right) dx + C \right]. \tag{6}$$

The formula in (6) need not be memorized. In a specific problem it is generally simpler to use the *method* by which we developed this formula. Begin by calculating the integrating factor given in Equation (4), $\rho = e^{\int P(x)\, dx}$. Then multiply each side of the differential equation by ρ, recognize the left-hand side of the resulting equation as the derivative of a product, integrate this equation, and finally solve for y. Moreover, given an initial condition $y(x_0) = y_0$, we can substitute $x = x_0$ and $y = y_0$ into (6) to solve for the value of C yielding the particular solution of (3) that satisfies this initial condition.

The integrating factor $\rho(x)$ is determined only to within a multiplicative constant. If we replace $\int P(x)\, dx$ by $\int P(x)\, dx + c$ in (4), the result is

$$\rho(x) = e^{(\int P(x)\, dx) + c} = e^c e^{\int P(x)\, dx}.$$

But the constant factor e^c does not affect the result of multiplying both sides of the differential equation in (3) by $\rho(x)$. Hence we may choose for $\int P(x)\, dx$ any convenient antiderivative of $P(x)$.

EXAMPLE 1 Solve the initial value problem

$$\frac{dy}{dx} - 3y = e^{2x}, \qquad y(0) = 3.$$

Solution   Here we have $P(x) = -3$ and $Q(x) = e^{2x}$, so the integrating factor is

$$\rho = e^{\int (-3)\, dx} = e^{-3x}.$$

Multiplication of each side of the given equation by e^{-3x} yields

$$e^{-3x}\frac{dy}{dx} - 3e^{-3x}y = e^{-x},$$

which we recognize as

$$\frac{d}{dx}(e^{-3x}y) = e^{-x}.$$

Hence integration with respect to x gives $e^{-3x}y = -e^{-x} + C$, so the general solution is

$$y = Ce^{3x} - e^{2x}.$$

Substitution of the initial condition, $(x_0, y_0) = (0, 3)$, produces the result $C = 4$. Thus the desired particular solution is

$$y = 4e^{3x} - e^{2x}.$$

EXAMPLE 2   Find the general solution of

$$(x^2 + 1)\frac{dy}{dx} + 3xy = 6x.$$

Solution   After division of each side of the equation by $x^2 + 1$, we recognize the result

$$\frac{dy}{dx} + \frac{3x}{x^2 + 1}y = \frac{6x}{x^2 + 1}$$

as a first order linear equation with $P(x) = 3x/(x^2 + 1)$ and $Q(x) = 6x/(x^2 + 1)$. Multiplication by

$$\rho = \exp\left(\int \frac{3x}{x^2 + 1}\, dx\right)$$
$$= \exp\left(\frac{3}{2}\ln(x^2 + 1)\right) = (x^2 + 1)^{3/2}$$

yields

$$(x^2 + 1)^{3/2}\frac{dy}{dx} + 3x(x^2 + 1)^{1/2}y = 6x(x^2 + 1)^{1/2},$$

and thus

$$D_x[(x^2 + 1)^{3/2}y] = 6x(x^2 + 1)^{1/2}.$$

Integration then yields

$$(x^2 + 1)^{3/2}y = \int 6x(x^2 + 1)^{1/2}\, dx = 2(x^2 + 1)^{3/2} + C.$$

Multiplication of both sides by $(x^2 + 1)^{-3/2}$ gives the general solution

$$y = 2 + C(x^2 + 1)^{-3/2}.$$

Frequently the integrations are not so simple as those in Examples 1 and 2. We may apply the fundamental theorem of calculus, however, to write an antiderivative of an arbitrary continuous function $f(x)$ in the form

$$\int_a^x f(t)\,dt.$$

EXAMPLE 3 Solve the initial value problem

$$x^2\frac{dy}{dx} + xy = \sin x, \qquad y(1) = 2.$$

Solution Division by x^2 gives

$$\frac{dy}{dx} + \frac{1}{x}y = \frac{\sin x}{x^2},$$

so the integrating factor is $\rho = e^{\int (1/x)\,dx} = e^{\ln x} = x$. Hence

$$x\frac{dy}{dx} + y = \frac{\sin x}{x};$$

$$D_x(xy) = \frac{\sin x}{x};$$

$$xy = \int_0^x \frac{\sin t}{t}\,dt + C.$$

The lower limit of integration $a = 0$ is permissible because $(\sin t)/t \longrightarrow 1$ as $t \longrightarrow 0$, and so the integral is not improper. Substitution of the initial condition $(x_0, y_0) = (1, 2)$ gives

$$C = 2 - \int_0^1 \frac{\sin t}{t}\,dt,$$

so

$$y = \frac{1}{x}\int_0^x \frac{\sin t}{t}\,dt + \frac{C}{x}$$

$$= \frac{1}{x}\int_0^x \frac{\sin t}{t}\,dt + \frac{2}{x} - \frac{1}{x}\int_0^1 \frac{\sin t}{t}\,dt$$

$$= \frac{2}{x} + \frac{1}{x}\int_1^x \frac{\sin t}{t}\,dt.$$

For example,

$$y(2) = 1 + \frac{1}{2}\int_1^2 \frac{\sin t}{t}\,dt.$$

In general, an integral such as this one would have to be approximated numerically (for example, by using Simpson's rule). In this case, however, we have the **sine integral** function

$$\mathrm{Si}(x) = \int_0^x \frac{\sin t}{t}\,dt,$$

which appears with sufficient frequency in applications that its values have been tabulated. A good set of tables of special functions is Abramowitz and Stegun, *Handbook of Mathematical Functions* (New

York: Dover, 1965). From Table 5.1 of this reference, we find that

$$y(2) = 1 + \tfrac{1}{2}[\text{Si}(2) - \text{Si}(1)]$$
$$\approx 1 + \tfrac{1}{2}(1.60541 - 0.94608) \approx 1.32967.$$

In the sequel we will see that it is the rule, rather than the exception, when a solution to a differential equation cannot be expressed in terms of elementary functions. We will study various devices for obtaining good approximations to the values of the nonelementary functions we encounter; in Chapter 6 we will discuss numerical integration of differential equations in some detail.

The derivation above of the solution in (6) of the linear first order equation in (3) bears a closer examination. Suppose that the functions $P(x)$ and $Q(x)$ are continuous on the (possibly unbounded) open interval I. Then the antiderivatives

$$\int P(x)\,dx \quad \text{and} \quad \int (Q(x)e^{\int P(x)\,dx})\,dx$$

exist on I. Our derivation of Eq. (6) shows that, *if $y = y(x)$ is a solution of Eq. (3) on I, then* $y(x)$ is given by the formula in (6) for some choice of the constant C. Conversely, you may verify by direct substitution (Problem 21) that the function $y(x)$ given in Equation (6) satisfies Eq. (3). Finally, given a point x_0 of I and any number y_0, there is (as previously noted) a unique value of C such that $y(x_0) = y_0$. Consequently we have the proved the following existence-uniqueness theorem.

THEOREM: THE LINEAR FIRST ORDER EQUATION

If the functions $P(x)$ and $Q(x)$ are continuous on the open interval I containing the point x_0, then the initial value problem

$$\frac{dy}{dx} + P(x)y = Q(x), \qquad y(x_0) = y_0$$

has a unique solution $y(x)$ on I, given by the formula in Equation (6) with an appropriate value of the constant C.

Remark 1: This theorem gives a solution on the *whole* interval I for a *linear* differential equation, in contrast with the basic existence-uniqueness theorem stated in Section 1.3, which guarantees only a solution on a possibly smaller interval J.

Remark 2: The theorem above tells us that *every* solution of Eq. (3) is included in the general solution given in (6). Thus a *linear* first order differential equation has *no* singular solutions.

EXAMPLE 4 If we write the equation

$$x\frac{dy}{dx} + y = 1$$

in the standard form shown in (3), we see that it has coefficient functions $P(x) = Q(x) = 1/x$ that are continuous on the intervals $x < 0$ and

$x > 0$, but discontinuous at $x = 0$. Because the equation as written is equivalent to $D_x(xy) = 1$, integration gives $xy = x + C$; thus

$$y = 1 + \frac{C}{x}.$$

Some typical **solution curves**—graphs of solutions—are shown in Fig. 1.32. The solution with $y(1) = 2$ is $y = 1 + (1/x)$ for $x > 0$, while the solution with $y(-1) = 2$ is $y = 1 - (1/x)$ for $x < 0$. The only solution that is continuous on the whole real line is the constant solution $y(x) \equiv 1$. We see also that, if $y_0 \neq 1$, there is *no* solution with $y(0) = y_0$. Note that, in accord with the theorem, all solutions are continuous except possibly at the discontinuity $x = 0$ of $P(x)$ and $Q(x)$, and that most solutions are singular at the singularity $x = 0$ of the differential equation.

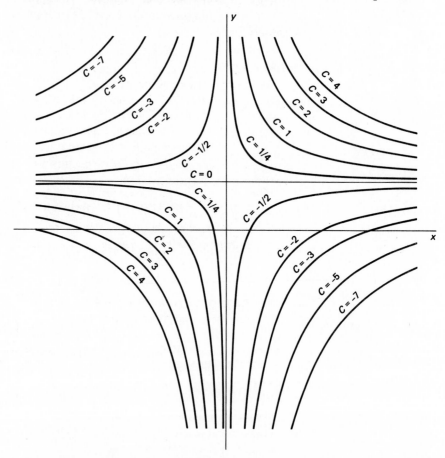

Figure 1.32 Graphs of some solutions for Example 4.

MIXTURE PROBLEMS

As a first application of linear first order equations, we consider a tank containing a solution—a mixture of solute and solvent—such as salt dissolved in water. There is both inflow and outflow, and we want to compute the *amount*

$x(t)$ of solute in the tank at time t, given the amount $x(0) = x_0$ at time $t = 0$. Suppose that solution with a concentration of c_i grams of solute per liter of solution flows into the tank at the constant rate of r_i liters per second, and that the solution in the tank—kept thoroughly mixed by stirring—flows out at the constant rate of r_o liters per second.

To set up a differential equation for $x(t)$, we estimate the change Δx in x during the brief time interval $[t, t + \Delta t]$. The amount of solute that flows into the tank during Δt seconds is $r_i c_i \Delta t$ grams. To check this, note how the cancellation of dimensions checks our computation:

$$\left(r_i \frac{\text{liters}}{\text{second}} \right) \left(c_i \frac{\text{grams}}{\text{liter}} \right) (\Delta t \text{ seconds})$$

yields a quantity measured in grams.

The amount of solute that flows out of the tank during the same time interval depends upon the concentration $c_o(t)$ in the tank at time t. But, as noted in Fig. 1.33, $c_o(t) = x(t)/V(t)$, where $V(t)$ denotes the volume (not constant unless $r_i = r_o$) of solution in the tank at time t. Then

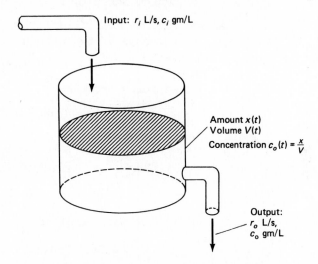

Input: r_i L/s, c_i gm/L

Amount $x(t)$
Volume $V(t)$
Concentration $c_o(t) = \frac{x}{V}$

Output:
r_o L/s,
c_o gm/L

Figure 1.33 The single-tank mixture problem.

$$\Delta x = \{\text{grams input}\} - \{\text{grams output}\}$$
$$\approx r_i c_i \, \Delta t - r_o c_o \, \Delta t$$
$$= r_i c_i \, \Delta t - r_o \frac{x}{V} \, \Delta t.$$

We now divide by Δt:

$$\frac{\Delta x}{\Delta t} \approx r_i c_i - r_o \frac{x}{V}.$$

Finally, we take the limit as $\Delta t \to 0$; if all the functions involved are continuous and x is differentiable, then the error in the approximation also approaches zero, and we obtain the differential equation

$$\frac{dx}{dt} = r_i c_i - \frac{r_o}{V} x. \tag{7}$$

If $V_0 = V(0)$, then $V(t) = V_0 + (r_i - r_o)t$, so Equation (7) is a linear first order differential equation for the amount $x(t)$ of solute in the tank.

Important: Equation (7) is *not* one you should commit to memory. It is the process we used to obtain that equation—examination of the behavior of the system over a short time interval $[t, t + \Delta t]$—that you should strive to understand, because it is a very useful tool for obtaining all sorts of differential equations.

EXAMPLE 5 Lake Erie has a volume of 458 km³, and its rate of inflow and outflow are both 175 km³/year. Suppose that at the time $t = 0$ (years), its pollutant concentration is 0.05%, and that thereafter the concentration of pollutants in the inflowing water is 0.01%. Assuming that the outflow is perfectly mixed lake water, how long will it take to reduce the pollution concentration in the lake to 0.02%?

Solution Here we have $V = 458$ (km³) and $r_i = r_o = 175$ (km³/year). Let $x(t)$ denote the volume of pollutants in the lake at time t. We are given

$$x(0) = (0.0005)(458) = 0.2290 \text{ (km}^3\text{)},$$

and we want to find when

$$x(t) = (0.0002)(458) = 0.0916 \text{ (km}^3\text{)}.$$

The change Δx in Δt years is

$$\Delta x \approx (0.0001)(175)\,\Delta t - \left(\frac{x}{458}\right)(175)\,\Delta t$$

$$= (0.0175 - 0.3821x)\,\Delta t,$$

so our differential equation is

$$\frac{dx}{dt} + (0.3821)x = 0.0175.$$

The integrating factor is $e^{(0.3821)t}$; it yields

$$D_t[e^{(0.3821)t}x] = (0.0175)e^{(0.3821)t};$$

$$[e^{(0.3821)t}]x = (0.0458)e^{(0.3821)t} + C.$$

Substitution of $x(0) = 0.2290$ into the last equation gives

$$C = 0.2290 - 0.0458 = 0.1832,$$

so the solution is

$$x(t) = 0.0458 + (0.1832)e^{-(0.3821)t}.$$

Finally, we solve the equation

$$0.0916 = 0.0458 + (0.1832)e^{-(0.3821)t}$$

for

$$t = \frac{-1}{0.3821} \ln \frac{0.0916 - 0.0458}{0.1832} \approx 3.63.$$

The answer to the problem, then, is after about 3.63 years.

EXAMPLE 6 A 120-gallon (gal) tank initially contains 90 lb of salt dissolved in 90 gal of water. Brine containing 2 lb/gal of salt flows into the tank at the rate of 4 gal/min, and the mixture flows out of the tank at the rate of 3 gal/min. How much salt does the tank contain when it is full?

Solution The interesting feature of this example is that, due to the differing rates in inflow and outflow, the volume of brine in the tank increases steadily with $V(t) = 90 + t$ (gallons). The change Δx in the amount x of salt in the tank from time t to time $t + \Delta t$ (minutes) is given by

$$\Delta x \approx (4)(2) \, \Delta t - 3\left(\frac{x}{90 + t}\right) \Delta t,$$

so our differential equation is

$$\frac{dx}{dt} + \frac{3}{90 + t} x = 8.$$

The integrating factor is

$$\exp \left(\int \frac{3 \, dt}{90 + t} \right) = e^{3 \ln (90 + t)} = (90 + t)^3,$$

which gives

$$D_t[(90 + t)^3 x] = 8(90 + t)^3;$$
$$(90 + t)^3 x = 2(90 + t)^4 + C.$$

Substitution of $x(0) = 90$ gives $C = -(90)^4$, so the amount of salt in the tank at time t is

$$x(t) = 2(90 + t) - \frac{(90)^4}{(90 + t)^3}.$$

The tank is full after 30 min, and when $t = 30$ we have

$$x(30) = 2(90 + 30) - \frac{(90)^4}{(120)^3} \approx 202 \text{ (lb)}$$

of salt.

1.5 Problems

Find general solutions of the differential equations in Problems 1–15. If an initial condition is given, find the corresponding particular solution. Throughout, primes denote derivatives with respect to x.

1. $xy' + y = 3xy$; $y(1) = 0$.
2. $xy' + 3y = 2x^5$; $y(2) = 1$.
3. $y' + y = e^x$; $y(0) = 1$.

4. $xy' - 3y = x^3$; $\quad y(1) = 10$.

5. $y' + 2xy = x$; $\quad y(0) = -2$.

6. $y' = (1 - y)\cos x$; $\quad y(\pi) = 2$.

7. $(1 + x)y' + y = \cos x$; $\quad y(0) = 1$.

8. $xy' = 2y + x^3 \cos x$.

9. $y' + y \cot x = \cos x$.

10. $y' = 1 + x + y + xy$; $\quad y(0) = 0$.

11. $xy' = 3y + x^4 \cos x$; $\quad y(2\pi) = 0$.

12. $y' = 2xy + 3x^2 \exp(x^2)$; $\quad y(0) = 5$.

13. $xy' + (2x - 3)y = 4x^4$.

14. $(x^2 + 4)y' + 3xy = x$; $\quad y(0) = 1$.

15. $(x^2 + 1)y' + 3x^3 y = 6x \exp\left(-\dfrac{3x^2}{2}\right)$; $\quad y(0) = 1$.

Solve the differential equations in Problems 16–18 by regarding y as the independent variable rather than x.

16. $(1 - 4xy^2)y' = y^3$

17. $(x + ye^y)y' = 1$.

18. $(1 + 2xy)y' = 1 + y^2$.

19. Express the general solution of $y' = 1 + 2xy$ in terms of the **error function**

$$\operatorname{erf}(x) = \frac{2}{\sqrt{\pi}} \int_0^x e^{-t^2}\, dt.$$

20. Express the solution of the initial value problem

$$2xy' = y + 2x \cos x, \qquad y(1) = 0$$

as an integral as in Example 3.

21. (a) Show that $y_c(x) = Ce^{-\int P(x)\, dx}$ is a general solution of $y' + P(x)y = 0$. (b) Show that

$$y_p(x) = e^{-\int P(x)\, dx}\left[\int \left(Q(x)e^{\int P(x)\, dx} \right) dx \right]$$

is a particular solution of $y' + P(x)y = Q(x)$. (c) If $y_c(x)$ is any general solution of $y' + P(x)y = 0$ and $y_p(x)$ is any particular solution of $y' + P(x)y = Q(x)$, then show that $y(x) = y_c(x) + y_p(x)$ is a general solution of $y' + P(x)y = Q(x)$.

22. (a) Find constants A and B such that $y_p(x) = A \sin x + B \cos x$ is a solution of $y' + y = 2 \sin x$. (b) Use the result of (a) and the method of Problem 21 to find the general solution of $y' + y = 2 \sin x$. (c) Solve the initial value problem $y' + y = 2 \sin x, y(0) = 1$.

23. A tank contains 1000 liters (L) of a solution consisting of 100 kg of salt dissolved in water. Pure water is pumped into the tank at the rate of 5 L/s, and the mixture—kept uniform by stirring—is pumped out at the same rate. How long will it be until only 10 kg of salt remain in the tank?

24. Consider a reservoir with a volume of 8 billion cubic feet (ft^3) and an initial pollutant concentration of 0.25%. There is a daily inflow of 500 million ft^3 of water with a pollutant concentration of 0.05% and an equal daily outflow of the well-mixed water of the reservoir. How long will it take to reduce the pollutant concentration in the reservoir to 0.10%?

25. Rework Example 5 for the case of Lake Ontario; the only differences are that this lake has a volume of 1636 km³ and an inflow-outflow rate of 209 km³/year.

26. A tank initially contains 60 gal of pure water. Brine containing 1 lb of salt per gallon enters the tank at 2 gal/min, and the (perfectly mixed) solution leaves the tank at 3 gal/min; the tank is empty after exactly 1 h. (a) Find the amount of salt in the tank after t minutes. (b) What is the maximum amount of salt ever in the tank?

27. A 400-gal tank initially contains 100 gal of brine containing 50 lb of salt. Brine containing 1 lb of salt per gallon enters the tank at the rate of 5 gal/s, and the mixed brine in the tank flows out at the rate of 3 gal/s. How much salt will the tank contain when it is full of brine?

28. Consider the *cascade* of two tanks shown in Fig. 1.34, with $V_1 = 100$ (gal) and $V_2 = 200$ (gal) the volumes of brine in the two tanks. Each tank initially contains 50 lb of salt. The three flow rates are each 5 gal/s, with pure water flowing into Tank 1. (a) Find the amount $x(t)$ of salt in Tank 1 at time t. (b) Suppose that $y(t)$ is the amount of salt in Tank 2 at time t. Show first that

$$\frac{dy}{dt} = \frac{5x}{100} - \frac{5y}{200}.$$

and then solve for $y(t)$, using the value of $x(t)$ found in part (a). (c) Finally, find the maximum amount of salt ever in Tank 2.

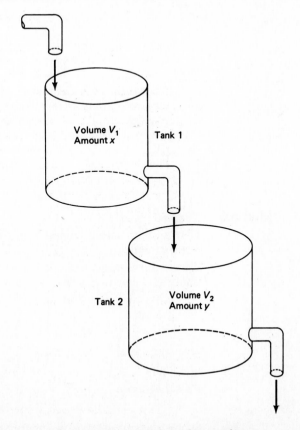

Figure 1.34 A cascade of two tanks.

29. Suppose that in the cascade shown in Fig. 1.34, Tank 1 initially contains 100 gal of pure ethyl alcohol and Tank 2 initially contains 100 gal of pure water. Pure water flows into Tank 1 at 10 gal/min, and the other two flow rates are also 10 gal/min. (a) Find the amounts $x(t)$ and $y(t)$ of alcohol in the two tanks. (b) Find the maximum amount of alcohol ever in Tank 2.

30. A multiple cascade is shown in Fig. 1.35. At time $t = 0$, Tank 0 contains 1 gal of alcohol and 1 gal of water, while each of the other tanks contains 2 gal of pure water. Pure water is pumped into Tank 0 at 1 gal/min, and the varying mixture in each tank is pumped into the one to its right at the same rate.

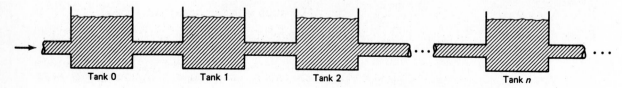

Tank 0   Tank 1   Tank 2   Tank n

Figure 1.35 A multiple cascade.

Assume, as usual, that the mixtures are kept perfectly uniform by stirring. Let $x_n(t)$ denote the amount of alcohol in Tank n at time t. (a) Show that $x_0(t) = e^{-t/2}$. (b) Show by induction on n that

$$x_n(t) = \frac{t^n e^{-t/2}}{n! \, 2^n} \quad \text{for } n > 0.$$

(c) Show that the maximum value of $x_n(t)$ for $n > 0$ is $M_n = n^n e^{-n}/n!$. (d) Conclude from **Stirling's approximation** $n! \approx (2\pi n)^{1/2} n^n e^{-n}$ that $M_n \approx (2\pi n)^{-1/2}$.

31. A 30-year-old woman accepts an engineering position with a starting salary of \$30,000 per year. Her salary $S(t)$ increases exponentially, with $S(t) = 30 e^{t/20}$ thousand dollars after t years. Meanwhile, 12% of her salary is deposited continuously in a retirement account, which accumulates interest at a continuous annual rate of 6%. (a) Estimate ΔA in terms of Δt to derive the differential equation for the amount $A(t)$ in her retirement account after t years. (b) Compute $A(40)$, the amount available for her retirement at age 70.

1.6
Substitution Methods

The first order differential equations we have solved in the previous sections have all been either separable or linear. But many applications involve differential equations that are neither separable nor linear. In this section we discuss substitution methods that sometimes can be used to transform a given differential equation into one that we already know how to solve.

Given a differential equation

$$\frac{dy}{dx} = f(x, y), \tag{1}$$

with dependent variable y and independent variable x, we can transform it into a new differential equation with the new dependent variable v by making the substitution

$$y = \phi(x, v). \tag{2}$$

Such a substitution may be suggested by the particular form of the differential equation, as in Example 1 below. At any rate, note that if the function ϕ has partial derivatives ϕ_x and ϕ_v with respect to x and v, then application of the chain rule—regarding v as an (unknown) function of x—yields

$$\frac{dy}{dx} = \phi_x(x, v) + \phi_v(x, v)\frac{dv}{dx}. \qquad (3)$$

Substitution of (2) and (3) in (1) gives

$$\phi_x(x, v) + \phi_v(x, v)\frac{dv}{dx} = f(x, \phi(x, v)). \qquad (4)$$

This is our *new* differential equation with *new* dependent variable v.

If $v = v(x)$ is a solution of Eq. (4), then $y = \phi(x, v(x))$ is a solution of the original Equation (1). The trick is to select a substitution $y = \phi(x, v)$ such that the transformed Equation (4) is one we can solve. Even when possible, this is not always easy; it may require a fair amount of ingenuity or trial-and-error.

Sometimes it is more convenient to begin with a relation $v = \psi(x, y)$ that can be solved for $y = \phi(x, v)$. Occasionally, as in the following example, the presence of a conspicuous combination of x and y in the differential equation will suggest a promising substitution.

EXAMPLE 1   Solve the differential equation

$$\frac{dy}{dx} = (x + y + 3)^2.$$

Solution   It seems reasonable to try the substitution

$$v = x + y + 3: \quad \text{that is,} \quad y = v - x - 3.$$

Then

$$\frac{dy}{dx} = \frac{dv}{dx} - 1,$$

so our transformed equation is

$$\frac{dv}{dx} = 1 + v^2.$$

This is a separable equation, and we have no difficulty in obtaining its solution

$$x = \int \frac{dv}{1 + v^2} = \tan^{-1} v + C.$$

So $v = \tan(x - C)$. Because $v = x + y + 3$, the general solution of the original equation $y' = (x + y + 3)^2$ is $x + y + 3 = \tan(x - C)$; that is,

$$y = \tan(x - C) - x - 3.$$

Example 1 illustrates the fact that any differential equation of the form

$$\frac{dy}{dx} = F(ax + by + c) \qquad (5)$$

can be transformed into a separable equation by use of the substitution $v = ax + by + c$; see Problem 36.

HOMOGENEOUS EQUATIONS

A **homogeneous** first order differential equation is one that can be written in the form

$$\frac{dy}{dx} = F\left(\frac{y}{x}\right).\tag{6}$$

If we make the substitutions

$$v = \frac{y}{x}, \qquad y = vx, \qquad \frac{dy}{dx} = v + x\frac{dv}{dx},\tag{7}$$

then Eq. (6) is transformed into the *separable* equation

$$x\frac{dv}{dx} = F(v) - v.$$

Thus any homogeneous first order differential equation can be reduced to an integration problem by means of the substitutions in (7).

EXAMPLE 2 Solve the differential equation

$$2xy\frac{dy}{dx} = 4x^2 + 3y^2.$$

Solution This equation is neither separable nor linear, but we recognize it as a homogeneous equation by writing it in the form

$$\frac{dy}{dx} = \frac{4x^2 + 3y^2}{2xy} = 2\left(\frac{x}{y}\right) + \frac{3}{2}\left(\frac{y}{x}\right).$$

The substitutions in (7) then take the form

$$y = vx, \quad \frac{dy}{dx} = v + x\frac{dv}{dx},$$

$$v = \frac{y}{x}, \quad \text{and} \quad \frac{1}{v} = \frac{x}{y}.$$

These yield

$$v + x\frac{dv}{dx} = \frac{2}{v} + \frac{3}{2}v,$$

and hence

$$x\frac{dv}{dx} = \frac{2}{v} + \frac{v}{2} = \frac{v^2 + 4}{2v}:$$

$$\int \frac{2v}{v^2 + 4}\,dv = \int \frac{1}{x}\,dx;$$

$$\ln(v^2 + 4) = \ln|x| + \ln C.$$

We apply the exponential function to each side of the last equation to obtain

$$v^2 + 4 = C|x|:$$

$$\frac{y^2}{x^2} + 4 = C|x|:$$

$$y^2 + 4x^2 = Cx^3,$$

because the sign of x can be absorbed by the arbitrary constant C.

EXAMPLE 3 Solve the initial value problem

$$x\frac{dy}{dx} = y + (x^2 - y^2)^{1/2}, \qquad y(1) = 0.$$

Solution We divide both sides by x and find that

$$\frac{dy}{dx} = \frac{y}{x} + \left[1 - \left(\frac{y}{x}\right)^2\right]^{1/2},$$

so we make the substitutions in (7); we get

$$v + x\frac{dv}{dx} = v + [1 - v^2]^{1/2};$$

$$\int \frac{1}{(1 - v^2)^{1/2}}\, dv = \int \frac{1}{x}\, dx;$$

$$\sin^{-1} v = \ln x + C.$$

There is no need to use $\ln |x|$ because $x > 0$ near $x = 1$ (part of the given initial condition). Also, $v = y/x = 0$ at the initial condition $(x_0, y_0) = (1, 0)$, so $C = \sin^{-1} 0 = 0$. Hence

$$v = \frac{y}{x} = \sin (\ln x),$$

and therefore $y = x \sin(\ln x)$ is the desired solution.

BERNOULLI EQUATIONS

A first order differential equation of the form

$$\frac{dy}{dx} + P(x)y = Q(x)y^n \tag{8}$$

is called a **Bernoulli equation**. If either $n = 0$ or $n = 1$, then (8) is linear. Otherwise, as we ask you to show in Problem 37, the substitution

$$v = y^{1-n} \tag{9}$$

transforms (8) into the *linear* equation

$$\frac{dv}{dx} + (1 - n)P(x)v = (1 - n)Q(x).$$

Rather than memorizing the form of this transformed equation, it is more efficient to make the substitution in (9) explicitly, as in the following examples.

EXAMPLE 4 If we rewrite the homogeneous equation $2xyy' = 4x^2 + 3y^2$ of Example 2 in the form

$$\frac{dy}{dx} - \frac{3}{2x}y = \frac{2x}{y},$$

we see that it is also a Bernoulli equation with $P(x) = -3/(2x)$, $Q(x) = 2x$, $n = -1$, and $1 - n = 2$. Hence we substitute

$$v = y^2, \qquad y = v^{1/2}, \quad \text{and} \quad \frac{dy}{dx} = \frac{dy}{dv}\frac{dv}{dx} = \frac{1}{2}v^{-1/2}\frac{dv}{dx}.$$

This gives

$$\frac{1}{2}v^{-1/2}\frac{dv}{dx} - \frac{3}{2x}v^{1/2} = 2xv^{-1/2}.$$

Multiplication by $2v^{1/2}$ produces the linear equation

$$\frac{dv}{dx} - \frac{3}{x}v = 4x$$

with integrating factor $\rho = e^{\int (-3/x)\,dx} = x^{-3}$. So we obtain

$$D_x(x^{-3}v) = \frac{4}{x^2};$$

$$x^{-3}v = -\frac{4}{x} + C;$$

$$x^{-3}y^2 = -\frac{4}{x} + C;$$

$$y^2 = -4x^2 + Cx^3.$$

EXAMPLE 5   The equation $x(dy/dx) + 6y = 3xy^{4/3}$ is neither separable nor linear nor homogeneous, but it is a Bernoulli equation with $n = \frac{4}{3}$, $1 - n = -\frac{1}{3}$. The substitutions

$$v = y^{-1/3}, \qquad y = v^{-3}, \quad \text{and} \quad \frac{dy}{dx} = \frac{dy}{dv}\frac{dv}{dx} = -3v^{-4}\frac{dv}{dx}$$

transform it into

$$-3xv^{-4}\frac{dv}{dx} + 6v^{-3} = 3xv^{-4}.$$

Division by $-3xv^{-4}$ yields the linear equation

$$\frac{dv}{dx} - \frac{2}{x}v = -1$$

with integrating factor $\rho = e^{\int (-2/x)\,dx} = x^{-2}$. This gives

$$D_x(x^{-2}v) = -\frac{1}{x^2};$$

$$x^{-2}v = \frac{1}{x} + C;$$

$$v = x + Cx^2;$$

and finally

$$y = \frac{1}{(x + Cx^2)^3}.$$

REDUCIBLE SECOND ORDER EQUATIONS

The general form of a second order differential equation is

$$F(x, y, y', y'') = 0. \tag{10}$$

If either the dependent variable y or the independent variable x is missing from a second order equation, then it is easily reduced to a first order equation by a change of variables.

If the dependent variable y is missing, so that the equation has the form

$$F(x, y', y'') = 0, \tag{11}$$

then we make the substitutions

$$p = y' = \frac{dy}{dx}, \qquad y'' = \frac{dp}{dx}. \tag{12}$$

These transform (11) into the first order equation

$$F\left(x, p, \frac{dp}{dx}\right) = 0$$

for p as a function of x. If we can find its general solution $p = p(x, C_1)$, then we get the general solution of the original second order equation in (11) by integrating:

$$y = \int y'(x)\, dx = \int p\, dx = \int p(x, C_1)\, dx + C_2.$$

Note that the solution above contains *two* arbitrary constants. This is to be expected—solving a second order equation would normally require two integrations.

EXAMPLE 6 Solve the equation $xy'' + 2y' = 6x$.

Solution The substitutions in (12) give

$$x\frac{dp}{dx} + 2p = 6x; \quad \text{that is,} \quad \frac{dp}{dx} + \frac{2}{x}p = 6.$$

Upon multiplying both sides of the linear equation (on the right) by its integrating factor x^2, we get

$$D_x(x^2 p) = 6x^2;$$
$$x^2 p = 2x^3 + C_1;$$
$$\frac{dy}{dx} = p = 2x + \frac{C_1}{x^2}.$$

Another integration produces the solution:

$$y = x^2 - \frac{C_1}{x} + C_2.$$

If the independent variable x is missing, so that the equation has the form

$$F(y, y', y'') = 0, \tag{13}$$

then we make the substitutions

$$p = y'(x), \qquad y'' = \frac{dp}{dx} = \frac{dp}{dy}\frac{dy}{dx} = p\frac{dp}{dy}. \tag{14}$$

These transform (13) into the first order equation

$$F\left(y, p, p\frac{dp}{dy}\right) = 0$$

for p as a function of y. If we can find its general solution $p = p(y, C_1)$, then we get the general solution of the original second order equation by integrating:

$$x = \int \frac{dx}{dy}\, dy = \int \frac{1}{p}\, dy = \int \frac{dy}{p(y, C_1)} + C_2.$$

For the same reason as in the first case, the final solution contains two arbitrary constants of integration. Note also that the solution expresses x as a function of y. Thus we obtain an implicit solution of (13) that is valid where $p = y'(x) \neq 0$.

EXAMPLE 7 Solve the equation $yy'' = (y')^2$ under the assumption that y and y' are known to be positive.

Solution The substitutions in (14) yield

$$yp\frac{dp}{dy} = p^2,$$

so

$$\int \frac{dp}{p} = \int \frac{dy}{y};$$

$$\ln p = \ln y + C \qquad \text{(because } y > 0 \text{ and } p > 0\text{)};$$

$$p = C_1 y$$

where $C_1 = e^C$. Thus

$$\frac{dx}{dy} = \frac{1}{p} = \frac{1}{C_1 y};$$

$$C_1 x = \int \frac{dy}{y} = \ln y + C_2.$$

Hence the general solution of $yy'' = (y')^2$ is

$$y = \exp\left(C_1 x - C_2\right) = A e^{Bx}$$

where $A = \exp(-C_2)$ and $B = C_1$.

THE HANGING CABLE

As an application, we now investigate the shape of a hanging cable made of a perfectly flexible yet inelastic material. We assume that the only *external* force acting on the cable (except for the supporting forces at its two ends) is the gravitational force of its own weight. We make the simple assumption that the weight of the cable is uniformly distributed along its length, and thus is *not* uniformly distributed horizontally.

Let w denote the density of the cable, measured in units such as pounds per foot of length. Fig. 1.36 shows the cable with lowest point P and with a coordinate system set up so that the y-axis passes through the lowest point of the cable and the x-axis is somewhere below that point.

Now consider a section PQ of the cable of length s. We plan to obtain a differential equation for the shape $y = f(x)$ of the hanging cable by balancing horizontal and vertical components of the forces acting on PQ. These forces are:

T_0: the horizontal tension pulling on the cable at P;

T: the tangential tension pulling on the cable at Q;

ws: the force of gravity pulling downward on the section PQ.

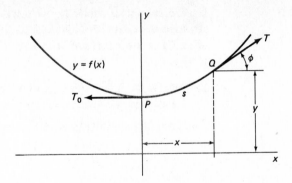

Figure 1.36 The hanging cable.

When we equate the horizontal and vertical components, we find that

$$T \cos \phi = T_0 \quad \text{and} \quad T \sin \phi = ws, \tag{15}$$

where, as indicated in Figure 1.36, ϕ is the angle the tangent to the cable at Q makes with the horizontal.

We divide the second equation above by the first to find that

$$\frac{dy}{dx} = \tan \phi = \frac{T \sin \phi}{T \cos \phi} = \frac{ws}{T_0}. \tag{16}$$

Because s is a function of x, this differential equation is not as simple as we might hope. But after we differentiate both sides, we find that

$$\frac{d^2 y}{dx^2} = \frac{w}{T_0} \frac{ds}{dx}.$$

But $ds/dx = [1 + (dy/dx)^2]^{1/2}$, and so

$$\frac{d^2 y}{dx^2} = \frac{w}{T_0} \left[1 + \left(\frac{dy}{dx} \right)^2 \right]^{1/2}. \tag{17}$$

This is the differential equation we must solve to find the shape $y = y(x)$ of the hanging cable.

Because the dependent variable y is missing, we substitute

$$p = \frac{dy}{dx} \quad \text{and} \quad \frac{dp}{dx} = \frac{d^2 y}{dx^2}$$

in Eq. (17). This yields the simpler equation

$$\frac{dp}{dx} = \frac{w}{T_0} (1 + p^2)^{1/2},$$

which we rewrite in the form

$$\frac{1}{(1 + p^2)^{1/2}} \frac{dp}{dx} = \frac{w}{T_0}.$$

Because $D_x \sinh^{-1} x = (1 + x^2)^{-1/2}$, integration of both sides of the equation above gives

$$\sinh^{-1} p = \frac{wx}{T_0} + C_1.$$

Now $p = 0$ when $x = 0$ because the cable has a horizontal tangent at its lowest point P. This tells us that $C_1 = \sinh^{-1} 0 = 0$, and thus that $\sinh^{-1} p = wx/T_0$. Therefore

$$\frac{dy}{dx} = p = \sinh\left(\frac{wx}{T_0}\right).$$

Finally, we integrate both sides of this last equation, and thus we find that

$$y = \frac{T_0}{w}\cosh\left(\frac{wx}{T_0}\right) + C_2.$$

If y_0 is the height of the point P above the x-axis, then $C_2 = y_0 - T_0/w$, and so the shape of the hanging cable is given by

$$y = \frac{T_0}{w}\cosh\left(\frac{wx}{T_0}\right) + y_0 - \frac{T_0}{w}. \tag{18}$$

We may choose the location of the x-axis so that $y_0 = T_0/w$, and then Eq. (18) takes the simple form

$$y = \frac{T_0}{w}\cosh\left(\frac{wx}{T_0}\right), \tag{19}$$

or, if we let $a = T_0/w$, the form $y = a\cosh(x/a)$. This curve is frequently called a *catenary*, from the Latin word *catena* (*chain*—a good physical approximation to a cable that is both perfectly flexible and inelastic).

Problems 46–54 deal with the relationship between the length S of the cable, the distance $2L$ between the two points from which it is suspended (at equal heights), and the dip or sag H of the cable at its middle. All these lengths are shown in Fig. 1.37.

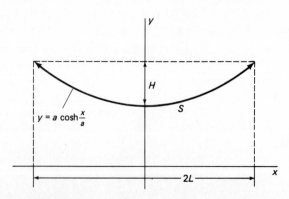

Figure 1.37 Parameters for Problems 46–54.

1.6 Problems

Find general solutions of the differential equations in Problems 1–35. Primes denote derivatives with respect to x throughout.

1. $(x + y)y' = x - y$.

2. $2xyy' = x^2 + 2y^2$.

3. $xy' = y + 2(xy)^{1/2}$.

4. $(x - y)y' = x + y$.

5. $x(x + y)y' = y(x - y)$.

6. $(x + 2y)y' = y$.

7. $xy^2y' = x^3 + y^3$.

8. $x^2y' = xy + x^2e^{y/x}$.

9. $x^2y' = xy + y^2$.

10. $xyy' = x^2 + 3y^2$.

11. $(x^2 - y^2)y' = 2xy$.

12. $xyy' = y^2 + x(4x^2 + y^2)^{1/2}$.

13. $xy' = y + (x^2 + y^2)^{1/2}$.

14. $yy' + x = (x^2 + y^2)^{1/2}$.

15. $x(x + y)y' + y(3x + y) = 0$.

16. $y' = (x + y + 1)^{1/2}$.

17. $y' = (2x + y)^2$.

18. $(x + y)y' = 1$.

19. $x^2y' + 2xy = 5y^3$.

20. $y^2y' + 2xy^3 = 6x$.

21. $y' = y + y^3$.

22. $x^2y' + 2xy = 5y^4$.

23. $xy' + 6y = 3xy^{4/3}$.

24. $2xy' + y^3 \cos x = 2xy$.

25. $y^2(xy' + y)(1 + x^4)^{1/2} = x$.

26. $y'' + 2y' = 0$.

27. $xy'' = y'$.

28. $yy'' + (y')^2 = 0$ (assume y and y' positive).

29. $y'' + 4y = 0$.

30. $xy'' + y' = 4x$.

31. $y'' = (y')^2$ (assume y and y' positive).

32. $x^2y'' + 3xy' = 2$.

33. $yy'' + (y')^2 = yy'$ (assume y and y' positive).

34. $y'' = (x + y')^2$.

35. $y'' = 2y(y')^3$.

36. Show that the substitution $v = ax + by + c$ transforms the differential equation $y' = F(ax + by + c)$ into a separable equation.

37. Suppose that $n \neq 0$ and $n \neq 1$. Show that the substitution $v = y^{1-n}$ transforms the Bernoulli equation $y' + P(x)y = Q(x)y^n$ into the linear equation

$$v' + (1 - n)P(x)v = (1 - n)Q(x).$$

38. Substitute $p = x'(t)$ to derive the general solution $x = A \cos \omega t + B \sin \omega t$ of the differential equation $x''(t) + \omega^2 x(t) = 0$.

39. Show that the substitution $v = \ln y$ transforms the differential equation $y' + P(x)y = Q(x)y \ln y$ into the linear equation $v' + P = Qv$.

40. Use the method of Problem 39 to solve the equation $xy' - 4x^2y + 2y \ln y = 0$.

41. Solve the equation

$$\frac{dy}{dx} = \frac{x - y - 1}{x + y + 3}$$

by finding h and k so that the substitutions $x = u + h$, $y = v + k$ transform it into the homogeneous equation

$$\frac{dv}{du} = \frac{u - v}{u + v}.$$

42. Use the method of Problem 41 to solve the equation

$$\frac{dy}{dx} = \frac{2y - x + 7}{4x - 3y - 18}.$$

43. The equation $dy/dx = A(x)y^2 + B(x)y + C(x)$ is called a **Riccati equation**. Suppose that one particular solution $y_1(x)$ of this equation is known. Show that the substitution

$$y = y_1 + \frac{1}{v}$$

transforms it into the *linear* equation

$$\frac{dv}{dx} + (B + 2Ay_1)v = -A.$$

Use the method of Problem 43 to solve the equations in Problems 44 and 45, given that $y_1 = x$ is a solution of each.

44. $y' + y^2 = 1 + x^2$.

45. $y' + 2xy = 1 + x^2 + y^2$.

46. Consider a hanging cable, of the sort to which Eq. (19) applies. Show that the length of the section lying above the interval $[0, x]$ is $s = (T_0/w) \sinh (wx/T_0)$.

47. Deduce from Eq. (19) that the sag of the cable at its middle is

$$H = \frac{T_0}{w} \left[\cosh \left(\frac{wL}{T_0} \right) - 1 \right].$$

48. Deduce from Eq. (15) that the tension in the cable satisfies the equation $T^2 = T_0^2 + w^2 s^2$.

49. Use the results of Problems 46 and 48 to show that the tension in the cable at the point (x, y) is $T = T_0 \cosh (wx/T_0) = wy$, the weight of a cable of the same density and of length y.

50. A high-voltage transmission line 205 ft long and weighing 1 lb/ft is strung between two towers 200 ft apart. (a) Use the result of Problem 46 to find the minimum tension T_0. (*Suggestion:* Use Newton's method to solve for $u = 100/T_0$.) (b) Use Problem 48 to find the maximum tension in the transmission line. (c) Use Problem 47 to find the sag H at its middle.

51. Deduce from Problems 46 and 47 that the length S, sag H, density w, and minimum tension T_0 of a hanging cable satisfy the equation $4wH^2 - wS^2 + 8HT_0 = 0$.

52. A 300-ft length of cable weighing 2 lb/ft is hung between two points at equal height. A 30-ft sag is then observed at the middle of the cable. (a) Use the result of Problem 51 to find the minimum tension T_0. (b) Find the distance $2L$ between the ends of the cable. (c) Find the maximum tension in the cable.

53. A 400-ft wire weighing 2 lb/ft can withstand a tension of 5000 lb. Use the results of Problems 48 and 51 to determine the minimum sag that can be allowed in hanging this wire between two points of equal height.

54. A cable is 105 ft long and weighs 2 lb/ft. It is hung with its ends at the same level, and the ends are 100 feet apart. Find the sag at the middle of the cable and the maximum and minimum tension in the cable.

1.7
Exact Equations
and Integrating
Factors

We have seen that a general solution $y(x)$ of a first order differential equation is often defined implicitly by an equation of the form

$$F(x, y(x)) = C \tag{1}$$

where C is a constant. On the other hand, given the identity in (1), we can recover the original differential equation by differentiating each side with respect to x. Provided that Eq. (1) implicitly defines y as a differentiable function of x, this gives the differential equation in the form $\partial F/\partial x + (\partial F/\partial y)(dy/dx) = 0$; that is,

$$M(x, y) + N(x, y)\frac{dy}{dx} = 0, \tag{2}$$

where $M(x, y) = F_x(x, y)$ and $N(x, y) = F_y(x, y)$.

It is sometimes convenient to rewrite Eq. (2) in the form

$$M(x, y)\, dx + N(x, y)\, dy = 0, \tag{3}$$

called its **differential** form. The general first order differential equation $y' = f(x, y)$ can be written in this form with $M = f(x, y)$ and $N = -1$. The discussion above shows that, *if* there exists a function $F(x, y)$ such that

$$\frac{\partial F}{\partial x} = M \quad \text{and} \quad \frac{\partial F}{\partial y} = N,$$

then the equation $F(x, y) = C$ implicitly defines a general solution of Eq. (3). In this case, Eq. (3) is called an **exact** differential equation—the differential $dF = F_x\, dx + F_y\, dy$ of $F(x, y)$ is exactly $M\, dx + N\, dy$.

A natural question is this: How can we determine whether the differential equation in (3) is exact? And if it is exact, how can we find the function $F(x, y)$ such that $F_x = M$ and $F_y = N$? To answer the first question, let us recall that if the mixed second order partial derivatives F_{xy} and F_{yx} are continuous on an open set in the xy-plane, then they are equal: $F_{xy} \equiv F_{yx}$. If Eq. (3) is exact and M and N have continuous partial derivatives, it then follows that

$$\frac{\partial M}{\partial y} = F_{xy} = F_{yx} = \frac{\partial N}{\partial x}.$$

Thus the equation

$$\frac{\partial M}{\partial y} = \frac{\partial N}{\partial x} \tag{4}$$

is a *necessary condition* that the differential equation $M\, dx + N\, dy = 0$ be exact. That is, if $M_y \neq N_x$, then the differential equation in question is not exact, so we need not attempt to find a function $F(x, y)$ such that $F_x = M$ and $F_y = N$: There is no such function.

EXAMPLE 1 The differential equation

$$y^3\, dx + 3xy^2\, dy = 0 \tag{5}$$

is exact because we can immediately see that the function $F(x, y) = xy^3$ has the property that $F_x(x, y) = y^3$ and $F_y(x, y) = 3xy^2$. Thus a general solution of (5) is $xy^3 = C$; if you prefer, $y = (C/x)^{1/3}$.

Suppose, however, that we divide each term of the differential equation of Example 1 by y^2 to obtain

$$y\, dx + 3x\, dy = 0. \tag{6}$$

This equation is not exact because $M = y$ and $N = 3x$, so that $M_y = 1 \neq 3 = N_x$. Hence the necessary condition in (4) is not satisfied.

We are confronted with a curious situation here: The differential equations in (5) and (6) are essentially equivalent—and they have the same solutions—yet one is exact and the other is not. This observation opens the way for an exciting possibility: that we may be able to convert an equation that is not exact into one that is, simply by multiplying each term by an appropriate factor. In brief, whether a given differential equation is exact or not has to do with the precise form $M \, dx + N \, dy = 0$ in which it is written. The following theorem tells us that, subject to differentiability conditions that usually are satisfied in practice, the necessary condition in (4) is also a *sufficient* condition for exactness. That is, if $M_y = N_x$, then the differential equation $M \, dx + N \, dy = 0$ is exact.

THEOREM: CRITERION FOR EXACTNESS

Suppose that the functions $M(x, y)$ and $N(x, y)$ are continuous and have continuous first order partial derivatives in the open rectangle $R: a < x < b$, $c < y < d$. Then the differential equation

$$M(x, y) \, dx + N(x, y) \, dy = 0 \tag{3}$$

is exact in R if and only if

$$\frac{\partial M}{\partial y} = \frac{\partial N}{\partial x} \tag{4}$$

at each point of R. That is, there exists a function $F(x, y)$ defined on R with $\partial F/\partial x = M$ and $\partial F/\partial y = N$ if and only if Eq. (4) holds on R.

Proof We have seen already that it is necessary for Eq. (4) to hold if Eq. (3) is to be exact. To prove the converse, we must show that if (4) holds, then we can construct a function $F(x, y)$ such that $\partial F/\partial x = M$ and $\partial F/\partial y = N$. Note first that, for *any* function $g(y)$, the function

$$F(x, y) = \int M(x, y) \, dx + g(y) \tag{7}$$

satisfies the condition $\partial F/\partial x = M$. (In (7), the notation $\int M(x, y) \, dx$ denotes an antiderivative of $M(x, y)$ with respect to x.) We want to choose $g(y)$ so that

$$N = \frac{\partial F}{\partial y} = \frac{\partial}{\partial y} \int M(x, y) \, dx + g'(y)$$

as well; that is, so that

$$g'(y) = N - \frac{\partial}{\partial y} \int M(x, y) \, dx. \tag{8}$$

To see that there *is* such a function of y, it suffices to show that the right-hand side in (8) is a function of y alone. We can then find $g(y)$ by integrating with respect to y. Because the right-hand side in (8) is defined on a rectangle, and hence on an interval as a function of x, it suffices to show that its derivative with respect to x is identically zero. But

$$\frac{\partial}{\partial x}\left(N - \frac{\partial}{\partial y}\int M(x, y)\, dx\right) = \frac{\partial N}{\partial x} - \frac{\partial}{\partial x}\frac{\partial}{\partial y}\int M(x, y)\, dx$$

$$= \frac{\partial N}{\partial x} - \frac{\partial}{\partial y}\frac{\partial}{\partial x}\int M(x, y)\, dx$$

$$= \frac{\partial N}{\partial x} - \frac{\partial M}{\partial y} = 0$$

by hypothesis. So we can, indeed, find the desired function $g(y)$ by integrating Eq. (8). We substitute this result in Eq. (7) to obtain

$$F(x, y) = \int M(x, y)\, dx + \int \left[N(x, y) - \frac{\partial}{\partial y}\int M(x, y)\, dx\right] dy \qquad (9)$$

as the desired function with $F_x = M$ and $F_y = N$.

Instead of remembering Eq. (9), it is usually better to solve an exact equation $M\, dx + N\, dy = 0$ by carrying out the process indicated by Eqs. (7) and (8). First we integrate $M(x, y)$ with respect to x and write

$$F(x, y) = \int M(x, y)\, dx + g(y),$$

thinking of the function $g(y)$ as an "arbitrary constant of integration" as far as the variable x is concerned. Then we determine $g(y)$ by imposing the condition that $\partial F/\partial y = N(x, y)$. This yields a general solution in the implicit form $F(x, y) = C$.

EXAMPLE 2   Solve the differential equation

$$(6xy - y^3)\, dx + (4y + 3x^2 - 3xy^2)\, dy = 0.$$

Solution   This equation is exact because

$$\frac{\partial M}{\partial y} = 6x - 3y^2 = \frac{\partial N}{\partial x}.$$

Integrating $\partial F/\partial x = M$ with respect to x, we get

$$F(x, y) = \int (6xy - y^3)\, dx = 3x^2y - xy^3 + g(y).$$

Then we differentiate with respect to y and set $\partial F/\partial y$ equal to N. This yields

$$3x^2 - 3xy^2 + g'(y) = 4y + 3x^2 - 3xy^2,$$

and it follows that $g'(y) = 4y$. Hence $g(y) = 2y^2 + C_1$, and thus

$$F(x, y) = 3x^2y - xy^3 + 2y^2 + C_1.$$

Therefore, a general solution of the given differential equation is defined implicitly by the equation $3x^2y - xy^3 + 2y^2 = C$ (we have absorbed the constant C_1 into the constant C).

INTEGRATING FACTORS

We sometimes see nonexact differential equations written in the form $M\,dx + N\,dy = 0$. As suggested in the discussion following Example 1, it may be possible to convert such an equation into an exact equation by multiplying its terms by a suitable integrating factor. For example, the equation $y\,dx - x\,dy = 0$ is not exact as it stands, but multiplication by $1/y^2$ gives

$$\frac{y\,dx - x\,dy}{y^2} = 0.$$

The left-hand side is the differential of $F(x, y) = x/y$, and thus we see that the general solution of $y\,dx - x\,dy = 0$ is $x/y = C$; if you prefer, $y = kx$.

The function $1/y^2$ used above is an example of an integrating factor. An **integrating factor** for the differential equation in the differential form

$$M(x, y)\,dx + N(x, y)\,dy = 0 \tag{10}$$

is a function $\rho(x, y)$ such that the equation

$$\rho(x, y)M(x, y)\,dx + \rho(x, y)N(x, y)\,dy = 0$$

is exact; that is,

$$\frac{\partial}{\partial y}(\rho M) = \frac{\partial}{\partial x}(\rho N). \tag{11}$$

For example, the separable equation $y\,dx + \sec x\,dy = 0$ is not exact. We actually separate the variables by multiplying the equation by the integrating factor $\rho(x, y) = 1/(y \sec x)$ to obtain the exact equation

$$\cos x\,dx + \frac{1}{y}\,dy = 0$$

with general solution $\sin x + \ln y = C$. This illustrates the fact that whenever we separate the variables in a separable equation, we are using an integrating factor.

Unfortunately, though it is known that every differential equation of the form in (10) has an integrating factor, no useful general method for finding one is known. Sometimes we can spot an integrating factor by recognizing the presence of propitious combinations such as those in the table shown in Fig. 1.38. This is the basis for the *method of grouping* illustrated next in Example 3.

$F(x, y)$	dF
xy	$y\,dx + x\,dy$
$\dfrac{x}{y}$	$\dfrac{y\,dx - x\,dy}{y^2}$
$\dfrac{1}{2}\ln(x^2 + y^2)$	$\dfrac{x\,dx + y\,dy}{x^2 + y^2}$
$\tan^{-1}(y/x)$	$\dfrac{-y\,dx + x\,dy}{x^2 + y^2}$

Figure 1.38 Important forms for the method of grouping.

EXAMPLE 3 Solve the equation

$$(x^2 + y^2 - y)\, dx + x\, dy = 0.$$

Solution This equation is not exact, nor is it susceptible to any of the previous methods of this chapter. But if we multiply each term by the integrating factor $p(x, y) = 1/(x^2 + y^2)$, we get the equation

$$dx + \frac{-y\, dx + x\, dy}{x^2 + y^2} = 0.$$

This equation can be integrated immediately, and its general solution is $x + \tan^{-1}(y/x) = C$; that is, $y = x \tan(C - x)$.

The integrating factor of Example 3 was obtained by a judicious guess. But *if* the equation $M\, dx + N\, dy = 0$ has an integrating factor p that is either a function of x alone or a function of y alone, then p can be found systematically. This situation is summarized in the table of Fig. 1.39.

Case	Integrating factor
$\frac{1}{N}(M_y - N_x) = f(x)$	$p(x) = e^{\int f(x)dx}$
$\frac{1}{M}(N_x - M_y) = g(y)$	$p(y) = e^{\int g(y)dy}$

Figure 1.39 Special integrating factors.

The meaning of the first line in Fig. 1.39 is that the equation $M\, dx + N\, dy = 0$ has an integrating factor that is a function of x alone if and only if $(M_y - N_x)/N$ is a function $f(x)$ of x alone, in which case $p(x) = \exp(\int f(x)\, dx)$ is an integrating factor of that equation. To see why, suppose that

$$\frac{1}{N}(M_y - N_x) = f(x),$$

and define $p(x)$ to be $\exp(\int f(x)\, dx)$. Then

$$\frac{\partial}{\partial x}(pN) = \frac{\partial p}{\partial x}N + p\frac{\partial N}{\partial x}$$

$$= \frac{p}{N}(M_y - N_x)N + p\frac{\partial N}{\partial x} \qquad \text{(because } p'(x) = p(x)f(x))$$

$$= p\frac{\partial M}{\partial y}$$

$$= \frac{\partial}{\partial y}(pM)$$

(because p is a function of x alone). Thus the equation $pM\, dx + pN\, dy = 0$ is exact, and we have verified that $p(x)$ is an integrating factor of $M\, dx + N\, dy = 0$. In Problem 33 we ask you to show that the condition $(M_y - N_x)/N = f(x)$ is also a necessary condition that $M\, dx + N\, dy = 0$ have an integrating factor that is a function of x alone.

EXAMPLE 4 Solve the differential equation

$$y^2 \cos x \, dx + (4 + 5y \sin x) \, dy = 0.$$

Solution With $M = y^2 \cos x$ and $N = 4 + 5y \sin x$, we find that

$$\frac{M_y - N_x}{N} = \frac{2y \cos x - 5y \cos x}{4 + 5y \sin x} = -\frac{3y \cos x}{4 + 5y \sin x}$$

is not a function of x alone, so the given equation has no integrating factor of the form $\rho(x)$. But

$$\frac{N_x - M_y}{M} = \frac{3y \cos x}{y^2 \cos x} = \frac{3}{y},$$

so the equation has the integrating factor

$$\rho(y) = e^{\int (3/y) \, dy} = e^{3 \ln y} = y^3.$$

After we multiply each term by y^3, we get the equation

$$y^5 \cos x \, dx + (4y^3 + 5y^4 \sin x) \, dy = 0,$$

which we easily verify to be exact. Integrating the coefficient of dx, we write

$$F(x, y) = \int y^5 \cos x \, dx = y^5 \sin x + g(y).$$

Then we set $\partial F / \partial y$ equal to the coefficient of dy to obtain

$$5y^4 \sin x + g'(y) = 4y^3 + 5y^4 \sin x.$$

Thus $g'(y) = 4y^3$, so we may choose $g(y) = y^4$. Therefore a general solution of the original differential equation is $y^5 \sin x + y^4 = C$.

EXAMPLE 5* Suppose that a flexible 4-ft rope starts with 3 ft of its length arranged in a heap right at the edge of a high horizontal table, with the remaining foot hanging (at rest) off the table. At time $t = 0$ the heap begins to unwind and the rope begins gradually to fall off the table, under the force of gravity pulling on the overhanging part. Under the assumption that frictional forces of all sorts are negligible, how long will it take for all the rope to fall off the table?

Solution We assume that some unspecified device prevents the entire heap from simply falling off the edge all at once. Let $x(t)$ denote the length of overhanging rope at time t, as shown in Fig. 1.40, and let $v(t)$ denote

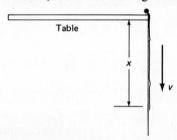

Table

x

v

Figure 1.40 The heap of rope unwinding off a table.

*We are indebted to Dr. Carol W. Penney for this interesting variant of a standard example (see Problem 36).

its velocity then, with the positive direction downward. Then

$$x(0) = 1 \quad \text{and} \quad v(0) = 0. \tag{12}$$

If the linear density of the rope is w (slugs/foot), then the mass of the over-hanging rope is $m = wx$ and the gravitational force acting upon it is $F = mg = wgx$. Then Newton's second law in the form

$$F = \frac{d}{dt}(mv)$$

yields

$$wgx = \frac{d}{dt}(wxv) = w\left(x\frac{dv}{dt} + v\frac{dx}{dt}\right).$$

Because $dx/dt = v$, we need to solve the differential equation

$$gx = x\frac{dv}{dt} + v^2 = x\frac{dv}{dx}\frac{dx}{dt} + v^2 = xv\frac{dv}{dx} + v^2;$$

that is,

$$\left(\frac{v^2}{x} - g\right) dx + v\, dv = 0. \tag{13}$$

With $M(x, v) = v^2/x - g$ and $N(x, v) = v$, we have $M_v = 2v/x$ but $N_x = 0$, so Eq. (13) is not exact. But

$$\frac{M_v - N_x}{N} = \frac{(2v/x) - 0}{v} = \frac{2}{x},$$

so it has the integrating factor

$$\rho(x) = e^{\int (2/x)\, dx} = e^{2 \ln x} = x^2.$$

Multiplication of the terms in (13) by x^2 gives the exact equation

$$(xv^2 - gx^2)\, dx + x^2v\, dv = 0.$$

Its general solution $\frac{1}{2}x^2v^2 - \frac{1}{3}gx^3 = C$ is readily found. The initial conditions in (12) imply that $C = -g/3$. With this value of C we solve for

$$v = \frac{dx}{dt} = \left(\frac{2g}{3}\right)^{1/2} \frac{(x^3 - 1)^{1/2}}{x},$$

taking the positive square root because $v \geqq 0$. Hence

$$t = \left(\frac{3}{2g}\right)^{1/2} \int \frac{x}{(x^3 - 1)^{1/2}}\, dx.$$

Because $x = 1$ when $t = 0$, the desired value of t when $x = 4$ is

$$T = \left(\frac{3}{2g}\right)^{1/2} \int_1^4 \frac{x}{(x^3 - 1)^{1/2}}\, dx.$$

This is a nonelementary improper integral, but the substitution $x^3 = \sec^2 u$ converts it into the proper integral

$$T = \left(\frac{2}{3g}\right)^{1/2} \int_0^{\cos^{-1}(1/8)} (\sec u)^{4/3}\, du.$$

Simpson's rule with $n = 100$ subintervals finally gives $T \approx 0.541$ seconds for the time required for all the rope to fall off the table.

1.7 Problems

In each of Problems 1–12, verify that the given differential equation is exact and then solve it.

1. $(2x + 3y)\, dx + (3x + 2y)\, dy = 0.$
2. $(4x - y)\, dx + (6y - x)\, dy = 0.$
3. $(3x^2 + 2y^2)\, dx + (4xy + 6y^2)\, dy = 0.$
4. $(2xy^2 + 3x^2)\, dx + (2x^2y + 4y^3)\, dy = 0.$
5. $\left(x^3 + \dfrac{y}{x}\right) dx + (y^2 + \ln x)\, dy = 0.$
6. $(1 + ye^{xy})\, dx + (2y + xe^{xy})\, dy = 0.$
7. $(\cos x + \ln y)\, dx + \left(\dfrac{x}{y} + e^y\right) dy = 0.$
8. $(x + \tan^{-1} y)\, dx + \dfrac{x + y}{1 + y^2}\, dy = 0.$
9. $(3x^2y^3 + y^4)\, dx + (3x^3y^2 + y^4 + 4xy^3)\, dy = 0.$
10. $(e^x \sin y + \tan y)\, dx + (e^x \cos y + x \sec^2 y)\, dy = 0.$
11. $\left(\dfrac{2x}{y} - \dfrac{3y^2}{x^4}\right) dx + \left(\dfrac{2y}{x^3} - \dfrac{x^2}{y^2} + \dfrac{1}{y^{1/2}}\right) dy = 0.$
12. $\dfrac{2x^{5/2} - 3y^{5/3}}{2x^{5/2}y^{2/3}}\, dx + \dfrac{3y^{5/3} - 2x^{5/2}}{3x^{3/2}y^{5/3}}\, dy = 0.$

Use the method of grouping to solve the differential equations in Problems 13–20.

13. $y\, dx - x\, dy = y^3\, dy.$
14. $x\, dy - y\, dx = x^2 e^x\, dx.$
15. $(y - 1)\, dx + (x - 2)\, dy = 0.$
16. $(xy^2 - x)\, dx + (x^2y - y)\, dy = 0.$
17. $3x^2y\, dx + (x^3 + e^y)\, dy = 0.$
18. $(2x + x^2 + y^2)\, dx + (2y - x^2 - y^2)\, dy = 0.$
19. $(x^2 + y^2 - y)\, dx + x\, dy = 0.$
20. $(y^2 + e^x \sin y)\, dx + (2xy + e^x \cos y)\, dy = 0.$

Solve each of the differential equations in Problems 21–28 by finding an integrating factor that is a function of a single variable.

21. $4y\, dx + x\, dy = 0.$
22. $(4x + 3y^3)\, dx + 3xy^2\, dy = 0.$
23. $2xy\, dx + (y^2 - x^2)\, dy = 0.$
24. $(4x^2 + 3 \cos y)\, dx - x \sin y\, dy = 0.$
25. $(y \ln y + ye^x)\, dx + (x + y \cos y)\, dy = 0.$
26. $(4xy^2 + y)\, dx + (6y^3 - x)\, dy = 0.$
27. $2x\, dx + x^2 \cot y\, dy = 0.$
28. $\left(1 + \dfrac{1}{x}\right) \tan y\, dx + \sec^2 y\, dy = 0.$

29. Solve $(7x^4y - 3y^8) \, dx + (2x^5 - 9xy^7) \, dy = 0$, given that there exists an integrating factor of the form $x^m y^n$.

30. Use the technique suggested in Problem 29 to solve the equation $(3y^2 + 10xy) \, dx + (5xy + 12x^2) \, dy = 0$.

Use the observation that

$$D_x \left[\tan^{-1} \frac{x}{y} \right] = \frac{y \, dx - x \, dy}{x^2 + y^2}$$

to solve the differential equations in Problems 31 and 32.

31. $[y + x(x^2 + y^2)^{1/2}] \, dx + [y(x^2 + y^2)^{1/2} - x] \, dy = 0$.

32. $\left(y + \dfrac{x}{x^2 + y^2} \right) dx + \left(\dfrac{y}{x^2 + y^2} - x \right) dy = 0$.

33. Suppose that $\rho(x)$ is a function of x alone and is an integrating factor of $M(x, y) \, dx + N(x, y) \, dy = 0$. Show that $(M_y - N_x)/N$ is a function of x alone.

34. Suppose that $(N_x - M_y)/M = g(y)$ is a function of y alone. Prove that $\rho(y) = \exp(\int g(y) \, dy)$ is an integrating factor of $M \, dx + N \, dy = 0$.

35. To solve the linear equation $y' + P(x)y = Q(x)$, rewrite it in the form $M \, dx + N \, dy = 0$ with $M = Py - Q$ and $N = 1$. Then show that this last differential equation has an integrating factor of the form $\rho(x)$, and use this fact to derive the solution

$$y = e^{-\int P(x) \, dx} \left[C + \int (Q(x) e^{\int P(x) \, dx}) \, dx \right].$$

36. In Example 5 suppose that the 3 ft of rope on the table, rather than being heaped in a pile at the edge of the table, are initially stretched out on the table perpendicular to its edge. At time $t = 0$, when 1 ft of rope is hanging from the table, the rope begins to slide off the table. (a) Using the notation in the text, derive the differential equation

$$Lx'' = gx; \quad \text{that is,} \quad x'' = 8x,$$

because $L = 4$ (ft) and $g = 32$ (ft/s^2). (b) Note that the independent variable t is missing from the differential equation derived in part (a), and that the initial conditions are $x(0) = 1$ and $x'(0) = 0$. Derive the solution $x(t) = \cosh(t\sqrt{8})$. (c) How long will it take until the last bit of rope leaves the table? (*Answer:* About 0.730 s.)

*1.8
Population Models

In Section 1.4 we introduced the exponential differential equation $dP/dt = kP$, with solution $P(t) = P_0 e^{kt}$, as a mathematical model for natural population growth that occurs as a result of constant birth and death rates. Here we present a more general population model that accommodates birth and death rates that are not necessarily constant. As before, however, our population function $P(t)$ will be a *continuous* approximation to the actual population, which of course grows by integral increments.

Suppose that the population changes only by the occurrence of births and deaths—there is no immigration or emigration. Let $B(t)$ and $D(t)$ denote, respectively, the numbers of births and deaths that have occurred (since $t = 0$) by time t. Then the *birth rate* $\beta(t)$ and *death rate* $\delta(t)$, in births or deaths per individual per unit of time, are defined as follows:

$$\beta(t) = \lim_{h \to 0} \frac{B(t+h) - B(t)}{hP(t)} = \frac{1}{P}\frac{dB}{dt},$$

$$\delta(t) = \lim_{h \to 0} \frac{D(t+h) - D(t)}{hP(t)} = \frac{1}{P}\frac{dD}{dt}. \qquad \left.\begin{array}{c}\\\\\\\end{array}\right\} \qquad (1)$$

Then

$$P'(t) = \lim_{h \to 0} \frac{P(t+h) - P(t)}{h}$$

$$= \lim_{h \to 0} \frac{[B(t+h) - B(t)] - [D(t+h) - D(t)]}{h}$$

$$= B'(t) - D'(t);$$

Thus

$$P'(t) = [\beta(t) - \delta(t)]P(t). \qquad (2)$$

We will also use the briefer notation $dP/dt = (\beta - \delta)P$. Eq. (2) is the **general population equation**. If β and δ are constant, it reduces to the natural growth equation with $k = \beta - \delta$. But it also includes the possibility that β and δ are variable functions of t. The birth and death rates need not be known in advance; they may well depend upon the unknown function $P(t)$.

LIMITED POPULATIONS

In situations as diverse as the human population of a nation and a fruit fly population in a closed container, it is often observed that the birth rate decreases as the population itself increases. The reasons may range from increased scientific or cultural sophistication to a limited food supply. Suppose, for example, that the birth rate β is a *linear* decreasing function of the population P, so that $\beta = \beta_0 - \beta_1 P$, where β_0 and β_1 are positive constants. If the death rate $\delta = \delta_0$ remains constant, then Eq. (2) takes the form $dP/dt = (\beta_0 - \beta_1 P - \delta_0)P$; that is,

$$\frac{dP}{dt} = kP(M - P) \qquad (3)$$

where $k = \beta_1$ and $M = (\beta_0 - \delta_0)/\beta_1$. We assume $\beta_0 > \delta_0$ so that $M > 0$.

Equation (3) is called the **logistic equation**. If we assume that $P < M$, it may be solved by separation of variables as follows.

$$\int \frac{dP}{P(M - P)} = \int k\, dt;$$

$$\frac{1}{M} \int \left(\frac{1}{P} + \frac{1}{M - P} \right) dP = \int k\, dt;$$

$$\ln \left(\frac{P}{M - P} \right) = kMt + C.$$

Exponentiation gives

$$\frac{P}{M - P} = Ae^{kMt}$$

where $A = e^c$. We substitute $t = 0$ into each side of this last equation to find that $A = P_0/(M - P_0)$. So

$$\frac{P}{M - P} = \frac{P_0 e^{kMt}}{M - P_0}.$$

This equation is easy to solve for

$$P(t) = \frac{MP_0}{P_0 + (M - P_0)e^{-kMt}}. \tag{4}$$

While we made the assumption that $P < M$ in order to derive Eq. (4), this restriction is unnecessary, because we can verify by direct substitution into Eq. (3) that $P(t)$ as given in (4) satisfies the logistic equation whether $P < M$ or $P \geqq M$.

If the initial population satisfies $P_0 < M$, then (4) shows that $P(t) < M$ for all $t \geqq 0$, and also that

$$\lim_{t \to \infty} P(t) = M. \tag{5}$$

Thus a population that satisfies the logistic equation is *not* like a naturally growing population; it does not grow without bound, but instead approaches the finite **limiting population** M as $t \to +\infty$. But because $dP/dt = kP(M - P) > 0$ in this case, we see that the population is steadily increasing. Moreover, differentiation with respect to t gives

$$\frac{d^2P}{dt^2} = \left[\frac{d}{dP}\left(\frac{dP}{dt}\right)\right]\left(\frac{dP}{dt}\right) = (kM - 2kP)[kP(M - P)].$$

Hence the graph of $P(t)$ has an inflection point where $P = M/2$. Therefore the graph of P has the shape of the lower curve in Fig. 1.41. The population increases at an increasing rate until $P = M/2$, and thereafter increases at a decreasing rate.

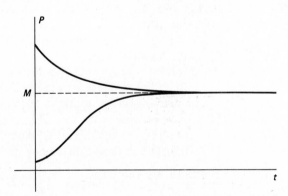

Figure 1.41   Two solutions of the logistic equation.

If $P_0 > M$, then a similar analysis (Problem 8) shows that $P(t)$ is a steadily decreasing function with a graph like the upper curve in Fig. 1.41.

In 1845 the Belgian demographer Verhulst used the 1790–1840 U.S. population data to predict the U.S. population through the year 1930, under his assumption that it would continue to satisfy the logistic equation. With

$P_0 = 3.9$ (in millions), $M = 197.3$ (in millions), and $k = 0.0001589$, Eq. (4) gives the remarkable results shown in the table of Fig. 1.42. Of course, the assumed limiting population of $M = 197.3$ millions has now been exceeded, so the population of the United States has not continued to satisfy the logistic equation in the decades following 1930.

Year	Actual United States population (millions)	Calculated population
1790	3.9	3.9
1800	5.3	5.3
1810	7.2	7.2
1820	9.6	9.7
1830	12.9	13.0
1840	17.1	17.4
1850	23.2	23.1
1860	31.4	30.2
1870	38.6	39.2
1880	50.2	49.9
1890	62.9	62.5
1900	76.0	76.6
1910	92.0	91.7
1920	106.5	107.1
1930	123.2	122.1

Figure 1.42 U.S. population data.

EXAMPLE 1 Suppose that in 1910 the population of a certain country was 50 million and was growing at the rate of 750,000 people per year at that time. Suppose also that in 1940 its population was 100 million and was then growing at the rate of 1 million per year. Assume that this population satisfies the logistic equation. Determine both the limiting population M and the predicted population for the year 2000.

Solution We substitute the two given pairs of data in Eq. (3) and find that

$$0.75 = 50k(M - 50),$$
$$1.00 = 100k(M - 100).$$

We solve simultaneously for $M = 200$ and $k = 0.0001$. Thus the limiting population of the country in question is 200 million. With these values of M and k and with $t = 0$ corresponding to the year 1940 (in which $P_0 = 100$), we find that, as a consequence of Eq. (4), the population in the year 2000 will be

$$P = \frac{(100)(200)}{100 + (200 - 100)e^{-(0.0001)(200)(60)}},$$

or about 153.7 million people.

We next describe some situations that illustrate the varied circumstances in which the logistic equation is a satisfactory mathematical model.

1. *Limited environment situation.* A certain environment can support a population of at most M individuals. It's then reasonable to expect the growth rate $\beta - \delta$ (the combined birth and death rates) to be proportional to $M - P$, because we may think of $M - P$ as the potential for further expansion. Then $\beta - \delta = k(M - P)$, so that

$$\frac{dP}{dt} = (\beta - \delta)P = kP(M - P).$$

The classic example of a limited environment situation is a fruit fly population in a closed container.

2. *Competition situation.* If the birth rate β is constant but the death rate δ is proportional to P, so that $\delta = \alpha P$, then

$$\frac{dP}{dt} = (\beta - \alpha P)P = kP(M - P).$$

This might be a reasonable working hypothesis in a study of a cannibalistic population, in which all deaths result from chance encounters between individuals. Of course, competition between individuals is not usually so deadly, nor its effects so immediate and decisive.

3. *Joint proportion situation.* Let $P(t)$ denote the number of individuals in a constant susceptible population M who are infected with a certain contagious and incurable disease. The disease in question is spread by chance encounters. Then $P'(t)$ should be proportional to the product of the number P of the individuals having the disease and the number $M - P$ of those not having it, so that $dP/dt = kP(M - P)$. Again we discover that the mathematical model is the logistic equation. The mathematical description of the spread of a rumor in a population of M individuals is identical.

EXAMPLE 2 Suppose that at time $t = 0$, half of a population of 100,000 persons have heard a certain rumor, and that the number of those who have heard it is then increasing at the rate of 1000 persons per day. How long will it take for this rumor to spread to 80% of the population?

Solution Let us work in units of thousands of persons. Then we take $M = 100$ for the total fixed population. We substitute $M = 100$, $P'(0) = 1$, and $P_0 = 50$ into the logistic equation, and thereby obtain

$$1 = k(50)(100 - 50), \quad \text{so that} \quad k = 0.0004.$$

If t denotes the number of days until 80 thousand people have heard the rumor, then Eq. (4) gives

$$80 = \frac{(50)(100)}{50 + (100 - 50)e^{-(0.04)t}},$$

so that t is approximately 34.66. Thus the rumor will have spread to 80% of the population in a little less than 35 days.

DOOMSDAY VERSUS EXTINCTION

Consider a population $P(t)$ of unsophisticated animals that rely solely on chance encounters to meet mates for reproductive purposes. It is reasonable to expect such encounters to occur at a rate that is proportional to the product of the number $P/2$ of males and the number $P/2$ of females, hence at a rate proportional to P^2. Hence we assume that births occur at the rate $dB/dt = kP^2$ (k constant), so that $\beta = kP$ by Eq. (1). If the death rate δ is constant, then the general population equation in (2) yields the differential equation

$$\frac{dP}{dt} = kP^2 - \delta P = kP(P - a), \tag{6}$$

where $a = \delta/k > 0$, as a mathematical model of the population. The solution of Eq. (6) depends upon whether the initial population $P_0 = P(0)$ is greater than or less than a.

Case 1: $P_0 > a$. From Eq. (6) we see that $P'(0) = kP_0(P_0 - a) > 0$, so $P(t)$ starts out increasing. Hence $P'(t)$ remains positive, so $P(t)$ continues to increase, and therefore $P(t) > a$ for all $t > 0$. We note that

$$\frac{1}{P(P - a)} = -\frac{1}{a}\left(\frac{1}{P} - \frac{1}{P - a}\right).$$

We separate variables in Eq. (6) and integrate as follows:

$$\int \frac{dP}{P(P - a)} = \int k\, dt;$$

$$-\int \left(\frac{1}{P} - \frac{1}{P - a}\right) dP = -\int ka\, dt;$$

$$\ln \frac{P}{P - a} = -kat + C_1.$$

Substitution of P_0 for P and 0 for t then gives

$$C_1 = \ln \frac{P_0}{P_0 - a} = \ln C$$

where $C = P_0/(P_0 - a) > 1$. Exponentiation then yields $P/(P - a) = Ce^{-kat}$ which we solve for

$$P(t) = \frac{Cae^{-kat}}{Ce^{-kat} - 1}. \tag{7}$$

Note that the denominator in (7) approaches zero as

$$t \longrightarrow T = \frac{\ln C}{ka} = \frac{1}{ka} \ln \frac{P_0}{P_0 - a} > 0.$$

Thus $\lim_{t \to T} P(t) = +\infty$. This is a *doomsday* situation.

Case 2: $P_0 < a$. In this case $P'(0) < 0$, and it follows that $P(t) < a$ for all $t > 0$. A similar separation of variables (Problem 14) now leads to

$$P(t) = \frac{Cae^{-kat}}{Ce^{-kat} + 1} \tag{8}$$

where $C = P_0/(a - P_0) > 0$. The difference is that the denominator in Eq. (8) remains greater than 1, and so $\lim_{t \to \infty} P(t) = 0$. This is an *extinction* situation.

Thus the population either explodes or is an endangered species threatened with extinction, depending upon its initial size. An approximation to this phenomenon is sometimes observed with actual animal populations, such as the alligator populations in certain areas of the southern United States.

1.8 Problems

1. Suppose that the fish population $P(t)$ in a lake is attacked by disease at time $t = 0$, with the results that the fish cease to reproduce (so that the birth rate is $\beta = 0$) and the death rate δ (deaths per week per fish) is thereafter proportional to $\sqrt{1/P}$. If there were initially 900 fish in the lake and 441 were left after 6 weeks, how long did it take all the fish in the lake to die?

2. Suppose that when a certain lake is stocked with fish, the birth and death rates β and δ are both inversely proportional to $\sqrt{P}$. (a) Show that $P(t) = (\frac{1}{2}kt + \sqrt{P_0})^2$ where k is a constant. (b) If $P_0 = 100$ and after 6 months there are 169 fish in the lake, how many will there be after 1 year?

3. Consider a prolific breed of rabbits whose birth and death rates, β and δ, are each proportional to the rabbit population $P = P(t)$, with $\beta > \delta$. (a) Show that

$$P(t) = \frac{P_0}{1 - kP_0 t}, \qquad k \text{ constant.}$$

Note that $P(t) \longrightarrow +\infty$ as $t \longrightarrow 1/(kP_0)$. This is doomsday. (b) Suppose that $P_0 = 2$ and that there are four rabbits after 3 months. When does doomsday occur?

4. Repeat both parts of Problem 3 in the case $\beta < \delta$. What now happens to the rabbit population in the long run?

5. Consider a human population $P(t)$ with constant birth and death rates β and δ, but also with I persons per year entering the country (immigration). Estimate the change ΔP during the short time interval Δt to derive the differential equation

$$\frac{dP}{dt} = kP + I \qquad (k = \beta - \delta).$$

Use the result of Problem 5 in Problems 6 and 7.

6. A certain city had a population of 1.5 million in 1980. Assume that it grows continuously at a 4% annual rate (this implies that $\beta - \delta = 0.04$) and also absorbs 50,000 newcomers per year. What will be its population in the year 2000?

7. Consider the U.S. population with $P_0 = 222$ million in 1980. (a) Compute the population in the year 2000, assuming a natural growth rate of 1% annually. (b) Rework part (a) with the additional assumption that immigration will remain at a constant 500,000 people per year.

8. Derive the solution in (4) of the logistic equation in (3) under the assumption that $P > M$, and show in this case that the graph of $P(t)$ looks like the upper curve in Fig. 1.41.

9. Suppose that as a certain salt dissolves in a solvent, the number $x(t)$ of grams of the salt in solution after t seconds satisfies the logistic equation $dx/dt = (0.8)x - (0.004)x^2$. (a) What is the maximum amount of the salt that will dissolve in this solvent? (b) If $x = 50$ when $t = 0$, how long will it take for an additional 50 g of the salt to dissolve?

10. Suppose that a community contains 15,000 people who are susceptible to a spreading contagious disease. At time $t = 0$ the number $N(t)$ of people who have the disease is 5000 and is increasing by 500 per day. How long will it take for another 5000 people to contract the disease? Assume that $N'(t)$ is proportional to the product of the numbers of those who have the disease and those who do not.

11. The data in the table of Fig. 1.43 are given for a certain population $P(t)$ that satisfies the logistic equation in (3). (a) What is the limiting population M? (*Suggestion:* Use the approximation

$$P'(t) \approx \frac{P(t + h) - P(t - h)}{2h}$$

Year	P (millions)
1909	49.26
1910	50.00
1911	50.76
.	.
.	.
.	.
1939	99.02
1940	100.00
1941	101.02

Figure 1.43 Population data for Problem 11.

with $h = 1$ to estimate the values of dP/dt when $P = 50$ and when $P = 100$. Then substitute these values into the logistic equation and solve for k and M.) (b) Use the values of k and M found in part (a) to determine when $P = 150$. (*Suggestion:* Take $t = 0$ to correspond to the year 1940.)

12. A population $P(t)$ of small rodents has birth rate $\beta = (0.001)P$ (births per month per rodent) and *constant* death rate δ. If $P(0) = 100$ and $P'(0) = 8$, how long (in months) will it take this population to double to 200 rodents? (*Suggestion:* First find the value of δ.)

13. Consider an animal population $P(t)$ with constant death rate $\delta = 0.01$ and with birth rate β proportional to P. Suppose that $P(0) = 200$ and $P'(0) = 2$. (a) When is $P = 1000$? (b) When does doomsday occur?

14. Derive the solution in (8) of Eq. (7) in the case $P(0) < a$.

15. A tumor may be regarded as a population of multiplying cells. It is found empirically that the "birth rate" of the cells in a tumor decreases exponentially with time, so that $\beta(t) = \beta_0 e^{-\alpha t}$ (where α and β_0 are positive constants), and

hence

$$\frac{dP}{dt} = \beta_0 e^{-\alpha t} P, \qquad P(0) = P_0.$$

Solve this initial value problem for

$$P(t) = P_0 \exp\left[\frac{\beta_0}{\alpha}(1 - e^{-\alpha t})\right].$$

Observe that $P(t)$ approaches the finite limiting population $P_0 \exp(\beta_0/\alpha)$ as $t \longrightarrow +\infty$.

16. For the tumor of Problem 15, suppose that at time $t = 0$ there are $P_0 = 10^6$ cells and that $P(t)$ is then increasing at the rate of 3×10^5 cells per month. After 6 months the tumor has doubled (in size and in number of cells). Solve numerically for α, and then find the limiting population of the tumor.

*1.9
Motion with Variable Acceleration

In Section 1.2 we discussed vertical motion near the surface of the earth under the influence of constant gravitational acceleration, $g \approx 32$ ft/s^2. Beginning with the acceleration $dv/dt = -g$, we integrated twice to obtain the equations

$$\frac{dy}{dt} = v = -gt + v_0$$

and

$$y = -\frac{1}{2}gt^2 + v_0 t + y_0$$

for the velocity v and height y at time t. We saw that these equations can be used to answer specific questions. For example, suppose that we want to predict the maximum height that will be attained by a bolt shot from a crossbow aimed directly upward from ground level ($y_0 = 0$). A modern crossbow can impart an initial velocity of about 288 ft/s to a lightweight bolt, so let us take $v_0 = 288$. We take $g = 32$ (exactly) for simplicity, and the equations of motion of the bolt take the form

$$v = -32t + 288;$$
$$y = -16t^2 + 288t.$$

Of course, we here assume that the crossbow is fired at time $t = 0$ seconds. Then $v = 0$ when $t = 9$, so the maximum height attained by the bolt will be

$$y_{\max} = y(9) = -(16)(9^2) + (288)(9) = 1296 \text{ (ft)}.$$

In this computation we have, however, ignored the effect of air resistance. This makes the mathematical analysis quite simple, but perhaps unrealistically so. Next we want to construct a more complete mathematical model, one that takes air resistance into account.

Empirical studies indicate that the force F_R of air resistance on a moving body with velocity v is approximately of the form $F_R = kv^p$, where $1 \leq p \leq 2$ and the value of k depends upon the size and shape of the body, as well as the density and viscosity of the air. Generally speaking, $p = 1$ for relatively low speeds and $p = 2$ for high speeds, while $1 < p < 2$ for intermediate speeds. But how slow "low speed" and fast "high speed" are depends upon the same factors that determine the value of the coefficient k.

Thus air resistance is a complicated physical phenomenon. But the simplifying assumption that F_R is *exactly* of the form given above, with either $p = 1$ or $p = 2$, yields a tractable mathematical model that exhibits the most important qualitative features of motion with resistance.

RESISTANCE PROPORTIONAL TO VELOCITY

So let us consider the vertical motion of a body with mass m near the surface of the earth, subject to two forces: a downward gravitational force F_G and a force F_R of air resistance that is proportional to velocity (so that $p = 1$) and of course directed opposite the direction of motion of the body. If we set up a coordinate system with the positive y-direction upward and $y = 0$ at ground level, then $F_G = -mg$ and

$$F_R = -kv, \qquad (1)$$

where k is a positive constant and $v = dy/dt$ is the velocity of the body. Note that the minus sign in Eq. (1) makes F_R positive (an upward force) if the body is falling (v is negative), while it makes F_R negative (a downward force) if the body is rising (v is positive). As indicated in Fig. 1.44, the net force acting on the body is then

$$F = F_R + F_G = -kv - mg,$$

and Newton's law of motion $F = m(dv/dt)$ yields the equation $m(dv/dt) = -kv - mg$. Thus

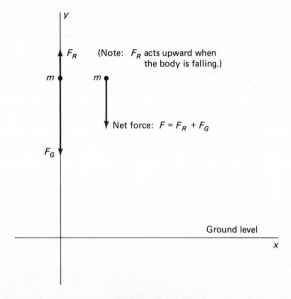

Figure 1.44 Vertical motion with air resistance.

$$\frac{dv}{dt} = -\rho v - g \tag{2}$$

where $\rho = k/m > 0$. The reader should verify that, if the positive y-axis were directed downward, then Eq. (2) would take the form $v' = -\rho v + g$.

Equation (2) is a separable first order differential equation, and its solution is

$$v(t) = \left(v_0 + \frac{g}{\rho}\right)e^{-\rho t} - \frac{g}{\rho}. \tag{3}$$

Here, $v_0 = v(0)$ is the initial velocity of the body. Note that

$$\lim_{t \to \infty} v(t) = -\frac{g}{\rho}. \tag{4}$$

Thus the speed of a body falling with air resistance does *not* increase indefinitely; instead, it approaches a *finite* limiting speed, or **terminal speed**,

$$v_\tau = \frac{g}{\rho} = \frac{mg}{k}. \tag{5}$$

This fact is what makes the parachute a practical invention; it even helps explain the occasional survival of people who fall without parachutes from high-flying airplanes.

We now rewrite Eq. (3) in the form

$$\frac{dy}{dt} = (v_0 + v_\tau)e^{-\rho t} - v_\tau. \tag{6}$$

Integration gives

$$y(t) = -\frac{1}{\rho}(v_0 + v_\tau)e^{-\rho t} - v_\tau t + C.$$

We substitute 0 for t and let $y_0 = y(0)$ denote the initial height of the body. Thus we find that $C = y_0 + (v_0 + v_\tau)/\rho$, and so

$$y(t) = y_0 - v_\tau t + \frac{1}{\rho}(v_0 + v_\tau)(1 - e^{-\rho t}). \tag{7}$$

Equations (6) and (7) give the velocity v and height y of a body moving vertically under the influence of gravity and air resistance. The formulas depend upon the initial height y_0 of the body, its initial velocity v_0, and what might be called the *drag coefficient* ρ, the constant such that the acceleration due to air resistance is $a_R = -\rho v$. The two equations also involve the terminal speed v_τ.

EXAMPLE 1 We consider again a bolt shot straight upward from a crossbow at ground level with initial velocity $v_0 = 288$ ft/s. But now let us also take into account air resistance, with $\rho = 0.04$. What will be the maximum height reached by the bolt, and when will that maximum be attained? When and with what speed will the bolt strike the ground?

Solution We substitute $y_0 = 0$, $v_0 = 288$, and $v_\tau = g/\rho = 800$ into Eqs. (3) and (7). We obtain

$$v = 1088e^{-t/25} - 800$$

and

$$y = 27200(1 - e^{-t/25}) - 800t.$$

To find the time required for the bolt to reach its maximum height (when $v = 0$), we solve the equation $1088e^{-t/25} - 800 = 0$ for $t = 25 \ln(34/25) \approx 7.687$ s. Its maximum height is then

$$y_{\max} \approx 27200(1 - e^{-0.3075}) - (800)(7.687) \approx 1050 \text{ (ft)}.$$

To find when the bolt strikes the ground ($y = 0$), we must solve the equation $27{,}200(1 - e^{-t/25}) - 800t = 0$, which can be written in the simpler form $34(1 - e^{-t/25}) - t = 0$. To solve the latter equation, we use the iterative formula of Newton's method:

$$t_{n+1} = t_n - \frac{f(t_n)}{f'(t_n)}. \tag{8}$$

Recall that this iteration successively improves an initial estimate t_0 of a solution of the equation $f(t) = 0$. (See, for example, Section 3-4 of Edwards and Penney, *Calculus and Analytic Geometry*, (Englewood Cliffs, N.J.: Prentice-Hall, 1982.)) With $f(t) = 34(1 - e^{-t/25}) - t$, Eq. (8) takes the form

$$t_{n+1} = t_n - \frac{34[1 - \exp(-t_n/25)] - t_n}{(34/25)\exp(-t_n/25) - 1}. \tag{9}$$

If we begin with $t_0 = 18$ s (the total time aloft in the case of no air resistance), the iteration in (9) yields

$$t_1 = 16.374115,$$
$$t_2 = 16.252254,$$
$$t_3 = 16.251529,$$

and

$$t_4 = 16.251529,$$

each given in seconds. Thus the bolt is in the air for about 16.25 s and hits the ground with velocity

$$v_{\text{imp}} = 1088 \exp(-0.65) - 800 \approx -232 \text{ (ft/s)}.$$

The effect of air resistance is to decrease the maximum height, the total time spent aloft, and the final impact speed. Note also that when the effect of air resistance is included, the bolt spends more time in descent (about 8.565 s) than in ascent (about 7.687 s).

For a person descending with the aid of a parachute, a typical value of ρ is 1.5, which corresponds to a terminal speed of $v_\tau \approx 21.3$ ft/s, or about 14.5 mi/h. With an unbuttoned overcoat flapping in the wind in place of a parachute, an unlucky skydiver might increase ρ to perhaps as much as 0.5, which gives a terminal speed of $v_\tau \approx 64$ ft/s, or about 44 mi/h. See Problems 10 and 11 for some parachute-jump computations.

RESISTANCE PROPORTIONAL TO SQUARE OF VELOCITY

Now let us assume that the force of air resistance is proportional to the *square* of the velocity:

$$F_R = \pm kv^2 \tag{10}$$

where k is a positive constant. Then Newton's second law gives

$$m\frac{dv}{dt} = F_G + F_R = \pm mg \pm kv^2;$$

that is,

$$\frac{dv}{dt} = \pm g \pm pv^2 \tag{11}$$

where $p = k/m > 0$. The choice of signs in Eq. (11) depends upon the direction of motion, as well as the choice of the direction of the positive y-axis. Consequently, we must discuss the two cases—upward and downward motion—separately.

Downward Motion

Suppose that the body is dropped from a given height with initial velocity $v(0) = 0$. In this case it is more convenient to choose the y-axis pointing *downward*, with $y_0 = y(0) = 0$, so $y(t)$ will denote the distance the body has fallen at time t. Then F_G is a downward force, thus positive, while F_R is an upward (negative) force. So we choose the signs in Eq. (11) to obtain

$$\frac{dv}{dt} = +g - pv^2 = g\left(1 - \frac{p}{g}v^2\right). \tag{12}$$

Recall that the hyperbolic tangent function may be defined as

$$\tanh u = \frac{\sinh u}{\cosh u} = \frac{e^u - e^{-u}}{e^u + e^{-u}}.$$

In Problem 13 we ask you to apply the integral

$$\int \frac{du}{1 - u^2} = \tanh^{-1} u + C$$

to solve Eq. (12) with $v(0) = 0$ for

$$v(t) = \sqrt{\frac{g}{p}} \tanh (t\sqrt{pg}). \tag{13}$$

Because $v = dy/dt$, another integration (see Problem 14) yields

$$y(t) = \frac{1}{p} \ln \cosh (t\sqrt{pg}) \tag{14}$$

for the distance the body has fallen at time t.

In the case of downward motion with resistance proportional to the velocity, $F_R = -kv$, we saw earlier that a body acquires a terminal speed $v_\tau = g/p$ (with $p = k/m$). Because

$$\lim_{t \to \infty} \tanh at = \lim_{t \to \infty} \frac{e^{at} - e^{-at}}{e^{at} + e^{-at}} = 1$$

if $a > 0$, it follows from Eq. (13) that, in the present case of downward motion with resistance $F_R = kv^2$, the body approaches a terminal speed

$$v_\tau = \sqrt{\frac{g}{p}} = \sqrt{\frac{mg}{k}}. \tag{15}$$

Upward Motion

Now let us consider a body that is projected straight upward from ground level $y_0 = 0$ with initial velocity $v(0) = v_0$ and with the positive y-axis now directed upward. Our differential equation now takes the form

$$\frac{dv}{dt} = -g - \rho v^2 \tag{16}$$

because F_G and F_R are each directed downward.

In Problem 16 we ask you to deduce that the velocity and height of the body at time t are given by

$$v = \sqrt{\frac{g}{\rho}} \tan\left(C - t\sqrt{\rho g}\right) \tag{17}$$

and

$$y = \frac{1}{\rho} \ln \frac{\cos\left(C - t\sqrt{\rho g}\right)}{\cos C} \tag{18}$$

where

$$C = \tan^{-1}\left(v_0 \sqrt{\frac{\rho}{g}}\right). \tag{19}$$

The motion of the body continues to be described by Eqs. (17) and (18) as long as the velocity given in (17) is still positive; that is, as long as the body is still moving upward. It reaches its maximum height when $t\sqrt{\rho g} = C$ (so that $v = 0$); hence Eq. (18) gives

$$y_{\max} = -\frac{1}{\rho} \ln \cos C. \tag{20}$$

Thereafter the body falls back toward the ground, with its (downward) velocity and distance fallen (t seconds after beginning to fall) given by Eqs. (13) and (14).

Let us now reconsider the crossbow bolt of Example 1, which we assumed shot straight upward from the ground with an initial velocity $v_0 = 288$ ft/s. Instead of air resistance proportional to velocity v, we assume air resistance proportional to v^2, with $\rho = 0.0002$ in Eqs. (12) and (16). In Problems 18 and 19 we ask you to verify the entries in the last line of the following table.

Air Resistance	Maximum Height (ft)	Ascent Time (s)	Descent Time (s)	Impact Speed (ft/s)
0	1296	9.000	9.000	288
$0.04v$	1050	7.687	8.565	232
$0.0002v^2$	1044	7.800	8.361	228

Comparison of the last two lines of data here suggests a qualitative similarity between the cases $F_R = kv$ and $F_R = kv^2$.

VARIABLE GRAVITATIONAL ACCELERATION

Unless a projectile in vertical motion remains in the immediate vicinity of the earth's surface, the gravitational acceleration acting upon it is not constant. According to Newton's law of gravitation, the gravitational force of attraction between two point masses M and m located at a distance r apart is given by

$$F = \frac{GMm}{r^2} \qquad (21)$$

where G is a certain empirical constant. The formula is also valid if either or both of the two masses are homogeneous spheres; in this case, the distance r is measured between the centers of the spheres.

Let M denote the mass of the earth and R its radius. We obtain the gravitational acceleration $a = g$ of a particle of mass m at the earth's surface by combining Eq. (21) above with the equation $F = ma$:

$$ma = mg = \frac{GMm}{r^2} = \frac{GMm}{R^2},$$

so that

$$g = \frac{GM}{R^2} \qquad \text{(about 32 ft/s}^2\text{).} \qquad (22)$$

We will use Eq. (22) from time to time to remove the necessity for the actual determination of G in our problems and examples. For instance, the mass M_1 of the moon is approximately $(0.0123)M$ and its radius R_1 is approximately $(0.2725)R$, so we can use Eq. (22) to compute the lunar gravitational acceleration g_1 at the surface of the moon:

$$g_1 = \frac{GM_1}{R_1^2} = G\frac{(0.0123)M}{(0.2725R)^2} \approx (0.1656)\frac{GM}{R^2}$$

$$= (0.1656)g \approx (0.1656)(32) \approx 5.3 \text{ ft/s}^2.$$

We now calculate the *escape velocity* from the earth—the minimum initial velocity with which an object must be launched straight upward from the earth's surface so that it will continue forever to move away from the earth. The condition that the upward motion will continue forever becomes the relation $v = dy/dt > 0$ for all $t \geqq 0$. We will use $y = y(t)$ to denote the distance from the object to the earth's *center* (because of the way in which Newton's law of gravitation is phrased). Our launch site, though, will be on the earth's *surface*, where we take $y_0 = y(0) = 6370$ km; that is, 6,370,000 m. The solution to the escape velocity problem will be the initial velocity $v_0 = v(0)$ that is just sufficient to insure that $v = v(t)$ is never zero or negative.

Now imagine the launched object as it is shown in Fig. 1.45, with mass m, at distance $y = y(t)$ from the center of the earth at time $t > 0$, and with velocity $v = v(t)$ then. The only force acting on the mass m as it ascends is the pull of gravity of the earth, given by Eq. (21), Newton's law of gravitation, as $F = -GMm/y^2$. The resulting acceleration of m is then given, with the *same* values of F and m, by Newton's second law of motion $F = ma$. We eliminate

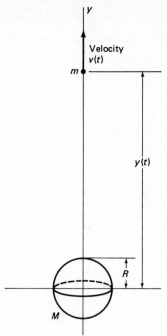

Figure 1.45 A mass m at a great distance from the earth.

F by equating the other sides of each equation:

$$ma = m\frac{dv}{dt} = -\frac{GMm}{y^2},$$

which we simplify to

$$\frac{dv}{dt} = -\frac{GM}{y^2}.$$

To solve this differential equation, we use the chain rule:

$$\frac{dv}{dt} = \frac{dv}{dy}\frac{dy}{dt} = v\frac{dv}{dy}.$$

Thus

$$v\frac{dv}{dy} = -\frac{GM}{y^2}.$$

That is,

$$D_y\left(\frac{1}{2}v^2\right) = D_y\left(\frac{GM}{y}\right).$$

Note that we think of y as the independent variable, rather than the more natural t. It follows immediately that

$$\frac{1}{2}v^2 = \frac{GM}{y} + C. \tag{23}$$

To evaluate the constant C, we use the fact that when $t = 0$, we have $v = v_0$ and $y = R$. Thus

$$\frac{1}{2}v_0^2 = \frac{GM}{R} + C;$$

we substitute the resulting value of C in Eq. (23) to obtain

$$v^2 = v_0{}^2 + 2GM\left(\frac{1}{y} - \frac{1}{R}\right).$$

In particular,

$$v^2 > v_0{}^2 - \frac{2GM}{R}.$$

Therefore v will remain positive provided that $v_0{}^2 \geqq 2GM/R$. With the aid of Eq. (22), we can now write a formula for the escape velocity for the earth:

$$v_0 = \sqrt{\frac{2GM}{R}} = \sqrt{2Rg}. \tag{24}$$

With $g = 9.8$ m/s² and $R = 6.37 \times 10^6$ m, this gives $v_0 \approx 11{,}174$ m/s, about 36,600 ft/s, about 24,995 mi/h, or about 6.94 mi/s.

The following example is a refinement of Example 2 in Section 1.2; we now take lunar gravity into account.

EXAMPLE 2 A lunar lander is free-falling toward the moon's surface at a speed of 1000 mi/h. Its retrorockets, when fired in free space, provide a deceleration of 33,000 mi/h². At what height above the lunar surface should the retrorockets be activated to insure a "soft" touchdown ($v = 0$ at impact)?

Solution We use units of kilomiles and hours and denote by $y(t)$ the lunar lander's distance from the center of the moon at time t. The moon's radius is about 1.08 kilomiles, so we want $v = 0$ when $y = 1.08$. The moon's *surface* gravitational acceleration is, as computed above, 5.3 ft/s²; this is approximately 13 kilomiles/h². Because gravitational acceleration is inversely proportional to the square of the distance, the lunar gravitational acceleration at distance y (from its center) is

$$\left(\frac{1.08}{y}\right)^2 (13) = \frac{15.16}{y^2}$$

kilomiles/h². We subtract this from the retrorocket acceleration of 33 kilomiles/h² to obtain

$$\frac{dv}{dt} = 33 - \frac{15.16}{y^2}$$

where $v = dy/dt$. We change the independent variable to y by using the chain rule to write

$$\frac{dv}{dt} = \frac{dv}{dy}\frac{dy}{dt} = v\frac{dv}{dy} = 33 - \frac{15.16}{y^2}.$$

Integration with respect to y gives

$$\frac{1}{2}v^2 = 33y + \frac{15.16}{y} + C.$$

The desired condition $v = 0$ when $y = 1.08$ implies that $C = -49.68$, so

$$\frac{1}{2}v^2 = 33y + \frac{15.16}{y} - 49.68.$$

Finally, we want to know the value of $y = y_0$ when $v_0 = -1$ (the initial velocity of the rocket, in kilomiles/h.) We substitute $v = -1$ into the last equation and simplify to obtain the quadratic equation $33y^2 - (50.18)y + 15.16 = 0$. The only roots of this equation are $y = 1.105$ and $y = 0.416$ (approximately). The second corresponds to a height *below* the lunar surface, so the first of the two is the solution we seek. And $1.105 - 1.080 = 0.025$ kilomiles, so the retrorockets should be activated at a distance of 25 miles above the lunar surface.

ROCKET PROPULSION

Suppose that the rocket of Fig. 1.46 blasts off straight upward from the surface of the earth at time $t = 0$. We want to compute its height y and velocity $v = dy/dt$ at time t. The rocket is propelled by exhaust gases that exit with constant speed c (relative to the rocket). Because of the combustion of its fuel, the mass $m = m(t)$ of the rocket is variable.

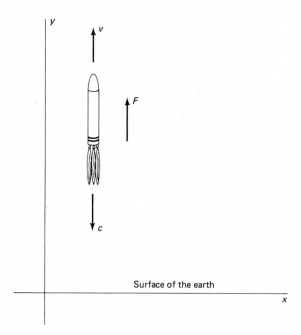

Figure 1.46 A rocket departing from the earth.

To derive the equation of motion of the rocket, we need Newton's second law in the form

$$\frac{dP}{dt} = F \tag{25}$$

where P is momentum (the product of mass and velocity) and F is the net external force (gravity, air resistance, and so on). If m is constant, then Eq. (25) takes the standard form $F = m(dv/dt)$.

Suppose that m changes to $m + \Delta m$ and v changes to $v + \Delta v$ during the short time interval from t to $t + \Delta t$. The change in the momentum of the *rocket itself* is

$$(m + \Delta m)(v + \Delta v) - mv = m\Delta v + v\Delta m + \Delta m \Delta v.$$

But the system also includes the exhaust gases expelled during this time interval, with mass $-\Delta m$ and velocity $v - c$. Hence the total change in momentum during the time interval Δt is

$$\Delta P \approx (m\Delta v + v\Delta m + \Delta m \Delta v) + (-\Delta m)(v - c)$$
$$= m\Delta v + c\Delta m + \Delta m \Delta v.$$

Dividing by Δt and taking the limit as $\Delta t \to 0$, we get

$$\frac{dP}{dt} = m\frac{dv}{dt} + c\frac{dm}{dt}$$

because Δm and Δv approach zero as $\Delta t \to 0$. We substitute this expression for dP/dt into Newton's law (25) and thereby obtain the **rocket propulsion equation**

$$m\frac{dv}{dt} + c\frac{dm}{dt} = F. \tag{26}$$

We shall assume that $F = F_G + F_R$, where $F_G = -mg$ is a constant force of gravity and $F_R = -kv$ is a force of resistance proportional to velocity. This gives

$$m\frac{dv}{dt} + c\frac{dm}{dt} = -mg - kv. \tag{27}$$

Now suppose that the fuel of the rocket is consumed at a constant rate β (burn rate) during the time interval $[0, \tau]$, during which time the mass of the rocket decreases from m_0 to m_1. Thus

$$m(0) = m_0, \qquad m(\tau) = m_1,$$
$$m(t) = m_0 - \beta t, \qquad \frac{dm}{dt} = -\beta \quad \text{for } t \leq \tau, \tag{28}$$

with burnout occurring at time $t = \tau$.

We proceed to compute the velocity at time $t \leq \tau$. If we substitute the expressions in (28) into Eq. (27), the result is

$$(m_0 - \beta t)\frac{dv}{dt} - \beta c = -(m_0 - \beta t)g - kv,$$

which we can rewrite in the form

$$\frac{dv}{dt} + \frac{k}{m_0 - \beta t}v = -g + \frac{\beta c}{m_0 - \beta t}. \tag{29}$$

This is a linear first order equation for v as a function of t, with integrating factor

$$\rho = \exp\left(\int \frac{k\,dt}{m_0 - \beta t}\right) = (m_0 - \beta t)^{-k/\beta}.$$

The details are somewhat tedious (Problem 31), but the result of integrating Eq. (29) is

$$v(t) = v_0 M^{k/\beta} + \frac{\beta c}{k}(1 - M^{k/\beta}) + \frac{gm_0}{\beta - k}(M - M^{k/\beta}), \tag{30}$$

where $v_0 = v(0)$ and

$$M = \frac{m_0 - \beta t}{m_0} = \frac{m(t)}{m_0}$$

is the **fractional mass** of the rocket at time t.

NO RESISTANCE

Let us simplify the situation by ignoring the force of resistance to the motion of the rocket. Then $k = 0$. In Problem 29 we ask you to take the limit of the right-hand side of Eq. (30) as $k \to 0$, using l'Hôpital's rule, to obtain

$$v(t) = v_0 - gt + c \ln \frac{m_0}{m_0 - \beta t} \tag{31}$$

for the case of no resistance. You can get the same result by setting $k = 0$ in the differential equation in (29) and then integrating (Problem 30). Because $m_0 - \beta\tau = m_1$, the velocity of the rocket at burnout $(t = \tau)$ is

$$v(\tau) = v_0 - g\tau + c \ln \left(\frac{m_0}{m_1}\right). \tag{32}$$

We can compute the height of the rocket at time $t \leq \tau$ by integrating Eq. (31):

$$y(t) = (v_0 + c \ln m_0)t - \frac{1}{2}gt^2 - c \int_0^t \ln(m_0 - \beta t)\, dt,$$

assuming that $y_0 = y(0) = 0$. Using the integration formula $\int \ln u\, du = u \ln u - u + C$, we find after some simplifications that

$$y(t) = (v_0 + c)t - \frac{1}{2}gt^2 - \frac{c}{\beta}(m_0 - \beta t) \ln \frac{m_0}{m_0 - \beta t}. \tag{33}$$

When we substitute $t = \tau$, we find the height at burnout to be

$$y(\tau) = (v_0 + c)\tau - \frac{1}{2}g\tau^2 - \frac{cm_1}{\beta} \ln \frac{m_0}{m_1}. \tag{34}$$

FREE SPACE

Suppose finally that the rocket is accelerating in free space, where there is neither gravity nor resistance, so that $g = k = 0$. With $g = 0$ in Eq. (32), we see that as the mass of the rocket decreases from m_0 to m_1, its increase in velocity is

$$\Delta v = v(\tau) - v_0 = c \ln \frac{m_0}{m_1}. \tag{35}$$

Note that Δv depends only upon the exhaust gas speed c and the initial-to-final mass ratio m_0/m_1, but does *not* depend upon the burn rate β. For example, if $c = 2$ mi/s and $m_0/m_1 = 10$, then $v = 2 \ln 10 \approx 4.61$ mi/s. Thus if a rocket initially consists predominately of fuel, then it can attain speeds significantly greater than the (relative) speed of its exhaust gases.

1.9 Problems

1. The acceleration of a certain sports car is proportional to the difference between 250 km/h and the velocity of the sports car. If this machine can accelerate from rest to 100 km/h in 10 seconds, how long will it take for the car to accelerate from rest to 200 km/h?

2. Suppose that a body moves through a resisting medium with resistance proportional to its velocity v, so that $dv/dt = -kv$. (a) Show that its velocity and position at time t are given by

$$v = v_0 e^{-kt} \quad \text{and} \quad x = x_0 + \left(\frac{v_0}{k}\right)(1 - e^{-kt}).$$

(b) Conclude that the body travels only a *finite* distance, and find that distance.

3. Suppose that a motorboat is moving at 40 ft/s when its motor suddenly quits, and that 10 s later the boat has slowed to 20 ft/s. Assume, as in Problem 2, that the resistance it encounters while coasting is proportional to its velocity. How far will the boat coast in all?

4. Consider a body that moves horizontally through a medium whose resistance is proportional to the *square* of the velocity, so that $dv/dt = -kv^2$. Show that

$$v(t) = \frac{v_0}{1 + v_0 kt}$$

and that

$$x(t) = x_0 + \frac{1}{k} \ln(1 + v_0 kt).$$

Note that, in contrast to the result of Problem 2, $x(t) \to \infty$ as $t \to +\infty$.

5. Assuming resistance proportional to the square of the velocity as in Problem 4, how far does the motorboat of Problem 3 coast in the first minute after its motor quits?

6. Assume that a body moving with velocity v encounters resistance of the form $dv/dt = -kv^{3/2}$. Show that

$$v(t) = \frac{4v_0}{(kt\sqrt{v_0} + 2)^2}$$

and

$$x(t) = x_0 + \sqrt{\frac{2v_0}{k}}\left(1 - \frac{2}{kt\sqrt{v_0} + 2}\right).$$

Conclude that under a $\frac{3}{2}$-power resistance a body travels only a finite distance before coasting to a stop.

7. Calculate $v(t)$ and $x(t)$ under the assumption that

$$\frac{dv}{dt} = -kv^{1+r} \quad \text{where} \quad 0 < r < 1.$$

Also compute the coasting distance.

8. A car starts from rest ($x_0 = v_0 = 0$) and travels along a straight road. Its engine provides a constant acceleration of 18 ft/s². The combination of air and road resistance provides a deceleration of 0.12 ft/s² for every foot per second of the car's velocity v. (a) Set up and solve a first order differential equation for $v(t)$. (b) Find v when $t = 10$ s and also the limiting velocity as $t \to +\infty$, both in miles per hour.

9. A motorboat weighs 32,000 lb, and its motor provides a thrust of 5000 lb. Assume that the water resistance is 100 pounds for each foot per second of the speed v of the boat. Then

$$1000 \frac{dv}{dt} = 5000 - 100v.$$

If the boat starts from rest, what is the maximum velocity that it can attain?

10. A woman bails out of an airplane at an altitude of 10,000 feet, falls freely for 20 s, then opens her parachute. How long will it take her to reach the ground? Assume that $\rho = 0.15$ without the parachute and that $\rho = 1.5$ with the parachute. (*Suggestion:* First determine her height and velocity when the parachute opens.)

11. According to a newspaper account, a paratrooper survived a training jump from 1200 feet when his parachute failed to open but provided some resistance by flapping unopened in the wind. Allegedly he hit the ground at 100 mi/h after falling for 8 s. Test the accuracy of this account. (*Suggestion:* Find ρ (in Eq. (2)) by assuming a terminal velocity of 100 mi/h. Then calculate the time required to fall 1200 ft.)

12. It is proposed to dispose of nuclear wastes—in drums with weight $W = 640$ lb and volume 8 ft³—by dropping them into the ocean ($v_0 = 0$). The force equation for a drum falling through water is

$$m \frac{dv}{dt} = -W + B + F_R,$$

where the bouyant force B is equal to the weight of the volume of water displaced by the drum (Archimedes' principle) and F_R is the force of water resistance, found empirically to be 1 lb per foot per second of the velocity of a drum. If the drums are likely to burst upon an impact of more than 75 ft/s, what is the maximum depth to which they can be dropped in the ocean without likelihood of bursting?

13. Derive Eq. (13).

14. Derive Eq. (14); note that

$$\int \tanh u \, du = \int \frac{\sinh u}{\cosh u} \, du = \ln \cosh u + C.$$

15. A motorboat starts from rest (initial velocity $v(0) = 0$). Its motor provides a constant acceleration of 4 ft/s², but water resistance causes a deceleration of $v^2/400$ ft/s². Find v when $t = 10$ s, and also find the *limiting velocity* as $t \longrightarrow +\infty$ (that is, the maximum possible speed of the boat).

16. Derive Eqs. (17)–(19) of the text.

17. If a ball is projected upward from the ground with initial velocity v_0, deduce from Eq. (20) that the maximum height it attains is

$$Y_{\max} = \frac{1}{2\rho} \ln \left(1 + \frac{\rho v_0^2}{g} \right).$$

18. Suppose that a crossbow bolt is shot straight upward with initial velocity 288 ft/s. Assume that air resistance is proportional to v^2, the square of the velocity, with $\rho = 0.0002$ in Eqs. (12) and (16). Find the maximum height the bolt attains and its time of ascent.

19. In a continuation of Problem 18, find the descent time of the bolt and the speed with which it strikes the ground.

20. Suppose that $\rho = 0.075$ in Eq. (12) for a paratrooper falling with parachute open. If he jumps from an altitude of 10,000 ft and opens his parachute immediately, what will be his terminal speed? How long will it take him to reach the ground?

21. Suppose that the paratrooper of Problem 20 freefalls for 30 s with $\rho = 0.00075$ before opening his parachute. How long will it now take him to reach the ground?

22. The mass of the sun is 329,320 times that of the earth, and its radius is 109 times the radius of the earth. (a) To what radius (in meters) would the earth have to be compressed in order for it to become a *black hole*—the escape velocity from its surface equal to the velocity $c = 3 \times 10^8$ m/s of light? (b) Repeat part (a) with the sun in place of the earth.

23. (a) Show that if a projectile is launched upward from the surface of the earth with initial velocity v_0 less that escape velocity, then the maximum distance from the center of the earth that the projectile will attain is

$$Y_{\max} = \frac{2gR^2}{2gR - v_0{}^2}$$

where g and R are the surface gravity and radius of the earth. (b) With what initial velocity v_0 (in miles per hour) must such a projectile be launched to yield a maximum height of 100 mi above the surface of the earth? (c) Find the maximum distance from the center of the earth, expressed in terms of earth radii, attained by a projectile launched from the surface of the earth with 90% of escape velocity.

24. (a) Suppose that a body is dropped ($v_0 = 0$) from a distance y_0 from the earth's center, so that its gravitational acceleration is $dv/dt = -k/y^2$ with $k = GM = R^2g$. Show that it reaches the height y at time

$$t = \sqrt{\frac{y_0}{2R^2g}}\left(\sqrt{yy_0 - y^2} + y_0 \cos^{-1}\sqrt{\frac{y}{y_0}}\right).$$

(*Suggestion:* Substitute $y = y_0 \cos^2 \theta$ to evaluate the antiderivative of $\sqrt{y/(y_0 - y)}$.) (b) If a body is dropped from a height of 1000 mi above the earth's surface, how long does it fall and with what speed will it strike the earth's surface?

25. Suppose that a projectile is fired straight upward from the surface of the earth with initial velocity v_0. Then its height $x(t)$ at time t satisfies the initial value problem

$$\frac{d^2x}{dt^2} = -\frac{gR^2}{(x + R)^2}; \qquad x(0) = 0, \qquad x'(0) = v_0.$$

Let us use the values $g = 32.15$ (ft/s²) or 0.006089 (mi/s²) for the gravitational acceleration of the earth at its surface and $R = 3960$ (mi) as the radius of the earth. (a) Substitute $dv/dt = v(dv/dx)$ and then integrate to obtain

$$v^2 = \frac{v_0{}^2R - (2gR - v_0{}^2)x}{R + x}$$

for the velocity of the projectile at altitude x. If $v_0 = 1$ (mi/s), what is the maximum height $x_{\max}$ attained by the projectile? (*Answer:* About 84 mi.) (b) Assume that $v_0{}^2 < 2gR$. Conclude from part (a) that the time required for the projectile to ascend to its maximum height $x_{\max}$ is

$$t_{\max} = \int_0^{x_{\max}} \sqrt{\frac{R+x}{\alpha^2 - \beta^2 x}}\, dx,$$

where $\alpha^2 = v_0^2 R$ and $\beta^2 = 2gR - v_0^2$. Make the rationalizing substitution $u^2 = (R+x)/(\alpha^2 - \beta^2 x)$, and then integrate to obtain

$$t_{\max} = \frac{2gR^2}{\beta^3}\left[\frac{\pi}{2} - \tan^{-1}\left(\frac{\beta}{v_0}\right) + \frac{\beta v_0}{2gR}\right].$$

In the case $v_0 = 1$ (mi/s), conclude that $t_{\max} \approx 169$ s.

26. A rocket has an initial weight of 25 tons, of which 20 tons consists of fuel mixture that burns at the rate of 1 ton/s. Its exhaust gas speed is 1 mi/s. It blasts off at time $t = 0$ with $y_0 = v_0 = 0$. Find its height and velocity (in miles per hour) at burnout. Ignore air resistance and use $g = 32$ ft/s².

27. For the rocket of **Problem 26**, how large must its exhaust speed be in order for it to get off the ground?

28. For a rocket in free space, show that Eq. (26) can be written in the form $dv/dm = -c/m$. Integrate this equation to obtain the velocity of the rocket as given in Eq. (35).

29. Derive Eq. (31) by taking the limit as $k \longrightarrow 0$ in Eq. (30).

30. Derive Eq. (31) by solving Eq. (29) with the value $k = 0$.

31. Derive Eq. (30) by solving Eq. (29) in the case $k > 0$.

32. The V-2 rocket of World War II had an initial weight of 28,300 lb (so $m_0 = 878.88$ slugs, using $g = 32.2$ ft/s²); 68.5% of its mass was fuel. This fuel burned uniformly for 70 s with an exhaust velocity of 1.25 mi/s. Assume an air resistance of $v/10$ lb (with v in ft/s). Find the velocity and height of the V-2 at burnout under the assumption that it is fired vertically upward. Begin by solving Eq. (29) with the numerical parameters given here.

Chapter 1 Summary and a Look Ahead

In this chapter we have discussed applications of and solution methods for several important types of first order differential equations, including those that are separable (Section 1.4), linear (Section 1.5), or exact (Section 1.7). In Section 1.6 we discussed substitution techniques that can sometimes be used to transform a given first order differential equation into one that is either separable, linear, or exact.

Lest it appear that these methods constitute a "grab bag" of special and unrelated techniques, it is important to note that they are all versions of a single idea. Given a differential equation

$$f(x, y, y') = 0, \tag{1}$$

we attempt to write it in the form

$$\frac{d}{dx}[G(x, y)] = 0. \tag{2}$$

It is precisely to attain the form in (2) that we multiply the terms in Eq. (1) by an appropriate integrating factor (even if all we are doing is separating the variables). But once we have found a function $G(x, y)$ such that (1) and (2) are equivalent, a general solution is defined implicitly by means of the equation

$$G(x, y) = C \qquad (3)$$

that one obtains by integrating (2).

Given a specific first order differential equation to be solved, we can attack it by means of the following steps:

Is it *separable*? If so, separate the variables and integrate (Section 1.4).

Is it *linear*? That is, can it be written in the form

$$\frac{dy}{dx} + P(x)y = Q(x)?$$

If so, multiply by the integrating factor $\rho = \exp(\int P\,dx)$ of Section 1.5.

Is it *exact*? That is, when the equation is written in the form $M\,dx + N\,dy = 0$, is $\partial M/\partial y = \partial N/\partial x$? Or can terms be grouped or an integrating factor found that produces an exact equation (Section 1.7)?

If the equation as it stands is not separable, linear, or exact, is there a plausible substitution that will make it so? For instance, is it homogeneous (Section 1.6)?

Many first order differential equations succumb to the line of attack outlined above. Nevertheless, many more do not. Because of the wide availability of computers—even inexpensive pocket computers—numerical techniques are being used with increasing frequency to *approximate* the solutions of differential equations that cannot easily be solved explicitly by the methods of this chapter.

To illustrate very briefly a standard technique of numerical approximation, let us consider the initial value problem

$$\frac{dy}{dx} = y, \qquad y(0) = 1. \qquad (4)$$

In order to *approximate* the exact solution $y(x) = e^x$ on the interval $0 \leq x \leq 1$, we begin with a subdivision of the interval into (say) 10 subintervals each of length $h = 0.1$, by means of the points

$$x_0 = 0.0,\ x_1 = 0.1,\ x_2 = 0.2, \ldots, x_{10} = 1.0.$$

We know that $y_0 = y(0) = 1$, and we want to find (for each $n = 1, 2, 3, \ldots,$ 10) an approximation y_n to the actual value $y(x_n)$ of the solution at x_n. If we have found y_n somehow, then the mean value theorem yields

$$\begin{aligned} y_{n+1} &= y(x_n) + y'(\overline{x_n})h \\ &\approx y(x_n) + y'(x_n)h \\ &= (1 + h)y(x_n), \end{aligned}$$

because our differential equation is $y' = y$. Thus each value of y should be approximately $1 + h$ times the preceding value of y, so we choose

$$y_{n+1} = (1 + h)y_n. \qquad (5)$$

Beginning with the initial value $y_0 = 1$, Eq. (5) is an iterative formula for computing successively the values $y_1, y_2, \ldots, y_{10}$. This will give us a table of *approximate* values of the solution of the initial value problem in (4).

Figure 1.47 is a listing of a simple BASIC program that carries out the iteration in (5). It can be run on any pocket computer that is programmable in BASIC and will display at each iteration the values of n, x_n, y_n, and $y(x_n) = \exp(x_n)$. Figure 1.48 shows a version of the same program, written for the IBM Personal Computer, fully documented, and formatted for ease of examination of what the program does. The REM (remark) statements are not executed by the computer; their purpose is explanation and spacing for the benefit of the reader.

```
100 X = 0
110 Y = 1
120 H = 0.1
130 FOR N = 1 TO 10
140 X = X + H
150 Y = (1 + H)*Y
160 PRINT N
170 PRINT X
180 PRINT Y
190 PRINT EXP (X)
200 NEXT N
210 END
```

Figure 1.47 A BASIC program for Euler's method for $y' = y, y(0) = 1$.

```
100 REM —— THIS PROGRAM APPLIES EULER'S METHOD TO APPROXIMATE
110 REM —— THE SOLUTION ON [0, 1] OF Y = Y', WITH SUBINTERVALS
120 REM —— OF LENGTH H = 0.1.  FOR EACH N = 1,2,———,10, THE
130 REM —— PROGRAM PRINTS THE VALUES OF N AND X, THE COMPUTED
140 REM —— (APPROXIMATE) VALUE OF Y, AND THE (ROUNDED) ACTUAL
150 REM —— VALUE Y(X) = EXP X.
160 REM
170 REM —— PROGRAM VARIABLES:
180 REM
190 REM —— X      THE INDEPENDENT VARIABLE
200 REM —— Y      THE APPROXIMATE VALUE OF Y(X)
210 REM —— H      THE LENGTH OF EACH SUBINTERVAL
220 REM
230 REM —— INITIAL VALUES:
240 REM
250          X = 0 : Y = 1 : H = 0.1
260 REM
270 REM —— LOOP TO COMPUTE NEXT X AND NEXT Y:
280 REM
290          FOR N = 1 TO 10
300              LET X = X + H : Y = (1 + H)*Y
310              LPRINT N, X, Y, EXP (X)
320          NEXT N
330 REM
340          END
```

Figure 1.48 The program of Figure 1.47 with comment lines and improved readability.

Figure 1.49 shows the output of this program as executed by the IBM Personal Computer. Note that $y_{10} = 2.593743$ is about 5% less than the actual value $y(1) = e \approx 2.718282$. A better approximation could be obtained

n	x_n	y_n	$y(x_n) = \exp(x_n)$
1	.1	1.1	1.105171
2	.2	1.21	1.221403
3	.3	1.331	1.349859
4	.4	1.4641	1.491825
5	.5	1.61051	1.648721
6	.6	1.771561	1.822119
7	.7	1.948717	2.013753
8	.8	2.143589	2.225541
9	.9	2.357948	2.459603
10	1	2.593743	2.718282

Figure 1.49 Output of the program of Figure 1.48.

by decreasing the value of h, and thereby increasing the number of subintervals. Such a table of approximate values can be used numerically or can be used to construct a curve that approximates the graph of the actual solution of the differential equation.

Chapter 6 is devoted to numerical solutions of differential equations, and the first two or three sections of Chapter 6 can well be studied at this time, before proceeding to higher order differential equations in Chapter 2.

Problems

For each of the following initial value problems, apply the iterative method described above to compute approximate values $y_1, y_2, \ldots, y_{10}$ of the solution on the interval $0 \leq x \leq 1$. In each case, apply the mean value theorem to derive first the indicated iterative formula for y_{n+1} in terms of x_n, y_n, and $h = 0.1$. Also compare your approximate values with the actual values of the solution.

1. $y' = -y, y(0) = 1$; $y_{n+1} = (1 - h)y_n$.
2. $y' = 2y, y(0) = 1$; $y_{n+1} = (1 + 2h)y_n$.
3. $y' = y^2, y(0) = 0.5$; $y_{n+1} = y_n + hy_n^2$.
4. $y' = 2xy, y(0) = 1$; $y_{n+1} = (1 + 2hx_n)y_n$.

Linear Equations
of Higher Order

2

Only in very special cases can an nth order differential equation of the form $G(x, y, y', y'', \ldots, y^{(n)}) = 0$ be solved exactly and explicitly. In this chapter we restrict our attention to *linear* equations of order $n > 1$. The general **nth order linear equation** has the form

$$a_n(x)\frac{d^n y}{dx^n} + a_{n-1}(x)\frac{d^{n-1}y}{dx^{n-1}} + \cdots + a_1(x)\frac{dy}{dx} + a_0(x)y = F(x). \qquad (1)$$

Unless otherwise noted, we will always assume that the coefficient functions $a_i(x)$ and $F(x)$ are continuous on some open interval I (perhaps unbounded) on which we wish to solve the differential equation, but they need not be linear functions.

If the function $F(x)$ on the right-hand side of (1) vanishes identically on I, then we call (1) a **homogeneous** linear equation; otherwise, it is **nonhomogeneous**. For example, the second order equation

$$x^2 y'' + 2xy' + 3y = \cos x$$

is nonhomogeneous; its *associated* homogeneous equation is

$$x^2 y'' + 2xy' + 3y = 0.$$

The homogeneous linear equation **associated** with Eq. (1) is

$$a_n(x)\frac{d^n y}{dx^n} + a_{n-1}(x)\frac{d^{n-1}y}{dx^{n-1}} + \cdots + a_1(x)\frac{dy}{dx} + a_0(x)y = 0. \qquad (2)$$

In order to solve an nth order differential equation we must, at least in principle, integrate n times (to get from $y^{(n)}$ to y). It is therefore natural to expect Eq. (2) to have a general solution involving n arbitrary constants $c_1, c_2, \ldots, c_n$ (of integration). Indeed, we will see in Section 2.2 that if $a_n(x)$ is nonzero on the interval I, then (2) has a general solution of the especially pleasant form

$$y = c_1 y_1 + c_2 y_2 + \cdots + c_n y_n, \tag{3}$$

a *linear combination* of n particular solutions $y_1, y_2, \ldots, y_n$. The general theory of homogeneous linear equations parallels the case of second order linear equations (the case $n = 2$) which is discussed below in this section.

With regard to the crucial matter of actually finding a general solution like (3), there is a substantial difference between the case $n = 1$ and the higher order cases $n \geq 2$. In Section 1.5 we saw that there exists a systematic procedure by which a general solution of any *first order* linear differential equation can always be found explicitly. By contrast, there does *not* exist a formula for computing the general solution of an arbitrary higher order linear equation with variable coefficients. Fortunately, many important applications involve only homogeneous equations with *constant* coefficients, and we will see in Section 2.3 how to solve such equations in a routine fashion.

Linear differential equations with constant coefficients frequently appear as mathematical models of mechanical systems and electrical circuits. For example, suppose that a mass m is attached both to a spring that exerts on it a force F_S and to a dashpot (shock absorber) that exerts a force F_R on the mass (see Fig. 2.1). Assume that the restoring force F_S of the spring is proportional to the displacement x (positive to the right, negative to the left) of the mass from equilibrium, and that the dashpot force F_R is proportional to the velocity $v = dx/dt$ of the mass. With the aid of Fig. 2.2, we also get the appropriate directions of action of these two forces:

$$F_S = -kx \quad \text{and} \quad F_R = -cv \quad (k, c > 0).$$

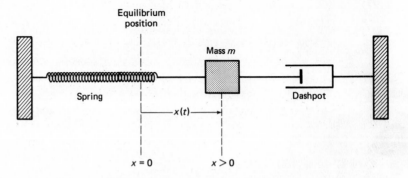

Figure 2.1 A mass-spring-dashpot system.

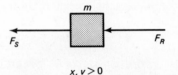

Figure 2.2 Directions of the forces acting on m.

The minus signs are correct—F_S and F_R are negative when x and v are positive. Newton's law $F = ma$ now gives

$$mx'' = F_S + F_R;\tag{4}$$

that is,

$$m\frac{d^2x}{dt^2} + c\frac{dx}{dt} + kx = 0.\tag{5}$$

This homogeneous second order linear equation governs the *free vibrations* of the mass; we will return to this problem in detail in Section 2.4.

If, in addition to F_S and F_R, the mass m is acted on by an external force $F(t)$—which must then be added to the right-hand side in Eq. (4)—the resulting equation is

$$m\frac{d^2x}{dt^2} + c\frac{dx}{dt} + kx = F(t).\tag{6}$$

This nonhomogeneous linear differential equation governs the *forced vibrations* of the mass under the influence of the external force $F(t)$.

As this example illustrates, an external force in a mechanical system typically corresponds to a nonhomogeneous term in the differential equation(s) describing the system. In Section 2.2 we will see that a general solution of a nonhomogeneous linear differential equation is the sum $y = y_c + y_p$ of (1) a general solution y_c of the associated homogeneous equation and (2) a single particular solution y_p of the given nonhomogeneous equation. We will take up the problem of finding y_p in Sections 2.5 and 2.7.

SECOND ORDER LINEAR EQUATIONS

Consider the general second order linear differential equation

$$A(x)y'' + B(x)y' + C(x)y = F(x)\tag{7}$$

where the coefficient functions A, B, C, and F are continuous on the open interval I. Here we will assume in addition that $A(x) \neq 0$ at each point of I, so we can divide each term in (7) by $A(x)$ and write it in the form

$$y'' + p(x)y' + q(x)y = f(x).\tag{8}$$

We will first discuss the associated homogeneous equation

$$y'' + p(x)y' + q(x)y = 0.\tag{9}$$

A particularly useful property of this *homogeneous* linear equation is the fact that the sum of any two solutions of (9) is again a solution, as is any constant multiple of a solution. This is the central idea of the following theorem.

THEOREM 1:   PRINCIPLE OF SUPERPOSITION

Let y_1 and y_2 be two solutions of the homogeneous linear Eq. (9) on the interval I. If c_1 and c_2 are constants, then the linear combination

$$y = c_1y_1 + c_2y_2\tag{10}$$

is also a solution of (9) on I.

Proof The conclusion follows almost immediately from the *linearity* of the operation of differentiation, which gives

$$y' = c_1 y_1' + c_2 y_2' \quad \text{and} \quad y'' = c_1 y_1'' + c_2 y_2''.$$

Then

$$
\begin{aligned}
y'' + py' + qy &= (c_1 y_1 + c_2 y_2)'' + p(c_1 y_1 + c_2 y_2)' + q(c_1 y_1 + c_2 y_2) \\
&= (c_1 y_1'' + c_2 y_2'') + p(c_1 y_1' + c_2 y_2') + q(c_1 y_1 + c_2 y_2) \\
&= c_1(y_1'' + py_1' + qy_1) + c_2(y_2'' + py_2' + qy_2) \\
&= (c_1)(0) + (c_2)(0) = 0
\end{aligned}
$$

because y_1 and y_2 are solutions. Thus $y = c_1 y_1 + c_2 y_2$ is also a solution.

EXAMPLE 1 We can see by inspection that $y_1 = \cos x$ and $y_2 = \sin x$ are two solutions of the equation $y'' + y = 0$. Theorem 1 tells us that any linear combination of these solutions, such as

$$y = 3y_1 - 2y_2 = 3\cos x - 2\sin x$$

is also a solution. We will see later that, conversely, *every* solution of $y'' + y = 0$ is a linear combination of these two particular solutions y_1 and y_2. Thus a general solution of $y'' + y = 0$ is $y = c_1 \cos x + c_2 \sin x$.

We gave above the linear equation $mx'' + cx' + kx = F(t)$ as a mathematical model of the motion of the mass of Fig. 2.1. Physical considerations suggest that the motion of the mass ought to be determined by its initial position and initial velocity. Hence, given any preassigned values of $x(0)$ and $x'(0)$, Eq. (6) ought to have a *unique* solution satisfying these initial conditions. More generally, in order to be a "good" mathematical model of a deterministic physical situation, a differential equation must have unique solutions satisfying any appropriate initial conditions. The following existence and uniqueness theorem (proved in Section 9.4) gives us this assurance for the general second order linear equation.

THEOREM 2: EXISTENCE AND UNIQUENESS

Suppose that the functions p, q, and f are continuous on the open interval I containing the point a. Then, given two numbers b_0 and b_1, Eq. (8)

$$y'' + p(x)y' + q(x)y = f(x)$$

has a unique (that is, one and only one) solution on the entire interval I that satisfies the initial conditions

$$y(a) = b_0, \qquad y'(a) = b_1. \tag{11}$$

Equation (8) and the conditions in (11) constitute a second order linear **initial value problem**. Theorem 2 tells us that any such initial value problem has a unique solution on the *whole* interval I where the coefficient functions in (8) are continuous. Recall from Section 1.3 that a *nonlinear* differential equation generally has a unique solution on only a smaller interval.

EXAMPLE 1 CONTINUED We saw above that $y = 3\cos x - 2\sin x$ is a solution (on the entire real line) of $y'' + y = 0$. It has the initial values

$y(0) = 3$, $y'(0) = -2$. Theorem 2 tells us that this is the *only* solution with these initial values. More generally, the solution

$$y = b_0 \cos x + b_1 \sin x$$

satisfies the *arbitrary* initial conditions $y(0) = b_0$, $y'(0) = b_1$; this illustrates the *existence* of such a solution, also as guaranteed by Theorem 2.

Example 1 suggests how, given a *homogeneous* second order linear equation, we might actually find the solution whose existence is assured by Theorem 2. First, we find two "essentially different" solutions y_1 and y_2; second, we attempt to impose on the general solution

$$y = c_1 y_1 + c_2 y_2 \tag{12}$$

the initial conditions $y(a) = b_0$, $y'(a) = b_1$. That is, we attempt to solve the simultaneous equations

$$\left. \begin{array}{l} c_1 y_1(a) + c_2 y_2(a) = b_0, \\ c_1 y_1'(a) + c_2 y_2'(a) = b_1 \end{array} \right\} \tag{13}$$

for the coefficients c_1 and c_2.

EXAMPLE 2 Verify that the functions $y_1 = e^x$ and $y_2 = xe^x$ are solutions of $y'' - 2y' + y = 0$, and then find a solution satisfying the initial conditions $y(0) = 3$, $y'(0) = 1$.

Solution The verification is routine; we omit it. We impose the given initial conditions on the general solution $y = c_1 e^x + c_2 xe^x$, for which

$$y' = (c_1 + c_2)e^x + c_2 xe^x,$$

to obtain the simultaneous equations

$$\begin{cases} y(0) = c_1 = 3, \\ y'(0) = c_1 + c_2 = 1. \end{cases}$$

The resulting solution is $c_1 = 3$, $c_2 = -2$. Hence the solution of the original initial value problem is $y = 3e^x - 2xe^x$.

In order for the procedure of Example 2 to succeed, the two solutions y_1 and y_2 must have the elusive property that the equations in (13) can always be solved for c_1 and c_2, no matter what the initial conditions b_0 and b_1 might be. The following definition tells precisely how different the two functions y_1 and y_2 must be.

DEFINITION: LINEAR INDEPENDENCE OF TWO FUNCTIONS

Two functions defined on an open interval are called **linearly independent** *provided that neither is a constant multiple of the other.*

Two functions are said to be **linearly dependent** if they are not linearly independent; that is, one of them *is* a constant multiple of the other. We can always determine whether two given functions f and g are linearly dependent on an interval I by noting at a glance whether either of the two quotients

f/g or g/f is a constant on I. Thus it is clear that the following pairs of functions are linearly independent on the entire real line:

$$\sin x \quad \text{and} \quad \cos x;$$
$$e^x \quad \text{and} \quad e^{-2x};$$
$$e^x \quad \text{and} \quad xe^x;$$
$$x + 1 \quad \text{and} \quad x^2;$$
$$x \quad \text{and} \quad |x|.$$

But the identically zero function and any other function g are linearly dependent on any interval because $0 = (0)(g(x))$. Also, the functions $f(x) = \sin 2x$ and $g(x) = \sin x \cos x$ are linearly dependent on any interval because $f(x) = 2g(x)$ for all x (a familiar trigonometric identity).

But does the homogeneous equation $y'' + py' + qy = 0$ always have two linearly independent solutions? Theorem 2 says yes! We need only choose y_1 and y_2 so that

$$y_1(a) = 1, \; y_1'(a) = 0 \quad \text{and} \quad y_2(a) = 0, \; y_2'(a) = 1.$$

It is then impossible that either $y_1 = ky_2$ or $y_2 = ky_1$ because $(k)(0) \neq 1$ for any constant k. Theorem 2 tells us that two such linearly independent solutions *exist*; actually finding them is a crucial matter that we will defer until later sections (beginning in Section 2.3).

We want to show, finally, that given *any* two linearly independent solutions y_1 and y_2 of the homogeneous equation

$$y'' + p(x)y' + q(x)y = 0, \tag{9}$$

every solution y of (9) can be expressed as a linear combination

$$y = c_1 y_1 + c_2 y_2 \tag{12}$$

of y_1 and y_2. This means that the function in (12) is a *general solution* of Eq. (9).

As suggested by the equations in (13), the determination of the constants c_1 and c_2 in (12) depends upon a certain 2×2 determinant of values of y_1, y_2, and their derivatives. Given two functions f and g, the **Wronskian** of f and g is the determinant

$$W = \begin{vmatrix} f & g \\ f' & g' \end{vmatrix} = fg' - f'g.$$

We write either $W(f, g)$ or $W(x)$, depending upon whether we wish to emphasize the two functions or the point x at which their Wronskian is to be evaluated. For example,

$$W(\cos x, \sin x) = \begin{vmatrix} \cos x & \sin x \\ -\sin x & \cos x \end{vmatrix} = \cos^2 x + \sin^2 x = 1$$

while

$$W(e^x, xe^x) = \begin{vmatrix} e^x & xe^x \\ e^x & e^x + xe^x \end{vmatrix} = e^{2x}.$$

These are examples of linearly *independent* pairs of solutions of differential equations (see Examples 1 and 2). Note that in each case the Wronskian is everywhere *nonzero*.

On the other hand, if the functions f and g are linearly dependent, with $f = kg$ (for example), then

$$W(f, g) = \begin{vmatrix} kg & g \\ kg' & g' \end{vmatrix} = kgg' - kg'g \equiv 0.$$

Thus the Wronskian of two linearly *dependent* functions is identically zero. In Section 2.2 we will prove that, if the two functions y_1 and y_2 are solutions of a homogeneous second order linear equation, then the strong converse stated in part (b) of the following theorem holds.

THEOREM 3: WRONSKIANS OF SOLUTIONS

Suppose that y_1 and y_2 are two solutions of the homogeneous second order linear equation (Eq. (9))

$$y'' + p(x)y' + q(x)y = 0$$

on an open interval I on which p and q are continuous.

(a) *If y_1 and y_2 are linearly dependent, then $W(y_1, y_2) \equiv 0$ on I;*
(b) *If y_1 and y_2 are linearly independent, then $W(y_1, y_2) \neq 0$ at each point of I.*

Thus, given two solutions of (9), there are just two possibilities: The Wronskian W is identically zero if the solutions are linearly dependent; the Wronskian is never zero if the solutions are linearly independent. The latter fact is what we need in order to show that $y = c_1 y_1 + c_2 y_2$ is a general solution of Eq. (9) if y_1 and y_2 are linearly independent solutions.

THEOREM 4: GENERAL SOLUTIONS

Let y_1 and y_2 be two linearly independent solutions of the homogeneous equation (Eq. (9))

$$y'' + p(x)y' + q(x)y = 0$$

with p and q continuous on the open interval I. If Y is any solution whatsoever of Eq. (9), then there exist numbers c_1 and c_1 such that

$$Y(x) = c_1 y_1(x) + c_2 y_2(x)$$

for all x in I.

In essence, Theorem 4 tells us that when we have found *two* linearly independent solutions of the homogeneous equation in (9), then we have found *all* of its solutions.

Proof Choose a point a of I, and consider the simultaneous equations

$$\left.\begin{array}{l} c_1 y_1(a) + c_2 y_2(a) = Y(a), \\ c_1 y_1'(a) + c_2 y_2'(a) = Y'(a). \end{array}\right\} \tag{14}$$

The determinant of the coefficients in this system of linear equations in the unknowns c_1 and c_2 is simply the Wronskian $W(y_1, y_2)$ evaluated at a. By Theorem 3, this determinant is nonzero, so by elementary algebra it follows that the equations in (14) can be solved for c_1 and c_2. With these values of c_1 and c_2, we define the solution

$$G(x) = c_1 y_1(x) + c_2 y_2(x)$$

of Eq. (9); then

$$G(a) = c_1 y_1(a) + c_2 y_2(a) = Y(a)$$

and

$$G'(a) = c_1 y_1'(a) + c_2 y_2'(a) = Y'(a).$$

Thus the two solutions Y and G have the same initial values at a, as do Y' and G'. By the uniqueness of a solution determined by such initial values (Theorem 2), it follows that Y and G agree on I. Thus we see that

$$Y(x) \equiv G(x) = c_1 y_1(x) + c_2 y_2(x),$$

as desired.

EXAMPLE 3 It is evident that $y_1 = e^{2x}$ and $y_2 = e^{-2x}$ are linearly independent solutions of

$$y'' - 4y = 0. \tag{15}$$

But $\cosh 2x$ and $\sinh 2x$ are also solutions of (15), so it follows from Theorem 4 that $\cosh 2x$ and $\sinh 2x$ can be expressed as linear combinations of $y_1 = e^{2x}$ and $y_2 = e^{-2x}$. Of course this is no surprise, because $\cosh 2x = \frac{1}{2}e^{2x} + \frac{1}{2}e^{-2x}$ and $\sinh 2x = \frac{1}{2}e^{2x} - \frac{1}{2}e^{-2x}$ by the definitions of the hyperbolic cosine and sine.

2.1 Problems

In each of Problems 1–16, a homogeneous second order linear differential equation, two functions y_1 and y_2, and a pair of initial conditions are given. First verify that y_1 and y_2 are solutions of the differential equation, and then find a particular solution of the form $y = c_1 y_1 + c_2 y_2$ that satisfies the given initial conditions.

1. $y'' - y = 0$; $y_1 = e^x, y_2 = e^{-x}$; $y(0) = 0, y'(0) = 5$.

2. $y'' - 9y = 0$; $y_1 = e^{3x}, y_2 = e^{-3x}$; $y(0) = -1, y'(0) = 15$.

3. $y'' + 4y = 0$; $y_1 = \cos 2x, y_2 = \sin 2x$; $y(0) = 3, y'(0) = 8$.

4. $y'' + 25y = 0$; $y_1 = \cos 5x, y_2 = \sin 5x$; $y(0) = 10, y'(0) = -10$.

5. $y'' - 3y' + 2y = 0$; $y_1 = e^x, y_2 = e^{2x}$; $y(0) = 1, y'(0) = 0$.

6. $y'' + y' - 6y = 0$; $y_1 = e^{2x}, y_2 = e^{-3x}$; $y(0) = 7, y'(0) = -1$.

7. $y'' + y' = 0$; $y_1 = 1, y_2 = e^{-x}$; $y(0) = -2, y'(0) = 8$.

8. $y'' - 3y' = 0$; $y_1 = 1, y_2 = e^{3x}$; $y(0) = 4, y'(0) = -2$.

9. $y'' + 2y' + y = 0$; $y_1 = e^{-x}, y_2 = xe^{-x}$; $y(0) = 2, y'(0) = -1$.

10. $y'' - 10y' + 25y = 0$; $y_1 = e^{5x}, y_2 = xe^{5x}$; $y(0) = 3, y'(0) = 13$.

11. $y'' - 2y' + 2y = 0$; $y_1 = e^x \cos x, y_2 = e^x \sin x$; $y(0) = 0, y'(0) = 5$.

12. $y'' + 6y' + 13y = 0$; $y_1 = e^{-3x} \cos 2x, y_2 = e^{-3x} \sin 2x$; $y(0) = 2, y'(0) = 0$.

13. $x^2 y'' - 2xy' + 2y = 0$; $y_1 = x, y_2 = x^2$; $y(1) = 3, y'(1) = 1.$

14. $x^2 y'' + 2xy' - 6y = 0$; $y_1 = x^2, y_2 = \dfrac{1}{x^3}$; $y(2) = 10, y'(2) = 15.$

15. $x^2 y'' - xy' + y = 0$; $y_1 = x, y_2 = x \ln x$; $y(1) = 7, y'(1) = 2.$

16. $x^2 y'' + xy' + y = 0$; $y_1 = \cos(\ln x), y_2 = \sin(\ln x)$; $y(1) = 2, y'(1) = 3.$

The following three problems illustrate the fact that the superposition principle does not generally hold for nonlinear equations.

17. Show that $y = 1/x$ is a solution of $y' + y^2 = 0$, but that if $c \neq 0, 1$, then $y = c/x$ is not a solution.

18. Show that $y = x^3$ is a solution of $yy'' = 6x^4$, but that if $c^2 \neq 1$, then $y = cx^3$ is not a solution.

19. Show that $y_1 = 1$ and $y_2 = x^{1/2}$ are solutions of $yy'' + (y')^2 = 0$, but that their sum $y = y_1 + y_2$ is not a solution.

Determine whether the pairs of functions in Problems 20–26 are linearly independent or linearly dependent on the real line.

20. $f(x) = \pi, g(x) = \cos^2 x + \sin^2 x.$

21. $f(x) = x^3, g(x) = x^2 |x|.$

22. $f(x) = 1 + x, g(x) = 1 + |x|.$

23. $f(x) = xe^x, g(x) = |x| e^x.$

24. $f(x) = \sin^2 x, g(x) = 1 - \cos 2x.$

25. $f(x) = e^x \sin x, g(x) = e^x \cos x.$

26. $f(x) = 2 \cos x + 3 \sin x, g(x) = 3 \cos x - 2 \sin x.$

27. Let y_p be a particular solution of the nonhomogeneous equation $y'' + py' + qy = f(x)$, and let y_c be a solution of its associated homogeneous equation. Show that $y = y_c + y_p$ is a solution of the given nonhomogeneous equation.

28. With $y_p \equiv 1$ and $y_c = c_1 \cos x + c_2 \sin x$ in the notation of Problem 27, find a solution of $y'' + y = 1$ satisfying the initial conditions $y(0) = -1 = y'(0).$

29. Show that $y_1 = x^2$ and $y_2 = x^3$ are two different solutions of $x^2 y'' - 4xy' + 6y = 0$, both satisfying the initial conditions $y(0) = 0 = y'(0)$. Explain why these facts do not contradict Theorem 2 (with respect to the guaranteed uniqueness).

30. (a) Show that $y_1 = x^3$ and $y_2 = |x^3|$ are linearly independent solutions on the real line of the equation $x^2 y'' - 3xy' + 3y = 0$. (b) Verify that $W(y_1, y_2)$ is identically zero. Why do these facts not contradict Theorem 3?

31. Show that $y_1 = \sin x^2$ and $y_2 = \cos x^2$ are linearly independent functions, but that their Wronskian vanishes at $x = 0$. Why does this imply that there is *no* differential equation of the form $y'' + p(x)y' + q(x)y = 0$, with p and q both continuous, having both y_1 and y_2 as solutions?

32. Let y_1 and y_2 be two solutions of $A(x)y'' + B(x)y' + C(x)y = 0$ on an open interval I where $A, B,$ and C are continuous and $A(x)$ is never zero. (a) Let $W = W(y_1, y_2)$. Show that

$$A(x) \frac{dW}{dx} = y_1(Ay_2'') - y_2(Ay_1'').$$

Then substitute for Ay_2'' and Ay_1'' from the original differential equation to show that

$$A(x)\frac{dW}{dx} = -B(x)W(x).$$

(b) Solve this first order equation to deduce **Abel's formula**

$$W(x) = K\exp\left(-\int \frac{B(x)}{A(x)}\,dx\right)$$

where K is a constant. (c) Why does Abel's formula imply that the Wronskian $W(y_1, y_2)$ is either zero everywhere or nonzero everywhere (as stated in Theorem 3)?

33. (a) Take as given the fact that $D_x e^{ix} = ie^{ix}$, where i is the complex number $\sqrt{-1}$. Show that $u = e^{ix}$ is a solution of $y'' + y = 0$. Why does it follow that e^{ix} is a linear combination of $\cos x$ and $\sin x$? (b) **Euler's formula** is the identity $e^{ix} = \cos x + i\sin x$. Deduce Euler's formula from the facts that $u(0) = 1$ and $u'(0) = i$.

2.2
General Solutions of Linear Equations

We now show that our discussion in Section 2.1 of second order linear equations generalizes in a very natural way to the general nth order *linear* differential equation of the form

$$P_0(x)y^{(n)} + P_1(x)y^{(n-1)} + \cdots + P_{n-1}(x)y' + P_n(x)y = F(x). \tag{1}$$

Unless otherwise noted, we will always assume that the coefficient functions $P_i(x)$ and $F(x)$ are continuous on some open interval I (perhaps unbounded) where we wish to solve the equation. Under the additional assumption that $P_0(x) \neq 0$ at each point x of I, we can divide each term in (1) by $P_0(x)$ to obtain an equation with leading coefficient 1, of the form

$$y^{(n)} + p_1(x)y^{(n-1)} + \cdots + p_{n-1}(x)y' + p_n(x)y = f(x). \tag{2}$$

The *homogeneous* linear equation *associated* with (2) is

$$y^{(n)} + p_1(x)y^{(n-1)} + \cdots + p_{n-1}(x)y' + p_n(x)y = 0. \tag{3}$$

Just as in the second order case, a *homogeneous* nth order linear differential equation has the nice property that any superposition, or *linear combination*, of solutions of the equation is again a solution. The proof of the following theorem is essentially the same—a routine verification—as that of Theorem 1 in Section 2.1.

THEOREM 1: PRINCIPLE OF SUPERPOSITION

Let $y_1, y_2, \ldots, y_n$ be n solutions of the homogeneous linear equation in (3) on the interval I. If $c_1, c_2, \ldots, c_n$ are constants, then the linear combination

$$y = c_1 y_1 + c_2 y_2 + \cdots + c_n y_n \tag{4}$$

is also a solution of (3) on I.

EXAMPLE 1 It is easy to verify that the three functions $y_1 = e^{3x}$, $y_2 = \cos 2x$, and $y_3 = \sin 2x$ are all solutions of the homogeneous third order equation

$$y^{(3)} - 3y'' + 4y' - 12y = 0$$

on the entire real line. Theorem 1 tells us that any linear combination of these solutions, such as

$$y = 7y_1 + 3y_2 - 2y_3 = 7e^{3x} + 3 \cos 2x - 2 \sin 2x,$$

is also a solution on the entire real line. We will see that, conversely, every solution of the differential equation above is a linear combination of the three particular solutions y_1, y_2, and y_3. Thus its general solution has the form

$$y = c_1 e^{3x} + c_2 \cos 2x + c_3 \sin 2x.$$

We saw in Section 2.1 that a particular solution of a *second order* linear differential equation is determined by *two* initial conditions. Similarly, a particular solution of an *n*th order linear differential equation is determined by *n* initial conditions. The following theorem, proved in Section 9.4, is the natural generalization of Theorem 2 in Section 2.1.

THEOREM 2: EXISTENCE AND UNIQUENESS

Suppose that the functions $p_1, p_2, \ldots, p_n$ are continuous on the open interval I containing the point a. Then, given n numbers $b_0, b_1, b_2, \ldots, b_{n-1}$, the equation (Eq. (2))

$$y^{(n)} + p_1(x)y^{(n-1)} + \cdots + p_{n-1}(x)y' + p_n(x)y = f(x)$$

has a unique (that is, one and only one) solution on the entire interval I that satisfies the n initial conditions

$$y(a) = b_0, y'(a) = b_1, \ldots, y^{(n-1)}(a) = b_{n-1}. \tag{5}$$

Equation (2) and the conditions in (5) constitute an *n*th order *initial value problem*. Theorem 2 tells us that any such initial value problem has a unique solution on the *whole* interval *I* where the coefficient functions in (2) are continuous. It tells us nothing, however, about how to find this solution. In Section 2.3 we will see how to construct explicit solutions of initial value problems in the *constant* coefficient case that occurs often in applications.

EXAMPLE 1 CONTINUED   We saw earlier that

$$y = 7e^{3x} + 3 \cos 2x - 2 \sin 2x$$

is a solution of

$$y^{(3)} - 3y'' + 4y' - 12y = 0$$

on the real line. It has the initial values $y(0) = 10$, $y'(0) = 17$, and $y''(0) = 51$. Theorem 2 assures us that this is the *only* solution with these initial values.

Note that Theorem 2 implies that the *trivial* solution $y(x) \equiv 0$ is the only solution of the *homogeneous* equation

$$y^{(n)} + p_1(x)y^{(n-1)} + \cdots + p_{n-1}(x)y' + p_n(x)y = 0 \tag{3}$$

that satisfies the *trivial* initial conditions

$$y(a) = y'(a) = \cdots = y^{(n-1)}(a) = 0.$$

EXAMPLE 2 It is easy to verify that $y_1 = x^2$ and $y_2 = x^3$ are two different solutions of $x^2 y'' - 4xy' + 6y = 0$, and that each satisfies the initial conditions $y(0) = y'(0) = 0$. Why does this not contradict the uniqueness part of Theorem 2? It is because the leading coefficient in this differential equation vanishes at $x = 0$, so the equation cannot be written in the form of Eq. (3) with coefficient functions *continuous* on an open interval containing the point $x = 0$.

On the basis of our knowledge of general solutions of second order linear equations, we anticipate that a general solution of the *homogeneous* nth order linear equation

$$y^{(n)} + p_1(x)y^{(n-1)} + \cdots + p_{n-1}(x)y' + p_n(x)y = 0 \qquad (3)$$

will be a linear combination

$$y = c_1 y_1 + c_2 y_2 + \cdots + c_n y_n, \qquad (4)$$

where $y_1, y_2, \ldots, y_n$ are particular solutions of Eq. (3). But these n particular solutions must be "sufficiently independent" that we can always choose the coefficients $c_1, c_2, \ldots, c_n$ in (4) to satisfy initial conditions of the form in (5). The question is this: What should be meant by *independence* of three or more functions?

Recall that *two* functions f_1 and f_2 are linearly *dependent* if one is a constant multiple of the other; that is, if either $f_1 = kf_2$ or $f_2 = kf_1$ for some constant k. If we rewrite these equations as

$$(1)f_1 + (-k)f_2 = 0 \quad \text{or} \quad (k)f_1 + (-1)f_2 = 0,$$

we see that the linear dependence of f_1 and f_2 implies that there exist two constants c_1 and c_2 *not both zero* such that

$$c_1 f_1 + c_2 f_2 = 0. \qquad (6)$$

Conversely, if c_1 and c_2 are not both zero, then (6) certainly implies that f_1 and f_2 are linearly dependent.

By analogy with Eq. (6), we say that n functions $f_1, f_2, \ldots, f_n$ are *linearly dependent* provided that some *nontrivial* linear combination

$$c_1 f_1 + c_2 f_2 + \cdots + c_n f_n$$

of them vanishes identically; "nontrivial" means that *not all* of the coefficients $c_1, c_2, \ldots, c_n$ are zero (although some of them may be).

DEFINITION: LINEAR DEPENDENCE OF FUNCTIONS

The n functions $f_1, f_2, \ldots, f_n$ are said to be **linearly dependent** *on the interval I provided that there exist constants $c_1, c_2, \ldots, c_n$ not all zero such that*

$$c_1 f_1 + c_2 f_2 + \cdots + c_n f_n = 0 \qquad (7)$$

on I; that is,

$$c_1 f_1(x) + c_2 f_2(x) + \cdots + c_n f_n(x) = 0$$

for all x in I.

If not all the coefficients in (7) are zero, then clearly we can solve for at least one of the functions as a linear combination of the others, and conversely. Thus the functions $f_1, f_2, \ldots, f_n$ are linearly dependent if and only if at least one of them is a linear combination of the others.

EXAMPLE 3 The functions $f_1(x) = \sin 2x, f_2(x) = \sin x \cos x,$ and $f_3(x) = e^x$ are linearly dependent on the real line because

$$(1)f_1 + (-2)f_2 + (0)f_3 = 0$$

(by the familiar trigonometric identity $\sin 2x = 2 \sin x \cos x$).

The n functions $f_1, f_2, \ldots, f_n$ are called **linearly independent** on the interval I provided they are not linearly dependent there. Equivalently, they are linearly independent on I provided that the identity

$$c_1 f_1 + c_2 f_2 + \cdots + c_n f_n = 0 \tag{7}$$

holds on I only in the trivial case

$$c_1 = c_2 = \cdots = c_n = 0;$$

that is, *no* nontrivial linear combination of these functions vanishes on I. Put another way, the functions $f_1, f_2, \ldots, f_n$ are linearly independent if no one of them is a linear combination of the others. (Why?)

Sometimes one can show that n given functions are linearly dependent by finding, as in Example 3, nontrivial values of the coefficients so that Eq. (7) holds. But in order to show that n given functions are linearly independent, we must prove that nontrivial values of the coefficients *cannot* be found, and this is seldom easy to do in any direct or obvious manner.

Fortunately, in the case of n solutions of a homogeneous nth order linear equations, there is a tool that makes the determination of their linear dependence or independence a routine matter. This tool is the Wronskian determinant, which we introduced (for the case $n = 2$) in Section 2.1. Suppose that the n functions $f_1, f_2, \ldots, f_n$ are each $n - 1$ times differentiable. Then their Wronskian is the $n \times n$ determinant

$$W = \begin{vmatrix} f_1 & f_2 & \cdots & f_n \\ f'_1 & f'_2 & \cdots & f'_n \\ \cdot & \cdot & & \cdot \\ \cdot & \cdot & & \cdot \\ \cdot & \cdot & & \cdot \\ f_1^{(n-1)} & f_2^{(n-1)} & \cdots & f_n^{(n-1)} \end{vmatrix}. \tag{8}$$

We write $W(f_1, f_2, \ldots, f_n)$ or $W(x)$, depending upon whether we wish to emphasize the functions or the point x at which their Wronskian is to be evaluated. The Wronskian is named after the Polish mathematician J. M. H. Wronski (1778–1853), most of whose other mathematical work is now forgotten.

We saw in Section 2.1 that the Wronskian of two linearly dependent functions vanishes identically. More generally, *the Wronskian of n linearly dependent functions $f_1, f_2, \ldots, f_n$ is identically zero.* To prove this, assume that

Eq. (7) holds on the interval I for some choice of the constants $c_1, c_2, \ldots, c_n$ not all zero. We then differentiate these equations $n - 1$ times in succession, obtaining the n equations

$$\left.\begin{aligned} c_1 f_1(x) + c_2 f_2(x) + \cdots + c_n f_n(x) &= 0, \\ c_1 f_1'(x) + c_2 f_2'(x) + \cdots + c_n f_n'(x) &= 0, \\ &\ \ \vdots \\ c_1 f_1^{(n-1)}(x) + c_2 f_2^{(n-1)}(x) + \cdots + c_n f_n^{(n-1)}(x) &= 0, \end{aligned}\right\} \tag{9}$$

which hold for all x in I. We recall from linear algebra that a system of n linear *homogeneous* equations in n unknowns has a nontrivial solution if and only if its determinant of coefficients vanishes. In (9) the unknowns are the constants $c_1, c_2, \ldots, c_n$ and the determinant of coefficients is simply the Wronskian $W(f_1, f_2, \ldots, f_n)$ evaluated at the typical point x of I. Because we know that the c_i are not all zero, it follows that $W(x) \equiv 0$, as we wanted to prove.

Therefore, in order to show that the functions $f_1, f_2, \ldots, f_n$ are *linearly independent* on the interval I, it suffices to show that their Wronskian is nonzero at just one point of I.

EXAMPLE 4 Show that the functions $y_1 = e^{3x}$, $y_2 = \cos 2x$, and $y_3 = \sin 2x$ (of Example 1) are linearly independent.

Solution Their Wronskian is

$$W = \begin{vmatrix} e^{3x} & \cos 2x & \sin 2x \\ 3e^{3x} & -2\sin 2x & 2\cos 2x \\ 9e^{3x} & -4\cos 2x & -4\sin 2x \end{vmatrix}$$

$$= e^{3x} \begin{vmatrix} -2\sin 2x & 2\cos 2x \\ -4\cos 2x & -4\sin 2x \end{vmatrix} - 3e^{3x} \begin{vmatrix} \cos 2x & \sin 2x \\ -4\cos 2x & -4\sin 2x \end{vmatrix}$$

$$+ 9e^{3x} \begin{vmatrix} \cos 2x & \sin 2x \\ -2\sin 2x & 2\cos 2x \end{vmatrix} = 26e^{3x}.$$

Because $W \neq 0$ everywhere, it follows that y_1, y_2, and y_3 are linearly independent on any open interval (including the whole real line).

EXAMPLE 5 Show first that the three solutions $y_1 = x$, $y_2 = x \ln x$, and $y_3 = x^2$ of the third order equation

$$x^3 y^{(3)} - 4x^3 y'' + 5xy' - 2y = 0 \tag{10}$$

are linearly independent on the open interval $x > 0$. Then find a particular solution of (10) that satisfies the initial conditions

$$y(1) = 3, \qquad y'(1) = 2, \qquad y''(1) = 1. \tag{11}$$

Solution Note that for $x > 0$, we could divide each term in (10) by x^3 to obtain a homogeneous linear equation of the standard form in (3). When we compute the Wronskian of the three given solutions, we find that

$$W = \begin{vmatrix} x & x \ln x & x^2 \\ 1 & 1 + \ln x & 2x \\ 0 & \dfrac{1}{x} & 2 \end{vmatrix} = x.$$

Thus $W \neq 0$ for $x > 0$, so y_1, y_2, and y_3 are linearly independent on the interval $x > 0$. To find the desired particular solution, we impose the initial conditions in (11) on

$$y(x) = c_1 x + c_2 x \ln x \qquad + c_3 x^2,$$
$$y'(x) = c_1 \quad + c_2(1 + \ln x) + 2c_3 x,$$
$$y''(x) = 0 \quad + \frac{c_2}{x} \qquad + 2c_3.$$

This yields the simultaneous equations

$$y(1) = c_1 \qquad + c_3 = 3,$$
$$y'(1) = c_1 + c_2 + 2c_3 = 2,$$
$$y''(1) = \qquad c_2 + 2c_3 = 1;$$

we solve to find $c_1 = 1$, $c_2 = -3$, and $c_3 = 2$. Thus our particular solution is

$$y = x - 3x \ln x + 2x^2.$$

Provided that $W(y_1, y_2, \ldots, y_n) \neq 0$, it turns out (Theorem 4) that we can always find values of the coefficients in the linear combination $y = c_1 y_1 + c_2 y_2 + \cdots + c_n y_n$ in order to satisfy any given initial conditions of the form in (5). The following theorem provides the necessary nonvanishing of W in the case of linearly independent solutions.

THEOREM 3: WRONSKIANS OF SOLUTIONS

Suppose that $y_1, y_2, \ldots, y_n$ are n solutions of the homogeneous nth order linear equation

$$y^{(n)} + p_1(x)y^{(n-1)} + \cdots + p_{n-1}(x)y' + p_n(x)y = 0 \qquad (3)$$

on an open interval I where each p_i is continuous. Let $W = W(y_1, y_2, \ldots, y_n)$.

(a) *If $y_1, y_2, \ldots, y_n$ are linearly dependent, then $W \equiv 0$ on I;*
(b) *If $y_1, y_2, \ldots, y_n$ are linearly independent, then $W \neq 0$ at each point of I.*

Thus there are just two possibilities: Either $W = 0$ everywhere on I, or $W \neq 0$ everywhere on I.

Proof We have already proven part (a). To prove part (b), it is sufficient to assume that $W(a) = 0$ for some point a of I, and show that this implies that the solutions $y_1, y_2, \ldots, y_n$ are linearly dependent. But $W(a)$ is simply the determinant of coefficients of the system of n homogeneous linear equations

$$
\left.\begin{aligned}
c_1 y_1(a) + c_2 y_2(a) \quad &+ \cdots + \quad c_n y_n(a) = 0, \\
c_1 y_1'(a) + c_2 y_2'(a) \quad &+ \cdots + \quad c_n y_n'(a) = 0, \\
&\ \vdots \\
c_1 y_1^{(n-1)}(a) + c_2 y_2^{(n-1)}(a) + &\cdots + c_n y_n^{(n-1)}(a) = 0
\end{aligned}\right\} \qquad (12)
$$

in the n unknowns $c_1, c_2, \ldots, c_n$. Because $W(a) = 0$, the basic fact from linear algebra quoted just following (9) above implies that the equations in (12) have a nontrivial solution. That is, the numbers $c_1, c_2, \ldots,$ and c_n cannot all be zero.

We now use these values to define the particular solution

$$
Y(x) = c_1 y_1(x) + c_2 y_2(x) + \cdots + c_n y_n(x) \qquad (13)
$$

of Eq. (3). Equations (12) then imply that Y satisfies the trivial initial conditions

$$
Y(a) = Y'(a) = \cdots = Y^{(n-1)}(a) = 0.
$$

Theorem 2 (uniqueness) therefore implies that $Y(x) \equiv 0$ on I. In view of (13) and the fact that $c_1, c_2, \ldots, c_n$ are not all zero, this is the desired conclusion that the solutions $y_1, y_2, \ldots, y_n$ are linearly dependent. This completes the proof of Theorem 3.

We can now show that every solution of a *homogeneous* nth order linear equation is a linear combination of n given linearly independent solutions. Using the fact from Theorem 3 that the Wronskian of n linearly independent solutions is nonzero, the proof of the following theorem is essentially the same as the proof of Theorem 4 of Section 2.1 (the case $n = 2$).

THEOREM 4: GENERAL SOLUTIONS

Let $y_1, y_2, \ldots, y_n$ be n linearly independent solutions of the homogeneous equation

$$
y^{(n)} + p_1(x)y^{(n-1)} + \cdots + p_{n-1}(x)y' + p_n(x)y = 0 \qquad (3)
$$

on an open interval I where the p_i are continuous. If Y is any solution whatsoever of Eq. (3), then there exist numbers $c_1, c_2, \ldots, c_n$ such that

$$
Y(x) = c_1 y_1(x) + c_2 y_2(x) + \cdots + c_n y_n(x)
$$

for all x in I.

Thus *every* solution of a homogeneous nth order linear differential equation is a linear combination

$$
y = c_1 y_1 + c_2 y_2 + \cdots + c_n y_n
$$

of *any* n given linearly independent solutions. On this basis we call such a linear combination a **general solution** of the differential equation.

NONHOMOGENEOUS EQUATIONS

We now consider the *nonhomogeneous* nth order linear differential equation

$$Ly = y^{(n)} + p_1(x)y^{(n-1)} + \cdots + p_{n-1}(x)y' + p_n(x)y = f(x) \qquad (2)$$

with associated homogeneous equation

$$Ly = y^{(n)} + p_1(x)y^{(n-1)} + \cdots + p_{n-1}(x)y' + p_n(x)y = 0. \qquad (3)$$

Here we introduce the symbol L to represent an **operator**; given an n times differentiable function y, L *operates* on y (as suggested in Fig. 2.3) to produce the linear combination

$$Ly = y^{(n)} + p_1 y^{(n-1)} + \cdots + p_{n-1}y' + p_n y \qquad (14)$$

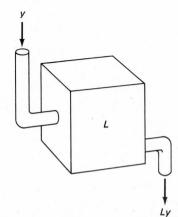

Figure 2.3 The idea of L "operating" on the function y.

of y and its first n derivatives. The principle of superposition (Theorem 1) means simply that the operator L is *linear*; that is,

$$L(c_1 y_1 + c_2 y_2) = c_1 L y_1 + c_2 L y_2 \qquad (15)$$

if c_1 and c_2 are constants.

Suppose that a single particular solution y_p of the nonhomogeneous equation in (2) is known and that Y is any other solution of (2). Then (15) implies that

$$L(Y - y_p) = LY - Ly_p = f - f = 0.$$

Thus $y_c = Y - y_p$ is a solution of the associated homogeneous equation in (3). Then

$$Y = y_c + y_p, \qquad (16)$$

and it follows from Theorem 4 that

$$y_c = c_1 y_1 + c_2 y_2 + \cdots + c_n y_n \qquad (17)$$

where $y_1, y_2, \ldots, y_n$ are linearly independent solutions of the associated *homogeneous* equation. We call y_c a **complementary function** of the nonhomogeneous equation and have thus proved that a *general solution* of the nonhomogeneous equation in (2) is the sum of its complementary function y_c and a single particular solution y_p of (2).

> **THEOREM 5:** SOLUTIONS OF NONHOMOGENEOUS EQUATIONS
>
> *Let y_p be a particular solution of the nonhomogeneous equation in (2) on an open interval I where the functions p_i and f are continuous. Let $y_1, y_2, \ldots,$ y_n be linearly independent solutions of the associated homogeneous equation in (3). If Y is any solution whatsoever of Eq. (2) on I, then there exist numbers $c_1, c_2, \ldots, c_n$ such that*
>
> $$Y(x) = c_1 y_1(x) + c_2 y_2(x) + \cdots + c_n y_n(x) + y_p(x) \qquad (18)$$
>
> *for all x in I.*

EXAMPLE 6 It is evident that $y_p(x) = 3x$ is a particular solution of the equation

$$y'' + 4y = 12x, \qquad (19)$$

and that $y_c = c_1 \cos 2x + c_2 \sin 2x$ is its complementary function. Find a solution of (19) that satisfies the initial conditions $y(0) = 5$, $y'(0) = 7$.

Solution The general solution of Eq. (19) is

$$y = c_1 \cos 2x + c_2 \sin 2x + 3x.$$

Now

$$y' = -2c_1 \sin 2x + 2c_2 \cos 2x + 3.$$

Hence the initial conditions give

$$y(0) = c_1 \qquad = 5,$$
$$y'(0) = 2c_2 + 3 = 7.$$

We find that $c_1 = 5$ and $c_2 = 2$. Thus the desired solution is

$$y = 5 \cos 2x + 2 \sin 2x + 3x.$$

2.2 Problems

In each of Problems 1–6, show directly that the given functions are linearly dependent on the real line. That is, find a nontrivial linear combination of the given functions that vanishes identically.

1. $f(x) = 2x, g(x) = 3x^2, h(x) = 5x - 8x^2$.
2. $f(x) = 5, g(x) = 2 - 3x^2, h(x) = 10 + 15x^2$.
3. $f(x) = 0, g(x) = \sin x, h(x) = e^x$.
4. $f(x) = 17, g(x) = 2 \sin^2 x, h(x) = 3 \cos^2 x$.
5. $f(x) = 17, g(x) = \cos^2 x, h(x) = \cos 2x$.
6. $f(x) = e^x, g(x) = \cosh x, h(x) = \sinh x$.

In each of Problems 7–12, use the Wronskian to prove that the given functions are linearly independent on the indicated interval.

7. $f(x) = 1, g(x) = x, h(x) = x^2$; the real line.

8. $f(x) = e^x, g(x) = e^{2x}, h(x) = e^{3x}$; the real line.

9. $f(x) = e^x, g(x) = \cos x, h(x) = \sin x$; the real line.

10. $f(x) = e^x, g(x) = x^{-2}, h(x) = x^{-2} \ln x$; $x > 0$.

11. $f(x) = x, g(x) = xe^x, h(x) = x^2 e^x$; the real line.

12. $f(x) = x, g(x) = \cos(\ln x), h(x) = \sin(\ln x)$; $x > 0$.

In each of Problems 13–20, a third order homogeneous linear equation and three linearly independent solutions are given. Find a particular solution satisfying the given initial conditions.

13. $y^{(3)} + 2y'' - y' - 2y = 0$; $y(0) = 1, y'(0) = 2, y''(0) = 0$; $y_1 = e^x$, $y_2 = e^{-x}, y_3 = e^{-2x}$.

14. $y^{(3)} - 6y'' + 11y' - 6y = 0$; $y(0) = 0, y'(0) = 0, y''(0) = 3$; $y_1 = e^x$, $y_2 = e^{2x}, y_3 = e^{3x}$.

15. $y^{(3)} - 3y'' + 3y' - y = 0$; $y(0) = 2, y'(0) = 0, y''(0) = 0$; $y_1 = e^x$, $y_2 = xe^x, y_3 = x^2 e^x$.

16. $y^{(3)} - 5y'' + 8y' - 4y = 0$; $y(0) = 1, y'(0) = 4, y''(0) = 0$; $y_1 = e^x$, $y_2 = e^{2x}, y_3 = xe^{2x}$.

17. $y^{(3)} + 9y' = 0$; $y(0) = 3, y'(0) = -1, y''(0) = 2$; $y_1 = 1, y_2 = \cos 3x$, $y_3 = \sin 3x$.

18. $y^{(3)} - 3y'' + 4y' - 2y = 0$; $y(0) = 1, y'(0) = 0, y''(0) = 0$; $y_1 = e^x$, $y_2 = e^x \cos x, y_3 = e^x \sin x$.

19. $x^3 y^{(3)} - 3x^2 y'' + 6xy' - 6y = 0$; $y(1) = 6, y'(1) = 14, y''(1) = 22$; $y_1 = x, y_2 = x^2, y_3 = x^3$.

20. $x^3 y^{(3)} + 6x^2 y'' + 4xy' - 4y = 0$; $y(1) = 1, y'(1) = 5, y''(1) = -11$; $y_1 = x, y_2 = x^{-2}, y_3 = x^{-2} \ln x$.

In each of Problems 21–24, a nonhomogeneous differential equation, a complementary solution y_c, and a particular solution y_p are given. Find a solution satisfying the given initial conditions.

21. $y'' + y = 3x$; $y(0) = 2, y'(0) = -2$; $y_c = c_1 \cos x + c_2 \sin x$; $y_p = 3x$.

22. $y'' - 4y = 12$; $y(0) = 0, y'(0) = 10$; $y_c = c_1 e^{2x} + c_2 e^{-2x}$; $y_p = -3$.

23. $y'' - 2y' - 3y = 6$; $y(0) = 3, y'(0) = 11$; $y_c = c_1 e^{-x} + c_2 e^{3x}$; $y_p = -2$.

24. $y'' - 2y' + 2y = 2x$; $y(0) = 4, y'(0) = 8, y_c = c_1 e^x \cos x + c_2 e^x \sin x$; $y_p = x + 1$.

25. Let $Ly = y'' + py' + qy$. Suppose that y_1 and y_2 are two functions such that $Ly_1 = f(x)$ and $Ly_2 = g(x)$. Show that their sum $y = y_1 + y_2$ satisfies the nonhomogeneous equation $Ly = f(x) + g(x)$.

26. a) Find by inspection particular solutions of the two nonhomogeneous equations $y'' + 2y = 4$ and $y'' + 2y = 6x$. (b) Use the method of Problem 25 to find a particular solution of the differential equation $y'' + 2y = 6x + 4$.

27. Prove directly that the functions $f_1(x) \equiv 1, f_2(x) = x$, and $f_3(x) = x^2$ are linearly independent on the whole real line. (*Suggestion:* Assume that

$c_1 + c_2 x + c_3 x^2 = 0$. Differentiate this equation twice. You now have three equations that must be satisfied by any x, including $x = 0$. Conclude that $c_1 = c_2 = c_3 = 0$.)

28. Generalize the method of Problem 27 to prove directly that the functions $f_0(x) \equiv 1, f_1(x) = x, f_2(x) = x^2, \ldots, f_n(x) = x^n$ are linearly independent on the real line.

29. Use the result of Problem 28 and the definition of linear independence to prove directly that, for any constant r, the functions $f_0(x) = e^{rx}, f_1(x) = xe^{rx}, \ldots, f_n(x) = x^n e^{rx}$ are linearly independent on the whole real line.

30. Verify that $y_1 = x$ and $y_2 = x^2$ are linearly independent solutions on the entire real line of the equation $x^2 y'' - 2xy' + 2y = 0$, but that $W(x, x^2)$ vanishes at $x = 0$. Why do these observations not contradict part (b) of Theorem 3?

31. This problem indicates why we can impose *only* n initial conditions on a solution of an nth order linear differential equation. (a) Given the equation $y'' + py' + qy = 0$, explain why the value of $y''(a)$ is determined by the values of $y(a)$ and $y'(a)$. (b) Prove that the equation $y'' - 2y' - 5y = 0$ has a solution satisfying the conditions $y(0) = 1, y'(0) = 0, y''(0) = C$ if and only if $C = 5$.

32. Prove that an nth order homogeneous linear differential equation satisfying the hypotheses of Theorem 2 has n *linearly independent* solutions $y_1, y_2, y_3, \ldots, y_n$. (*Suggestion:* Let y_i be the unique solution such that $y_i^{(i-1)}(a) = 1$ and $y_i^{(k)}(a) = 0$ if $k \neq i - 1$.)

33. Suppose that the three numbers $r_1, r_2,$ and r_3 are distinct. Show that the three functions $\exp(r_1 x), \exp(r_2 x),$ and $\exp(r_3 x)$ are linearly independent by showing that their Wronskian

$$W = \exp([r_1 + r_2 + r_3]x) \cdot \begin{vmatrix} 1 & 1 & 1 \\ r_1 & r_2 & r_3 \\ r_1^2 & r_2^2 & r_3^2 \end{vmatrix}$$

is nonzero for all x.

34. Assume as known that the **Vandermonde determinant**

$$V = \begin{vmatrix} 1 & 1 & \cdots & 1 \\ r_1 & r_2 & \cdots & r_n \\ r_1^2 & r_2^2 & \cdots & r_n^2 \\ \cdot & \cdot & & \cdot \\ \cdot & \cdot & & \cdot \\ \cdot & \cdot & & \cdot \\ r_1^{n-1} & r_2^{n-1} & \cdots & r_n^{n-1} \end{vmatrix}$$

is nonzero if the numbers $r_1, r_2, \ldots, r_n$ are distinct. Prove by the method of Problem 33 that the functions

$$f_i(x) = \exp(r_i x), \qquad 1 \leq i \leq n$$

are linearly independent.

35. According to Problem 32 in Section 2.1, the Wronskian $W(y_1, y_2)$ of two solutions of the second order equation

$$y'' + p_1(x)y' + p_2(x)y = 0$$

is given by Abel's formula

$$W(x) = K \exp\left(-\int p_1(x)\,dx\right)$$

for some constant K. It can be shown that the Wronskian of n solutions y_1, $y_2, \ldots, y_n$ of the nth order equation

$$y^{(n)} + p_1(x)y^{(n-1)} + \cdots + p_{n-1}(x)y' + p_n(x)y = 0$$

satisfies the same identity. Prove this for the case $n = 3$ as follows:

(a) The derivative of a determinant of functions is the sum of the determinants obtained by separately differentiating the rows of the original determinant. Conclude that

$$W' = \begin{vmatrix} y_1 & y_2 & y_3 \\ y_1' & y_2' & y_3' \\ y_1^{(3)} & y_2^{(3)} & y_3^{(3)} \end{vmatrix}.$$

(b) Substitute for $y_1^{(3)}$, $y_2^{(3)}$, and $y_3^{(3)}$ from the differential equation $y^{(3)} + p_1 y'' + p_2 y' + p_3 y = 0$, and then show that $W' = -p_1 W$. Integration now gives Abel's formula.

2.3
Homogeneous Equations with Constant Coefficients

In the first two sections of this chapter we saw that a general solution of an nth order homogeneous linear equation is a linear combination of n linearly independent particular solutions, but we said little about how to actually find even a single solution. The solution of a linear differential equation with *variable* coefficients ordinarily requires infinite series methods (Chapter 3) or numerical methods (Chapter 6). But we can now show how to find, explicitly and in a rather straightforward way, n linearly independent solutions of a given nth order homogeneous linear equation if it has constant coefficients. The general such equation may be written in the form

$$a_n y^{(n)} + a_{n-1} y^{(n-1)} + \cdots + a_2 y'' + a_1 y' + a_0 y = 0, \tag{1}$$

where the coefficients $a_0, a_1, a_2, \ldots, a_n$ are real constants with $a_n \neq 0$.

We first look for a *single* solution of Eq. (1), and begin with the observation that

$$\frac{d^k}{dx^k}(e^{rx}) = r^k e^{rx}, \tag{2}$$

so any derivative of e^{rx} is a constant multiple of e^{rx}. Hence, if we substituted $y = e^{rx}$ in Eq. (1), each term would be a constant multiple of e^{rx}, with the constant coefficients depending upon r and the coefficients a_i. This suggests that we try to find r so that all these multiples of e^{rx} will have sum zero, in which case $y = e^{rx}$ will be a solution of Eq. (1).

For instance, if we substitute $y = e^{rx}$ in the equation $y'' - 5y' + 6y = 0$, we obtain

$$r^2 e^{rx} - 5re^{rx} + 6e^{rx} = 0;$$

thus $(r - 2)(r - 3)e^{rx} = 0$. Hence $y = e^{rx}$ will be a solution if either $r = 2$ or $r = 3$. So, in searching for a single solution, we actually have found *two* solutions: $y_1 = e^{2x}$ and $y_2 = e^{3x}$.

To carry out this technique in the general case, we substitute $y = e^{rx}$ in Eq. (1), and with the aid of (2) we find the result to be

$$a_n r^n e^{rx} + a_{n-1} r^{n-1} e^{rx} + \cdots + a_2 r^2 e^{rx} + a_1 r e^{rx} + a_0 e^{rx} = 0;$$

that is,

$$e^{rx}(a_n r^n + a_{n-1} r^{n-1} + \cdots + a_2 r^2 + a_1 r + a_0) = 0.$$

Because e^{rx} is never zero, we see that $y = e^{rx}$ will be a solution of Eq. (1) precisely when r is a root of the equation

$$a_n r^n + a_{n-1} r^{n-1} + \cdots + a_2 r^2 + a_1 r + a_0 = 0. \tag{3}$$

This equation is called the **characteristic equation**, or **auxiliary equation**, of the differential equation in (1). Our problem, then, is reduced to the solution of this purely algebraic equation.

According to the fundamental theorem of algebra, every nth degree polynomial—such as the one in (3)—has n zeros, though not necessarily distinct and not necessarily real. Finding the exact values of these zeros may be difficult or even impossible; the quadratic formula is sufficient for second degree equations, but for equations of high degree we may need to spot a fortuitous factorization, or resort to numerical methods (such as Newton's method for finding real zeros or Bairstow's method for finding complex zeros—see any numerical analysis text).

Whatever the method we use, let us suppose that we have solved the characteristic equation. Then we can always write a general solution of the differential equation. The situation is slightly more complicated in the case of repeated roots or complex roots of Eq. (3), so let us first examine the simplest case—in which the characteristic equation has n *distinct* (no two equal) *real* roots $r_1, r_2, \ldots, r_n$. Then the functions

$$e^{r_1 x}, e^{r_2 x}, \ldots, e^{r_n x}$$

are all solutions of Eq. (1), and (by Problem 34 in Section 2.2) these n solutions are linearly independent (on the entire real line). In summary, we have proved Theorem 1.

THEOREM 1:   DISTINCT REAL ROOTS

If the n roots $r_1, r_2, \ldots, r_n$ of the characteristic equation in (3) are real and distinct, then

$$y = c_1 e^{r_1 x} + c_2 e^{r_2 x} + \cdots + c_n e^{r_n x} \tag{4}$$

is a general solution of Eq. (1).

EXAMPLE 1   Find a general solution of $y^{(3)} - y'' - 6y' = 0$.

Solution   The characteristic equation of this differential equation is $r^3 - r^2 - 6r = 0$, which we solve by factoring: $r(r - 3)(r + 2) = 0$. The three roots are $0, 3$, and -2. They are real and distinct, and therefore—because $e^0 = 1$—a general solution of the given differential equation is

$$y = c_1 + c_2 e^{3x} + c_3 e^{-2x}.$$

REPEATED ROOTS

If the roots of the characteristic equation in (3) are *not* distinct—there are repeated roots—then we cannot produce n linearly independent solutions of Eq. (1) by the method of Theorem 1. For example, if the roots are 1, 2, 2, and 2, we obtain only the *two* functions e^x and e^{2x}. The problem, then, is to produce the missing linearly independent solutions. For this purpose it is convenient to adopt the operator notation introduced near the conclusion of Section 2.2. Equation (1) corresponds to the operator equation $Ly = 0$ where L is the operator

$$L = a_n \frac{d^n}{dx^n} + a_{n-1} \frac{d^{n-1}}{dx^{n-1}} + \cdots + a_2 \frac{d^2}{dx^2} + a_1 \frac{d}{dx} + a_0. \tag{5}$$

We also denote by $D = d/dx$ the operation of differentiation with respect to x, so that

$$Dy = y', \qquad D^2 y = y'', \qquad D^3 y = y^{(3)},$$

and so on. In terms of D, the operator L in (5) may be written

$$L = a_n D^n + a_{n-1} D^{n-1} + \cdots + a_2 D^2 + a_1 D + a_0, \tag{6}$$

and we will find it useful to think of the right-hand side in (6) as a (formal) nth degree polynomial in the "variable" D; it is a **polynomial operator**.

A first degree polynomial operator has the form $D - a$ where a is a real number. It operates on the function $y = y(x)$ to produce

$$(D - a)y = Dy - ay = y' - ay.$$

The important fact about such operators is that any two of them *commute*:

$$(D - a)(D - b)y = (D - b)(D - a)y \tag{7}$$

for any twice differentiable function $y = y(x)$. The proof of the formula in (7) is the following computation:

$$\begin{aligned}
(D - a)(D - b)y &= (D - a)(y' - by) \\
&= D(y' - by) - a(y' - by) = y'' - (b + a)y' + aby \\
&= y'' - (a + b)y' + bay = D(y' - ay) - b(y' - ay) \\
&= (D - b)(y' - ay) = (D - b)(D - a)y.
\end{aligned}$$

Let us now consider the possibility that the characteristic equation

$$a_n r^n + a_{n-1} r^{n-1} + \cdots + a_1 r + a_0 = 0 \tag{3}$$

has *repeated* roots. For example, suppose that Eq. (3) has only two distinct roots, r_0 of multiplicity 1 and r_1 of multiplicity $k > 1$. Then (3) can be rewritten in the form

$$(r - r_1)^k (r - r_0) = (r - r_0)(r - r_1)^k = 0. \tag{8}$$

Similarly, the operator L in (6) can be written as

$$L = (D - r_1)^k (D - r_0) = (D - r_0)(D - r_1)^k, \tag{9}$$

the order of the factors making no difference because of the formula in (7).

Two solutions of the differential equation $Ly = 0$ are certainly $y_0 = e^{r_0 x}$ and $y_1 = e^{r_1 x}$. This is, however, not sufficient; we need $k + 1$ linearly independent solutions in order to construct a general solution, because the equation is of order $k + 1$. To find the missing $k - 1$ solutions, we note that

$$Ly = (D - r_0)[(D - r_1)^k y] = 0.$$

Consequently *every* solution of the kth order equation

$$(D - r_1)^k y = 0 \tag{10}$$

will also be a solution of the original equation $Ly = 0$. Hence our problem is reduced to that of finding the general solution of the differential equation in (10).

The fact that $e^{r_1 x}$ is one solution of (10) suggests that we try the substitution

$$y(x) = u(x)e^{r_1 x}, \tag{11}$$

where $u(x)$ is a function yet to be determined. Observe that

$$(D - r_1)[ue^{r_1 x}] = (Du)e^{r_1 x} + r_1 u e^{r_1 x} - r_1 u e^{r_1 x},$$

so

$$(D - r_1)[ue^{r_1 x}] = (Du)e^{r_1 x}. \tag{12}$$

It therefore follows by induction on k that

$$(D - r_1)^k[ue^{r_1 x}] = (D^k u)e^{r_1 x} \tag{13}$$

for any function $u(x)$. Hence $y = ue^{r_1 x}$ will be a solution of (10) if and only if $D^k u = u^{(k)} = 0$. But this is so if and only if

$$u(x) = c_1 + c_2 x + c_3 x^2 + \cdots + c_k x^{k-1},$$

a polynomial of degree at most $k - 1$. Hence our desired solution of (10) is

$$y = ue^{r_1 x} = (c_1 + c_2 x + \cdots + c_k x^{k-1})e^{r_1 x}.$$

In particular, we see here the additional solutions $xe^{r_1 x}, x^2 e^{r_1 x}, \ldots, x^{k-1}e^{r_1 x}$ of our original differential equation $Ly = 0$.

The analysis above can be carried out with the operator $D - r_0$ replaced with an arbitrary polynomial operator. When this is done, the result is a proof of the following theorem.

THEOREM 2: REPEATED ROOTS

If the characteristic equation in (3) has a repeated root r of multiplicity k, then the part of a general solution of the differential equation in (1) corresponding to r is of the form

$$(c_1 + c_2 x + c_3 x^2 + \cdots + c_k x^{k-1})e^{rx}. \tag{14}$$

We may observe that, according to Problem 29 in Section 2.2, the k functions $e^{rx}, xe^{rx}, x^2 e^{rx}, \ldots,$ and $x^{k-1}e^{rx}$ involved in (14) are linearly independent on the real line. Thus a root of multiplicity k corresponds to k linearly independent solutions of the differential equation.

EXAMPLE 2 Find a general solution of $y^{(4)} + 3y^{(3)} + 3y'' + y' = 0$.

Solution The characteristic equation of this differential equation is

$$r^4 + 3r^3 + 3r^2 + r = r(r + 1)^3 = 0.$$

It has the simple root $r_1 = 0$, which contributes $y_1 = c_1$ to the general solution. It has also the triple ($k = 3$) root $r_2 = -1$, which contributes $y_2 = (c_2 + c_3 x + c_4 x^2)e^{-x}$ to the solution. Hence a general solution of the differential equation is

$$y = c_1 + (c_2 + c_3 x + c_4 x^2)e^{-x}.$$

COMPLEX ROOTS

Because we have assumed that the coefficients of the differential equation and its characteristic equation are real, any complex (nonreal) roots will occur in complex conjugate pairs $a \pm bi$ where a and b are real and $i = \sqrt{-1}$. This raises the question as to what might be meant by an exponential such as $\exp([a + bi]x)$.

To answer this question, we recall from elementary calculus the Taylor series for the exponential function:

$$e^t = \sum_{n=0}^{\infty} \frac{t^n}{n!} = 1 + t + \frac{t^2}{2!} + \frac{t^3}{3!} + \frac{t^4}{4!} + \cdots.$$

If we substitute $t = ix$ in this series, we get

$$e^{ix} = \sum_{n=0}^{\infty} \frac{(ix)^n}{n!}$$

$$= 1 + ix - \frac{x^2}{2!} - \frac{ix^3}{3!} + \frac{x^4}{4!} + \frac{ix^5}{5!} - \cdots$$

$$= \left(1 - \frac{x^2}{2!} + \frac{x^4}{4!} - \cdots\right) + i\left(x - \frac{x^3}{3!} + \frac{x^5}{5!} - \cdots\right).$$

Because the two real series in the last line are the Taylor series for $\cos x$ and $\sin x$, respectively, this implies that

$$e^{ix} = \cos x + i \sin x. \tag{15}$$

This result is known as *Euler's formula*. Because of it we *define* the exponential function e^z, for $z = x + iy$ an arbitrary complex number, to be

$$e^z = e^{x+iy} = e^x e^{iy} = e^x (\cos y + i \sin y). \tag{16}$$

Thus it appears that complex roots of the characteristic equation will lead to complex-valued solutions of the differential equation. A **complex-valued function** F of the real variable x associates with each real number x (in its domain of definition) the complex number

$$y = F(x) = f(x) + ig(x). \tag{17}$$

The real-valued functions f and g are called the **real** and **imaginary** parts, respectively, of F. If they are differentiable, we define the **derivative** F' of F to be

$$F'(x) = f'(x) + ig'(x). \tag{18}$$

Thus we simply differentiate the real and imaginary parts of F separately. Similarly, we say that the complex-valued function $y = F(x)$ **satisfies** the differential equation in (1) provided that its real and imaginary parts separately satisfy that differential equation.

The particular complex-valued functions of interest here are of the form $F(x) = e^{rx}$ where $r = a \pm bi$. We note from Euler's formula that

$$e^{(a+bi)x} = e^{ax}(\cos bx + i \sin bx) \qquad (19a)$$

and

$$e^{(a-bi)x} = e^{ax}(\cos bx - i \sin bx). \qquad (19b)$$

The most important property of e^{rx} is that

$$D_x(e^{rx}) = re^{rx} \qquad (20)$$

even if r should be a complex number. The proof of this assertion is a straightforward computation based on the definitions and formulas given above:

$$D_x(e^{rx}) = D_x(e^{ax}\cos bx) + iD_x(e^{ax}\sin bx)$$
$$= [ae^{ax}\cos bx - be^{ax}\sin bx] + i[ae^{ax}\sin bx + be^{ax}\cos bx]$$
$$= (a + bi)(e^{ax}\cos bx + ie^{ax}\sin bx) = re^{rx}.$$

It follows from (20) that, when r is complex (just as when r is real), e^{rx} will be a solution of the differential equation in (1) if and only if r is a root of its characteristic equation. If the complex conjugate pair of roots $r_1 = a + bi$ and $r_2 = a - bi$ are simple (nonrepeated), then the corresponding part of a general solution of (1) is

$$C_1 e^{(a+bi)x} + C_2 e^{(a-bi)x} = C_1 e^{ax}(\cos bx + i \sin bx)$$
$$+ C_2 e^{ax}(\cos bx - i \sin bx)$$
$$= e^{ax}(c_1 \cos bx + c_2 \sin bx)$$

where $c_1 = C_1 + C_2$ and $c_2 = (C_1 - C_2)i$. Thus the conjugate pair of roots $a \pm bi$ leads to the linearly independent *real-valued* solutions $e^{ax}\cos bx$ and $e^{ax}\sin bx$. This yields the following result.

THEOREM 3 :   COMPLEX ROOTS

If the characteristic equation in (3) has an unrepeated pair of complex conjugate roots $a \pm bi$ (with $b \neq 0$), then the corresponding part of a general solution of (1) is of the form

$$e^{ax}(c_1 \cos bx + c_2 \sin bx). \qquad (21)$$

EXAMPLE 3 The characteristic equation of $y'' + b^2 y = 0$ $(b > 0)$ is $r^2 + b^2 = 0$, with roots $\pm bi$. So Theorem 3 gives the general solution

$$y = c_1 \cos bx + c_2 \sin bx.$$

EXAMPLE 4 Find the particular solution of $y'' - 4y' + 5y = 0$ for which $y(0) = 1$ and $y'(0) = 5$.

Solution The characteristic equation is $r^2 - 4r + 5 = 0$, with roots $2 + i$ and $2 - i$. Hence a general solution is

$$y = e^{2x}(c_1 \cos x + c_2 \sin x).$$

Then

$$y' = 2e^{2x}(c_1 \cos x + c_2 \sin x) + e^{2x}(-c_1 \sin x + c_2 \cos x),$$

so the initial conditions give

$$y(0) = c_1 = 1 \quad \text{and} \quad y'(0) = 2c_1 + c_2 = 5.$$

It follows that $c_2 = 3$, and so the desired particular solution is

$$y = e^{2x}(\cos x + 3 \sin x).$$

EXAMPLE 5 Find a general solution of $y^{(4)} + k^4 y = 0$ $(k > 0)$.

Solution The characteristic equation is

$$r^4 + k^4 = (r^2 + k^2 i)(r^2 - k^2 i) = 0,$$

and its four roots are $\pm k\sqrt{\pm i}$. Now $i = e^{i\pi/2}$ and $-i = e^{3i\pi/2}$, so

$$\sqrt{i} = (e^{i\pi/2})^{1/2} = e^{i\pi/4} = \frac{1 + i}{\sqrt{2}}$$

and

$$\sqrt{-i} = (e^{3i\pi/2})^{1/2} = e^{3i\pi/4} = \frac{-1 + i}{\sqrt{2}}.$$

Thus the four (distinct) roots of the characteristic equation are $r = \pm(1 \pm i)a$ where $a = k/\sqrt{2}$. These two pairs of complex conjugate roots, $a \pm ai$ and $-a \pm ai$, give a general solution

$$y = e^{ax}(c_1 \cos ax + c_2 \sin ax) + e^{-ax}(c_3 \cos ax + c_4 \sin ax)$$

of $y^{(4)} + k^4 y = 0$, where $a = k/\sqrt{2}$.

In Example 5 we employed the **polar form**

$$x + iy = re^{i\theta} \tag{22}$$

of a complex number. The relation between the real and imaginary parts x, y and the **modulus** r and **argument** θ is indicated in Fig. 2.4. One consequence of (22) is that the nonzero complex number $x + iy$ has the two square roots

$$\pm(x + iy)^{1/2} = \pm(re^{i\theta})^{1/2} = \pm r^{1/2} e^{i\theta/2}. \tag{23}$$

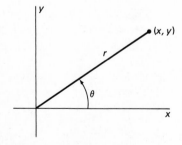

Figure 2.4 Modulus and argument of the complex number $x + iy$.

REPEATED COMPLEX ROOTS

Theorem 2 holds for repeated complex roots. If the conjugate pair $a \pm bi$ has multiplicity k, then the corresponding part of the general solution is of the form

$$(A_1 + A_2 x + \cdots + A_k x^{k-1})e^{(a+bi)x} + (B_1 + B_2 x + \cdots + B_k x^{k-1})e^{(a-bi)x}$$

$$= \sum_{p=0}^{k-1} x^p e^{ax}(c_i \cos bx + d_i \sin bx). \tag{24}$$

It can be shown that the $2k$ functions

$$x^p e^{ax} \cos bx, \; x^p e^{ax} \sin bx, \quad 0 \leq p \leq k - 1$$

that appear in (24) are linearly independent.

EXAMPLE 6   Find a general solution of $(D^2 + 2D + 4)^2 y = 0$.

Solution   The characteristic equation $(r^2 + 2r + 4)^2 = 0$ has as its roots the conjugate pair $-1 \pm i\sqrt{3}$ of multiplicity 2. Hence (24) gives a general solution

$$y = e^{-x}(c_1 \cos x\sqrt{3} + d_1 \sin x\sqrt{3}) + xe^{-x}(c_2 \cos x\sqrt{3} + d_2 \sin x\sqrt{3}).$$

In applications we are seldom presented in advance with a factorization as convenient as the one in Example 6. Often the most difficult part of solving a homogeneous linear equation is finding the roots of its characteristic equation. Sometimes we can find one root r_1, either by guessing (if it is an integer, as is the root $r_1 = 1$ of the equation $r^3 + 4r - 5 = 0$), or by a numerical method as in the following example. We can then carry out a division to obtain a partial factorization.

EXAMPLE 7   Find a general solution of $y^{(3)} + y' + y = 0$.

Solution   The characteristic equation is $p(r) = r^3 + r + 1 = 0$. Because $p(-1) = -1$ and $p(0) = +1$, there is a root between -1 and 0. We begin with the initial estimate $r_0 = -0.5$ and use the iteration

$$r_{n+1} = r_n - \frac{p(r_n)}{p'(r_n)} = r_n - \frac{r_n^3 + r_n + 1}{3r_n^2 + 1}$$

of Newton's method. This technique yields the approximate root -0.68233. The simple BASIC program shown in Fig. 2.5 may be used to find this root to the degree of accuracy determined by the stopping criterion $E = 0.0001$; if your computer is capable of handling numbers with more significant digits, then you may, of course, obtain greater accuracy by choosing E much closer to zero.

```
10     R = -0.5
20     E = 0.0001
30     F = R*R*R + R + 1
40     D = 3*R*R + 1
50     S = R - F/D
60     PRINT "ESTIMATE: ";R;" VALUE OF P: ";F
70     IF ABS(R - S) < E THEN GOTO 110
90     R = S
100    GOTO 30
110    PRINT "FINAL ESTIMATE: ";S
120    END                              Figure 2.5
```

The long division

$$(r + 0.68233))\overline{r^3 + r + 1}$$

then produces as its quotient the quadratic factor

$$r^2 - (0.68233)r + 1.46557.$$

With the aid of the quadratic formula, we find that the zeros of the latter are approximately $0.34116 \pm (1.16154)i$. These roots give an approximate general solution

$$y = c_1 e^{-(0.68233)x} + e^{(0.34116)x}[c_2 \cos (1.16154)x + c_3 \sin (1.16154)x].$$

2.3 Problems

Find a general solution of each of the differential equations given in Problems 1–20.

1. $y'' - 4y = 0.$

2. $2y'' - 3y' = 0.$

3. $y'' + 3y' - 10y = 0.$

4. $2y'' - 7y' + 3y = 0.$

5. $y'' + 6y' + 9y = 0.$

6. $y'' + 5y' + 5y = 0.$

7. $4y'' - 12y' + 9y = 0.$

8. $y'' - 6y' + 13y = 0.$

9. $y'' + 8y' + 25y = 0.$

10. $5y^{(4)} + 3y^{(3)} = 0.$

11. $y^{(4)} - 8y^{(3)} + 16y'' = 0.$

12. $y^{(4)} - 3y^{(3)} + 3y'' - y' = 0.$

13. $9y^{(3)} + 12y'' + 4y' = 0.$

14. $y^{(4)} + 3y'' - 4y = 0.$

15. $y^{(4)} - 8y'' + 16y = 0.$

16. $y^{(4)} + 18y'' + 81y = 0.$

17. $6y^{(4)} + 11y'' + 4y = 0.$

18. $y^{(4)} = 16y.$

19. $y^{(3)} + y'' - y' - y = 0.$

20. $y^{(4)} + 2y^{(3)} + 3y'' + 2y' + y = 0.$

Solve each of the initial value problems given in Problems 21–26.

21. $y'' - 4y' + 3y = 0;$ $y(0) = 7, y'(0) = 11.$

22. $9y'' + 6y' + 4y = 0;$ $y(0) = 3, y'(0) = 4.$

23. $y'' - 6y' + 25y = 0;$ $y(0) = 3, y'(0) = 1.$

24. $2y^{(3)} - 3y'' - 2y' = 0;$ $y(0) = 1, y'(0) = -1, y''(0) = 3.$

25. $3y^{(3)} + 2y'' = 0;$ $y(0) = -1, y'(0) = 0, y''(0) = 1.$

26. $y^{(3)} + 10y'' + 25y' = 0;$ $y(0) = 3, y'(0) = 4, y''(0) = 5.$

Find a general solution of each of the equations in Problems 27–30. First find a small integral root of the characteristic equation by inspection; then factor by division.

27. $y^{(3)} + 3y'' - 4y = 0.$

28. $2y^{(3)} - y'' - 5y' - 2y = 0.$

29. $y^{(3)} + 27y = 0.$

30. $y^{(4)} - y^{(3)} + y'' - 3y' - 6y = 0.$

Find general solutions of the equations in Problems 31–33. First find numerically (as in Example 7) a root of the characteristic equation.

31. $y^{(3)} - 3y'' + y = 0.$

32. $y^{(3)} + 3y' + 5y = 0.$

33. $y^{(4)} + 4y' - 7y = 0.$

In each of Problems 34–36, one solution of the differential equation is given. Find the general solution.

34. $3y^{(3)} - 2y'' + 12y' - 8y = 0;\quad y = e^{2x/3}.$

35. $6y^{(4)} + 5y^{(3)} + 25y'' + 20y' + 4y = 0;\quad y = \cos 2x.$

36. $9y^{(3)} + 11y'' + 4y' - 14y = 0;\quad y = e^{-x}\sin x.$

Problems 37–41 pertain to the solution of differential equations with complex coefficients.

37. (a) Use Euler's formula to show that every complex number can be written in the form $re^{i\theta}$, where $r \geq 0$ and $-\pi < \theta \leq \pi$. (b) Express the numbers 4, -2, $3i$, $1 + i$, and $-1 + i\sqrt{3}$ in the form $re^{i\theta}$. (c) The two square roots of $re^{i\theta}$ are $\pm\sqrt{r}\,e^{i\theta/2}$. Find the square roots of the numbers $2 - 2i\sqrt{3}$ and $-2 + 2i\sqrt{3}$.

38. Use the quadratic formula to solve the following equations. Note in each case that the roots are not complex conjugates. (a) $x^2 + ix + 2 = 0$. (b) $x^2 - 2ix + 3 = 0$.

39. Find a general solution of $y'' - 2iy' + 3y = 0$.

40. Find a general solution of $y'' - iy' + 6y = 0$.

41. Find a general solution of $y'' = (-2 + 2i\sqrt{3})y$.

2.4

Mechanical Vibrations

The motion of a mass attached to a spring serves as a relatively simple exemplar of the vibrations that occur in more complex mechanical systems. For many such systems the analysis of these vibrations is a problem in the solution of linear differential equations with constant coefficients.

We consider a body of mass m attached to one end of an ordinary spring that resists compression as well as stretching; the other end of the spring is attached to a fixed wall, as shown in Fig. 2.6. Assume that the body rests on a frictionless horizontal plane, so that it can move only back and forth as the spring stretches and compresses. Denote by x the distance of the body from its **equilibrium position**—its position when the spring is unstretched. We take $x > 0$ when the spring is stretched and $x < 0$ when it is compressed.

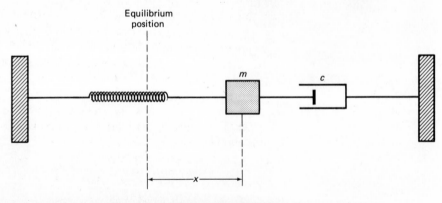

Figure 2.6 A mass-spring-dashpot system.

According to Hooke's law, the restorative force F_S that the spring exerts on the mass is proportional to the distance that the spring has been stretched or compressed. Because this is the same as the displacement of the mass m from its equilibrium position, it follows that

$$F_S = -kx. \tag{1}$$

The positive constant of proportionality k is called the **spring constant**. Note that F_S and x have opposite signs: $F_S < 0$ when $x > 0$, $F_S > 0$ when $x < 0$.

Figure 2.6 also shows the mass attached to a dashpot—a device, like a shock absorber, that provides a force directed opposite to the instantaneous direction of motion of the mass m. We assume the dashpot is designed so that this force F_R is proportional to the velocity $v = dx/dt$ of the mass, so that

$$F_R = -cv = -c\frac{dx}{dt}. \tag{2}$$

The positive constant c is the **damping constant** of the dashpot. More generally, we may regard (2) as specifying frictional forces in our system (including air resistance to the motion of m).

If, in addition to the forces F_S and F_R, the mass is subjected to a given **external force** $F_E = F(t)$, then the total force acting on the mass is $F = F_S + F_R + F_E$. Using Newton's law

$$F = ma = m\frac{d^2x}{dt^2} = mx'',$$

we obtain the second order linear differential equation

$$mx'' + cx' + kx = F(t) \tag{3}$$

that governs the motion of the mass.

If there is no dashpot (and we ignore all frictional forces), then we set $c = 0$ in (3) and call the motion **undamped**; it is **damped** motion if $c > 0$. If there is no external force we replace $F(t)$ by 0 in (3). We refer to the motion as **free** in this case and **forced** in the case $F(t) \neq 0$. Thus the homogeneous equation

$$mx'' + cx' + kx = 0 \tag{4}$$

describes free motion of a mass on a spring with dashpot but with no external forces applied. We will defer discussion of forced motion until Section 2.8.

For an alternative example, we might attach the mass to the lower end of a spring that is suspended vertically from a fixed support, as in Fig. 2.7. In this case the weight $W = mg$ of the mass would stretch the spring a distance s_0 determined by Eq. (1) with $F_S = -W$ and $x = s_0$. That is, $mg = ks_0$, so that $s_0 = mg/k$. This gives the **static** equilibrium position of the mass. If y denotes the displacement of the mass in motion, measured downward from its static equilibrium position, then we ask you to show in Problem 9 that y satisfies Eq. (3); specifically, that

$$my'' + cy' + ky = F(t) \tag{5}$$

if we include damping and external forces.

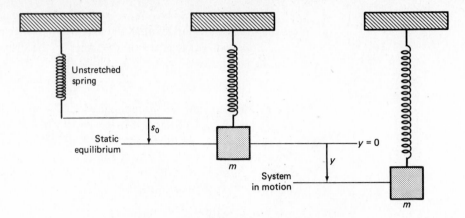

Figure 2.7 A mass suspended vertically from a spring.

The importance of the differential equation that appears in Eqs. (3) and (5) stems from the fact that it describes the motion of many other simple mechanical systems. For example, a **simple pendulum** consists of a mass m swinging back and forth in a vertical plane on the end of a string (or, better, a *massless rod*) of length L, as shown in Fig. 2.8. We may specify the position of the mass at time t by giving the counterclockwise angle $\theta = \theta(t)$ that the string or rod makes with the vertical at time t. To analyze the motion of the mass m, we will apply the law of the conservation of mechanical energy, according to which the sum of the kinetic energy and the potential energy of m remains constant.

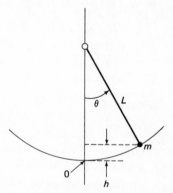

Figure 2.8 The simple pendulum.

The distance along the circular arc from 0 to m is $s = L\theta$, so the velocity of the mass is $v = ds/dt = L(d\theta/dt)$, and therefore its kinetic energy is

$$T = \frac{1}{2}mv^2 = \frac{1}{2}m\left(\frac{ds}{dt}\right)^2 = \frac{1}{2}mL^2\left(\frac{d\theta}{dt}\right)^2.$$

We next choose as reference point the lowest point O reached by the mass (see Fig. 2.8). Then its potential energy V is the product of its weight mg and its vertical height $h = L(1 - \cos\theta)$ above O, so that $V = mgL(1 - \cos\theta)$. The fact that the sum of T and V is a constant C therefore gives

$$\frac{1}{2}mL^2\left(\frac{d\theta}{dt}\right)^2 + mgL(1 - \cos\theta) = C.$$

We differentiate each side of this identity with respect to t to obtain

$$mL^2\left(\frac{d\theta}{dt}\right)\left(\frac{d^2\theta}{dt^2}\right) + mgL(\sin\theta)\frac{d\theta}{dt} = 0,$$

and so

$$\frac{d^2\theta}{dt^2} + \frac{g}{L}\sin\theta = 0 \tag{6}$$

after removal of the common factor $mL^2(d\theta/dt)$.

Now recall that $\sin\theta \approx \theta$ when θ is small; in particular, θ and $\sin\theta$ agree to two decimal places when θ is at most $\pi/12$ (that is, 15°). In a typical pendulum clock, for example, θ would certainly never exceed 15°. It therefore seems reasonable to simplify our mathematical model of the simple pendulum by replacing $\sin\theta$ with θ in Eq. (6). If we also insert a term $c\theta'$ to account for the frictional resistance of the surrounding medium, the result is an equation of the form of Eq. (4):

$$\theta'' + c\theta' + k\theta = 0 \tag{7}$$

where $k = g/L$. Note that this equation is independent of the mass m on the end of the rod. We might, however, expect the effects of the discrepancy between θ and $\sin\theta$ to accumulate over a period of time, so that Eq. (7) might not describe accurately the actual motion of the pendulum over a long period of time.

In the remainder of this section, we first analyze free undamped motion and then free damped motion.

FREE UNDAMPED MOTION

If we have only a mass on a spring, with neither damping nor external force, then Eq. (3) takes the simpler form

$$mx'' + kx = 0. \tag{8}$$

It is convenient to define

$$\omega_0 = \sqrt{\frac{k}{m}}, \tag{9}$$

and rewrite Eq. (8) as

$$x'' + \omega_0^2 x = 0. \tag{8'}$$

The general solution of Eq. (8′) is

$$x(t) = A\cos\omega_0 t + B\sin\omega_0 t. \tag{10}$$

To analyze the motion described by this solution, we choose constants C and α so that

$$C = (A^2 + B^2)^{1/2}, \quad \cos\alpha = \frac{A}{C}, \quad \text{and} \quad \sin\alpha = \frac{B}{C}, \tag{11}$$

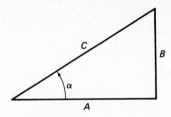

Figure 2.9 The angle α.

as indicated in Fig. 2.9. Note that, while tan $\alpha = B/A$, the angle α is *not* given by the principal branch of the inverse tangent function (which gives values only in the interval $(-\pi/2, \pi/2)$). Instead, α is the angle between 0 and 2π whose cosine and sine have the signs given in (11).

In any event, from (10) and (11) we get

$$x(t) = C\left(\frac{A}{C}\cos \omega_0 t + \frac{B}{C}\sin \omega_0 t\right)$$

$$= C(\cos \alpha \cos \omega_0 t + \sin \alpha \sin \omega_0 t);$$

with the aid of the cosine addition formula, we find that

$$x(t) = C\cos(\omega_0 t - \alpha). \tag{12}$$

Thus the mass oscillates to-and-fro about its equilibrium position with

Amplitude	C,
Circular frequency	ω_0,

and

Phase angle	α.

Such motion is called **simple harmonic motion**. A typical graph of $x(t)$ is shown in Fig. 2.10. If time t is measured in seconds, the circular frequency ω_0 has dimensions of radians per second (rad/s). The **period** of the motion is the time required for the system to complete one full oscillation, so is given by

$$T = \frac{2\pi}{\omega_0} \tag{13}$$

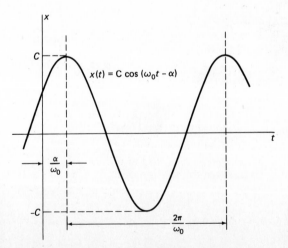

Figure 2.10 Simple harmonic motion.

seconds; its **frequency** is

$$\frac{1}{T} = \frac{\omega_0}{2\pi} \tag{14}$$

in hertz (Hz), which measures the number of complete cycles per second. Note that frequency is measured in cycles per second, while circular frequency has the dimensions of radians per second.

If the initial position $x(0) = x_0$ and initial velocity $x'(0) = v_0$ of the mass are given, we first determine the values of the coefficients A and B in (10), and then find the amplitude C and phase angle α by carrying out the transformation of $x(t)$ to the form in (12), as indicated above.

EXAMPLE 1 A body that weighs 16 lb is attached to the end of a spring which is stretched 2 ft by a force of 100 lb. It is set in motion with initial position $x_0 = 0.5$ (ft) and initial velocity $v_0 = -10$ (ft/s). (Note that these data indicate that the body is displaced to the right and moving to the left at time $t = 0$.) Find the position function of the body, as well as the amplitude, frequency, period of oscillation, and phase angle of its motion.

Solution We take $g = 32$ ft/s². The mass of the body is then $m = W/g = 0.5$ (slugs). The spring constant is $k = 100/2 = 50$ (lb/ft), so Eq. (8) yields $\frac{1}{2}x'' + 50x = 0$; that is, $x'' + 100x = 0$. Consequently, the circular frequency will be $\omega_0 = 10$ (rad/s). So the body will oscillate with

$$\text{Frequency:} \quad \frac{10}{2\pi} \approx 1.59 \text{ Hz}$$

and

$$\text{period:} \quad \frac{2\pi}{10} \approx 0.63 \text{ s}.$$

We now impose the initial conditions $x(0) = 0.5$ and $x'(0) = -10$ in the general solution $x = A \cos 10t + B \sin 10t$, and it follows that $A = 0.5$ and $B = -1$. So the position function of the body is

$$x(t) = \tfrac{1}{2} \cos 10t - \sin 10t.$$

Hence its amplitude of motion is

$$C = [(\tfrac{1}{2})^2 + (1)^2]^{1/2} = \tfrac{1}{2}\sqrt{5} \approx 1.12 \text{ (ft)}.$$

To find the phase angle we write

$$x = \frac{\sqrt{5}}{2}\left(\frac{1}{\sqrt{5}} \cos 10t - \frac{2}{\sqrt{5}} \sin 10t\right) = \frac{\sqrt{5}}{2} \cos(10t - \alpha).$$

Thus we require $\cos \alpha = 1/\sqrt{5} > 0$ and $\sin \alpha = -2/\sqrt{5} < 0$. Hence α is the fourth-quadrant angle

$$\alpha = 2\pi - \tan^{-1}\left(\frac{2/\sqrt{5}}{1/\sqrt{5}}\right) \approx 5.1760 \qquad \text{(rad)}.$$

In the form in which the amplitude and phase angle are made explicit, the position function is

$$x(t) \approx \frac{\sqrt{5}}{2} \cos (10t - 5.1760).$$

FREE DAMPED MOTION

With damping but no external force, the differential equation we have been studying takes the form $mx'' + cx' + kx = 0$; alternatively,

$$x'' + 2px' + \omega_0^2 x = 0, \tag{15}$$

where $\omega_0 = \sqrt{k/m}$ is the corresponding *undamped* circular frequency and

$$p = \frac{c}{2m} > 0. \tag{16}$$

The characteristic equation $r^2 + 2pr + \omega_0^2 = 0$ of (15) has roots

$$r_1, r_2 = -p \pm (p^2 - \omega_0^2)^{1/2} \tag{17}$$

that depend upon the sign of

$$p^2 - \omega_0^2 = \frac{c^2}{4m^2} - \frac{k}{m} = \frac{c^2 - 4km}{4m^2}.$$

The **critical damping** c_{CR} is given by $c_{CR} = \sqrt{4km}$, and we distinguish three cases, according as $c > c_{CR}$, $c = c_{CR}$, or $c < c_{CR}$.

Overdamped Case: $c > c_{CR}$ ($c^2 > 4km$)

Because c is relatively large in this case, we are dealing with a strong resistance in comparison with a relatively weak spring or a small mass. Then (17) gives distinct real roots r_1 and r_2, both of which are negative. The position function has the form

$$x(t) = c_1 \exp (r_1 t) + c_2 \exp (r_2 t). \tag{18}$$

It is easy to see that $x(t) \rightarrow 0$ as $t \rightarrow +\infty$ and that the body settles to its equilibrium position without any oscillations (Problem 25). Figure 2.11 shows some typical graphs of the position function for the overdamped case; we chose x_0 a fixed positive number and illustrated the effects of changing the initial velocity v_0. In every case the would-be oscillations are damped out.

Critically Damped Case: $c = c_{CR}$ ($c^2 = 4km$)

In this case (17) gives equal roots $r_1 = r_2 = -p$ of the characteristic equation, so the general solution is

$$x(t) = e^{-pt}(c_1 + c_2 t). \tag{19}$$

Because $e^{-pt} > 0$ and $c_1 + c_2 t$ has at most one positive zero, the body passes through its equilibrium position at most once, and it is clear that $x(t) \rightarrow 0$ as $t \rightarrow +\infty$. Some graphs of the motion in the critically damped case appear in Fig. 2.12, and they resemble those of the overdamped case (Fig. 2.11). In

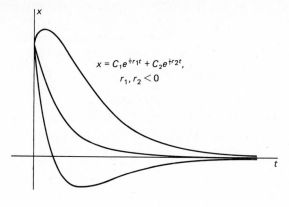

$$x = C_1 e^{+r_1 t} + C_2 e^{+r_2 t},$$
$$r_1, r_2 < 0$$

Figure 2.11 Overdamped motion.

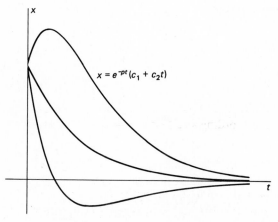

$$x = e^{-pt}(c_1 + c_2 t)$$

Figure 2.12 Critically damped motion.

the critically damped case, the resistance of the dashpot is just large enough to damp out any oscillations, but even a slight decrease in resistance will bring us to the remaining case, the one that shows the most dramatic behavior.

Underdamped Case: $c < c_{CR}$ ($c^2 < 4km$)

The characteristic equation now has the two complex conjugate roots $-p \pm i(\omega_0^2 - p^2)^{1/2}$, and the general solution is

$$x(t) = e^{-pt}(c_1 \cos \omega_1 t + c_2 \sin \omega_1 t), \tag{20}$$

where

$$\omega_1 = (\omega_0^2 - p^2)^{1/2} = \frac{\sqrt{4km - c^2}}{2m}. \tag{21}$$

Using the cosine addition formula as in the derivation of Eq. (12), we may rewrite (20) as

$$x(t) = C e^{-pt} \cos (\omega_1 t - \alpha), \tag{22}$$

where $C = (c_1^2 + c_2^2)^{1/2}$ and $\tan \alpha = c_2/c_1$.

The solution in (22) represents exponentially damped oscillations of the body about its equilibrium position. The graph of $x(t)$ lies between the curves $x = Ce^{-pt}$ and $x = -Ce^{-pt}$ and touches them when $\omega_1 t - \alpha$ is an integral multiple of π. The motion is not actually periodic, but it nevertheless is useful to call ω_1 its **circular frequency**, $T_1 = 2\pi/\omega_1$ its **pseudoperiod** of oscillation, and Ce^{-pt} its **time-varying amplitude**. Most of these quantities are shown in the typical graph of underdamped motion shown in Fig. 2.13. Note from (21) that in this case ω_1 is less than the undamped circular frequency ω_0, so T_1 is larger than the period T of oscillation of the same mass without damping on the same spring. Thus the damping of the dashpot has at least three effects: (1) It exponentially damps the oscillations, in accord with the time-varying amplitude; (2) It slows the motion; that is, the dashpot decreases the frequency of the motion; and (3) It delays the motion—this is the effect of the phase angle in Equation (22).

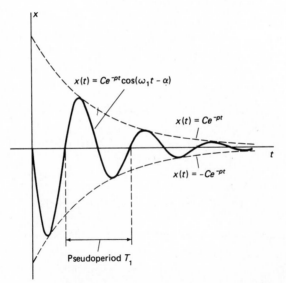

Figure 2.13 Underdamped motion.

EXAMPLE 2 The mass and spring of Example 1 are now attached also to a dashpot that provides 6 lb of resistance for each foot per second of velocity. The mass is set in motion with the same initial position $x(0) = 0.5$ (ft) and the same initial velocity $x'(0) = -10$ (ft/s). Find the position function of the mass, its new frequency and pseudoperiod, its phase angle, and the amplitudes of its first four (local) maxima and minima.

Solution Rather than memorizing the various formulas given above, it is far better practice in a particular case to set up the differential equation and then solve it directly. Recall that $m = 1/2$ and $k = 50$; we are now given $c = 6$ in fps units. Hence Eq. (3) is $\frac{1}{2}x'' + 6x' + 50x = 0$; that is, $x'' + 12x' + 100x = 0$. The roots of the characteristic equation

$r^2 + 12r + 100 = 0$ are

$$\frac{-12 \pm \sqrt{144 - 400}}{2} = -6 \pm 8i,$$

so the general solution is

$$x(t) = e^{-6t}(A \cos 8t + B \sin 8t). \tag{23}$$

The new circular frequency is $\omega_1 = 8$ (rad/s) and the pseudoperiod and new frequency are

$$T_1 = \frac{2\pi}{8} \approx 0.79 \quad \text{(s)}$$

and

$$\frac{1}{T_1} = \frac{8}{2\pi} \approx 1.27 \text{ (Hz)}$$

(as opposed to 0.63 and 1.59 Hz, respectively, in the undamped case).

From (23) we compute

$$x'(t) = e^{-6t}(-8A \sin 8t + 8B \cos 8t) - 6e^{-6t}(A \cos 8t + B \sin 8t).$$

Our initial conditions therefore give the equations

$$x(0) = A = \tfrac{1}{2}$$

and

$$x'(0) = -6A + 8B = -10,$$

so $A = \tfrac{1}{2}$ and $B = -\tfrac{7}{8}$. Thus

$$x(t) = e^{-6t}(\tfrac{1}{2} \cos 8t - \tfrac{7}{8} \sin 8t),$$

so with $C = [(\tfrac{1}{2})^2 + (\tfrac{7}{8})^2]^{1/2} = \sqrt{65}/8$ we have

$$x(t) = \frac{\sqrt{65}}{8} e^{-6t} \left(\frac{4}{\sqrt{65}} \cos 8t - \frac{7}{\sqrt{65}} \sin 8t \right).$$

We require $\cos \alpha = 4/\sqrt{65} > 0$ and $\sin \alpha = -7/\sqrt{65} < 0$, so α is the fourth-quadrant angle

$$\alpha = 2\pi - \tan^{-1}(\tfrac{7}{4}) \approx 5.2315 \text{ (rad.)}$$

Finally,

$$x(t) \approx \frac{\sqrt{65}}{8} e^{-6t} \cos(8t - 5.2315). \tag{24}$$

The local maxima and minima of $x(t)$ occur when

$$0 = x'(t) \approx \frac{\sqrt{65}}{8}[-6e^{-6t} \cos(8t - 5.2315) - 8e^{-6t} \sin(8t - 5.2315)],$$

and thus when

$$\tan(8t - 5.2315) \approx -0.75.$$

Because $\tan^{-1}(0.75) \approx -0.6435$, we want the first four positive values of t such that $8t - 5.2315$ is equal to -0.6435 plus an integral multiple of π. These values of t and the corresponding values of x computed using (24) are as follows:

t (s)	0.1808	0.5735	0.9662	1.3589
x (ft)	-0.2725	0.0258	-0.0024	0.0002

We see that the oscillations are damped out very rapidly, with their amplitude decreasing by a factor of about 10 every half-cycle. See Problems 28 and 29 for a more general discussion of this phenomenon.

2.4 Problems

1. Determine the period and frequency of the simple harmonic motion of a 4-kg mass on the end of a spring with spring constant 16 N/m.

2. Determine the period and frequency of the simple harmonic motion of a body weighing 24 lb on the end of a spring with spring constant 48 lb/ft.

3. A mass of 3 kg is attached to the end of a spring that is stretched 20 cm by a force of 15 N. It is set in motion with initial position $x_0 = 0$ and initial velocity $v_0 = -10$ m/s. Find the amplitude, period, and frequency of the resulting motion.

4. A body weighing 8 lb is attached to the end of a spring that is stretched 1 in. by a force of 3 lb. At time $t = 0$ the body is pulled 1 ft to the right (spring stretched) and set in motion with an initial velocity of 5 ft/s to the left.
   (a) Find $x(t)$ in the form $C \cos(\omega_0 t + \alpha)$.
   (b) Find the amplitude and period of the motion of the body.

In Problems 5–8, assume that the differential equation of a simple pendulum of length L is $L\theta'' + g\theta = 0$, where $g = GM/R^2$ is the gravitational acceleration at the location of the pendulum (at distance R from the center of the earth; M denotes the mass of the earth).

5. Two pendulums are of lengths L_1 and L_2, and—when located at the respective distances R_1 and R_2 from the center of the earth—have periods p_1 and p_2. Show that
$$\frac{p_1}{p_2} = \frac{R_1\sqrt{L_1}}{R_2\sqrt{L_2}}.$$

6. A certain pendulum clock keeps perfect time in Paris, where the radius of the earth is $R = 3956$ (mi). But this clock loses 2 min 40 s per day at a location on the equator. Use the result of Problem 5 to find the amount of the equatorial bulge of the earth.

7. A pendulum of length 100.10 in., located at a point at sea level where the radius of the earth is $R = 3960$ (mi), has the same period as does a pendulum of length 100.00 in. atop a nearby mountain. Use the result of Problem 5 to find the height of the mountain.

8. Most grandfather clocks have pendulums with lengths that are adjustable. One such clock loses 10 mins per day when the length of the pendulum is 30 in. With what length pendulum will this clock keep perfect time?

9. Derive Equation (5) describing the motion of a mass attached to the bottom of a vertically suspended spring. (*Suggestion:* First denote by $x(t)$ the displace-

ment of the mass below the unstretched position of the spring; set up the differential equation for x. Then substitute $y = x - s_0$ in this differential equation.)

10. Consider a floating cylindrical buoy with radius r, height h, and density $\rho \leq 0.5$ (recall that the density of water is 1 g/cm^3). The buoy is initially suspended at rest with its bottom at the top surface of the water and is released at time $t = 0$. Thereafter it is acted upon by two forces: a downward gravitational force equal to its weight $mg = \rho\pi r^2 hg$ and an upward force of buoyancy equal to the weight $\pi r^2 xg$ of water displaced, where $x = x(t)$ is the depth of the bottom of the buoy beneath the surface at time t. Conclude that the buoy undergoes simple harmonic motion about the equilibrium position $x_e = \rho h$ with period $p = 2\pi\sqrt{\rho h/g}$. Compute p and the amplitude of the motion if $\rho = 0.5$ g/cm^3, $h = 200$ cm, and $g = 980$ cm/s^2.

11. A cylindrical buoy weighing 100 lb floats in water with its axis vertical (as in Problem 10). When depressed slightly and released, it oscillates up and down four times every 10 s. Assume that friction is negligible. Find the radius of the buoy.

12. Assume that the earth is a solid sphere of uniform density, with mass M and radius $R = 3960$ (mi). For a particle of mass m *within* the earth at distance r from the center of the earth, the gravitational force attracting m toward the center is $F_r = -GM_r m/r^2$, where M_r is the mass of the part of the earth within a sphere of radius r. (a) Show that $F_r = -GMmr/R^3$. (b) Now suppose that a hole is drilled straight through the center of the earth, thus connecting two antipodal points on its surface. Let a particle of mass m be dropped at time $t = 0$ into this hole with initial speed zero, and let $r(t)$ be its distance from the center of the earth at time t. Conclude from Newton's second law and part (a) that $r''(t) = -k^2 r(t)$, where $k^2 = GM/R^3 = g/R$. (c) Take $g = 32.2$ ft/s^2, and conclude from part (b) that the particle undergoes simple harmonic motion back and forth between the ends of the hole, with a period of about 84 min. (d) Look up (or derive) the period of a satellite that just skims the surface of the earth; compare the value with the results in (c). How do you explain the coincidence? Or *is* it a coincidence? (e) With what speed (in miles per hour) does the particle pass through the center of the earth? (f) Look up or derive the orbital velocity of a satellite that just skims the surface of the earth; compare the value with the result in (e). How do you explain the coincidence? Or *is* it a coincidence?

The remaining problems in this section deal with free damped motion. In Problems 13–17, a mass m is attached both to a spring (with given spring constant k) and a dashpot (with given damping constant c). The mass is set in motion with initial position x_0 and initial velocity v_0. Find the position function x(t) and determine whether the motion is overdamped, critically damped, or underdamped. If it is underdamped, write x(t) in the form $Ce^{-pt}\cos(\omega_1 t - \alpha)$.

13. $m = \frac{1}{2}, c = 3, k = 4$; $x_0 = 2, v_0 = 0$.

14. $m = 3, c = 30, k = 63$; $x_0 = 2, v_0 = 2$.

15. $m = 1, c = 8, k = 16$; $x_0 = 5, v_0 = -10$.

16. $m = 2, c = 12, k = 50$; $x_0 = 0, v_0 = -8$.

17. $m = 4, c = 20, k = 169$; $x_0 = 4, v_0 = 16$.

18. A 12-lb weight is attached both to a vertically suspended spring that it stretches 6 in. and to a dashpot that provides 3 lb of resistance for every foot per second

of velocity. (a) If the weight is pulled down 1 ft below its static equilibrium position and then released from rest at time $t = 0$, find its position function $x(t)$. (b) Find the frequency, time-varying amplitude, and phase angle of the motion.

19. This problem deals with a highly simplified model of a car of weight 3200 lb. Assume that the suspension system of the car acts like a single spring and its shock absorbers like a single dashpot, so that its vertical vibrations satisfy Eq. (4) with appropriate values of the coefficients. (a) Find the stiffness coefficient k of the spring if the car undergoes free vibrations at 80 cycles per minute (cycles/min) when its shock absorbers are disconnected. (b) With the shock absorbers connected the car is set into vibration by driving it over a bump, and the resulting damped vibrations have a frequency of 78 cycles/min. After how long will the time-varying amplitude be 1% of its original value?

Each of Problems 20–30 deals with a mass-spring-dashpot system having position function $x(t)$ satisfying Eq. (4). We write $x_0 = x(0)$ and $v_0 = x'(0)$ and recall that $p = c/2m$, $\omega_0^2 = k/m$, and $\omega_1^2 = \omega_0^2 - p^2$. The system is critically damped, overdamped, or underdamped, as specified in each problem.

20. (Critically damped) Show in this case that
$$x(t) = (x_0 + v_0 t + p x_0 t) e^{-pt}.$$

21. (Critically damped) Deduce from Problem 20 that the mass passes through $x = 0$ at some instant $t > 0$ if and only if x_0 and $v_0 + p x_0$ have opposite signs.

22. (Critically damped) Deduce from Problem 20 that $x(t)$ has a local maximum or minimum at some instant $t > 0$ if and only if v_0 and $v_0 + p x_0$ have the same sign.

23. (Overdamped) Show in this case that
$$x(t) = \frac{1}{2\gamma}[(v_0 - r_2 x_0) \exp(r_1 t) - (v_0 - r_1 x_0) \exp(r_2 t)],$$
where $r_1, r_2 = -p \pm (p^2 - \omega_0^2)^{1/2}$ and $\gamma = (r_1 - r_2)/2 > 0$.

24. (Overdamped) If $x_0 = 0$ deduce from Problem 23 that
$$x(t) = \frac{v_0}{\gamma} e^{-pt} \sinh \gamma t.$$

25. (Overdamped) Prove that in this case the mass can pass through its equilibrium position $x = 0$ at most once.

26. (Underdamped) Show that in this case
$$x(t) = e^{-pt}\left[x_0 \cos \omega_1 t + \frac{v_0 + p x_0}{\omega_1} \sin \omega_1 t \right].$$

27. (Underdamped) If the damping constant c is small in comparison with $\sqrt{8mk}$, apply the binomial series to show that
$$\omega_1 \approx \omega_0 \left(1 - \frac{c^2}{8mk}\right).$$

28. (Underdamped) Show that the local maxima and minima of $x(t) = Ce^{-pt} \cos(\omega_1 t - \alpha)$ occur where
$$\tan(\omega_1 t - \alpha) = \frac{-p}{\omega_1}.$$
Conclude that $t_2 - t_1 = 2\pi/\omega_1$ if two consecutive maxima occur at times t_1 and t_2.

29. (Underdamped) Let x_1 and x_2 be two consecutive local maximum values of $x(t)$. Deduce from the result of Problem 28 that

$$\ln \frac{x_1}{x_2} = \frac{2\pi p}{\omega_1}.$$

The constant $\Delta = 2\pi p/\omega_1$ is called the **logarithmic decrement** of the oscillation. Note also that $c = m\omega_1 \Delta/\pi$ because $p = c/2m$.

    Note: The result of Problem 29 provides an accurate method for measuring the *viscosity* of a fluid, which is an important parameter in fluid dynamics, but which is not easily measured directly. According to Stokes' drag law, a spherical body of radius a moving at a (relatively slow) speed v through a fluid of viscosity μ experiences a resistive force $F_R = 6\pi\mu av$. Thus if a spherical mass on a spring is immersed in the fluid and set in motion, this drag resistance damps its oscillations with damping constant $c = 6\pi a\mu$. The frequency ω_1 and logarithmic decrement Δ of the oscillations can be measured by direct observation. The final formula in Problem 29 then gives c and hence the viscosity of the fluid.

30. (Underdamped) A body weighing 100 lb is oscillating attached to a spring and a dashpot. Its first two maximum displacements of 6.73 in. and 1.46 in. are observed to occur at times 0.34 s and 1.17 s, respectively. Compute the damping constant (in pound-seconds per foot) and spring constant (in pounds per foot).

2.5
Nonhomogeneous Equations and the Method of Undetermined Coefficients

We learned in Section 2.3 how to solve homogeneous linear equations with constant coefficients, but we saw in Section 2.4 that an external force in a simple mechanical system contributes a nonhomogeneous term to its differential equation. The general nonhomogeneous nth order linear equation with constant coefficients has the form

$$a_n y^{(n)} + a_{n-1} y^{(n-1)} + \cdots + a_1 y' + a_0 y = f(x). \tag{1}$$

By Theorem 5 in Section 2.2, a general solution of (1) is of the form

$$y = y_c + y_p \tag{2}$$

where the complementary function y_c is a general solution of the associated homogeneous equation

$$a_n y^{(n)} + a_{n-1} y^{(n-1)} + \cdots + a_1 y' + a_0 y = 0, \tag{3}$$

and y_p is a single particular solution of (1). Thus our remaining task is to find y_p.

    The **method of undetermined coefficients** is a straightforward way of doing this when the given function $f(x)$ in (1) is sufficiently simple that we can make an intelligent guess as to the general form of y_p. For example, suppose that $f(x)$ is a polynomial of degree m. Then, because the derivatives of a polynomial are themselves polynomials of lower degree, it is reasonable to suspect a particular solution

$$y_p = A_m x^m + A_{m-1} x^{m-1} + \cdots + A_1 x + A_0$$

that is also a polynomial of degree m, but with as yet undetermined coefficients. We may, therefore, substitute this form of y_p into (1), and then—by equating coefficients of like powers of x on the two sides of the resulting equation—attempt to determine the coefficients $A_0, A_1, \ldots, A_m$ so that y_p will, indeed, be a particular solution.

Similarly, suppose that $f(x) = a \cos kx + b \sin kx$. Then it is reasonable to suspect a particular solution of the same form:

$$y_p = A \cos kx + B \sin kx,$$

a linear combination with undetermined coefficients A and B. The reason is that any derivative of such a linear combination of $\cos kx$ and $\sin kx$ has the same form. We may therefore substitute this form of y_p in (1), and then—by equating coefficients of $\cos kx$ and $\sin kx$ on the two sides of the resulting equation—attempt to determine the coefficients A and B so that y_p will, indeed, be a particular solution.

It turns out that this approach does succeed whenever all the derivatives of $f(x)$ have the same form as $f(x)$ itself. Before describing the method in full generality, we illustrate it with several preliminary examples.

EXAMPLE 1   Find a particular solution of $y'' + 3y' + 4y = 3x + 2$.

Solution   Here $f(x) = 3x + 2$, a polynomial of degree 1, so our guess is $y_p = Ax + B$. Then $y_p' = A$ and $y_p'' = 0$, so y_p will satisfy the differential equation provided that

$$(0) + 3(A) + 4(Ax + B) = 3x + 2.$$

This is so if and only if $4A = 3$ and $3A + 4B = 2$. These two equations yield $A = \frac{3}{4}$ and $B = -\frac{1}{16}$, so we have found the particular solution $y_p = \frac{3}{4}x - \frac{1}{16}$.

EXAMPLE 2   Find a particular solution of $y'' - 4y = 2e^{3x}$.

Solution   Any derivative of e^{3x} is a constant multiple of e^{3x}, so it is reasonable to try $y_p = Ae^{3x}$. Then $y_p'' = 9Ae^{3x}$, so the differential equation will be satisfied provided that

$$9Ae^{3x} - 4(Ae^{3x}) = 2e^{3x};$$

that is, $5A = 2$, so that $A = \frac{2}{5}$. Thus our particular solution is $y_p = \frac{2}{5}e^{3x}$.

EXAMPLE 3   Find a particular solution of $3y'' + y' - 2y = 2 \cos x$.

Solution   Our first guess might be $A \cos x$, but the presence of y' on the left-hand side signals that we may need a term involving $\sin x$ as well. So we try

$$y_p = A \cos x + B \sin x,$$
$$y_p' = -A \sin x + B \cos x,$$
$$y_p'' = -A \cos x - B \sin x.$$

Then substitution of y_p into the differential equation yields

$$3(-A \cos x - B \sin x) + (-A \sin x + B \cos x)$$
$$- 2(A \cos x + B \sin x) = 2 \cos x.$$

We equate the coefficient of $\cos x$ on the left-hand side with that of $\cos x$ on the right and do the same with $\sin x$ to get the equations

$$-5A + B = 2,$$
$$-A - 5B = 0,$$

with solution $A = -\frac{5}{13}$, $B = \frac{1}{13}$. Hence our particular solution is $y_p = -\frac{5}{13} \cos x + \frac{1}{13} \sin x$.

The following example, which superficially resembles Example 2, indicates that the method of undetermined coefficients is not always quite so simple as it appears.

EXAMPLE 4 Find a particular solution of $y'' - 4y = 2e^{2x}$.

Solution If we try $y_p = Ae^{2x}$, we find that

$$y_p'' - 4y_p = 4Ae^{2x} - 4Ae^{2x} = 0 \neq 2e^{2x};$$

so, no matter what A is, Ae^{2x} cannot satisfy the given nonhomogeneous equation. In fact, the above computation shows that Ae^{2x} satisfies instead the associated *homogeneous* equation. Therefore, we should begin with a trial function whose derivative involves both e^{2x} *and something else* that can cancel upon substitution into the differential equation to leave the e^{2x} term that we need. A reasonable guess is $y_p = Axe^{2x}$, with $y_p' = Ae^{2x} + 2Axe^{2x}$ and $y_p'' = 4Ae^{2x} + 4Axe^{2x}$. Substitution yields

$$(4Ae^{2x} + 4Axe^{2x}) - 4(Axe^{2x}) = 2e^{2x}.$$

The terms involving xe^{2x} obligingly cancel, leaving only $4Ae^{2x} = 2e^{2x}$, so that $A = \frac{1}{2}$. Thus our particular solution is $y_p = \frac{1}{2}xe^{2x}$.

Our initial difficulty in Example 4 resulted from the fact that $f(x) = 2e^{2x}$ satisfies the associated homogeneous equation. Rule 1, given shortly, tells what to do when we do not have this difficulty, and Rule 2 tells what to do when we do have it.

The method of undetermined coefficients applies whenever the function $f(x)$ in Eq. (1) is a linear combination of (finite) products of functions of the following three types:

1. A polynomial in x;
2. An exponential function e^{rx}; $\qquad$ (4)
3. $\cos kx$ or $\sin kx$.

Any such function, for example $f(x) = (3 - 4x^2)e^{5x} - 4x^3 \cos 10x$, has the crucial property that only *finitely* many linearly independent functions appear as terms (summands) in $f(x)$ and its derivatives of all orders. In Rules 1 and 2 we assume that $Ly = f(x)$ is a nonhomogeneous linear equation with constant coefficients, and that $f(x)$ is a function of this kind.

RULE 1: METHOD OF UNDETERMINED COEFFICIENTS

Suppose that no *term appearing either in $f(x)$ or in any of its derivatives satisfies the associated homogeneous equation $Ly = 0$. Then take as a trial solution for y_p a linear combination of all linearly independent such terms. Determine the coefficients by substitution of this trial solution into the nonhomogeneous equation.*

Note that this rule is not a theorem requiring proof; it is merely a procedure to be followed in searching for a particular solution y_p. If we succeed in finding y_p, then nothing more need be said. It can be proved, however, that this procedure will always succeed under the conditions specified.

In practice we check the supposition made in Rule 1 by first using the characteristic equation to find the complementary function y_c, and then write a list of all the terms appearing in $f(x)$ and its successive derivatives. If none of the terms in this list duplicates a term in y_c, then we're in business!

EXAMPLE 5 Find a particular solution of

$$y'' + 4y = 3x^3. \tag{5}$$

Solution The (familiar) complementary function of this equation is $y_c = c_1 \cos 2x + c_2 \sin 2x$. The function $f(x) = 3x^3$ and its derivatives are constant multiples of the linearly independent functions x^3, x^2, x, and 1. Because none of these appears in y_c, we try

$$y_p = Ax^3 + Bx^2 + Cx + D,$$
$$y_p' = 3Ax^2 + 2Bx + C,$$
$$y_p'' = 6Ax + 2B.$$

Substitution in (5) gives

$$y_p'' + 4y_p = (6Ax + 2B) + 4(Ax^3 + Bx^2 + Cx + D)$$
$$= 4Ax^3 + 4Bx^2 + (6A + 4C)x + (2B + D)$$
$$= 3x^3.$$

We equate coefficients of like powers of x to get the equations

$$4A = 3, \qquad 4B = 0,$$
$$6A + 4C = 0, \qquad 2B + D = 0$$

with solution $A = \frac{3}{4}$, $B = 0$, $C = -\frac{9}{8}$, and $D = 0$. Hence a particular solution of (5) is $y_p = \frac{3}{4}x^3 - \frac{9}{8}x$.

EXAMPLE 6 Solve the initial value problem

$$y'' - 3y' + 2y = 3e^{-x} - 10\cos 3x; \tag{6}$$
$$y(0) = 1, \qquad y'(0) = 2.$$

Solution The characteristic equation $r^2 - 3r + 2 = 0$ has roots $r = 1$ and $r = 2$, so the complementary function is $y_c = c_1 e^x + c_2 e^{2x}$. The terms involved in $f(x) = 3e^{-x} - 10\cos 3x$ and its derivatives are e^{-x}, $\cos 3x$, and $\sin 3x$. Because none of these appears in y_c, we try

$$y_p = Ae^{-x} + B\cos 3x + C\sin 3x,$$
$$y_p' = -Ae^{-x} - 3B\sin 3x + 3C\cos 3x,$$
$$y_p'' = Ae^{-x} - 9B\cos 3x - 9C\sin 3x.$$

After we substitute these expressions into (6) and collect coefficients, we get

$$y_p'' - 3y_p' + 2y_p = 6Ae^{-x} + (-7B - 9C)\cos 3x + (9B - 7C)\sin 3x$$
$$= 3e^{-x} - 10\cos 3x.$$

We equate the coefficients of the terms involving e^{-x}, $\cos 3x$, and $\sin 3x$ to obtain

$$6A = 3,$$
$$-7B - 9C = -10,$$
$$9B - 7C = 0$$

with solution $A = \frac{1}{2}$, $B = \frac{7}{13}$, and $C = \frac{9}{13}$. This gives the particular solution

$$y_p = \tfrac{1}{2}e^{-x} + \tfrac{7}{13}\cos 3x + \tfrac{9}{13}\sin 3x$$

which, however, does not have the required initial values.

To satisfy the initial conditions, we begin with the *general* solution

$$y = y_c + y_p$$
$$= c_1 e^x + c_2 e^{2x} + \tfrac{1}{2}e^{-x} + \tfrac{7}{13}\cos 3x + \tfrac{9}{13}\sin 3x$$

with derivative

$$y' = c_1 e^x + 2c_2 e^{2x} - \tfrac{1}{2}e^{-x} - \tfrac{21}{13}\sin 3x + \tfrac{27}{13}\cos 3x.$$

The initial conditions lead to the equations

$$y(0) = c_1 + c_2 + \tfrac{1}{2} + \tfrac{7}{13} = 1,$$
$$y'(0) = c_1 + 2c_2 - \tfrac{1}{2} + \tfrac{27}{13} = 2$$

with solution $c_1 = -\frac{1}{2}$, $c_2 = \frac{6}{13}$. The desired particular solution is therefore

$$y = -\tfrac{1}{2}e^x + \tfrac{6}{13}e^{2x} + \tfrac{1}{2}e^{-x} + \tfrac{7}{13}\cos 3x + \tfrac{9}{13}\sin 3x.$$

EXAMPLE 7 Find the general form of a particular solution of

$$y^{(3)} + 9y' = x\sin x + x^2 e^{2x}. \tag{7}$$

Solution The characteristic equation $r^3 + 9r = 0$ has roots $r = 0, -3i$, and $3i$. So the complementary function is $y_c = c_1 + c_2\cos 3x + c_3\sin 3x$. The derivatives of the right-hand side in (7) involve the terms $\cos x$, $\sin x$, $x\cos x$, $x\sin x$, e^{2x}, xe^{2x}, and $x^2 e^{2x}$. Because there is no duplication with the terms of the complementary function, the trial function takes the form

$$y_p = A \cos x + B \sin x + Cx \cos x + Dx \sin x$$
$$+ Ee^{2x} + Fxe^{2x} + Gx^2e^{2x}.$$

Upon substituting y_p in (7) and equating coefficients, we would get seven equations determining the seven coefficients A, B, C, D, E, F, and G.

Now we turn our attention to the situation in which Rule 1 does not apply: Some of the terms involved in $f(x)$ and its derivatives satisfy the associated homogeneous equation. For instance, suppose that we want to find a particular solution of

$$(D - r)^3 y = (2x - 3)e^{rx}. \tag{8}$$

Proceeding as in Rule 1, our first guess would be

$$y_p = Ae^{rx} + Bxe^{rx}. \tag{9}$$

This form of y_p will not be adequate because the complementary function of (8) is

$$y_c = c_1 e^{rx} + c_2 xe^{rx} + c_3 x^2 e^{rx}, \tag{10}$$

so substitution of (9) in the left-hand side of (8) would yield zero rather than $(2x - 3)e^{rx}$.

To see how to amend our first guess, we observe that

$$(D - r)^2[(2x - 3)e^{rx}] = [D^2(2x - 3)]e^{rx} = 0$$

by Eq. (13) in Section 2.3. If y is *any* solution of Eq. (8) and we apply the operator $(D - r)^2$ to both sides, we therefore see that y is also a solution of the equation $(D - r)^5 y = 0$. The general solution of this *homogeneous* equation can be written as

$$y = \underbrace{c_1 e^{rx} + c_2 xe^{rx} + c_3 x^2 e^{rx}}_{y_c} + \underbrace{Ax^3 e^{rx} + Bx^4 e^{rx}}_{y_p}.$$

Thus *every* solution of our original equation in (8) is the sum of a complementary function and a *particular solution* of the form

$$y_p = Ax^3 e^{rx} + Bx^4 e^{rx}. \tag{11}$$

Note that (11) may be obtained by multiplying each term of our first guess in (9) by the least power of x (in this case, x^3) that suffices to eliminate duplication between the terms of the resulting trial solution y_p and the complementary function y_c given in (10). This procedure succeeds in the general case.

To simplify the general statement, we observe that, in order to find a particular solution of the nonhomogeneous linear differential equation

$$Ly = f_1(x) + f_2(x), \tag{12}$$

it suffices to find *separately* particular solutions Y_1 and Y_2 of the two equations

$$Ly = f_1(x) \quad \text{and} \quad Ly = f_2(x), \tag{13}$$

respectively. For linearity then gives

$$L[Y_1 + Y_2] = LY_1 + LY_2 = f_1(x) + f_2(x),$$

and therefore $y_p = Y_1 + Y_2$ is a particular solution of (2). This is the **principle of superposition** for nonhomogeneous linear equations.

Now our problem is to find a particular solution of the equation $Ly = f(x)$, where $f(x)$ is a linear combination of products of the elementary functions listed in (4). Thus $f(x)$ can be written as a sum of terms each of the form

$$P_m(x)e^{rx}\{\cos kx \quad \text{or} \quad \sin kx\} \tag{14}$$

where $P_m(x)$ is a polynomial in x. Note that any derivative of such a term is of the same form, but with *both* sines and cosines appearing. The procedure by which we arrived above at the particular solution in (11) of Eq. (8) can be generalized to show that the following procedure is always successful.

RULE 2: METHOD OF UNDETERMINED COEFFICIENTS

If the function $f(x)$ is of the form in (14), take as the trial solution

$$y_p = x^s[(A_0 + A_1x + \cdots + A_mx^m)e^{rx}\cos kx$$
$$+ (B_0 + B_1x + \cdots + B_mx^m)e^{rx}\sin kx], \tag{15}$$

where s is the smallest nonnegative integer such that no term in y_p duplicates a term in the complementary function y_c. Then determine the coefficients in (15) by substituting y_p into the nonhomogeneous equation.

In practice we seldom need to deal with a function $f(x)$ exhibiting the full generality in (14). The table of Fig. 2.14 lists the form of y_p in various common cases, corresponding to the possibilities $m = 0$, $r = 0$, and $k = 0$.

$f(x)$	y_p
$P_m(x) = b_0 + b_1x + b_2x^2 + \ldots + b_mx^m$	$x^s(A_0 + A_1x + A_2x^2 + \ldots + A_mx^m)$
$a\cos kx + b\sin kx$	$x^s(A\cos kx + B\sin kx)$
$e^{rx}(a\cos kx + b\sin kx)$	$x^se^{rx}(A\cos kx + B\sin kx)$
$P_m(x)e^{rx}$	$x^s(A_0 + A_1x + A_2x^2 + \ldots + A_mx^m)e^{rx}$
$P_m(x)(a\cos kx + b\sin kx)$	$x^s[(A_0 + A_1x + \ldots + A_mx^m)\cos kx$ $+ (B_0 + B_1x + \ldots + B_mx^m)\sin kx]$

Figure 2.14 Substitutions in the method of undetermined coefficients.

On the other hand, it is not uncommon to have $f(x) = f_1(x) + f_2(x)$, where $f_1(x)$ and $f_2(x)$ are different functions of the sort listed in the table of Fig. 2.14. In this event, we take as y_p the sum of the indicated particular functions, choosing s *separately* for each part to eliminate duplication with the complementary function. This procedure is illustrated in the following examples.

EXAMPLE 8 Find a particular solution of

$$y^{(3)} + y'' = 3e^x + 4x^2. \tag{16}$$

Solution The characteristic equation $r^3 + r^2 = 0$ has roots $r = 0, 0$, and -1, so the complementary function is $y_c = c_1 + c_2 x + c_3 e^{-x}$. As a first step toward our particular solution, we form the sum

$$(Ae^x) + (B + Cx + Dx^2).$$

The part Ae^x corresponding to $3e^x$ does not duplicate any part of the complementary function, but the part $B + Cx + Dx^2$ must be multiplied by x^2 to eliminate duplication. Next,

$$y_p = Ae^x + Bx^2 + Cx^3 + Dx^4,$$
$$y_p' = Ae^x + 2Bx + 3Cx^2 + 4Dx^3,$$
$$y_p'' = Ae^x + 2B + 6Cx + 12Dx^2,$$

and

$$y_p{}^{(3)} = Ae^x + 6C + 24Dx.$$

Substitution of these derivatives in (16) yields

$$2Ae^x + (2B + 6C) + (6C + 24D)x + 12Dx^2 = 3e^x + 4x^2.$$

The system of equations

$$2A = 3, \qquad 2B + 6C = 0,$$
$$6C + 24D = 0, \qquad 12D = 4$$

has the solution $A = \frac{3}{2}$, $B = 4$, $C = -\frac{4}{3}$, and $D = \frac{1}{3}$. Hence the desired particular solution is

$$y_p = \tfrac{3}{2}e^x + 4x^2 - \tfrac{4}{3}x^3 + \tfrac{1}{3}x^4.$$

EXAMPLE 9 Determine the appropriate form for a particular solution of $y'' + 6y' + 13y = e^{-3x} \cos 2x$.

Solution The characteristic equation $r^2 + 6r + 13 = 0$ has roots $-3 \pm 2i$, so the complementary function is

$$y_c = e^{-3x}(c_1 \cos 2x + c_2 \sin 2x).$$

This is the same form as a first attempt $e^{-3x}(A \cos 2x + B \sin 2x)$ at a particular solution, so we must multiply by x to eliminate duplication. Hence we would take $y_p = e^{-3x}(Ax \cos 2x + Bx \sin 2x)$.

EXAMPLE 10 Determine the appropriate form for a particular solution of the fifth order equation

$$(D - 2)^3(D^2 + 9)y = x^2 e^{2x} + x \sin 3x.$$

Solution The characteristic equation $(r - 2)^3(r^2 + 9) = 0$ has roots $r = 2$, $2, 2, 3i$, and $-3i$ so the complementary function is

$$y_c = c_1 e^{2x} + c_2 x e^{2x} + c_3 x^2 e^{2x} + c_4 \cos 3x + c_5 \sin 3x.$$

As a first step toward the form of a particular solution, we examine the sum

$$[(A + Bx + Cx^2)e^{2x}] + [(D + Ex)\cos 3x + (F + Gx)\sin 3x].$$

To eliminate duplication with terms of y_c, the first part—corresponding to x^2e^{2x}—must be multiplied by x^3, and the second part—corresponding to $x \sin 3x$—must be multiplied by x. Hence we would take

$$y_p = (Ax^3 + Bx^4 + Cx^5)e^{2x} + (Dx + Ex^2)\cos 3x$$
$$+ (Fx + Gx^2)\sin 3x.$$

Finally, let us point out the kind of situation in which the method of undetermined coefficients cannot be used. Consider, for instance, the equation

$$y'' + y = \tan x, \tag{17}$$

which at first glance may appear similar to those considered in the examples above. Not so; the function $f(x) = \tan x$ has *infinitely many* linearly independent derivatives

$$\sec^2 x, \ 2 \sec^2 x \tan x, \ 4 \sec^2 x \tan^2 x + 2 \sec^4 x, \ldots.$$

Therefore we do not have available a *finite* linear combination to use as a trial solution. In Section 2.7 (Variation of Parameters) we will discuss an alternative method that can be applied to nonhomogeneous equations such as (17).

2.5 Problems

In each of Problems 1–20, find a particular solution y_p of the given equation.

1. $y'' + 16y = e^{3x}$.

2. $y'' - y' - 2y = 3x + 4$.

3. $y'' - y' - 6y = 2 \sin 3x$.

4. $4y'' + 4y' + y = 3xe^x$.

5. $y'' + y' + y = \sin^2 x$.

6. $2y'' + 4y' + 7y = x^2$.

7. $y'' - 4y = \sinh x$.

8. $y'' - 4y = \cosh 2x$.

9. $y'' + 2y' - 3y = 1 + xe^x$.

10. $y'' + 9y = 2 \cos 3x + 3 \sin 3x$.

11. $y^{(3)} + 4y' = 3x - 1$.

12. $y^{(3)} + y' = 2 - \sin x$.

13. $y'' + 2y' + 5y = e^x \sin x$.

14. $y^{(4)} - 2y'' + y = xe^x$.

15. $y^{(5)} + 5y^{(4)} - y = 17$.

16. $y'' + 9y = 2x^2 e^{3x} + 5$.

17. $y'' + y = \sin x + x \cos x$.

18. $y^{(4)} - 5y'' + 4y = e^x - xe^{2x}$.

19. $y^{(5)} + 2y^{(3)} + 2y'' = 3x^2 - 1$.

20. $y^{(3)} - y = e^x + 7$.

In each of Problems 21–30, set up the appropriate form of a particular solution y_p, but do not determine the values of the coefficients.

21. $y'' - 2y' + 2y = e^x \sin x$.

22. $y^{(5)} - y^{(3)} = e^x + 2x^2 - 5$.

23. $y'' + 4y = 3x \cos 2x$.

24. $y^{(3)} - y'' - 12y' = x - 2xe^{-3x}$.

25. $y'' + 3y' + 2y = x(e^{-x} - e^{-2x})$.

26. $y'' - 6y' + 13y = xe^{3x} \sin 2x$.

27. $y^{(4)} + 5y'' + 4y = \sin x + \cos 2x$.

28. $y^{(4)} + 9y'' = (x^2 + 1)\sin 3x$.

29. $(D - 1)^3(D^2 - 4)y = xe^x + e^{2x} + e^{-2x}$.

30. $y^{(4)} - 2y'' + y = x^2 \cos x$.

Solve the initial value problems in Problems 31–40.

31. $y'' + 4y = 2x$; $y(0) = 1, y'(0) = 2$.

32. $y'' + 3y' + 2y = e^x$; $y(0) = 0, y'(0) = 3$.

33. $y'' + 9y = \sin 2x$; $y(0) = 1, y'(0) = 0$.

34. $y'' + y = \cos x$; $y(0) = 1, y'(0) = -1$.

35. $y'' - 2y' + 2y = x + 1$; $y(0) = 3, y'(0) = 0$.

36. $y^{(4)} - 4y'' = x^2$; $y(0) = y'(0) = 1, y''(0) = y^{(3)}(0) = -1$.

37. $y^{(3)} - 2y'' + y' = 1 + xe^x$; $y(0) = y'(0) = 0, y''(0) = 1$.

38. $y'' + 2y' + 2y = \sin 3x$; $y(0) = 2, y'(0) = 0$.

39. $y^{(3)} + y'' = x + e^{-x}$; $y(0) = 1, y'(0) = 0, y''(0) = 1$.

40. $y^{(4)} - y = 5$; $y(0) = y'(0) = y''(0) = y^{(3)}(0) = 0$.

41. (a) Write

$$\cos 3x + i \sin 3x = e^{3ix} = (\cos x + i \sin x)^3$$

by Euler's formula, expand, and equate real and imaginary parts to derive the identity $\cos^3 x = \frac{1}{4} \cos 3x + \frac{3}{4} \cos x$ and a similar identity for $\sin^3 x$. (b) Use the result of part (a) to find a general solution of $y'' + 4y = \cos^3 x$.

Use trigonometric identities to find general solutions of the equations in Problems 42–44.

42. $y'' + y' + y = \sin x \sin 3x$.

43. $y'' + 9y = \sin^4 x$.

44. $y'' + y = x \cos^3 x$.

Let $Ly = f(x)$ be a nonhomogeneous linear differential equation with constant coefficients. If M is a linear differential operator with constant coefficients such that $M[f(x)] \equiv 0$, then M is called an **annihilator** *of $f(x)$. In this case every solution of $Ly = f(x)$ satisfies the homogeneous equation $MLy = 0$. If the general solution of $MLy = 0$ is written in the form $y = y_c + y_p$, where y_c is a general solution of $Ly = 0$, then y_p will be an appropriate form of a particular solution of the original nonhomogeneous equation $Ly = f(x)$. This method was used in the text, with $L = (D - r)^3$ and $M = (D - r)^2$, to discover the particular solution (11) of Eq. (8). Use this annihilator method in each of Problems 45–50 to discover the form of a particular solution of the given equation.*

45. $y'' - 5y' = x^2$;   take $M = D^3$.

46. $y'' - 5y' = e^{5x}$;   take $M = D - 5$.

47. $y'' - 5y' = xe^{5x}$;   take $M = (D - 5)^2$.

48. $y'' + y = \cos x$;   take $M = D^2 + 1$.

49. $y^{(4)} + 8y'' + 16y = \sin 2x$;   take $M = D^2 + 4$.

50. $(D^2 - 2D + 2)^2 y = e^x \cos x$;   take $M = D^2 - 2D + 2$.

2.6
Reduction of Order and Euler-Cauchy Equations

In Section 2.3 we discussed fully the solution of homogeneous linear differential equations with constant coefficients and mentioned that, unfortunately, there exists no similar procedure for finding routinely a general solution of a higher order linear equation with variable coefficients. Sometimes, however, we can find *one* solution of such an equation by inspection. For example, consider the equation

$$x^2y'' - 5xy' + 9y = 0. \tag{1}$$

Because the exponent matches the order of differentiation in each term, it is plausible that there is a solution of the form $y_1 = x^r$. Upon substitution of y_1 in (1), we obtain

$$x^2 \cdot r(r-1)x^{r-2} - 5x \cdot rx^{r-1} + 9x^r = 0,$$

so y_1 will indeed be a solution of (1) provided that

$$r(r-1) - 5r + 9 = 0;$$

that is, if $(r-3)^2 = 0$. Thus $y_1 = x^3$ is one solution of the equation in (1), but this approach does not yield the second solution we need to form a general solution.

We discuss here the method of **reduction of order**, which enables us to use one known solution y_1 of a second order homogeneous linear differential equation to find a second linearly independent solution y_2. Consider the second order equation

$$y'' + p(x)y' + q(x)y = 0 \tag{2}$$

on an open interval I on which p and q are continuous functions. Suppose that we know one solution y_1 of (2). By Theorem 2 in Section 2.1, there exists a second linearly independent solution y_2; our problem is to find y_2. Equivalently, we would like to find the quotient

$$v(x) = \frac{y_2(x)}{y_1(x)}. \tag{3}$$

Once we know $v(x)$, y_2 will be given by

$$y_2(x) = v(x)y_1(x). \tag{4}$$

We begin by substituting the expression in (4) in Eq. (2), using the derivatives

$$y_2' = vy_1' + v'y_1,$$
$$y_2'' = vy_1'' + 2v'y_1' + v''y_1. \tag{5}$$

We get

$$[vy_1'' + 2v'y_1' + v''y_1] + p[vy_1' + v'y_1] + qvy_1 = 0,$$

and rearrangement gives

$$v[y_1'' + py_1' + qy_1] + v''y_1 + 2v'y_1' + pv'y_1 = 0.$$

But the bracketed terms in this last equation vanish because y_1 is a solution of (2). This leaves the equation

$$y_1v'' + (2y_1' + py_1)v' = 0. \tag{6}$$

The key to the success of this method is that Eq. (6) is *linear* in v'. Thus the substitution in (4) has reduced the second order linear equation in (2) to the first order (in v') linear equation in (6). If we write $u = v'$ and assume that

y_1 never vanishes on I, then (6) yields

$$u' + \left(2\frac{y_1'}{y_1} + p(x)\right)u = 0. \tag{7}$$

An integrating factor for (7) is

$$\rho = \exp\left(\int \left(2\frac{y_1'}{y_1} + p(x)\right) dx\right)$$

$$= \exp\left(2\ln|y_1| + \int p(x)\,dx\right);$$

thus

$$\rho = y_1{}^2 e^{\int p(x)\,dx}.$$

We now integrate the equation in (7) to obtain

$$uy_1{}^2 e^{\int p(x)\,dx} = C,$$

so

$$v' = u = \frac{C}{y_1{}^2}e^{-\int p(x)\,dx}.$$

Another integration now gives

$$\frac{y_2}{y_1} = v = C\int \frac{e^{-\int p(x)\,dx}}{y_1{}^2}\,dx + K.$$

With the particular choices $C = 1$ and $K = 0$, we get

$$y_2 = y_1 \int \frac{e^{-\int p(x)\,dx}}{y_1{}^2}\,dx. \tag{8}$$

This formula provides a second solution $y_2(x)$ of Eq. (2) on any interval where y_1 is never zero. Note that, because an exponential never vanishes, y_2 is a nonconstant multiple of y_1, so y_1 and y_2 are linearly independent solutions.

It can be shown that if y_1 is a known solution of an nth order homogeneous linear equation, then the substitution $y_2 = vy_1$ reduces this nth order equation to a homogeneous linear equation of order $n - 1$ in $u = v'$; hence the terminology *reduction of order*. Unfortunately, the resulting equation of order $n - 1$ will generally have variable coefficients, so if $n > 2$, it may be as difficult to solve as the original equation of order n. Consequently the method of reduction of order is useful primarily for solving second order equations.

EXAMPLE 1 We noted above that $y_1 = x^3$ is one solution of the equation

$$x^2 y'' - 5xy' + 9y = 0. \tag{1}$$

To apply the formula in (8), we first rewrite (1) in the form

$$y'' - \frac{5}{x}y' + \frac{9}{x^2}y = 0 \qquad (x > 0).$$

Then $p(x) = -5/x$ and $e^{-\int p(x)\,dx} = e^{5\ln x} = x^5$, so the formula in (8) gives a second solution

$$y_2 = x^3 \int \frac{x^5}{(x^3)^2}\,dx = x^3 \ln x,$$

and a general solution of (1) for $x > 0$ is $y = c_1 x^3 + c_2 x^3 \ln x$.

Sometimes it is just as convenient to substitute $y_2 = vy_1$ directly into the differential equation as to apply the formula in (8). If we substitute

$$y_2 = x^3 v, \qquad y_2' = 3x^2 v + x^3 v',$$
$$y_2'' = 6xv + 6x^2 v' + x^3 v''$$

in (1), we get

$$(6x^3 v + 6x^4 v' + x^5 v'') - 5(3x^3 v + x^4 v') + 9x^3 v = 0;$$
$$x^4 v' + x^5 v'' = 0;$$
$$D_x[xv'] = 0;$$
$$v' = \frac{C}{x};$$
$$v = C \ln x.$$

Thus we obtain the same independent solution $y_2 = x^3 v = x^3 \ln x$ (if we choose $C = 1$).

EXAMPLE 2 Given the solution $y_1 = x$, find a general solution of

$$(x^2 - 1)y'' - 2xy' + 2y = 0 \qquad (x^2 < 1).$$

Solution First divide each term by $x^2 - 1$ to obtain

$$y'' - \frac{2x}{x^2 - 1}y' + \frac{2}{x^2 - 1}y = 0$$

with $p(x) = -2x/(x^2 - 1)$. Hence

$$e^{-\int p(x)\,dx} = \exp\left(\int \frac{2x}{x^2 - 1}\,dx\right)$$
$$= \exp\left(\ln|x^2 - 1|\right) = 1 - x^2,$$

because $x^2 < 1$. Therefore, the formula in (8) yields

$$y_2 = x \int \frac{1 - x^2}{x^2}\,dx$$
$$= x \int (x^{-2} - 1)\,dx = -1 - x^2.$$

Thus a general solution is $y = c_1 x + c_2(1 + x^2)$. It is interesting to observe that we needed the condition $x^2 < 1$ both for the preliminary division and for the first integration; we also needed the condition $x \neq 0$ to apply the formula in (8). Yet you may verify without difficulty that the general solution above is valid for all x.

EULER-CAUCHY EQUATIONS

Equation (1) above is an example of an Euler-Cauchy equation. The general nth order **Euler-Cauchy equation** is the equation

$$a_n x^n y^{(n)} + a_{n-1} x^{n-1} y^{(n-1)} + \cdots + a_2 x^2 y'' + a_1 xy' + a_0 y = 0 \qquad (9)$$

where $a_0, a_1, \ldots, a_n$ are constants with $a_n \neq 0$. It is sometimes called an **equidimensional** equation because the exponent of each coefficient matches the order of the derivative; this implies that the substitution $y = x^r$ will yield terms all of the same degree. If we make this substitution in (9) and then divide by x^r, we get the polynomial equation

$$a_n r(r-1)\cdots(r-n+1) + \cdots + a_2 r(r-1) + a_1 r + a_0 = 0. \qquad (10)$$

If the nth degree equation in (10) has n distinct roots $r_1, r_2, \ldots, r_n$, then we get n linearly independent solutions and a general solution

$$y = c_1 x^{r_1} + c_2 x^{r_2} + \cdots + c_n x^{r_n} \qquad (11)$$

of Eq. (9) for $x > 0$. For $x < 0$, x^r is not defined for all r. This is not a difficulty; it is easy to verify that the form of Eq. (9) is unchanged by the substitution $t = -x$, so we get the solution for $x < 0$ by replacing x by $-x = |x|$ in the solution for $x > 0$. Hence if the roots of Eq. (10) are distinct, we have a general solution:

$$y = c_1 |x|^{r_1} + c_2 |x|^{r_2} + \cdots + c_n |x|^{r_n} \qquad (12)$$

for all x.

To discuss the second order Euler-Cauchy equation in detail, let us divide each term by the leading coefficient to obtain

$$x^2 y'' + p_0 x y' + q_0 y = 0. \qquad (13)$$

Equation (10) now takes the form

$$r(r-1) + p_0 r + q_0 = 0 \qquad (14)$$

with roots

$$r_1, r_2 = \frac{-(p_0 - 1) \pm \sqrt{(p_0 - 1)^2 - 4q_0}}{2} \qquad (15)$$

The usual three cases occur.

Distinct Real Roots: $r_1 \neq r_2$

As indicated above, a general solution of (13) for $x > 0$ is

$$y = c_1 x^{r_1} + c_2 x^{r_2}. \qquad (16)$$

Equal Real Roots: $r_1 = r_2$

In this case we begin with the single solution

$$y_1 = x^{r_1} \quad \text{where} \quad r_1 = -\tfrac{1}{2}(p_0 - 1).$$

Because

$$\exp\left(-\int \frac{p_0}{x}\, dx\right) = e^{-p_0 \ln x} = x^{-p_0},$$

the formula in (8) gives the second solution

$$y_2 = x^{r_1} \int \frac{x^{-p_0}}{x^{-(p_0 - 1)}}\, dx$$

$$= x^{r_1} \int \frac{1}{x}\, dx = x^{r_1} \ln x,$$

so a general solution for $x > 0$ is

$$y = c_1 x^{r_1} + c_2 x^{r_1} \ln x. \tag{17}$$

Complex Conjugate Roots: $r_1 = a + bi, r_2 = a - bi$

Then

$$x^{(a \pm bi)} = e^{(a \pm bi) \ln x} = e^{a \ln x} e^{\pm i (b \ln x)}$$

$$= x^a [\cos (b \ln x) \pm i \sin (b \ln x)].$$

The real and imaginary parts are linearly independent solutions, so for $x > 0$ a general solution of (13) in this case is

$$y = c_1 x^a \cos (b \ln x) + c_2 x^a \sin (b \ln x). \tag{18}$$

EXAMPLE 3   Find a general solution of $2x^2 y'' + xy' - 15y = 0$.

Solution   Equation (10) is

$$2r(r - 1) + r - 15 = (r - 3)(2r + 5) = 0$$

with distinct roots $r_1 = 3$ and $r_2 = -\frac{5}{2}$, so a general solution is

$$y = c_1 x^3 + \frac{c_2}{x^{5/2}}.$$

EXAMPLE 4   Find a general solution of $x^2 y'' + 7xy' + 13y = 0$.

Solution   Equation (10) is

$$r(r - 1) + 7r + 13 = r^2 + 6r + 13 = 0$$

with complex conjugate roots $-3 \pm 2i$, so the general solution given in (18) is

$$y = c_1 \frac{\cos (2 \ln x)}{x^3} + c_2 \frac{\sin (2 \ln x)}{x^3}.$$

Finally, suppose that $n > 2$ and that the roots of Eq. (10) are not distinct. In this event we can discover the missing solutions by transforming Eq. (9) with the aid of the substitution $x = e^t$. This will always produce a homogeneous linear equation with constant coefficients, which can then be solved using the methods of Section 2.3. The following example illustrates this technique.

EXAMPLE 5   Find a general solution of

$$x^3 \frac{d^3 y}{dx^3} + 6x^2 \frac{d^2 y}{dx^2} + 7x \frac{dy}{dx} + y = 0.$$

Solution   We must preface the substitution $x = e^t$ by some computations. First, $dx/dt = e^t = x$. This implies that

$$\frac{dy}{dx} = \frac{dy}{dt} \frac{dt}{dx} = \frac{1}{e^t} \frac{dy}{dt} = \frac{1}{x} \frac{dy}{dt},$$

so

$$x \frac{dy}{dx} = \frac{dy}{dt}. \tag{19}$$

Next,

$$\frac{d^2y}{dx^2} = \frac{1}{x}\frac{d}{dx}\left(\frac{dy}{dt}\right) + \frac{dy}{dt}\frac{d}{dx}\left(\frac{1}{x}\right)$$

$$= \frac{1}{x}\frac{d^2y}{dx^2}\frac{dt}{dx} - \frac{1}{x^2}\frac{dy}{dt}$$

$$= \frac{1}{x^2}\left(\frac{d^2y}{dt^2} - \frac{dy}{dt}\right),$$

so

$$x^2\frac{d^2y}{dx^2} = \frac{d^2y}{dt^2} - \frac{dy}{dt}. \tag{20}$$

Another similar computation gives

$$x^3\frac{d^3y}{dx^3} = \frac{d^3y}{dt^3} - 3\frac{d^2y}{dt^2} + 2\frac{dy}{dt}. \tag{21}$$

Now substitution of (19), (20), and (21) in the given equation yields

$$\left(\frac{d^3y}{dt^3} - 3\frac{d^2y}{dt^2} + 2\frac{dy}{dt}\right) + 6\left(\frac{d^2y}{dt^2} - \frac{dy}{dt}\right) + 7\frac{dy}{dt} + y = 0;$$

that is,

$$\frac{d^3y}{dt^3} + 3\frac{d^2y}{dt^2} + 3\frac{dy}{dt} + y = 0.$$

The characteristic equation

$$r^3 + 3r^2 + 3r + 1 = (r+1)^3 = 0$$

has the root -1 of multiplicity 3. Because $t = \ln x$, a general solution is therefore

$$y = c_1e^{-t} + c_2te^{-t} + c_3t^2e^{-t}$$

$$= \frac{c_1}{x} + \frac{c_2\ln x}{x} + \frac{c_3(\ln x)^2}{x}.$$

As Example 5 may suggest, it can be shown in general that a root r of multiplicity k of Eq. (10) corresponds to the k linearly independent solutions

$$x^r(\ln x)^m, \quad m = 0, 1, 2, \ldots, k - 1.$$

2.6 Problems

In each of Problems 1–10, a differential equation and one solution y_1 are given. Apply the formula in (8) to find a second linearly independent solution y_2.

1. $y'' - 4y' + 4y = 0$; $y_1 = e^{2x}$.

2. $x^2y'' + 3xy' + y = 0$; $y_1 = \dfrac{1}{x}$.

3. $4x^2y'' + y = 0$; $y_1 = \sqrt{x}$.

4. $xy'' + (x - 1)y' - y = 0$; $y_1 = e^{-x}$.

5. $xy'' - 2(x + 1)y' + 4y = 0$; $y_1 = e^{2x}$.

6. $(x^2 + 1)y'' - 2xy' + 2y = 0$; $y_1 = x$.

7. $x^2y'' - 2xy' + (x^2 + 2)y = 0$; $y_1 = x\cos x$.

8. $xy'' - y' + 4x^3 y = 0$; $y_1 = \cos x^2$.

9. $(x + x^2)y'' - (2x + 1)y' + 2y = 0$; $y_1 = x^2$.

10. $(x^4 - x^2)y'' - (3x^3 - x)y' + 8y = 0$; $y_1 = x^4$.

In each of Problems 11–17, a differential equation and one solution y_1 are given. Substitute $y_2 = vy_1$ in the equation to find a second linearly independent solution y_2.

11. $4y'' - 4y' + y = 0$; $y_1 = e^{x/2}$.

12. $xy'' - 3y' = 0$; $y_1 = 1$.

13. $x^2 y'' + xy' - 9y = 0$; $y_1 = x^3$.

14. $(x + 1)y'' - (x + 2)y' + y = 0$; $y_1 = e^x$.

15. $x^2 y'' - x(x + 2)y' + (x + 2)y = 0$; $y_1 = x$.

16. $(1 + x^3)y'' - 3x^2 y' + 3xy = 0$; $y_1 = x$.

17. $x^2 y'' - 4xy' + (x^2 + 6)y = 0$; $y_1 = x^2 \sin x$.

18. Note that $y_1 = x$ is one solution of Legendre's equation of order 1, $(1 - x^2)y'' - 2xy' + 2y = 0$, and use the method of reduction of order to derive the second solution,

$$y_2 = 1 - \frac{x}{2} \ln \frac{1 + x}{1 - x}$$

for $-1 < x < 1$.

19. Verify that $y_1 = x^{-1/2} \cos x$ is one solution of Bessel's equation of order $\frac{1}{2}$, $x^2 y'' + xy' + (x^2 - \frac{1}{4})y = 0$ $(x > 0)$, and derive the second solution $y_2 = x^{-1/2} \sin x$.

20. Show that $y_1 = 3x^2 - 1$ is one solution of Legendre's equation of order 2, $(1 - x^2)y'' - 2xy' + 6y = 0$. Then apply the formula in (8) to derive the second solution,

$$y_2 = y_1 \ln \frac{1 + x}{1 - x} - 6x.$$

Note that the integral

$$\int \frac{dx}{(3x^2 - 1)^2(1 - x^2)}$$

can be evaluated by the method of partial fractions.

Find general solutions of the Euler-Cauchy equations in Problems 21–29.

21. $x^2 y'' + xy' = 0$.

22. $x^2 y'' + xy' - y = 0$.

23. $x^2 y'' + 2xy' - 12y = 0$.

24. $x^2 y'' - 3xy' + 4y = 0$.

25. $4x^2 y'' + 8xy' - 3y = 0$.

26. $9x^2 y'' + 3xy' - 8y = 0$.

27. $x^2 y'' - xy' + 2y = 0$.

28. $x^2 y'' + 7xy' + 25y = 0$.

29. $2x^2 y'' + 5y = 0$.

30. Verify Eq. (21) of the text.

Apply the method of Example 5 to find general solutions of the Euler-Cauchy equations in Problems 31–36.

31. $x^3 y^{(3)} - x^2 y'' + xy' = 0$.

32. $x^3 y^{(3)} + 6x^2 y'' + 4xy' = 0$.

33. $x^3 y^{(3)} + 3x^2 y'' + xy' = 0$.

34. $x^3 y^{(3)} - 2xy' + y = 0$.

35. $x^3 y^{(3)} + 3x^2 y'' + xy' + y = 0$.

36. $x^3 y^{(3)} + 2x^2 y'' + y = 0$.

2.7

Variation of Parameters

In Section 2.5 we presented the method of undetermined coefficients which, when it is applicable, is usually the simplest method of finding a particular solution of a nonhomogeneous linear differential equation with constant coefficients. We pointed out there, however, that this method cannot succeed with an equation such as $y'' + y = \tan x$, because the function $f(x) = \tan x$ has infinitely many linearly independent derivatives. In addition, the method of undetermined coefficients can be used only with equations having constant coefficients.

We discuss here the method of **variation of parameters** which—in principle (that is, if the integrals that appear can be evaluated)—can always be used to find a particular solution of the nonhomogeneous linear differential equation

$$y^{(n)} + p_{n-1}(x)y^{(n-1)} + \cdots + p_1(x)y' + p_0(x)y = f(x), \tag{1}$$

provided that we know the general solution

$$y_c = c_1 y_1 + c_2 y_2 + \cdots + c_n y_n \tag{2}$$

of the associated homogeneous equation

$$y^{(n)} + p_{n-1}(x)y^{(n-1)} + \cdots + p_1(x)y' + p_0(x)y = 0. \tag{3}$$

Here, in brief, is the basic idea of the method of variation of parameters. Suppose that we replace the constants, or *parameters*, $c_1, c_2, \ldots, c_n$ in the complementary function in (2) by variables: functions $u_1, u_2, \ldots, u_n$ of x. We ask whether it is possible to choose these functions in such a way that the combination

$$y_p(x) = u_1(x)y_1(x) + u_2(x)y_2(x) + \cdots + u_n(x)y_n(x) \tag{4}$$

is a particular solution of the nonhomogeneous Eq. (1). It turns out that this *is* always possible.

The method is essentially the same for all orders $n \geq 2$. We will describe it in detail for the case $n = 2$, and then state the generalization. So we begin with the second order nonhomogeneous equation

$$L[y] = y'' + P(x)y' + Q(x)y = f(x) \tag{5}$$

with complementary function

$$y_c = c_1 y_1 + c_2 y_2 \tag{6}$$

on some open interval I where the functions P and Q are continuous. We want to find functions $u_1(x)$ and $u_2(x)$ such that

$$y_p = u_1 y_1 + u_2 y_2 \tag{7}$$

is a particular solution of (5).

One condition on the two functions u_1 and u_2 is that $L[y_p] = f(x)$. Because two conditions are required to determine two functions, we are free to impose an additional condition of our choice. We will do this in a way that simplifies the computations as much as possible. But first, to impose the condition $L[y_p] = f(x)$, we must compute the derivatives y_p' and y_p''. The product rule gives

$$y_p' = (u_1 y_1' + u_2 y_2') + (u_1' y_1 + u_2' y_2).$$

Our first imposed condition will be that the second sum here must vanish:

$$u_1' y_1 + u_2' y_2 = 0. \tag{8}$$

Then

$$y_p' = u_1 y_1' + u_2 y_2', \tag{9}$$

and the product rule gives

$$y_p'' = (u_1 y_1'' + u_2 y_2'') + (u_1' y_1' + u_2' y_2'). \tag{10}$$

But both y_1 and y_2 satisfy the homogeneous equation $y'' + Py' + Qy = 0$ associated with the nonhomogeneous equation in (5), so

$$y_i'' = -Py_i' - Qy_i \tag{11}$$

for $i = 1, 2$. It therefore follows from (10) that

$$y_p'' = (u_1' y_1' + u_2' y_2') - P(u_1 y_1' + u_2 y_2') - Q(u_1 y_1 + u_2 y_2).$$

In view of Eqs. (7) and (9), this means that

$$y_p'' = (u_1' y_1' + u_2' y_2') - Py_p' - Qy_p;$$

that is, that

$$L[y_p] = u_1' y_1' + u_2' y_2'. \tag{12}$$

The requirement that y_p satisfy the nonhomogeneous equation in (5)—that is, that $L[y_p] = f(x)$—therefore implies that

$$u_1' y_1' + u_2' y_2' = f(x). \tag{13}$$

Finally, Eqs. (8) and (13) determine the functions u_1 and u_2 that we need. Collecting them together, we have a system

$$\left. \begin{array}{l} u_1' y_1 + u_2' y_2 = 0, \\ u_1' y_1' + u_2' y_2' = f(x) \end{array} \right\} \tag{14}$$

of two linear equations in the two *derivatives* u_1' and u_2'. Note that the determinant of coefficients in (14) is simply the Wronskian $W(y_1, y_2)$. Once we have solved the equations in (14) for the derivatives u_1' and u_2', we integrate each to obtain the functions u_1 and u_2 such that $y_p = u_1 y_1 + u_2 y_2$ is the desired particular solution of Eq. (5).

In the general case of the nth order nonhomogeneous equation in (1), a generalization of the procedure described above yields the system

$$\left. \begin{array}{l} u_1' y_1 + u_2' y_2 + \cdots + u_n' y_n = 0, \\ u_1' y_1' + u_2' y_2' + \cdots + u_n' y_n' = 0, \\ u_1' y_1'' + u_2' y_2'' + \cdots + u_n' y_n'' = 0, \\ \qquad\qquad\qquad\qquad \vdots \\ u_1' y_1^{(n-1)} + u_2' y_2^{(n-1)} + \cdots + u_n' y_n^{(n-1)} = f(x) \end{array} \right\} \tag{15}$$

of n linear equations in the *derivatives* of the n functions $u_1, u_2, \ldots, u_n$. The determinant of coefficients in (15) is the Wronskian $W = W(x)$ of the solutions $y_1, y_2, \ldots, y_n$. If $W_i(x)$ denotes the determinant obtained from $W(x)$ upon replacing its ith column

$$\langle y_i, y_i', \ldots, y_i^{(n-1)} \rangle^T$$

with the column

$$\langle 0, 0, \ldots, 0, 1 \rangle^T,$$

then Cramer's rule for the solution of a linear system of equations yields

$$u_i' = \frac{W_i(x)f(x)}{W(x)}; \tag{16}$$

finally, integration yields

$$u_i(x) = \int \frac{W_i(x)f(x)}{W(x)} \, dx \tag{17}$$

for $i = 1, 2, \ldots, n$. Because we are looking only for a single particular solution, the choice of constants of integration in (17) is not material. On substituting (17) in the formula in (4), we get the general **variation of parameters formula**

$$y_p(x) = \sum_{i=1}^{n} y_i(x) \int \frac{W_i(x)f(x)}{W(x)} \, dx \tag{18}$$

for a particular solution of our original nth order nonhomogeneous linear differential equation in (1). In the case $n = 2$, this formula reduces to the formula

$$y_p(x) = -y_1(x) \int \frac{y_2(x)f(x)}{W(x)} \, dx + y_2(x) \int \frac{y_1(x)f(x)}{W(x)} \, dx \tag{18a}$$

for a particular solution in terms of two linearly independent solutions $y_1(x)$ and $y_2(x)$ of the associated homogeneous equation.

Remark: In using the formulas just given, it is important to remember that we began with the nonhomogeneous equation in (1) written in standard form with leading coefficient 1. Hence we always begin by dividing each term of the given differential equation by its leading coefficient. Moreover, rather than memorizing (17) or (18), it is ordinarily better practice to set up the system of equations in (15) and then solve them explicitly for the derivatives $u_1', u_2', \ldots, u_n'$. The equations in (15) are easy to remember by virtue of the fact that their determinant of coefficients is the Wronskian $W = W(y_1, y_2, \ldots, y_n)$.

EXAMPLE 1 Find a particular solution of $y'' + y = \tan x$.

Solution The complementary function is $y_c = c_1 \cos x + c_2 \sin x$, so

$$y_1 = \cos x, \qquad y_2 = \sin x,$$
$$y_1' = -\sin x, \qquad y_2' = \cos x.$$

Hence the equations in (15) with $n = 2$ are

$$u_1'(\cos x) + u_2'(\sin x) = 0,$$
$$u_1'(-\sin x) + u_2'(\cos x) = \tan x.$$

We easily solve these equations for

$$u_1' = -\sin x \tan x = \frac{-\sin^2 x}{\cos x} = \cos x - \sec x,$$

$$u_2' = \cos x \tan x = \sin x.$$

Hence we take

$$u_1 = \int (\cos x - \sec x)\, dx = \sin x - \ln|\sec x + \tan x|$$

and

$$u_2 = \int \sin x \, dx = -\cos x.$$

Thus our particular solution is

$$y_p = u_1 y_1 + u_2 y_2$$
$$= (\sin x - \ln|\sec x + \tan x|)(\cos x) + (-\cos x)(\sin x);$$

that is, $y_p(x) = -(\cos x)\ln|\sec x + \tan x|$.

EXAMPLE 2   Find a particular solution of the nonhomogeneous equation

$$x^3 y^{(3)} + x^2 y'' - 6xy' + 6y = 30x.$$

Note that the associated homogeneous (Euler-Cauchy) equation has linearly independent solutions $y_1 = x$, $y_2 = x^3$, and $y_3 = x^{-2}$.

Solution   We first write the equation in standard form:

$$y^{(3)} + \frac{1}{x}y'' - \frac{6}{x^2}y' + \frac{6}{x^3}y = 30x^{-2},$$

and thereby note that $f(x) = 30x^{-2}$. Inserting the derivatives of y_1, y_2, and y_3 into the equations in (15) with $n = 3$, we obtain the system

$$(x)u_1' + (x^3)u_2' + (x^{-2})u_3' = 0,$$
$$(1)u_1' + (3x^2)u_2' + (-2x^{-3})u_3' = 0,$$
$$(0)u_1' + (6x)u_2' + (6x^{-4})u_3' = 30x^{-2}.$$

The determinant of coefficients is the Wronskian

$$W = \begin{vmatrix} x & x^3 & x^{-2} \\ 1 & 3x^2 & -2x^{-3} \\ 0 & 6x & 6x^{-4} \end{vmatrix} = \frac{30}{x}.$$

Cramer's rule then gives

$$u_1' = \frac{x}{30}\begin{vmatrix} 0 & x^3 & x^{-2} \\ 0 & 3x^2 & -2x^{-3} \\ 30x^{-2} & 6x & 6x^{-4} \end{vmatrix} = -\frac{5}{x},$$

$$u_2' = \frac{x}{30}\begin{vmatrix} x & 0 & x^{-2} \\ 1 & 0 & -2x^{-3} \\ 0 & 30x^{-2} & 6x^{-4} \end{vmatrix} = \frac{3}{x^3},$$

and

$$u_3' = \frac{x}{30}\begin{vmatrix} x & x^3 & 0 \\ 1 & 3x^2 & 0 \\ 0 & 6x & 30x^{-2} \end{vmatrix} = 2x^2.$$

Integrating these derivatives, we get

$$u_1 = 5 \ln x, \qquad u_2 = -\frac{3}{2x^2}, \qquad u_3 = \frac{2}{3}x^3.$$

Thus our particular solution (for $x > 0$) is

$$y_p = (-5 \ln x)(x) + \left(-\frac{3}{2x^2}\right)(x^3) + \left(\frac{2}{3}x^3\right)(x^{-2}) = -\frac{5}{6}x - 5x \ln x.$$

The integral formulas in (17) are most useful precisely when the integrals are nonelementary, as in the following example.

EXAMPLE 3 Solve the initial value problem

$$y'' + y = \frac{1}{\sqrt{2\pi x}}, \qquad y(\pi) = y'(\pi) = 0.$$

Solution Here $y_1 = \cos x$ and $y_2 = \sin x$, so

$$W = \begin{vmatrix} \cos x & \sin x \\ -\sin x & \cos x \end{vmatrix} = \cos^2 x + \sin^2 x = 1;$$

$$W_1 = \begin{vmatrix} 0 & \sin x \\ 1 & \cos x \end{vmatrix} = -\sin x,$$

$$W_2 = \begin{vmatrix} \cos x & 0 \\ -\sin x & 1 \end{vmatrix} = \cos x.$$

Hence the formulas in (16) give

$$u_1' = -\frac{\sin x}{\sqrt{2\pi x}}, \qquad u_2' = \frac{\cos x}{\sqrt{2\pi x}}.$$

For u_1 and u_2 we take the specific antiderivatives

$$u_1(x) = -\frac{1}{\sqrt{2\pi}} \int_0^x \frac{\sin t}{\sqrt{t}} \, dt = -S_2(x),$$

$$u_2(x) = \frac{1}{\sqrt{2\pi}} \int_0^x \frac{\cos t}{\sqrt{t}} \, dt = C_2(x).$$

The *Fresnel integrals* $S_2(x)$ and $C_2(x)$ are nonelementary functions, but are tabulated in Table 7.7 of Abramowitz and Stegun, *Handbook of Mathematical Functions* (New York: Dover, 1965).
   The general solution of $y'' + y = 1/\sqrt{2\pi x}$ is therefore

$$y = y_c + y_p = Ay_1 + By_2 + u_1 y_1 + u_2 y_2$$
$$= A \cos x + B \sin x - S_2(x) \cos x + C_2(x) \sin x,$$

and its derivative is

$$y' = -A \sin x + B \cos x - \frac{\sin x \cos x}{\sqrt{2\pi x}} + \frac{\cos x \sin x}{\sqrt{2\pi x}}$$
$$+ S_2(x) \sin x + C_2(x) \cos x;$$

that is, $y' = -A \sin x + B \cos x + S_2(x) \sin x + C_2(x) \cos x$. The initial conditions $y(\pi) = y'(\pi) = 0$ now give $A = S_2(\pi)$ and $B = -C_2(\pi)$.

Thus the solution of the initial value problem is

$$y = [S_2(\pi) - S_2(x)] \cos x - [C_2(\pi) - C_2(x)] \sin x.$$

From the tables mentioned earlier, we find that $S_2(\pi) \approx 0.714$ and $C_2(\pi) \approx 0.529$. Because

$$S_2(\pi) - S_2(x) = \frac{1}{\sqrt{2\pi}} \int_x^\pi \frac{\sin t}{\sqrt{t}} \, dt$$

(and similarly for $C_2(\pi) - C_2(x)$), an interesting alternative form of our solution is

$$y = \frac{\cos x}{\sqrt{2\pi}} \int_x^\pi \frac{\sin t}{\sqrt{t}} \, dt - \frac{\sin x}{\sqrt{2\pi}} \int_x^\pi \frac{\cos t}{\sqrt{t}} \, dt;$$

thus

$$y(x) = \frac{1}{\sqrt{2\pi}} \int_x^\pi \frac{\sin (t - x)}{\sqrt{t}} \, dt,$$

with the aid of a familiar trigonometric identity. If a table of values of the Fresnel integrals were not available, we could use this formula to approximate the value $y(x)$ for any given x by numerical integration (for instance, by using Simpson's rule).

2.7 Problems

In each of Problems 1–18, use the method of variation of parameters to find a particular solution of the given differential equation.

1. $y'' + 3y' + 2y = 4e^x.$

2. $y'' - 2y' - 8y = 3e^{-2x}.$

3. $y'' - 4y' + 4y = 2e^{2x}.$

4. $y'' - 4y = \sinh 2x.$

5. $y'' + 4y = \cos 3x.$

6. $y'' + 9y = \sin 3x.$

7. $y'' + 9y = 2 \sec 3x.$

8. $y'' + y = \csc^2 x.$

9. $y'' + 4y = \sin^2 x.$

10. $y'' - 4y = xe^x.$

11. $y'' - 2y' + y = x^{-2}e^x.$

12. $x^2 y'' - 4xy' + 6y = x^3.$

13. $x^2 y'' - 3xy' + 4y = x^4.$

14. $4x^2 y'' - 4xy' + 3y = 8x^{4/3}$

15. $x^2 y'' + xy' + y = \ln x.$

16. $y^{(3)} - y'' - 2y' = x^2.$

17. $y^{(3)} + 3y'' + 3y' + y = e^{-x}.$

18. $y^{(3)} + 4y' = \cot 2x.$

In Problems 19–24, find particular solutions involving nonelementary indefinite integrals (as in Example 3).

19. $y'' - y = x^{-2}e^x.$

20. $y'' + y = x^{1/2}.$

21. $y'' + 4y = \exp(-x^2).$

22. $y^{(3)} + y'' + y' + y = \dfrac{1}{x}.$

23. $y^{(3)} - y'' = \ln x.$

24. $y^{(4)} - y = \tanh x.$

25. Find a particular solution of the equation

$$x^3 y^{(3)} + 5x^2 y'' + 2xy' - 2y = x^4,$$

given that its complementary function is $y_c = c_1 x + c_2 x^{-1} + c_3 x^{-2}$.

26. Find a particular solution of the equation

$$24x^3 y^{(3)} + 46x^2 y'' + 7xy' - y = 24x^3,$$

given that its complementary function is $y_c = c_1 x^{1/2} + c_2 x^{1/3} + c_3 x^{1/4}$.

27. Find a particular solution of the equation

$$(x^2 - 1)y'' - 2xy' + 2y = x^2 - 1;$$

recall from Example 2 in Section 2.6 that its complementary function is $y_c = c_1 x + c_2(1 + x^2)$.

28. Find a particular solution of the equation

$$x^2 y'' + xy' + (x^2 - \tfrac{1}{4})y = x^{3/2} \cos x;$$

recall from Problem 19 in Section 2.6 that its complementary function is $y_c = x^{-1/2}(c_1 \cos x + c_2 \sin x)$.

29. Express the solution of the initial value problem

$$y'' - y = \frac{1}{x}, \qquad y(1) = y'(1) = 1$$

in terms of integrals of the form

$$\int_1^x \frac{1}{t} e^{\pm t} \, dt.$$

30. Express the solution of the initial value problem

$$y'' - 3y' + 2y = \sin x^2; \qquad y(0) = 2, \qquad y'(0) = 3$$

in terms of integrals of the form

$$\int_0^x e^{-at} \sin t^2 \, dt.$$

31. Generalize the method of Example 3 to derive the formula

$$y(x) = \int_a^x f(t) \sin (x - t) \, dt$$

for the solution of the initial value problem $y'' + y = f(x)$, $y(a) = 0 = y'(a)$.

32. Use variation of parameters to derive the particular solution

$$y_p = \tfrac{1}{2} \int_0^x f(t)[e^{x-t} - e^{t-x}] \, dt$$

of the equation $y'' - y = f(x)$.

---

*2.8

Forced Oscillations and Resonance

In Section 2.4 we derived the differential equation

$$mx'' + cx' + kx = F(t) \tag{1}$$

that governs the one-dimensional motion of a mass m that is attached to a spring (with constant k) and a dashpot (with constant c) and is also acted on by an external force $F(t)$. Machines with rotating components commonly involve mass-spring systems (or their equivalents) in which the external force is simple harmonic:

$$F(t) = F_0 \cos \omega t \quad \text{or} \quad F(t) = F_0 \sin \omega t, \tag{2}$$

where the constant F_0 is the amplitude of the periodic force and ω is its circular frequency.

For an example of how a rotating machine component can provide a simple harmonic force, consider the cart with a rotating vertical flywheel shown in Fig. 2.15. The cart has mass $m - m_0$, not including the flywheel of mass m_0. The centroid of the flywheel is off-center at distance a from its center, and its angular speed is ω radians per second. The cart is attached to a spring (with constant k) as shown. Assume that the centroid of the cart itself is directly beneath the center of the flywheel, and denote by $x(t)$ its displacement from its equilibrium position (where the spring is unstretched). Figure 2.15 helps us to see that the displacement $\bar{x}$ of the centroid of the combined cart plus flywheel is given by

$$\bar{x} = \frac{(m - m_0)x + m_0(x + a \cos \omega t)}{m} = x + \frac{m_0 a}{m} \cos \omega t.$$

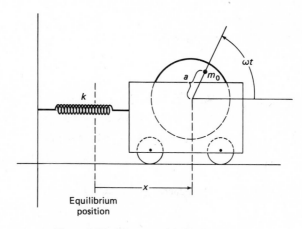

Figure 2.15 The cart-with-flywheel system.

Let us ignore friction and apply Newton's second law $m\bar{x}'' = -kx$, because the force exerted by the spring is $-kx$. We substitute for $\bar{x}$ in the last equation to obtain

$$mx'' - m_0 a\omega^2 \cos \omega t = -kx;$$

that is,

$$mx'' + kx = m_0 a\omega^2 \cos \omega t. \tag{3}$$

Thus the cart with its rotating flywheel acts like a mass on a spring under the influence of a simple harmonic external force with amplitude $F_0 = m_0 a\omega^2$. Such a system is a reasonable model of a front-loading washing machine with the clothes being washed loaded off-center. This illustrates the practical importance of analyzing solutions of Eq. (1) with external forces as in (2).

UNDAMPED FORCED OSCILLATIONS

To study undamped oscillations under the influence of the external force $F(t) = F_0 \cos \omega t$, we set $c = 0$ in Eq. (1), and thereby begin with the equation

$$mx'' + kx = F_0 \cos \omega t \tag{4}$$

whose complementary function is $x_c = c_1 \cos \omega_0 t + c_2 \sin \omega_0 t$. Here, $\omega_0 = \sqrt{k/m}$ is the (circular) **natural frequency** of the mass-spring system. Let us assume initially that the external and natural frequencies are *unequal*: $\omega \neq \omega_0$. We substitute $x_p = A \cos \omega t$ in (4) to find a particular solution (no sine term is needed in x_p because there is no term involving x' on the left-hand side in (4)). This gives

$$-m\omega^2 A \cos \omega t + kA \cos \omega t = F_0 \cos \omega t,$$

so

$$A = \frac{F_0}{k - m\omega^2} = \frac{F_0/m}{\omega_0^2 - \omega^2}, \tag{5}$$

and thus

$$x_p = \frac{F_0/m}{\omega_0^2 - \omega^2} \cos \omega t. \tag{6}$$

Therefore the general solution $x = x_c + x_p$ is given by

$$x(t) = c_1 \cos \omega_0 t + c_2 \sin \omega_0 t + \frac{F_0/m}{\omega_0^2 - \omega^2} \cos \omega t, \tag{7}$$

where the constants c_1 and c_2 are determined by the initial values $x(0)$ and $x'(0)$. Equivalently, as in Eq. (12) of Section 2.4, we can rewrite (7) as

$$x(t) = C \cos (\omega_0 t - \alpha) + \frac{F_0/m}{\omega_0^2 - \omega^2} \cos \omega t, \tag{8}$$

so we see that the resulting motion is a superposition of two oscillations, one with natural circular frequency ω_0, the other with the external circular frequency ω.

BEATS

If we impose the initial conditions $x(0) = x'(0) = 0$ on the solution in (7), we find that $c_1 = -F_0/m(\omega_0^2 - \omega^2)$ and $c_2 = 0$, so the particular solution is

$$x = \frac{F_0/m}{\omega_0^2 - \omega^2} (\cos \omega t - \cos \omega_0 t). \tag{9}$$

The trigonometric identity $2 \sin A \sin B = \cos (A - B) - \cos (A + B)$, applied with $A = \frac{1}{2}(\omega_0 + \omega)t$ and $B = \frac{1}{2}(\omega_0 - \omega)t$, enables us to rewrite (9) as

$$x = \frac{2F_0}{m(\omega_0^2 - \omega^2)} \sin \frac{1}{2}(\omega_0 - \omega)t \sin \frac{1}{2}(\omega_0 + \omega)t. \tag{10}$$

Suppose now that $\omega \approx \omega_0$, so that $\omega_0 + \omega$ is large in comparison with $|\omega_0 - \omega|$. Then $\sin \frac{1}{2}(\omega_0 + \omega)t$ is a *rapidly* varying function, while $\sin \frac{1}{2}(\omega_0 - \omega)t$ is a *slowly* varying function. We may therefore interpret (10) as a rapid oscillation with circular frequency $\frac{1}{2}(\omega_0 + \omega)$, $x = A(t) \sin \frac{1}{2}(\omega_0 + \omega)t$, but with a slowly varying amplitude

$$A(t) = \frac{2F_0}{m(\omega_0{}^2 - \omega^2)} \sin \frac{1}{2}(\omega_0 - \omega)t.$$

The graph of $x(t)$ is shown in Fig. 2.16. An oscillation such as this, with a slowly varying periodic amplitude, exhibits the phenomenon of *beats*. For example, if two horns not exactly attuned to one another simultaneously play their middle C, one at $\omega_0/2\pi = 258$ Hz and the other at $\omega/2\pi = 254$ Hz, then one hears a beat—an audible variation in the *amplitude* of the combined sound—with a frequency of

$$\frac{(\omega_0 - \omega)/2}{2\pi} = \frac{1}{2}(258 - 254) = 2 \text{ Hz}.$$

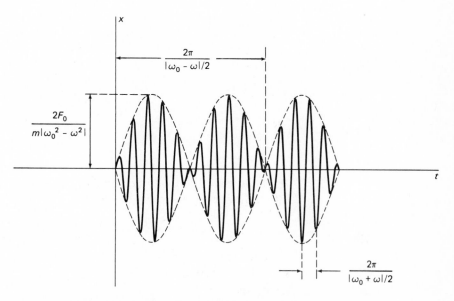

Figure 2.16 Superposition of differing frequencies produces beats.

RESONANCE

Looking at Eq. (6), we see that the amplitude A of x_p is large when the natural and external frequencies ω_0 and ω are approximately equal. It is sometimes useful to rewrite (5) in the form

$$A = \frac{F_0}{k - m\omega^2} = \frac{F_0/k}{1 - (\omega/\omega_0)^2} = \pm \frac{\rho F_0}{k}, \tag{11}$$

where F_0/k is the **static displacement** of a spring with constant k due to a *constant* force F_0, and the **amplification factor** ρ is defined to be

$$\rho = \frac{1}{|1 - (\omega/\omega_0)^2|}. \tag{12}$$

It is clear that $\rho \longrightarrow +\infty$ as $\omega \longrightarrow \omega_0$. This is the phenomenon of **resonance**—the increase without bound (as $\omega \longrightarrow \omega_0$) in the amplitude of the oscillations of an undamped system with natural frequency ω_0 in response to an external force with frequency ω.

We have been assuming that $\omega \neq \omega_0$. What sort of catastrophe should one expect if ω and ω_0 are precisely equal? Then Eq. (4), upon division of each term by m, becomes

$$x'' + \omega_0{}^2 x = \frac{F_0}{m} \cos \omega_0 t. \tag{13}$$

Because $\cos \omega_0 t$ is a term of the complementary function, the method of undetermined coefficients calls for us to try $x_p = t(A \cos \omega_0 t + B \sin \omega_0 t)$. We substitute this in (13), and thereby find that $A = 0$ and $B = F_0/2m\omega_0$. Hence the particular solution is

$$x_p(t) = \frac{F_0}{2m\omega_0} t \sin \omega_0 t. \tag{14}$$

The graph of $x_p(t)$ in Fig. 2.17 shows vividly how the amplitude of the oscillation theoretically would increase without bound in this case of *pure resonance*, $\omega = \omega_0$. We may interpret the phenomenon as reinforcement of the natural vibrations of the system by externally impressed vibrations at the same frequency.

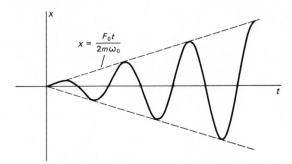

Figure 2.17   Pure resonance.

For a numerical example, suppose that $m = 4$ slugs and that $k = 400$ lb/ft in the cart with flywheel of Fig. 2.15. Then the natural frequency is $\omega_0 = \sqrt{k/m} = 10$ rad/s, or $10/2\pi \approx 1.59$ Hz. We would therefore expect oscillations of very large amplitude to occur if the flywheel revolves at about $(1.59)(60) \approx 95$ revolutions per minute (rpm).

In practice, a mechanical system with very little damping can be destroyed by resonance vibrations. A spectacular example can occur when a column of soldiers marches in step over a bridge. Any complicated structure such as a bridge has many natural frequencies of vibration. If the frequency of the soldiers' cadence is approximately equal to one of the natural frequencies of the structure, then—just as in our simple example of a mass on a spring—resonance will occur. Indeed, the resulting resonance vibrations can be of such large amplitude that the bridge will collapse. This has actually happened —for example, the collapse of Broughton Bridge near Manchester, England, in 1831—and it is the reason for the now-standard practice of breaking cadence when crossing a bridge. Resonance may have been involved in the 1981 Kansas

City disaster in which a hotel balcony (called a *skywalk*) collapsed with dancers on it. The collapse of a building in an earthquake is sometimes due to resonance vibrations caused by the ground oscillating at one of the natural frequencies of the structure. On occasion an airplane has crashed because of resonant wing oscillations caused by vibrations of the engines. It is reported that for some of the first commercial jet aircraft, the natural frequency of the vertical vibrations of the airplane during turbulence was almost exactly that of the mass-spring system consisting of the pilot's head (mass) and spine (spring). Resonance occurred, causing pilots to have difficulty in reading the instruments. The newer wide-bodied jets have different natural frequencies, so that this resonance problem no longer occurs.

The avoidance of destructive resonance vibrations is a constant factor in the design of mechanical structures and systems of all types. Often the most important step in determining the natural frequency of vibration of a system is the formulation of its differential equation. In addition to Newton's law $F = ma$, the principle of conservation of energy is sometimes useful for this purpose (as in the derivation of the pendulum equation in Section 2.4). The following kinetic and potential energy formulas are often useful:

1. Kinetic energy: $T = \frac{1}{2}mv^2$ for translation of a mass with velocity v;
2. Kinetic energy: $T = \frac{1}{2}I\omega^2$ for rotation of a body of moment of inertia I with angular velocity ω;
3. Potential energy: $V = \frac{1}{2}kx^2$ for a spring with constant k stretched or compressed a distance x;
4. Potential energy: $V = mgh$ for the gravitational potential energy of a mass m at height h above the reference level (the level at which $V = 0$).

EXAMPLE 1   Find the natural frequency of a mass m on a spring (with constant k) if, instead of sliding without friction, it is a uniform disk of radius a that rolls without slipping, as shown in Fig. 2.18.

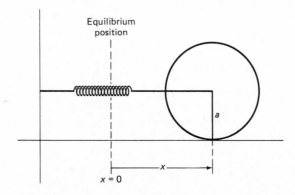

Figure 2.18   The rolling disk.

Solution   With the notation above, the principle of conservation of energy gives

$$\tfrac{1}{2}mv^2 + \tfrac{1}{2}I\omega^2 + \tfrac{1}{2}kx^2 = E$$

where E is a constant (the total mechanical energy of the system). We note that $v = a\omega$, and recall that $I = ma^2/2$ for a uniform circular disk. Thus we may simplify the above equation to

$$\tfrac{3}{4}mv^2 + \tfrac{1}{2}kx^2 = E.$$

Differentiation $(v = x', v' = x'')$ now gives $\tfrac{3}{2}mx'x'' + kxx' = 0$. We divide each term by $\tfrac{3}{2}mx'$ to obtain

$$x'' + \frac{2k}{3m}x = 0.$$

Thus the natural circular frequency is $\omega_0 = \sqrt{2k/3m}$, which is $\sqrt{\tfrac{2}{3}}$ times the frequency in the previous situation of sliding without friction.

EXAMPLE 2   Assume that a car weighing 1600 lb oscillates vertically as if it were a mass $m = 50$ slugs on a single spring (with constant $k = 4800$ lb/ft), attached to a single dashpot (with constant $c = 200$ lb-s/ft). Suppose that this car with the dashpot *disconnected* is driven along a washboard road surface with an amplitude of 2 in. and a wavelength of $L = 30$ ft (see Fig. 2.19). At what car speed (in miles per hour) will resonance vibrations occur?

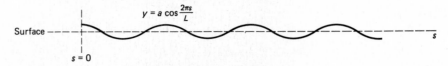

Figure 2.19   The washboard road.

Solution   We think of the car as a unicycle, as pictured in Fig. 2.20. Let $x(t)$ denote the upward displacement of the mass m from its equilibrium position; we ignore the force of gravity, because it merely displaces the equilibrium position as in Problem 9 of Section 2.4. We write the equation

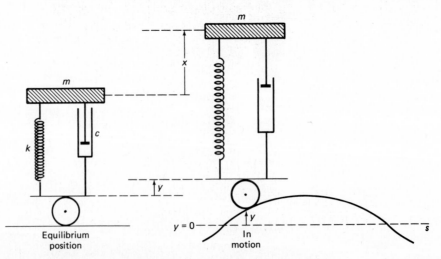

Figure 2.20   The "unicycle model" of a car.

of the road surface as

$$y = a \cos \frac{2\pi s}{L} \qquad \left(a = \frac{1}{6} \text{ ft}, L = 30 \text{ ft}\right).$$ (15)

When the car is in motion, the spring is stretched by the amount $x - y$, so Newton's second law, $F = ma$, gives $mx'' = -k(x - y)$; that is,

$$mx'' + kx = ky.$$ (16)

If the velocity of the car is v, then $s = vt$ in (15), so (16) takes the form

$$mx'' + kx = ka \cos \frac{2\pi vt}{L}.$$ (16')

This is the differential equation governing the vertical oscillations of the car. In comparing it with Eq. (4), we see that we have forced oscillations with circular frequency $\omega = 2\pi v/L$. Resonance will occur when $\omega = \omega_0 = \sqrt{k/m}$. We insert our numerical data to find the speed of the car at resonance:

$$v = \frac{L}{2\pi} \sqrt{\frac{k}{m}} = \frac{30}{2\pi} \sqrt{\frac{4800}{50}} \text{ (ft/s)},$$

or about 32 mi/h.

DAMPED FORCED OSCILLATIONS

In real physical systems there is always some damping, from frictional effects if nothing else. The complementary function x_c of the equation

$$mx'' + cx' + kx = F_0 \cos \omega t$$ (17)

is given by Eq. (18), (19), or (20) in Section 2.4, depending upon whether $c > c_{CR}$, $c = c_{CR}$, or $c < c_{CR}$. The specific form is not important here. What is important is that, in any case, these formulas show that $x_c(t) \to 0$ as $t \to +\infty$. Thus x_c is a **transient solution** of (17)—one that dies out with the passage of time, leaving only the particular solution x_p.

The method of undetermined coefficients indicates that we should substitute $x_p = A \cos \omega t + B \sin \omega t$ in (17). When we do so, collect terms, and equate coefficients of $\cos \omega t$ and $\sin \omega t$, we obtain the two equations

$$\left. \begin{array}{r} (k - m\omega^2)A + c\omega B = F_0, \\ -c\omega A + (k - m\omega^2)B = 0 \end{array} \right\}$$ (18)

that we solve without difficulty for

$$A = \frac{(k - m\omega^2)F_0}{(k - m\omega^2)^2 + (c\omega)^2}, \qquad B = \frac{(c\omega)F_0}{(k - m\omega^2)^2 + (c\omega)^2}.$$

To simplify the notation, it is convenient to introduce the quantity

$$\rho = \frac{k}{\sqrt{(k - m\omega^2)^2 + (c\omega)^2}}$$ (19)

and the angle α of Fig. 2.21. Then we find that

$$A = \rho \frac{F_0}{k} \cos \alpha, \qquad B = \rho \frac{F_0}{k} \sin \alpha.$$

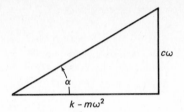

Figure 2.21 The angle α.

Hence our particular solution is

$$x_p = A \cos \omega t + B \sin \omega t$$

$$= \rho \frac{F_0}{k}(\cos \omega t \cos \alpha + \sin \omega t \sin \omega \alpha);$$

more concisely,

$$x_p = \rho \frac{F_0}{k} \cos (\omega t - \alpha). \tag{20}$$

Thus we get a **steady periodic solution** that remains after the transient solution has died away. This steady-state solution has amplitude $\rho(F_0/k)$, circular frequency ω, and phase angle α given by

$$\alpha = \tan^{-1} \frac{c\omega}{k - m\omega^2}, \qquad 0 \leq \alpha \leq \pi. \tag{21}$$

Note that α lies in the first or second quadrant, so the formula above does not involve the principal value of the inverse tangent function. If a calculator gives a negative value, we must add π to that value to obtain the actual value of α.

The **amplification factor** ρ, defined in (19), is the amount by which the static displacement F_0/k must be multiplied to get the amplitude of the steady periodic oscillation. Note that when $c > 0$, the amplitude always remains finite (unlike the undamped case). The amplitude may reach a maximum for some value of ω; this is *practical* resonance. To see when it occurs for various values of the constants m, c, and k, it is useful to express ρ in terms of the dimensionless ratios

$$\tilde{\omega} = \frac{\omega}{\omega_0} = \frac{\omega}{\sqrt{k/m}}, \quad \tilde{c} = \frac{c}{c_{CR}} = \frac{c}{\sqrt{4km}}.$$

Then (19) is equivalent to

$$\rho = \frac{1}{\sqrt{(1 - \tilde{\omega}^2)^2 + 4\tilde{c}^2\tilde{\omega}^2}}. \tag{19'}$$

Figure 2.22 shows the graph of ρ versus $\tilde{\omega}$ for various values of $\tilde{c}$. It can be shown that if $c \geq c_{CR}/\sqrt{2}$, then ρ steadily decreases as ω increases, but if $c < c_{CR}/\sqrt{2}$, then ρ reaches a maximum value—practical resonance—at some value of ω less than ω_0, and then approaches 0 as $\omega \to +\infty$. It follows that an underdamped system typically will undergo forced oscillations whose amplitude is

1. Large if ω is close to the critical resonance frequency;
2. Close to F_0/k if ω is very small;
3. Very small if ω is very large.

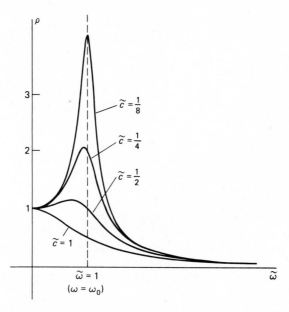

Figure 2.22   Resonance.

EXAMPLE 3   Find the transient and steady periodic solutions of

$$x'' + 2x' + 2x = 20 \cos 2t, \quad x(0) = x'(0) = 0.$$

Solution   Instead of applying the general formulas derived above, it is better in a concrete problem to work the problem directly. The roots of the characteristic equation $r^2 + 2r + 2 = 0$ are $-1 \pm i$, so the complementary function is $x_c = e^{-t}(c_1 \cos t + c_2 \sin t)$. When we substitute $x_p = A \cos 2t + B \sin 2t$ in the given equation, collect coefficients, and equate coefficients of $\cos 2t$ and $\sin 2t$, we obtain the equations

$$-2A + 4B = 20,$$
$$-4A - 2B = 0,$$

with solution $A = -2$, $B = 4$. Hence the general solution is

$$x = e^{-t}(c_1 \cos t + c_2 \sin t) - 2 \cos 2t + 4 \sin 2t.$$

At this point we impose the initial conditions $x(0) = x'(0) = 0$, and find easily that $c_1 = 2, c_2 = -6$. Therefore the transient solution x_{tr} and the steady periodic solution x_{sp} are given by

$$x_{tr} = e^{-t}(2 \cos t - 6 \sin t)$$

and

$$x_{sp} = -2 \cos 2t + 4 \sin 2t$$

$$= 2\sqrt{5}\left(-\frac{1}{\sqrt{5}} \cos 2t + \frac{2}{\sqrt{5}} \sin 2t\right).$$

The latter can be written in the form $x_{sp} = 2\sqrt{5}\,[\cos{(2t - \alpha)}]$ where $\alpha = \pi - \tan^{-1}(2) \approx 2.0344$.

Finally, we indicate how to include damping in the analysis of the vibrations of the car of Example 2 (though we leave the detailed computations to the problems). We use without proof the fact that when $c > 0$, the differential equation is

$$mx'' + cx' + kx = cy' + ky. \qquad (22)$$

With $y = a \sin{\omega t}$ for the road surface, this equation becomes

$$mx'' + cx' + kx = E_0 \cos{\omega t} + F_0 \sin{\omega t}, \qquad (23)$$

where $E_0 = c\omega a$ and $F_0 = ka$. Substituting the trial solution $x_{sp} = A \cos{\omega t} + B \sin{\omega t}$ in (23) to determine the coefficients A and B, we arrive finally at the steady periodic solution

$$x_{sp} = \frac{\sqrt{E_0^2 + F_0^2}}{\sqrt{(k - m\omega^2)^2 + (c\omega)^2}} \cos{(\omega t - \alpha - \beta)} \qquad (24)$$

where α is defined in (21) and $\beta = \tan^{-1}(F_0/E_0)$. Upon substitution of $E_0 = c\omega a$ and $F_0 = ka$, we see that the amplitude of the steady periodic oscillations of the car is

$$C = \frac{a\sqrt{k^2 + (c\omega)^2}}{\sqrt{(k - m\omega^2)^2 + (c\omega)^2}}. \qquad (25)$$

Velocity v (mph)	Amplitude C (in.)
7.5	2.12
15.0	2.54
22.5	3.59
30.0	5.34
37.5	3.62
45.0	2.01
52.5	1.31
60.0	0.95
75.0	0.60
90.0	0.43

Because $\omega = 2\pi v/L$ when the car is moving with velocity v, this gives C as a function of v. The formula in (25) was used, with the numerical parameters given in Example 2, to calculate the entries in the table shown in Fig. 2.23. As the car accelerates gradually from rest, it initially oscillates with amplitude slightly over 2 in. (the road surface amplitude). Maximum resonance oscillations with amplitude over 5 in. occur around 30 mi/h, but then subside to more tolerable levels at high speeds.

Figure 2.23 Effect of the washboard road on the car.

2.8 Problems

In each of Problems 1–6, express the solution of the given initial value problem as a sum of two oscillations (as in Eq. (8)).

1. $x'' + 9x = 10 \cos{2t}; \quad x(0) = x'(0) = 0.$

2. $x'' + 4x = 5 \sin{3t}; \quad x(0) = x'(0) = 0.$

3. $x'' + 100x = 15 \cos{5t} + 20 \sin{5t}; \quad x(0) = 25, \, x'(0) = 0.$

4. $x'' + 25x = 10 \cos{4t}; \quad x(0) = 0, \, x'(0) = 10.$

5. $mx'' + kx = F_0 \cos \omega t$ with $\omega \neq \omega_0$; $x(0) = x_0, x'(0) = 0$.

6. $mx'' + kx = F_0 \sin \omega t$ with $\omega = \omega_0$; $x(0) = 0, x'(0) = v_0$.

In each of Problems 7–15, find the steady periodic solution in the form $x_{sp} = C\cos(\omega t - \alpha)$. If initial conditions are given, also find the transient solution.

7. $x'' + 4x' + 4x = 10 \cos 3t$.

8. $x'' + 3x' + 5x = -4 \cos 5t$.

9. $2x'' + 2x' + x = 3 \sin 10t$.

10. $x'' + 3x' + 3x = 8 \cos 10t + 6 \sin 10t$.

11. $x'' + 4x' + 5x = 10 \cos 3t$; $x(0) = x'(0) = 0$.

12. $x'' + 6x' + 13x = 10 \sin 5t$; $x(0) = x'(0) = 0$.

13. $x'' + 2x' + 6x = 3 \cos 10t$; $x(0) = 10, x'(0) = 0$.

14. $x'' + 8x' + 25x = 5 \cos t + 13 \sin t$; $x(0) = 5, x'(0) = 0$.

15. A mass weighing 100 lb is attached to the end of a spring that is stretched 1 in. by a force of 100 lb. A force $F_0 \cos \omega t$ acts on the mass. At what frequency (in Hertz) will resonance oscillations occur? Neglect damping.

16. A front-loading washing machine is mounted on a thick rubber pad that acts like a spring; the weight of the machine depresses the pad exactly $\frac{1}{4}$ in. When its rotor spins at ω radians per second, the rotor exerts a vertical force $F_0 \cos \omega t$ pounds on the machine. At what speed (in revolutions per minute) will resonance vibrations occur? Neglect damping.

17. See Fig. 2.24, which shows a mass m on the end of a pendulum (of length L) also attached to a horizontal spring (with constant k). Assume small oscillations of m so that the spring remains essentially horizontal, and neglect damping. Find the natural circular frequency ω_0 in terms of L, k, m, and the gravitational constant g.

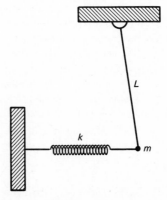

Figure 2.24 The pendulum-and-spring system of Problem 17.

18. A mass m hangs on the end of a cord around a pulley of radius a and moment of inertia I, as shown in Fig. 2.25. The rim of the pulley is attached to a spring (with constant k). Assume small oscillations so that the spring remains essentially horizontal, and neglect friction. Find the natural circular frequency of the system in terms of m, a, k, I, and g.

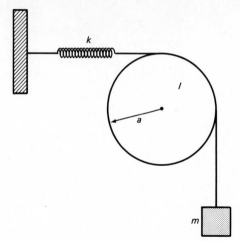

Figure 2.25 The mass-spring-pulley system of Problem 18.

19. A building consists of two floors. The first floor is attached rigidly to the ground, and the second floor is of mass m and weighs 16 tons (32,000 lb). The elastic frame of the building behaves as a spring that resists horizontal displacements of the second floor; it requires a horizontal force of 5 tons to displace the second floor a distance of 1 ft. Assume that in an earthquake the ground oscillates horizontally with amplitude A_0 and circular frequency ω, resulting in an external horizontal force $F(t) = mA_0\omega^2 \sin \omega t$ on the second floor. (a) What is the natural frequency (in Hertz) of oscillations of the second floor? (b) If the ground undergoes one oscillation every 2.25 s with an amplitude of 3 in., what is the amplitude of the resulting forced oscillations of the second floor?

20. A mass on a spring without damping is acted upon by the external force $F(t) = F_0 \cos^3 \omega t$. Show that there are *two* values of ω for which resonance occurs, and find both.

21. Derive the steady periodic solution of $mx'' + cx' + kx = F_0 \sin \omega t$. In particular, show that it is what one would expect—the same as the formula in (20) with the same values of ρ and ω, except with $\sin(\omega t - \alpha)$ in place of $\cos(\omega t - \alpha)$.

22. Derive the steady periodic solution in (24) of Eq. (23)—with both sine and cosine forces—by superposition of the steady periodic solutions separately corresponding to $E_0 \cos \omega t$ and $F_0 \sin \omega t$ (see Problem 21).

23. Recall that the amplification factor ρ is given in terms of the impressed frequency ω by

$$\rho = k[(k - m\omega^2)^2 + (c\omega)^2]^{-1/2}.$$

(a) If $c \geq c_{CR}/\sqrt{2}$, where $c_{CR} = \sqrt{4km}$, show that ρ steadily decreases as ω increases. (b) If $c < c_{CR}/\sqrt{2}$, show that ρ attains a maximum value (practical resonance) when

$$\omega = \omega_m = \sqrt{\frac{k}{m} - \frac{c^2}{2m^2}} < \omega_0 = \sqrt{\frac{k}{m}}.$$

24. Consider the car discussed in this section—with $m = 50$ slugs, $c = 200$ lb-s/ft, $k = 4800$ lb/ft—traveling with velocity v on a washboard road surface described

by $y = \frac{1}{6} \sin(2\pi s/30)$. Find the velocity v_m (in miles per hour) at which practical resonance occurs and the amplitude (in inches) of the oscillations of the car at this critical speed. See Problem 23.

25. One end of a spring is attached to a moving support, as in Fig. 2.26. A mass m is attached both to the other end and to a dashpot (with damping constant c). Show that the differential equation of motion of the mass is

$$mx'' + cx' + kx = mA\omega^2 \cos \omega t.$$

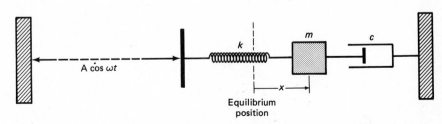

Figure 2.26   The mechanical system of Problem 25.

26. As indicated by the cart-with-flywheel example discussed in this section, an unbalanced rotating machine part typically results in a force having amplitude proportional to the square of the frequency ω. (a) Show that the amplitude of the steady periodic solution of the differential equation of Problem 25 is $\rho(mA/k)$, where the amplification factor is

$$\rho = k\omega^2[(k - m\omega^2)^2 + (c\omega)^2]^{-1/2}.$$

(b) Suppose that $c^2 < 2mk$. Show that the maximum amplitude occurs at the frequency ω_m given by

$$\omega_m{}^2 = \frac{k}{m}\left(\frac{2mk}{2mk - c^2}\right).$$

Thus the resonance frequency in this case is *larger* (in contrast to the result of Problem 23) than the natural frequency $\omega_0 = \sqrt{k/m}$. (*Suggestion:* Maximize the *square* of ρ.)

*2.9
Electrical Circuits

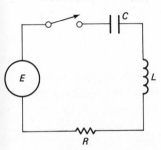

Figure 2.27   The series RLC circuit.

Here we examine the RLC circuit that is a basic building block in more complicated electrical circuits and networks. As shown in Fig. 2.27, it consists of

A **resistor** with a resistance of R *ohms,*
An **inductor** with an inductance of L *henries,* and
A **capacitor** with a capacitance of C *farads*

in series with a source of electromotive force (such as a battery or a generator) that supplies a voltage of $E(t)$ *volts* at time t. If the switch shown in the circuit of Fig. 2.27 is closed, this results in a current of $I(t)$ *amperes* in the circuit and a charge of $Q(t)$ *coulombs* on the capacitor at time t. The relation between the functions Q and I is

$$\frac{dQ}{dt} = I. \tag{1}$$

We will always use mks electrical units, in which time is measured in seconds.

According to elementary principles of electricity, the **voltage drops** across the three circuit elements are those shown in the table of Fig. 2.28. We

Circuit element	Voltage drop
Inductor	$L\frac{dI}{dt}$
Resistor	RI
Capacitor	$\frac{1}{C}Q$

Figure 2.28 Table of voltage drops.

can analyze the behavior of the series circuit of Fig. 2.27 with the aid of this table and one of Kirchoff's laws:

The (algebraic) sum of the voltage drops across
the elements in a simple loop of an electrical
circuit is equal to the applied voltage.

As a consequence, the current and charge in the simple RLC circuit of Fig. 2.27 satisfy the basic circuit equation

$$L\frac{dI}{dt} + RI + \frac{1}{C}Q = E(t). \tag{2}$$

If we substitute (1) in (2), we get the second order linear differential equation

$$LQ'' + RQ' + \frac{1}{C}Q = E(t) \tag{3}$$

for the charge $Q(t)$, under the assumption that the voltage $E(t)$ is known.

In most practical problems it is the current I rather than the charge Q that is of primary interest, so we differentiate each side of Eq. (3) and substitute $I = Q'$ to obtain

$$LI'' + RI' + \frac{1}{C}I = E'(t). \tag{4}$$

We do *not* assume here a prior familiarity with electrical circuits. It suffices to regard the resistor, inductor, and capacitor in an electrical circuit as "black boxes" that are calibrated by the constants R, C, and L above. A battery or generator is described by the voltage $E(t)$ that it supplies. When the switch is open, no current flows in the circuit; when it is closed, there is a current $I(t)$ in the circuit and a charge $Q(t)$ on the capacitor. All we need to know about these constants and varying quantities is that they satisfy Eqs.

(1)–(4)—our mathematical model for the RLC circuit. We can then learn a good deal about electricity by studying the mathematical model.

Note that the equations in (3) and (4) have precisely the same form as the equation

$$mx'' + cx' + kx = F(t) \tag{5}$$

of a mass-spring-dashpot system with external force $F(t)$. The table in Fig. 2.29 details this important **mechanical-electrical analogy**. As a consequence, most of the results derived in Section 2.8 for mechanical systems can be applied at once to electrical circuits. The fact that the same differential equation serves as a mathematical model for such different physical systems is a striking illustration of the unifying role of mathematics in the investigation of natural phenomena. More concretely, the correspondences in Fig. 2.29 can be used to construct an electrical model of a given mechanical system, using inexpensive and readily available circuit elements. The performance of the mechanical system can then be predicted by means of accurate and simple measurements in the electrical circuit. This is especially useful when the actual mechanical system would be expensive to construct and when measurements of displacements and velocities would be inconvenient or inaccurate. This idea is the basis of *analog computers*—electrical models of mechanical systems.

Mechanical system	Electrical system
Mass m	Inductance L
Damping constant c	Resistance R
Spring constant k	Reciprocal capacitance $1/C$
Position x	Charge Q (using (3)) (or Current I using (4))
Force F	Electromotive force E (or its derivative E')

Figure 2.29 Mechanical-electrical analogies.

In the typical case of an alternating current voltage $E(t) = E_0 \sin \omega t$, Eq. (4) takes the form

$$LI'' + RI' + \frac{1}{C}I = \omega E_0 \cos \omega t. \tag{6}$$

As in a mass-spring-dashpot system with a simple harmonic external force, the solution of Eq. (6) is the sum of a **transient current** I_{tr} that approaches zero as $t \longrightarrow +\infty$ (under the assumption that the coefficients in (5) are all positive, so the characteristic roots have negative real parts), and a **steady periodic current** I_{sp}; thus

$$I = I_{tr} + I_{sp}. \tag{7}$$

Recall from Section 2.8 (Eqs. (19)–(21) there) that the steady periodic solution of Eq. (5) with $F(t) = F_0 \cos \omega t$ is

$$x_{sp} = \frac{F_0 \cos(\omega t - \alpha)}{\sqrt{(k - m\omega^2)^2 + (c\omega)^2}}$$

where

$$\alpha = \tan^{-1} \frac{c\omega}{k - m\omega^2}, \qquad 0 \leq \alpha \leq \pi.$$

If we make the substitutions L for m, R for c, $1/C$ for k, and ωE_0 for F_0, we get the steady periodic current

$$I_{sp} = \frac{E_0 \cos(\omega t - \alpha)}{\sqrt{R^2 + \left(\omega L - \dfrac{1}{\omega C}\right)^2}} \tag{8}$$

with the phase angle given by

$$\alpha = \tan^{-1} \frac{\omega RC}{1 - LC\omega^2}, \qquad 0 \leq \alpha \leq \pi. \tag{9}$$

The quantity

$$Z = \sqrt{R^2 + \left(\omega L - \frac{1}{\omega C}\right)^2} \quad \text{(ohms)} \tag{10}$$

is called the **impedance** of the circuit. Then the steady periodic current

$$I_{sp} = \frac{E_0}{Z} \cos(\omega t - \alpha) \tag{11}$$

has amplitude

$$I_0 = \frac{E_0}{Z}, \tag{12}$$

reminiscent of Ohm's law $I = E/R$.

Equation (11) gives the steady periodic current as a cosine function, whereas the input voltage $E(t) = E_0 \sin \omega t$ was a sine function. To convert I_{sp} to a sine function, we first introduce the **reactance**

$$S = \omega L - \frac{1}{\omega C}. \tag{13}$$

Then $Z = \sqrt{R^2 + S^2}$, and we see from Eq. (9) that α is as in Fig. 2.30, with delay angle $\delta = \alpha - \pi/2$. Equation (11) now yields

$$I_{sp} = \frac{E_0}{Z}(\cos \alpha \cos \omega t + \sin \alpha \sin \omega t)$$

$$= \frac{E_0}{Z}\left(-\frac{S}{Z}\cos \omega t + \frac{R}{Z}\sin \omega t\right)$$

$$= \frac{E_0}{Z}(\cos \delta \sin \omega t - \sin \delta \cos \omega t).$$

Therefore

$$I_{sp} = \frac{E_0}{Z}\sin(\omega t - \delta), \tag{14}$$

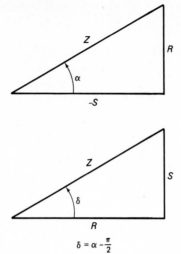

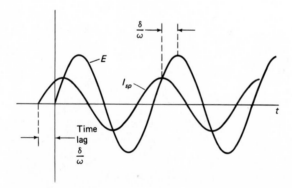

$$\delta = \alpha - \frac{\pi}{2}$$

Figure 2.30 Reactance and delay angle.

where

$$\delta = \tan^{-1}\frac{S}{R} = \tan^{-1}\frac{LC\omega^2 - 1}{\omega RC} \; \epsilon \; [0, \pi]. \qquad (15)$$

This finally gives the **time lag** δ/ω (in seconds) of the steady periodic current I_{sp} behind the input voltage. See Fig. 2.31.

Figure 2.31 Time lag of current behind imposed voltage.

When we want to find the transient current, we are usually given the initial values $I(0)$ and $Q(0)$. So we first must find $I'(0)$. To do so, we substitute $t = 0$ in Eq. (2) to obtain the equation

$$LI'(0) + RI(0) + \frac{1}{C}Q(0) = E(0) \qquad (16)$$

to determine $I'(0)$ in terms of the initial values of current, charge, and voltage.

EXAMPLE 1 Consider an RLC circuit with $R = 50$ ohms, $L = 0.1$ henries (H), and $C = 5 \times 10^{-4}$ farads (F). At time $t = 0$, when both $I(0)$ and $Q(0)$ are zero, the circuit is connected to a 110-V, 60-Hz alternating current generator. Find the current in the circuit and the lag of the steady periodic current behind the voltage.

Solution A frequency of 60 Hz means that $\omega = (2\pi)(60)$ rad/s, approximately 377 rad/s. So we take $E(t) = 110 \sin 377t$, and use equality in place of the symbol for approximate equality in the rest of this discussion. The differential equation in (6) takes the form

$$(0.1)I'' + 50I' + 2000I = (377)(110)\cos 377t.$$

We substitute the given values of R, L, C, and $\omega = 377$ in (10) to find that the impedance is $Z = 59.58$ ohms, so the steady periodic amplitude is

$$I_0 = \frac{110 \text{ (volts)}}{59.58 \text{ (ohms)}} = 1.846 \text{ amperes (A)}.$$

With the same data, Eq. (15) gives the sine phase angle:

$$\delta = \tan^{-1}(0.648) = 0.575.$$

Thus the time lag of current behind voltage is

$$\frac{\delta}{\omega} = \frac{0.575}{377} = 0.0015 \text{ s},$$

and the steady periodic current is $I_{sp} = (1.846) \sin (377t - 0.575)$.

The characteristic equation $(0.1)r^2 + 50r + 2000 = 0$ has roots $r_1 \approx -44$ and $r_2 \approx -456$. With these approximations, the general solution is

$$I(t) = c_1 e^{-44t} + c_2 e^{-456t} + (1.846) \sin (377t - 0.575),$$

with derivative

$$I'(t) = -44c_1 e^{-44t} - 456c_2 e^{-456t} + 696 \cos(377t - 0.575).$$

Because $I(0) = Q(0) = 0$, Eq. (16) gives $I'(0) = 0$ as well. With these initial values substituted, we obtain the equations

$$I(0) = c_1 + c_2 - 1.004 = 0,$$

$$I'(0) = -44c_1 - 456c_2 + 584 = 0;$$

the solution is $c_1 = -0.307$, $c_2 = 1.311$. Thus the transient solution is

$$I_{tr} = (-0.307)e^{-44t} + (1.311)e^{-456t},$$

and it dies out very rapidly, indeed.

EXAMPLE 2 Suppose that the RLC circuit of Example 1, still with $I(0) = Q(0) = 0$, is connected at time $t = 0$ to a battery supplying a constant 110 volts. Now find the current in the circuit.

Solution We now have $E(t) = 110$, so Eq. (16) gives

$$I'(0) = \frac{E(0)}{L} = \frac{110}{0.1} = 1100 \text{ A/s},$$

and the differential equation is

$$(0.1)I'' + 50I' + 2000I = E'(t) = 0.$$

Its general solution is the complementary function we found in Example 1: $I(t) = c_1 e^{-44t} + c_2 e^{-456t}$. We solve the equations

$$I(0) = c_1 + c_2 = 0,$$

$$I'(0) = -44c_1 - 345c_2 = 1100$$

for $c_1 = -c_2 = 2.671$. Therefore $I(t) = (2.671)(e^{-44t} - e^{-456t})$. Note that $I \to 0$ as $t \to +\infty$, even though the voltage is constant.

ELECTRICAL RESONANCE

Consider again the current differential equation in (6) corresponding to a sinusoidal input voltage $E(t) = E_0 \sin \omega t$. We have seen that the amplitude of its steady periodic current is

$$I_0 = \frac{E_0}{Z} = \frac{E_0}{\sqrt{R^2 + \left(\omega L - \frac{1}{\omega C}\right)^2}}. \tag{17}$$

For typical values of the constants R, L, C, and E_0, the graph of I_0 as a function of ω resembles the one shown in Fig. 2.32. It reaches a maximum value at $\omega_m = 1/\sqrt{LC}$ and then approaches zero as $\omega \to +\infty$; the critical frequency ω_m is the **resonance frequency** of the circuit.

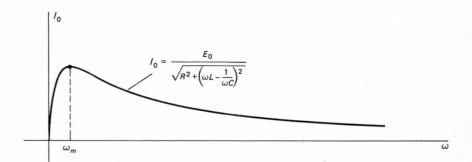

Figure 2.32   Effect of frequency on I_0.

In Section 2.8 we emphasized the importance of avoiding resonance in most mechanical systems (the seismograph is an example of a mechanical system in which resonance is *sought*). By contrast, many common electrical devices could not function properly without taking advantage of the phenomenon of resonance. The radio is a familiar example. A highly simplified model of its tuning circuit is the RLC circuit we have discussed. Its inductance L and resistance R are constant, but its capacitance C is varied as one operates the tuning dial (thereby varying the effective area of the plates of its variable capacitor).

Suppose that we want to pick up a particular radio station that is broadcasting at frequency ω, and thereby (in effect) provides an input voltage $E(t) = E_0 \sin \omega t$ to the tuning circuit of the radio. The resulting steady periodic current I_{sp} in the tuning circuit drives its amplifier, and in turn its loudspeaker, with the volume of sound that we hear roughly proportional to the amplitude I_0 of I_{sp}. In order to hear our preferred station (of frequency ω) the loudest (and simultaneously tune out stations broadcasting at other frequencies), we therefore want to choose C to maximize I_0. But examine Eq. (17), thinking of ω as constant and with C the only variable. We see at a glance—no calculus required—that I_0 is maximal when

$$\omega L - \frac{1}{\omega C} = 0;$$

that is, when

$$C = \frac{1}{L\omega^2}. \tag{18}$$

So we merely turn the dial to set the capacitance at this value.

This is the way that the old crystal radios worked, but modern AM radios have a more sophisticated design. A *pair* of variable capacitors are used: The first controls the frequency selected as described above; the second controls the frequency of a signal that the radio itself generates, kept close to to 455 kilohertz (kHz) above the desired frequency. The resulting *beat* frequency of 455 kHz, known as the *intermediate frequency*, is then amplified in several stages. This technique has the advantage that the several RLC circuits used in the amplification stages easily can be designed to resonate at 455 kHz and reject other frequencies, resulting in more selectivity of the receiver as well as better amplification of the desired signal.

2.9 Problems

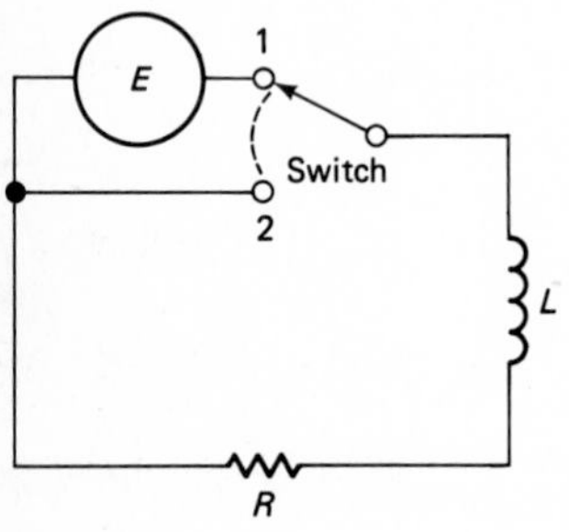

Problems 1–6 deal with the LR circuit of Fig. 2.33, a series circuit containing an inductor with an inductance of L henries, a resistor with a resistance of R ohms, and a source of electromotive force (emf), but no capacitor. In this case Eq. (2) reduces to the linear first order equation

$$LI' + RI = E(t).$$

Figure 2.33 The circuit for the first six problems.

1. In the circuit of Fig. 2.33, suppose that $L = 5$ H, $R = 25$ ohms, and that the source E of emf is a battery supplying 100 V to the circuit. Suppose also that the switch has been in position 1 for a long time, so that a steady current of 4 A is flowing in the circuit. At time $t = 0$, the switch is thrown to position 2, so that $I(0) = 4$ and $E = 0$ for $t \geq 0$. Find $I(t)$.

2. Given the same circuit as in Problem 1, suppose that the switch is initially in position 2, but is thrown to position 1 at time $t = 0$, so that $I(0) = 0$ and $E = 100$ for $t \geq 0$. Find $I(t)$, and show that $I(t)$ approaches 4 as $t \longrightarrow +\infty$.

3. The battery in Problem 2 is replaced with an alternating current generator that supplies a voltage of $E(t) = 100 \cos 60t$ volts. With everything else the same, now find $I(t)$.

4. In the circuit of Fig. 2.33, with the switch in position 1, suppose that $L = 2$, $R = 40$, $E = 100e^{-10t}$, and $I(0) = 0$. Find the maximum current in the circuit for $t \geq 0$.

5. In the circuit of Fig. 2.33, with the switch in position 1, suppose that $L = 2$, $R = 20$, $I(0) = 0$, and $E = 100e^{-10t} \cos 60t$. Find $I(t)$.

6. In the circuit of Fig. 2.33, with the switch in position 1, take $L = 1$, $R = 10$, and $E = 30 \cos 60t + 40 \sin 60t$. (a) Substitute $I_{sp} = A \cos 60t + B \sin 60t$ and then determine A and B to find the steady-state current I_{sp} in the circuit. (b) Write the solution in the form $I_{sp} = C \cos(\omega t - \alpha)$.

Problems 7–10 deal with the RC circuit shown in Fig. 2.34 containing a resistor (R ohms), a capacitor (C farads), a switch, a source E of emf, but no inductor. Substitution of $L = 0$ in Eq. (3) gives the linear first order differential equation

$$R\frac{dQ}{dt} + \frac{1}{C}Q = E(t)$$

for the change $Q = Q(t)$ on the capacitor at time t. Note that $I(t) = Q'(t)$.

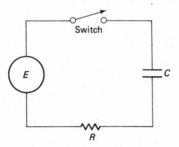

Switch

E

C

R

Figure 2.34 The circuit for Problems 7–10.

7. (a) Find the charge $Q(t)$ and current $I(t)$ in the RC circuit if $E(t) = E_0$ (a constant voltage supplied by a battery) and the switch is closed at time $t = 0$, so that $Q(0) = 0$.
   (b) Show that $\lim\limits_{x \to +\infty} Q(t) = E_0 C$ and that $\lim\limits_{x \to +\infty} I(t) = 0$.

8. Suppose that in the circuit of Fig. 2.34, we have $R = 10$, $C = 0.02$, $Q(0) = 0$, and $E(t) = 100e^{-5t}$ (volts). (a) Find $Q(t)$ and $I(t)$. (b) What is the maximum charge on the capacitor for $t \geq 0$, and when does it occur?

9. Suppose that in the circuit of Fig. 2.34, $R = 200$, $C = 2.5 \times 10^{-4}$, $Q(0) = 0$, and $E(t) = 100 \cos 120t$. (a) Find $Q(t)$ and $I(t)$. (b) What is the amplitude of the steady-state current?

10. An emf of voltage $E(t) = E_0 \cos \omega t$ is applied to the RC circuit of Fig. 2.34 at time $t = 0$, and $Q(0) = 0$. Substitute $Q_{sp} = A \cos \omega t + B \sin \omega t$ in the differential equation to show that the steady periodic charge on the capacitor is

$$Q_{sp} = \frac{E_0 C}{(1 + \omega^2 R^2 C^2)^{1/2}} \cos(\omega t - \beta)$$

where $\beta = \tan^{-1}(\omega R C)$.

In each of Problems 11–16, the parameters of an RLC circuit with input voltage E(t) are given. Substitute $I_{sp} = A \cos \omega t + B \sin \omega t$ in Eq. (4), using the appropriate values of ω, to find the steady periodic current in the form $I_{sp} = I_0 \sin(\omega t - \delta)$.

11. $R = 30$ ohms, $L = 10$ H, $C = 0.02$ F; $E(t) = 50 \sin 2t$ volts.
12. $R = 200$ ohms, $L = 5$ H, $C = 0.001$ F; $E(t) = 100 \sin 10t$ volts.
13. $R = 20$ ohms, $L = 10$ H, $C = 0.01$ F; $E(t) = 200 \cos 5t$ volts.
14. $R = 50$ ohms, $L = 5$ H, $C = 0.005$ F; $E(t) = 300 \cos 100t + 400 \sin 100t$ volts.

15. $R = 100$ ohms, $L = 2$ H, $C = 5 \times 10^{-6}$ F; $E(t) = 110 \sin 60\pi t$ volts.

16. $R = 25$ ohms, $L = 0.2$ H, $C = 5 \times 10^{-4}$ F; $E(t) = 120 \cos 377t$ volts.

In each of Problems 17–22, an RLC circuit with input voltage $E(t)$ is described. Find the current $I(t)$ given the initial current (in amperes) and charge on the capacitor (in coulombs).

17. $R = 16$ ohms, $L = 2$ H, $C = 0.02$ F; $E(t) = 100$ V; $I(0) = 0$, $Q(0) = 5$.

18. $R = 60$ ohms, $L = 2$ H, $C = 0.0025$ F; $E(t) = 100e^{-t}$ V; $I(0) = Q(0) = 0$.

19. $R = 60$ ohms, $L = 2$ H, $C = 0.0025$ F; $E(t) = 100e^{-10t}$ V; $I(0) = 0$, $Q(0) = 1$.

20. The circuit and input voltage of Problem 11 with $I(0) = 0$ and $Q(0) = 0$.

21. The circuit and input voltage of Problem 13 with $I(0) = 0$ and $Q(0) = 0.3$.

22. The circuit and input voltage of Problem 15 with $I(0) = 0$ and $Q(0) = 0$.

23. Consider an LC circuit—that is, an RLC circuit with $R = 0$—with input voltage $E(t) = E_0 \sin \omega t$. Show that unbounded oscillations of current occur for a certain resonance frequency; express this frequency in terms of L and C.

24. It was stated in the text that, if R, L, and C are positive, then any solution of $LI'' + RI' + I/C = 0$ is a transient solution—it approaches zero as $t \to +\infty$. Prove this.

25. Prove that the amplitude I_0 of the steady periodic solution of Eq. (6) is maximal at frequency $\omega = 1/\sqrt{LC}$.

*2.10
Endpoint Problems and Eigenvalues

You are now familiar with the fact that a solution of a second order linear differential equation is uniquely determined by two initial conditions. In particular, the only solution of the initial value problem

$$y'' + p(x)y' + q(x)y = 0;$$
$$y(a) = 0, \qquad y'(a) = 0 \tag{1}$$

is the trivial solution $y(x) \equiv 0$. Most of Chapter 2 has been based, directly or indirectly, on the uniqueness of solutions of linear initial value problems (as guaranteed by Theorem 2 in Section 2.2).

In this section we will see that the situation is radically different for a problem such as

$$y'' + p(x)y' + q(x)y = 0;$$
$$y(a) = 0, \qquad y(b) = 0. \tag{2}$$

The difference between the problems in Eqs. (1) and (2) is that in (2) the two conditions are imposed at two *different* points a and b with (say) $a < b$. In (2) we are to find a solution of the differential equation on the interval (a, b) that satisfies the conditions $y(a) = 0$ and $y(b) = 0$ at the endpoints of the interval. Such a problem is called an **endpoint** or **boundary value** problem. The following two examples illustrate the sorts of complications that can arise in endpoint problems.

EXAMPLE 1 Consider the endpoint problem

$$y'' + 3y = 0; \qquad y(0) = 0, \qquad y(\pi) = 0. \tag{3}$$

The general solution of the differential equation is

$$y = A \cos \sqrt{3}\, x + B \sin \sqrt{3}\, x.$$

The endpoint conditions give $y(0) = A = 0$ and $y(\pi) = B \sin \pi\sqrt{3} \approx (-0.7458)B = 0$, so $B = 0$ as well. Thus the only solution of the endpoint problem in (3) is the trivial solution $y(x) \equiv 0$; there is no surprise here.

EXAMPLE 2 Consider the endpoint problem

$$y'' + 4y = 0; \qquad y(0) = 0, \qquad y(\pi) = 0. \tag{4}$$

The general solution of the differential equation is

$$y = A \cos 2x + B \sin 2x.$$

The endpoint conditions yield $y(0) = A = 0$ and

$$y(\pi) = A \cos 2\pi + B \sin 2\pi = B \cdot 0 = 0.$$

No matter what the value of B, the condition $y(\pi) = 0$ is automatically satisfied because $\sin 2\pi = 0$. Thus the function $y = B \sin 2x$ satisfies the endpoint problem in (4) for *every* value of B. This is an example of an endpoint problem having infinitely many nontrivial solutions!

Rather than being the extreme cases, these two examples illustrate the typical situation for an endpoint problem as in (2): It may have no nontrivial solution, or it may have infinitely many nontrivial solutions. Note that the problems in (3) and (4) can both be written in the form

$$y'' + p(x)y' + \lambda q(x)y = 0; \qquad y(a) = 0, \qquad y(b) = 0, \tag{5}$$

with $p(x) \equiv 0, q(x) \equiv 1, a = 0$, and $b = \pi$. The number λ is a parameter in the problem (nothing to do with the parameters that were varied in Section 2.7). If we take $\lambda = 3$, we get the equations in (3); with $\lambda = 4$, we obtain the equations in (4). Examples 1 and 2 show that the situation in an endpoint problem containing a parameter can (and generally will) depend strongly upon the specific numerical value of the parameter.

An endpoint problem containing a parameter λ, such as the endpoint problem in (5), is called an **eigenvalue problem**. The question we ask in an eigenvalue problem is this: For what (real) values of the parameter λ does there exist a *nontrivial* solution of the endpoint problem? Such a value of λ is called an **eigenvalue** or **characteristic value** of the problem. Thus we saw in Example 2 that $\lambda = 4$ is an eigenvalue of the eigenvalue problem

$$y'' + \lambda y = 0; \qquad y(0) = 0, \qquad y(\pi) = 0. \tag{6}$$

In Example 1 we saw that $\lambda = 3$ is not an eigenvalue of this problem.

Suppose that λ_* is an eigenvalue of the problem in (5), and that $y_*(x)$ is a nontrivial solution of the problem with this value of λ inserted; that is, that

$$y_*'' + p(x)y_*' + \lambda_* q(x)y_* = 0$$

and
$$y_*(a) = 0, \qquad y_*(b) = 0.$$

Then we call y_* an **eigenfunction** associated with the eigenvalue λ_*. Thus we saw in Example 2 that $y_* = \sin 2x$ is an eigenfunction associated with the eigenvalue $\lambda_* = 4$, as is any constant multiple of $\sin 2x$.

More generally, note that the problem in (5) is *homogeneous* in the sense that any constant multiple of an eigenfunction is again an eigenfunction; indeed, one associated with the same eigenvalue. That is, if $y = y_*(x)$ satisfies the problem in (5) with $\lambda = \lambda_*$, then so does any constant multiple $cy_*(x)$. It can be proved under mild restrictions on the coefficient functions p and q that any two eigenfunctions associated with the same eigenvalue must be linearly dependent.

EXAMPLE 3 Determine the eigenvalues and associated eigenfunctions of the problem

$$y'' + \lambda y = 0; \qquad y(0) = 0, \quad y(L) = 0 \qquad (L > 0). \qquad (7)$$

Solution   We must consider all possible (real) values of λ—positive, zero, and negative.

If $\lambda = 0$, then the equation is simply $y'' = 0$, and its general solution is $y = Ax + B$. Then the endpoint conditions $y(0) = 0 = y(L)$ immediately imply that $A = B = 0$, so the only solution in this case is the trivial function $y(x) \equiv 0$. Therefore $\lambda = 0$ is *not* an eigenvalue of the problem in (7).

If $\lambda < 0$, let us then write $\lambda = -\alpha^2$ (with $\alpha > 0$) to be specific. Then the differential equation takes the form $y'' - \alpha^2 y = 0$, and its general solution is

$$y = c_1 e^{\alpha x} + c_2 e^{-\alpha x} = A \cosh \alpha x + B \sinh \alpha x,$$

where $A = c_1 + c_2$ and $B = c_1 - c_2$. (Recall that $\cosh x = (e^{\alpha x} + e^{-\alpha x})/2$ and $\sinh x = (e^{\alpha x} - e^{-\alpha x})/2$.) The condition $y(0) = 0$ then gives

$$y(0) = A \cosh 0 + B \sinh 0 = A = 0,$$

so that $y = B \sinh \alpha x$. But now the second endpoint condition, $y(L) = 0$, gives $y(L) = B \sinh \alpha L = 0$. This implies that $B = 0$, because $\alpha \neq 0$, and $\sinh x = 0$ only for $x = 0$ (examine the graphs of $\sinh x$ and $\cosh x$ in Fig. 2.35). Thus the only solution of the problem in (7) in the case $\lambda < 0$ is the trivial solution $y \equiv 0$, and we may therefore conclude that the problem has *no* negative eigenvalues.

The only remaining possibility is that $\lambda = \alpha^2 > 0$ with $\alpha > 0$. In this case the differential equation is $y'' + \alpha^2 y = 0$, with general solution $y = A \cos \alpha x + B \sin \alpha x$. The condition that $y(0) = 0$ implies that $A = 0$, so $y = B \sin \alpha x$. The condition $y(L) = 0$ then gives $y(L) = B \sin \alpha L = 0$. Can this occur if $B \neq 0$? Yes, but only provided that αL is a (positive) integral multiple of π:

$$\alpha L = \pi, 2\pi, 3\pi, \ldots, n\pi, \ldots;$$

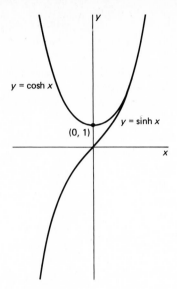

Figure 2.35 The hyperbolic sine and cosine graphs.

that is, if

$$\lambda = \alpha^2 = \frac{\pi^2}{L^2}, \frac{4\pi^2}{L^2}, \frac{9\pi^2}{L^2}, \ldots, \frac{n^2\pi^2}{L^2}, \ldots$$

Thus we have discovered that the problem in (7) has an *infinite sequence* of positive eigenvalues,

$$\lambda_n = \frac{n^2\pi^2}{L^2}, \quad n = 1, 2, 3, \ldots. \tag{8}$$

With $B = 1$, the eigenfunction associated with the eigenvalue λ_n is

$$y_n(x) = \sin \frac{n\pi x}{L}, \qquad n = 1, 2, 3, \ldots. \tag{9}$$

Example 3 illustrates the general situation. According to a theorem whose precise statement we will defer until Section 8.1, under the assumption that $q(x) > 0$ on the interval $[a, b]$, any eigenvalue problem of the form in (7) has a divergent increasing sequence

$$\lambda_1 < \lambda_2 < \lambda_3 < \cdots < \lambda_n < \cdots \longrightarrow +\infty$$

of eigenvalues, each with an associated eigenfunction. This is also true of the following slightly more general type of eigenvalue problem, in which the endpoint conditions involve values of the derivative y' as well as values of y:

$$y'' + p(x)y' + \lambda q(x)y = 0;$$
$$a_1 y(a) + a_2 y'(a) = 0, \qquad b_1 y(b) + b_2 y'(b) = 0, \tag{10}$$

where a_1, a_2, b_1, and b_2 are given constants. With $a_1 = 1 = b_2$ and $a_2 = 0 = b_1$, we get the problem of the following example (in which $p(x) \equiv 0$ and $q(x) \equiv 1$, as in the previous examples).

EXAMPLE 4 Determine the eigenvalues and eigenfunctions of the problem

$$y'' + \lambda y = 0; \qquad y(0) = 0, \qquad y'(L) = 0. \tag{11}$$

Solution Virtually the same argument as that used in Example 3 shows that the only possible eigenvalues are positive, so we take $\lambda = \alpha^2 > 0$ to be specific. Then the differential equation is $y'' + \alpha^2 y = 0$, with general solution $y = A \cos \alpha x + B \sin \alpha x$. The condition $y(0) = 0$ immediately gives $A = 0$, so $y = B \sin \alpha x$, and $y' = B\alpha \cos \alpha x$. The second endpoint condition, $y'(L) = 0$, now gives

$$y'(L) = B\alpha \cos \alpha L = 0.$$

This will be so with $B \neq 0$ provided that αL is an odd positive multiple of $\pi/2$:

$$\alpha L = \frac{\pi}{2}, \frac{3\pi}{2}, \ldots, \frac{(2n-1)\pi}{2}, \ldots;$$

that is, if

$$\lambda = \frac{\pi^2}{4L^2}, \frac{9\pi^2}{4L^2}, \ldots, \frac{(2n-1)^2\pi^2}{4L^2}, \ldots.$$

Thus the nth eigenvalue λ_n and associated eigenfunction of the problem in (11) are given by

$$\lambda_n = \frac{(2n-1)^2\pi^2}{4L^2} \quad \text{and} \quad y_n = \sin \frac{(2n-1)\pi x}{2L} \quad \text{for } n = 1, 2, 3, \ldots. \tag{12}$$

A general procedure for determining the eigenvalues of the problem in (10) can be outlined as follows. We first write the general solution of the differential equation in the form

$$y = Ay_1(x, \lambda) + By_2(x, \lambda).$$

We write $y_i(x, \lambda)$ because y_1 and y_2 will depend upon λ, as in Examples 3 and 4, in which

$$y_1 = \cos \alpha x = \cos \sqrt{\lambda}\, x \quad \text{and} \quad y_2 = \sin \alpha x = \sin \sqrt{\lambda}\, x.$$

Then we impose the two endpoint conditions, noting that each is linear in y and y', and hence also linear in A and B. When we collect coefficients of A and B in the resulting pair of equations, we therefore get a system of the form

$$\left. \begin{array}{l} \alpha_1(\lambda)A + \beta_1(\lambda)B = 0, \\ \alpha_2(\lambda)A + \beta_2(\lambda)B = 0. \end{array} \right\} \tag{13}$$

Now λ is an eigenvalue if and only if (13) has a nontrivial solution (one with A and B not both zero). But such a homogeneous system of linear equations has a nontrivial solution if and only if the determinant of its coefficients vanishes. We therefore conclude that the eigenvalues of the problem in (10) are the (real) solutions of the equation

$$D(\lambda) = \alpha_1(\lambda)\beta_2(\lambda) - \alpha_2(\lambda)\beta_1(\lambda) = 0. \tag{14}$$

This can be a formidable equation to solve and may require a numerical approximation technique such as Newton's method.

Much of the interest in eigenvalue problems is due to their very diverse physical applications. The remainder of this section is devoted to one such application. Numerous additional applications are included in Chapters 7 and 8 (on partial differential equations and boundary value problems).

THE WHIRLING STRING

Who of us has not wondered about the shape of a quickly spinning jump rope? Let us consider the shape assumed by a tightly stretched flexible string of length L and constant linear density ρ (mass per unit length) if it is rotated or whirled (like a jump rope) with constant angular speed ω (in radians per second) about its equilibrium position along the x-axis. We assume that the portion of the string to one side of any point exerts a constant tension force T on the portion of the string to the other side of the point, with the direction of T tangential to the string. We further assume that, as the string whirls around the x-axis, each point moves in a circle centered at that point's equilibrium position on the x-axis. Thus the string is elastic, so that as it whirls it also stretches to assume a curved shape. Denote by $y(x)$ the displacement of the string from the axis of rotation. Finally, we assume that the deflection of the string is so slight that $\sin \theta \approx \tan \theta \approx y'(x)$ in Fig. 2.36(c).

We plan to derive a differential equation for $y(x)$ by application of Newton's law $F = ma$ to the piece of string of mass $\rho \Delta x$ corresponding to the

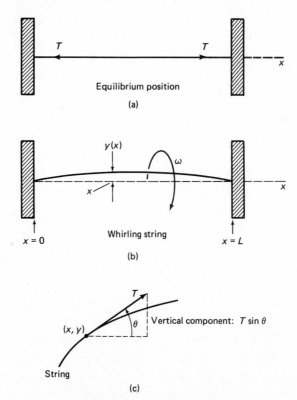

Figure 2.36 The whirling string.

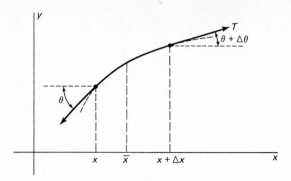

Figure 2.37 Forces on a short segment of the whirling string.

interval $[x, x + \Delta x]$. The only forces acting on this piece are the tension forces at its two ends. From Fig. 2.37 we see that the net vertical force in the positive y-direction is

$$F = T \sin (\theta + \Delta\theta) - T \sin \theta$$
$$\approx T \tan (\theta + \Delta\theta) - T \tan \theta,$$

so that

$$F \approx Ty'(x + \Delta x) - Ty'(x). \tag{15}$$

Next we recall from elementary calculus or physics the formula $a = r\omega^2$ for the (inward) centripetal acceleration of a body in uniform circular motion (r is the radius of the circle and ω is the angular speed of the body). Here we have $r = y$, so the vertical acceleration of our piece of string is $a = -\omega^2 y$, the minus sign because the inward direction is the negative y-direction. Because $m = \rho\Delta x$, substitution of this and (15) in $F = ma$ yields

$$Ty'(x + \Delta x) - Ty'(x) \approx -\rho\omega^2 y\Delta x,$$

so that

$$T\frac{y'(x + \Delta x) - y(x)}{\Delta x} \approx -\rho\omega^2 y.$$

We now take the limit as $\Delta x \to 0$ to get the differential equation of motion of the string:

$$Ty'' + \rho\omega^2 y = 0. \tag{16}$$

If we write

$$\lambda = \frac{\rho\omega^2}{T} \tag{17}$$

and impose the condition that the ends of the string are fixed, we get finally the eigenvalue problem

$$y'' + \lambda y = 0; \quad y(0) = 0, \quad y(L) = 0 \tag{7}$$

that we considered in Example 3. We found there that the eigenvalues of the problem in (7) are

$$\lambda_n = \frac{n^2\pi^2}{L^2}, \qquad n = 1, 2, 3, \ldots \tag{8}$$

with the eigenfunction $y_n = \sin(n\pi x/L)$ associated with λ_n.

But what does all this mean in terms of our whirling string? It means that unless λ in (17) is one of the eigenvalues in (8), then the only solution of the problem in (7) is the trivial solution $y(x) \equiv 0$. In this case the string remains in its equilibrium position with zero deflection. But if we equate (17) and (8) and solve for the value ω_n corresponding to λ_n,

$$\omega_n = \sqrt{\frac{\lambda_n T}{\rho}} = \frac{n\pi}{L}\sqrt{\frac{T}{\rho}} \tag{18}$$

for $n = 1, 2, 3, \ldots$, we get a sequence of **critical speeds** of angular rotation. Only at these critical angular speeds can the string whirl up out of its equilibrium position. At angular speed ω_n it assumes a shape of the form $y_n = c_n \sin(n\pi x/L)$; our mathematical model is not sufficiently complete to determine the coefficient c_n.

Suppose that we start the string rotating at speed

$$\omega < \omega_1 = \frac{\pi}{L}\sqrt{\frac{T}{\rho}},$$

but gradually increase its speed of rotation. So long as $\omega < \omega_1$, the string remains in its undeflected position $y \equiv 0$. But when $\omega = \omega_1$, the string pops into a whirling position $y = c \sin(\pi x/L)$. And when ω is increased further, the string pops back into its undeflected position along the axis of rotation!

2.10 Problems

The eigenvalues in each of Problems 1–5 are all nonnegative. First determine whether $\lambda = 0$ is an eigenvalue; then find the positive eigenvalues and associated eigenfunctions.

1. $y'' + \lambda y = 0$; $\quad y'(0) = 0, y(1) = 0$.

2. $y'' + \lambda y = 0$; $\quad y'(0) = 0, y'(\pi) = 0$.

3. $y'' + \lambda y = 0$; $\quad y(-\pi) = 0, y(\pi) = 0$.

4. $y'' + \lambda y = 0$; $\quad y'(-\pi) = 0, y'(\pi) = 0$.

5. $y'' + \lambda y = 0$; $\quad y(-2) = 0, y'(2) = 0$.

6. Consider the eigenvalue problem

$$y'' + \lambda y = 0; \qquad y'(0) = 0, \qquad y(1) + y'(1) = 0.$$

All the eigenvalues are nonnegative, so write $\lambda = \alpha^2$ with $\alpha \geq 0$. (a) Show that $\lambda = 0$ is not an eigenvalue. (b) Show that $y = A \cos \alpha x + B \sin \alpha x$ satisfies the endpoint conditions if and only if $B = 0$ and α is a positive root of the equation $\tan z = 1/z$. These roots $\{\alpha_n\}_1^\infty$ are the abscissas of the points of intersection of the curves $y = \tan z$ and $y = 1/z$, as indicated in Fig. 2.38. Thus the eigenvalues and eigenfunctions of this problem are the numbers $\{\alpha_n^2\}_1^\infty$ and the functions $\{\cos \alpha_n x\}_1^\infty$, respectively.

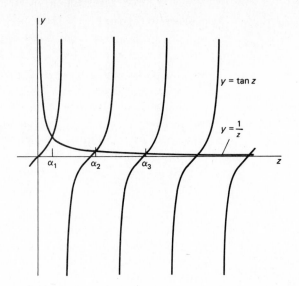

Figure 2.38 The eigenvalues are determined by the intersections of the graphs of $y = \tan z$ and $y = 1/z$.

7. Consider the eigenvalue problem

$$y'' + \lambda y = 0; \qquad y(0) = 0, \qquad y(1) + y'(1) = 0;$$

all its eigenvalues are nonnegative. (a) Show that $\lambda = 0$ is not an eigenvalue. (b) Show that the eigenfunctions are the functions $\{\sin \alpha_n x\}_1^\infty$, where α_n is the nth positive root of the equation $\tan z = -z$. (c) Draw a sketch indicating the roots $\{\alpha_n\}_1^\infty$ as the points of intersection of the curves $y = z$ and $y = -\tan z$.

8. Consider the eigenvalue problem

$$y'' + \lambda y = 0; \qquad y(0) = 0, \qquad y(1) = y'(1);$$

all its eigenvalues are nonnegative. (a) Show that $\lambda = 0$ is an eigenvalue with associated eigenfunction $y_0(x) = x$. (b) Show that the remaining eigenfunctions are given by $y_n(x) = \sin \alpha_n x$, where α_n is the nth positive root of the equation $\tan z = z$. Draw a sketch showing these roots.

9. Prove that the eigenvalue problem of Example 4 has no negative eigenvalues.

10. Prove that the eigenvalue problem

$$y'' + \lambda y = 0; \qquad y(0) = 0, \qquad y(1) + y'(1) = 0$$

has no negative eigenvalues. (*Suggestion:* Show graphically that the only root of the equation $\tanh z = -z$ is $z = 0$.)

11. Use a method similar to that suggested in Problem 10 to prove that the eigenvalue problem in Problem 6 has no negative eigenvalues.

12. Consider the eigenvalue problem

$$y'' + \lambda y = 0; \qquad y(-\pi) = y(\pi), \qquad y'(-\pi) = y'(\pi),$$

which is not of the type in (10) because the two endpoint conditions are not "separated" between the two endpoints. (a) Show that $\lambda_0 = 0$ is an eigenvalue with eigenfunction $y_0(x) \equiv 1$. (b) Show that there are no negative eigenvalues. (c) Show that the nth positive eigenvalue is n^2, and that it has two linearly independent associated eigenfunctions, $\cos nx$ and $\sin nx$.

13. Consider the eigenvalue problem

$$y'' + 2y' + \lambda y = 0; \qquad y(0) = y(1) = 0.$$

(a) Show that $\lambda = 1$ is not an eigenvalue. (b) Show that there is no eigenvalue λ such that $\lambda < 1$. (c) Show that the nth positive eigenvalue is $\lambda_n = n^2\pi^2 + 1$, with associated eigenfunction $y_n(x) = e^{-x} \sin (n\pi x)$.

14. Consider the eigenvalue problem

$$y'' + 2y' + \lambda y = 0; \qquad y(0) = 0, \qquad y'(1) = 0.$$

Show that the eigenvalues are all positive, and that the nth positive eigenvalue is $\lambda_n = \alpha_n^2 + 1$ with associated eigenfunction $y_n(x) = e^{-x} \sin \alpha_n x$, where α_n is the nth positive root of $\tan z = -z$.

15. Consider the eigenvalue problem

$$x^2 y'' + xy' + \lambda y = 0; \qquad y(1) = 0, \qquad y(e) = 0,$$

in which the differential equation is an Euler-Cauchy equation. Show that the eigenvalues are all positive, the nth one being $\lambda_n = n^2\pi^2$, with associated eigenfunction $y_n(x) = \sin (n\pi \ln x)$.

16. Consider the eigenvalue problem

$$x^2 y'' - 3xy' + \lambda y = 0; \qquad y(1) = 0, \qquad y(e) = 0.$$

Show that the eigenvalues all exceed 4, and that the nth one is $\lambda_n = n^2\pi^2 + 4$, with associated eigenfunction $y_n(x) = x^2 \sin (n\pi \ln x)$.

Power Series Solutions of Linear Equations

3

3.1
Introduction and Review of Power Series

In Section 2.3 we saw that solving a homogeneous linear differential equation with constant coefficients can be reduced to the algebraic problem of finding the roots of its characteristic equation. There is no similar procedure for solving linear differential equations with *variable* coefficients, at least not routinely and in finitely many steps. With the exception of special types, such as the Euler-Cauchy equations of Section 2.8, and the occasional equation that can be solved by inspection (perhaps followed by reduction of order), linear equations with variable coefficients generally require the power series techniques of this chapter.

These techniques suffice for many of the nonelementary differential equations that appear most frequently in applications. Perhaps the most important (because of its applications in such areas as acoustics, heat flow, and nuclear reactor design) is **Bessel's equation** of order n:

$$x^2 y'' + xy' + (x^2 - n^2)y = 0.$$

Legendre's equation of order n is important in many applications; it has the form

$$(1 - x^2)y'' - 2xy' + n(n + 1)y = 0.$$

In this section we introduce the **power series method** in its simplest form and, along the way, state (without proof) several theorems that constitute a review of the basic facts about power series. Recall first that a **power series** in (powers of) $x - a$ is an infinite series of the form

199

$$\sum_{n=0}^{\infty} c_n(x-a)^n = c_0 + c_1(x-a) + c_2(x-a)^2 + \cdots + c_n(x-a)^n + \cdots. \quad (1)$$

If $a = 0$, this is a power series in x:

$$\sum_{n=0}^{\infty} c_n x^n = c_0 + c_1 x + c_2 x^2 + \cdots + c_n x^n + \cdots. \quad (2)$$

We will confine our review mainly to power series in x, but every general property of power series in x can be converted to a general property of power series in $x - a$ by replacement of x by $x - a$.

The power series in (2) **converges** on the interval I provided that the limit

$$\sum_{n=0}^{\infty} c_n x^n = \lim_{N \to \infty} \sum_{n=0}^{N} c_n x^n \quad (3)$$

exists for all x in I. In this case the sum

$$f(x) = \sum_{n=0}^{\infty} c_n x^n \quad (4)$$

is defined on I, and we call the series $\sum c_n x^n$ a **power series representation** of the function f on I. The following power series representations of elementary functions should be familiar to you from introductory calculus.

$$e^x = \sum_{n=0}^{\infty} \frac{x^n}{n!} = 1 + x + \frac{x^2}{2!} + \frac{x^3}{3!} + \cdots; \quad (5)$$

$$\cos x = \sum_{n=0}^{\infty} (-1)^n \frac{x^{2n}}{(2n)!} = 1 - \frac{x^2}{2!} + \frac{x^4}{4!} - \cdots; \quad (6)$$

$$\sin x = \sum_{n=0}^{\infty} (-1)^n \frac{x^{2n+1}}{(2n+1)!} = x - \frac{x^3}{3!} + \frac{x^5}{5!} - \cdots; \quad (7)$$

$$\cosh x = \sum_{n=0}^{\infty} \frac{x^{2n}}{(2n)!} = 1 + \frac{x^2}{2!} + \frac{x^4}{4!} + \cdots; \quad (8)$$

$$\sinh x = \sum_{n=0}^{\infty} \frac{x^{2n+1}}{(2n+1)!} = x + \frac{x^3}{3!} + \frac{x^5}{5!} + \cdots; \quad (9)$$

$$\ln(1+x) = \sum_{n=1}^{\infty} (-1)^{n+1} \frac{x^n}{n} = x - \frac{x^2}{2} + \frac{x^3}{3} - \cdots; \quad (10)$$

$$\frac{1}{1-x} = \sum_{n=0}^{\infty} x^n = 1 + x + x^2 + x^3 + \cdots; \quad (11)$$

and

$$(1+x)^\alpha = 1 + \alpha x + \frac{\alpha(\alpha-1)}{2!} x^2 + \frac{\alpha(\alpha-1)(\alpha-2)}{3!} x^3 + \cdots. \quad (12)$$

In compact summation notation, we observe the usual conventions that $0! = 1$ and that $c_0 x^0 = c_0$ for all x (including $x = 0$). The series in (5)–(9) converge to the indicated functions for all x, while the series in (10)–(12) converge if $|x| < 1$, but diverge if $|x| > 1$. The series in (12), with α an arbitrary real number, is the **binomial series**; the series in (11) is the **geometric series**.

Power series such as those listed above are often derived as Taylor series. The **Taylor series** with **center** $x = a$ of the function f is the power series

$$\sum_{n=0}^{\infty} \frac{f^{(n)}(a)}{n!}(x-a)^n = f(a) + f'(a)(x-a) + \frac{f''(a)}{2!}(x-a)^2 + \cdots \quad (13)$$

in powers of $x - a$, under the hypothesis that f is infinitely differentiable at a (so that the coefficients in (13) are all defined). If the Taylor series of f converges to $f(x)$ for all x in some open interval containing a, then we say that the function f is **analytic** at $x = a$. For example, every polynomial is analytic everywhere, and every rational function is analytic wherever its denominator is nonzero. More generally, if the two functions f and g are both analytic at $x = a$, then so are their sum $f + g$ and their product $f \cdot g$, as is their quotient f/g wherever g is nonzero.

For instance, the function $\tan x = (\sin x)/(\cos x)$ is analytic at $x = 0$ because $\cos 0 = 1 \neq 0$ and the functions $\sin x$ and $\cos x$ are analytic (by virtue of their convergent power series representations in Eqs. (6) and (7)). It is rather awkward to compute the Taylor series of the function $\tan x$ using (13) because of the way in which its successive derivatives grow in complexity (try it!). Fortunately, power series may be manipulated algebraically in much the same way as polynomials. For example, if

$$f(x) = \sum_{n=0}^{\infty} a_n x^n \quad \text{and} \quad g(x) = \sum_{n=0}^{\infty} b_n x^n, \tag{14}$$

then

$$f(x) + g(x) = \sum_{n=0}^{\infty} (a_n + b_n) x^n \tag{15}$$

and

$$f(x)g(x) = \sum_{n=0}^{\infty} c_n x^n$$
$$= a_0 b_0 + (a_0 b_1 + a_1 b_0)x + (a_0 b_2 + a_1 b_1 + a_2 b_0)x^2 + \cdots \tag{16}$$

where $c_n = a_0 b_n + a_1 b_{n-1} + \cdots + a_n b_0$. The series in (15) is the result of **termwise addition**, while the series in (16) is the result of **formal multiplication**—multiplying each term of the first series by each term of the second and then collecting coefficients of like powers of x. (Thus the processes strongly resemble addition and multiplication of ordinary polynomials.) The series in (15) and (16) converge to $f(x) + g(x)$ and $f(x)g(x)$, respectively, on any open interval on which both the series in (14) converge. For example,

$$\sin x \cos x = \left(x - \frac{1}{6}x^3 + \frac{1}{120}x^5 - \cdots \right)\left(1 - \frac{1}{2}x^2 + \frac{1}{24}x^4 - \cdots \right)$$

$$= x + \left(-\frac{1}{6} - \frac{1}{2} \right)x^3 + \left(\frac{1}{24} + \frac{1}{12} + \frac{1}{120} \right)x^5 + \cdots$$

$$= x - \frac{4}{6}x^3 + \frac{16}{120}x^5 - \cdots$$

$$= \frac{1}{2}\left[(2x) - \frac{(2x)^3}{3!} + \frac{(2x)^5}{5!} - \cdots \right] = \frac{1}{2}\sin 2x$$

for all x.

Similarly, the quotient of two power series can be computed by long division, as illustrated by the computation shown in Fig. 3.1. This division of the Taylor series for $\cos x$ into that for $\sin x$ yields the first few terms of the series

$$\tan x = x + \frac{1}{3}x^3 + \frac{2}{15}x^5 + \frac{17}{315}x^7 + \cdots. \tag{17}$$

$$x + \frac{x^3}{3} + \frac{2x^5}{15} + \frac{17x^7}{315} + \cdots$$

$$1 - \frac{x^2}{2} + \frac{x^4}{24} - \frac{x^6}{720} + \cdots \overline{\smash{\big)}\, x - \frac{x^3}{6} + \frac{x^5}{120} - \frac{x^7}{5040} + \cdots}$$

$$x - \frac{x^3}{2} + \frac{x^5}{24} - \frac{x^7}{720} + \cdots$$

$$\frac{x^3}{3} - \frac{x^5}{30} + \frac{x^7}{840} + \cdots$$

$$\frac{x^3}{3} - \frac{x^5}{6} + \frac{x^7}{72} - \cdots$$

$$\frac{2x^5}{15} - \frac{4x^7}{315} + \cdots$$

$$\frac{2x^5}{15} - \frac{x^7}{15} + \cdots$$

$$\frac{17x^7}{315} - \cdots$$

Figure 3.1 Obtaining the series for $\tan x$ by division of series.

Division of power series is more treacherous than multiplication; the series thus obtained for f/g may fail to converge at some points where the series for f and g both converge. For instance, the sine and cosine series converge for all x, but the tangent series in (17) converges only if $|x| < \pi/2$.

The **power series method** for solving a differential equation consists of substituting the power series

$$y = \sum_{n=0}^{\infty} c_n x^n \tag{18}$$

in the differential equation and then attempting to determine what the coefficients $c_0, c_1, c_2, \ldots$ must be in order that the power series will satisfy the differential equation. This is much like the method of undetermined coefficients, but now we have infinitely many coefficients somehow to determine. This method is not always successful, but when it is we obtain an infinite series representation of a solution, in contrast to the "closed form" solutions that our previous methods have yielded.

Before we can substitute the power series in (18) in a differential equation, we must first know what to substitute for the derivatives $y', y'', \ldots$. The following theorem (stated without proof) tells us that the derivative y' of $y = \sum c_n x^n$ is obtained by the simple procedure of writing the sum of the derivatives of the individual terms of y.

THEOREM 1: TERMWISE DIFFERENTIATION OF POWER SERIES

If the power series representation

$$f(x) = \sum_{n=0}^{\infty} c_n x^n = c_0 + c_1 x + c_2 x^2 + c_3 x^3 + \cdots \tag{19}$$

of the function f converges on the open interval I, then f is differentiable on

I, and

$$f'(x) = \sum_{n=1}^{\infty} nc_n x^{n-1} = c_1 + 2c_2 x + 3c_3 x^2 + \cdots \qquad (20)$$

at each point of I.

For example, differentiation of the geometric series

$$\frac{1}{1-x} = \sum_{n=0}^{\infty} x^n = 1 + x + x^2 + x^3 + \cdots \qquad (11)$$

gives the series

$$\frac{1}{(1-x)^2} = \sum_{n=1}^{\infty} nx^{n-1} = 1 + 2x + 3x^2 + 4x^3 + \cdots .$$

The process of determining the coefficients in the series $y = \sum c_n x^n$ so that it will satisfy a given differential equation depends also upon Theorem 2. This theorem—stated without proof—tells us that if two power series represent the same function, then they are the same series. In particular, the Taylor series in (13) is the only power series (in powers of x) that represents the function $f(x)$.

THEOREM 2: IDENTITY PRINCIPLE

If

$$\sum_{n=0}^{\infty} a_n x^n = \sum_{n=0}^{\infty} b_n x^n$$

for every point x in some open interval I, then $a_n = b_n$ for all $n \geqq 0$.

In particular, if $\sum a_n x^n = 0$ for all x in some open interval, it follows from Theorem 2 that $a_n = 0$ for all n.

EXAMPLE 1 Solve the equation $y' + 2y = 0$.

Solution We substitute the series

$$y = \sum_{n=0}^{\infty} c_n x^n \quad \text{and} \quad y' = \sum_{n=1}^{\infty} nc_n x^{n-1},$$

and obtain

$$\sum_{n=1}^{\infty} nc_n x^{n-1} + 2 \sum_{n=0}^{\infty} c_n x^n = 0. \qquad (21)$$

To compare coefficients here, we need the general term in each to be the term containing x^n. To accomplish this, we shift the index of summation in the first sum. To see how to do this, note that

$$\sum_{n=1}^{\infty} nc_n x^{n-1} = c_1 + 2c_2 x + 3c_3 x^2 + \cdots = \sum_{n=0}^{\infty} (n+1)c_{n+1} x^n.$$

Thus we can replace n by $n + 1$ if, at the same time, we start counting one step lower; that is, at $n = 0$ rather than at $n = 1$. This is a shift of $+1$ in the index of summation. The result of making this shift in (21) is the identity

$$\sum_{n=0}^{\infty} (n+1)c_{n+1} x^n + 2 \sum_{n=0}^{\infty} c_n x^n = 0;$$

that is,

$$\sum_{n=0}^{\infty} [(n+1)c_{n+1} + 2c_n]x^n = 0.$$

If this is true on some open interval, then it follows from the identity principle that $(n+1)c_{n+1} + 2c_n = 0$ for all $n \geq 0$; consequently,

$$c_{n+1} = -\frac{2c_n}{n+1} \tag{22}$$

for all $n \geq 0$. Equation (22) is a **recursion formula** from which we can successively compute $c_1, c_2, c_3, \ldots$ in terms of c_0; the latter will turn out to be the arbitrary constant we expect to find in a solution of a first order differential equation.

With $n = 0$, (22) gives

$$c_1 = -\frac{2c_0}{1}.$$

With $n = 1$, (22) gives

$$c_2 = -\frac{2c_1}{2} = +\frac{2^2 c_0}{1 \cdot 2} = +\frac{2^2 c_0}{2!}.$$

With $n = 2$, (22) gives

$$c_3 = -\frac{2c_2}{3} = -\frac{2^3 c_0}{1 \cdot 2 \cdot 3} = -\frac{2^3 c_0}{3!}.$$

By now it should be clear that after n such steps, we will have

$$c_n = (-1)^n \frac{2^n c_0}{n!}, \qquad n \geq 1.$$

(This is easy to prove by induction on n.) Consequently, our solution takes the form

$$y = \sum_{n=0}^{\infty} c_n x^n = \sum_{n=0}^{\infty} (-1)^n \frac{2^n c_0}{n!} x^n$$

$$= c_0 \sum_{n=0}^{\infty} \frac{(-2x)^n}{n!} = c_0 e^{-2x}.$$

In the final step we have used the familiar exponential series in (5) to identify our power series solution as the same solution $y = c_0 e^{-2x}$ we could have obtained immediately by the method of separation of variables.

We know that the power series obtained in Example 1 converges for all x because it is an exponential series. More commonly, a power series solution is not recognizable in terms of the familiar elementary functions. When we get an unfamiliar power series solution, we need a way of ascertaining where it converges. After all, $y = \sum c_n x^n$ is merely an *assumed* form of the solution. The procedure illustrated in Example 1 for determining the coefficients $\{c_n\}$ is only a formal process and may or may not be valid. Its validity—in applying Theorem 1 to compute y' and in applying Theorem 2 to obtain a recursion formula for the coefficients—depends on the convergence of the initially *unknown* series $y = \sum c_n x^n$. Hence this formal process is justified only if in the end we can show that the power series obtained converges on some open interval. If so, it then represents a solution of the differential equation on that interval. The following theorem (which we state without proof) may be used for this purpose.

THEOREM 3: RADIUS OF CONVERGENCE

Given the power series $\sum c_n x^n$, suppose that the limit

$$\rho = \lim_{n \to +\infty} \left| \frac{c_n}{c_{n+1}} \right| \tag{23}$$

exists (ρ is finite) or is infinite (in this case, we will write $\rho = \infty$). Then:

(a) *If $\rho = 0$, the series diverges for all $x \neq 0$.*
(b) *If $0 < \rho < \infty$, then $\sum c_n x^n$ converges if $|x| < \rho$ and diverges if $|x| > \rho$.*
(c) *If $\rho = \infty$, the series converges for all x.*

The number ρ in (23) is called the **radius of convergence** of the power series $\sum c_n x^n$. For instance, for the power series obtained in Example 1, we have

$$\rho = \lim_{n \to +\infty} \left| \frac{(-1)^n 2^n c_0/n!}{(-1)^{n+1} 2^{n+1} c_0/(n+1)!} \right| = \lim_{n \to +\infty} \frac{n+1}{2} = \infty,$$

and consequently the series we obtained in Example 1 converges for all x. Even if the limit in (23) fails to exist, there always will exist a number ρ such that exactly one of the alternatives (a), (b), or (c) in Theorem 3 holds. This number may be difficult to find, but for the power series we will consider in this chapter, (23) will be quite sufficient for computation of their radii of convergence.

EXAMPLE 2   Solve the equation $(x - 3)y' + 2y = 0$.

Solution   As before, we substitute

$$y = \sum_{n=0}^{\infty} c_n x^n \quad \text{and} \quad y' = \sum_{n=1}^{\infty} nc_n x^{n-1}$$

to obtain

$$(x - 3) \sum_{n=1}^{\infty} nc_n x^{n-1} + 2 \sum_{n=0}^{\infty} c_n x^n = 0,$$

so that

$$\sum_{n=1}^{\infty} nc_n x^n - 3 \sum_{n=1}^{\infty} nc_n x^{n-1} + 2 \sum_{n=0}^{\infty} c_n x^n = 0.$$

In the first sum we can replace $n = 1$ by $n = 0$ with no effect on the sum. In the second sum we shift the index of summation by $+1$. This yields

$$\sum_{n=0}^{\infty} nc_n x^n - 3 \sum_{n=0}^{\infty} (n + 1)c_{n+1} x^n + 2 \sum_{n=0}^{\infty} c_n x^n = 0;$$

that is,

$$\sum_{n=0}^{\infty} [nc_n - 3(n + 1)c_{n+1} + 2c_n]x^n = 0.$$

The identity principle then gives

$$nc_n - 3(n+1)c_{n+1} + 2c_n = 0,$$

from which we obtain the recursion formula

$$c_{n+1} = \frac{n+2}{3(n+1)}c_n.$$

We apply this formula with $n = 0$, $n = 1$, and $n = 2$ in turn, and find that

$$c_1 = \frac{2}{3}c_0, \qquad c_2 = \frac{3}{3\cdot 2}c_1 = \frac{3}{3^2}c_0,$$

and

$$c_3 = \frac{4}{3\cdot 3}c_2 = \frac{4}{3^3}c_0.$$

This is almost enough to make the pattern evident; it is not difficult to show by induction on n that

$$c_n = \frac{n+1}{3^n}c_0 \quad \text{if } n \geqq 1.$$

Hence our proposed power series solution is

$$y = c_0 \sum_{n=0}^{\infty} \frac{n+1}{3^n}x^n. \tag{24}$$

Its radius of convergence is

$$\rho = \lim_{n \to +\infty} \left| \frac{c_n}{c_{n+1}} \right| = \lim_{n \to +\infty} \frac{3n+3}{n+2} = 3.$$

Thus the series in (24) converges if $-3 < x < 3$, but diverges if $|x| > 3$. In this particular example we can explain why. An elementary solution (obtained by separation of variables) of our differential equation is $y = 1/(3 - x)^2$. If we differentiate termwise the geometric series

$$\frac{1}{3 - x} = \frac{\frac{1}{3}}{1 - (x/3)} = \frac{1}{3} \sum_{n=0}^{\infty} \frac{x^n}{3^n},$$

we get a constant multiple of the series in (24). Thus this series (with the arbitrary constant c_0 appropriately chosen) represents the solution $y = 1/(3 - x)^2$ on the interval $-3 < x < 3$, and the singularity at $x = 3$ is the reason why the radius of convergence of our power series solution turned out to be $\rho = 3$.

EXAMPLE 3 Solve the equation $x^2 y' = y - x - 1$.

Solution We make the usual substitutions $y = \sum c_n x^n$ and $y' = \sum nc_n x^{n-1}$, which yield

$$x^2 \sum_{n=1}^{\infty} nc_n x^{n-1} = -1 - x + \sum_{n=0}^{\infty} c_n x^n,$$

so that

$$\sum_{n=1}^{\infty} nc_n x^{n+1} = -1 - x + \sum_{n=0}^{\infty} c_n x^n.$$

Because of the presence of the terms -1 and $-x$ on the right-hand side, we need to split off the first two terms, $c_0 + c_1 x$, of the series on the right for comparison. If we also shift the index of summation in the series on the left by -1 (replace $n = 1$ with $n = 2$ and n with $n - 1$), we get

$$\sum_{n=2}^{\infty} (n - 1)c_{n-1}x^n = -1 - x + c_0 + c_1 x + \sum_{n=2}^{\infty} c_n x^n.$$

Because the left-hand side contains neither a constant term nor a term containing x to the first power, the identity principle now yields $c_0 = 1$, $c_1 = 1$, and $c_n = (n - 1)c_{n-1}$ for $n \geq 2$. It follows that

$$c_2 = 1 \cdot c_1 = 1!, \; c_3 = 2 \cdot c_2 = 2!, \; c_4 = 3 \cdot c_3 = 3!,$$

and, in general, that $c_n = (n - 1)!$ for $n \geq 2$. Thus we obtain the power series

$$y = 1 + x + \sum_{n=2}^{\infty} (n - 1)!x^n.$$

But the radius of convergence of this series is

$$\rho = \lim_{n \to +\infty} \frac{(n - 1)!}{n!} = \lim_{n \to +\infty} \frac{1}{n} = 0,$$

so it converges only for $x = 0$. What does this mean? Simply that the given differential equation does not have a (convergent) power series solution of the assumed form $y = \sum c_n x^n$. This example serves as a warning that the simple act of writing $y = \sum c_n x^n$ involves an assumption that may be false.

EXAMPLE 4 Solve the equation $y'' + y = 0$.

Solution If we assume a solution of the form

$$y = \sum_{n=0}^{\infty} c_n x^n,$$

we find that

$$y' = \sum_{n=1}^{\infty} nc_n x^{n-1} \quad \text{and} \quad y'' = \sum_{n=2}^{\infty} n(n - 1)c_n x^{n-2}.$$

Substitution for y and y'' in the differential equation then yields

$$\sum_{n=2}^{\infty} n(n - 1)c_n x^{n-2} + \sum_{n=0}^{\infty} c_n x^n = 0.$$

We shift the index of summation in the left sum by $+2$ (by replacing $n = 2$ with $n = 0$ and n with $n + 2$). This gives

$$\sum_{n=0}^{\infty} (n + 2)(n + 1)c_{n+2}x^n + \sum_{n=0}^{\infty} c_n x^n = 0.$$

The identity principle now leads to the identity $(n + 2)(n + 1)c_{n+2} + c_n = 0$, and thus we obtain the recursion formula

$$c_{n+2} = -\frac{c_n}{(n + 1)(n + 2)} \tag{25}$$

for $n \geq 0$. It is evident that this formula will determine the coefficients c_n with even subscripts in terms of c_0, and those of odd subscript in terms of c_1; c_0 and c_1 are not predetermined, and thus will be the two arbitrary constants we expect to find in a general solution of a second order equation.

When we apply the recursion formula in (25) with $n = 0, 2,$ and 4 in turn, we get

$$c_2 = -\frac{c_0}{2!}, \qquad c_4 = \frac{c_0}{4!}, \quad \text{and} \quad c_6 = -\frac{c_0}{6!}$$

Taking $n = 1, 3,$ and 5 in turn, we find that

$$c_3 = -\frac{c_1}{3!}, \qquad c_5 = \frac{c_1}{5!}, \quad \text{and} \quad c_7 = -\frac{c_1}{7!}.$$

Again, the pattern is clear; we leave it for you to show (by induction) that for $k \geq 1$,

$$c_{2k} = \frac{(-1)^k c_0}{(2k)!} \quad \text{and} \quad c_{2k+1} = \frac{(-1)^k c_1}{(2k+1)!}.$$

Thus we get the power series solution

$$y = c_0\left(1 - \frac{x^2}{2!} + \frac{x^4}{4!} - \frac{x^6}{6!} + \cdots\right)$$

$$+ c_1\left(x - \frac{x^3}{3!} + \frac{x^5}{5!} - \frac{x^7}{7!} + \cdots\right)$$

—that is, $y = c_0 \cos x + c_1 \sin x$. Note that we have no problem with the radius of convergence here; the Taylor series for the sine and cosine functions converge for all x.

The solution of Example 4 can bear further comment. Suppose that we had never heard of the sine and cosine functions, let alone their Taylor series. We would then have discovered the two power series solutions

$$C(x) = \sum_{n=0}^{\infty} (-1)^n \frac{x^{2n}}{(2n)!} = 1 - \frac{x^2}{2!} + \frac{x^4}{4!} - \cdots \tag{26}$$

and

$$S(x) = \sum_{n=0}^{\infty} (-1)^n \frac{x^{2n+1}}{(2n+1)!} = x - \frac{x^3}{3!} + \frac{x^5}{5!} - \cdots \tag{27}$$

of the differential equation $y'' + y = 0$. It is clear that $C(0) = 1$ and that $S(0) = 0$. After verifying that the two series in (26) and (27) converge for all x, we can differentiate them term by term to find that

$$C'(x) = -S(x) \quad \text{and} \quad S'(x) = C(x). \tag{28}$$

Consequently $C'(0) = 0$ and $S'(0) = 1$. Thus with the aid of the power series method (all the while knowing nothing about the sine and cosine functions), we have discovered that $y = C(x)$ is the unique solution of $y'' + y = 0$ that satisfies the initial conditions $y(0) = 1$ and $y'(0) = 0$, and that $y = S(x)$ is the solution that satisfies the initial conditions $y(0) = 0$ and $y'(0) = 1$. It follows that $C(x)$ and $S(x)$ are linearly independent, and—recognizing the importance of the differential equation $y'' + y = 0$—we can agree to call $C(x)$ the *cosine* function and $S(x)$ the *sine* function. Indeed, all the usual properties of these two functions can be established, using only their initial values (at $x = 0$) and the derivatives in (28); there is no need to refer to triangles or even to angles. (Can you use the series in (26) and (27) to show that $[C(x)]^2 + [S(x)]^2 = 1$ for all x?) This demonstrates that the cosine and sine functions are fully determined by the differential equation $y'' + y = 0$ of which they are the natural linearly independent solutions.

This is by no means an uncommon situation. Many important special functions of mathematics occur in the first instance as power series solutions of differential equations, and thus are in practice *defined* by means of these power series. In the remaining sections of this chapter we will see numerous examples of such functions.

3.1 Problems

In each of Problems 1–10, find a power series solution of the given differential equation. Determine the radius of convergence of the resulting series, and use the series in (5)–(12) to identify the series solution in terms of familiar elementary functions. (Of course, no one can prevent you from checking your work by also solving the equations by the methods of earlier chapters!)

1. $y' = y$.
2. $y' = 4y$.
3. $2y' + 3y = 0$.
4. $y' + 2xy = 0$.
5. $y' = x^2 y$.
6. $(x - 2)y' + y = 0$.
7. $(2x - 1)y' + 2y = 0$.
8. $2(x + 1)y' = y$.
9. $(x - 1)y' + 2y = 0$.
10. $2(x - 1)y' = 3y$.

In each of Problems 11–14, use the method of Example 4 to find two linearly independent power series solutions of the given differential equation. Determine the radius of convergence of each series, and identify the general solution in terms of familiar elementary functions.

11. $y'' = y$.
12. $y'' = 4y$.
13. $y'' + 9y = 0$.
14. $y'' + y = x$.

Show (as in Example 3) that the power series method fails to yield a power series solution of the form $y = \sum c_n x^n$ for the differential equations in Problems 15–18.

15. $xy' + y = 0$.
16. $2xy' = y$.
17. $x^2 y' + y = 0$.
18. $x^3 y' = 2y$.

In each of Problems 19–22, first derive a recursion formula giving c_n for $n \geqq 2$ in terms of c_0 or c_1 (or both). Then apply the given initial conditions to find the values of c_0 and c_1. Next determine c_n (in terms of n, as in the text), and, finally, identify the particular solution in terms of familiar elementary functions.

19. $y'' + 4y = 0$; $y(0) = 0, y'(0) = 3$.
20. $y'' - 4y = 0$; $y(0) = 2, y'(0) = 0$.
21. $y'' - 2y' + y = 0$; $y(0) = 0, y'(0) = 1$.
22. $y'' + y' - 2y = 0$; $y(0) = 1, y'(0) = -2$.
23. Show that the equation

$$x^2 y'' + x^2 y' + y = 0$$

has no power series solution of the form $y = \sum c_n x^n$.

24. Establish the binomial series in (12) by means of the following steps. (a) Show that $y = (1 + x)^\alpha$ satisfies the initial value problem $(1 + x)y' = \alpha y$, $y(0) = 1$. (b) Show that the power series method gives the binomial series in (12) as the solution of the initial value problem in part (a), and that this series converges if $|x| < 1$. (c) Explain why the validity of the binomial series given in (12) follows from (a) and (b).

3.2
Series Solutions Near Ordinary Points

The power series method introduced in Section 3.1 can be applied to linear equations of any order, but its most important applications are to homogeneous second order linear differential equations of the form

$$A(x)y'' + B(x)y' + C(x)y = 0, \tag{1}$$

where the coefficients A, B, and C are analytic functions of x. Indeed, in most applications these coefficient functions are simple polynomials.

We saw in Example 3 of Section 3.1 that the series method does not always yield a series solution. To discover when it does succeed, it is best to rewrite Eq. (1) in the form

$$y'' + P(x)y' + Q(x)y = 0 \tag{2}$$

where $P = B/A$ and $Q = C/A$. Note that $P(x)$ and $Q(x)$ will generally fail to be analytic at points where $A(x)$ vanishes. For instance, consider the equation

$$xy'' + y' + xy = 0. \tag{3}$$

The coefficient functions in (3) are all analytic everywhere. But in the form of (2) it is the equation

$$y'' + \frac{1}{x}y' + y = 0 \tag{4}$$

with $P(x) = 1/x$ not analytic at $x = 0$.

The point $x = a$ is called an **ordinary point** of Eq. (2)—and of the equivalent Eq. (1)—provided that the functions $P(x)$ and $Q(x)$ are both analytic at $x = a$. Otherwise, $x = a$ is a **singular point**. Thus the only singular point of Eqs. (3) and (4) is $x = 0$. Recall that a quotient of analytic functions is analytic wherever the denominator is nonzero. It follows that, if $A(a) \neq 0$ in Eq. (1) with analytic coefficients, then $x = a$ is an ordinary point. If $A(x)$, $B(x)$, and $C(x)$ are *polynomials* with no common factors, then $x = a$ is an ordinary point if and only if $A(a) \neq 0$.

EXAMPLE 1   The point $x = 0$ is an ordinary point of the equation

$$xy'' + (\sin x)y' + x^2y = 0,$$

despite the fact that $A(x) = x$ vanishes at $x = 0$. The reason is that

$$P(x) = \frac{\sin x}{x} = \frac{1}{x}\left(x - \frac{x^3}{3!} + \frac{x^5}{5!} - \cdots\right)$$

$$= 1 - \frac{x^2}{3!} + \frac{x^4}{5!} - \cdots$$

is nevertheless analytic at $x = 0$ because the division by x gives a convergent power series.

Theorem 2 of Section 2.1 implies that Eq. (2) has two linearly independent solutions on any open interval where the coefficient functions $P(x)$ and $Q(x)$ are continuous. The basic fact for our present purpose is that near an *ordinary* point a, these solutions will be power series in powers of $x - a$. A proof of the following theorem can be found in Chapter 3 of Coddington, *An Introduction to Ordinary Differential Equations* (Englewood Cliffs, N.J.: Prentice-Hall, 1961).

THEOREM: SOLUTIONS NEAR AN ORDINARY POINT

Suppose that a is an ordinary point of the equation

$$A(x)y'' + B(x)y' + C(x)y = 0; \tag{1}$$

that is, the functions $P = B/A$ and $Q = C/A$ are analytic at $x = a$. Then Eq. (1) has two linearly independent solutions, each of the form

$$y = \sum_{n=0}^{\infty} c_n(x - a)^n. \tag{5}$$

The radius of convergence of any such series solution is at least as large as the distance from a to the nearest (real or complex) singular point of Eq. (1). The coefficients in the series in (5) can be determined by its substitution in Eq. (1).

EXAMPLE 2   Determine the radius of convergence guaranteed by the above theorem of a series solution of

$$(x^2 + 9)y'' + xy' + x^2y = 0 \tag{6}$$

in powers of x. Repeat for a series in powers of $x - 4$.

Solution   This example illustrates the fact that we must take into account complex singular points as well as real ones. Because $P(x) = x/(x^2 + 9)$ and $Q(x) = x^2/(x^2 + 9)$, the only singular points of the equation in (6) are $+3i$ and $-3i$. The distance (in the complex plane) of each from 0 is 3, so a series solution of the form $\sum c_n x^n$ has radius of convergence at least 3. The distance of each singular point from 4 is 5, so a series solution of the form $\sum c_n(x - 4)^n$ has radius of convergence at least 5. See Fig. 3.2.

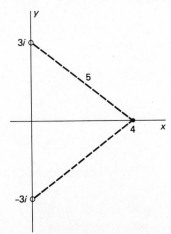

Figure 3.2   Radius of convergence as distance to nearest singularity.

EXAMPLE 3 Find the general solution in powers of x of
$$(x^2 - 4)y'' + 3xy' + y = 0. \tag{7}$$
Then find the particular solution with $y(0) = 4$, $y'(0) = 1$.

Solution The only singular points of (7) are $+2$ and -2, so the series we get will have radius of convergence at least 2. Substitution of
$$y = \sum_{n=0}^{\infty} c_n x^n, \qquad y' = \sum_{n=1}^{\infty} n c_n x^{n-1},$$
and
$$y'' = \sum_{n=2}^{\infty} n(n-1) c_n x^{n-2}$$
in Eq. (7) yields
$$\sum_{n=2}^{\infty} n(n-1) c_n x^n - 4 \sum_{n=2}^{\infty} n(n-1) c_n x^{n-2} + 3 \sum_{n=1}^{\infty} n c_n x^n + \sum_{n=0}^{\infty} c_n x^n = 0.$$

We can begin the first and third summations at $n = 0$ as well, because no nonzero terms are thereby introduced. We shift the index of summation in the second sum by $+2$, replacing n with $n + 2$ and using the initial value $n = 0$. This gives
$$\sum_{n=0}^{\infty} n(n-1) c_n x^n - 4 \sum_{n=0}^{\infty} (n+2)(n+1) c_{n+2} x^n + 3 \sum_{n=0}^{\infty} n c_n x^n$$
$$+ \sum_{n=0}^{\infty} c_n x^n = 0.$$

After collecting coefficients of c_n and c_{n+2}, we obtain
$$\sum_{n=0}^{\infty} [(n^2 + 2n + 1) c_n - 4(n+2)(n+1) c_{n+2}] x^n = 0.$$

The identity principle yields $(n+1)^2 c_n - 4(n+2)(n+1) c_{n+2} = 0$, which leads to the recursion formula
$$c_{n+2} = \frac{(n+1) c_n}{4(n+2)} \tag{8}$$
for $n \geq 0$. With $n = 0$, 2, and 4 in turn, we get
$$c_2 = \frac{c_0}{4 \cdot 2}, \qquad c_4 = \frac{3 c_2}{4 \cdot 4} = \frac{3 c_0}{4^2 \cdot 2 \cdot 4},$$
and
$$c_6 = \frac{5 c_4}{4 \cdot 6} = \frac{3 \cdot 5 c_0}{4^3 \cdot 2 \cdot 4 \cdot 6}.$$

Continuing in this fashion, we evidently would get
$$c_{2n} = \frac{3 \cdot 5 \cdots (2n-1)}{4^n \cdot 2 \cdot 4 \cdots (2n)} c_0.$$

With the common notation
$$(2n+1)!! = 1 \cdot 3 \cdot 5 \cdots (2n+1) = \frac{(2n+1)!}{2^n \cdot n!}$$

and the observation that $2 \cdot 4 \cdot 6 \cdots (2n) = 2^n \cdot n!$, we finally obtain

$$c_{2n} = \frac{(2n-1)!!}{2^{3n} \cdot n!} c_0. \tag{9}$$

(We also used the fact that $4^n \cdot 2^n = 2^{3n}$.)

With $n = 1, 3,$ and 5 in (8), we get

$$c_3 = \frac{2c_1}{4 \cdot 3}, \qquad c_5 = \frac{4c_3}{4 \cdot 5} = \frac{2 \cdot 4 c_1}{4^2 \cdot 3 \cdot 5},$$

and

$$c_7 = \frac{6c_5}{4 \cdot 7} = \frac{2 \cdot 4 \cdot 6 c_1}{4^3 \cdot 3 \cdot 5 \cdot 7}$$

It is apparent that the pattern is

$$c_{2n+1} = \frac{2 \cdot 4 \cdot 6 \cdots (2n)}{4^n \cdot 3 \cdot 5 \cdots (2n+1)} c_1 = \frac{n!}{2^n \cdot (2n+1)!!} c_1. \tag{10}$$

The formula in (9) gives the coefficients of even subscript in terms of c_0, and the formula in (10) gives the coefficients of odd subscript in terms of c_1. After we separately collect the terms of the series of even and odd degree, we get the general solution

$$y = c_0 \left(1 + \sum_{n=1}^{\infty} \frac{(2n-1)!!}{2^{3n}n!} x^{2n} \right) + c_1 \left(x + \sum_{n=1}^{\infty} \frac{n!}{2^n(2n+1)!!} x^{2n+1} \right). \tag{11'}$$

Alternatively,

$$y = c_0 (1 + \tfrac{1}{8}x^2 + \tfrac{3}{128}x^4 + \tfrac{5}{1024}x^6 + \cdots)$$
$$+ c_1 (x + \tfrac{1}{6}x^3 + \tfrac{1}{30}x^5 + \tfrac{1}{140}x^7 + \cdots). \tag{11''}$$

Because $y(0) = c_0$ and $y'(0) = c_1$, the given initial conditions imply that $c_0 = 4$ and $c_1 = 1$. Using these values in (11''), the first several terms of the particular solution satisfying $y(0) = 4$ and $y'(0) = 1$ are

$$y = 4 + x + \tfrac{1}{2}x^2 + \tfrac{1}{6}x^3 + \tfrac{3}{32}x^4 + \tfrac{1}{30}x^5 + \cdots. \tag{12}$$

If in Example 3 we had sought a particular solution with given initial values $y(a)$ and $y'(a)$, we would have needed the general solution in the form

$$y = \sum_{n=0}^{\infty} c_n (x-a)^n; \tag{13}$$

that is, in powers of $x - a$ rather than in powers of x. For only with a solution of the form in (13) is it true that $y(a) = c_0$ and $y'(a) = c_1$ determine the arbitrary constants c_0 and c_1 in terms of the initial values of y and y'. Consequently, in order to solve an initial value problem, we need a series expansion of the general solution centered at the point where the initial conditions are specified.

EXAMPLE 4 Solve the initial value problem

$$(t^2 - 2t - 3)\frac{d^2 y}{dt^2} + 3(t - 1)\frac{dy}{dt} + y = 0; \tag{14}$$

$$y(1) = 4, \qquad y'(1) = 1.$$

Solution We need a general solution of the form $\sum c_n(t - 1)^n$. But instead of substituting this series in (14) to determine the coefficients, it simplifies the computations if we first make the substitution $x = t - 1$, so that we wind up looking for a series of the form $\sum c_n x^n$ after all. To transform Eq. (14) into one with the new independent variable x, we note that

$$t^2 - 2t - 3 = (x + 1)^2 - 2(x + 1) - 3 = x^2 - 4,$$

$$\frac{dy}{dt} = \frac{dy}{dx}\frac{dx}{dt} = \frac{dy}{dx} = y',$$

and

$$\frac{d^2y}{dt^2} = \left[\frac{d}{dx}\left(\frac{dy}{dt}\right)\right]\frac{dx}{dt} = \frac{d}{dx}(y') = y'',$$

where primes denote differentiation with respect to x. Hence we transform Eq. (14) into

$$(x^2 - 4)y'' + 3xy' + y = 0$$

with initial conditions $y = 4$ and $y' = 1$ at $x = 0$ (corresponding to $t = 1$). This is the initial value problem we solved in Example 3, so the particular solution in (12) is available. We substitute $t - 1$ for x in (12), and thereby obtain the desired particular solution

$$y(t) = 4 + (t - 1) + \tfrac{1}{2}(t - 1)^2 + \tfrac{1}{6}(t - 1)^3$$
$$+ \tfrac{3}{32}(t - 1)^4 + \tfrac{1}{30}(t - 1)^5 + \cdots.$$

This series converges if $-1 < t < 3$. (Why?) A series such as this can be used to estimate numerical values of the solution. For instance,

$$y(0.8) = 4 + (-0.2) + \frac{(-0.2)^2}{2} + \frac{(-0.2)^3}{6} + \frac{3(-0.2)^4}{32}$$

$$+ \frac{(-0.2)^5}{30} + \cdots$$

$$\approx 4 - 0.2 + 0.02 - 0.00133 + 0.00015 - 0.00001 + \cdots,$$

so that $y(0.8) \approx 3.8188$.

The computation above illustrates the fact that series solutions of differential equations are useful not only for establishing general properties of a solution, but also for numerical computations when an expression of the solution in terms of familiar elementary functions is not available.

The formula in (8) in Example 3 is an example of a **two-term** recurrence formula; it expresses each coefficient in the series in terms of *one* of the preceding coefficients. A **many-term** recurrence formula expresses each coefficient in the series in terms of two or more preceding coefficients. In the case of a many-term recurrence formula, it is generally inconvenient or even impossible to find a formula that gives the typical c_n in terms of n. The following example shows what we can do with a three-term recurrence formula.

EXAMPLE 5 Find two linearly independent solutions of

$$y'' - xy' - x^2y = 0. \tag{15}$$

Solution We make the usual substitution of the power series $y = \sum c_n x^n$. This results in the equation

$$\sum_{n=2}^{\infty} n(n-1)c_n x^{n-2} - \sum_{n=1}^{\infty} nc_n x^n - \sum_{n=0}^{\infty} c_n x^{n+2} = 0.$$

We can start the second term at $n = 0$ without changing anything else. In order to make each sum include x^n in its general term, we shift the index of summation in the first sum by $+2$ (replace n by $n + 2$), and we shift it by -2 in the third sum (replace n by $n - 2$). These shifts yield

$$\sum_{n=0}^{\infty} (n+2)(n+1)c_{n+2} x^n - \sum_{n=0}^{\infty} nc_n x^n - \sum_{n=2}^{\infty} c_{n-2} x^n = 0.$$

The common range of these three summations is $n \geq 2$, so we must separate the terms corresponding to $n = 0$ and $n = 1$ in the first two sums before collecting coefficients of x^n. This gives

$$2c_2 + 6c_3 x - c_1 x + \sum_{n=2}^{\infty} [(n+2)(n+1)c_{n+2} - nc_n - c_{n-2}]x^n = 0.$$

The identity principle now implies that $2c_2 = 0$, $6c_3 - c_1 = 0$, and

$$(n+2)(n+1)c_{n+2} - nc_n - c_{n-2} = 0$$

for $n \geq 2$. Hence we get $c_2 = 0$, $c_3 = \frac{1}{6}c_1$, and the three-term recursion formula

$$c_{n+2} = \frac{nc_n + c_{n-2}}{(n+2)(n+1)} \tag{16}$$

for $n \geq 2$. In particular,

$$c_4 = \frac{2c_2 + c_0}{12}, \qquad c_5 = \frac{3c_3 + c_1}{20},$$

$$c_6 = \frac{4c_4 + c_2}{30}, \qquad c_7 = \frac{5c_5 + c_3}{42}, \tag{17}$$

$$c_8 = \frac{6c_6 + c_4}{56}.$$

Thus all values of c_n for $n \geq 4$ are given in terms of the arbitrary constants c_0 and c_1 because $c_2 = 0$ and $c_3 = c_1/6$.

To get our first solution y_1 of Eq. (15), we choose $c_0 = 1$ and $c_1 = 0$, so that $c_2 = c_3 = 0$. Then the formulas in (17) yield

$$c_4 = \tfrac{1}{12}, \qquad c_5 = 0, \qquad c_6 = \tfrac{1}{90}, \qquad c_7 = 0, \qquad c_8 = \tfrac{3}{1120};$$

thus

$$y_1 = 1 + \tfrac{1}{12}x^4 + \tfrac{1}{90}x^6 + \tfrac{3}{1120}x^8 + \cdots. \tag{18}$$

Because $c_1 = c_3 = 0$, it is clear from (16) that this series contains only terms of even degree.

To obtain our second linearly independent solution y_2 of (15), we take $c_0 = 0$ and $c_1 = 1$, so that $c_2 = 0$ and $c_3 = \frac{1}{6}$. Then the formulas in (17) yield

$$c_4 = 0, \qquad c_5 = \tfrac{3}{40}, \qquad c_6 = 0, \qquad c_7 = \tfrac{13}{1008},$$

so that

$$y_2 = x + \tfrac{1}{6}x^3 + \tfrac{3}{40}x^5 + \tfrac{13}{1008}x^7 + \cdots. \tag{19}$$

Because $c_0 = c_2 = 0$, it is clear from (16) that this series contains only terms of odd degree. The solutions $y_1(x)$ and $y_2(x)$ are linearly independent because $y_1(0) = 1$ and $y_1'(0) = 0$, while $y_2(0) = 0$ and $y_2'(0) = 1$. The general solution of Eq. (15) is a linear combination of the power series in (18) and (19). Equation (15) has no singular points, so the power series representing y_1 and y_2 converge for all x.

*THE LEGENDRE EQUATION

The **Legendre equation** of order α is the second order linear differential equation

$$(1 - x^2)y'' - 2xy' + \alpha(\alpha + 1)y = 0, \tag{20}$$

where the real number α satisfies the inequality $\alpha > -1$. This differential equation has extensive applications, ranging from numerical integration formulas (such as Gaussian quadrature) to the problem of determining the steady state temperature within a solid spherical ball when the temperature at points of its boundary is known. The only singular points of the Legendre equation are at $+1$ and -1, so it has two linearly independent solutions that can be expressed as power series in powers of x with radius of convergence at least 1. The substitution $y = \sum c_m x^m$ in (20) leads (see Problem 29) to the recursion formula

$$c_{m+2} = -\frac{(\alpha - m)(\alpha + m + 1)}{(m + 1)(m + 2)} c_m \tag{21}$$

for $m \geqq 0$. We are using m as the index of summation because we have another role for n to play.

In terms of the arbitrary constants c_0 and c_1, the formula in (21) yields

$$c_2 = -\frac{\alpha(\alpha + 1)}{2!} c_0,$$

$$c_3 = -\frac{(\alpha - 1)(\alpha + 2)}{3!} c_1,$$

$$c_4 = \frac{\alpha(\alpha - 2)(\alpha + 1)(\alpha + 3)}{4!} c_0,$$

$$c_5 = \frac{(\alpha - 1)(\alpha - 3)(\alpha + 2)(\alpha + 4)}{5!} c_1.$$

One can show without much trouble that, for $m > 0$,

$$c_{2m} = (-1)^m \frac{\alpha(\alpha - 2)(\alpha - 4) \cdots (\alpha - 2m + 2)(\alpha + 1)(\alpha + 3) \cdots (\alpha + 2m - 1)}{(2m)!} c_0$$

$$\tag{22}$$

and

$$c_{2m+1} = (-1)^m \frac{(\alpha - 1)(\alpha - 3) \cdots (\alpha - 2m + 1)(\alpha + 2)(\alpha + 4) \cdots (\alpha + 2m)}{(2m + 1)!} c_1.$$

$$\tag{23}$$

Alternatively,

$$c_{2m} = (-1)^m a_{2m} c_0 \quad \text{and} \quad c_{2m+1} = (-1)^m a_{2m+1} c_1,$$

where a_{2m} and a_{2m+1} denote the fractions in (22) and (23), respectively. With this notation, we get two linearly independent power series solutions

$$y_1 = c_0 \sum_{m=0}^{\infty} (-1)^m a_{2m} x^{2m} \quad \text{and} \quad y_2 = c_1 \sum_{m=0}^{\infty} (-1)^m a_{2m+1} x^{2m+1} \qquad (24)$$

of Legendre's equation of order α.

Now suppose that $\alpha = n$, a nonnegative *integer*. If $\alpha = n$ is even, we see from (22) that $a_{2m} = 0$ when $2m > n$. In this case y_1 is a *polynomial* of degree n and y_2 is a (nonterminating) infinite series. If $\alpha = n$ is odd, we see from (23) that $a_{2m+1} = 0$ when $2m + 1 > n$. In this case, y_2 is a *polynomial* of degree n and y_1 is a nonterminating infinite series. Thus in either case, one of the two solutions in (24) is a polynomial of degree n and the other is a nonterminating series.

With an appropriate choice (made separately for each n) of the arbitrary constant c_0 (n even) or c_1 (n odd), the nth degree polynomial solution of Legendre's equation of order n,

$$(1 - x^2)y'' - 2xy' + n(n + 1)y = 0, \qquad (25)$$

is denoted by $P_n(x)$ and is called the **Legendre polynomial** of degree n. It is customary (for a reason indicated in Problem 30) to choose the arbitrary constant so that the coefficient of x^n in $P_n(x)$ is $(2n)!/[2^n(n!)^2]$. It then turns out that

$$P_n(x) = \sum_{k=0}^{N} \frac{(-1)^k (2n - 2k)!}{2^n k! (n - k)! (n - 2k)!} x^{n-2k} \qquad (26)$$

where $N = [\![n/2]\!]$, the integral part of $n/2$. The first six Legendre polynomials are

$$P_0(x) = 1, \qquad\qquad P_1(x) = x,$$
$$P_2(x) = \tfrac{1}{2}(3x^2 - 1), \qquad P_3(x) = \tfrac{1}{2}(5x^3 - 3x),$$
$$P_4(x) = \tfrac{1}{8}(35x^4 - 30x^2 + 3), \qquad P_5(x) = \tfrac{1}{8}(63x^5 - 70x^3 + 15x).$$

3.2 Problems

Find general solutions in powers of x of the differential equations in Problems 1–15. State the recurrence relation and the guaranteed radius of convergence in each case.

1. $(x^2 - 1)y'' + 4xy' + 2y = 0.$
2. $(x^2 + 2)y'' + 4xy' + 2y = 0.$
3. $y'' + xy' + y = 0.$
4. $(x^2 + 1)y'' + 6xy' + 4y = 0.$
5. $(x^2 - 3)y'' + 2xy' = 0.$
6. $(x^2 - 1)y'' - 6xy' + 12y = 0.$
7. $(x^2 + 3)y'' - 7xy' + 16y = 0.$
8. $(2 - x^2)y'' - xy' + 16y = 0.$
9. $(x^2 - 1)y'' + 8xy' + 12y = 0.$
10. $3y'' + xy' - 4y = 0.$
11. $5y'' - 2xy' + 10y = 0.$
12. $y'' - x^2y' - 3xy = 0.$
13. $y'' + x^2y' + 2xy = 0.$
14. $y'' + xy = 0$ (an Airy equation).
15. $y'' + x^2y = 0.$

Solve the initial value problems in Problems 16–20. First make a substitution of the form $t = x - a$, *and then find a solution* $\sum c_n t^n$ *of the transformed differential equation. State the interval of values of x for which the theorem in this section guarantees convergence.*

16. $y'' + (x - 1)y' + y = 0$; $y(1) = 2$, $y'(1) = 0$.

17. $(2x - x^2)y'' - 6(x - 1)y' - 4y = 0$; $y(1) = 0$, $y'(1) = 1$.

18. $(x^2 - 6x + 10)y'' - 4(x - 3)y' + 6y = 0$; $y(3) = 2$, $y'(3) = 0$.

19. $(4x^2 + 16x + 17)y'' = 8y$; $y(-2) = 1$, $y'(-2) = 0$.

20. $(x^2 + 6x)y'' + (3x + 9)y' - 3y = 0$; $y(-3) = 0$, $y'(-3) = 2$.

In each of Problems 21–24, find a three-term recursion formula for solutions of the form $y = \sum c_n x^n$. *Then find the first three nonzero terms in each of two linearly independent solutions.*

21. $y'' + (1 + x)y = 0$.

22. $(x^2 - 1)y'' + 2xy' + 2xy = 0$.

23. $y'' + x^2 y' + x^2 y = 0$.

24. $(1 + x^3)y'' + x^4 y = 0$.

25. Solve the initial value problem

$$y'' + xy' + (2x^2 + 1)y = 0; \quad y(0) = 1, \, y'(0) = -1..$$

Determine sufficiently many terms to compute $y(\tfrac{1}{2})$ accurate to four decimal places.

In Problems 26–28, find the first three nonzero terms in each of two linearly independent solutions of the form $y = \sum c_n x^n$. *Substitute known Taylor series for the analytic functions and retain enough terms to compute the necessary coefficients.*

26. $y'' + e^{-x}y = 0$.

27. $(\cos x)y'' + y = 0$.

28. $xy'' + (\sin x)y' + xy = 0$.

29. Derive the recursion formula in (21) for the Legendre equation.

30. Follow the steps outlined below to establish **Rodrigues' formula**

$$P_n(x) = \frac{1}{n!2^n} \frac{d^n}{dx^n}(x^2 - 1)^n$$

for the *n*th degree Legendre polynomial.

(a) Show that $v = (x^2 - 1)^n$ satisfies the differential equation ·

$$(1 - x^2)v' + 2nxv = 0.$$

Differentiate each side of this equation to obtain

$$(1 - x^2)v'' + 2(n - 1)xv' + 2nv = 0.$$

(b) Differentiate each side of the last equation *n* times in succession to obtain

$$(1 - x^2)v^{(n+2)} - 2xv^{(n+1)} + n(n + 1)v^{(n)} = 0.$$

Thus $u = v^{(n)} = D^n(1 - x^2)^n$ satisfies Legendre's equation of order *n*.

(c) Show that the coefficient of x^n in u is $(2n)!/n!$; then state why this proves Rodrigues' formula. (Note that the coefficient of x^n in $P_n(x)$ is $(2n)!/[2^n(n!)^2]$.)

31. The **Hermite equation** of order α is

$$y'' - 2xy' + 2\alpha y = 0.$$

(a) Derive the two power series solutions

$$y_1 = 1 + \sum_{n=1}^{\infty} (-1)^m \frac{2^m \alpha(\alpha - 2) \cdots (\alpha - 2m + 2)}{(2m)!} x^{2m}$$

and

$$y_2 = x + \sum_{n=1}^{\infty} (-1)^m \frac{2^m(\alpha - 1)(\alpha - 3) \cdots (\alpha - 2m + 1)}{(2m + 1)!} x^{2m+1}.$$

Show that y_1 is a polynomial if α is an even integer, while y_2 is a polynomial if α is an odd integer.

(b) The **Hermite polynomial** of degree n is denoted by $H_n(x)$. It is the nth degree polynomial solution of Hermite's equation, multiplied by a suitable constant so that the coefficient of x^n is 2^n. Show that the first six Hermite polynomials are

$$H_0(x) = 1, \qquad\qquad H_1(x) = 2x,$$

$$H_2(x) = 4x^2 - 2, \qquad H_3(x) = 8x^3 - 12x,$$

$$H_4(x) = 16x^4 - 48x^2 + 12, \qquad H_5(x) = 32x^5 - 160x^3 + 120x.$$

(c) The general formula for the Hermite polynomials is

$$H_n(x) = (-1)^n e^{x^2} \frac{d^n}{dx^n} (e^{-x^2}).$$

Verify that this does in fact give an nth degree polynomial.

3.3
Regular Singular Points

We now investigate the solution of the homogeneous second order linear equation

$$A(x)y'' + B(x)y' + C(x)y = 0 \qquad\qquad (1)$$

near a singular point. Recall that if the functions A, B, and C are polynomials having no common factors, then the singular points of (1) are simply those points where $A(x)$ vanishes. For instance, $x = 0$ is the only singular point of the Bessel equation of order n,

$$x^2 y'' + xy' + (x^2 - n^2)y = 0,$$

while the Legendre equation of order n,

$$(1 - x^2)y'' - 2xy' + n(n + 1)y = 0,$$

has the two singular points $x = -1$ and $x = 1$. It turns out that some of the features of the solutions of such equations of the most importance for applications are largely determined by their behavior near their singular points.

We will restrict our attention to the case in which $x = 0$ is a singular point of Eq. (1). A differential equation having $x = a$ as a singular point is easily transformed by the substitution $t = x - a$ into one having a corresponding singular point at 0. For example, let us substitute $t = x - 1$ in the Legendre equation above. Because

$$y' = \frac{dy}{dx} = \frac{dy}{dt}\frac{dt}{dx} = \frac{dy}{dt},$$

$$y'' = \frac{d^2 y}{dx^2} = \left[\frac{d}{dt}\left(\frac{dy}{dt}\right)\right]\frac{dt}{dx} = \frac{d^2 y}{dt^2},$$

and $1 - x^2 = 1 - (t + 1)^2 = -2t - t^2$, we get the equation

$$-t(t + 2)\frac{d^2y}{dt^2} - 2(t + 1)\frac{dy}{dt} + n(n + 1)y = 0.$$

This new equation has the singular point $t = 0$ corresponding to $x = 1$ in the original equation; it has also the singular point $t = -2$ corresponding to $x = -1$.

A differential equation having a singular point at 0 ordinarily will not have power series solutions of the form $y = \sum c_n x^n$, so the straightforward method of Section 3.2 fails in this case. To see the form that a solution of such an equation might take, we assume that Eq. (1) has analytic coefficient functions and rewrite it as

$$y'' + P(x)y' + Q(x)y = 0 \tag{2}$$

where $P = B/A$ and $Q = C/A$. Recall that $x = 0$ is an ordinary point (rather than a singular point) of the equation in (2) if the functions $P(x)$ and $Q(x)$ are analytic at $x = 0$; that is, if $P(x)$ and $Q(x)$ have convergent power series expansions in powers of x on some open interval containing $x = 0$. Now it can be proved that each of the functions $P(x)$ and $Q(x)$ *either* is analytic at 0 *or* approaches ∞ as $x \rightarrow 0$. Consequently $x = 0$ is a singular point of (2) provided that either $P(x)$ or $Q(x)$ (or both) approaches ∞ as $x \rightarrow 0$. For instance, if we rewrite the Bessel equation above in the form

$$y'' + \frac{1}{x}y' + \left(1 - \frac{n^2}{x^2}\right)y = 0,$$

we see that $P(x) = 1/x$ and $Q(x) = 1 - (n/x)^2$ each approach infinity as $x \rightarrow 0$.

Later we will see that the power series method can be generalized to apply near the singular point $x = 0$ of Eq. (2) provided that $P(x)$ approaches infinity no more rapidly than $1/x$, and $Q(x)$ no more rapidly than $1/x^2$, as $x \rightarrow 0$. This is a way of saying that $P(x)$ and $Q(x)$ have only *weak* singularities at $x = 0$. To state it more precisely, we rewrite Eq. (2) in the form

$$y'' + \frac{p(x)}{x}y' + \frac{q(x)}{x^2}y = 0, \tag{3}$$

where

$$p(x) = xP(x) \quad \text{and} \quad q(x) = x^2 Q(x). \tag{4}$$

DEFINITION: REGULAR SINGULAR POINT

The singular point $x = 0$ of Eq. (3) is a **regular singular point** *if the functions $p(x)$ and $q(x)$ are both analytic at $x = 0$. Otherwise it is an* **irregular singular point**.

In particular, the singular point $x = 0$ is a *regular* singular point if $p(x)$ and $q(x)$ are both polynomials. For instance, we see that $x = 0$ is a regular singular point of Bessel's equation of order n by writing that equation in the form

$$y'' + \frac{1}{x} y' + \frac{x^2 - n^2}{x^2} = 0,$$

noting that $p(x) = 1$ and $q(x) = x^2 - n^2$ are both polynomials in x.

By contrast, consider the equation $2x^3 y'' + (1 + x)y' + 3xy = 0$, which has the singular point $x = 0$. If we write this equation in the form of (3), we get

$$y'' + \frac{(1 + x)/(2x^2)}{x} y' + \frac{\frac{3}{2}}{x^2} y = 0.$$

Because $p(x) = 1/(2x^2) + 1/(2x)$ is not a polynomial (although $q(x) = \frac{3}{2}$ is), $x = 0$ is an irregular singular point. We will not discuss the solution of differential equations near irregular singular points; this is a considerably more advanced topic than the solution of differential equations near regular singular points.

It may happen that when we begin with a differential equation in the general form in (1) and rewrite it in the form in (3), the functions $p(x)$ and $q(x)$ as given in (4) are indeterminate forms at $x = 0$. In this case the situation is determined by the limits

$$p_0 = p(0) = \lim_{x \to 0} p(x) = \lim_{x \to 0} xP(x) \tag{5}$$

and

$$q_0 = q(0) = \lim_{x \to 0} q(x) = \lim_{x \to 0} x^2 Q(x). \tag{6}$$

If $p_0 = 0 = q_0$ then $x = 0$ may be an ordinary point. Otherwise, if the limits in both (5) and (6) exist and are *finite*, then $x = 0$ is a regular singular point. If either limit fails to exist or is infinite, then $x = 0$ is an irregular singular point.

Remark: The most common case in applications, for the differential equation written in the form

$$y'' + \frac{p(x)}{x} y' + \frac{q(x)}{x^2} y = 0, \tag{3}$$

is that the functions $p(x)$ and $q(x)$ are *polynomials*. In this case $p_0 = p(0)$ and $q_0 = q(0)$ are simply the constant terms of these polynomials, so the limits in (5) and (6) need not be evaluated.

EXAMPLE 1 To investigate the nature of the point $x = 0$ for the differential equation

$$x^4 y'' + (x^2 \sin x)y' + (1 - \cos x)y = 0,$$

we first rewrite it in the form in (3):

$$y'' + \frac{(\sin x)/x}{x} y' + \frac{(1 - \cos x)/x^2}{x^2} y = 0.$$

Then l'Hôpital's rule gives the values

$$p_0 = \lim_{x \to 0} \frac{\sin x}{x} = \lim_{x \to 0} \frac{\cos x}{1} = 1$$

and

$$q_0 = \lim_{x \to 0} \frac{1 - \cos x}{x^2} = \lim_{x \to 0} \frac{\sin x}{2x} = \frac{1}{2}$$

for the limits in (5) and (6). While not both are zero, we see that $x = 0$ is a regular singular point because each limit is finite. Alternatively, we could write

$$p(x) = \frac{\sin x}{x} = \frac{1}{x}\left(x - \frac{x^3}{3!} + \frac{x^5}{5!} - \cdots\right)$$

$$= 1 - \frac{x^2}{3!} + \frac{x^4}{5!} - \cdots$$

and

$$q(x) = \frac{1 - \cos x}{x^2} = \frac{1}{x^2}\left[1 - \left(1 - \frac{x^2}{2!} + \frac{x^4}{4!} - \frac{x^6}{6!} + \cdots\right)\right]$$

$$= \frac{1}{2!} - \frac{x^2}{4!} + \frac{x^4}{6!} - \cdots.$$

These (convergent) power series show explicitly that $p(x)$ and $q(x)$ are analytic, and moreover that $p_0 = p(0) = 1$ and $q_0 = q(0) = \frac{1}{2}$.

We now approach the task of actually finding solutions of a second order differential equation near the regular singular point $x = 0$. The simplest such equation is the Euler-Cauchy equation

$$x^2 y'' + p_0 x y' + q_0 y = 0, \tag{7}$$

which we initially rewrite in the form

$$y'' + \frac{p_0}{x} y' + \frac{q_0}{x^2} y = 0 \tag{7'}$$

with $p(x) = p_0$ and $q(x) = q_0$. In Section 2.6 we saw that $y = x^r$ is a solution of Eq. (7) if r is a root of the quadratic equation

$$r(r - 1) + p_0 r + q_0 = 0. \tag{8}$$

In the general case, in which $p(x)$ and $q(x)$ are power series rather than constants, it is a reasonable conjecture that our differential equation might have a solution of the form

$$y = x^r \sum_{n=0}^{\infty} c_n x^n = \sum_{n=0}^{\infty} c_n x^{n+r} \tag{9}$$

$$= c_0 x^r + c_1 x^{r+1} + c_2 x^{r+2} + \cdots$$

—the product of x^r and a power series. This turns out to be a very fruitful conjecture; according to the theorem stated below (following Example 2), every equation of the form in (1) having $x = 0$ as a regular singular point does, indeed, have at least one such solution. This fact is the basis for the *method of Frobenius*, named for the German mathematician Georg Frobenius (1848–1917), who discovered this method in the 1870s.

An infinite series of the form in (9) is called a **Frobenius series**. Note that a Frobenius series is generally *not* a power series. For instance, with $r = -\frac{1}{2}$

the series in (9) takes the form

$$y = c_0 x^{-1/2} + c_1 x^{1/2} + c_2 x^{3/2} + \cdots;$$

it is not a series of *integral* powers of x.

To investigate the possible existence of Frobenius series solutions, we begin with the equation

$$x^2 y'' + xp(x)y' + q(x)y = 0 \tag{10}$$

obtained by multiplying the equation in (3) by x^2. If $x = 0$ is a regular singular point, then $p(x)$ and $q(x)$ are analytic at $x = 0$, so

$$\left. \begin{aligned} p(x) &= p_0 + p_1 x + p_2 x^2 + \cdots, \\ q(x) &= q_0 + q_1 x + q_2 x^2 + \cdots. \end{aligned} \right\} \tag{11}$$

Suppose that Eq. (10) has the Frobenius series solution

$$y = \sum_{n=0}^{\infty} c_n x^{n+r}. \tag{12}$$

We may (and always do) assume that $c_0 \neq 0$ because the series must have a first nonzero term. Termwise differentiation in (12) leads to

$$y' = \sum_{n=0}^{\infty} c_n (n + r) x^{n+r-1} \tag{13}$$

and

$$y'' = \sum_{n=0}^{\infty} c_n (n + r)(n + r - 1) x^{n+r-2}. \tag{14}$$

Substitution of the series in (11)–(14) in (10) now yields

$$\begin{aligned} &[r(r-1)c_0 x^r + (r+1)rc_1 x^{r+1} + \cdots] \\ &+ [p_0 x + p_1 x^2 + \cdots][rc_0 x^{r-1} + (r+1)c_1 x^r + \cdots] \\ &+ [q_0 + q_1 x + \cdots][c_0 x^r + c_1 x^{r+1} + \cdots] = 0. \end{aligned} \tag{15}$$

The lowest power of x that appears in (15) is x^r. If (15) is to be satisfied identically, the coefficient $r(r-1)c_0 + p_0 r c_0 + q_0 c_0$ of x^r must vanish. Because $c_0 \neq 0$, it follows that r must satisfy the quadratic equation

$$r(r-1) + p_0 r + q_0 = 0 \tag{16}$$

of precisely the same form as that obtained with the Euler-Cauchy equation. Equation (16) is called the **indicial equation** of the differential equation in (10), and its two roots (possibly equal) are the **exponents** of the differential equation (at the regular singular point $x = 0$).

Our derivation of (16) shows that *if* the Frobenius series $y = x^r \sum c_n x^n$ is to be a solution of the differential equation in (10), *then* the exponent r must be one of the roots r_1 and r_2 of the indicial equation in (16). If $r_1 \neq r_2$ it follows that there are two possible Frobenius series solutions, while if $r_1 = r_2$ there is only one possible Frobenius series solution; the second solution cannot be a Frobenius series. The exponents r_1 and r_2 in the possible Frobenius series solutions are determined (using the indicial equation) by the values $p_0 = p(0)$ and $q_0 = q(0)$ that we have discussed. In practice, particularly when the coefficients in the differential equation in the original form in (1) are polynomials, the simplest way of finding p_0 and q_0 is often to write the equation in the form

$$y'' + \frac{p_0 + p_1 x + \cdots}{x} y' + \frac{q_0 + q_1 x + \cdots}{x^2} y = 0. \tag{17}$$

Then inspection of the series that appear in the two numerators reveals the constants p_0 and q_0.

EXAMPLE 2 Find the exponents in the possible Frobenius series solutions of the equation

$$2x^2(1 + x)y'' + 3x(1 + x)^3 y' - (1 - x^2)y = 0.$$

Solution We divide each term by $2x^2(1 + x)$ to recast the differential equation in the form

$$y'' + \frac{(\frac{3}{2})(1 + 2x + x^2)}{x} y' - \frac{(\frac{1}{2})(1 - x)}{x^2} y = 0,$$

and thus see that $p_0 = \frac{3}{2}$ and $q_0 = -\frac{1}{2}$. Hence the indicial equation is

$$r(r - 1) + \tfrac{3}{2} r - \tfrac{1}{2} = r^2 + \tfrac{1}{2} r - \tfrac{1}{2} = (r + 1)(r - \tfrac{1}{2}) = 0,$$

with roots $r_1 = \frac{1}{2}$ and $r_2 = -1$. The two possible Frobenius series solutions are then of the forms

$$y_1 = x^{1/2} \sum_{n=0}^{\infty} a_n x^n \quad \text{and} \quad y_2 = x^{-1} \sum_{n=0}^{\infty} b_n x^n.$$

Once the coefficients r_1 and r_2 are known, the coefficients in a Frobenius series solution are determined by substitution of the series in (12)–(14) in the differential equation, essentially the same method as was used to determine coefficients in power series solutions in Section 3.2. If the exponents r_1 and r_2 are complex conjugates, then there always exist two independent Frobenius series solutions. We will restrict our attention here to the case in which r_1 and r_2 are both real. We will also seek only solutions for $x > 0$. Once such a solution has been found, we need only replace x^{r_i} by $|x|^{r_i}$ to obtain a solution for $x < 0$. The following theorem is proved in Chapter 4 of Coddington's *An Introduction to Ordinary Differential Equations*.

THEOREM: FROBENIUS SERIES SOLUTIONS

Suppose that $x = 0$ is a regular singular point of the equation

$$x^2 y'' + xp(x)y' + q(x)y = 0. \tag{10}$$

Let $\rho > 0$ denote the minimum of the radii of convergence of the power series

$$p(x) = \sum_{n=0}^{\infty} p_n x^n \quad \text{and} \quad q(x) = \sum_{n=0}^{\infty} q_n x^n.$$

Let r_1 and r_2 be the roots, with $r_1 \geqq r_2$, of the indicial equation $r(r - 1) + p_0 r + q_0 = 0$. Then:

(a) For $x > 0$, there exists a solution of the form

$$y_1 = x^{r_1} \sum_{n=0}^{\infty} a_n x^n \quad (a_0 \neq 0) \tag{18}$$

of Eq. (10) corresponding to the larger root r_1.

(b) If $r_1 - r_2$ is neither zero nor a positive integer, then there exists a second linearly independent solution for $x > 0$ of the form

$$y_2 = x^{r_2} \sum_{n=0}^{\infty} b_n x^n \qquad (b_0 \neq 0) \tag{19}$$

corresponding to the smaller root r_2.

The radii of convergence of the power series in (18) and (19) are each at least ρ. The coefficients in these series can be determined by substituting the series in the differential equation

$$x^2 y'' + x p(x) y' + q(x) y = 0. \tag{10}$$

We have already seen that if $r_1 = r_2$, then there can exist only one Frobenius series solution. It turns out that, if $r_1 - r_2$ is a positive integer, there may or may not exist a second Frobenius series solution of the form in (19) corresponding to the smaller root r_2. These exceptional cases are discussed in Section 3.4. The examples below illustrate the process of determining the coefficients in those Frobenius series solutions that are guaranteed by the theorem.

EXAMPLE 3 Find the Frobenius series solutions of

$$2x^2 y'' + 3x y' - (x^2 + 1) y = 0. \tag{20}$$

Solution First we divide each term by $2x^2$ to put the equation in the form in (17):

$$y'' + \frac{\frac{3}{2}}{x} y' + \frac{-(\frac{1}{2}) - (\frac{1}{2})x^2}{x^2} y = 0. \tag{21}$$

We now see that $x = 0$ is a regular singular point, and that $p_0 = \frac{3}{2}$ and $q_0 = -\frac{1}{2}$. Because $p(x) = \frac{3}{2}$ and $q(x) = -(\frac{1}{2}) - (\frac{1}{2})x^2$ are polynomials, the Frobenius series we obtain will converge for all $x > 0$. The indicial equation is

$$r(r - 1) + \tfrac{3}{2} r - \tfrac{1}{2} = (r - \tfrac{1}{2})(r + 1) = 0,$$

so the exponents are $r_1 = \frac{1}{2}$ and $r_2 = -1$. They do not differ by an integer, so the theorem guarantees the existence of two linearly independent Frobenius series solutions. Rather than separately substituting $y_1 = x^{1/2} \sum a_n x^n$ and $y_2 = x^{-1} \sum b_n x^n$ in the differential equation in (20), it is more efficient to begin by substituting $y = x^r \sum c_n x^n$. We will then get a recursion formula that depends on r. With the value $r_1 = \frac{1}{2}$ it becomes a recursion formula for the series for y_1, while with $r_2 = -1$ it becomes a recursion formula for the series for y_2.

When we substitute

$$y = \sum_{n=0}^{\infty} c_n x^{n+r}, \qquad y' = \sum_{n=0}^{\infty} (n + r) c_n x^{n+r-1},$$

and

$$y'' = \sum_{n=0}^{\infty} (n + r)(n + r - 1) c_n x^{n+r-2}$$

in (20)—the original differential equation, rather than (21)—we get

$$2 \sum_{n=0}^{\infty} (n + r)(n + r - 1)c_n x^{n+r} + 3 \sum_{n=0}^{\infty} (n + r)c_n x^{n+r}$$

$$- \sum_{n=0}^{\infty} c_n x^{n+r+2} - \sum_{n=0}^{\infty} c_n x^{n+r} = 0. \qquad (22)$$

At this stage there are several ways to proceed. A good standard practice is to shift indices so that each exponent will be the same as the smallest one present. In the case above, we shift the index of summation in the third sum by -2 to bring its exponent down from $n + r + 2$ to $n + r$. This gives

$$2 \sum_{n=0}^{\infty} (n + r)(n + r - 1)c_n x^{n+r} + 3 \sum_{n=0}^{\infty} (n + r)c_n x^{n+r}$$

$$- \sum_{n=2}^{\infty} c_{n-2} x^{n+r} - \sum_{n=0}^{\infty} c_n x^{n+r} = 0. \qquad (23)$$

The common range of summation is $n \geq 2$, so we must treat $n = 0$ and $n = 1$ separately. Following our standard practice, the terms corresponding to $n = 0$ will always give the indicial equation

$$[2r(r - 1) + 3r - 1]c_0 = 2(r^2 + \tfrac{1}{2}r - \tfrac{1}{2})c_0 = 0.$$

The terms corresponding to $n = 1$ yield

$$[2(r + 1)r + 3(r + 1) - 1]c_1 = (2r^2 + 5r + 2)c_1 = 0.$$

Because the coefficient $2r^2 + 5r + 2$ of c_1 is nonzero whether $r = \tfrac{1}{2}$ or $r = -1$, it follows that

$$c_1 = 0 \qquad (24)$$

in either case.

The coefficient of x^{n+r} in (23) is

$$2(n + r)(n + r - 1)c_n + 3(n + r)c_n - c_{n-2} - c_n = 0.$$

We solve for c_n and simplify to obtain the recursion formula

$$c_n = \frac{c_{n-2}}{2(n + r)^2 + (n + r) - 1} \quad \text{for } n \geq 2. \qquad (25)$$

The case $r_1 = \tfrac{1}{2}$: We now write a_n in place of c_n and substitute $r = \tfrac{1}{2}$ in (25). This gives the recursion formula

$$a_n = \frac{a_{n-2}}{2n^2 + 3n} \quad \text{for } n \geq 2. \qquad (26)$$

With this formula we can determine the coefficients in the first Frobenius solution y_1. In view of (24) we see that $a_n = 0$ whenever n is odd. With $n = 2, 4,$ and 6 in (26), we get

$$a_2 = \frac{a_0}{14}, \qquad a_4 = \frac{a_2}{44} = \frac{a_0}{616}, \quad \text{and} \quad a_6 = \frac{a_4}{90} = \frac{a_0}{55,440}.$$

Hence our first Frobenius series solution is

$$y_1 = a_0 x^{1/2} \left(1 + \frac{x^2}{14} + \frac{x^4}{616} + \frac{x^6}{55,440} + \cdots \right).$$

The case $r_2 = -1$: We may now write b_n in place of c_n and sub-stitute $r = -1$ in (25). This gives the recursion formula

$$b_n = \frac{b_{n-2}}{2n^2 - 3n} \quad \text{for} \quad n \geq 2. \tag{27}$$

Again (24) implies that $b_n = 0$ for n odd. With $n = 2, 4,$ and 6 in (27), we get

$$b_2 = \frac{b_0}{2}, \qquad b_4 = \frac{b_2}{20} = \frac{b_0}{40}, \quad \text{and} \quad b_6 = \frac{b_4}{54} = \frac{b_0}{2160}.$$

Hence our second Frobenius series solution is

$$y_2 = b_0 x^{-1}\left(1 + \frac{1}{2}x^2 + \frac{x^4}{40} + \frac{x^6}{2160} + \cdots\right).$$

EXAMPLE 4 Find a Frobenius series solution of Bessel's equation of order zero:

$$x^2 y'' + xy' + x^2 y = 0. \tag{28}$$

Solution In the form of (17) the equation becomes

$$y'' + \frac{1}{x}y' + \frac{x^2}{x^2}y = 0.$$

Hence $x = 0$ is a regular singular point with $p(x) = 1$ and $q(x) = x^2$, so our series will converge for all $x > 0$. Because $p_0 = 1$ and $q_0 = 0$, the indicial equation is $r(r - 1) + r = r^2 = 0$. Thus we obtain only the single exponent $r = 0$, and so there is only one Frobenius series solution

$$y = x^0 \sum_{n=0}^{\infty} c_n x^n = \sum_{n=0}^{\infty} c_n x^n$$

of (28); it is in fact a power series.

As usual we substitute $y = \sum c_n x^n$ in (28); the result is

$$\sum_{n=0}^{\infty} n(n-1)c_n x^n + \sum_{n=0}^{\infty} nc_n x^n + \sum_{n=0}^{\infty} c_n x^{n+2} = 0.$$

We combine the first two sums and shift the index of summation in the third by -2 to obtain

$$\sum_{n=0}^{\infty} n^2 c_n x^n + \sum_{n=2}^{\infty} c_{n-2} x^n = 0.$$

The term corresponding to x^0 gives $0 = 0$: no information. The term corresponding to x^1 gives $c_1 = 0$, and the term for x^n yields the recursion formula

$$c_n = -\frac{c_{n-2}}{n^2} \quad \text{for } n \geq 2. \tag{29}$$

Because $c_1 = 0$, we see that $c_n = 0$ whenever n is odd. Substituting $n = 2, 4,$ and 6 in (29), we get

$$c_2 = -\frac{c_0}{2^2}, \qquad c_4 = -\frac{c_2}{4^2} = \frac{c_0}{2^2 4^2}, \quad \text{and} \quad c_6 = -\frac{c_4}{6^2} = -\frac{c_0}{2^2 4^2 6^2}.$$

Evidently the pattern is

$$c_{2n} = \frac{(-1)^n c_0}{2^2 \cdot 4^2 \cdots (2n)^2} = \frac{(-1)^n c_0}{2^{2n}(n!)^2}.$$

The choice $c_0 = 1$ gives us one of the most important special functions in mathematics, the **Bessel function of order zero of the first kind**, denoted by $J_0(x)$. Thus

$$J_0(x) = \sum_{n=0}^{\infty} \frac{(-1)^n x^{2n}}{2^{2n}(n!)^2} = 1 - \frac{x^2}{4} + \frac{x^4}{64} - \frac{x^6}{2304} + \cdots. \tag{30}$$

In this example we have not been able to find a second linearly independent solution of Bessel's equation of order zero. We will derive that solution in Section 3.4; it will not be a Frobenius series.

Recall that, if $r_1 - r_2$ is a positive integer, the theorem of this section guarantees only the existence of the Frobenius series solution corresponding to the larger exponent r_1. The following example illustrates the fortunate case in which the series method nevertheless yields a second Frobenius series solution. The case in which the second solution is not a Frobenius series will be discussed in Section 3.4.

EXAMPLE 5   Find the Frobenius series solutions of

$$xy'' + 2y' + xy = 0. \tag{31}$$

Solution   In standard form the equation becomes

$$y'' + \frac{2}{x} y' + \frac{x^2}{x^2} y = 0,$$

so we see that $x = 0$ is a regular singular point with $p_0 = 2$ and $q_0 = 0$. The indicial equation $r(r - 1) + 2r = r(r + 1) = 0$ has roots $r_1 = 0$ and $r_2 = -1$, which differ by an integer. In this case when $r_1 - r_2$ is an integer, it is best to depart from the standard procedure of Example 3, and begin our work with the *smaller* exponent. As you will see below, the recursion formula will then tell us whether or not a second Frobenius series solution exists. If it does exist, then our computations will simultaneously yield *both* Frobenius series solutions. If the second solution does not exist, we begin anew with the larger exponent $r = r_1$ to obtain the one Frobenius series solution guaranteed by the theorem of this section.

Hence we begin by substituting

$$y = x^{-1} \sum_{n=0}^{\infty} c_n x^n = \sum_{n=0}^{\infty} c_n x^{n-1}$$

in (31). This gives

$$\sum_{n=0}^{\infty} (n-1)(n-2)c_n x^{n-2} + 2\sum_{n=0}^{\infty} (n-1)c_n x^{n-2} + \sum_{n=0}^{\infty} c_n x^n = 0.$$

We combine the first two sums and shift the index by -2 in the third to obtain

$$\sum_{n=0}^{\infty} n(n-1)c_n x^{n-2} + \sum_{n=2}^{\infty} c_{n-2} x^{n-2} = 0. \tag{32}$$

The cases $n = 0$ and $n = 1$ reduce to

$$0 \cdot c_0 = 0 \quad \text{and} \quad 0 \cdot c_1 = 0.$$

Hence we have *two* arbitrary constants c_0 and c_1, and therefore can expect to find a general solution incorporating two linearly independent Frobenius series solutions. If, for $n = 1$, we had obtained an equation such as $0 \cdot c_1 = 3$, which can be satisfied for *no* choice of c_1, this would have told us that no second Frobenius series solution could exist.

Now knowing that all is well, from (32) we read the recursion formula

$$c_n = -\frac{c_{n-2}}{n(n-1)} \quad \text{for } n \geq 2. \tag{33}$$

The first few values of n give

$$c_2 = -\frac{1}{2 \cdot 1} c_0, \qquad\qquad c_3 = -\frac{1}{3 \cdot 2} c_1,$$

$$c_4 = -\frac{1}{4 \cdot 3} c_2 = \frac{c_0}{4!}, \qquad c_5 = -\frac{1}{5 \cdot 4} c_3 = \frac{c_1}{5!},$$

$$c_6 = -\frac{1}{6 \cdot 5} c_4 = -\frac{c_0}{6!}, \qquad c_7 = -\frac{1}{7 \cdot 6} c_5 = -\frac{c_1}{7!};$$

evidently the pattern is

$$c_{2n} = \frac{(-1)^n c_0}{(2n)!}, \qquad c_{2n+1} = \frac{(-1)^n c_1}{(2n+1)!}$$

for $n \geq 1$. Therefore our solution is

$$\begin{aligned}
y &= x^{-1} \sum_{n=0}^{\infty} c_n x^n \\
&= \frac{c_0}{x}\left(1 - \frac{x^2}{2!} + \frac{x^4}{4!} - \cdots\right) + \frac{c_1}{x}\left(x - \frac{x^3}{3!} + \frac{x^5}{5!} - \cdots\right) \\
&= \frac{c_0}{x} \sum_{n=0}^{\infty} \frac{(-1)^n x^{2n}}{(2n)!} + \frac{c_1}{x} \sum_{n=0}^{\infty} \frac{(-1)^n x^{2n+1}}{(2n+1)!};
\end{aligned}$$

thus

$$y = \frac{1}{x}(c_0 \cos x + c_1 \sin x).$$

We have thus found a general solution expressed as the sum of two Frobenius series solutions $(c_0/x) \cos x$ and $(c_1/x) \sin x$.

3.3 Problems

In each of Problems 1–8, determine whether $x = 0$ is an ordinary point, a regular singular point, or an irregular singular point. If it is a regular singular point, find the exponents of the differential equation at $x = 0$.

1. $xy'' + (x - x^3)y' + (\sin x)y = 0.$ **2.** $xy'' + x^2 y' + (e^x - 1)y = 0.$

3. $x^2 y'' + (\cos x)y' + xy = 0.$ **4.** $3x^3 y'' + 2x^2 y' + (1 - x^2)y = 0.$

5. $x(1 + x)y'' + 2y' + 3xy = 0.$

6. $x^2(1 - x^2)y'' + 2xy' - 2y = 0.$

7. $x^2y'' + (6 \sin x)y' + 6y = 0.$

8. $(6x^2 + 2x^3)y'' + 21xy' + 9(x^2 - 1)y = 0.$

If $x = a \neq 0$ is a singular point of a second order linear differential equation, the substitution $t = x - a$ transforms it into a differential equation having $t = 0$ as a singular point. We then attribute to the original equation at $x = a$ the behavior of the new equation at $t = 0$. Classify (as regular or irregular) the singular points of the differential equations in Problems 9–16.

9. $(1 - x)y'' + xy' + x^2y = 0.$

10. $(1 - x)^2y'' + (2x - 2)y' + y = 0.$

11. $(1 - x^2)y'' - 2xy' + 12y = 0.$

12. $(x - 2)^3y'' + 3(x - 2)^2y' + x^3y = 0.$

13. $(x^2 - 4)y'' + (x - 2)y' + (x + 2)y = 0.$

14. $(x^2 - 9)^2y'' + (x^2 + 9)y' + (x^2 + 4)y = 0.$

15. $(x - 2)^2y'' - (x^2 - 4)y' + (x + 2)y = 0.$

16. $x^3(1 - x)y'' + (3x + 2)y' + xy = 0.$

Find two linearly independent Frobenius series solutions (for $x > 0$) of each of the differential equations in Problems 17–26.

17. $4xy'' + 2y' + y = 0.$ **18.** $2xy'' + 3y' - y = 0.$

19. $2xy'' - y' - y = 0.$ **20.** $3xy'' + 2y' + 2y = 0.$

21. $2x^2y'' + xy' - (1 + 2x^2)y = 0.$ **22.** $2x^2y'' + xy' - (3 - 2x^2)y = 0.$

23. $6x^2y'' + 7xy' - (x^2 + 2)y = 0.$ **24.** $3x^2y'' + 2xy' + x^2y = 0.$

25. $2xy'' + (1 + x)y' + y = 0.$ **26.** $2xy'' + (1 - 2x^2)y' - 4xy = 0.$

Use the method of Example 5 to find two linearly independent Frobenius series solutions of each of the differential equations in Problems 27–31.

27. $xy'' + 2y' + 9xy = 0.$ **28.** $xy'' + 2y' - 4xy = 0.$

29. $4xy'' + 8y' + xy = 0.$ **30.** $xy'' - y' + 4x^2y = 0.$

31. $4x^2y'' - 4xy' + (3 - 4x^2)y = 0.$

In each of Problems 32–34, find the first three nonzero terms of each of two linearly independent Frobenius series solutions.

32. $2x^2y'' + x(x + 1)y' - (2x + 1)y = 0.$

33. $(2x^2 + 5x^3)y'' + (3x - x^2)y' - (1 + x)y = 0.$

34. $2x^2y'' + (\sin x)y' - (\cos x)y = 0.$

35. Note that $x = 0$ is an irregular point of the equation

$$x^2y'' + (3x - 1)y' + y = 0.$$

(a) Show that $y = x^r \sum c_n x^n$ can satisfy this equation only if $r = 0$. (b) Substitute $y = \sum c_n x^n$ to derive the "formal" solution $y = \sum n! x^n$. What is the radius of convergence of this series?

36. (a) Suppose that A and B are nonzero constants. Show that the equation $x^2 y'' + Ay' + By = 0$ has at most one solution of the form $y = x^r \sum c_n x^n$. (b) Repeat part (a) with the equation $x^3 y'' + Axy' + By = 0$. (c) Show that the equation $x^3 y'' + Ax^2 y' + By = 0$ has no Frobenius series solution. (*Suggestion:* In each case substitute $y = x^r \sum c_n x^n$ in the given equation to determine the possible values of r.)

37. (a) Use the method of Frobenius to derive the solution $y_1 = x$ of the equation $x^3 y'' - xy' + y = 0$.
(b) Derive by reduction of order the second solution $y_2 = xe^{-1/x}$. Does y_2 have a Frobenius series representation?

38. Apply the method of Frobenius to Bessel's equation of order $\frac{1}{2}$,

$$x^2 y'' + xy' + (x^2 - \tfrac{1}{4})y = 0,$$

to derive its general solution for $x > 0$,

$$y = c_0 \frac{\cos x}{\sqrt{x}} + c_1 \frac{\sin x}{\sqrt{x}}.$$

39. (a) Show that Bessel's equation of order 1,

$$x^2 y'' + xy' + (x^2 - 1)y = 0,$$

has exponents $r_1 = 1$ and $r_2 = -1$ at $x = 0$, and that the Frobenius series solution corresponding to $r_1 = 1$ is

$$J_1(x) = \frac{x}{2} \sum_{n=0}^{\infty} \frac{(-1)^n x^{2n}}{n!(n+1)!2^{2n}}.$$

(b) Show that there is no Frobenius solution corresponding to the smaller exponent $r_2 = -1$; that is, show that it is impossible to determine the coefficients in $y_2 = x^{-1} \sum c_n x^n$.

40. Consider the equation $x^2 y'' + xy' + (1 - x)y = 0$. (a) Show that its exponents are $\pm i$, so it has complex-valued Frobenius series solutions $y_+ = x^i \sum p_n x^n$ and $y_- = x^{-i} \sum q_n x^n$ with $p_0 = q_0 = 1$. (b) Show that the recursion formula is

$$c_n = \frac{c_{n-1}}{n^2 + 2rn}.$$

Apply it with $r = i$ to obtain $p_n = c_n$, $r = -i$ to obtain $q_n = c_n$. Conclude that p_n and q_n are complex conjugates: $p_n = a_n + ib_n$ and $q_n = a_n - ib_n$, where the numbers $\{a_n\}$ and $\{b_n\}$ are real. (c) Deduce from (b) that the differential equation of this problem has real-valued solutions of the form

$$y_1 = A(x) \cos (\ln x) - B(x) \sin (\ln x),$$
$$y_2 = A(x) \sin (\ln x) + B(x) \cos (\ln x)$$

where $A(x) = \sum a_n x^n$ and $B(x) = \sum b_n x^n$.

---

*3.4

Method of Frobenius —The Exceptional Cases

We continue our discussion of the equation

$$y'' + \frac{p(x)}{x}y' + \frac{q(x)}{x^2}y = 0 \tag{1}$$

where $p(x)$ and $q(x)$ are analytic at $x = 0$, and $x = 0$ is a regular singular point. If the roots r_1 and r_2 of the indicial equation

$$\phi(r) = r(r-1) + p_0 r + q_0 = 0 \tag{2}$$

do not differ by an integer, then the theorem of Section 3.3 guarantees that Eq. (1) has two linearly independent Frobenius series solutions. We consider now the more complex situation in which $r_1 - r_2$ is an integer. If $r_1 = r_2$, then there is only one exponent available, and thus there can be only one Frobenius series solution. But we saw in Example 5 of Section 3.3 that if $r_1 = r_2 + N$, with N a positive integer, it is possible that a second Frobenius series solution exists. We will also see that it is possible that such a solution does not exist. In fact, the second solution involves $\ln x$ when it is not a Frobenius series. As you will see in Examples 3 and 4, these exceptional cases occur in the solution of Bessel's equation, which for applications is the most important second order linear differential equation with variable coefficients.

THE NONLOGARITHMIC CASE WITH $r_1 = r_2 + N$

In Section 3.3 we derived the indicial equation by substituting the power series $p(x) = \sum p_n x^n$ and $q(x) = \sum q_n x^n$ and the Frobenius series

$$y = x^r \sum_{n=0}^{\infty} c_n x^n = \sum_{n=0}^{\infty} c_n x^{n+r} \qquad (c_0 \neq 0) \tag{3}$$

in the differential equation in the form

$$x^2 y'' + xp(x)y' + q(x)y = 0. \tag{4}$$

The result of this substitution, after collection of the coefficients of like powers of x, is an equation of the form

$$\sum_{n=0}^{\infty} F_n(r)x^{n+r} = 0 \tag{5}$$

with the coefficients depending on r. It turns out that the coefficient of x^r is

$$F_0(r) = [r(r-1) + p_0 r + q_0]c_0 = \phi(r)c_0, \tag{6}$$

which gives the indicial equation because $c_0 \neq 0$ by assumption; also, for $n \geq 1$, the coefficient of x^{n+r} is of the form

$$F_n(r) = \phi(r+n)c_n + L_n(r; c_0, c_1, \ldots, c_{n-1}). \tag{7}$$

Here L_n is a certain linear combination of $c_0, c_1, \ldots, c_{n-1}$. Although the exact formula is not necessary for our purposes, it happens that

$$L_n = \sum_{k=0}^{n-1} [(r+k)p_{n-k} + q_{n-k}]c_k. \tag{8}$$

Because all the coefficients in (5) must vanish for our Frobenius series to be a solution of (4), it follows that the exponent r and the coefficients $c_0, c_1, \ldots, c_n$ must satisfy the equation

$$\phi(r+n)c_n + L_n(r; c_0, c_1, \ldots, c_{n-1}) = 0. \tag{9}$$

This is a *recursion formula* for c_n in terms of $c_0, c_1, \ldots, c_{n-1}$.

Suppose now that $r_1 = r_2 + N$ with N a *positive* integer. If we use the larger exponent r_1 in (9), then the coefficient $\phi(r_1 + n)$ of c_n will be nonzero for every $n \geq 1$ because $\phi(r) = 0$ only when $r = r_1$ and when $r = r_2 < r_1$. Once

$c_0, c_1, \ldots, c_{n-1}$ have been determined, we therefore can solve Eq. (9) for c_n and continue to compute successive coefficients in the Frobenius series solution corresponding to the exponent r_1.

But when we use the smaller exponent r_2, there is a potential difficulty in computing c_N. For in this case $\phi(r_2 + N) = 0$, so Eq. (9) becomes

$$0 \cdot c_N + L_N(r_2; c_0, c_1, \ldots, c_{N-1}) = 0. \tag{10}$$

At this stage $c_0, c_1, \ldots, c_{N-1}$ have already been determined. If it happens that $L_N(r_2; c_0, c_1, \ldots, c_{N-1}) = 0$, then we can choose c_N arbitrarily and continue to determine the remaining coefficients in a second Frobenius series solution. But if it happens that $L_N(r_2; c_0, c_1, \ldots, c_{N-1}) \neq 0$, then Eq. (10) is not satisfied with any choice of c_N; in this case there cannot exist a Frobenius series solution corresponding to the smaller exponent r_2. Examples 1 and 2 illustrate these two possibilities.

EXAMPLE 1   Consider the equation

$$x^2 y'' + (6x + x^2)y' + xy = 0. \tag{11}$$

Here $p_0 = 6$ and $q_0 = 0$, so the indicial equation is

$$\phi(r) = r(r-1) + 6r = r^2 + 5r = 0 \tag{12}$$

with roots $r_1 = 0$ and $r_2 = -5$; the roots differ by the integer $N = 5$. We substitute the Frobenius series $y = \sum c_n x^{n+r}$ and get

$$\sum_{n=0}^{\infty} (n+r)(n+r-1)c_n x^{n+r} + 6 \sum_{n=0}^{\infty} (n+r)c_n x^{n+r}$$

$$+ \sum_{n=0}^{\infty} (n+r)c_n x^{n+r+1} + \sum_{n=0}^{\infty} c_n x^{n+r+1} = 0.$$

When we combine the first two and also the last two sums, and in the latter shift the index by -1, the result is

$$\sum_{n=0}^{\infty} [(n+r)^2 + 5(n+r)]c_n x^{n+r} + \sum_{n=1}^{\infty} (n+r)c_{n-1} x^{n+r} = 0.$$

The terms corresponding to $n = 0$ give the indicial equation in (12), while for $n \geq 1$ we get the equation

$$[(n+r)^2 + 5(n+r)]c_n + (n+r)c_{n-1} = 0, \tag{13}$$

which in this example corresponds to the general equation in (9). Note that the coefficient of c_n is $\phi(n+r)$.

We now follow the recommendations in Section 3.3 for the case $r_1 = r_2 + N$; we begin with the smaller root $r_2 = -5$. With $r = -5$, Eq. (13) reduces to

$$n(n-5)c_n + (n-5)c_{n-1} = 0. \tag{14}$$

If $n \neq 5$, we can solve this equation for c_n to get the recursion formula

$$c_n = -\frac{c_{n-1}}{n} \quad \text{for } n \neq 5. \tag{15}$$

This yields

$$c_1 = -c_0, \qquad c_2 = -\frac{c_1}{2} = \frac{c_0}{2},$$
$$c_3 = -\frac{c_2}{3} = \frac{c_0}{6}, \quad \text{and} \quad c_4 = -\frac{c_3}{4} = \frac{c_0}{24}. \qquad \Bigg\} \qquad (16)$$

In the case $r_1 = r_2 + N$, it is always the coefficient c_N that requires special consideration. Here $N = 5$, and for $n = 5$ Eq. (14) takes the form $0 \cdot c_5 + 0 = 0$. Hence c_5 is a second arbitrary constant, and we can compute additional coefficients, using the recursion formula in (15):

$$c_6 = -\frac{c_5}{6}, \quad c_7 = -\frac{c_6}{7} = \frac{c_5}{6 \cdot 7}, \quad c_8 = -\frac{c_7}{8} = -\frac{c_5}{6 \cdot 7 \cdot 8}, \qquad (17)$$

and so on.

When we combine the results in (16) and (17), we get

$$y = x^{-5} \sum_{n=0}^{\infty} c_n x^n$$

$$= c_0 x^{-5} \left(1 - x + \frac{x^2}{2} - \frac{x^3}{6} + \frac{x^4}{24} \right)$$

$$+ c_5 x^{-5} \left(x^5 - \frac{x^6}{6} + \frac{x^7}{6 \cdot 7} - \frac{x^8}{6 \cdot 7 \cdot 8} + \cdots \right)$$

in terms of the two arbitrary constants c_0 and c_5. Thus we have found the two Frobenius series solutions

$$y_1 = x^{-5} \left(1 - x + \frac{x^2}{2} - \frac{x^3}{6} + \frac{x^4}{24} \right)$$

and

$$y_2 = 1 + \sum_{n=1}^{\infty} \frac{(-1)^n x^n}{6 \cdot 7 \cdots (n+5)} = 1 + \sum_{n=1}^{\infty} \frac{(-1)^n 120 x^n}{(n+5)!}$$

of Eq. (11).

EXAMPLE 2 Determine whether or not the equation

$$x^2 y'' - xy' + (x^2 - 8)y = 0 \qquad (18)$$

has two linearly independent Frobenius series solutions.

Solution Here $p_0 = -1$ and $q_0 = -8$, so the indicial equation is

$$\phi(r) = r(r-1) - r - 8 = r^2 - 2r - 8 = 0$$

with roots $r_1 = 4$ and $r_2 = -2$ differing by $N = 6$. On substitution of $y = \sum c_n x^{n+r}$ in (18), we get

$$\sum_{n=0}^{\infty} (n+r)(n+r-1)c_n x^{n+r} - \sum_{n=0}^{\infty} (n+r)c_n x^{n+r}$$

$$+ \sum_{n=0}^{\infty} c_n x^{n+r+2} - 8 \sum_{n=0}^{\infty} c_n x^{n+r} = 0.$$

If we shift the index by -2 in the third sum and combine the other three sums, we get

$$\sum_{n=0}^{\infty} [(n + r)^2 - 2(n + r) - 8]c_n x^{n+r} + \sum_{n=2}^{\infty} c_{n-2} x^{n+r} = 0.$$

The coefficient of x^r gives the indicial equation, and the coefficient of x^{r+1} gives $[(r + 1)^2 - 2(r + 1) - 8]c_1 = 0$. Because the coefficient of c_1 is nonzero both for $r = 4$ and for $r = -2$, it follows that $c_1 = 0$ in each case. For $n \geq 2$ we get the equation

$$[(n + r)^2 - 2(n + r) - 8]c_n + c_{n-2} = 0, \tag{19}$$

which corresponds in this example to the general equation in (9); note that the coefficient of c_n is $\phi(n + r)$.

We work first with the smaller root $r = r_2 = -2$. Then Eq. (19) becomes

$$n(n - 6)c_n + c_{n-2} = 0 \tag{20}$$

for $n \geq 2$. For $n \neq 6$ we can solve for the recursion formula

$$c_n = -\frac{c_{n-2}}{n(n - 6)} \qquad (n \geq 2, \ n \neq 6) \tag{21}$$

Because $c_1 = 0$, this formula gives

$$c_2 = \frac{c_0}{8}, \qquad\qquad c_3 = 0,$$

$$c_4 = \frac{c_2}{8} = \frac{c_0}{64}, \quad \text{and} \quad c_5 = 0.$$

Now Eq. (20) with $n = 6$ reduces to

$$0 \cdot c_6 + \frac{c_0}{64} = 0.$$

But $c_0 \neq 0$ by assumption, and hence there is no way to choose c_6 so that this equation holds. Thus there is *no* Frobenius series solution corresponding to the smaller root $r_2 = -2$.

To find the single Frobenius series solution corresponding to the larger root $r_1 = 4$, we substitute $r = 4$ in (19) to obtain the recursion formula

$$c_n = -\frac{c_{n-2}}{n(n + 6)} \qquad (n \geq 2). \tag{22}$$

This gives

$$c_2 = -\frac{c_0}{2 \cdot 8}, \qquad c_4 = -\frac{c_2}{4 \cdot 10} = \frac{c_0}{2 \cdot 4 \cdot 8 \cdot 10}.$$

The general pattern is

$$c_{2n} = \frac{(-1)^n c_0}{2 \cdot 4 \cdots (2n) \cdot 8 \cdot 10 \cdots (2n + 6)} = \frac{(-1)^n 48 c_0}{2^{2n+3} n! (n + 3)!}.$$

This yields the Frobenius series solution

$$y_1 = x^4 \left(1 + \sum_{n=1}^{\infty} \frac{(-1)^n 48 x^{2n}}{2^{2n+3} n! (n + 3)!} \right)$$

of Eq. (18).

THE LOGARITHMIC CASES

We now investigate the general form of the second solution of the equation

$$y'' + \frac{p(x)}{x}y' + \frac{q(x)}{x^2}y = 0, \tag{1}$$

under the assumption that its exponents r_1 and $r_2 = r_1 - N$ differ by the integer $N \geq 0$. We assume that we have already found the Frobenius series solution

$$y_1 = x^{r_1} \sum_{n=0}^{\infty} a_n x^n \qquad (a_0 \neq 0) \tag{23}$$

for $x > 0$ corresponding to the larger exponent, r_1. Let us write $P(x)$ for $p(x)/x$ and $Q(x)$ for $q(x)/x^2$. Recall from Section 2.6 (reduction of order) that a second solution y_2 is given by

$$y_2 = y_1 \int \frac{e^{-\int P(x)\,dx}}{y_1^2}\,dx \tag{24}$$

on any interval on which y_1 is nonzero.

Because the indicial equation has roots r_1 and $r_2 = r_1 - N$, it can be factored easily:

$$r^2 + (p_0 - 1)r + q_0 = (r - r_1)(r - r_1 + N)$$
$$= r^2 + (N - 2r_1)r + (r_1^2 - r_1 N) = 0,$$

so we see that

$$p_0 - 1 = N - 2r_1;$$

that is,

$$-p_0 - 2r_1 = -1 - N. \tag{25}$$

In preparation for use of the reduction of order formula in (24), we write

$$P(x) = \frac{p_0 + p_1 x + p_2 x^2 + \cdots}{x} = \frac{p_0}{x} + p_1 + p_2 x + \cdots.$$

Then

$$e^{-\int P(x)\,dx} = \exp\left(-\int \left(\frac{p_0}{x} + p_1 + p_2 x + \cdots\right) dx\right)$$

$$= \exp\left(-p_0 \ln x + p_1 x + \frac{1}{2}p_2 x^2 + \cdots\right)$$

$$= x^{-p_0}\exp\left(p_1 x + \frac{1}{2}p_1 x^2 + \cdots\right),$$

so that

$$e^{-\int P(x)\,dx} = x^{-p_0}(1 + A_1 x + A_2 x^2 + \cdots). \tag{26}$$

In the last step we have used the fact that a composition of analytic functions is analytic and therefore has a power series representation; the initial coefficient of that series in (26) is 1 because $e^0 = 1$.

We now substitute (23) and (26) in (24); with the choice $a_0 = 1$ in (23), this yields

$$y_2 = y_1 \int \frac{x^{-p_0}(1 + A_1 x + A_2 x^2 + \cdots)}{x^{2r_1}(1 + a_1 x + a_2 x^2 + \cdots)^2}\,dx.$$

We square the denominator and simplify:

$$y_2 = x_1 \int \frac{x^{-p_0 - 2r_1}(1 + A_1 x + A_2 x^2 + \cdots)}{(1 + B_1 x + B_2 x^2 + \cdots)}\, dx$$

$$= y_1 \int x^{-1-N}(1 + C_1 x + C_2 x^2 + \cdots)\, dx \tag{27}$$

(we used long division and (25) in the last step). We now consider separately the cases $N = 0$ and $N > 0$. We want to ascertain the general form of y_2 without keeping track of specific coefficients.

Equal Exponents: $r_1 = r_2$

With $N = 0$, (27) gives

$$y_2 = y_1 \int \left(\frac{1}{x} + C_1 + C_2 x + \cdots\right) dx$$

$$= y_1 \ln x + y_1\left(C_1 x + \frac{1}{2}C_2 x^2 + \cdots\right)$$

$$= y_1 \ln x + x^{r_1}(1 + a_1 x + \cdots)\left(C_1 x + \frac{1}{2}C_2 x^2 + \cdots\right)$$

$$= y_1 \ln x + x^{r_1}(b_0 x + b_1 x^2 + b_2 x^3 + \cdots)$$

Consequently, in the case of equal exponents, the general form of y_2 is

$$y_2 = y_1 \ln x + x^{r_1+1} \sum_{n=0}^{\infty} b_n x^n. \tag{28}$$

Note the logarithmic term; it is always present when $r_1 = r_2$.

Positive Integral Difference: $r_1 = r_2 + N$

With $N > 0$, (27) gives

$$y_2 = y_1 \int x^{-1-N}(1 + C_1 x + C_2 x^2 + \cdots + C_N x^N + \cdots)\, dx$$

$$= y_1 \int \left(\frac{C_N}{x} + \frac{1}{x^{N+1}} + \frac{C_1}{x^N} + \cdots\right) dx$$

$$= C_N y_1 \ln x + y_1\left(\frac{x^{-N}}{-N} + \frac{C_1 x^{-N+1}}{-N+1} + \cdots\right)$$

$$= C_N y_1 \ln x + x^{r_2+N}\left(\sum_{n=0}^{\infty} a_n x^n\right)x^{-N}\left(-\frac{1}{N} + \frac{C_1 x}{-N+1} + \cdots\right),$$

so that

$$y_2 = C_N y_1 \ln x + x^{r_2} \sum_{n=0}^{\infty} b_n x^n \tag{29}$$

where $b_0 = -a_0/N \neq 0$. This gives the general form of y_2 in the case of exponents differing by a positive integer. Note the coefficient C_N that appears in (29) but not in (28). If it happens that $C_N = 0$, then there is no logarithmic term; if so, Eq. (1) has a second Frobenius series solution (as in Example 1).

In our derivation of the formulas in (28) and (29)—which exhibit the general form of the second solution in the cases $r_1 = r_2$ and $r_1 - r_2 = N$, respectively—we have said nothing about the radii of convergence of the various power series that appear. The following theorem is a summation of the discussion above and also tells where the series in (28) and (29) converge. As in the theorem of Section 3.3, we restrict our attention to solutions for $x > 0$. Once a solution for $x > 0$ has been found, we need only replace x^{r_i} by $|x|^{r_i}$ and $\ln x$ by $\ln|x|$ to obtain a solution for $x < 0$.

THEOREM: THE EXCEPTIONAL CASES

Suppose that $x = 0$ is a regular singular point of the equation

$$x^2 y'' + xp(x)y' + q(x)y = 0. \tag{4}$$

Let $\rho > 0$ denote the minimum of the radii of convergence of the power series

$$p(x) = \sum_{n=0}^{\infty} p_n x^n \quad \text{and} \quad q(x) = \sum_{n=0}^{\infty} q_n x^n.$$

Let r_1 and r_2 be the roots, with $r_1 \geqq r_2$, of the indicial equation

$$r(r - 1) + p_0 r + q_0 = 0.$$

(a) If $r_1 = r_2$, then (4) has two solutions y_1 and y_2 of the forms

$$y_1 = x^{r_1} \sum_{n=0}^{\infty} a_n x^n \qquad (a_0 \neq 0) \tag{30a}$$

and

$$y_2 = y_1 \ln x + x^{r_2+1} \sum_{n=0}^{\infty} b_n x^n. \tag{30b}$$

(b) If $r_1 - r_2 = N$, a positive integer, then (4) has two solutions y_1 and y_2 of the forms

$$y_1 = x^{r_1} \sum_{n=0}^{\infty} a_n x^n \qquad (a_0 \neq 0) \tag{31a}$$

and

$$y_2 = Cy_1 \ln x + x^{r_2} \sum_{n=0}^{\infty} b_n x^n. \tag{31b}$$

In (31b), $b_0 \neq 0$ but C may be either zero or nonzero, so the logarithmic term may or may not actually be present in this case. The radii of convergence of the power series shown above are all at least ρ. The coefficients in these series [and the constant C in (31b)] may be determined by direct substitution of the series in the differential equation in (4).

EXAMPLE 3 We will illustrate the case $r_1 = r_2$ of the theorem by deriving the second solution of Bessel's equation of order zero,

$$x^2 y'' + xy' + x^2 y = 0, \tag{32}$$

for which $r_1 = r_2 = 0$. In Example 4 of Section 3.3 we found the first solution

$$y_1 = J_0(x) = \sum_{n=0}^{\infty} \frac{(-1)^n x^{2n}}{2^{2n}(n!)^2}. \tag{33}$$

According to (30b) the second solution will have the form

$$y_2 = y_1 \ln x + \sum_{n=1}^{\infty} b_n x^n. \tag{34}$$

The first two derivatives of y_2 are

$$y_2' = y_1' \ln x + \frac{y_1}{x} + \sum_{n=1}^{\infty} nb_n x^{n-1}$$

and

$$y_2'' = y_1'' \ln x + \frac{2y_1'}{x} - \frac{y_1}{x^2} + \sum_{n=2}^{\infty} n(n-1)b_n x^{n-2}.$$

We substitute these in Eq. (32) and use the fact that $J_0(x)$ also satisfies this equation to obtain

$$0 = x^2 y_2'' + x y_2' + x^2 y_2$$
$$= [x^2 y_1'' + x y_1' + x^2 y_1] \ln x + 2x y_1'$$
$$+ \sum_{n=2}^{\infty} n(n-1)b_n x^n + \sum_{n=1}^{\infty} nb_n x^n + \sum_{n=1}^{\infty} b_n x^{n+2},$$

and it follows that

$$0 = 2 \sum_{n=1}^{\infty} \frac{(-1)^n 2n x^{2n}}{2^{2n}(n!)^2} + b_1 x$$

$$+ 2^2 b_2 x^2 + \sum_{n=3}^{\infty} (n^2 b_n + b_{n-2}) x^n. \tag{35}$$

The only term involving x in (35) is $b_1 x$, so $b_1 = 0$. But $n^2 b_n + b_{n-2} = 0$ if n is odd, so it follows that all the coefficients of odd subscript in y_2 vanish.

Now we examine the coefficients with even subscript in (35). First we see that

$$b_2 = -2 \frac{(-1)(2)}{2^2 2^2 (1!)^2} = \frac{1}{4}. \tag{36}$$

For $n \geq 2$, we read the recursion formula

$$(2n)^2 b_{2n} + b_{2n-2} = -\frac{(2)(-1)^n(2n)}{2^{2n}(n!)^2} \tag{37}$$

from (35). Nonhomogeneous recursion formulas such as (37) are typical of the exceptional cases of the method of Frobenius, and their solution often requires a bit of ingenuity. The usual strategy depends on detecting the most conspicuous dependence of b_{2n} on n. We note the presence of $2^{2n}(n!)^2$ on the right-hand side in (37); in conjunction with the coefficient $(2n)^2$ on the left-hand side, we are induced to think of b_{2n} as *something* divided by $2^{2n}(n!)^2$. Noting also the alternation of sign, we make the substitution

$$b_{2n} = \frac{(-1)^{n+1}c_{2n}}{2^{2n}(n!)^2},\tag{38}$$

in the expectation that the recursion formula for c_{2n} will be simpler than the one for b_{2n}. We chose $(-1)^{n+1}$ rather than $(-1)^n$ because $b_2 = \frac{1}{4} > 0$; with $n = 1$ in (38), we get $c_2 = 1$. Substitution of (38) in (37) gives

$$(2n)^2\frac{(-1)^{n+1}c_{2n}}{2^{2n}(n!)^2} + \frac{(-1)^n c_{2n-2}}{2^{2n-2}[(n-1)!]^2} = \frac{(-2)(-2)^n(2n)}{2^{2n}(n!)^2},$$

which boils down to the extremely simple recursion formula

$$c_{2n} = c_{2n-2} + \frac{1}{n}.$$

Thus

$$c_4 = c_2 + \frac{1}{2} = 1 + \frac{1}{2},$$

$$c_6 = c_4 + \frac{1}{3} = 1 + \frac{1}{2} + \frac{1}{3},$$

$$c_8 = c_6 + \frac{1}{4} = 1 + \frac{1}{2} + \frac{1}{3} + \frac{1}{4},$$

and so on. Evidently

$$c_{2n} = 1 + \frac{1}{2} + \frac{1}{3} + \cdots + \frac{1}{n} = H_n,\tag{39}$$

where by H_n we denote the nth partial sum of the harmonic series $\sum(1/n)$.

Finally, keeping in mind that the coefficients of odd subscript are all zero, we substitute (38) and (39) in (34) to obtain the second solution

$$y_2 = J_0(x)\ln x + \sum_{n=1}^{\infty}\frac{(-1)^{n+1}H_n x^{2n}}{2^{2n}(n!)^2}\tag{40}$$

$$= J_0(x)\ln x + \frac{x^2}{4} - \frac{3x^4}{128} + \frac{11x^6}{13,824} - \cdots$$

of Bessel's equation of order zero. The power series in (40) converges for all x. The most commonly used linearly independent [of $J_0(x)$] second solution is

$$Y_0(x) = \frac{2}{\pi}(\gamma - \ln 2)y_1 + \frac{2}{\pi}y_2;$$

that is,

$$Y_0(x) = \frac{2}{\pi}\left[\left(\gamma + \ln\frac{x}{2}\right)J_0(x) + \sum_{n=1}^{\infty}\frac{(-1)^{n+1}H_n x^{2n}}{2^{2n}(n!)^2}\right],\tag{41}$$

where γ denotes Euler's constant:

$$\gamma = \lim_{n\to\infty}(H_n - \ln n) \approx 0.57722.\tag{42}$$

This particular combination $Y_0(x)$ is chosen because of its nice behavior as $x \to +\infty$; it is called the **Bessel function of order zero of the second kind.**

EXAMPLE 4 As an alternative to the method of substitution, we illustrate the case $r_1 - r_2 = N$ by employing the technique of reduction of order to derive a second solution of Bessel's equation of order 1,

$$x^2 y'' + xy' + (x^2 - 1)y = 0; \tag{43}$$

the associated indicial equation has roots $r_1 = 1$ and $r_2 = -1$. According to Problem 39 in Section 3.3, one solution of (43) is

$$y_1 = J_1(x) = \frac{x}{2} \sum_{n=0}^{\infty} \frac{(-1)^n x^{2n}}{2^n n! (n+1)!} \tag{44}$$

$$= \frac{x}{2} - \frac{x^3}{16} + \frac{x^5}{384} - \frac{x^7}{18,432} + \cdots.$$

With $P(x) = 1/x$ from (43), the reduction of order formula in (24) yields

$$y_2 = y_1 \int \frac{dx}{x y_1^2}$$

$$= y_1 \int \frac{dx}{x(x/2 - x^3/16 + x^5/384 - x^7/18,432 + \cdots)^2}$$

$$= y_1 \int \frac{4\,dx}{x^3(1 - x^2/8 + x^4/192 - x^6/9216 + \cdots)^2}$$

$$= 4y_1 \int \frac{dx}{x^3(1 - x^2/4 + 5x^4/192 - 7x^6/4608 + \cdots)}$$

$$= 4y_1 \int \frac{1}{x^3}\left(1 + \frac{x^2}{4} + \frac{7x^4}{192} + \frac{19x^6}{4608} + \cdots\right) dx$$

$$= 4y_1 \int \left(\frac{1}{4x} + \frac{1}{x^3} + \frac{7x}{192} + \frac{19x^3}{4608} + \cdots\right) dx$$

$$= y_1 \ln x + 4y_1\left(-\frac{1}{2x^2} + \frac{7x^2}{384} + \frac{19x^4}{18,432} + \cdots\right).$$

Thus

$$y_2 = y_1 \ln x - \frac{1}{x} + \frac{x}{8} + \frac{x^3}{32} - \frac{11x^5}{4608} + \cdots. \tag{45}$$

Note that the technique of reduction of order readily yields the first several terms of the series, but it does not provide a recursion formula that can be used to determine the general term of the series.

With a computation similar to that shown in Example 3—but more complicated (see Problem 21)—the method of substitution can be used to derive the solution

$$y_3 = y_1 \ln x - \frac{1}{x} + \sum_{n=1}^{\infty} \frac{(-1)^n (H_n + H_{n-1}) x^{2n-1}}{2^{2n} n! (n-1)!}, \tag{46}$$

where H_n is defined in (39) for $n \geq 1$; $H_0 = 0$. The reader can verify that the terms shown in (45) agree with

$$y_2 = \frac{3}{4} J_1(x) + y_3. \tag{47}$$

The most commonly used linearly independent (of J_1) solution of Bessel's equation of order 1 is the combination

$$Y_1(x) = \frac{2}{\pi}(\gamma - \ln 2)y_1 + \frac{2}{\pi}y_3$$

$$= \frac{2}{\pi}\left[\left(\gamma + \ln\frac{x}{2}\right)J_1(x) - \frac{1}{x} + \sum_{n=1}^{\infty}\frac{(-1)^n(H_n + H_{n-1})x^{2n-1}}{2^{2n}n!(n-1)!}\right].$$

Examples 3 and 4 illustrate two methods of finding the solution in the logarithmic cases—direct substitution and reduction of order. A third alternative is outlined in Problem 19.

3.4 Problems

In each of Problems 1–8, apply the method of Example 1 to find two linearly independent series solutions.

1. $xy'' + (3 - x)y' - y = 0.$
2. $xy'' + (5 - x)y' - y = 0.$
3. $xy'' + (5 + 3x)y' + 3y = 0.$
4. $5xy'' + (30 + 3x)y' + 3y = 0.$
5. $xy'' - (4 + x)y' + 3y = 0.$
6. $2xy'' - (6 + 2x)y' + y = 0.$
7. $x^2y'' + (2x + 3x^2)y' - 2y = 0.$
8. $x(1 - x)y'' - 3y' + 2y = 0.$

In each of Problems 9–14, first find the first four nonzero terms in a Frobenius series solution of the given differential equation. Then use the reduction of order technique (as in Example 4) to find the logarithmic term and the first three nonzero terms in a second linearly independent series solution.

9. $xy'' + y' - xy = 0.$
10. $x^2y'' - xy' + (x^2 + 1)y = 0.$
11. $x^2y'' + (x^2 - 3x)y' + 4y = 0.$
12. $x^2y'' + x^2y' - 2y = 0.$
13. $x^2y'' + (2x^2 - 3x)y' + 3y = 0.$
14. $x^2y'' + x(1 + x)y' - 4y = 0.$

15. Begin with

$$J_0(x) = 1 - \frac{x^2}{4} + \frac{x^4}{64} - \frac{x^6}{2304} + \cdots ;$$

by reduction of order, derive the second solution

$$y_2 = J_0(x)\ln x + \frac{x^2}{4} - \frac{3x^4}{128} + \frac{11x^6}{13{,}824} - \cdots$$

of Bessel's equation of order zero.

16. Find two linearly independent Frobenius series solutions of Bessel's equation of order $\frac{3}{2}$,

$$x^2y'' + xy' + (x^2 - \tfrac{9}{4})y = 0.$$

17. (a) Verify that $y_1 = xe^x$ is one solution of

$$x^2y'' - x(1 + x)y' + y = 0.$$

(b) Note that $r_1 = r_2 = 1$. Substitute

$$y_2 = y_1 \ln x + \sum_{n=1}^{\infty} b_n x^{n+1}$$

in the differential equation to deduce that $b_1 = -1$ and that

$$nb_n - b_{n-1} = -\frac{1}{n!} \quad \text{for } n \geq 2.$$

(c) Substitute $b_n = +c_n/n!$ in this recursion formula, and conclude from the result that $c_n = -H_n$. Thus the second solution is

$$y_2 = xe^x \ln x - \sum_{n=1}^{\infty} \frac{H_n x^{n+1}}{n!}.$$

18. Consider the equation $xy'' - y = 0$, which has exponents $r_1 = 1$ and $r_2 = 0$ at $x = 0$. (a) Derive the Frobenius series solution

$$y_1 = \sum_{n=1}^{\infty} \frac{x^n}{n!\,(n-1)!}.$$

(b) Substitute

$$y_2 = Cy_1 \ln x + \sum_{n=0}^{\infty} b_n x^n$$

in the equation $xy'' - y = 0$ to derive the recursion formula

$$n(n+1)b_{n+1} - b_n = -\frac{2n+1}{(n+1)!\,n!}C.$$

Conclude from this result that a second solution is

$$y_2 = y_1 \ln x + 1 - \sum_{n=1}^{\infty} \frac{H_n + H_{n-1}}{n!\,(n-1)!}x^n.$$

19. Suppose that the differential equation

$$L[y] = x^2 y'' + xp(x)y' + q(x)y = 0 \tag{49}$$

has equal exponents $r_1 = r_2$ at the regular singular point $x = 0$, so that its indicial equation is

$$\phi(r) = (r - r_1)^2 = 0.$$

Let $c_0 = 1$ and define $c_n(r)$ for $n \geq 1$ by using Eq. (9); that is,

$$c_n(r) = -\frac{L_n(r; c_0, c_1, \ldots, c_{n-1})}{\phi(r+n)}. \tag{50}$$

Then define the function $y(x, r)$ of x and r to be

$$y(x, r) = \sum_{n=0}^{\infty} c_n(r)x^{n+r}. \tag{51}$$

(a) Deduce from the discussion preceding Eq. (9) that

$$L[y(x, r)] = x^r(r - r_1)^2. \tag{52}$$

Hence deduce that

$$y_1 = y(x, r_1) = \sum_{n=0}^{\infty} c_n(r_1)x^{n+r_1} \tag{53}$$

is one solution of Eq. (49). (b) Differentiate Eq. (52) with respect to r to show that

$$L[y_r(x, r_1)] = \frac{\partial}{\partial r}[x^r(r - r_1)^2]\Big|_{r=r_1} = 0.$$

Deduce that $y_2 = y_r(x, r_1)$ is a second solution of Eq. (49). (c) Differentiate Eq. (53) with respect to r to show that

$$y_2 = y_1 \ln x + x^{r_1} \sum_{n=1}^{\infty} c_n'(r_1)x^n. \tag{54}$$

20. Use the method of Problem 19 to derive both the solutions in (33) and (40) of Bessel's equation of order zero. The following steps outline this computation.
(a) Take $c_0 = 1$; show that Eq. (50) reduces in this case to

$$(r + 1)^2 c_1(r) = 0$$

and

$$c_n(r) = -\frac{c_{n-2}(r)}{(n + r)^2} \quad \text{for } n \geq 2. \Bigg\} \tag{55}$$

(b) Next show that $c_1(0) = c_1'(0) = 0$, and then deduce from (55) that $c_n(0) = c_n'(0) = 0$ for n odd. Hence you need to compute $c_n(0)$ and $c_n'(0)$ only for n even.
(c) Deduce from (55) that

$$c_{2n}(r) = \frac{(-1)^n}{(r + 2)^2(r + 4)^2 \cdots (r + 2n)^2}. \tag{56}$$

With $r = r_1 = 0$ in (53), this gives $J_0(x)$.
(d) Differentiate (56) to show that

$$c_{2n}'(0) = \frac{(-1)^{n+1} H_n}{2^{2n}(n!)^2}.$$

Substitution of this result in (54) gives the second solution in (40).

21. Derive the logarithmic solution in (46) of Bessel's equation of order 1 by the method of substitution. The following steps outline this computation.
(a) Substitute

$$y_2 = C J_1(x) \ln x + x^{-1}\left(1 + \sum_{n=1}^{\infty} b_n x^n\right)$$

in Bessel's equation to obtain

$$-b_1 + x + \sum_{n=2}^{\infty} [(n^2 - 1)b_{n+1} + b_{n-1}]x^n$$

$$+ C\left[x + \sum_{n=1}^{\infty} \frac{(-1)^n(2n + 1)x^{2n+1}}{2^{2n}(n + 1)! \, n!}\right] = 0. \tag{57}$$

(b) Deduce from Eq. (57) that $C = -1$ and that $b_n = 0$ for n odd.
(c) Next deduce the recursion formula

$$[(2n + 1)^2 - 1]b_{2n+2} + b_{2n} = \frac{(-1)^n(2n + 1)}{2^{2n}(n + 1)! \, n!} \tag{58}$$

for $n \geq 1$. Note that if b_2 is chosen arbitrarily, then b_{2n} is determined for all $n > 1$.
(d) Take $b_2 = \frac{1}{4}$ and substitute

$$b_{2n} = \frac{(-1)^{n+1} c_{2n}}{2^{2n} n! \, (n - 1)!}$$

in (58) to obtain

$$c_{2n+2} - c_{2n} = \frac{1}{n + 1} + \frac{1}{n}.$$

(e) Note that $c_2 = 1 = H_1 + H_0$ and deduce that $c_{2n} = H_n + H_{n-1}$.

3.5

Bessel's Equation   We have already seen several cases of Bessel's equation of order $p \geq 0$,

$$x^2 y'' + xy' + (x^2 - p^2)y = 0. \tag{1}$$

Its solutions are now called Bessel functions of order p. Such functions first appeared in the 1730s in the work of Daniel Bernoulli and Euler on the oscillations of a vertically suspended chain. The equation itself appears in a 1764 article by Euler on the vibrations of a circular drumhead, and Fourier used Bessel functions in his classical treatise on heat (1822). But their general properties were first studied systematically in an 1824 memoir by F. W. Bessel, who was investigating the motion of planets. The standard source of information on Bessel functions is G. N. Watson's *A Treatise on the Theory of Bessel Functions* (2nd ed., Cambridge: Cambridge University Press, 1944). Its 36 pages of references, which cover only the period up to 1922, give some idea of the vast literature of this subject.

Bessel's equation in (1) has indicial equation $r^2 - p^2 = 0$, with roots $r = \pm p$. If we substitute $y = \sum c_m x^{m+r}$ in (1), we find in the usual manner that $c_1 = 0$ and that

$$[(m + r)^2 - p^2]c_m + c_{m-2} = 0 \tag{2}$$

for $m \geq 2$. The verification of (2) is left to the reader (see Problem 6).

The Case $r = p > 0$

If we use $r = p$ and write a_m in place of c_m, then (2) yields the recursion formula

$$a_m = \frac{-a_{m-2}}{m(2p + m)}. \tag{3}$$

Because $a_1 = 0$, it follows that $a_m = 0$ for all odd m. The first few even coefficients are

$$a_2 = \frac{-a_0}{2(2p + 2)} = -\frac{a_0}{2^2(p + 1)},$$

$$a_4 = \frac{-a_2}{4(2p + 4)} = \frac{a_0}{2^4 \cdot 2(p + 1)(p + 2)},$$

$$a_6 = \frac{-a_4}{6(2p + 6)} = -\frac{a_0}{2^6 \cdot 2 \cdot 3(p + 1)(p + 2)(p + 3)}.$$

The general pattern is

$$a_{2m} = \frac{(-1)^m a_0}{2^{2m} m!(p + 1)(p + 2) \cdots (p + m)},$$

so with the larger root $r = p$ we get the solution

$$y_1 = a_0 \sum_{m=1}^{\infty} \frac{(-1)^m x^{2m+p}}{2^{2m} m!(p + 1)(p + 2) \cdots (p + m)}. \tag{4}$$

If $p = 0$ this is the only Frobenius series solution; with $a_0 = 1$ it is the function $J_0(x)$ we have seen before.

The Case $r = -p < 0$

If we use $r = -p$ and write b_m in place of c_m, Eq. (2) takes the form

$$m(m - 2p)b_m + b_{m-2} = 0 \qquad (5)$$

for $m \geq 2$, while $b_1 = 0$. We see that there is a potential difficulty if it happens that $2p$ is a positive integer—that is, if p is either a positive integer or an odd integral multiple of $\frac{1}{2}$. For then when $m = 2p$, Eq. (5) is simply $0 \cdot b_m + b_{m-2} = 0$. Thus if $b_{m-2} \neq 0$, no value of b_m will satisfy this equation.

But if p is an odd integral multiple of $\frac{1}{2}$, we can circumvent this difficulty. For suppose that $p = k/2$ where k is an odd positive integer. Then we need only choose $b_m = 0$ for all odd values of m. The crucial step is the kth step,

$$k(k - k)b_k + b_{k-2} = 0;$$

and this equation will hold because $b_k = b_{k-2} = 0$.

Hence if p is *not* a positive integer, we take $b_m = 0$ for m odd and define the coefficients of even subscript in terms of b_0 by means of the recursion formula

$$b_m = \frac{-b_{m-2}}{m(m - 2p)}, \qquad m \geq 2. \qquad (6)$$

In comparing (6) with (3), we see that (6) will lead to the same result as that in (4), except with p replaced by $-p$. Thus in this case we obtain the solution

$$y_2 = b_0 \sum_{m=1}^{\infty} \frac{(-1)^m x^{2m-p}}{2^{2m} m! (-p + 1)(-p + 2) \cdots (-p + m)}. \qquad (7)$$

The series in (4) and (7) converge for all $x > 0$ because $x = 0$ is the only singular point of Bessel's equation. If $p > 0$, the leading term in y_1 is $a_0 x^p$, while the leading term in y_2 is $b_0 x^{-p}$. Hence $y_1(0) = 0$, while $y_2(x) \to \pm\infty$ as $x \to 0$, so it is clear that y_1 and y_2 are linearly independent solutions of Bessel's equation of order $p > 0$.

THE GAMMA FUNCTION

The formulas in (4) and (7) can be simplified by use of the **gamma function** $\Gamma(x)$, which is defined for $x > 0$ by

$$\Gamma(x) = \int_0^{\infty} e^{-t} t^{x-1} \, dt. \qquad (8)$$

It is not difficult to show that this improper integral converges for each $x > 0$. The gamma function is a generalization for $x > 0$ of the factorial function $n!$, which is defined only if n is a nonnegative integer. To see the way in which $\Gamma(x)$ is a generalization of $n!$, we note first that

$$\Gamma(1) = \int_0^{\infty} e^{-t} \, dt = \lim_{b \to \infty} \left[-e^t \right]_0^b = 1. \qquad (9)$$

Then we integrate by parts:

$$\Gamma(x + 1) = \lim_{b \to \infty} \int_0^b e^{-t} t^x \, dt$$

$$= \lim_{b \to \infty} \left(\left[-e^{-t} t^x \right]_0^b + \int_0^b x e^{-t} t^{x-1} \, dt \right)$$

$$= x \lim_{b \to \infty} \int_0^b e^{-t} t^{x-1} \, dt;$$

that is,
$$\Gamma(x+1) = x\Gamma(x). \qquad (10)$$

This is the most important property of the gamma function.

If we combine (9) and (10), we see that
$$\Gamma(2) = 1\Gamma(1) = 1!, \qquad \Gamma(3) = 2\Gamma(2) = 2!, \qquad \Gamma(4) = 3\Gamma(3) = 3!,$$

and in general that
$$\Gamma(n+1) = n! \quad \text{for } n \geq 0. \qquad (11)$$

An important special value of the gamma function is
$$\Gamma(\tfrac{1}{2}) = \int_0^\infty e^{-t} t^{-1/2} \, dt = 2 \int_0^\infty e^{-u^2} \, du = \sqrt{\pi}, \qquad (12)$$

where we have substituted u^2 for t in the integral; the fact that
$$\int_0^\infty e^{-u^2} \, du = \frac{\sqrt{\pi}}{2}$$

is known, but is far from obvious.

Although $\Gamma(x)$ is defined in (8) only for $x > 0$, we can use the recursion formula in (10) to define $\Gamma(x)$ whenever x is neither zero nor a negative integer. If $-1 < x < 0$, then
$$\Gamma(x) = \frac{\Gamma(x+1)}{x};$$

the right-hand term is defined because $0 < x + 1 < 1$. The same formula may then be used to extend the definition of $\Gamma(x)$ to the open interval $(-2, -1)$, then to the open interval $(-3, -2)$, and so on. The graph of the gamma function thus extended is shown in Fig. 3.3. The student who would like to pursue

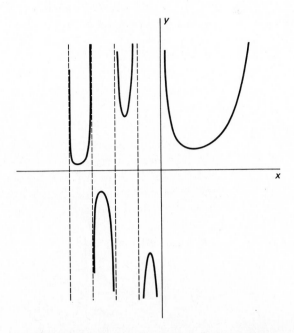

Figure 3.3 Graph of the extended gamma function.

this fascinating topic further should consult Artin's *The Gamma Function* (New York: Holt, Rinehart & Winston, 1964). In only 39 pages, this is one of the finest expositions in the entire literature of mathematics.

BESSEL FUNCTIONS OF THE FIRST KIND

If we choose $a_0 = 1/[2^p\Gamma(p + 1)]$ in (4) and note that

$$\Gamma(p + m + 1) = (p + m)(p + m - 1) \cdots (p + 2)(p + 1)\Gamma(p + 1)$$

by repeated application of (10), we can write the **Bessel function of the first kind of order** p very concisely with the aid of the gamma function:

$$J_p(x) = \sum_{m=0}^{\infty} \frac{(-1)^m}{m!\,\Gamma(p + m + 1)}\left(\frac{x}{2}\right)^{2m+p}. \tag{13}$$

Similarly, if p is not an integer we choose $b_0 = 1/[2^{-p}\Gamma(-p + 1)]$ in (7) to obtain the linearly independent second solution

$$J_{-p}(x) = \sum_{m=0}^{\infty} \frac{(-1)^m}{m!\,\Gamma(-p + m + 1)}\left(\frac{x}{2}\right)^{2m-p} \tag{14}$$

of Bessel's equation of order p. If p is not an integer, we have the general solution

$$y = c_1 J_p(x) + c_2 J_{-p}(x) \tag{15}$$

for $x > 0$; x^p must be replaced by $|x|^p$ in Eqs. (13)–(15) to get the correct solutions for $x < 0$.

If $p = n$, a nonnegative integer, then Eq. (13) gives

$$J_n(x) = \sum_{m=0}^{\infty} \frac{(-1)^m}{m!\,(m + n)!}\left(\frac{x}{2}\right)^{2m+n} \tag{16}$$

for the Bessel functions of the first kind of integral order. Thus

$$J_0(x) = \sum_{m=0}^{\infty} \frac{(-1)^m x^{2m}}{2^{2m}(m!)^2} = 1 - \frac{x^2}{2^2} + \frac{x^4}{2^2 4^2} - \frac{x^6}{2^2 4^2 6^2} + \cdots \tag{17}$$

and

$$J_1(x) = \sum_{m=0}^{\infty} \frac{(-1)^m x^{2m+1}}{2^{2m+1} m!\,(m + 1)!} = \frac{x}{2} - \frac{1}{2!}\left(\frac{x}{2}\right)^3 + \frac{1}{2!\,3!}\left(\frac{x}{2}\right)^5 - \cdots. \tag{18}$$

The graphs of $J_0(x)$ and $J_1(x)$ are shown in Fig. 3.4. In a general way they resemble damped cosine and sine oscillations, respectively. Indeed, if you examine the series in (17), you can see part of the reason why $J_0(x)$ and $\cos x$ should be similar—only minor changes in the denominators in (17) are needed to produce the Taylor series for $\cos x$. As suggested by Fig. 3.4, the zeros of

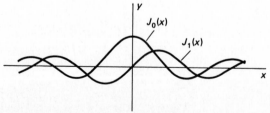

Figure 3.4 The Bessel functions $J_0(x)$ and $J_1(x)$.

the functions $J_0(x)$ and $J_1(x)$ are *interlaced*—between any two consecutive zeros of $J_0(x)$ there is precisely one zero of $J_1(x)$ (see Problem 24) and vice versa. The first four zeros of $J_0(x)$ are approximately 2.4048, 5.5201, 8.6537, and 11.7915. For n large, the nth zero of $J_0(x)$ is approximately $(n - \frac{1}{4})\pi$, while the nth zero of $J_1(x)$ is approximately $(n + \frac{1}{4})\pi$. Thus the interval between consecutive zeros of either $J_0(x)$ or of $J_1(x)$ is approximately π—another similarity with $\cos x$ and $\sin x$.

It turns out that $J_p(x)$ is an elementary function if the order p is half an odd integer. For instance, on substitution of $p = \frac{1}{2}$ and $p = -\frac{1}{2}$ in Eqs. (13) and (14), respectively, the results can be recognized (Problem 2) as

$$J_{1/2}(x) = \sqrt{\frac{2}{\pi x}} \sin x \quad \text{and} \quad J_{-1/2}(x) = \sqrt{\frac{2}{\pi x}} \cos x. \tag{19}$$

BESSEL FUNCTIONS OF THE SECOND KIND

The methods of Section 3.4 must be used to find linearly independent second solutions of integral order. A very complicated generalization of Example 3 in that section gives the formula

$$Y_n(x) = \frac{2}{\pi}\left(\gamma + \ln\frac{x}{2}\right)J_n(x) - \frac{1}{\pi}\sum_{m=0}^{n-1}\frac{2^{n-2m}(n - m - 1)!}{m!\,x^{n-2m}}$$

$$- \frac{1}{\pi}\sum_{m=0}^{\infty}\frac{(-1)^m(H_m + H_{m+n})}{m!\,(m + n)!}\left(\frac{x}{2}\right)^{n+2m}, \tag{20}$$

with the notation used there. If $n = 0$ the first sum in (20) is taken to be zero. Here, $Y_n(x)$ is called the **Bessel function of the second kind of integral order** $n \geq 0$.

The general solution of Bessel's equation of order n is

$$y = c_1 J_n(x) + c_2 Y_n(x). \tag{21}$$

It is important to note that $Y_n(x) \to -\infty$ as $x \to 0$ (see Fig. 3.5). Hence $c_2 = 0$ in (21) if $y(x)$ is continuous at $x = 0$.

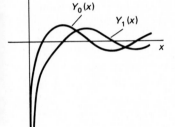

Figure 3.5 The Bessel functions $Y_0(x)$ and $Y_1(x)$.

BESSEL FUNCTION IDENTITIES

Bessel functions are analogous to trigonometric functions in that they satisfy a large number of standard identities of frequent utility, especially in the evaluation of integrals involving Bessel functions. Differentiation of

$$J_p(x) = \sum_{m=0}^{\infty}\frac{(-1)^m}{m!\,\Gamma(p + m + 1)}\left(\frac{x}{2}\right)^{2m+p} \tag{13}$$

in the case that p is a nonnegative integer gives

$$\frac{d}{dx}[x^p J_p(x)] = \frac{d}{dx}\sum_{m=0}^{\infty}\frac{(-1)^m x^{2m+2p}}{2^{2m+p}m!\,(p + m)!}$$

$$= \sum_{m=0}^{\infty}\frac{(-1)^m x^{2m+2p-1}}{2^{2m+p-1}m!\,(p + m - 1)!}$$

$$= x^p\sum_{m=0}^{\infty}\frac{(-1)^m x^{2m+p-1}}{2^{2m+p-1}m!\,(p + m - 1)!},$$

and thus we have shown that

$$\frac{d}{dx}[x^p J_p(x)] = x^p J_{p-1}(x).\tag{22}$$

Similarly,

$$\frac{d}{dx}[x^{-p} J_p(x)] = -x^{-p} J_{p+1}(x).\tag{23}$$

If we carry out the differentiations in (22) and (23) and then divide the resulting identities by x^p and x^{-p}, respectively, we obtain (see Problem 8) the identities

$$J_p'(x) = J_{p-1}(x) - \frac{p}{x} J_p(x)\tag{24}$$

and

$$J_p'(x) = \frac{p}{x} J_p(x) - J_{p+1}(x).\tag{25}$$

Thus we may express the derivatives of Bessel functions in terms of Bessel functions themselves. Subtraction of (25) from (24) gives the recursion formula

$$J_{p+1}(x) = \frac{2p}{x} J_p(x) - J_{p-1}(x),\tag{26}$$

which can be used to express Bessel functions of high order in terms of Bessel functions of lower orders. In the form

$$J_{p-1}(x) = \frac{2p}{x} J_p(x) - J_{p+1}(x),\tag{27}$$

it can be used to express Bessel functions of large negative order in terms of Bessel functions of numerically smaller negative orders.

The identities in (22)–(27) above hold wherever they are meaningful—that is, whenever no Bessel functions of negative integral order appear. In particular, they hold for all nonintegral values of p.

EXAMPLE 1   With $p = 0$, the identity in (22) gives

$$\int x J_0(x)\, dx = x J_1(x) + C.$$

Similarly, with $p = 0$, the identity in (23) gives

$$\int J_1(x)\, dx = -J_0(x) + C.$$

EXAMPLE 2   Using first $p = 2$ and then $p = 1$ in (26), we get

$$J_3(x) = \frac{4}{x} J_2(x) - J_1(x) = \frac{4}{x}\left[\frac{2}{x} J_1(x) - J_0(x)\right] - J_1(x),$$

so that

$$J_3(x) = \frac{4}{x} J_1(x) + \left(\frac{8}{x^2} - 1\right) J_0(x).$$

With similar manipulations every Bessel function of positive integral order can be expressed in terms of $J_0(x)$ and $J_1(x)$.

EXAMPLE 3 To evaluate $\int x J_2(x)\, dx$, we first note that $\int x^{-1} J_2(x)\, dx = -x^{-1} J_1(x) + C$ by (23) with $p = 1$. We therefore write

$$\int x J_2(x)\, dx = \int x^2 [x^{-1} J_2(x)]\, dx$$

and integrate by parts with

$$u = x^2, \qquad\qquad dv = x^{-1} J_2(x)\, dx,$$
$$du = 2x\, dx, \quad \text{and} \quad v = -x^{-1} J_1(x).$$

This gives

$$\int x J_2(x)\, dx = -x J_1(x) + 2 \int J_1(x)\, dx$$
$$= -x J_1(x) - 2 J_0(x) + C,$$

with the aid of the second result of Example 1.

THE PARAMETRIC BESSEL EQUATION

The **parametric Bessel equation of order n** is

$$x^2 y'' + x y' + (\alpha^2 x^2 - n^2) y = 0, \tag{28}$$

with α a positive parameter. As we will see in Chapter 8, this equation appears in the solution of Laplace's equation in polar coordinates. It is easy to see (Problem 9) that the substitution $t = \alpha x$ transforms (28) into the (standard) Bessel equation

$$t^2 \frac{d^2 y}{dt^2} + t \frac{dy}{dt} + (t^2 - n^2) y = 0 \tag{29}$$

with general solution $y = c_1 J_n(t) + c_2 Y_n(t)$. Hence the general solution of (28) is

$$y = c_1 J_n(\alpha x) + c_2 Y_n(\alpha x). \tag{30}$$

Now consider the eigenvalue problem

$$\left. \begin{array}{c} x^2 y'' + x y' + (\lambda x^2 - n^2) y = 0, \\ y(L) = 0 \end{array} \right\} \tag{31}$$

on the interval $[0, L]$. We seek the *positive* values of λ for which there exists a nontrivial solution of (31) that is *continuous* on $[0, L]$. If we write $\lambda = \alpha^2$, then the differential equation in (31) is that in (28), so its general solution is given in (30). Because $Y_n(x) \to -\infty$ as $x \to 0$ while $J_n(0)$ is finite, the continuity of $y(x)$ requires that $c_2 = 0$. Thus $y(x) = c_1 J_n(\alpha x)$. The endpoint condition $y(L) = 0$ now implies that $z = \alpha L$ must be a (positive) root of the equation

$$J_n(z) = 0. \tag{32}$$

For $n > 1$, $J_n(x)$ oscillates rather like $J_1(x)$ in Fig. 3.4 and hence has an infinite sequence of positive roots $\gamma_{n1}, \gamma_{n2}, \gamma_{n3}, \ldots$. It follows that the kth positive eigenvalue of the problem in (31) is

$$\gamma_k = \alpha_k^2 = \frac{\gamma_{nk}^2}{L^2}, \tag{33}$$

and that its associated eigenfunction is

$$y_k(x) = J_n\left(\frac{\gamma_{nk}x}{L}\right). \tag{34}$$

The roots γ_{nk} of Eq. (32) for $n \le 8$ and $k \le 20$ are tabulated in Table 9.5 of Abramowitz and Stegun's *Handbook of Mathematical Functions* (New York: Dover, 1965).

3.5 Problems

1. Differentiate termwise the series for $J_0(x)$ to show directly that $J_0'(x) = -J_1(x)$.

2. (a) Deduce from Eqs. (10) and (12) that
$$\Gamma\left(n + \frac{1}{2}\right) = \frac{1 \cdot 3 \cdots (2n - 1)}{2^n}\sqrt{\pi}.$$

   (b) Use the result of part (a) to verify the formulas in Eq. (19) for $J_{1/2}(x)$ and $J_{-1/2}(x)$.

3. (a) Suppose that m is a positive integer. Show that
$$\Gamma\left(m + \frac{2}{3}\right) = \frac{2 \cdot 5 \cdots (3m - 1)}{3^m}\Gamma\left(\frac{2}{3}\right).$$

   (b) Conclude from part (a) and the formula in (13) that
$$J_{-1/3}(x) = \frac{(x/2)^{-1/3}}{\Gamma(\frac{2}{3})}\left(1 + \sum_{m=1}^{\infty} \frac{(-1)^m 3^m x^{2m}}{2^{2m}m! \cdot 2 \cdot 5 \cdots (3m - 1)}\right).$$

4. Apply the formulas in (19), (26), and (27) to show that
$$J_{3/2}(x) = \sqrt{\frac{2}{\pi x^3}}(\sin x - x \cos x)$$

   and
$$J_{-3/2}(x) = -\sqrt{\frac{2}{\pi x^3}}(\cos x + x \sin x).$$

5. Express $J_4(x)$ in terms of $J_0(x)$ and $J_1(x)$.

6. Derive the recursion formula in Eq. (2) for Bessel's equation.

7. Verify the identity in Eq. (23) by termwise differentiation.

8. Deduce the identities in Eqs. (24) and (25) from those in Eqs. (22) and (23).

9. Verify that the substitution $t = \alpha x$ transforms the parametric Bessel equation in (28) into the equation in (29).

10. Show that
$$4J_p''(x) = J_{p-2}(x) - 2J_p(x) + J_{p+2}(x).$$

Any integral of the form $\int x^m J_n(x)\, dx$ can be evaluated in terms of Bessel functions and the indefinite integral $\int J_0(x)\, dx$. The latter integral cannot be simplified further, but the function $\int_0^x J_0(t)\, dt$ is tabulated in Table 11.1 of Abramowitz and Stegun. Use the identities in Eqs. (22) and (23) to evaluate the integrals in Problems 11–19.

11. $\int x^2 J_0(x)\, dx$.   12. $\int x^3 J_0(x)\, dx$.   13. $\int x^4 J_0(x)\, dx$.

14. $\int x J_1(x)\, dx$.   15. $\int x^2 J_1(x)\, dx$.   16. $\int x^3 J_1(x)\, dx$.

17. $\int x^4 J_1(x)\, dx$.   18. $\int J_2(x)\, dx$.   19. $\int J_3(x)\, dx$.

20. Prove that

$$J_0(x) = \frac{1}{\pi} \int_0^\pi \cos(x \sin \theta) \, d\theta$$

by showing that the right-hand side satisfies Bessel's equation of order zero and has the value $J_0(0)$ when $x = 0$. Explain why this constitutes a proof.

21. Prove that

$$J_1(x) = \frac{1}{\pi} \int_0^\pi \cos(\theta - x \sin \theta) \, d\theta$$

by showing that the right-hand side satisfies Bessel's equation of order 1 and that its derivative has the value $J_1'(0)$ when $x = 0$. Explain why this constitutes a proof.

22. It can be shown that

$$J_n(x) = \frac{1}{\pi} \int_0^\pi \cos(n\theta - x \sin \theta) \, d\theta.$$

With $n \geq 2$, show that the right-hand side satisfies Bessel's equation of order n and also agrees with the values $J_n(0)$ and $J_n'(0)$. Explain why this does *not* suffice to prove the assertion above.

23. Deduce from Problem 20 that

$$J_0(x) = \frac{1}{2\pi} \int_0^{2\pi} e^{ix \sin \theta} \, d\theta.$$

(*Suggestion:* Show first that

$$\int_0^{2\pi} e^{ix \sin \theta} \, d\theta = \int_0^\pi (e^{ix \sin \theta} + e^{-ix \sin \theta}) \, d\theta;$$

then use Euler's formula.)

24. Use the identities in Eqs. (22) and (23) and Rolle's theorem to prove that between any two consecutive zeros of $J_n(x)$ there is precisely one zero of $J_{n+1}(x)$.

*3.6
Applications
of Bessel Functions

The importance of Bessel functions stems not only from the frequent appearance of Bessel's equation in applications, but also from the fact that the solutions of many other second order linear differential equations can be expressed in terms of Bessel functions. To see how this comes about, we begin with Bessel's equation of order p in the form

$$z^2 \frac{d^2 w}{dz^2} + z \frac{dw}{dz} + (z^2 - p^2)w = 0, \tag{1}$$

and substitute

$$w = x^{-\alpha} y, \qquad z = kx^\beta. \tag{2}$$

Then a routine though lengthy transformation of (1) yields the equation

$$x^2 y'' + (1 - 2\alpha)xy' + (\alpha^2 - \beta^2 p^2 + \beta^2 k^2 x^{2\beta})y = 0;$$

that is,

$$x^2 y'' + Axy' + (B + Cx^q)y = 0, \tag{3}$$

where the constants A, B, C, and q are given by

$$A = 1 - 2\alpha, \quad B = \alpha^2 - \beta^2 p^2, \quad C = \beta^2 k^2, \quad \text{and} \quad q = 2\beta. \qquad (4)$$

It is a simple matter to solve the equations in (4) for

$$\alpha = \frac{1 - A}{2} \qquad \beta = \frac{q}{2},$$

$$k = \frac{2\sqrt{C}}{q}, \quad \text{and} \quad p = \frac{\sqrt{(1 - A)^2 - 4B}}{q}. \qquad (5)$$

Under the assumption that the square roots in (5) are real, it follows that the general solution of (3) is $y = x^\alpha w(z) = x^\alpha w(kx^\beta)$, where w is the general solution of the equation in (1). This establishes the following result.

> **THEOREM: SOLUTIONS IN BESSEL FUNCTIONS**
>
> If $C > 0$, $q \neq 0$, and $(1 - A)^2 \geq 4B$, then the general solution of Eq. (3) is
>
> $$y = x^\alpha[c_1 J_p(kx^\beta) + c_2 J_{-p}(kx^\beta)] \qquad (6)$$
>
> where α, β, k, and p are given by the equations in (5). If p is an integer, then J_{-p} is to be replaced by Y_p.

EXAMPLE 1 Solve the equation

$$4x^2 y'' + 8xy' + (x^4 - 3)y = 0. \qquad (7)$$

Solution To compare the equation above with that in (3), we rewrite it as

$$x^2 y'' + 2xy' + (-\tfrac{3}{4} + \tfrac{1}{4}x^4)y = 0$$

and see that $A = 2$, $B = -\tfrac{3}{4}$, $C = \tfrac{1}{4}$, and $q = 4$. Then the equations in (5) give $\alpha = -\tfrac{1}{2}$, $\beta = 2$, $k = \tfrac{1}{4}$, and $p = \tfrac{1}{2}$. Thus the general solution in (6) of Eq. (7) is

$$y = x^{-1/2}[c_1 J_{1/2}(\tfrac{1}{4}x^2) + c_2 J_{-1/2}(\tfrac{1}{4}x^2)].$$

EXAMPLE 2 Solve the Airy equation

$$y'' + 9xy = 0. \qquad (8)$$

Solution First we rewrite the equation in the form

$$x^2 y'' + 9x^3 y = 0.$$

This is a special case of the equation in (3) with $A = B = 0$, $C = +9$, and $q = 3$. It follows from the equations in (5) that $\alpha = \tfrac{1}{2}$, $\beta = \tfrac{3}{2}$, $k = 2$, and $p = \tfrac{1}{3}$. Thus the general solution of Eq. (8) is

$$y = x^{1/2}[c_1 J_{1/3}(2x^{3/2}) + c_2 J_{-1/3}(2x^{3/2})].$$

BUCKLING OF A VERTICAL COLUMN

For a practical application, we now consider the problem of determining when a uniform vertical column will buckle under its own weight. We take $x = 0$ at the free top end of the column and $x = L$ at its bottom; we assume that the bottom is rigidly imbedded in the ground (or in concrete). See Fig. 3.6. Denote

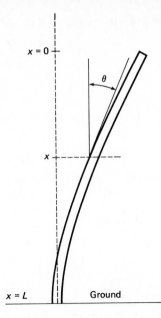

θ

x

$x = L$ Ground

Figure 3.6 The buckling
column.

the angular deflection of the column at the point x by $\theta(x)$. From the theory
of elasticity it follows that

$$EI\frac{d^2\theta}{dx^2} + g\rho x\theta = 0, \tag{9}$$

where E is the Young's modulus of the material of the column, I is its cross-
sectional moment of inertia, ρ is the linear density of the column, and g is
gravitational acceleration. The boundary conditions corresponding to the
situation as described are

$$\theta'(0) = 0, \qquad \theta(L) = 0. \tag{10}$$

We will accept (9) and (10) as an appropriate statement of the problem and
attempt to solve it in this form. With

$$\lambda = \gamma^2 = \frac{g\rho}{EI}, \tag{11}$$

we have the eigenvalue problem

$$\theta'' + \gamma^2 x\theta = 0;$$
$$\theta'(0) = 0, \qquad \theta(L) = 0. \tag{12}$$

The column can buckle only if there is a nontrivial solution of (12); otherwise
the column will remain in its undeflected vertical position.

The differential equation in (12) is an Airy equation. It has the form of
Eq. (3) with $A = 0$, $B = 0$, $C = \gamma^2$, and $q = 3$. The equations in (5) give
$\alpha = \frac{1}{2}$, $\beta = \frac{3}{2}$, $k = 2\gamma/3$, and $p = \frac{1}{3}$. So the general solution is

$$\theta(x) = x^{1/2}[c_1 J_{1/3}(\tfrac{2}{3}\gamma x^{3/2}) + c_2 J_{-1/3}(\tfrac{2}{3}\gamma x^{3/2})]. \tag{13}$$

In order to apply the initial conditions, we substitute $p = \pm\frac{1}{3}$ in

$$J_p(x) = \sum_{m=0}^{\infty} \frac{(-1)^m}{m!\,\Gamma(p + m + 1)}\left(\frac{x}{2}\right)^{2m+p},$$

and find after some simplification that

$$\theta(x) = \frac{c_1\gamma^{1/3}}{3^{1/3}\Gamma(\frac{4}{3})}\left(x - \frac{\gamma^2 x^4}{12} + \frac{\gamma^4 x^7}{504} - \cdots\right)$$
$$+ \frac{c_2 3^{1/3}}{\gamma^{1/3}\Gamma(\frac{2}{3})}\left(1 - \frac{\gamma^2 x^3}{6} + \frac{\gamma^4 x^6}{180} - \cdots\right).$$

From this it is clear that the endpoint condition $\theta'(0) = 0$ implies that $c_1 = 0$, so

$$\theta(x) = c_2 x^{1/2} J_{-1/3}(\tfrac{2}{3}\gamma x^{3/2}). \tag{14}$$

The endpoint condition $\theta(L) = 0$ now gives

$$J_{-1/3}(\tfrac{2}{3}\gamma L^{3/2}) = 0. \tag{15}$$

Thus the column will buckle only if $z = \frac{2}{3}\gamma L^{3/2}$ is a root of the equation $J_{-1/3}(z) = 0$. The graph of

$$J_{-1/3}(z) = \frac{(z/2)^{-1/3}}{\Gamma(\frac{2}{3})}\left(1 + \sum_{m=1}^{\infty} \frac{(-1)^m 3^m z^{2m}}{2^{2m}(m!\, 2\cdot 5 \cdots (3m-1)}\right) \tag{16}$$

(see Problem 3 in Section 3.5) is shown in Fig. 3.7. Its zeros $\{z_n\}_1^\infty$ are [ignoring the leading factor in (16)] the roots of the equation

$$F(z) = 1 + \sum_{m=1}^{\infty} \frac{(-1)^m 3^m z^{2m}}{2^{2m}(m!)2\cdot 5 \cdots (3m-1)} = 0. \tag{17}$$

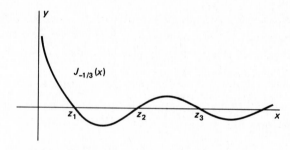

Figure 3.7 The graph of $J_{-1/3}(x)$.

If $|z|$ is not too large, the series in (17) is an alternating series with terms rapidly approaching zero. Hence the values of $F(z)$ are not difficult to compute. Indeed, if the series is truncated, the error in computing $F(z)$ is numerically less than the first term dropped.

Some typical values of $F(z)$ are given in the table of Fig. 3.8. From these data it is apparent that the least positive root z_1 of $F(z) = 0$ is between $z = 1.5$ and $z = 2.0$. Figure 3.9 shows a BASIC program for an IBM Personal Computer, which we used to compute z_1. This program employs the iteration

$$x_{n+1} = x_n - \frac{F(x_n)}{F'(x_n)} \tag{18}$$

of Newton's method, beginning with the initial guess $x_0 = 1.5$. The subroutine starting at line 300 computes the value of $F(x)$ by summing the series in (17), accurate to nine decimal places. The subroutine starting at line 400 similarly

z	$F(z)$
0.0	+1.00000
0.5	+0.90799
1.0	+0.65226
1.5	+0.28900
2.0	−0.10257
2.5	−0.43830
3.0	−0.64819

Figure 3.8 Values of $F(z)$.

```
100 REM––PROGRAM COMPUTES ROOT OF J SUB MINUS 1/3
110 REM
120        DEFDBL D, F, S, T, X, Y
130 REM
140 REM––NEWTON'S METHOD ITERATION:
150 REM
160        X = 1.5 : N = 1
170        GOSUB 300 :            REM––GET F = F(X)
180        GOSUB 400 :            REM––GET D = D(X) = F'(X)
190        Y = X − F/D
200        LPRINT USING "##          #.######"; N, Y
210        IF ABS (X − Y) < 1E-08 THEN GOTO 999
220        X = Y : N = N + 1
230        GOTO 170 :             REM––DO IT AGAIN
240 REM
300 REM––SUBROUTINE COMPUTES VALUE OF FUNCTION F(X)
305 REM
310 REM––T IS TYPICAL TERM, S IS PARTIAL SUM OF SERIES
315 REM
320        T = 1 : S = 1 : M = 1
330        T = −T*3*X*X/(4*M*(3*M − 1) )
340        S = S + T :   F = S
350        IF ABS(T) < 1E-10 THEN GOTO 380
360        M = M + 1
370        GOTO 330
380        RETURN
390 REM
400 REM––SUBROUTINE COMPUTES VALUE OF DERIVATIVE D(X) = F'(X)
405 REM
410 REM––T IS TYPICAL TERM, S IS PARTIAL SUM OF SERIES
415 REM
420        T = −3*X/4 : S = −3*X/4 : M = 1
430        T = −T*3*X*X/(4*M*(3*M + 2) )
440        S = S + T : D = S
450        IF ABS (T) < 1E-10 THEN GOTO 480
460        M = M + 1
470        GOTO 430
480        RETURN
490 REM
999        END
```

Figure 3.9 Program to compute roots of $J_{-1/3}(z) = 0$.

computes the value of the derivative

$$D(x) = F'(x)$$

$$= -\frac{3x}{4} + \sum_{m=1}^{\infty} \frac{(-1)^{m+1} 3^{m+1} x^{2m+1}}{2^{2m+1} \cdot (m!) \cdot 2 \cdot 5 \cdots (3m + 2)}. \tag{19}$$

The output of the program was $x_1 = 1.868914$, $x_2 = 1.866350$, and $x_3 = x_4 = 1.866351$. Thus the smallest positive root of $J_{-1/3}(z) = 0$ is $z = 1.86635$, rounded to five decimal places.

The shortest length L_1 for which the column will buckle under its own weight is

$$L_1 = \left(\frac{3z_1}{2\gamma}\right)^{2/3} = \left[\frac{3z_1}{2}\left(\frac{EI}{\rho g}\right)^{1/2}\right]^{2/3}.$$

If we substitute $z_1 \approx 1.86635$ and $\rho = \delta A$, where δ is the volumetric density of the material of the column and A is its cross-sectional area, we finally get

$$L_1 \approx (1.986)\left(\frac{EI}{g\delta A}\right)^{1/3} \tag{20}$$

for the critical buckling length. For example, with a steel column or rod for which $E = 2.8 \times 10^7$ lb/in.^{2} and $g\delta = 0.28$ lb/in.^{3}, the formula in (20) gives the results shown in the table in Fig. 3.10.

Cross section of rod	Shortest buckling length L_1
Circular with $r = \frac{1}{2}$ in.	30 ft 6 in.
Circular with $r = 1\frac{1}{2}$ in.	63 ft 5 in.
Annular with $r_{inner} = 1\frac{1}{4}$ in. and $r_{outer} = 1\frac{1}{2}$ in.	75 ft 7 in.

Figure 3.10

We have used the familiar formulas $A = \pi r^2$ and $I = \frac{1}{4}\pi r^4$ for a circular disk. The data in the table show why flagpoles are hollow.

3.6 Problems

In each of Problems 1–12, express the general solution of the given differential equation in terms of Bessel functions.

1. $x^2 y'' - xy' + (1 + x^2)y = 0$.
2. $xy'' + 3y' + xy = 0$.
3. $xy'' - y' + 36x^3 y = 0$.
4. $x^2 y'' - 5xy' + (8 + x)y = 0$.
5. $36x^2 y'' + 60xy' + (9x^3 - 5)y = 0$.
6. $16x^2 y'' + 24xy' + (1 + 144x^3)y = 0$.
7. $x^2 y'' + 3xy' + (1 + x^2)y = 0$.
8. $4x^2 y'' - 12xy' + (15 + 16x)y = 0$.
9. $16x^2 y'' - (5 - 144x^3)y = 0$.
10. $2x^2 y'' + 3xy' - 2(14 - x^5)y = 0$.
11. $y'' + x^4 y = 0$.
12. $y'' + 4x^3 y = 0$.
13. Apply the theorem in this section to show that the general solution of

$$xy'' + 2y' + xy = 0$$

is $y = x^{-1}(A \cos x + B \sin x)$.
14. (a) Show that the substitution $y = (-1/u)(du/dx)$ transforms the Ricatti equation $dy/dx = x^2 + y^2$ into $u'' + x^2 u = 0$. (b) Show that the general

solution of $y' = x^2 + y^2$ is

$$y = x\frac{J_{3/4}(x^2/2) - cJ_{-3/4}(x^2/2)}{cJ_{1/4}(x^2/2) + J_{-1/4}(x^2/2)}.$$

(*Suggestion:* Apply the identities in (24) and (25) of Section 3.5.)

15. Verify that the substitutions in (2) in Bessel's equation (Eq. (1)) yield Eq. (3).

*3.7
Appendix
on Infinite Series
and the Atom

Here we illustrate the applications of infinite series methods by briefly outlining one of the great triumphs of modern physics—the interpretation of atomic spectra. During the nineteenth century it was discovered that when a gas is rendered luminous in an electric discharge tube, the light that it emits consists of specific, sharply defined wavelengths that are characteristic of the particular gas used. For example, when the light emitted by hydrogen is analyzed with a spectroscope, we find that its *spectrum* consists of a sequence of sharply defined lines (on the photographic plate), beginning with a red line at wavelength $\lambda = 6563$ angstroms (Å) (1 Å is defined to be 10^{-10} m) and apparently converging toward a limiting line at 3646 Å; see Fig. 3.11. In 1885 J. J. Balmer observed that the wavelengths λ_n of these lines are given by the formula

$$\frac{1}{\lambda_n} = R\left(\frac{1}{2^2} - \frac{1}{n^2}\right) \tag{1}$$

for $n = 3, 4, 5, \ldots$, where $R \approx 109{,}678$ cm^{-1} is the **Rydberg constant**. Thus the formula in (1) fits the **Balmer series** of visible spectral lines of hydrogen. (There are other series of spectral lines outside the visible range, but we shall not discuss them here.)

During the early twentieth century there gradually emerged, from the work of physicists such as Rutherford and Bohr, an understanding that the light emitted by an element originates in the electrons in its individual atoms. In particular, it was found that an electron of an atom can exist in any one of several *states* with well-defined energy levels. When an electron jumps from one state to another with lower energy, it emits a photon of light; the frequency $v = c/\lambda$ of the light (where c denotes the velocity of light) is given by

$$\Delta E = hv, \tag{2}$$

where ΔE is the energy lost by the electron and h is **Planck's constant**. The problem, then, was to construct a mathematical model of the atom consistent with the formulas in (1) and (2).

In the modern quantum theory of atomic structure, the electron in a hydrogen atom does not travel in a well-defined orbit about the nucleus (the proton) of the atom. Instead, it occupies a state called an **orbital**, which may be visualized as a cloud of negative electricity surrounding the nucleus. The density of this "electron cloud" at a particular point is a measure of the probability that, at a given instant, the electron is near that point.

More precisely, an orbital is described by a *probability amplitude function* $\psi(x, y, z)$. This function has the property that the probability $P(x, y, z; \Delta V)$ of

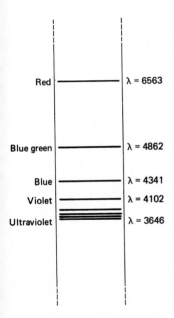

Figure 3.11 The hydrogen spectrum.

Red — $\lambda = 6563$

Blue green — $\lambda = 4862$

Blue — $\lambda = 4341$

Violet — $\lambda = 4102$

Ultraviolet — $\lambda = 3646$

finding the electron in a small volume element ΔV surrounding the point (x, y, z) is given approximately by

$$P(x, y, z; \Delta V) \approx [\psi(x, y, z)]^2 \, \Delta V. \tag{3}$$

The error in this approximation is small in comparison with ΔV; that is,

$$\lim_{\Delta V \to 0} \frac{P(x, y, z, \Delta V)}{\Delta V} = [\psi(x, y, z)]^2. \tag{4}$$

If an electron with mass m has constant total energy E—the sum of its kinetic energy and its potential energy $V(x, y, z)$—then its probability amplitude function must satisfy the *time-independent Schrödinger equation*

$$\nabla^2 \psi + \frac{8\pi^2 m}{h^2}(E - V)\psi = 0. \tag{5}$$

Here $\nabla^2 \psi$ denotes the **Laplacian**, defined as follows:

$$\nabla^2 \psi = \frac{\partial^2 \psi}{\partial x^2} + \frac{\partial^2 \psi}{\partial y^2} + \frac{\partial^2 \psi}{\partial z^2}.$$

We want to investigate the possibility of *spherically symmetric* orbitals for which $\psi = \psi(r)$ is a function of the radial coordinate $r = \sqrt{x^2 + y^2 + z^2}$ alone, considering the nucleus of the hydrogen atom (the proton) to be fixed at the origin. In this case

$$\frac{\partial \psi}{\partial x} = \frac{d\psi}{dr} \frac{\partial r}{\partial x} = \frac{x}{r} \frac{d\psi}{dr},$$

so

$$\frac{\partial^2 \psi}{\partial x^2} = \frac{\partial}{\partial x}\left(\frac{x}{r} \frac{d\psi}{dr}\right)$$

$$= \frac{1}{r} \frac{d\psi}{dr} - \frac{x}{r^2} \frac{d\psi}{dr} \frac{\partial r}{\partial x} + \frac{x}{r} \frac{d^2\psi}{dr^2} \frac{\partial r}{\partial x}$$

and thus

$$\frac{\partial^2 \psi}{\partial x^2} = \frac{1}{r} \frac{d\psi}{dr} - \frac{x^2}{r^3} \frac{d\psi}{dr} + \frac{x^2}{r^2} \frac{d^2\psi}{dr^2},$$

with similar formulas for $\partial^2 \psi / \partial y^2$ and $\partial^2 \psi / \partial z^2$, x being replaced by y and z, respectively. Adding these formulas for the second derivatives of ψ, we get

$$\nabla^2 \psi = \frac{3}{r} \frac{d\psi}{dr} - \frac{x^2 + y^2 + z^2}{r^3} \frac{d\psi}{dr} + \frac{x^2 + y^2 + z^2}{r^2} \frac{d^2\psi}{dr^2},$$

so

$$\nabla^2 \psi = \frac{d^2\psi}{dr^2} + \frac{2}{r} \frac{d\psi}{dr}.$$

Hence the Schrödinger equation in (5) becomes

$$\frac{d^2\psi}{dr^2} + \frac{2}{r} \frac{d\psi}{dr} + \frac{8\pi^2 m}{h^2}(E - V)\psi = 0 \tag{6}$$

in the radially symmetric case.

The proton and the electron have charges $+q$ and $-q$, so in appropriate electrical units the potential energy of the electron orbiting the fixed proton is the *Coulomb potential*

$$V = -\frac{q^2}{r}.$$ (7)

Then the total energy of an electron at rest at infinity ($r = \infty$) will be zero, and the total energy of an electron bound to the nucleus will be negative,

$$E = -\alpha^2.$$ (8)

On substitution of (7) and (8) in (6) we obtain the ordinary differential equation

$$\frac{d^2\psi}{dr^2} + \frac{2}{r}\frac{d\psi}{dr} + \frac{8\pi^2 m}{h^2}\left(-\alpha^2 - \frac{q^2}{r}\right)\psi = 0.$$ (9)

When r is very large, Eq. (9) is approximately the equation

$$\frac{d^2\psi}{dr^2} - \frac{8\pi^2\alpha^2 m}{h^2}\psi = 0,$$

whose solution is of the form

$$u = A \exp\left(\sqrt{\frac{8\pi^2\alpha^2 m}{h^2}}\, r\right) + B \exp\left(-\sqrt{\frac{8\pi^2\alpha^2 m}{h^2}}\, r\right);$$

that is,

$$u = Ae^{-x/2} + Be^{+x/2},$$ (10)

where

$$x = \frac{4\pi\alpha r}{h}\sqrt{2m}.$$ (11)

Now let us impose the physically motivated condition that we are looking for a solution $\psi(r)$ that is continuous everywhere with $\psi(r) \to 0$ as $r \to \infty$ (so the probability of finding the electron far from the nucleus is negligible). If $\psi = u(x)$ is to be given approximately by (10) when x is large, it is necessary that $B = 0$, so that $\psi \approx Ae^{-x/2}$.

This preliminary analysis suggests that we attempt to simplify Eq. (9) by splitting the factor $e^{-x/2}$ out of its solution. In Problem 1 we ask you to show that the substitution

$$\psi = e^{-x/2}v(x), \qquad x = \frac{4\pi\alpha r}{h}\sqrt{2m}$$ (12)

transforms (9) into the differential equation

$$x\frac{d^2v}{dx^2} + (2 - x)\frac{dv}{dx} + (p - 1)v = 0,$$ (13)

where $p = (\pi q^2\sqrt{2m})/\alpha h$. This equation is closely related to **Laguerre's equation** of order p:

$$x\frac{d^2y}{dx^2} + (1 - x)\frac{dy}{dx} + py = 0.$$ (14)

Indeed, differentiation of (14) gives

$$xy^{(3)} + (2 - x)y'' + (p - 1)y' = 0,$$

so $v(x) = y'(x)$ is a solution of (13) if $y(x)$ is a solution of Laguerre's equation. So let us examine Laguerre's equation. First we write it in the form

$$y'' + \frac{1 - x}{x}y' + \frac{p}{x}y = 0.$$

Clearly this equation has $x = 0$ as a regular singular point, and its indicial equation is

$$r(r - 1) + (1)r + 0 = 0;$$

that is, $r^2 = 0$. From the theorem in Section 3.4 we therefore see that Laguerre's equation has two linearly independent solutions of the forms

$$y_1(x) = \sum_{k=1}^{\infty} a_k x^k$$

and

$$y_2(x) = y_1(x) \ln x + \sum_{k=1}^{\infty} b_k x^k.$$

But we are searching for a solution that is continuous at $x = 0$, and this condition excludes the singular solution.

In Problem 2 we ask you to verify that substitution of the power series $y = \sum a_k x^k$ in Laguerre's equation yields the recursion formula

$$a_k = -\frac{p - k + 1}{k^2} a_{k-1}. \tag{15}$$

From this it follows by induction that

$$a_k = (-1)^k \frac{p \cdot (p - 1) \cdots (p - k + 1)}{(k!)^2} a_0.$$

Consequently

$$y(x) = a_0 \sum_{k=0}^{\infty} (-1)^k \frac{p \cdot (p - 1) \cdots (p - k + 1)}{(k!)^2} x^k, \tag{16}$$

so the solution of Eq. (13) that we seek is

$$v(x) = y'(x) = \sum_{k=0}^{\infty} b_k x^k \tag{17}$$

where

$$b_k = (-1)^{k+1} a_0 \frac{p \cdot (p - 1) \cdots (p - k)}{(k!)(k + 1)!}. \tag{18}$$

But what is the point to all this? In order to determine what frequency of light photon is emitted when the hydrogen atom's electron jumps from one energy level to a lower one, we need to determine the possible values of the total energy constant $E = -\alpha^2$ that appears in Eq. (9). Equivalently, what are the possible values of $p = (\pi q^2 \sqrt{2m})/\alpha h$?

We will find the answer to this question by imposing the condition that

$$\psi = \frac{v(x)}{e^{x/2}} \longrightarrow 0 \quad \text{as} \quad x \longrightarrow \infty. \tag{19}$$

If p is a *nonnegative integer*, then it follows from (18) that $v(x)$ is a polynomial, so the condition in (19) is satisfied. Otherwise, the series in (17) for $v(x)$ is infinite, with

$$\frac{b_{k+1}}{b_k} = \frac{k+1-p}{(k+1)(k+2)}.$$

Now

$$e^{x/2} = \sum_{n=0}^{\infty} c_k x^k$$

where $c_k = 1/(k! 2^k)$, so

$$\frac{c_{k+1}}{c_k} = \frac{1}{2(k+1)}.$$

Hence

$$\frac{b_{k+1}/b_k}{c_{k+1}/c_k} = \frac{2(k+1)(k+1-p)}{(k+1)(k+2)} \longrightarrow 2$$

as $k \to \infty$. It follows that

$$\frac{b_{k+1}}{b_k} > \frac{3}{2} \frac{c_{k+1}}{c_k}$$

for k sufficiently large, say for $k \geqq K$. Hence

$$\frac{b_{K+k}}{c_{K+k}} > \left(\frac{3}{2}\right)^k \frac{b_K}{c_K} > 1$$

for k sufficiently large. Thus, if p is *not* a nonnegative integer, then the terms of $v = \sum b_k x^k$ from some point on are positive and larger than those of $e^{x/2} = \sum c_k x^k$, so the condition in (19) cannot be satisfied. We get a nontrivial solution only if p is a *positive integer* because, if $p = 0$, then $y(x) = a_0$, so that $v(x) = y'(x) \equiv 0$.

Thus we finally see that the only values of p for which Eq. (9) has a nontrivial solution satisfying our conditions are the positive integers

$$p = \frac{\pi q^2 \sqrt{2m}}{\alpha h} = n = 1, 2, 3, \ldots.$$

The nth value of α is

$$\alpha_n = \frac{\pi q^2 \sqrt{2m}}{nh},$$

and the energy of the electron in its nth state is

$$E_n = -\alpha_n^2 = -\frac{2\pi^2 m q^4}{n^2 h^2}. \tag{20}$$

If the electron falls from energy level E_n to energy level E_k, where $n > k$, its energy loss is

$$\Delta E = \frac{2\pi^2 m q^4}{h^2} \left(\frac{1}{k^2} - \frac{1}{n^2}\right). \tag{21}$$

Because $\Delta E = h\nu = hc/\lambda$ by Eq. (2), we see that the wavelength λ of the photon emitted is given by

$$\frac{1}{\lambda} = \frac{\Delta E}{hc} = \frac{2\pi^2 m q^4}{ch^3} \left(\frac{1}{k^2} - \frac{1}{n^2}\right). \tag{22}$$

In comparing this result with the formula in (1), we see that the Balmer series of visible spectral lines results from electron transitions from the state n to the state $k = 2$ for $n = 3, 4, 5, \ldots$. We see also that the Rydberg constant is given by

$$R = \frac{2\pi^2 m q^4}{ch^3}. \tag{23}$$

Indeed, when the known values of $c, m, q,$ and h are substituted in (23), the result is the empirically measured value $R \approx 109{,}678 \text{ cm}^{-1}$. It is the spectacular agreement of theory with experiment that is the great triumph.

The polynomial solution of Laguerre's equation,

$$xy'' + (1 - x)y' + ny = 0 \qquad (n \text{ is an integer}),$$

that we found above is called the nth **Laguerre polynomial** $L_n(x)$. In Problem 3 we ask you to deduce from the formula in (16) with $a_0 = 1$ that

$$L_n(x) = \sum_{k=0}^{n} (-1)^k \binom{n}{k} \frac{x^k}{k!}. \tag{24}$$

Thus $L_0(x) = 1$, $L_1(x) = 1 - x$, $L_2(x) = 1 - 2x + \frac{1}{2}x^2$, $L_3(x) = 1 - 3x + \frac{3}{2}x^2 - \frac{1}{6}x^3$, and so on.

3.7 Problems

1. Verify that the substitution given by the equations in (12) transforms the differential equation in (9) into Eq. (13).

2. Verify the recursion formula in (15) for the Laguerre polynomial of order p.

3. Verify that if $p = n$, a nonnegative integer, then the infinite series in (16) with $a_0 = 1$ reduces to the Laguerre polynomial as given in Eq. (24).

The Laplace Transform

4

4.1
Laplace Transforms and Inverse Transforms

In Chapter 2 we saw that linear differential equations with constant coefficients have numerous applications and can be solved systematically. There are common situations, however, in which the alternative method of this chapter is preferable. For example, recall the differential equations

$$mx'' + cx' + kx = F(t)$$

and

$$LI'' + RI' + \frac{1}{C}I = E'(t)$$

corresponding to a mass-spring-dashpot system and a series RLC circuit, respectively. It often happens in practice that the forcing term, $F(t)$ or $E'(t)$, has discontinuities—for instance, when the voltage supplied to an electrical circuit is periodically turned off and on. In this case the methods of Chapter 2 can be quite awkward, and the Laplace transform method is more convenient.

The differentiation operator D can be viewed as a transformation which, when applied to the function $f(t)$, yields the new function $D\{f(t)\} = f'(t)$. The Laplace transformation $\mathcal{L}$ involves the operation of integration, and yields the new function $\mathcal{L}\{f(t)\} = F(s)$ of a new independent variable s. The situation is diagrammed in Fig. 4.1. After learning in this section how to compute the Laplace transform $F(s)$ of a function $f(t)$, we will see in Section 4.2 that the Laplace transformation converts a *differential* equation in the unknown function $f(t)$ into an *algebraic* equation in $F(s)$. Because algebraic equations are generally easier to solve than are differential equations, this is one method that simplifies the problem of finding the solution $f(t)$.

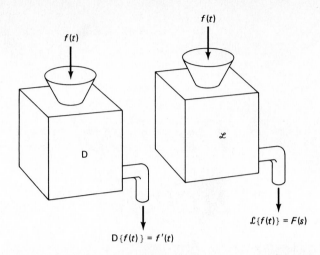

Figure 4.1   Transformation of a function: $\mathcal{L}$ in analogy with D.

DEFINITION: THE LAPLACE TRANSFORM

Given a function $f(t)$ defined for all $t \geqq 0$, the **Laplace transform** *of f is the function F of s defined as follows:*

$$\mathcal{L}\{f(t)\} = F(s) = \int_0^\infty e^{-st} f(t)\, dt \tag{1}$$

for all values of s for which the improper integral converges.

Recall that an **improper integral** over an infinite interval is defined as a limit of integrals over finite intervals; that is,

$$\int_a^\infty g(t)\, dt = \lim_{b \to \infty} \int_a^b g(t)\, dt. \tag{2}$$

If the limit in (2) exists, then we say that the improper integral **converges**. Otherwise, it **diverges** or fails to exist. Note that the integrand of the improper integral in (1) contains the parameter s in addition to the variable of integration t. Therefore, when the integral in (1) converges, it converges not merely to a number, but to a *function F of s*. As in the following examples, it is typical for the improper integral in the definition of $\mathcal{L}\{f(t)\}$ to converge for some values of s and diverge for others.

EXAMPLE 1   With $f(t) \equiv 1$ for $t \geqq 0$, the definition of the Laplace transform in (1) gives

$$\mathcal{L}\{1\} = \int_0^\infty e^{-st}\, dt = \left[-\frac{1}{s} e^{-st} \right]_0^\infty$$

$$= \lim_{b \to \infty} \left[-\frac{1}{s} e^{-bs} + \frac{1}{s} \right],$$

and therefore

$$\mathcal{L}\{1\} = \frac{1}{s} \quad \text{for } s > 0. \tag{3}$$

As in (3), it's good practice to specify the domain of the Laplace transform—in problems as well as in examples. Also, in this computation we have used the common abbreviation

$$\left[g(t) \right]_a^\infty = \lim_{b \to \infty} \left[g(t) \right]_a^b. \tag{4}$$

Remark: The limit we computed above would not exist if $s < 0$, for then $(1/s)e^{-bs}$ becomes unbounded as $b \to \infty$. Hence $\mathcal{L}\{1\}$ is defined only for $s > 0$. This is typical of Laplace transforms; the domain of a transform is normally of the form $s > a$ for some number a.

EXAMPLE 2 With $f(t) = e^{at}$ for $t \geq 0$, we obtain

$$\mathcal{L}\{e^{at}\} = \int_0^\infty e^{-st} e^{at} \, dt$$

$$= \int_0^\infty e^{-(s-a)t} \, dt = \left[-\frac{e^{-(s-a)t}}{s-a} \right]_0^\infty.$$

If $s - a > 0$, then $e^{-(s-a)t} \to 0$ as $t \to +\infty$, so it follows that

$$\mathcal{L}\{e^{at}\} = \frac{1}{s-a} \quad \text{for } s > a. \tag{5}$$

Note here that the improper integral giving $\mathcal{L}\{e^{at}\}$ diverges if $s \leq a$. It is worth noting also that the formula in (5) holds if a is a complex number. For then, with $a = \alpha + i\beta$,

$$e^{-(s-a)t} = e^{i\beta t} e^{-(s-\alpha)t} \longrightarrow 0$$

as $t \to +\infty$, provided that $s > \alpha = \mathcal{R}e(a)$; recall that $e^{i\beta t} = \cos \beta t + i \sin \beta t$.

EXAMPLE 3 Suppose that $f(t) = t^a$ for $t \geq 0$, where a is real and $a > -1$. Then

$$\mathcal{L}\{t^a\} = \int_0^\infty e^{-st} t^a \, dt. \tag{6}$$

One could attack this integral using integration by parts (at least if a is an integer), but instead we use the gamma function, which was defined in Section 3.5 for $x > 0$ by the formula

$$\Gamma(x) = \int_0^\infty e^{-t} t^{x-1} \, dt.$$

For an elementary discussion of $\Gamma(x)$, see the subsection on the gamma function in Section 3.5.

If we substitute $u = st$, $t = u/s$, and $dt = du/s$ in (6), we get

$$\mathcal{L}\{t^a\} = \frac{1}{s^{a+1}} \int_0^\infty e^{-u} u^a \, du = \frac{\Gamma(a+1)}{s^{a+1}} \tag{7}$$

for all $s > 0$ (so that $u = st > 0$). Because $\Gamma(n + 1) = n!$ if n is a nonnegative integer, we see that

$$\mathcal{L}\{t^n\} = \frac{n!}{s^{n+1}} \quad \text{for } s > 0. \tag{8}$$

For instance,

$$\mathcal{L}\{t\} = \frac{1}{s^2}, \qquad \mathcal{L}\{t^2\} = \frac{2}{s^3}, \quad \text{and} \quad \mathcal{L}\{t^3\} = \frac{6}{s^4}.$$

As in Problems 1 and 2, these formulas can be derived immediately from the definition, without use of the gamma function.

It is not necessary for us to proceed much further in the computation of Laplace transforms directly from the definition. Once we know the Laplace transforms of several functions, we can combine them to obtain transforms of other functions. The reason is that the Laplace transformation is a *linear* operation.

THEOREM 1: LINEARITY OF THE LAPLACE TRANSFORMATION

If a and b are constants, then

$$\mathcal{L}\{af(t) + bg(t)\} = a\mathcal{L}\{f(t)\} + b\mathcal{L}\{g(t)\} \tag{9}$$

for all s such that the Laplace transforms of the functions f and g both exist.

The proof of Theorem 1 follows immediately from the linearity of the operations of taking limits and of integration.

$$\mathcal{L}\{af(t) + bg(t)\} = \int_0^\infty e^{-st}[af(t) + bg(t)] \, dt$$

$$= \lim_{c \to \infty} \int_0^b e^{-st}[af(t) + bg(t)] \, dt$$

$$= a\left(\lim_{c \to \infty} \int_0^c e^{-st}f(t) \, dt\right) + b\left(\lim_{c \to \infty} \int_0^c e^{-st}g(t) \, dt\right)$$

$$= a\mathcal{L}\{f(t)\} + b\mathcal{L}\{g(t)\}.$$

EXAMPLE 4   The formulas in (7)–(9) yield

$$\mathcal{L}\{3t^2 + 4t^{3/2}\} = 3\frac{2!}{t^3} + \frac{4\Gamma(\frac{5}{2})}{t^{5/2}}$$

$$= \frac{6}{t^3} + 3\sqrt{\frac{\pi}{t^5}},$$

using the fact that

$$\Gamma(\tfrac{5}{2}) = \tfrac{3}{2}\Gamma(\tfrac{3}{2}) = \tfrac{3}{2}\cdot\tfrac{1}{2}\Gamma(\tfrac{1}{2}) = \tfrac{3}{4}\sqrt{\pi},$$

because $\Gamma(x + 1) = x\Gamma(x)$ and $\Gamma(\tfrac{1}{2}) = \sqrt{\pi}$.

EXAMPLE 5   Recall that $\cosh kt = (e^{kt} + e^{-kt})/2$. If $k > 0$, then Theorem 1 and Example 2 together give

$$\mathcal{L}\{\cosh kt\} = \frac{1}{2}\mathcal{L}\{e^{kt}\} + \frac{1}{2}\mathcal{L}\{e^{-kt}\}$$

$$= \frac{1}{2}\left(\frac{1}{s-k} + \frac{1}{s+k}\right);$$

that is,

$$\mathcal{L}\{\cosh kt\} = \frac{s}{s^2 - k^2} \quad \text{for } s > k > 0. \tag{10}$$

Similarly,

$$\mathcal{L}\{\sinh kt\} = \frac{k}{s^2 - k^2} \quad \text{for } s > k > 0. \tag{11}$$

Because $\cos kt = (e^{ikt} + e^{-ikt})/2$, the formula in (5) (with $a = ik$) yields

$$\mathcal{L}\{\cos kt\} = \frac{1}{2}\left(\frac{1}{s - ik} + \frac{1}{s + ik}\right) = \frac{1}{2} \cdot \frac{2s}{s^2 - (ik)^2},$$

and thus

$$\mathcal{L}\{\cos kt\} = \frac{s}{s^2 + k^2} \quad \text{for } s > 0. \tag{12}$$

(The domain follows from $s > \mathcal{R}e(ik) = 0$.) Similarly,

$$\mathcal{L}\{\sin kt\} = \frac{k}{s^2 + k^2} \quad \text{for } s > 0. \tag{13}$$

EXAMPLE 6 Applying linearity, the formula in (12), and a familiar trigonometric identity, we get

$$\mathcal{L}\{3e^{2s} + 2\sin^2 3t\} = \mathcal{L}\{3e^{2t} + 1 - \cos 6t\}$$

$$= \frac{3}{s - 2} + \frac{1}{s} - \frac{s}{s^2 + 36}$$

$$= \frac{3s^3 + 144s - 72}{s(s - 2)(s^2 + 36)} \quad \text{for } s > 0.$$

According to Theorem 3, no two different functions that are both continuous for all $t \geq 0$ can have the same Laplace transform. Thus if $F(s)$ is the transform of some continuous function $f(t)$, then $f(t)$ is uniquely determined. This observation allows us to make the following definition: If $F(s) = \mathcal{L}\{f(t)\}$, then we call $f(t)$ the **inverse Laplace transform** of $F(s)$ and write

$$f(t) = \mathcal{L}^{-1}\{F(s)\}. \tag{14}$$

Thus

$$\mathcal{L}^{-1}\left\{\frac{1}{s^3}\right\} = \frac{1}{2}t^2, \qquad \mathcal{L}^{-1}\left\{\frac{1}{s + 2}\right\} = e^{-2t},$$

$$\mathcal{L}^{-1}\left\{\frac{2}{s^2 + 9}\right\} = \frac{2}{3}\sin 3t,$$

and so on.

NOTATION: FUNCTIONS AND THEIR TRANSFORMS

Throughout this chapter we denote functions of t by lowercase letters. The transform of a function will always be denoted by the same letter capitalized. Thus F(s) is the Laplace transform of f(t), and x(t) is the inverse Laplace transform of X(s).

A table of Laplace transforms serves a purpose similar to that of a table of known integrals. The table in Fig. 4.2 lists the transforms derived in this section; many additional transforms can be derived from these few, using various general properties of the Laplace transformation (which we will discuss in subsequent sections).

$f(t)$	$F(s)$
1	$\dfrac{1}{s}$ $\quad (s > 0)$
t	$\dfrac{1}{s^2}$ $\quad (s > 0)$
t^n $(n \geqslant 0)$	$\dfrac{n!}{s^{n+1}}$ $\quad (s > 0)$
t^a $(a > -1)$	$\dfrac{\Gamma(a+1)}{s^{a+1}}$ $\quad (s > 0)$
e^{at}	$\dfrac{1}{s-a}$ $\quad (s > a)$
$\cos kt$	$\dfrac{s}{s^2 + k^2}$ $\quad (s > 0)$
$\sin kt$	$\dfrac{k}{s^2 + k^2}$ $\quad (s > 0)$
$\cosh kt$	$\dfrac{s}{s^2 - k^2}$ $\quad (s > |k|)$
$\sinh kt$	$\dfrac{k}{s^2 - k^2}$ $\quad (s > |k|)$
$u(t - a)$	$\dfrac{e^{-as}}{s}$ $\quad (s > 0)$

Figure 4.2 A short table of Laplace transforms.

As we remarked at the beginning of this section, we need to be able to handle certain types of discontinuous functions. The function $f(t)$ is said to be **piecewise continuous** on the bounded interval $a \leqq t \leqq b$ provided that $[a, b]$ can be subdivided into finitely many abutting subintervals so that:

1. f is continuous in the interior of each of these subintervals; and
2. $f(t)$ has a finite one-sided limit as t approaches each endpoint of each such subinterval from its interior.

We say that f is piecewise continuous for $t \geqq 0$ if it is piecewise continuous on every bounded subinterval of the nonnegative real axis. Thus a piecewise continuous function has only simple discontinuities (if any) at isolated points. At such points the value of the function experiences a finite jump, as indicated in Fig. 4.3. The **jump in $f(t)$ at the point c** is defined to be $f(c+) - f(c-)$, where

$$f(c+) = \lim_{\epsilon \to 0^+} f(c + \epsilon)$$

and

$$f(c-) = \lim_{\epsilon \to 0^+} f(c - \epsilon).$$

Perhaps the simplest piecewise continuous function is the **unit step func-**

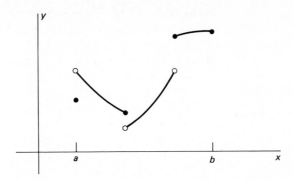

Figure 4.3 Graph of a piecewise continuous function; the solid dots indicate values of the function at discontinuities.

tion, shown in Fig. 4.4. It is defined as follows:

$$u(t) = \begin{cases} 0 & \text{for } t < 0, \\ 1 & \text{for } t \geq 0. \end{cases} \tag{15}$$

Because $u(t) = 1$ for $t \geq 0$ and because the Laplace transform involves only the values of a function for $t \geq 0$, we see immediately that

$$\mathcal{L}\{u(t)\} = \frac{1}{s} \qquad (s > 0) \tag{16}$$

The unit step function $u_a(t) = u(t - a)$ is shown in Fig. 4.5. Its jump occurs at $t = a$ rather than at $t = 0$; equivalently,

$$u_a(t) = u(t - a) = \begin{cases} 0 & \text{for } t < a, \\ 1 & \text{for } t \geq a. \end{cases} \tag{17}$$

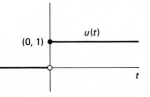

Figure 4.4 Graph of the unit step function.

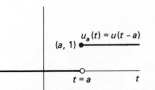

Figure 4.5 The unit step function $u_a(t)$—jump at $t = a$.

EXAMPLE 7 Find $\mathcal{L}\{u_a(t)\}$ if $a > 0$.

Solution We begin with the definition of the Laplace transform. We obtain

$$\mathcal{L}\{u_a(t)\} = \int_0^\infty e^{-st} u_a(t)\, dt = \int_a^\infty e^{-st}\, dt = \lim_{b \to \infty} \left[-\frac{e^{-st}}{s} \right]_a^b;$$

consequently,

$$\mathcal{L}\{u_a(t)\} = \frac{e^{-as}}{s} \qquad (s > 0,\ a > 0). \tag{18}$$

It is a familiar fact from calculus that the integral $\int_a^b g(t)\, dt$ exists if g is piecewise continuous on the bounded interval $[a, b]$. Hence if f is piecewise continuous for $t \geq 0$, it follows that the integral $\int_0^b e^{-st} f(t)\, dt$ exists for all $b < \infty$. But in order for $F(s)$—the limit of this last integral as $b \to \infty$—to exist, we need some condition on the rate of growth of $f(t)$ as $t \to \infty$. The function f is said to be of **exponential order** as $t \to \infty$ if there exist nonnegative constants M, c, and T such that

$$|f(t)| \leq Me^{ct} \quad \text{for } t \geq T. \tag{19}$$

Thus a function is of exponential order provided that it grows no more rapidly (as $t \to \infty$) than a constant multiple of some exponential function with a linear exponent. The particular values of M, c, and T are not so important. What is important is that some such values exist so that the condition in (19) is satisfied.

Every polynomial $p(t)$ is of exponential order, with $M = c = 1$ in (19); this follows from the fact that $p(t)/e^t \to 0$ as $t \to \infty$. The function e^{t^2} is an example of one that is *not* of exponential order: Because

$$\lim_{t \to \infty} \frac{e^{t^2}}{e^{ct}} = \lim_{t \to \infty} e^{t^2 - ct} = +\infty,$$

the condition in (19) cannot hold for any (finite) value of M. The condition in (19) merely says that $f(t)/e^{ct}$ is *bounded* for t sufficiently large. In particular, any bounded function—such as $\cos kt$ or $\sin kt$—is of exponential order.

THEOREM 2:   EXISTENCE OF LAPLACE TRANSFORMS

If the function f is piecewise continuous for $t \geq 0$ and is of of exponential order as $t \to \infty$, then its Laplace transform $F(s) = \mathcal{L}\{f(t)\}$ exists. More precisely, if f is piecewise continuous and satisfies the condition in (19), then $F(s)$ exists for all $s > c$.

Proof   First we note that we can take $T = 0$ in (19). For by piecewise continuity, $|f(t)|$ is bounded on $[0, T]$. Increasing M in (19) if necessary, we can therefore assume that $|f(t)| \leq M$ if $0 \leq t \leq T$. Because $e^{ct} \geq 1$ for $t \geq 0$, it then follows that $|f(t)| \leq Me^{ct}$ for all $t \geq 0$.

By a standard theorem on convergence of improper integrals—the fact that absolute convergence implies convergence—it suffices for us to prove that the integral $\int_0^\infty |e^{-st}f(t)| \, dt$ exists for $s > c$. To do this, it suffices in turn to show that the value of the integral $\int_0^b |e^{-st}f(t)| \, dt$ remains bounded as $b \to \infty$. But the fact that $|f(t)| \leq Me^{ct}$ for all $t \geq 0$ implies that

$$\int_0^b |e^{-st}f(t)| \, dt \leq \int_0^b |e^{-st}Me^{ct}| \, dt = M \int_0^b e^{-(s-c)t} \, dt$$

$$\leq M \int_0^\infty e^{-(s-c)t} \, dt = \frac{M}{s-c}$$

if $s > c$. This proves Theorem 2. We have shown, moreover, that

$$|F(s)| \leq \int_0^\infty |e^{-st}f(t)| \, dt \leq \frac{M}{s-c} \tag{20}$$

if $s > c$. When we take limits as $s \to \infty$, we get the following additional result.

If $f(t)$ satisfies the hypotheses of Theorem 2, then

$$\lim_{s \to \infty} F(s) = 0. \tag{21}$$

The condition in (21) severely limits the functions that can be Laplace transforms. For instance, the function $s/(s + 1)$ cannot be the Laplace transform of any "reasonable" function because its limit as $s \to \infty$ is 1, not 0. More generally, a rational function—a quotient of two polynomials—can be (and is, as we shall see) a Laplace transform only if the degree of its numerator is less than the degree of its denominator.

On the other hand, the hypotheses of Theorem 2 are sufficient, but not necessary, conditions for existence of the Laplace transform of $f(t)$. For example, the function $f(t) = t^{-1/2} = 1/\sqrt{t}$ fails to be piecewise continuous (at 0), but nevertheless (Example 3 with $a = -\frac{1}{2} > -1$) its Laplace transform

$$\mathcal{L}\{t^{-1/2}\} = \frac{\Gamma(\frac{1}{2})}{s^{1/2}} = \sqrt{\frac{\pi}{s}}$$

both exists and violates the condition in (20), which would imply that $sF(s)$ remains bounded as $s \to \infty$.

The remainder of this chapter will be devoted largely to techniques for solving a differential equation by first finding the Laplace transform of its solution. It is then vital for us to know that this uniquely determines the solution of the differential equation; that is, that the function of s we have found has only one inverse Laplace transform that could be the desired solution. The following theorem is proved in Chapter 6 of Churchill's *Operational Mathematics* (New York: McGraw-Hill, 3rd ed., 1972).

THEOREM 3: UNIQUENESS OF INVERSE LAPLACE TRANSFORMS

Suppose that the functions $f(t)$ and $g(t)$ satisfy the hypotheses of Theorem 2, so that their Laplace transforms $F(s)$ and $G(s)$ exist. If $F(s) = G(s)$ for all $s > c$ (for some c), then $f(t) = g(t)$ wherever f and g are both continuous.

Thus two piecewise continuous functions of exponential order with the same Laplace transform can differ only at their isolated points of discontinuity. This is of no importance in most practical applications, so we may regard inverse Laplace transforms as being essentially unique. In particular, two solutions of a differential equation must both be continuous, and hence must be the same solution if they have the same Laplace transform.

Laplace transforms have an interesting history. The integral in the definition of the Laplace transform probably appeared first in the work of Euler. It is customary in mathematics to name a technique or theorem for the next person after Euler to discover it (or else there would be several hundred different examples of "Euler's theorem"). In this case, the next person was the French mathematician Pierre Simon de Laplace (1749–1827), who employed such integrals in his work on probability theory. The so-called operational techniques for solving differential equations, which are based on Laplace transforms, were not exploited by Laplace. Indeed, they were discovered and popularized by practicing engineers—notably the English electrical engineer Oliver Heaviside (1850–1925). These techniques were successfully and widely applied before they had been rigorously justified, and around the beginning of this century their validity was the subject of considerable controversy.

4.1 Problems

Apply the definition in (1) to find directly the Laplace transforms of the functions described (by formula or graph) in Problems 1–10.

1. $f(t) = t$.

2. $f(t) = t^2$.

3. $f(t) = e^{3t+1}$.

4. $f(t) = \cos t$.

5. $f(t) = \sinh t$.

6. $f(t) = \sin^2 t$.

7.

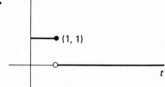

Figure 4.6 Graph of the function for Problem 7.

8.

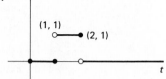

Figure 4.7 Graph of the function for Problem 8.

9.

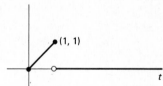

Figure 4.8 Graph of the function for Problem 9.

10.

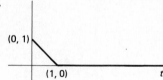

Figure 4.9 Graph of the function for Problem 10.

Use the transforms in Fig. 4.2 to find the Laplace transforms of the functions in Problems 11–22. A preliminary integration by parts may be necessary.

11. $f(t) = \sqrt{t} + 3t$.

12. $f(t) = 3t^{5/2} - 4t^3$.

13. $f(t) = t - 2e^{3t}$.

14. $f(t) = t^{3/2} + e^{-10t}$.

15. $f(t) = 1 + \cosh 5t$.

16. $f(t) = \sin 2t + \cos 2t$.

17. $f(t) = \cos^2 2t$.

18. $f(t) = \sin 3t \cos 3t$.

19. $f(t) = (1 + t)^3$.

20. $f(t) = te^t$.

21. $f(t) = t \cos 2t$.

22. $f(t) = \sinh^2 3t$.

Use the transforms in Fig. 4.2 to find the inverse Laplace transforms of the functions in Problems 23–32.

23. $F(s) = \dfrac{3}{s^4}$.

24. $F(s) = s^{-3/2}$.

25. $F(s) = \dfrac{1}{s} - \dfrac{2}{s^{5/2}}$.

26. $F(s) = \dfrac{1}{s+5}$.

27. $F(s) = \dfrac{3}{s-4}.$

28. $F(s) = \dfrac{3s+1}{s^2+4}.$

29. $F(s) = \dfrac{5-3s}{s^2+9}.$

30. $F(s) = \dfrac{9+s}{4-s^2}.$

31. $F(s) = \dfrac{10s-3}{25-s^2}.$

32. $F(s) = 2s^{-1}e^{-3s}.$

33. Derive the transform of $\sin kt$ by the method used in the text to derive the formula in (12).

34. Derive the transform of $\sinh kt$ by the method used in the text to derive the formula in (10).

35. Use the tabulated integral $\int e^{ax} \cos bx \, dx$ to obtain $\mathcal{L}\{\cos kt\}$ directly from the definition of the Laplace transform.

36. Show that the function $f(t) = \sin(e^{t^2})$ is of exponential order as $t \to \infty$, but that its derivative is not.

37. Let $f(t) = 1$ for $0 \le t \le a$, $f(t) = 0$ for $t > a$ (where $a > 0$). Express f in terms of unit step functions to show that $\mathcal{L}\{f(t)\} = s^{-1}(1 - e^{-as})$.

38. Let $f(t) = 1$ if $a \le t \le b$, $f(t) = 0$ if either $t < a$ or $t > b$ (where $0 < a < b$). Express f in terms of unit step functions to show that $\mathcal{L}\{f(t)\} = s^{-1}(e^{-as} - e^{-bs})$.

39. The unit staircase function is defined as follows:

$$f(t) = n \quad \text{if} \quad n - 1 < t \le n, \, n = 1, 2, 3, \dots.$$

(a) Sketch the graph of f to see why its name is appropriate.

(b) Show that

$$f(t) = \sum_{n=0}^{\infty} u(t - n)$$

for all $t > 0$. (c) Assume that the Laplace transform of the infinite series in (b) can be taken termwise. Apply the geometric series to obtain the result

$$\mathcal{L}\{f(t)\} = \dfrac{1}{s(1 - e^{-s})}.$$

40. (a) The graph of the function f is shown in Fig. 4.10. Show that f can be written in the form

$$f(t) = \sum_{n=0}^{\infty} (-1)^n u(t - n).$$

(b) Use the method of Problem 39 to show that

$$\mathcal{L}\{f(t)\} = \dfrac{1}{s(1 + e^{-s})}.$$

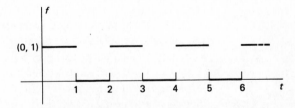

Figure 4.10 Graph of the function for Problem 40.

41. The graph of the square wave function $g(t)$ is shown in Fig. 4.11. Express g in terms of the function f of Problem 40 and hence deduce that

$$\mathcal{L}\{g(t)\} = \frac{1 - e^{-s}}{s(1 + e^{-s})} = \frac{1}{s} \tanh \frac{s}{2}.$$

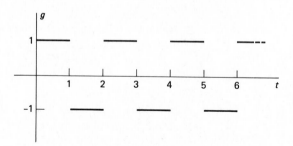

Figure 4.11 Graph of the function for Problem 41.

4.2
Transformation of Initial Value Problems

We now discuss the application of Laplace transforms to solve a linear differential equation with constant coefficients such as

$$ax''(t) + bx'(t) + cx(t) = f(t), \tag{1}$$

with given initial conditions $x(0) = x_0$ and $x'(0) = x_0'$. By the linearity of the Laplace transformation, we can transform Eq. (1) by separately taking the Laplace transform of each term in the equation. The transformed equation is

$$a\mathcal{L}\{x''(t)\} + b\mathcal{L}\{x'(t)\} + c\mathcal{L}\{x(t)\} = \mathcal{L}\{f(t)\}; \tag{2}$$

it involves the transforms of the derivatives x' and x'' of the unknown function $x(t)$. The key to the method is the following theorem, which tells us how to express the transform of the *derivative* of a function in terms of the transform of the function itself.

THEOREM 1: TRANSFORMS OF DERIVATIVES

Suppose that the function $f(t)$ is continuous and piecewise smooth for $t \geqq 0$ and is of exponential order as $t \to +\infty$, so that there exist nonnegative constants M, c, and T such that

$$|f(t)| \leqq Me^{ct} \quad \text{for } t \geqq T. \tag{3}$$

Then $\mathcal{L}\{f'(t)\}$ exists for $s > c$, and

$$\mathcal{L}\{f'(t)\} = s\mathcal{L}\{f(t)\} - f(0) = sF(s) - f(0). \tag{4}$$

The function f is called **piecewise smooth** on the bounded interval $[a, b]$ if it is piecewise continuous on $[a, b]$ and differentiable except at finitely many points, with $f'(t)$ being piecewise continuous on $[a, b]$. We may assign arbitrary values to $f'(t)$ at the isolated points at which f is not differentiable. We say that f is piecewise smooth for $t \geqq 0$ if it is piecewise smooth on every bounded subinterval of the nonnegative real axis. Figure 4.12 indicates how "corners" on the graph of f correspond to discontinuities in its derivative f'.

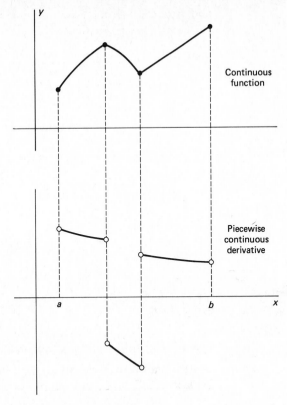

Figure 4.12 Discontinuities of f' corresponding to "corners" on the graph of f.

The main idea of the proof of Theorem 1 is exhibited best by the case when $f'(t)$ is continuous (not merely piecewise continuous) for $t \geqq 0$. Then, beginning with the definition of $\mathcal{L}\{f'(t)\}$ and integrating by parts, we get

$$\mathcal{L}\{f'(t)\} = \int_0^\infty e^{-st} f'(t)\, dt$$

$$= \left[e^{-st} f(t) \right]_0^\infty + s \int_0^\infty e^{-st} f(t)\, dt.$$

Because of (3), the integrated term $e^{-st}f(t)$ approaches 0 (when $s > c$) as $t \longrightarrow +\infty$, and its value at the lower limit $t = 0$ contributes $-f(0)$ to the evaluation of the expression above. The integral that remains above is simply $\mathcal{L}\{f(t)\}$; by Theorem 2 in Section 4.1, the integral converges when $s > c$. Thus $\mathcal{L}\{f'(t)\}$ exists when $s > c$, and its value is that given in Eq. (4). We will defer the case in which $f'(t)$ has isolated discontinuities to the end of this section.

In order to transform Eq. (1), we need the transform of the second derivative as well. If we assume that $g(t) = f'(t)$ satisfies the hypotheses of Theorem 1, then that theorem yields

$$\mathcal{L}\{f''(t)\} = \mathcal{L}\{g'(t)\} = s\mathcal{L}\{g(t)\} - g(0)$$

$$= s\mathcal{L}\{f'(t)\} - f'(0) = s[s\mathcal{L}\{f(t)\} - f(0)] - f'(0),$$

and thus
$$\mathcal{L}\{f''(t)\} = s^2 F(s) - s f(0) - f'(0). \tag{5}$$

A repetition of this calculation gives
$$\mathcal{L}\{f'''(t)\} = s\mathcal{L}\{f''(t)\} - f''(0)$$
$$= s^3 F(s) - s^2 f(0) - s f'(0) - f''(0). \tag{6}$$

After finitely many such steps we obtain the following extension of Theorem 1.

COROLLARY: TRANSFORMS OF HIGHER DERIVATIVES

Suppose that the functions $f, f', f'', \ldots, f^{(n-1)}$ are continuous and piece-wise smooth for $t \geq 0$, and that each of these functions satisfies the condition in (3) with the same values of M and c. Then $\mathcal{L}\{f^{(n)}(t)\}$ exists when $s > c$, and
$$\mathcal{L}\{f^{(n)}(t)\} = s^n \mathcal{L}\{f(t)\} - s^{n-1} f(0) - s^{n-2} f'(0) - \cdots - f^{(n-1)}(0)$$
$$= s^n F(s) - s^{n-1} f(0) - \cdots - s f^{(n-2)}(0) - f^{(n-1)}(0). \tag{7}$$

EXAMPLE 1   Solve the initial value problem
$$x'' - x' - 6x = 0; \qquad x(0) = 2, \qquad x'(0) = -1.$$

Solution   With the given initial values the formulas in (4) and (5) yield
$$\mathcal{L}\{x'(t)\} = s\mathcal{L}\{x(t)\} - x(0) = sX(s) - 2$$
and
$$\mathcal{L}\{x''(t)\} = s^2 \mathcal{L}\{x(t)\} - sx(0) - x'(0) = s^2 X(s) - 2s + 1,$$

where (according to our convention about notation) $X(s)$ denotes the Laplace transform of the (unknown) function $x(t)$. Hence the transformed equation is
$$[s^2 X(s) - 2s + 1] - [sX(s) - 2] - 6[X(s)] = 0,$$
which we quickly simplify to
$$(s^2 - s - 6)X(s) - 2s + 3 = 0.$$
Thus
$$X(s) = \frac{2s - 3}{s^2 - s - 6} = \frac{2s - 3}{(s - 3)(s + 2)}.$$

By the elementary method of partial fractions, there exist constants A and B such that
$$\frac{2s - 3}{(s - 3)(s + 2)} = \frac{A}{s - 3} + \frac{B}{s + 2},$$

and this leads to the identity $2s - 3 = A(s + 2) + B(s - 3)$. If we substitute $s = 3$, we find that $A = \frac{3}{5}$; substitution of $s = -2$ shows that $B = \frac{7}{5}$. Hence
$$\mathcal{L}\{x(t)\} = \frac{\frac{3}{5}}{s - 3} + \frac{\frac{7}{5}}{s + 2}.$$

Because $\mathcal{L}^{-1}\{1/(s-a)\} = e^{at}$, it follows that

$$x(t) = \tfrac{3}{5}e^{3t} + \tfrac{7}{5}e^{-2t}$$

is the solution of the original initial value problem. Note that we did not first find the general solution of the differential equation. The Laplace transform method directly yields the desired particular solution, automatically taking into account—via Theorem 1 and its corollary—the given initial conditions.

EXAMPLE 2 Solve the initial value problem

$$x'' + 4x = \sin 3t; \qquad x(0) = 0 = x'(0).$$

Solution Because both initial values are zero, the formula in (5) yields $\mathcal{L}\{x''(t)\} = s^2 X(s)$. We read the transform of $\sin 3t$ from the table on page 270, and thereby get the transformed equation

$$s^2 X(s) + 4X(s) = \frac{3}{s^2 + 9},$$

and so

$$X(s) = \frac{3}{(s^2 + 4)(s^2 + 9)}.$$

The method of partial fractions calls for

$$\frac{3}{(s^2 + 4)(s^2 + 9)} = \frac{As + B}{s^2 + 4} + \frac{Cs + D}{s^2 + 9}.$$

The fact that there are no terms of odd degree on the left-hand side suggests that $A = C = 0$, and we then solve easily—by multiplying each side by $(s^2 + 4)(s^2 + 9)$ and then equating coefficients of like powers of s—for $B = \tfrac{3}{5}$ and $D = -\tfrac{3}{5}$. Hence

$$\mathcal{L}\{x(t)\} = \frac{3}{10} \cdot \frac{2}{s^2 + 4} - \frac{1}{5} \cdot \frac{3}{s^2 + 9}.$$

Because $\mathcal{L}\{\sin 2t\} = 2/(s^2 + 4)$ and $\mathcal{L}\{\sin 3t\} = 3/(s^2 + 9)$, it follows that

$$x(t) = \tfrac{3}{10}\sin 2t - \tfrac{1}{5}\sin 3t.$$

Note that the Laplace transform method again gives the solution directly, without the necessity of first finding the complementary function and a particular solution of the original nonhomogeneous differential equation. Thus nonhomogeneous equations are solved in exactly the same manner as are homogeneous equations.

If we begin with the general linear second order equation

$$ax'' + bx' + cx = f(t) \tag{8}$$

with constant coefficients, the transformed equation is

$$a[s^2 X(s) - sx(0) - x'(0)] + b[sX(s) - x(0)] + cX(s) = F(s). \tag{9}$$

Note that (9) is an *algebraic* equation—indeed, a linear equation—in the "unknown" $X(s)$. This is the source of the power of the Laplace transform method: Linear differential equations are transformed into readily solved algebraic equations. If we solve (9) for $X(s)$, we get

$$X(s) = \frac{F(s)}{Z(s)} + \frac{[ax(0)]s + [ax'(0) + bx(0)]}{Z(s)}, \tag{10}$$

where

$$Z(s) = as^2 + bs + c. \tag{11}$$

If Eq. (8) describes the behavior of a physical system under the influence of an external force $f(t)$, then $Z(s)$ depends only on the system itself. For instance, $Z(s) = ms^2 + cs + k$ for the familiar mass-spring-dashpot (mass m, spring constant k, damping constant c) system. Then Eq. (10) presents $\mathcal{L}\{s(t)\}$ as the sum of a term depending only on the external force and one depending only on the initial conditions. In the case of an underdamped system, these two terms are the transforms of the steady periodic solution and the transient solution, respectively. The only potential difficulty in finding these solutions is in finding the inverse Laplace transform of the right-hand side in (10). Much of the remainder of this chapter is devoted to techniques for finding transforms and inverse transforms. In particular, we seek those methods that are sufficiently powerful to enable us to solve problems (unlike Examples 1 and 2) that cannot be solved readily by the methods of Chapter 2.

EXAMPLE 3   Show that

$$\mathcal{L}\{te^{at}\} = \frac{1}{(s-a)^2}.$$

Solution   If $f(t) = te^{at}$, then $f(0) = 0$ and $f'(t) = e^{at} + ate^{at}$. Hence Theorem 1 gives

$$\mathcal{L}\{e^{at} + ate^{at}\} = \mathcal{L}\{f'(t)\} = s\mathcal{L}\{f(t)\} = s\mathcal{L}\{te^{at}\}.$$

It follows by linearity of the transform that

$$\mathcal{L}\{e^{at}\} + a\mathcal{L}\{te^{at}\} = s\mathcal{L}\{te^{at}\}.$$

Hence

$$\mathcal{L}\{te^{at}\} = \frac{\mathcal{L}\{e^{at}\}}{s-a} = \frac{1}{(s-a)^2} \tag{12}$$

because $\mathcal{L}\{e^{at}\} = 1/(s-a)$.

EXAMPLE 4   Find $\mathcal{L}\{t \sin kt\}$.

Solution   Let $f(t) = t \sin kt$. Then $f(0) = 0$ and

$$f'(t) = \sin kt + kt \cos kt.$$

The derivative involves the new function $t \cos kt$, so we note that $f'(0) = 0$ and differentiate again. The result is

$$f''(t) = 2k \cos kt - k^2t \sin kt.$$

But $\mathcal{L}\{f''(t)\} = s^2\mathcal{L}\{f(t)\}$ by the formula in (5) for the transform of the second derivative and $\mathcal{L}\{\cos kt\} = s/(s^2 + k^2)$, so we have

$$\frac{2ks}{s^2 + k^2} - k^2 \mathcal{L}\{t \sin kt\} = s^2 \mathcal{L}\{t \sin kt\}.$$

Finally we solve this equation for

$$\mathcal{L}\{t \sin kt\} = \frac{2ks}{(s^2 + k^2)^2}. \tag{13}$$

This procedure is considerably more pleasant than the alternative of evaluating the integral

$$\mathcal{L}\{t \sin kt\} = \int_0^\infty te^{-st} \sin kt \, dt.$$

Examples 3 and 4 exploit the fact that, if $f(0) = 0$, then differentiation of f corresponds to multiplication of its transform by s. It is reasonable to expect the inverse operation of integration (antidifferentiation) to correspond to division of the transform by s.

THEOREM 2: TRANSFORMS OF INTEGRALS

If $f(t)$ is a piecewise continuous function for $t \geq 0$ and satisfies the condition of exponential order $|f(t)| \leq Me^{ct}$ for $t \geq T$, then

$$\mathcal{L}\left\{\int_0^t f(\tau) \, d\tau\right\} = \frac{1}{s} \mathcal{L}\{f(t)\} = \frac{F(s)}{s} \tag{14}$$

for $s > c$. Equivalently,

$$\mathcal{L}^{-1}\left\{\frac{F(s)}{s}\right\} = \int_0^t f(\tau) \, d\tau. \tag{15}$$

Proof Because $f(t)$ is piecewise continuous, the fundamental theorem of calculus implies that

$$g(t) = \int_0^t f(\tau) \, d\tau$$

is continuous and that $g'(t) = f(t)$ where f is continuous; thus $g(t)$ is continuous and piecewise smooth for $t \geq 0$. Furthermore,

$$|g(t)| \leq \int_0^t |f(\tau)| \, d\tau \leq M \int_0^t e^{c\tau} \, d\tau$$

$$= \frac{M}{c}(e^{ct} - 1) < \frac{M}{c} e^{ct},$$

so $g(t)$ is of exponential order. Hence we can apply Theorem 1 to g; this gives

$$\mathcal{L}\{f(t)\} = \mathcal{L}\{g'(t)\} = s\mathcal{L}\{g(t)\} - g(0).$$

Now $g(0) = 0$, so division by s yields

$$\mathcal{L}\left\{\int_0^t f(\tau) \, d\tau\right\} = \mathcal{L}\{g(t)\} = \frac{\mathcal{L}\{f(t)\}}{s},$$

which completes the proof.

EXAMPLE 5   Find the inverse Laplace transform of

$$\frac{1}{s^2(s-a)}.$$

Solution   In effect, Eq. (15) means that we can delete a factor of s from the denominator, find the inverse transform of the resulting simpler expression, and finally integrate from 0 to t (to "correct" for the missing factor s). Thus

$$\mathcal{L}^{-1}\left\{\frac{1}{s(s-a)}\right\} = \int_0^t \mathcal{L}^{-1}\left\{\frac{1}{s-a}\right\} d\tau$$

$$= \int_0^t e^{a\tau}\, d\tau = \frac{1}{a}(e^{at} - 1).$$

We now repeat the technique to obtain

$$\mathcal{L}^{-1}\left\{\frac{1}{s^2(s-a)}\right\} = \int_0^t \mathcal{L}^{-1}\left\{\frac{1}{s(s-a)}\right\} d\tau = \int_0^t \frac{1}{a}(e^{a\tau} - 1)\, d\tau$$

$$= \left[\frac{1}{a}\left(\frac{1}{a}e^{a\tau} - \tau\right)\right]_0^t = \frac{1}{a^2}(e^{at} - at - 1).$$

This technique is often a more convenient way than the method of partial fractions for finding an inverse transform of a fraction of the from $p(s)/[s^n q(s)]$.

Proof of Theorem 1

We conclude this section with the proof of Theorem 1 in the general case in which f' is merely piecewise continuous. We need to prove that the limit

$$\lim_{b\to\infty} \int_0^b e^{-st} f'(t)\, dt$$

exists and also need to find its value. With b fixed, let $t_1, t_2, \ldots, t_{k-1}$ be the points interior to the interval $[0, b]$ at which f' is discontinuous. Let $t_0 = 0$ and $t_k = b$. Then we can integrate by parts on each interval (t_{n-1}, t_n) where f' is continuous. This yields

$$\int_0^b e^{-st} f'(t)\, dt = \sum_{n=1}^{k} \int_{t_{n-1}}^{t_n} e^{-st} f'(t)\, dt$$

$$= \sum_{n=1}^{k} \left(\left[e^{-st} f(t)\right]_{t_{n-1}}^{t_n} + s \int_{t_{n-1}}^{t_n} e^{-st} f(t)\, dt\right)$$

$$= -f(0) - \sum_{n=1}^{k-1} e^{(-st_n)} j_f(t_n) + e^{-sb} f(b) + s \int_0^b e^{-st} f(t)\, dt, \quad (16)$$

where

$$j_f(t_n) = f(t_n^+) - f(t_n^-) \tag{17}$$

is the jump in $f(t)$ at $t = t_n$. But because f is continuous, each jump is zero: $j_f(t_n) = 0$ for all n. Moreover, if $b > c$, then $e^{-sb} f(b) \to 0$ as $b \to \infty$. There-

fore when we take the limit in (16) as $b \longrightarrow \infty$, we get the desired result:
$\mathcal{L}\{f'(t)\} = s\mathcal{L}\{f(t)\} - f(0)$.

Extension of Theorem 1

Suppose that the original function $f(t)$ is itself only piecewise continuous (instead of continuous), with its (finite jump) discontinuities located at the points $t_1, t_2, t_3, \ldots$. Assuming that $\mathcal{L}\{f'(t)\}$ exists, when we take the limit in (16) as $b \longrightarrow \infty$, we get

$$\mathcal{L}\{f'(t)\} = sF(s) - f(0) - \sum_{n=1}^{\infty} e^{-(st_n)} j_f(t_n). \tag{18}$$

EXAMPLE 6   Let $f(t) = 1 + [\![t]\!]$ be the unit staircase function; its graph is shown in Fig. 4.13. Then $f(0) = 1$, $f'(t) \equiv 0$, and $j_f(n) = 1$ for each $n = 1, 2, 3, \ldots$. Hence (18) yields

$$0 = sF(s) - 1 - \sum_{n=1}^{\infty} e^{-ns},$$

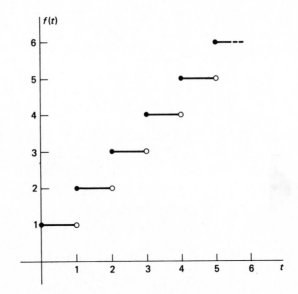

Figure 4.13   The unit staircase function of Example 6.

so the Laplace transform of $f(t)$ is

$$F(s) = \frac{1}{s} \sum_{n=0}^{\infty} e^{-ns} = \frac{1}{s(1 - e^{-s})}.$$

In the last step, we used the formula for the sum of a geometric series,

$$\sum_{n=1}^{\infty} x^n = \frac{1}{1 - x},$$

with $x = e^{-s} < 1$.

4.2 Problems

Use Laplace transforms to solve the initial value problems in Problems 1–10.

1. $x'' + 4x = 0$; $x(0) = 5, x'(0) = 0$.

2. $x'' + 9x = 0$; $x(0) = 3, x'(0) = 4$.

3. $x'' - x' - 2x = 0$; $x(0) = 0, x'(0) = 2$.

4. $x'' + 8x' + 15x = 0$; $x(0) = 2, x'(0) = -3$.

5. $x'' + x = \sin 2t$; $x(0) = 0 = x'(0)$.

6. $x'' + 4x = \cos t$; $x(0) = 0 = x'(0)$.

7. $x'' + x = \cos 3t$; $x(0) = 1, x'(0) = 0$.

8. $x'' + 9x = 1$; $x(0) = 0 = x'(0)$.

9. $x'' + 4x' + 3x = 1$; $x(0) = x'(0) = 0$.

10. $x'' + 3x' + 2x = t$; $x(0) = 0, x'(0) = 2$.

Apply Theorem 2 to find the inverse Laplace transforms of the functions in Problems 11–18.

11. $F(s) = \dfrac{1}{s(s - 3)}$.

12. $F(s) = \dfrac{3}{s(s + 5)}$.

13. $F(s) = \dfrac{1}{s(s^2 + 4)}$.

14. $F(s) = \dfrac{2s + 1}{s(s^2 + 9)}$.

15. $F(s) = \dfrac{1}{s^2(s^2 + 1)}$.

16. $F(s) = \dfrac{1}{s(s^2 - 9)}$.

17. $F(s) = \dfrac{1}{s^2(s^2 - 1)}$.

18. $F(s) = \dfrac{1}{s(s + 1)(s + 2)}$.

19. Apply Theorem 1 to derive $\mathcal{L}\{\sin kt\}$ from the formula for $\mathcal{L}\{\cos kt\}$.

20. Apply Theorem 1 to derive $\mathcal{L}\{\cosh kt\}$ from the formula for $\mathcal{L}\{\sinh kt\}$.

21. (a) Apply Theorem 1 to show that

$$\mathcal{L}\{t^n e^{at}\} = \frac{n}{s - a}\, \mathcal{L}\{t^{n-1} e^{at}\}.$$

(b) Deduce that $\mathcal{L}\{t^n e^{at}\} = n!/(s - a)^{n+1}$ for $n = 1, 2, 3, \ldots$.

Apply Theorem 1 as in Example 4 to derive the Laplace transforms in Problems 22–24.

22. $\mathcal{L}\{t \cos kt\} = \dfrac{s^2 - k^2}{(s^2 + k^2)^2}$.

23. $\mathcal{L}\{t \sinh kt\} = \dfrac{2ks}{(s^2 - k^2)^2}$.

24. $\mathcal{L}\{t \cosh kt\} = \dfrac{s^2 + k^2}{(s^2 - k^2)^2}$.

25. Apply the results in Example 4 and Problem 22 to show that

$$\mathcal{L}^{-1}\left\{\frac{1}{(s^2 + k^2)^2}\right\} = \frac{1}{2k^3}(\sin kt - kt \cos kt).$$

Apply the extension of Theorem 1 in (18) to derive the Laplace transforms given in Problems 26–31.

26. $\mathcal{L}\{u(t - a)\} = s^{-1} e^{-as}$ for $a > 0$.

27. If $f(t) = 1$ on the interval $[a, b]$ (where $0 < a < b$) and $f(t) = 0$ otherwise, then $\mathcal{L}\{f(t)\} = s^{-1}(e^{-as} - e^{-bs})$.

28. If $f(t) = (-1)^{\llbracket t \rrbracket}$ is the square wave function shown in Fig. 4.14, then

$$\mathcal{L}\{f(t)\} = \frac{1}{s}\tanh\frac{s}{2}.$$

(*Suggestion:* Use the geometric series.)

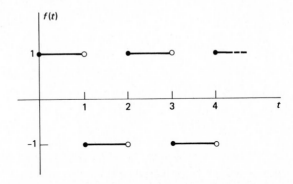

Figure 4.14 The square wave function of Problem 28.

29. If $f(t)$ is the unit on-off function shown in Fig. 4.15, then

$$\mathcal{L}\{f(t)\} = \frac{1}{s(1 + e^{-s})}.$$

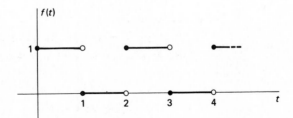

Figure 4.15 The off-on function of Problem 29.

30. If $g(t)$ is the triangular wave function shown in Fig. 4.16, then

$$\mathcal{L}\{g(t)\} = \frac{1}{s^2}\tanh\frac{s}{2}.$$

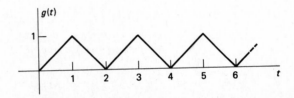

Figure 4.16 The triangular wave function of Problem 30.

31. If $f(t)$ is the sawtooth function shown in Fig. 4.17, then

$$\mathcal{L}\{f(t)\} = \frac{1}{s^2} - \frac{e^{-s}}{s(1 - e^{-s})}.$$

(*Suggestion:* Note that $f'(t) \equiv 1$.)

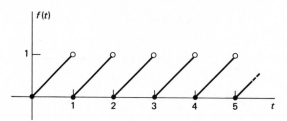

Figure 4.17 The sawtooth function of Problem 31.

4.3

Translation and Partial Fractions

As illustrated by Examples 1 and 2 in Section 4.2, the solution of a linear differential equation with constant coefficients can often be reduced to the matter of finding the inverse Laplace transform of a rational function of the form

$$R(s) = \frac{P(s)}{Q(s)} \tag{1}$$

where the degree of $P(s)$ is less than that of $Q(s)$. The technique for finding $\mathcal{L}^{-1}\{R(s)\}$ is based on the same method of partial fractions that we use in elementary calculus to integrate rational functions. The following two rules describe the **partial fraction decomposition** of $R(s)$, in terms of the factorization of the denominator $Q(s)$ into linear factors and irreducible quadratic factors corresponding to the real and complex zeros, respectively, of $Q(s)$.

RULE 1: LINEAR FACTOR PARTIAL FRACTIONS

The part of the partial fraction decomposition of $R(s)$ corresponding to the linear factor $s - a$ of multiplicity n is a sum of n partial fractions, having the form

$$\frac{A_1}{(s-a)^n} + \frac{A_2}{(s-a)^{n-1}} + \cdots + \frac{A_n}{s-a}, \tag{2}$$

where $A_1, A_2, \ldots,$ and A_n are constants.

RULE 2: QUADRATIC FACTOR PARTIAL FRACTIONS

The part of the partial fraction decomposition corresponding to the irreducible quadratic factor $(s - a)^2 + b^2$ of multiplicity n is a sum of n partial fractions, having the form

$$\frac{A_1 s + B_1}{[(s-a)^2 + b^2]^n} + \cdots + \frac{A_n s + B_n}{(s-a)^2 + b^2}, \tag{3}$$

where $A_1, A_2, \ldots, A_n, B_1, B_2, \ldots, B_n$ are constants.

Finding $\mathcal{L}^{-1}\{R(s)\}$ involves two steps: First we must find the partial fraction decomposition of $R(s)$, and then we must find the inverse Laplace transform of each of the individual partial fractions of the types that appear in (2) and (3). The latter step is based on the following elementary property of Laplace transforms.

THEOREM: TRANSLATION ON THE s-AXIS

If $F(s) = \mathcal{L}\{f(t)\}$ exists for $s > c$, then $\mathcal{L}\{e^{at}f(t)\}$ exists for $s > a + c$, and

$$\mathcal{L}\{e^{at}f(t)\} = F(s - a). \tag{4}$$

Equivalently,

$$\mathcal{L}^{-1}\{F(s - a)\} = e^{at}f(t). \tag{5}$$

Thus the translation $s \longrightarrow s - a$ in the transform corresponds to multiplication of the original function of t by e^{at}.

Proof If we simply replace s by $s - a$ in the definition of $F(s) = \mathcal{L}\{f(t)\}$, we obtain

$$F(s - a) = \int_0^\infty e^{-(s-a)t}f(t)\, dt$$

$$= \int_0^\infty e^{-st}[e^{at}f(t)]\, dt = \mathcal{L}\{e^{at}f(t)\}.$$

This is (4), and it is clear that (5) is the same.

If we apply the translation theorem to the formulas for the Laplace transforms of t^n, $\cos kt$, and $\sin kt$ that we already know—multiplying each of these functions by e^{at} and replacing s by $s - a$ in the transforms—we get the following additions to the table in Fig. 4.2 of Section 4.1.

$f(t)$	$F(s)$		
$e^{at}t^n$	$\dfrac{n!}{(s - a)^{n+1}},$	$s > a$	(6)
$e^{at}\cos kt$	$\dfrac{s - a}{(s - a)^2 + k^2},$	$s > a$	(7)
$e^{at}\sin kt$	$\dfrac{k}{(s - a)^2 + k^2},$	$s > a$	(8)

For ready reference, all the Laplace transforms derived in this chapter are listed in the table of transforms that appears at the end of the chapter.

EXAMPLE 1 Consider a mass-spring-dashpot system with $m = \frac{1}{2}$, $k = 17$, and $c = 3$ in mks units. As usual, let $x(t)$ denote the displacement of the mass m from its equilibrium position. If the mass is set in motion with with $x(0) = 3$ and $x'(0) = 1$, find $x(t)$ for the resulting damped free oscillations.

Solution The differential equation of motion is $\frac{1}{2}x'' + 3x' + 17x = 0$, so we need to solve the initial value problem

$$x'' + 6x' + 34x = 0; \qquad x(0) = 3, \qquad x'(0) = 1.$$

We take the Laplace transform of each term of the differential equation; because (obviously) $\mathcal{L}\{0\} = 0$, we get the equation

$$[s^2 X(s) - 3s - 1] + 6[sX(s) - 3] + 34X(s) = 0,$$

which we solve for

$$X(s) = \frac{3s + 19}{s^2 + 6s + 34}$$

$$= 3\frac{s + 3}{(s + 3)^2 + 25} + 2\frac{5}{(s + 3)^2 + 25}.$$

Applying the formulas in (7) and (8) with $a = -3$ and $k = 5$, we now see that $x(t) = e^{-3t}(3 \cos 5t + 2 \sin 5t)$.

The following example illustrates a useful technique for finding the partial fraction coefficients in the case of nonrepeated linear factors.

EXAMPLE 2 To find the inverse Laplace transform of

$$R(s) = \frac{s^2 + 1}{s^3 - 2s^2 - 8s},$$

we note that the denominator factors as $Q(s) = s(s + 2)(s - 4)$, so

$$\frac{s^2 + 1}{s^3 - 2s^2 - 8s} = \frac{A}{s} + \frac{B}{s + 2} + \frac{C}{s - 4}.$$

Multiplication of each term of this equation by $Q(s)$ yields

$$s^2 + 1 = A(s + 2)(s - 4) + Bs(s - 4) + Cs(s + 2).$$

When we successively substitute the three zeros $s = 0$, $s = -2$, and $s = 4$ of the denominator $Q(s)$ in this equation, we get the results

$$-8A = 1, \qquad 12B = 5, \quad \text{and} \quad 24C = 17.$$

Thus $A = -\frac{1}{8}$, $B = \frac{5}{12}$, and $C = \frac{17}{24}$, so

$$\frac{s^2 + 1}{s^3 - 2s^2 - 8s} = \frac{-\frac{1}{8}}{s} + \frac{\frac{5}{12}}{s + 2} + \frac{\frac{17}{24}}{s - 4},$$

and hence

$$\mathcal{L}^{-1}\left\{\frac{s^2 + 1}{s^3 - 2s^2 - 8s}\right\} = -\frac{1}{8} + \frac{5}{12}e^{-2t} + \frac{17}{24}e^{4t}.$$

The following example illustrates a differentiation technique for finding the partial fraction coefficients in the case of repeated linear factors.

EXAMPLE 3 Solve the initial value problem

$$y'' + 4y' + 4y = t^2; \qquad y(0) = y'(0) = 0.$$

Solution The transformed equation is

$$s^2 Y(s) + 4sY(s) + 4Y(s) = \frac{2}{s^3}.$$

Thus

$$Y(s) = \frac{2}{s^3(s+2)^2}$$

$$= \frac{A}{s^3} + \frac{B}{s^2} + \frac{C}{s} + \frac{D}{(s+2)^2} + \frac{E}{s+2}. \tag{9}$$

To find A, B, and C, we multiply both sides by s^3 to obtain

$$\frac{2}{(s+2)^2} = A + Bs + Cs^2 + s^3F(s), \tag{10}$$

where $F(s) = D(s+2)^{-2} + E(s+2)^{-1}$ is the sum of the two partial fractions corresponding to $(s+2)^2$. Substitution of $s = 0$ in (10) yields $A = \frac{1}{2}$. To find B and C, we differentiate Eq. (9) twice to obtain

$$\frac{-4}{(s+2)^3} = B + 2Cs + 3s^2F(s) + s^3F'(s) \tag{11}$$

and

$$\frac{12}{(s+2)^4} = 2C + 6sF(s) + 6s^2F'(s) + s^3F''(s). \tag{12}$$

Now substitution of $s = 0$ in (11) yields $B = -\frac{1}{2}$, and substitution of $s = 0$ in (12) yields $C = \frac{3}{8}$.

To find D and E, we multiply each side in Eq. (9) by $(s+2)^2$ to get

$$\frac{2}{s^3} = D + E(s+2) + (s+2)^2G(s) \tag{13}$$

where $G(s) = A/s^3 + B/s^2 + C/s$, and then differentiate to obtain

$$\frac{-6}{s^4} = E + 2(s+2)G(s) + (s+2)^2G'(s). \tag{14}$$

Substitution of $s = -2$ in (13) and (14) now yields $D = -\frac{1}{4}$ and $E = -\frac{3}{8}$. Thus

$$Y(s) = \frac{\frac{1}{2}}{s^3} - \frac{\frac{1}{2}}{s^2} + \frac{\frac{3}{8}}{s} - \frac{\frac{1}{4}}{(s+2)^2} - \frac{\frac{3}{8}}{s+2},$$

so the solution of our initial value problem is

$$y(t) = \tfrac{1}{4}t^2 - \tfrac{1}{2}t + \tfrac{3}{8} - \tfrac{1}{4}te^{-2t} - \tfrac{3}{8}e^{-2t}.$$

The following three examples illustrate techniques for dealing with quadratic factors in partial fraction decompositions.

EXAMPLE 4   Consider the mass-spring-dashpot system as in Example 1, but with initial conditions $x(0) = x'(0) = 0$ and with the imposed external force $F(t) = 15\sin 2t$. Find the resulting transient motion and steady periodic motion of the mass.

Solution   The initial value problem we need to solve is

$$x'' + 6x' + 34x = 30\sin 2t; \qquad x(0) = x'(0) = 0.$$

The transformed equation is

$$s^2 X(s) + 6s X(s) + 34 X(s) = \frac{60}{s^2 + 4}.$$

Hence

$$X(s) = \frac{60}{(s^2 + 4)[(s + 3)^2 + 25]}$$

$$= \frac{As + B}{s^2 + 4} + \frac{Cs + D}{[(s + 3)^2 + 25]}.$$

When we multiply both sides by the common denominator, we get

$$60 = (As + B)[(s + 3)^2 + 25] + (Cs + D)(s^2 + 4). \qquad (15)$$

To find A and B, we substitute the zero $s = 2i$ of the quadratic factor $s^2 + 4$ in (15); the result is

$$60 = (2iA + B)[(2i + 3)^2 + 25],$$

which we simplify to

$$60 = (-24A + 30B) + (60A + 12B)i.$$

We now equate real parts and imaginary parts on each side of this equation to obtain the two linear equations

$$-24A + 30B = 60 \quad \text{and} \quad 20A + 12B = 0,$$

which are readily solved for $A = -\frac{10}{29}$ and $B = \frac{50}{29}$.

To find C and D, we substitute the zero $s = -3 + 5i$ of the quadratic factor $[(s + 3)^2 + 25]$ in (15) and get

$$60 = [C(-3 + 5i) + D][(-3 + 5i)^2 + 4],$$

which we simplify to

$$60 = (186C - 12D) + (30C - 30D)i.$$

Again we equate real parts and imaginary parts; this yields the two linear equations

$$186C - 12D = 60 \quad \text{and} \quad 30C - 30D = 0,$$

and we readily find their solution to be $C = D = \frac{10}{29}$.

With these values of the coefficients A, B, C, and D, our partial fraction decomposition of $X(s)$ is

$$X(s) = \frac{1}{29}\left(\frac{-10s + 50}{s^2 + 4} + \frac{10s + 10}{(s + 3)^2 + 25}\right)$$

$$= \frac{1}{29}\left(\frac{-10(s) + (25)(2)}{s^2 + 4} + \frac{10(s + 3) - (4)(5)}{(s + 3)^2 + 25}\right).$$

After we compute the inverse Laplace transforms, we get the position function

$$x(t) = \frac{5}{29}(-2 \cos 2t + 5 \sin 2t) + \frac{2}{29} e^{-3t}(5 \cos 5t - 2 \sin 5t).$$

The terms of circular frequency 2 constitute the steady periodic forced oscillations of the mass, while the exponentially damped terms of circular frequency 5 constitute its transient motion. Note that the transient motion is nonzero even though both initial conditions are zero.

The following two inverse Laplace transforms are used in inverting partial fractions that correspond to the case of repeated quadratic factors:

$$\mathcal{L}^{-1}\left\{\frac{s}{(s^2 + k^2)^2}\right\} = \frac{1}{2k}t \sin kt \tag{15}$$

and

$$\mathcal{L}^{-1}\left\{\frac{1}{(s^2 + k^2)^2}\right\} = \frac{1}{2k^3}(\sin kt - kt \cos kt). \tag{16}$$

These follow from Example 4 and Problem 25 in Section 4.2, respectively. Because of the presence in (15) and (16) of the terms $t \sin kt$ and $t \cos kt$, a repeated quadratic factor ordinarily signals the phenomenon of resonance in an undamped mechanical or electrical system.

EXAMPLE 5 Use Laplace transforms to solve the initial value problem

$$x'' + \omega_0^2 x = F_0 \sin \omega t; \qquad x(0) = 0 = x'(0)$$

that determines the undamped forced oscillations of a mass on a spring.

Solution When we transform the differential equation, we get the equation

$$s^2 X(s) + \omega_0^2 X(s) = \frac{F_0 \omega}{s^2 + \omega^2},$$

so

$$X(s) = \frac{F_0 \omega}{(s^2 + \omega^2)(s^2 + \omega_0^2)}$$

If $\omega \neq \omega_0$, we find without difficulty that

$$X(s) = \frac{F_0 \omega}{\omega^2 - \omega_0^2}\left(\frac{1}{s^2 + \omega_0^2} - \frac{1}{s^2 + \omega^2}\right),$$

so it follows that

$$x(t) = \frac{F_0 \omega}{\omega^2 - \omega_0^2}\left(\frac{1}{\omega_0} \sin \omega_0 t - \frac{1}{\omega} \sin \omega t\right).$$

But if $\omega = \omega_0$, we have

$$X(s) = \frac{F_0 \omega_0}{(s^2 + \omega_0^2)^2},$$

so (16) yields the resonance solution

$$x(t) = \frac{F_0}{2\omega_0^2}(\sin \omega_0 t - \omega_0 t \cos \omega_0 t).$$

EXAMPLE 6 Solve the initial value problem

$$\begin{cases} y^{(4)} + 2y'' + y = 4te^t; \\ y(0) = y'(0) = y''(0) = y'''(0) = 0. \end{cases}$$

Solution First we observe that $\mathcal{L}\{y''(t)\} = s^2 Y(s)$, $\mathcal{L}\{y^{(4)}(t)\} = s^4 Y(s)$, and $\mathcal{L}\{te^t\} = 1/(s - 1)^2$. Hence the transformed equation is

$$(s^4 + 2s^2 + 1)Y(s) = \frac{4}{(s - 1)^2}.$$

Thus our problem is to find the inverse transform of

$$Y(s) = \frac{4}{(s - 1)^2(s^2 + 1)^2}$$

$$= \frac{A}{(s - 1)^2} + \frac{B}{s - 1} + \frac{Cs + D}{(s^2 + 1)^2} + \frac{Es + F}{s^2 + 1}. \qquad (17)$$

If we multiply by the common denominator $(s - 1)^2(s^2 + 1)^2$ we get the equation

$$A(s^2 + 1)^2 + B(s - 1)(s^2 + 1)^2 + Cs(s - 1)^2$$
$$+ D(s - 1)^2 + Es(s - 1)^2(s^2 + 1) + F(s - 1)^2(s^2 + 1) = 4. \qquad (18)$$

Upon substituting $s = 1$ we find that $A = 1$.

Equation (18) is an identity that holds for all values of s. To find the values of the remaining coefficients, we substitute successively the values $s = 0$, $s = -1$, $s = 2$, $s = -2$, and $s = 3$ in (18). This yields the system

$$\left.\begin{array}{r}
-B \qquad\qquad + D \qquad\quad + \quad F = 3, \\
-8B - \quad 4C + 4D - \quad 8E + \quad 8F = 0, \\
25B + \quad 2C + \quad D + \quad 10E + \quad 5F = -21, \\
-75B - 18C + 9D - \quad 90E + 45F = -21, \\
200B + 12C + 4D + 120E + 40F = -96
\end{array}\right\} \qquad (19)$$

of five linear equations in B, C, D, E, and F. With the aid of a calculator programmed to solve linear systems, we find that $B = -2$, $C = 2$, $D = 0$, $E = 2$, and $F = 1$.

We now substitute the coefficients we have found in (17), and thus obtain

$$Y(s) = \frac{1}{(s - 1)^2} - \frac{2}{s - 1} + \frac{2s}{(s^2 + 1)^2} + \frac{2s + 1}{s^2 + 1}.$$

Recalling (15), the translation property, and the familiar transforms of $\cos t$ and $\sin t$, we see finally that the solution of our initial value problem is

$$y(t) = (t - 2)e^t + 2\cos t + \sin t + 4t\sin t.$$

4.3 Problems

Apply the translation theorem to find the Laplace transforms of the functions in Problems 1–4.

1. $f(t) = t^4 e^{\pi t}$.

2. $f(t) = t^{3/2} e^{-4t}$.

3. $f(t) = e^{-2t} \sin 3\pi t$.

4. $f(t) = e^{-t/2} \cos 2\left(t - \dfrac{\pi}{8}\right)$.

Apply the translation theorem to find the inverse Laplace transforms of the functions in Problems 5–10.

5. $F(s) = \dfrac{3}{2s - 4}$.

6. $F(s) = \dfrac{s - 1}{(s + 1)^3}$.

7. $F(s) = \dfrac{1}{s^2 + 4s + 4}$.

8. $F(s) = \dfrac{s + 2}{s^2 + 4s + 5}$.

9. $F(s) = \dfrac{3s + 5}{s^2 - 6s + 25}$.

10. $F(s) = \dfrac{2s - 3}{9s^2 - 12s + 20}$.

Use partial fractions to find the inverse Laplace transforms of the functions in Problems 11–22.

11. $F(s) = \dfrac{1}{s^2 - 4}$.

12. $F(s) = \dfrac{5s - 6}{s^2 - 3s}$.

13. $F(s) = \dfrac{5 - 2s}{s^2 + 7s + 10}$.

14. $F(s) = \dfrac{5s - 4}{s^3 - s^2 - 2s}$.

15. $F(s) = \dfrac{1}{s^3 - 5s^2}$.

16. $F(s) = \dfrac{1}{(s^2 + s - 6)^2}$.

17. $F(s) = \dfrac{1}{s^4 - 16}$.

18. $F(s) = \dfrac{s^3}{(s - 4)^4}$.

19. $F(s) = \dfrac{s^2 - 2s}{s^4 + 5s^2 + 4}$.

20. $F(s) = \dfrac{1}{s^4 - 8s^2 + 16}$.

21. $F(s) = \dfrac{s^2 + 3}{(s^2 + 2s + 2)^2}$.

22. $F(s) = \dfrac{2s^3 - s^2}{(4s^2 - 4s + 5)^2}$.

Use the factorization

$$s^4 + 4a^4 = (s^2 - 2as + 2a^2)(s^2 + 2as + 2a^2)$$

to derive the inverse Laplace transforms listed in Problems 23–26.

23. $\mathcal{L}^{-1}\left\{\dfrac{s^3}{s^4 + 4a^4}\right\} = \cosh at \cos at$.

24. $\mathcal{L}^{-1}\left\{\dfrac{s}{s^4 + 4a^4}\right\} = \dfrac{1}{2a^2} \sinh at \sin at$.

25. $\mathcal{L}^{-1}\left\{\dfrac{s^2}{s^4 + 4a^4}\right\} = \dfrac{1}{2a}(\cosh at \sin at + \sinh at \cos at)$.

26. $\mathcal{L}^{-1}\left\{\dfrac{1}{s^4 + 4a^4}\right\} = \dfrac{1}{4a^3}(\cosh at \sin at - \sinh at \cos at)$.

Use Laplace transforms to solve the initial value problems in Problems 27–38.

27. $x'' + 6x' + 25x = 0$; $x(0) = 2, x'(0) = 3$.

28. $x'' - 6x' + 8x = 2$; $x(0) = x'(0) = 0$.

29. $x'' - 4x = 3t$; $x(0) = 0 = x'(0)$.

30. $x'' + 4x' + 8x = e^{-t}$; $x(0) = x'(0) = 0$.

31. $x''' + x'' - 6x' = 0$; $x(0) = 0, x'(0) = x''(0) = 1$.

32. $x^{(4)} - x = 0$; $x(0) = 1, x'(0) = x''(0) = x'''(0) = 0$.

33. $x^{(4)} + x = 0$; $x(0) = x'(0) = x''(0) = 0, x'''(0) = 1$.

34. $x^{(4)} + 13x'' + 36x = 0$; $x(0) = x''(0) = 0, x'(0) = 2, x'''(0) = -13$.

35. $x^{(4)} + 8x'' + 16x = 0$; $x(0) = x'(0) = x''(0) = 0, x'''(0) = 1$.

36. $x^{(4)} + 2x'' + x = e^{2t}$; $x(0) = x'(0) = x''(0) = x'''(0) = 0$.

37. $x'' + 4x' + 13x = te^{-t}$; $x(0) = 0$, $x'(0) = 2$.

38. $x'' + 6x' + 18x = \cos 2t$; $x(0) = 1$, $x'(0) = -1$.

4.4

Derivatives, Integrals, and Products of Transforms

The Laplace transform of the (initially unknown) solution of a differential equation is sometimes recognizable as the product of the transforms of two *known* functions. For example, when we transform the initial value problem

$$\begin{cases} x'' + x = \cos t; \\ x(0) = x'(0) = 0, \end{cases}$$

we get

$$X(s) = \frac{s}{(s^2 + 1)^2} = \frac{s}{s^2 + 1} \cdot \frac{1}{s^2 + 1} = \mathcal{L}\{\cos t\} \cdot \mathcal{L}\{\sin t\}.$$

This strongly suggests that there ought to be a way of combining the two functions $\sin t$ and $\cos t$ to obtain a function $x(t)$ whose transform is the *product* of *their* transforms. But obviously $x(t)$ is *not* simply the product of $\cos t$ and $\sin t$, because

$$\mathcal{L}\{\cos t \sin t\} = \mathcal{L}\{\tfrac{1}{2} \sin 2t\} = \frac{1}{s^2 + 4} \neq \frac{s}{(s^2 + 1)^2}.$$

Thus $\mathcal{L}\{\cos t \sin t\} \neq \mathcal{L}\{\cos t\} \cdot \mathcal{L}\{\sin t\}$.

Theorem 1 tells us that the function

$$h(t) = \int_0^t f(\tau)g(t - \tau)\, d\tau \tag{1}$$

has the desired property that

$$\mathcal{L}\{h(t)\} = H(s) = F(s) \cdot G(s). \tag{2}$$

The new function of t defined as the integral in (1) depends only on f and g and is called the *convolution* of f and g. It is denoted by $f * g$, the idea being that it is a new type of product of f and g, so tailored that its transform is the product of the transforms of f and g. Thus the **convolution** $f * g$ of the piecewise continuous functions f and g is defined for $t \geq 0$ to be

$$(f * g)(t) = \int_0^t f(\tau)g(t - \tau)\, d\tau. \tag{3}$$

We also write $f(t) * g(t)$ when convenient. In terms of this convolution product, Theorem 1 says that $\mathcal{L}\{f * g\} = \mathcal{L}\{f\} \cdot \mathcal{L}\{g\}$.

If we make the substitution $u = t - \tau$ in the integral in (3), we see that

$$f(t) * g(t) = \int_0^t f(\tau)g(t - \tau)\, d\tau$$

$$= \int_t^0 f(t - u)g(u)(-du) = \int_0^t g(u)f(t - u)\, du = g(t) * f(t).$$

Thus the convolution is *commutative*: $f * g = g * f$.

EXAMPLE 1 The convolution of $\cos t$ and $\sin t$ is

$$(\cos t) * (\sin t) = \int_0^t \cos \tau \sin (t - \tau) \, d\tau.$$

We apply the trigonometric identity

$$\cos A \sin B = \tfrac{1}{2}[\sin (A + B) - \sin (A - B)]$$

to obtain

$$(\cos t) * (\sin t) = \int_0^t \tfrac{1}{2}[\sin t - \sin (2\tau - t)] \, d\tau$$

$$= \tfrac{1}{2}\left[\tau \sin t + \tfrac{1}{2} \cos (2\tau - t) \right]_{\tau=0}^t ;$$

that is,

$$(\cos t) * (\sin t) = \tfrac{1}{2} t \sin t.$$

And we recall from Example 4 in Section 4.2 that the Laplace transform of $\tfrac{1}{2} t \sin t$ is indeed $s/(s^2 + 1)^2$.

The following theorem is proved at the end of this section.

THEOREM 1:   THE CONVOLUTION PROPERTY

Suppose that $f(t)$ and $g(t)$ are piecewise continuous for $t \geq 0$ and that $|f(t)|$ and $|g(t)|$ are bounded by Me^{ct} as $t \to +\infty$. Then the Laplace transform of the convolution $f(t) * g(t)$ exists when $s > c$; moreover,

$$\mathcal{L}\{f(t) * g(t)\} = \mathcal{L}\{f(t)\} \cdot \mathcal{L}\{g(t)\} \tag{4}$$

and

$$\mathcal{L}^{-1}\{F(s) \cdot G(s)\} = f(t) * g(t). \tag{5}$$

The following example illustrates the fact that convolution often provides a convenient alternative to the use of partial fractions for finding inverse transforms.

EXAMPLE 2 With $f(t) = \sin 2t$ and $g(t) = e^t$, convolution yields

$$\mathcal{L}^{-1}\left\{ \frac{2}{(s - 1)(s^2 + 4)} \right\} = (\sin 2t) * e^t$$

$$= \int_0^t e^{t-\tau} \sin 2\tau \, d\tau$$

$$= e^t \int_0^t e^{-\tau} \sin 2\tau \, d\tau$$

$$= e^t \left[\frac{e^{-\tau}}{5}(-\sin 2\tau - 2 \cos 2\tau) \right]_0^t,$$

and so

$$\mathcal{L}^{-1}\left\{ \frac{2}{(s - 1)(s^2 + 4)} \right\} = \frac{2}{5} e^t - \frac{1}{5} \sin 2t - \frac{2}{5} \cos 2t.$$

DIFFERENTIATION OF TRANSFORMS

According to Theorem 1 in Section 4.2, if $f(0) = 0$ then differentiation of $f(t)$ corresponds to multiplication of its transform $F(s)$ by s. The following theorem (proved at the end of this section) tells us that differentiation of the transform $F(s)$ corresponds to multiplication of the original function $f(t)$ by $-t$.

THEOREM 2: DIFFERENTIATION OF TRANSFORMS

If $f(t)$ is piecewise continuous for $t \geq 0$ and $|f(t)| \leq Me^{ct}$ as $t \to +\infty$, then

$$\mathcal{L}\{-tf(t)\} = F'(s) \tag{6}$$

for $s > c$. Equivalently,

$$f(t) = \mathcal{L}^{-1}\{F(s)\} = -\frac{1}{t}\mathcal{L}^{-1}\{F'(s)\}. \tag{7}$$

Repeated application of (6) yields

$$\mathcal{L}\{t^n f(t)\} = (-1)^n F^{(n)}(s) \tag{8}$$

for $n = 1, 2, 3, \ldots$.

EXAMPLE 3   Find $\mathcal{L}\{t^2 \sin kt\}$.

Solution   The formula in (8) gives

$$\mathcal{L}\{t^2 \sin kt\} = (-1)^2 \frac{d^2}{ds^2}\left(\frac{k}{s^2 + k^2}\right)$$

$$= \frac{d}{ds}\left[\frac{-2ks}{(s^2 + k^2)^2}\right] = \frac{6ks^2 - 2k^3}{(s^2 + k^2)^3}.$$

The form of the differentiation property in (7) is often helpful in finding an inverse transform when the *derivative* of the transform is easier to work with than the transform itself.

EXAMPLE 4   Find $\mathcal{L}^{-1}\{\tan^{-1} 1/s\}$.

Solution   The derivative of $\tan^{-1}(1/s)$ is a simple rational function, so we apply (7):

$$\mathcal{L}^{-1}\left\{\tan^{-1}\frac{1}{s}\right\} = -\frac{1}{t}\mathcal{L}^{-1}\left\{\frac{d}{ds}\tan^{-1}\frac{1}{s}\right\} = -\frac{1}{t}\mathcal{L}^{-1}\left\{\frac{-1/s^2}{1 + (1/s)^2}\right\}$$

$$= -\frac{1}{t}\mathcal{L}^{-1}\left\{\frac{-1}{s^2 + 1}\right\} = -\frac{1}{t}(-\sin t).$$

Therefore

$$\mathcal{L}^{-1}\left\{\tan^{-1}\frac{1}{s}\right\} = \frac{\sin t}{t}.$$

The formula in (8) can be applied to transform a linear differential equation having polynomial, rather than constant, coefficients. The result will be a differential equation involving the transform; whether this procedure leads to success depends, of course, on whether we can solve the new equation more readily than the old one.

EXAMPLE 5 Let $x(t)$ be the solution of Bessel's equation of order zero, $tx'' + x' + tx = 0$, such that $x(0) = 1$ and $x'(0) = 0$. Because

$$\mathcal{L}\{x'(t)\} = sX(s) - 1 \quad \text{and} \quad \mathcal{L}\{x''(t)\} = s^2 X(s) - s,$$

and because x and x'' are each multiplied by t, application of (7) yields the transformed equation

$$-\frac{d}{ds}[s^2 X(s) - s] + [sX(s) - 1] - \frac{d}{ds}[X(s)] = 0.$$

The result of differentiation and simplification is the differential equation $(s^2 + 1)X'(s) + sX(s) = 0$. This equation is separable—

$$\frac{X'(s)}{X(s)} = \frac{-s}{s^2 + 1};$$

its general solution is

$$X(s) = \frac{C}{\sqrt{s^2 + 1}}.$$

In Problem 39 we outline the argument that $C = 1$. In Section 3.5 we denoted this solution by $x = J_0(t)$. Because $X(s) = \mathcal{L}\{J_0(t)\}$, it follows that

$$\mathcal{L}\{J_0(t)\} = \frac{1}{\sqrt{s^2 + 1}}. \tag{10}$$

INTEGRATION OF TRANSFORMS

Differentiation of $F(s)$ corresponds to multiplication of $f(t)$ by t (together with a change of sign). It is therefore natural to expect that integration of $F(s)$ will correspond to division of $f(t)$ by t. The following theorem (proved at the end of this section) confirms this, provided that the resulting quotient $f(t)/t$ remains "nice" as $t \to 0$; that is, provided that

$$\lim_{t \to 0^+} \frac{f(t)}{t} \quad \text{exists and is finite.} \tag{11}$$

THEOREM 3: INTEGRATION OF TRANSFORMS

Suppose that $f(t)$ is piecewise continuous for $t \geq 0$, that $f(t)$ satisfies the condition in (11), and that $|f(t)| \leq Me^{ct}$ as $t \to +\infty$. Then

$$\mathcal{L}\left\{\frac{f(t)}{t}\right\} = \int_s^\infty F(\sigma)\, d\sigma \tag{12}$$

when $s > c$. Equivalently,

$$f(t) = \mathcal{L}^{-1}\{F(s)\} = t\mathcal{L}^{-1}\left\{\int_s^\infty F(\sigma)\, d\sigma\right\}. \tag{13}$$

EXAMPLE 6 Find $\mathcal{L}\{(\sinh t)/t\}$.

Solution We first verify that

$$\lim_{t \to 0} \frac{\sinh t}{t} = \lim_{t \to 0} \frac{e^t - e^{-t}}{2t} = \lim_{t \to 0} \frac{e^t + e^{-t}}{2} = 1,$$

with the aid of l'Hôpital's rule. Then the formula in (12), with $f(t) =$ sinh t, yields

$$\mathcal{L}\left\{\frac{\sinh t}{t}\right\} = \int_s^\infty \mathcal{L}\{\sinh t\} \, d\sigma = \int_s^\infty \frac{d\sigma}{\sigma^2 - 1}$$

$$= \frac{1}{2}\int_s^\infty \left(\frac{1}{\sigma - 1} - \frac{1}{\sigma + 1}\right) d\sigma = \frac{1}{2}\left[\ln\frac{\sigma - 1}{\sigma + 1}\right]_s^\infty.$$

Therefore

$$\mathcal{L}\left\{\frac{\sinh t}{t}\right\} = \frac{1}{2}\ln\frac{s - 1}{s + 1},$$

because ln $1 = 0$.

The form of the integration property in (13) is often helpful in finding an inverse transform when the indefinite *integral* of the transform is easier to handle than the transform itself.

EXAMPLE 7 Find $\mathcal{L}^{-1}\{2s/(s^2 - 1)^2\}$.

Solution We could use partial fractions, but it is much simpler to apply (13). This gives

$$\mathcal{L}^{-1}\left\{\frac{2s}{(s^2 - 1)^2}\right\} = t\mathcal{L}^{-1}\left\{\int_s^\infty \frac{2\sigma \, d\sigma}{(\sigma^2 - 1)^2}\right\}$$

$$= t\mathcal{L}^{-1}\left\{\left[\frac{-1}{\sigma^2 - 1}\right]_s^\infty\right\} = t\mathcal{L}^{-1}\left\{\frac{1}{s^2 - 1}\right\},$$

and therefore

$$\mathcal{L}^{-1}\left\{\frac{2s}{(s^2 - 1)^2}\right\} = t \sinh t.$$

*PROOFS OF THEOREMS

Proof of Theorem 1

The transforms $F(s)$ and $G(s)$ exist when $s > c$ by Theorem 2 in Section 4.1. The definition of the Laplace transform gives

$$G(s) = \int_0^\infty e^{-su}g(u) \, du$$

$$= \int_{-\tau}^\infty e^{-s(t-\tau)}g(t - \tau) \, dt \qquad (u = t - \tau),$$

and, therefore,

$$G(s) = e^{s\tau}\int_0^\infty e^{-st}g(t - \tau) \, dt,$$

because we may *define* $f(t)$ and $g(t)$ to be zero for $t < 0$. Then

$$F(s)G(s) = G(s) \int_0^\infty e^{-s\tau} f(\tau)\, d\tau$$

$$= \int_0^\infty e^{-s\tau} f(\tau) G(s)\, d\tau$$

$$= \int_0^\infty e^{-s\tau} f(\tau) \left(e^{s\tau} \int_0^\infty e^{-st} g(t-\tau)\, dt \right) d\tau$$

$$= \int_0^\infty \left(\int_0^\infty e^{-st} f(\tau) g(t-\tau)\, dt \right) d\tau.$$

Now our hypotheses on f and g imply that the order of integration may be reversed. (The proof of this requires a discussion of uniform convergence of improper integrals, and can be found in Chapter 2 of Churchill's *Operational Mathematics* (New York: McGraw-Hill, 3rd ed., 1972)). Hence

$$F(s)G(s) = \int_0^\infty \left(\int_0^\infty e^{-st} f(\tau) g(t-\tau)\, d\tau \right) dt$$

$$= \int_0^\infty e^{-st} \left(\int_0^t f(\tau) g(t-\tau)\, d\tau \right) dt$$

$$= \int_0^\infty e^{-st} [f(t) * g(t)]\, dt,$$

and therefore

$$F(s)G(s) = \mathcal{L}\{f(t) * g(t)\}.$$

We replaced the upper limit of the inner integral by t because $g(t-\tau) = 0$ whenever $\tau > t$. This completes the proof of Theorem 1.

Proof of Theorem 2

Because

$$F(s) = \int_0^\infty e^{-st} f(t)\, dt,$$

differentiation under the integral sign yields

$$F'(s) = \frac{d}{ds} \int_0^\infty e^{-st} f(t)\, dt$$

$$= \int_0^\infty \frac{d}{ds} [e^{-st} f(t)]\, dt = \int_0^\infty e^{-st} [-t f(t)]\, dt;$$

thus

$$F'(s) = \mathcal{L}\{-t f(t)\},$$

which is Eq. (6). We obtain (7) by applying $\mathcal{L}^{-1}$ and then dividing by $-t$. The validity of differentiation under the integral sign depends upon uniform convergence of the resulting integral; this is discussed in Chapter 2 of the book by Churchill just mentioned.

Proof of Theorem 3

By definition,

$$F(\sigma) = \int_0^\infty e^{-\sigma t} f(t)\, dt.$$

So integration of $F(\sigma)$ from s to $+\infty$ gives

$$\int_s^\infty F(\sigma)\, d\sigma = \int_s^\infty \left(\int_0^\infty e^{-\sigma t} f(t)\, dt \right) d\sigma.$$

Under the hypotheses of the theorem, the order of integration may be reversed (cf. Churchill's book again); it follows that

$$\int_s^\infty F(\sigma)\, d\sigma = \int_0^\infty \left(\int_s^\infty e^{-\sigma t} f(t)\, d\sigma \right) dt$$

$$= \int_0^\infty \left[\frac{e^{-\sigma t}}{-t} \right]_{\sigma=s}^\infty f(t)\, dt = \int_0^\infty e^{-st} \frac{f(t)}{t}\, dt$$

$$= \mathcal{L} \left\{ \frac{f(t)}{t} \right\}.$$

This verifies (12), and (13) follows upon first applying $\mathcal{L}^{-1}$ and then multiplying by t.

4.4 Problems

Find the convolution $f(t) * g(t)$ in each of Problems 1–6.

1. $f(t) = t, g(t) = 1.$ **2.** $f(t) = t, g(t) = e^{at}.$

3. $f(t) = \sin t, g(t) = \sin t.$ **4.** $f(t) = t^2, g(t) = \cos t.$

5. $f(t) = e^{at}, g(t) = e^{at}.$ **6.** $f(t) = e^{at}, g(t) = e^{bt}$ $(a \neq b).$

Apply the convolution theorem to find the inverse Laplace transforms of the functions in Problems 7–14.

7. $F(s) = \dfrac{1}{s(s - 3)}.$ **8.** $F(s) = \dfrac{1}{s(s^2 + 4)}.$

9. $F(s) = \dfrac{1}{(s^2 + 9)^2}.$ **10.** $F(s) = \dfrac{1}{s^2(s^2 + k^2)}.$

11. $F(s) = \dfrac{s^2}{(s^2 + 4)^2}.$ **12.** $F(s) = \dfrac{1}{s(s^2 + 4s + 5)}.$

13. $F(s) = \dfrac{s}{(s - 3)(s^2 + 1)}.$ **14.** $F(s) = \dfrac{s}{s^4 + 5s^2 + 4}.$

In each of Problems 15–22, apply either Theorem 2 or Theorem 3 to find the Laplace transform of $f(t)$.

15. $f(t) = t \sin 3t.$ **16.** $f(t) = t^2 \cos 2t.$

17. $f(t) = te^{2t} \cos 3t.$ **18.** $f(t) = te^{-t} \sin^2 t.$

19. $f(t) = \dfrac{\sin t}{t}.$ **20.** $f(t) = \dfrac{1 - \cos 2t}{t}.$

21. $f(t) = \dfrac{e^{3t} - 1}{t}.$ **22.** $f(t) = \dfrac{e^t - e^{-t}}{t}.$

Find the inverse transforms of the functions in Problems 23–28.

23. $F(s) = \ln \dfrac{s-2}{s+2}$.

24. $F(s) = \ln \dfrac{s^2+1}{s^2+4}$.

25. $F(s) = \ln \dfrac{s^2+1}{(s+2)(s-3)}$.

26. $F(s) = \tan^{-1} \dfrac{3}{s+2}$.

27. $F(s) = \ln\left(1 + \dfrac{1}{s^2}\right)$.

28. $F(s) = \dfrac{s}{(s^2+1)^3}$.

In each of Problems 29–34, transform the given differential equation to find a nontrivial solution such that $x(0) = 0$.

29. $tx'' + (t-2)x' + x = 0$.

30. $tx'' + (3t-1)x' + 3x = 0$.

31. $tx'' - (4t+1)x' + 2(2t+1)x = 0$.

32. $tx'' + 2(t-1)x' - 2x = 0$.

33. $tx'' - 2x' + tx = 0$.

34. $tx'' + (4t-2)x' + (13t-4)x = 0$.

35. Apply the convolution theorem to show that

$$\mathcal{L}^{-1}\left\{\frac{1}{(s-1)\sqrt{s}}\right\} = \frac{2e^t}{\sqrt{\pi}} \int_0^{\sqrt{t}} e^{-u^2}\, du = e^t \operatorname{erf}\sqrt{t}\,.$$

(*Suggestion:* Substitute $u = \sqrt{\tau}$.)

In each of Problems 36–39, apply the convolution theorem to derive the indicated solution $x(t)$ of the given differential equation with initial conditions $x(0) = x'(0) = 0$.

36. $x'' + 4x = f(t);$ $x(t) = \frac{1}{2}\int_0^t f(t-\tau)\sin 2\tau\, d\tau$.

37. $x'' + 2x' + x = f(t);$ $x(t) = \int_0^t \tau e^{-\tau} f(t-\tau)\, d\tau$.

38. $x'' + 4x' + 13x = f(t);$ $x(t) = \frac{1}{3}\int_0^t f(t-\tau)e^{-2\tau}\sin 3\tau\, d\tau$.

Termwise Inverse Transformation of Series

In Chapter 2 of Churchill's *Operational Mathematics*, the following theorem is proved. Suppose that $f(t)$ is continuous for $t \geq 0$, that $f(t)$ is of exponential order as $t \rightarrow +\infty$, and that

$$F(s) = \sum_{n=0}^{\infty} \frac{a_n}{s^{n+k+1}}$$

where $0 \leq k < 1$ and the series converges absolutely for $s > c$. Then

$$f(t) = \sum_{n=0}^{\infty} \frac{a_n t^{n+k}}{\Gamma(n+k+1)}.$$

Apply this result in Problems 39–41.

39. In Example 5 it was shown that

$$\mathcal{L}\{J_0(t)\} = \frac{C}{\sqrt{s^2+1}} = \frac{C}{s}\left(1 + \frac{1}{s^2}\right)^{-1/2}.$$

Expand by the binomial series and then compute the inverse transformation term by term to obtain

$$J_0(t) = C \sum_{n=0}^{\infty} \frac{(-1)^n t^{2n}}{2^{2n}(n!)^2}.$$

Finally note that $J_0(0) = 1$ implies that $C = 1$.

40. Expand the function $s^{-1/2}e^{-1/s}$ in powers of s^{-1} to show that

$$\mathcal{L}^{-1}\left\{\frac{1}{\sqrt{s}}e^{-1/s}\right\} = \frac{1}{\sqrt{\pi t}}\cos 2\sqrt{t}.$$

41. Show that $\mathcal{L}^{-1}\left\{\frac{1}{s}e^{-1/s}\right\} = J_0(2\sqrt{t})$.

---

*4.5
Periodic and Piecewise Continuous Forcing Functions

Mathematical models of mechanical or electrical systems often involve functions with discontinuities corresponding to external forces that are turned abruptly on or off. One such simple on-off function is the **unit step function** at $t = a$; its formula is

$$u_a(t) = u(t - a) = \begin{cases} 0 & \text{if } t < a; \\ 1 & \text{if } t \geq a. \end{cases} \tag{1}$$

In Example 7 of Section 4.1 we saw that if $a \geq 0$, then

$$\mathcal{L}\{u(t - a)\} = \frac{e^{-as}}{s}. \tag{2}$$

Because $\mathcal{L}\{u(t)\} = 1/s$, the formula in (2) implies that multiplication of the transform of $u(t)$ by e^{-as} corresponds to the translation $t \rightarrow t - a$ in the original independent variable. The following theorem tells us that this fact, when properly interpreted, is a general property of the Laplace transformation.

THEOREM 1: TRANSLATION ON THE T-AXIS

If $\mathcal{L}\{f(t)\} = F(s)$ exists for $s > c$, then

$$\mathcal{L}\{u(t - a)f(t - a)\} = e^{-as}F(s) \tag{3a}$$

and

$$\mathcal{L}^{-1}\{e^{-as}F(s)\} = u(t - a)f(t - a) \tag{3b}$$

when $s > c + a$.

Note that

$$u(t - a)f(t - a) = \begin{cases} 0 & \text{if } t < a, \\ f(t - a) & \text{if } t \geq a. \end{cases} \tag{4}$$

Thus Theorem 1 implies that $\mathcal{L}^{-1}\{e^{-as}F(s)\}$ is the function whose graph for $t \geq a$ is the translation a units to the right of the graph of $f(t)$. Note that the part of the graph of $f(t)$ to the left of $x = 0$ (if any) is cut off and not translated; see Fig. 4.18.

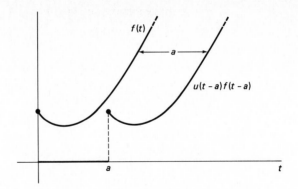

Figure 4.18 Translation of $f(t)$ a units to the right.

Proof of Theorem 1

From the definition of $\mathcal{L}\{f(t)\}$, we get

$$e^{-as}F(s) = e^{-as} \int_0^\infty e^{-s\tau} f(\tau)\, d\tau = \int_0^\infty e^{-s(\tau+a)} f(\tau)\, d\tau.$$

The substitution $t = \tau + a$ then yields

$$e^{-as}F(s) = \int_a^\infty e^{-st} f(t - a)\, dt.$$

From (4) we see that this is the same as

$$e^{-as}F(s) = \int_0^\infty e^{-st} u(t - a) f(t - a)\, dt = \mathcal{L}\{u(t - a)f(t - a)\},$$

because $u(t - a)f(t - a) = 0$ for $t < a$. We have therefore completed the proof of Theorem 1.

EXAMPLE 1 With $f(t) = \frac{1}{2}t^2$, Theorem 1 gives

$$\mathcal{L}^{-1}\left\{\frac{e^{-as}}{s^3}\right\} = u(t - a)\frac{1}{2}(t - a)^2 = \begin{cases} 0 & \text{if } t < a, \\ \frac{1}{2}(t - a)^2 & \text{if } t \geq a. \end{cases}$$

EXAMPLE 2 Find $\mathcal{L}\{g(t)\}$ if

$$g(t) = \begin{cases} 0 & \text{if } t < 3, \\ t^2 & \text{if } t \geq 3. \end{cases}$$

Solution Before applying Theorem 1, we must first write $g(t)$ in the form $u(t - 3)f(t - 3)$. The function $f(t)$ whose translation 3 units to the right agrees (for $t \geq 3$) with $g(t) = t^2$ is $f(t) = (t + 3)^2$ because $f(t - 3) = t^2$. But then

$$F(s) = \mathcal{L}\{t^2 + 6t + 9\} = \frac{2}{s^3} + \frac{6}{s^2} + \frac{9}{s},$$

so now Theorem 1 yields

$$\mathcal{L}\{g(t)\} = \mathcal{L}\{u(t-3)f(t-3)\}$$

$$= e^{-3s}F(s) = e^{-3s}\left(\frac{2}{s^3} + \frac{6}{s^2} + \frac{9}{s}\right).$$

EXAMPLE 3   Find $\mathcal{L}\{f(t)\}$ if

$$f(t) = \begin{cases} \cos 2t & \text{if } 0 \leq t < 2\pi, \\ 0 & \text{if } t \geq 2\pi. \end{cases}$$

Solution   We note first that

$$f(t) = [1 - u(t - 2\pi)]\cos 2t$$

$$= \cos 2t - u(t - 2\pi)\cos 2(t - 2\pi)$$

because of the periodicity of the cosine function. Hence Theorem 1 gives

$$\mathcal{L}\{f(t)\} = \mathcal{L}\{\cos 2t\} - e^{-2\pi s}\mathcal{L}\{\cos 2t\} = \frac{s(1 - e^{-2\pi s})}{s^2 + 4}.$$

EXAMPLE 4   A mass that weighs 32 lb (mass $m = 1$ slug) is attached to the free end of a long light spring that is stretched 1 ft by a force of 4 lb ($k = 4$ lb/ft). The mass is initially at rest in its equilibrium position. Beginning at time $t = 0$ (seconds), an external force $F(t) = \cos 2t$ is applied to the mass, but at time $t = 2\pi$ this force is turned off (abruptly discontinued), and the mass is allowed to continue its motion unimpeded. Find the resulting position function $x(t)$ of the mass.

Solution   We need to solve the initial value problem

$$\begin{cases} x'' + 4x = f(t); \\ x(0) = x'(0) = 0, \end{cases}$$

where $f(t)$ is the function of Example 3. The transformed equation is

$$(s^2 + 4)X(s) = F(s) = \frac{s(1 - e^{-2\pi s})}{s^2 + 4},$$

so

$$X(s) = \frac{s}{(s^2 + 4)^2} - e^{-2\pi s}\frac{s}{(s^2 + 4)^2}.$$

Because

$$\mathcal{L}^{-1}\left\{\frac{s}{(s^2 + 4)^2}\right\} = \frac{1}{4}t \sin 2t$$

by (22) in Section 4.3, it follows that

$$x(t) = \tfrac{1}{4}t \sin 2t - u(t - 2\pi)\cdot\tfrac{1}{4}(t - 2\pi)\sin 2(t - 2\pi).$$

If we separate the cases $t < 2\pi$ and $t \geq 2\pi$, we find that the position function may be written in the form

$$x(t) = \begin{cases} \dfrac{1}{4}t \sin 2t & \text{if } t < 2\pi, \\[2ex] \dfrac{\pi}{2}\sin 2t & \text{if } t \geq 2\pi. \end{cases}$$

As indicated by the graph of $x(t)$ shown in Fig. 4.19, the mass oscillates with circular frequency $\omega = 2$ and with linearly increasing amplitude until the force is removed at time $t = 2\pi$. Thereafter, the mass continues to oscillate with the same frequency but with constant amplitude $\pi/2$. The force $F(t) = \cos 2t$ would produce pure resonance if continued indefinitely, but we see that its effect ceases immediately when it is turned off.

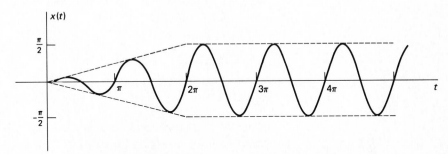

Figure 4.19   Graph of the function $x(t)$ of Example 4.

If we were to attack Example 4 with the methods of Chapter 2, we would need to solve one problem for the interval $0 \leq t < 2\pi$, and then a new problem with different initial conditions for the interval $t > 2\pi$. In such a situation the Laplace transform method enjoys the distinct advantage of not requiring the solution of different problems on different intervals.

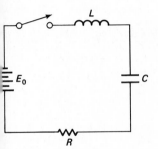

Figure 4.20   The series RLC circuit of Example 5.

EXAMPLE 5   Consider the RLC circuit shown in Fig. 4.20, with $R = 110$ ohms, $L = 1$ H, $C = 0.001$ F, and a battery supplying $E_0 = 90$ V. Initially there is no current in the circuit and no charge on the capacitor. At time $t = 0$ the switch is closed and left closed for one second. At time $t = 1$ it is opened and left open. Find the resulting current in the circuit.

Solution   We recall from Section 2.10 the basic series circuit equation

$$L\frac{di}{dt} + Ri + \frac{1}{C}q = e(t); \tag{5}$$

we use lowercase letters for current and charge and reserve uppercase letters for transforms. With the given circuit elements, Eq. (5) is

$$\frac{di}{dt} + 110i + 1000q = e(t), \tag{6}$$

where $e(t) = 90[1 - u(t-1)]$, corresponding to the closing and opening of the switch. In Section 2.10 our strategy was to differentiate each side of Eq. (5), then apply the relation

$$i = \frac{dq}{dt} \tag{7}$$

to obtain the second order equation

$$L\frac{d^2i}{dt^2} + R\frac{di}{dt} + \frac{1}{C}i = e'(t).$$

Here we do not use that method, because $e'(t) = 0$ except at $t = 1$, while the jump from $e(t) = 90$ when $t < 1$ to $e(t) = 0$ when $t > 1$ would seem to require that $e'(1) = -\infty$. Thus $e'(t)$ appears to have an "infinite discontinuity" at $t = 1$. This phenomenon will be discussed in Section 4.6. For now, we will simply note that it is an odd situation and circumvent it rather than attempt to deal with it here.

In order to avoid the possible problem at $t = 1$, we observe that the initial value $q(0) = 0$ and the relation in (7) yield, upon integration,

$$q(t) = \int_0^t i(\tau) \, d\tau. \tag{8}$$

We substitute (8) in Eq. (5) to obtain

$$L\frac{di}{dt} + Ri + \frac{1}{C} \int_0^t i(\tau) \, d\tau = e(t). \tag{9}$$

This is the **integrodifferential equation** of a series RLC circuit; it involves both the integral and the derivative of the unknown function $i(t)$. The Laplace transform method works well with such an equation.

In the present example, Eq. (9) is

$$\frac{di}{dt} + 110i + 1000 \int_0^t i(\tau) \, d\tau = 90[1 - u(t - 1)]. \tag{10}$$

Because

$$\mathcal{L}\left\{ \int_0^t i(\tau) \, d\tau \right\} = \frac{I(s)}{s}$$

by Theorem 2 in Section 4.2 on transforms of integrals, the transformed equation is

$$sI(s) + 110I(s) + 1000\frac{I(s)}{s} = \frac{90}{s}(1 - e^{-s}).$$

We solve this equation for $I(s)$ to obtain

$$I(s) = \frac{90(1 - e^{-s})}{s^2 + 110s + 1000}.$$

But

$$\frac{90}{s^2 + 110s + 1000} = \frac{1}{s + 10} - \frac{1}{s + 100},$$

so we have

$$I(s) = \frac{1}{s + 10} - \frac{1}{s + 100} - e^{-s}\left(\frac{1}{s + 10} - \frac{1}{s + 100} \right).$$

We now apply Theorem 1 with $f(t) = e^{-10t} - e^{-100t}$; thus the inverse transform is

$$i(t) = e^{-10t} - e^{-100t} - u(t - 1)[e^{-10(t-1)} - e^{-100(t-1)}].$$

After we separate the cases $t < 1$ and $t \geq 1$, we find that the current in the circuit is given by

$$i(t) = \begin{cases} e^{-10t} - e^{-100t} & \text{if } t < 1, \\ (1 - e^{10})e^{-10t} - (1 - e^{100})e^{-100t} & \text{if } t \geq 1. \end{cases}$$

The portion $e^{-10t} - e^{-100t}$ of the solution would describe the current if the switch were left closed for all t rather than being open for $t \geq 1$.

TRANSFORMS OF PERIODIC FUNCTIONS

Periodic forcing functions in practical mechanical or electrical systems often are more complicated than pure sines or cosines. The nonconstant function $f(t)$ defined for $t \geq 0$ is said to be **periodic** if there is a number $p > 0$ such that

$$f(t + p) = f(t) \tag{11}$$

for all $t \geq 0$. The least positive value of p (if any) for which (11) holds is called the **period** of f. Such a function is shown in Fig. 4.21. The following theorem simplifies the computation of the Laplace transform of a periodic function.

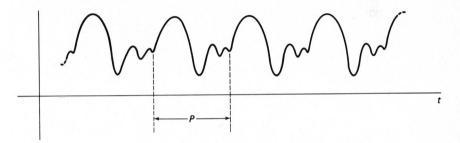

Figure 4.21 A function with period p.

THEOREM 2: TRANSFORMS OF PERIODIC FUNCTIONS

Let $f(t)$ be periodic with period p and piecewise continuous for $t \geq 0$. Then the transform $F(s) = \mathcal{L}\{f(t)\}$ exists for $s > 0$, and is given by

$$F(s) = \frac{1}{1 - e^{-ps}} \int_0^p e^{-st} f(t) \, dt. \tag{12}$$

Proof The definition of the Laplace transform gives

$$F(s) = \int_0^\infty e^{-st} f(t) \, dt = \sum_{n=0}^\infty \int_{np}^{(n+1)p} e^{-st} f(t) \, dt.$$

The substitution $t = \tau + np$ in the nth integral following the summation sign yields

$$\int_{np}^{(n+1)p} e^{-st} f(t) \, dt = \int_0^p e^{-s(\tau+np)} f(\tau + np) \, d\tau = e^{-nps} \int_0^p e^{-s\tau} f(\tau) \, d\tau$$

because $f(\tau + np) = f(\tau)$ by periodicity. Thus

$$F(s) = \sum_{n=0}^\infty e^{-nps} \int_0^p e^{-s\tau} f(\tau) \, d\tau$$

$$= (1 + e^{-ps} + e^{-2ps} + \cdots) \int_0^p e^{-s\tau} f(\tau) \, d\tau.$$

Consequently,

$$F(s) = \frac{1}{1 - e^{-ps}} \int_0^p e^{-s\tau} f(\tau) \, d\tau.$$

We used the geometric series

$$\frac{1}{1 - x} = 1 + x + x^2 + x^3 + \cdots,$$

with $x = e^{-ps} < 1$ (for $s > 0$), to sum the series in the final step. Thus we have derived the formula in (12).

The principal advantage of Theorem 2 is that it enables us to find the Laplace transform of a periodic function without the necessity of an explicit evaluation of an improper integral.

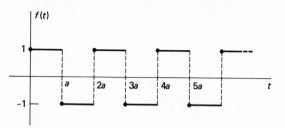

Figure 4.22 The square wave function of Example 6.

EXAMPLE 6 Figure 4.22 shows the graph of the square wave function $f(t) = (-1)^{\llbracket at \rrbracket}$ of period $p = 2a$; $\llbracket x \rrbracket$ denotes the greatest integer not exceeding x. By Theorem 2 the Laplace transform of $f(t)$ is

$$F(s) = \frac{1}{1 - e^{-2as}} \int_0^{2a} e^{-st} f(t) \, dt$$

$$= \frac{1}{1 - e^{-2as}} \left(\int_0^a e^{-st} \, dt + \int_a^{2a} (-1)e^{-st} \, dt \right)$$

$$= \frac{1}{1 - e^{-2as}} \left(\left[-\frac{1}{s} e^{-st} \right]_0^a - \left[-\frac{1}{s} e^{-st} \right]_a^{2a} \right)$$

$$= \frac{(1 - e^{-as})^2}{s(1 - e^{-2as})} = \frac{1 - e^{-as}}{s(1 + e^{-as})}.$$

Therefore

$$F(s) = \frac{1 - e^{-as}}{s(1 + e^{-as})} \tag{13a}$$

$$= \frac{e^{as/2} - e^{-as/2}}{s(e^{as/2} + e^{-as/2})} = \frac{1}{s} \tanh \frac{as}{2}. \tag{13b}$$

EXAMPLE 7 Figure 4.23 shows the graph of a triangular wave function $g(t)$ of period $p = 2a$. Because the derivative $g'(t)$ is the square wave function of Example 6, it follows from the formula in (13b) and Theorem 2 in Section 4.2 that the transform of this triangular wave function is

$$G(s) = \frac{F(s)}{s} = \frac{1}{s^2} \tanh \frac{as}{2}. \tag{14}$$

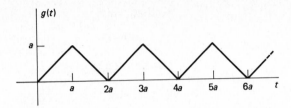

Figure 4.23 The triangular wave function of Example 7.

EXAMPLE 8 Consider a mass-spring-dashpot system with $m = 1$, $c = 4$, and $k = 20$ in appropriate units. Suppose that the system is initially at rest at equilibrium ($x(0) = x'(0) = 0$) and that the mass is acted on by the external force $f(t)$ whose graph is shown in Fig. 4.24: the square wave with amplitude 20 and period 2π. Find the position function $x(t)$.

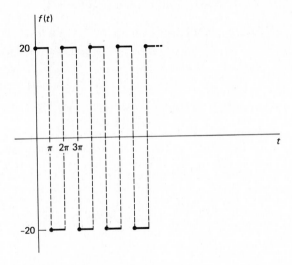

Figure 4.24 The external force function of Example 8.

Solution The initial value problem is

$$\begin{cases} x'' + 4x' + 20x = f(t); \\ x(0) = 0 = x'(0). \end{cases}$$

The transformed equation is

$$s^2 X(s) + 4s X(s) + 20X(s) = F(s). \tag{15}$$

From Example 6 we see that the transform of $f(t)$ is

$$F(s) = \frac{20}{s} \cdot \frac{1 - e^{-\pi s}}{1 + e^{-\pi s}}$$

$$= \frac{20}{s}(1 - e^{-\pi s})(1 - e^{-\pi s} + e^{-2\pi s} - e^{-3\pi s} + \cdots)$$

$$= \frac{20}{s}(1 - 2e^{-\pi s} + 2e^{-2\pi s} - 2e^{-3\pi s} + \cdots),$$

so that

$$F(s) = \frac{20}{s} + \frac{40}{s} \sum_{n=1}^{\infty} (-1)^n e^{-n\pi s}.$$ (16)

Substitution of (16) in (15) yields

$$X(s) = \frac{F(s)}{s^2 + 4s + 16}$$

$$= \frac{20}{s[(s + 2)^2 + 16]} + 2 \sum_{n=1}^{\infty} (-1)^n \frac{20e^{-n\pi s}}{s[(s + 2)^2 + 16]}.$$ (17)

From the formula in (22) of Section 4.3, we get

$$\mathcal{L}^{-1}\left\{\frac{20}{(s + 2)^2 + 16}\right\} = 5e^{-2t} \sin 4t,$$

so by Theorem 2 in Section 4.2 we have

$$g(t) = \mathcal{L}^{-1}\left\{\frac{20}{s[(s + 2)^2 + 16]}\right\} = \int_0^t 5e^{-2\tau} \sin 4\tau \, d\tau.$$

Using a tabulated formula for $\int e^{at} \sin bt \, dt$, we get

$$g(t) = 1 - e^{-2t}(\cos 4t + \tfrac{1}{2} \sin 4t) = 1 - h(t)$$ (18)

where

$$h(t) = e^{-2t}(\cos 4t + \tfrac{1}{2} \sin 4t).$$ (19)

Now we apply Theorem 1 to find the inverse transform of the expression in Eq. (17). The result is

$$x(t) = g(t) + 2 \sum_{n=1}^{\infty} (-1)^n u(t - n\pi) g(t - n\pi),$$ (20)

and we note that for any fixed value of t the sum above is finite. Moreover,

$$g(t - n\pi) = 1 - e^{-2(t-n\pi)}[\cos 4(t - n\pi) + \tfrac{1}{2} \sin 4(t - n\pi)]$$

$$= 1 - e^{2n\pi} e^{-2t}(\cos 4t + \tfrac{1}{2} \sin 4t).$$

Therefore

$$g(t - n\pi) = 1 - e^{2n\pi} h(t).$$ (21)

Hence if $0 < t < \pi$, then

$$x(t) = 1 - h(t).$$

If $\pi < t < 2\pi$, then

$$x(t) = [1 - h(t)] - 2[1 - e^{2\pi} h(t)]$$

$$= -1 + h(t) - 2h(t)[1 - e^{2\pi}].$$

If $2\pi < t < 3\pi$, then

$$x(t) = [1 - h(t)] - 2[1 - e^{2\pi} h(t)] + 2[1 - e^{4\pi} h(t)]$$

$$= 1 + h(t) - 2h(t)[1 - e^{2\pi} + e^{4\pi}].$$

The general expression for $n\pi < t < (n + 1)\pi$ is

$$x(t) = h(t) + (-1)^n - 2h(t)[1 - e^{2\pi} + \cdots + (-1)^n e^{2n\pi}]$$

$$= h(t) + (-1)^n - 2h(t)\frac{1 + (-1)^n e^{2(n+1)\pi}}{1 + e^{2\pi}},$$ (22)

which we obtained with the aid of the familiar formula for the sum of a finite geometric progression. A rearrangement of (22) finally gives, with the aid of (19),

$$x(t) = \frac{e^{2\pi} - 1}{e^{2\pi} + 1} e^{-2t}\left(\cos 4t + \frac{1}{2}\sin 4t\right) + (-1)^n$$

$$- \frac{(2)(-1)^n e^{2\pi}}{e^{2\pi} + 1} e^{-2(t - n\pi)}\left(\cos 4t + \frac{1}{2}\sin 4t\right) \tag{23}$$

for $n\pi < t < (n + 1)\pi$. The first term in (23) is the transient solution

$$x_{tr} \approx (0.9963)e^{-2t}(\cos 4t + \tfrac{1}{2}\sin 4t)$$

$$\approx (1.1139)e^{-2t}\cos(4t - 0.4636). \tag{24}$$

The last two terms in (23) give the steady periodic solution x_{sp}. To investigate it, we write $\tau = t - n\pi$ for t in the interval $n\pi < t < (n + 1)\pi$. Then

$$x_{sp}(t) = (-1)^n\left[1 - \frac{2e^{2\pi}}{e^{2\pi} + 1} e^{-2\tau}\left(\cos 4\tau + \frac{1}{2}\sin 4\tau\right)\right]$$

$$\approx (-1)^n[1 - (2.2319)e^{-2\tau}\cos(4\tau - 0.4636)]. \tag{25}$$

Figure 4.25 shows the graph of $x_{sp}(t)$. Its most interesting feature is the appearance of *periodically damped* oscillations with a frequency four times that of the imposed force $f(t)$. In Chapter 7 on Fourier series we will see why a periodic external force sometimes excites oscillations at a higher frequency than the imposed frequency.

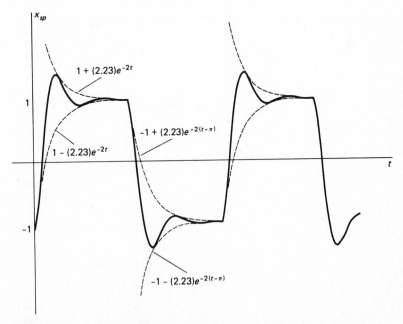

Figure 4.25 The steady periodic solution for Example 8; note the "periodically damped" oscillations with frequency four times that of the imposed force.

4.5 Problems

Find the inverse Laplace transform $f(t)$ of each of the functions given in Problems 1–10, and sketch the graph of $f(t)$.

1. $F(s) = \dfrac{e^{-3s}}{s^2}.$

2. $F(s) = \dfrac{e^{-s} - e^{-3s}}{s^2}.$

3. $F(s) = \dfrac{e^{-s}}{s + 2}.$

4. $F(s) = \dfrac{e^{-s} - e^{2-2s}}{s - 1}.$

5. $F(s) = \dfrac{e^{-\pi s}}{s^2 + 1}.$

6. $F(s) = \dfrac{se^{-s}}{s^2 + \pi^2}.$

7. $F(s) = \dfrac{1 - e^{-2\pi s}}{s^2 + 1}.$

8. $F(s) = \dfrac{s(1 - e^{-2s})}{s^2 + \pi^2}.$

9. $F(s) = \dfrac{s(1 + e^{-3s})}{s^2 + \pi^2}.$

10. $F(s) = \dfrac{2s(e^{-\pi s} - e^{-2\pi s})}{s^2 + 4}.$

Find the Laplace transform of each of the functions defined in Problems 11–22.

11. $f(t) = 2$ if $0 \leq t < 3$; $f(t) = 0$ if $t \geq 3$.

12. $f(t) = 3$ if $1 \leq t \leq 4$; $f(t) = 0$ if $t < 1$ or $t > 4$.

13. $f(t) = \sin t$ if $0 \leq t \leq 2\pi$; $f(t) = 0$ if $t > 2\pi$.

14. $f(t) = \cos \pi t$ if $0 \leq t \leq 2$; $f(t) = 0$ if $t > 2$.

15. $f(t) = \sin t$ if $0 \leq t \leq 3\pi$; $f(t) = 0$ if $t > 3\pi$.

16. $f(t) = \sin 2t$ if $\pi \leq t \leq 2\pi$; $f(t) = 0$ if $t < \pi$ or if $t > 2\pi$.

17. $f(t) = \sin \pi t$ if $2 \leq t \leq 3$; $f(t) = 0$ if $t < 2$ or if $t > 3$.

18. $f(t) = \cos \dfrac{\pi t}{2}$ if $3 \leq t \leq 5$; $f(t) = 0$ if $t < 3$ or $t > 5$.

19. $f(t) = 0$ if $t < 1$; $f(t) = t$ if $t \geq 1$.

20. $f(t) = t$ if $t \leq 1$; $f(t) = 1$ if $t > 1$.

21. $f(t) = t$ if $t \leq 1$; $f(t) = 2 - t$ if $1 \leq t \leq 2$; $f(t) = 0$ if $t > 2$.

22. $f(t) = t^3$ if $1 \leq t \leq 2$; $f(t) = 0$ if $t < 1$ or if $t > 2$.

23. Apply Theorem 2 with $p = 1$ to verify that $\mathcal{L}\{1\} = 1/s$

24. Apply Theorem 2 to verify that $\mathcal{L}\{\cos kt\} = s/(s^2 + k^2)$.

25. Apply Theorem 1 to show that the Laplace transform of the square wave function of Fig. 4.26 is

$$\mathcal{L}\{f(t)\} = \frac{1}{s(1 + e^{-as})}.$$

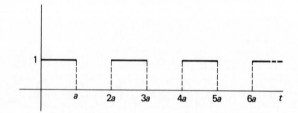

Figure 4.26 The square wave function of Problem 25.

26. Apply Theorem 2 to show that the Laplace transform of the sawtooth function $f(t)$ of Fig. 4.27 is

$$F(s) = \frac{1}{as^2} - \frac{e^{-as}}{s(1 - e^{-as})}.$$

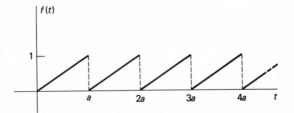

Figure 4.27 The sawtooth function of Problem 26.

27. Let $g(t)$ be the staircase function of Fig. 4.28. Show that $g(t) = (t/a) - f(t)$, where $f(t)$ is the sawtooth function of Fig. 4.27, and hence deduce that

$$\mathcal{L}\{g(t)\} = \frac{e^{-as}}{s(1 - e^{-as})}.$$

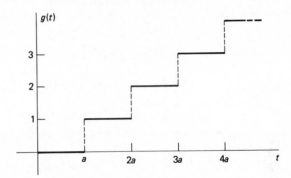

Figure 4.28 The staircase function of Problem 27.

28. Suppose that $f(t)$ is a periodic function of period $2a$ with $f(t) = t$ if $0 \leq t < a$ and $f(0) = 0$ if $a \leq t < 2a$. Find $\mathcal{L}\{f(t)\}$.

29. Suppose that $f(t)$ is the half-wave rectification of $\sin kt$, shown in Fig. 4.29. Show that

$$\mathcal{L}\{f(t)\} = \frac{k}{(s^2 + k^2)(1 - e^{-\pi s/k})}.$$

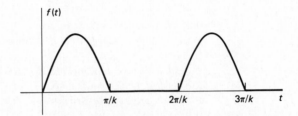

Figure 4.29 The half-wave rectification of $\sin kt$.

30. Let $g(t) = u(t - \pi/k)f(t - \pi/k)$, where $f(t)$ is the function of Problem 29. Note that $h(t) = f(t) + g(t)$ is the full-wave rectification of $\sin kt$ shown in Fig. 4.30. Hence deduce from Problem 29 that

$$\mathcal{L}\{h(t)\} = \frac{k}{s^2 + k^2} \coth \frac{\pi s}{2k}.$$

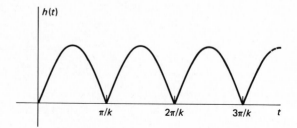

Figure 4.30   The full-wave rectification of $\sin kt$.

In each of Problems 31–35, the values of mass m, spring constant k, dashpot resistance c, and force f(t) are given for a mass-spring-dashpot system with external forcing function. Solve the initial value problem

$$\begin{cases} mx'' + cx' + kx = f(t); \\ x(0) = 0 = x'(0) \end{cases}$$

with the given data.

31. $m = 1, k = 4, c = 0$; $f(t) = 1$ if $0 \le t < \pi, f(t) = 0$ if $t > \pi$.
32. $m = 1, k = 4, c = 5$; $f(t) = 1$ if $0 \le t < 2, f(t) = 0$ if $t > 2$.
33. $m = 1, k = 9, c = 0$; $f(t) = \sin t$ if $0 \le t \le 2\pi, f(t) = 0$ if $t > 2\pi$.
34. $m = 1, k = 1, c = 0$; $f(t) = t$ if $0 \le t < 1, f(t) = 0$ if $t \ge 1$.
35. $m = 1, k = 4, c = 4$; $f(t) = t$ if $0 \le t < 2, f(t) = 0$ if $t \ge 2$.

In each of Problems 36–40, the values of the elements of an RLC circuit are given. Solve the initial value problem

$$\begin{cases} L\frac{di}{dt} + Ri + \frac{1}{C} \int_0^t i(\tau)\, d\tau = e(t); \\ \\ i(0) = 0, \end{cases}$$

with the given impressed voltage e(t).

36. $L = 0, R = 100, C = 10^{-3}$; $e(t) = 100$ if $0 \le t < 1$, $e(t) = 0$ if $t \ge 1$.
37. $L = 1, R = 0, C = 10^{-4}$; $e(t) = 100$ if $0 \le t < 2\pi$, $e(t) = 0$ if $t \ge 2\pi$.
38. $L = 1, R = 0, C = 10^{-4}$; $e(t) = 100 \sin 10t$ if $0 \le t < \pi$, $e(t) = 0$ if $t \ge \pi$.
39. $L = 1, R = 150, C = 2 \times 10^{-4}$; $e(t) = 100t$ if $0 \le t < 1$, $e(t) = 0$ if $t > 1$.
40. $L = 1, R = 100, C = 4 \times 10^{-4}$; $e(t) = 50t$ if $0 \le t < 1$, $e(t) = 0$ if $t > 1$.

In each of Problems 41 and 42, a mass-spring-dashpot system with external force f(t) is described. Under the assumption that x(0) = 0 = x'(0), use the method of Example 8 to find the transient and steady periodic motions of the mass.

41. $m = 1, k = 4, c = 0$; $f(t)$ is a square wave function with amplitude 4 and period 2π.

42. $m = 1, k = 10, c = 2;$ $f(t)$ is a square wave function with amplitude 10 and period 2π.

Consider a force $f(t)$ that acts only during a very short time interval $a \leqq t \leqq b$, with $f(t) = 0$ outside this interval. A typical example would be the *impulsive force* of a bat striking a ball—the impact is almost instantaneous. A quick surge of voltage (resulting from a lightning bolt, for instance) is an analogous electrical phenomenon. In such a situation it often happens that the principal effect of the force depends only on the value of the integral

$$p = \int_a^b f(t) \, dt \tag{1}$$

and does not depend otherwise on precisely how $f(t)$ varies with t. The number p in (1) is called the **impulse** of the force $f(t)$ over the interval $[a, b]$.

In the case of a force $f(t)$ that acts on a particle of mass m in linear motion, integration of Newton's law

$$f(t) = mv'(t) = \frac{d}{dt}[mv(t)]$$

yields

$$p = \int_a^b \frac{d}{dt}[mv(t)] \, dt = mv(b) - mv(a). \tag{2}$$

Thus the impulse of the force is equal to the change in momentum of the particle. So if change in momentum is the only effect with which we are concerned, we need know only the impulse of the force; we need know neither the precise function $f(t)$ nor even the precise time interval during which it acts. This is fortunate, because in a situation such as that of a batted ball, we are unlikely to have such detailed information about the impulsive force that acts on the ball.

Our strategy for handling such a situation is to set up a reasonable mathematical model in which the unknown force $f(t)$ is replaced with a simple and explicit force that has the same impulse. Suppose for simplicity that $f(t)$ has impulse 1 and acts during some brief time interval beginning at time $t = a \geqq 0$. Then we can select a fixed number $\epsilon > 0$ that approximates the length of this time interval and replace $f(t)$ with the specific function

$$d_{a, \epsilon}(t) = \begin{cases} \dfrac{1}{\epsilon} & \text{if } a \leqq t < a + \epsilon, \\ 0 & \text{otherwise.} \end{cases} \tag{3}$$

This is a function of t, with a and ϵ being parameters that specify the time interval $[a, a + \epsilon]$. If $b \geqq a + \epsilon$, then we see (as in Fig. 4.31) that the impulse of $d_{a, \epsilon}(t)$ over $[a, b]$ is

$$p = \int_a^b d_{a, \epsilon}(t) \, dt = \int_a^{a+\epsilon} \frac{1}{\epsilon} \, dt = 1.$$

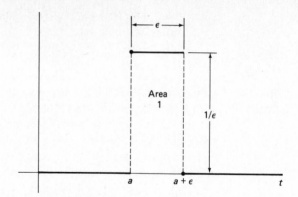

Figure 4.31 The impulse function $d_{a,\epsilon}(t)$.

Thus $d_{a,\epsilon}(t)$ has a *unit* impulse, whatever the number $\epsilon > 0$ may be. Essentially the same computation gives

$$\int_0^\infty d_{a,\epsilon}(t)\,dt = 1. \tag{4}$$

Because the precise time interval during which the force acts seems unimportant, it is tempting to think of an *instantaneous impulse* that occurs precisely at the instant $t = a$. We might try to model such an instantaneous unit impulse by taking the limit as $\epsilon \to 0$, thereby defining

$$\delta_a(t) = \lim_{\epsilon \to 0} d_{a,\epsilon}(t) \tag{5}$$

where $a \geqq 0$. If we can also take the limit under the integral sign in (4), then it will follow that

$$\int_0^\infty \delta_a(t)\,dt = 1. \tag{6}$$

But the limit in (5) gives

$$\delta_a(t) = \begin{cases} +\infty & \text{if } t = a, \\ 0 & \text{if } t \neq a. \end{cases} \tag{7}$$

Obviously no function can satisfy both (6) and (7)—if a function is zero except at a single point, then its integral is 0, not 1. Nevertheless the symbol $\delta_a(t)$ is very useful. Whatever it may mean, it is called the **Dirac delta function** at a, after the British theoretical physicist P. A. M. Dirac (1902–1982), who in the early 1930s introduced a "function" allegedly enjoying the properties in (6) and (7).

The following computation motivates the meaning that we will attach here to the symbol $\delta_a(t)$. If $g(t)$ is a continuous function, then the mean value theorem for integrals implies that

$$\int_a^{a+\epsilon} g(t)\,dt = \epsilon g(\bar{t})$$

for some point $\bar{t}$ in $[a, a + \epsilon]$. It follows that

$$\lim_{\epsilon \to 0} \int_0^\infty g(t) d_{a,\,\epsilon}(t)\, dt = \lim_{\epsilon \to 0} \int_a^{a+\epsilon} g(t) \cdot \frac{1}{\epsilon}\, dt \tag{8}$$

$$= \lim_{\epsilon \to 0} g(\bar{t}) = g(a)$$

by continuity of g at $t = a$. If $\delta_a(t)$ *were* a function in the strictest sense of the definition, and if we could interchange the limit and the integral in (8), we therefore could conclude that

$$\int_0^\infty g(t) \delta_a(t)\, dt = g(a). \tag{9}$$

We take the equation in (9) as the *definition* (!) of the symbol $\delta_a(t)$. Although we call it the delta function, it is not a genuine function; instead, it specifies the *operation*

$$\int_0^\infty \dots \delta_a(t)\, dt$$

which—when applied to a continuous function $g(t)$—sifts out or selects the value $g(a)$ of this function at the point $a \geq 0$. This idea is shown schematically in Fig. 4.32. Note that we will use the symbol $\delta_a(t)$ only in the context of integrals such as that in (9), or when it will appear subsequently in such an integral.

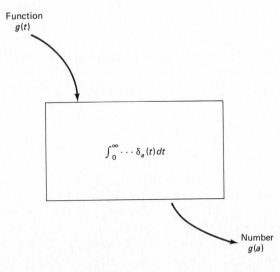

Function
$g(t)$

$\int_0^\infty \dots \delta_a(t)\, dt$

Number
$g(a)$

Figure 4.32 Diagram of how the delta function "sifts out" $g(a)$.

For instance, if we take $g(t) = e^{-st}$ in (9), the result is

$$\int_0^\infty e^{-st} \delta_a(t)\, dt = e^{-as} \tag{10}$$

We therefore *define* the Laplace transform of the delta function to be

$$\mathcal{L}\{\delta_a(t)\} = e^{-as} \qquad (a \geq 0). \tag{11}$$

If we write

$$\delta(t) = \delta_0(t) \quad \text{and} \quad \delta(t-a) = \delta_a(t), \tag{12}$$

then (11) with $a = 0$ gives

$$\mathcal{L}\{\delta(t)\} = 1. \tag{13}$$

Note that if $\delta(t)$ were an actual function, then (13) would contradict the corollary to Theorem 2 in Section 4.1. There is no problem here; $\delta(t)$ is not a function, and (13) is our *definition* of $\mathcal{L}\{\delta(t)\}$.

Now, finally, suppose that we are given a mechanical system whose response $x(t)$ to the external force $f(t)$ is determined by the differential equation

$$Ax'' + Bx' + Cx = f(t). \tag{14}$$

To investigate the response of this system to a unit impulse at the instant $t = a$, it seems reasonable to replace $f(t)$ with $\delta_a(t)$ and begin with the equation

$$Ax'' + Bx' + Cx = \delta_a(t). \tag{15}$$

But what is meant by a solution of such an equation? We will call $x(t)$ a solution of (15) provided that

$$x(t) = \lim_{\epsilon \to 0} x_\epsilon(t), \tag{16}$$

where $x_\epsilon(t)$ is a solution of

$$Ax'' + Bx' + Cx = d_{a,\epsilon}(t). \tag{17}$$

Because

$$d_{a,\epsilon}(t) = \frac{1}{\epsilon}[u_a(t) - u_{a+\epsilon}(t)] \tag{18}$$

is an ordinary function, the equation in (17) makes sense. For simplicity suppose the initial conditions to be $x(0) = x'(0) = 0$. When we transform Eq. (17), writing $X_\epsilon = \mathcal{L}\{x_\epsilon\}$, we get the equation

$$(As^2 + Bs + C)X(s) = \frac{1}{\epsilon}\left(\frac{e^{-as}}{s} - \frac{e^{-(a+\epsilon)s}}{s}\right)$$

$$= (e^{-as})\frac{1 - e^{-s\epsilon}}{s\epsilon}.$$

If we take the limit in the last equation as $\epsilon \to 0$, and note that

$$\lim_{\epsilon \to 0} \frac{1 - e^{-s\epsilon}}{s\epsilon} = 1$$

by l'Hôpital's rule, we get the equation

$$(As^2 + Bs + C)X(s) = e^{-as}. \tag{19}$$

Note that this is precisely the same result that we would obtain if we transformed the equation in (15), using the fact that $\mathcal{L}\{\delta_a(t)\} = e^{-as}$.

On this basis it is reasonable to solve a differential equation involving a delta function by employing the Laplace transform method exactly as if $\delta_a(t)$ were an ordinary function. It is important to verify that the solution so obtained agrees with the one defined in (16), but this depends upon a highly technical analysis of the limiting procedures involved; we consider it beyond the scope of the present discussion. The method is valid in all the examples of this section and will produce correct results in the subsequent problem set.

EXAMPLE 1 A mass $m = 1$ is attached to a spring with constant $k = 4$; there is no dashpot. The mass is released from rest with $x(0) = 3$. At the instant $t = 2\pi$ the mass is struck with a hammer, providing an impulse $p = 8$. Determine the motion of the mass.

Solution We need to solve the initial value problem

$$\begin{cases} x'' + 4x = 8\delta_{2\pi}(t); \\ x(0) = 3, \quad x'(0) = 0. \end{cases}$$

We apply the Laplace transform to get

$$s^2 X(s) - 3s + 4X(s) = 8e^{-2\pi s},$$

so

$$X(s) = \frac{3s}{s^2 + 4} + \frac{8e^{-2\pi s}}{s^2 + 4}.$$

Recalling the transforms of sine and cosine, as well as the theorem on translations on the t-axis (Theorem 1 in Section 4.5), we see that the inverse transform is

$$x(t) = 3 \cos 2t + 4u(t - 2\pi) \sin (t - 2\pi).$$

Because $3 \cos 2t + 4 \sin 2t = 5 \cos (2t - \alpha)$ with $\alpha = \tan^{-1} (\frac{4}{3}) \approx 0.6747$, separation of the cases $t < 2\pi$ and $t > 2\pi$ gives

$$x(t) = \begin{cases} 3 \cos 2t & \text{if } t \leq 2\pi, \\ 5 \cos (2t - 0.6747) & \text{if } t \geq 2\pi. \end{cases}$$

The resulting motion is shown in Fig. 4.33. Note that the impulse at $t = 2\pi$ results in a discontinuity in the velocity at $t = 2\pi$.

It is useful to regard the delta function $\delta_a(t)$ as the derivative of the unit step function $u_a(t)$. To see why this is reasonable, consider the continuous

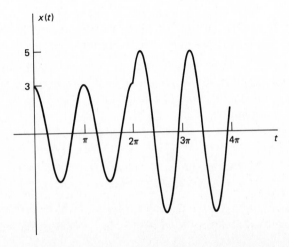

Figure 4.33 The motion of the mass of Example 1.

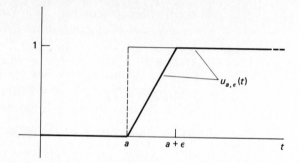

Figure 4.34 Approximation of $u_a(t)$ by $u_{a,\epsilon}(t)$.

approximation $u_{a,\epsilon}(t)$ to $u_a(t)$ shown in Fig. 4.34. We readily verify that

$$\frac{d}{dt}u_{a,\epsilon}(t) = d_{a,\epsilon}(t).$$

Because

$$u_a(t) = \lim_{\epsilon \to 0} u_{a,\epsilon}(t)$$

and

$$\delta_a(t) = \lim_{\epsilon \to 0} d_{a,\epsilon}(t),$$

an interchange of limits and derivatives yields

$$\frac{d}{dt}u_a(t) = \lim_{\epsilon \to 0} \frac{d}{dt}u_{a,\epsilon}(t) = \lim_{\epsilon \to 0} d_{a,\epsilon}(t),$$

and therefore

$$\frac{d}{dt}u_a(t) = \delta_a(t) = \delta(t-a). \tag{20}$$

We may regard this as the *formal definition* of the derivative of the unit step function, although $u_a(t)$ is not differentiable in the ordinary sense at $t = a$.

EXAMPLE 2   We return to the RLC circuit of Example 5 in Section 4.5, with $R = 110$ ohms, $L = 1$ H, $C = 0.001$ F, and a battery supplying $E_0 = 90$ V. Suppose that the circuit is initially passive—no current and no charge. At time $t = 0$ the switch is closed, and at time $t = 1$ it is opened and left open. Find the resulting current $i(t)$ in the circuit.

Solution   In Section 4.5 we circumvented the discontinuity in the voltage by employing the integrodifferential form of the circuit equation. Now that delta functions are available, we may begin with the ordinary circuit equation

$$Li'' + Ri' + \frac{1}{C}i = e'(t).$$

In this example we have $e(t) = 90 - 90u(t-1) = 90 - 90u_1(t)$, so $e'(t) = -90\delta(t-1)$. Hence we want to solve the initial value problem

$$\begin{cases} i'' + 110i' + 1000i = -90\delta(t-1); \\ i(0) = 0, \quad i'(0) = 90. \end{cases} \tag{21}$$

The fact that $i'(0) = 90$ comes from substitution of $t = 0$ in the equation

$$Li'(t) + Ri(t) + \frac{1}{C}q(t) = e(t)$$

with the numerical values $i(0) = q(0) = 0$ and $e(0) = 90$.

When we transform the problem in (21), we get the equation

$$s^2I(s) - 90 + 110sI(s) + 1000I(s) = -90e^{-s}.$$

Hence

$$I(s) = \frac{90(1 - e^{-s})}{s^2 + 110s + 1000}.$$

This is precisely the same transform $I(s)$ we found in Example 5 of Section 4.5, so inversion of $I(s)$ yields the same solution $i(t)$ as recorded there.

EXAMPLE 3 Consider a mass on a spring with $m = k = 1$ and $x(0) = x'(0) = 0$. At each of the instants $t = 0, \pi, 2\pi, \ldots, n\pi, \ldots$, the mass is struck a hammer blow with a unit impulse. Determine the resulting motion.

Solution We need to solve the initial value problem

$$\begin{cases} x'' + x = \sum\limits_{n=0}^{\infty} \delta_{n\pi}(t); \\ x(0) = 0 = x'(0). \end{cases}$$

Because $\mathcal{L}\{\delta_{n\pi}(t)\} = e^{-n\pi s}$, the transformed equation is

$$s^2X(s) + X(s) = \sum\limits_{n=0}^{\infty} e^{-n\pi s},$$

so

$$X(s) = \sum\limits_{n=0}^{\infty} \frac{e^{-n\pi s}}{s^2 + 1}.$$

We compute the inverse Laplace transform term by term; the result is

$$x(t) = \sum\limits_{n=0}^{\infty} u(t - n\pi) \sin(t - n\pi).$$

Because $\sin(t - n\pi) = (-1)^n \sin t$ and $u(t - n\pi) = 0$ for $t < n\pi$, we see that if $n\pi < t < (n+1)\pi$, then

$$x(t) = \sin t - \sin t + \sin t - \cdots + (-1)^n \sin t;$$

that is,

$$x(t) = \begin{cases} \sin t & \text{if } n \text{ is even,} \\ 0 & \text{if } n \text{ is odd.} \end{cases}$$

Hence $x(t)$ is the half-wave rectification of $\sin t$ shown in Fig. 4.35. The physical explanation is that the first hammer blow starts the mass moving to the right; just as it returns to the origin, the second hammer blow stops it dead; it remains motionless until the third hammer blow starts it moving again, and so on. Of course, if the hammer blows are not perfectly synchronized then the motion of the mass will be quite different.

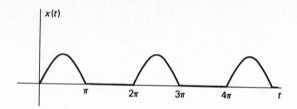

Figure 4.35   The half-wave rectification of sin t.

SYSTEMS ANALYSIS AND DUHAMEL'S PRINCIPLE

Consider a physical system in which the *output* or *response* $x(t)$ to the *input* function $f(t)$ is described by the differential equation

$$ax'' + bx' + cx = f(t), \tag{22}$$

where the constant coefficients a, b, and c are determined by the physical parameters of the system and are independent of $f(t)$. The mass-spring-dashpot system and the series RLC circuit are familiar examples of this general situation.

For simplicity we assume that the system is initially passive: $x(0) = x'(0) = 0$. Then the transform of Eq. (22) is

$$as^2 X(s) + bs X(s) + c X(s) = F(s),$$

so

$$X(s) = \frac{F(s)}{as^2 + bs + c} = W(s)F(s). \tag{23}$$

The function

$$W(s) = \frac{1}{as^2 + bs + c} \tag{24}$$

is called the **transfer function** of the system. Thus the transform of the response to the input $f(t)$ is the product of $W(s)$ and the transform $F(s)$.

The function

$$w(t) = \mathcal{L}^{-1}\{W(s)\} \tag{25}$$

is called the **weight function** of the system. From (24) we see by convolution that

$$x(t) = \int_0^t w(\tau)f(t - \tau)\, d\tau. \tag{26}$$

This formula is **Duhamel's principle** for the system. What is important is that the weight function $w(t)$ is determined completely by the parameters of the system. Once $w(t)$ has been determined, the integral in (26) gives the response of the system to an arbitrary input function $f(t)$.

EXAMPLE 4   Consider a mass-spring-dashpot system (initially passive) which responds to the external force $f(t)$ in accord with the equation $x'' + 6x' + 10x = f(t)$. Then

$$W(s) = \frac{1}{s^2 + 6s + 10} = \frac{1}{(s + 3)^2 + 1},$$

so the weight function is $w(t) = e^{-3t} \sin t$. Then Duhamel's principle implies that the response $x(t)$ to the force $f(t)$ is

$$x(t) = \int_0^t e^{-3\tau} (\sin \tau) f(t - \tau) d\tau.$$

Note that

$$W(s) = \frac{1}{as^2 + bs + c} = \frac{\mathcal{L}\{\delta(t)\}}{as^2 + bs + c}.$$

Consequently, it follows from (23) that the weight function is simply the response of the system to the delta function input $\delta(t)$. For this reason $w(t)$ is sometimes called the **unit impulse response**. A response that is usually easier to measure in practice is the response $h(t)$ to the unit step function $u(t)$; $h(t)$ is the **unit step response**. Because $\mathcal{L}\{u(t)\} = 1/s$, we see from (23) that the transform of $h(t)$ is

$$H(s) = \frac{W(s)}{s}.$$

It follows from the formula for transforms of integrals that

$$h(t) = \int_0^t w(\tau) \, d\tau, \quad \text{so that} \quad w(t) = h'(t). \tag{27}$$

Thus the weight function, or unit impulse response, is the derivative of the unit step response. Substitution of (27) in Duhamel's principle gives

$$x(t) = \int_0^t h'(\tau) f(t - \tau) \, d\tau \tag{28}$$

for the response of the system to the input $f(t)$.

To describe a typical application of (28), suppose that we are given a complex series circuit containing many inductors, resistors, and capacitors. Assume that its circuit equation is a linear equation of the form in (22), though with i in place of x. What if the coefficients a, b, and c are unknown, perhaps only because they are too complicated to compute? We would still want to know the current $i(t)$ corresponding to any input $f(t) = e'(t)$. We connect the circuit to a linearly increasing voltage $e(t) = t$, so that $f(t) = e'(t) = 1 = u(t)$, and measure the response $h(t)$ with an ammeter. We then compute the derivative $h'(t)$, either numerically or graphically. Then according to (28), the output current $i(t)$ corresponding to the input voltage $e(t)$ will be given by

$$i(t) = \int_0^t h'(\tau) e'(t - \tau) \, d\tau$$

(using the fact that $f(t) = e'(t)$).

In conclusion we remark that around 1950, after engineers and physicists had been using delta functions widely and fruitfully for about 20 years without rigorous justification, the French mathematician Laurent Schwartz developed a rigorous mathematical theory of *generalized functions* that supplied the missing logical foundation for delta function techniques. Every piecewise continuous ordinary function is a generalized function, but the delta function is an example of a generalized function that is not an ordinary function.

4.6 Problems

Solve the initial value problems in Problems 1–8.

1. $x'' + 4x = \delta(t); \quad x(0) = x'(0) = 0.$
2. $x'' + 4x = \delta(t) + \delta(t - \pi); \quad x(0) = x'(0) = 0.$
3. $x'' + 4x' + 4x = 1 + \delta(t - 2); \quad x(0) = x'(0) = 0.$
4. $x'' + 2x' + x = t + \delta(t); \quad x(0) = 0, x'(0) = 1.$
5. $x'' + 2x' + 2x = 2\delta(t - \pi); \quad x(0) = 0 = x'(0).$
6. $x'' + 9x = \delta(t - 3\pi) + \cos 3t; \quad x(0) = x'(0) = 0.$
7. $x'' + 4x' + 5x = \delta(t - \pi) + \delta(t - 2\pi); \quad x(0) = 0, x'(0) = 2.$
8. $x'' + 2x' + x = \delta(t) - \delta(t - 2); \quad x(0) = 2 = x'(0).$

Apply Duhamel's principle to write an integral formula for the solution of each of the initial value problems in Problems 9–12.

9. $x'' + 4x = f(t); \quad x(0) = x'(0) = 0.$
10. $x'' + 6x' + 9x = f(t); \quad x(0) = x'(0) = 0.$
11. $x'' + 6x' + 8x = f(t); \quad x(0) = x'(0) = 0.$
12. $x'' + 4x' + 8x = f(t); \quad x(0) = x'(0) = 0.$

13. This problem deals with a particle of mass m, initially at rest at the origin, that receives an impulse p at time $t = 0$.
    (a) Find the solution $x_\epsilon(t)$ of the problem
    $$mx'' = pd_{0,\,\epsilon}(t); \qquad x(0) = 0 = x'(0).$$
    (b) Show that $\lim\limits_{\epsilon \to 0} x_\epsilon(t)$ agrees with the solution of the problem
    $$mx'' = p\delta(t); \qquad x(0) = x'(0) = 0.$$
    (c) Show that $mv = p$ for $t > 0$ $(v = x')$.

14. Verify that $u'(t - a) = \delta(t - a)$ by solving the problem
    $$x' = \delta(t - a); \qquad x(0) = 0$$
    to obtain $x(t) = u(t - a)$.

15. This problem deals with a mass m on a spring (with constant k) that receives an impulse $p_0 = mv_0$ at time $t = 0$. Show that the initial value problems
    $$mx'' + kx = 0; \qquad x(0) = 0, \qquad x'(0) = v_0$$
    and
    $$mx'' + kx = p_0\delta(t); \qquad x(0) = 0, \qquad x'(0) = 0$$
    have the same solution. Thus the effect of $p_0\delta(t)$ is, indeed, to impart to the particle an initial momentum p_0.

16. This is a generalization of Problem 15. Show that the problems
    $$ax'' + bx' + cx = f(t); \qquad x(0) = 0, \qquad x'(0) = v_0$$
    and
    $$ax'' + bx' + cx = f(t) + av_0\delta(t); \qquad x(0) = x'(0) = 0$$
    have the same solution for $t > 0$. Thus the effect of the term $av_0\delta(t)$ is to supply the initial condition $x'(0) = v_0$.

17. Consider an initially passive RC circuit (no inductance) with a battery supplying E_0 volts.
    (a) If the switch to the battery is closed at time $t = a$ and opened at time $t = b$ $> a$ (and left open thereafter), show that the current $i(t)$ in the circuit

satisfies the initial value problem

$$\begin{cases} Ri' + \dfrac{1}{C}i = E_0\delta(t - a) - E_0\delta(t - b); \\ i(0) = 0. \end{cases}$$

(b) Solve this problem if $R = 100$ ohms, $C = 10^{-4}$ F, $E_0 = 100$ V, $a = 1$ (s), and $b = 2$ (s). Show that $i(t) > 0$ if $1 < t < 2$, and that $i(t) < 0$ if $t > 2$.

18. Consider an initially passive LC circuit (no resistance) with a battery supplying E_0 volts.

(a) If the switch is closed at time $t = 0$ and opened at time $t = a$, show that the current $i(t)$ in the circuit satisfies the initial value problem

$$\begin{cases} Li'' + \dfrac{1}{C}i = E_0\delta(t) - E_0\delta(t - a); \\ i(0) = i'(0) = 0. \end{cases}$$

(b) If $L = 1$ H, $C = 10^{-2}$ F, $E_0 = 10$ V, and $a = \pi$ (s), show that

$$i(t) = \begin{cases} \sin 10t & \text{if } t < \pi, \\ 0 & \text{if } t > \pi. \end{cases}$$

Thus the current oscillates through five cycles and then stops abruptly when the switch is opened.

19. Consider the LC circuit of Problem 18(b), except suppose that the switch is alternately closed and opened at times $t = 0, (0.1)\pi, (0.2)\pi, \ldots$.

(a) Show that $i(t)$ satisfies the initial value problem

$$\begin{cases} i'' + 100i = 10 \displaystyle\sum_{n=0}^{\infty} (-1)^n\delta(t - [0.1]n\pi), \\ i(0) = i'(0) = 1. \end{cases}$$

(b) Solve this problem to show that $i(t)$ is the half-wave rectification of $\sin 10t$.

20. Consider an RLC circuit in series with a battery, with $L = 1$ H, $R = 60$ ohms, $C = 10^{-3}$ F, and $E_0 = 10$ V.

(a) Suppose that the switch is alternately closed and opened at times $t = 0, (0.1)\pi, (0.2)\pi, \ldots$. Show that $i(t)$ satisfies the initial value problem

$$\begin{cases} i'' + 60i' + 1000i = 10 \displaystyle\sum_{n=0}^{\infty} (-1)^n\delta(t - [0.1]n\pi), \\ i(0) = i'(0) = 0. \end{cases}$$

(b) Solve this problem to show that if $(0.1)n\pi < t < (0.1)(n + 1)\pi$, then

$$i(t) = \left(\frac{e^{3n\pi+3\pi} - 1}{e^{3\pi} - 1}\right)e^{-30t}\sin 10t.$$

21. Consider a mass $m = 1$ on a spring with constant $k = 1$, initially at rest, but struck with a hammer at each of the instants $t = 0, 2\pi, 4\pi, \ldots$. Suppose that each hammer blow imparts an impulse of $+1$. Show that the position $x(t)$ of the mass satisfies the initial value problem

$$\begin{cases} x'' + x = \displaystyle\sum_{n=0}^{\infty} \delta(t - 2n\pi), \\ x(0) = x'(0) = 0. \end{cases}$$

Solve this problem to show that if $2n\pi < t < 2(n + 1)\pi$, then $x(t) = (n + 1)\sin t$. Thus resonance occurs because the mass is struck each time it passes through the origin moving to the right—in contrast to Example 3, in which the mass was struck each time it returned to the origin.

Table of Laplace Transforms

This table summarizes the general properties of Laplace transforms and the Laplace transforms of particular functions derived in this chapter.

Function	Transform
$f(t)$	$F(s)$
$af(t) + bg(t)$	$aF(s) + bG(s)$
$f'(t)$	$sF(s) - f(0)$
$f''(t)$	$s^2 F(s) - sf(0) - f'(0)$
$f^{(n)}(t)$	$s^n F(s) - s^{n-1} f(0) - \cdots - f^{(n-1)}(0)$
$\int_0^t f(\tau)\, d\tau$	$\dfrac{F(s)}{s}$
$e^{at} f(t)$	$F(s - a)$
$u(t - a)f(t - a)$	$e^{-as} F(s)$
$\int_0^t f(\tau) g(t - \tau)\, d\tau$	$F(s)G(s)$
$tf(t)$	$-F'(s)$
$t^n f(t)$	$(-1)^n F^{(n)}(s)$
$\dfrac{f(t)}{t}$	$\int_s^\infty F(\sigma)\, d\sigma$
$f(t)$, period p	$\dfrac{1}{1 - e^{-ps}} \int_0^p e^{-st} f(t)\, dt$
1	$\dfrac{1}{s}$
t	$\dfrac{1}{s^2}$
t^n	$\dfrac{n!}{s^{n+1}}$
$\dfrac{1}{\sqrt{\pi t}}$	$\dfrac{1}{\sqrt{s}}$
t^a	$\dfrac{\Gamma(a + 1)}{s^{a+1}}$
e^{at}	$\dfrac{1}{s - a}$
$t^n e^{at}$	$\dfrac{n!}{(s - a)^{n+1}}$
$\cos kt$	$\dfrac{s}{s^2 + k^2}$
$\sin kt$	$\dfrac{k}{s^2 + k^2}$
$\cosh kt$	$\dfrac{s}{s^2 - k^2}$
$\sinh kt$	$\dfrac{k}{s^2 - k^2}$
$\dfrac{1}{2k^3}(\sin kt - kt \cos kt)$	$\dfrac{1}{(s^2 + k^2)^2}$
$\dfrac{t}{2k} \sin kt$	$\dfrac{s}{(s^2 + k^2)^2}$
$\dfrac{1}{2k}(\sin kt + kt \cos kt)$	$\dfrac{s^2}{(s^2 + k^2)^2}$
$u(t - a)$	$\dfrac{e^{-as}}{s}$
$\delta(t - a)$	e^{-as}
$(-1)^{[\![at]\!]}$ (square wave)	$\dfrac{1}{s} \tanh \dfrac{as}{2}$
$\left[\!\left[\dfrac{t}{a} \right]\!\right]$ (staircase)	$\dfrac{e^{-as}}{s(1 - e^{-as})}$

Linear Systems
of Differential
Equations

5

5.1
Introduction
to Systems

In the preceding chapters we have discussed methods for solving an ordinary differential equation that involves only one dependent variable. Many applications, however, require the use of two or more dependent variables, each a function of a single independent variable (typically time). Such problems lead naturally to a *system* of simultaneous ordinary differential equations. We will usually denote the independent variable by t and the dependent variables (the unknown functions of t) by $x_1, x_2, x_3, \ldots$, or by $x, y, z, \ldots$. Primes will denote differentiation with respect to t.

We will restrict our attention to systems in which the number of equations is the same as the number of dependent variables (unknown functions). For instance, a system of two first order differential equations in the dependent variables x and y has the general form

$$\left. \begin{array}{l} f(t, x, y, x', y') = 0 \\ g(t, x, y, x', y') = 0 \end{array} \right\}, \tag{1}$$

where the functions f and g are given. A solution of this system is a *pair* $x(t)$, $y(t)$ of functions of t that satisfy both equations identically over some interval of values of t.

For an example of a second order system, consider a particle of mass m that moves in space under the influence of a force field $\mathbf{F}$ that depends upon time t, the position $(x(t), y(t), z(t))$ of the particle, and its velocity $(x'(t), y'(t), z'(t))$. Applying Newton's law $m\mathbf{a} = \mathbf{F}$ componentwise, we get the system

$$mx'' = F_1(t, x, y, z, x', y', z'),$$
$$my'' = F_2(t, x, y, z, x', y', z'), \tag{2}$$
$$mz'' = F_3(t, x, y, z, x', y', z'),$$

where F_1, F_2, and F_3 are the components of the vector function $\mathbf{F}$.

The following three examples illustrate further how systems of differential equations arise naturally in scientific problems.

EXAMPLE 1 Consider the system of two masses and two springs shown in Fig. 5.1, with a given external force $f(t)$ acting on the right-hand mass m_2. We denote by $x(t)$ the displacement (to the right) of the mass m_1 from its static position (when the system is motionless and in equilibrium and $f(t) = 0$), and by $y(t)$ the displacement of the mass m_2 from its static position. Thus the two springs are neither stretched nor compressed when x and y are zero.

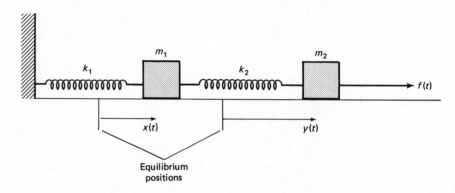

Figure 5.1 The mass-and-spring system of Example 1.

In the configuration in Fig. 5.1, the first spring is stretched x units and the second by $y - x$ units. We apply Newton's law of motion to the two "free body diagrams" shown in Fig. 5.2; we thereby obtain the system

$$m_1 x'' = -k_1 x + k_2(y - x),$$
$$m_2 y'' = -k_2(y - x) + f(t) \tag{3}$$

of differential equations that the position functions $x(t)$ and $y(t)$ must satisfy. For instance, if $m_1 = 2$, $m_2 = 1$, $k_1 = 4$, $k_2 = 2$, and $f(t) =$

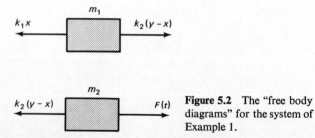

Figure 5.2 The "free body diagrams" for the system of Example 1.

40 sin 3t in appropriate physical units, the system in (3) then reduces to

$$2x'' = -6x + 2y,$$
$$y'' = 2x - 2y + 40 \sin 3t. \tag{4}$$

EXAMPLE 2 Consider two brine tanks connected as shown in Fig. 5.3. Tank 1 contains $x(t)$ pounds of salt in 100 gal of brine and Tank 2 contains $y(t)$ pounds of salt in 200 gal of brine. The brine in each tank is kept

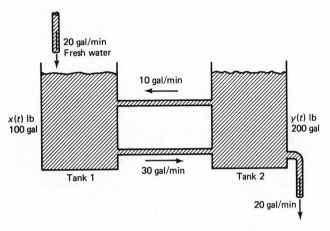

Figure 5.3 The two brine tanks of Example 2.

uniform by stirring, and brine is pumped from each tank to the other at the rates indicated in the figure. In addition, fresh water flows into Tank 1 at 20 gal/min, and the brine in Tank 2 flows out at 20 gal/min (so the total volume of brine in the two tanks remains constant). The salt concentrations in the two tanks are $x/100$ pounds per gallon and $y/200$ pounds per gallon, respectively. When we compute the rates of change of the amounts of salt in the two tanks, we therefore get the system of differential equations that $x(t)$ and $y(t)$ must satisfy:

$$x' = -(30)\left(\frac{x}{100}\right) + (10)\left(\frac{y}{200}\right) = -\frac{3}{10}x + \frac{1}{20}y,$$
$$y' = (30)\left(\frac{x}{100}\right) - (10)\left(\frac{y}{200}\right) - (20)\left(\frac{y}{200}\right) = \frac{3}{10}x - \frac{3}{20}y. \tag{5}$$

EXAMPLE 3 Consider the electrical network shown in Fig. 5.4, where $I_1(t)$ denotes the current in the indicated direction through the inductor and $I_2(t)$ denotes the current through the resistor R_2. The current through the resistor R is $I = I_1 - I_2$ in the direction indicated. We recall Kirchoff's voltage law to the effect that the (algebraic) sum of the voltage drops around any closed loop of such a network is zero. As in Section 2.9, the voltage drops across the three types of circuit elements are those shown in Fig. 5.5. We apply Kirchoff's law to the left-hand loop of the network

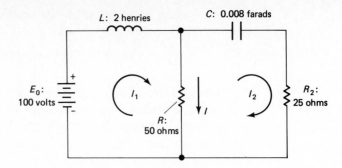

Figure 5.4 The electrical network of Example 3.

Circuit element	Voltage drop
Inductor	$L \dfrac{dI}{dt}$
Resistor	RI
Capacitor	$\dfrac{1}{C} Q$

Figure 5.5 Voltage drops across common circuit elements.

to obtain

$$2 \frac{dI_1}{dt} + 50(I_1 - I_2) - 100 = 0, \tag{6}$$

because the voltage drop from the negative to the positive pole of the battery is negative. The right-hand loop yields the equation

$$125 Q_2 + 25 I_2 + 50(I_2 - I_1) = 0 \tag{7}$$

where $Q_2(t)$ is the charge on the capacitor. Because $dQ_2/dt = I_2$, differentiation of each side of Eq. (7) yields

$$-50 \frac{dI_1}{dt} + 75 \frac{dI_2}{dt} + 125 I_2 = 0. \tag{8}$$

Dividing Eqs. (6) and (8) by the factors 2 and -25 respectively, we get the system

$$\frac{dI_1}{dt} + 25 I_1 - 25 I_2 = 50,$$

$$2 \frac{dI_1}{dt} - 3 \frac{dI_2}{dt} - 5 I_2 = 0 \tag{9}$$

of differential equations that the currents $I_1(t)$ and $I_2(t)$ must satisfy.

FIRST ORDER SYSTEMS

Consider a system of differential equations that can be solved for the highest order derivatives of the dependent variables that appear, as explicit functions of t and of lower order derivatives of the dependent variables. For instance, in the case of a system of two second order equations, our assumption is that it can be written in the form

$$x_1'' = f_1(t, x_1, x_2, x_1', x_2'),$$
$$x_2'' = f_2(t, x_1, x_2, x_1', x_2'), \tag{10}$$

It is of both practical and theoretical importance that any such higher order system can be transformed into an equivalent system of *first order* equations.

To describe how such a transformation is accomplished, we consider first the "system" consisting of the single nth order equation

$$x^{(n)} = f(t, x, x', \ldots, x^{(n-1)}). \tag{11}$$

We introduce the dependent variables $x_1, x_2, \ldots, x_n$ defined as follows:

$$x_1 = x, x_2 = x', x_3 = x'', \ldots, x_n = x^{(n-1)}. \tag{12}$$

Note that $x_1' = x' = x_2$, $x_2' = x'' = x_3$, and so on. Hence the substitutions of (12) in Eq. (11) yield the system

$$x_1' = x_2,$$
$$x_2' = x_3,$$
$$\cdot$$
$$\cdot \tag{13}$$
$$\cdot$$
$$x_{n-1}' = x_n,$$
$$x_n' = f(t, x_1, x_2, \ldots, x_n)$$

of n *first order* equations. Evidently this system is equivalent to the original nth order equation in (11), in the sense that $x(t)$ is a solution of (11) if and only if the functions $x_1(t), x_2(t), \ldots, x_n(t)$ defined in (12) satisfy the system of equations in (13).

For instance, the third order equation $x''' + 3x'' + 2x' - 5x = \sin 2t$ is equivalent to the system

$$x_1' = x_2,$$
$$x_2' = x_3,$$
$$x_3' = 5x_1 - 2x_2 - 3x_3 + \sin 2t$$

where $x_1 = x$. It may appear that this system offers little advantage because we could use the methods of Chapter 2 to solve the original (linear) third order equation. But suppose that we were confronted with the nonlinear equation $x'' = x^3 + (x')^3$, to which none of our earlier methods can be applied. The corresponding first order system is

$$x_1' = x_2,$$
$$x_2' = (x_1)^3 + (x_2)^3, \tag{14}$$

and we will see in Chapter 6 that there exist effective numerical techniques for approximating the solution of essentially any first order system. So in this case the transformation to a first order system *is* advantageous. From a practical viewpoint, large systems of higher order differential equations typically are solved numerically with the aid of a computer, and the first step is to transform such a system into a first order system for which a standard computer program is available.

EXAMPLE 4 The system

$$2x'' = -6x + 2y, \tag{4}$$
$$y'' = \quad 2x - 2y + 40 \sin 3t$$

of second order equations was derived in Example 1. Transform this system into an equivalent first order system.

Solution Motivated by the equations in (12), we define

$$x_1 = x, \qquad x_2 = x' = x_1', \qquad y_1 = y, \qquad y_2 = y' = y_1'.$$

Then the system in (4) yields the system

$$
\begin{aligned}
x_1' &= x_2, \\
2x_2' &= -6x_1 + 2y_1, \\
y_1' &= y_2, \\
y_2' &= 2x_1 - 2y_1 + 40 \sin 3t
\end{aligned}
\tag{15}
$$

of four first order equations in the dependent variables x_1, x_2, y_1, and y_2.

In addition to practical advantages for numerical computation, the general theory of systems and systematic solution techniques are more easily and more concisely described for first order systems than for higher order systems. Throughout most of this chapter, we will concentrate our attention on *linear* first order systems of the form

$$
\begin{aligned}
x_1' &= p_{11}(t)x_1 + p_{12}(t)x_2 + \cdots + p_{1n}(t)x_n + f_1(t), \\
x_2' &= p_{21}(t)x_1 + p_{22}(t)x_2 + \cdots + p_{2n}(t)x_n + f_2(t), \\
&\;\;\vdots \\
x_n' &= p_{n1}(t)x_1 + p_{n2}(t)x_2 + \ldots + p_{nn}(t)x_n + f_n(t).
\end{aligned}
\tag{16}
$$

We say that this system is **homogeneous** if the functions $f_1, f_2, \ldots, f_n$ are all identically zero; otherwise it is **nonhomogeneous**. Thus the linear system in (5) is homogeneous, while the linear system in (15) is nonhomogeneous. The system in (14) is nonlinear because the right-hand side of the second equation is not a linear function of the dependent variables x_1 and x_2.

A **solution** of the system in (16) is an n-tuple of functions $x_1(t)$, $x_2(t)$, $\ldots$, $x_n(t)$ that (on some interval) identically satisfy each of the equations in (16). We will see that the general theory of a system of n linear first order equations has many similarities with the general theory of a single nth order linear differential equation. The theorem below (proved in Section 9.4) is analogous to Theorem 2 (existence and uniqueness of the solution of an nth order equation) in Section 2.2. It tells us that if the coefficient functions p_{ij} and f_j in (16) are continuous, then the system has a unique solution satisfying given initial conditions.

THEOREM: EXISTENCE AND UNIQUENESS FOR LINEAR SYSTEMS

Suppose that the functions $p_{11}, p_{12}, \ldots, p_{nn}$ and the functions $f_1, f_2, \ldots,$ f_n are continuous on the open interval I containing the point a. Then, given n numbers $b_1, b_2, \ldots, b_n$, the system in (16) has a unique solution on the entire interval I that satisfies the n initial conditions

$$x_1(a) = b_1, \, x_2(a) = b_2, \ldots, x_n(a) = b_n. \tag{17}$$

Thus n initial conditions are needed to determine a solution of a system of n linear first order equations, and we therefore expect a general solution of such a system to involve n arbitrary constants. For instance, we saw in Example 4 that the second order system in (4)—which describes the position functions $x(t)$ and $y(t)$ of the two masses in Example 1—is equivalent to the system of *four* first order linear equations in (15). Hence four initial conditions would be needed to determine the subsequent motions of the two masses in Example 1; typical initial values would be the initial positions $x(0)$ and $y(0)$ and the initial velocities $x'(0)$ and $y'(0)$. On the other hand, we found that the amounts $x(t)$ and $y(t)$ of salt in the two tanks of Example 2 are described by the system in (5) of *two* first order linear equations. Hence the two initial values $x(0)$ and $y(0)$ should suffice to determine the solution. Given a higher order system, we often must transform it into an equivalent first order system to discover how many initial conditions are needed to determine a unique solution; the theorem above tells us that the number of such conditions is precisely the same as the number of equations in the equivalent first order system.

THE LAPLACE TRANSFORM METHOD*

Laplace transforms are frequently used in engineering problems to solve linear systems in which the coefficients are all constants. When initial conditions are specified, the Laplace transform reduces such a linear system of differential equations to a system of linear algebraic equations in which the unknowns are the transforms of the solution functions. As the following example illustrates, the technique for a system is essentially the same as for a single linear differential equation with constant coefficients.

EXAMPLE 5   Solve the system

$$\begin{aligned} 2x'' &= -6x + 2y, \\ y'' &= \quad 2x - 2y + 40 \sin 3t \end{aligned} \tag{4}$$

subject to the initial conditions

$$x(0) = x'(0) = y(0) = y'(0) = 0. \tag{18}$$

Thus the force $f(t) = 40 \sin 3t$ is suddenly applied to the second mass of Example 1 at time $t = 0$ when the system is at rest in its equilibrium position.

*This subsection can be omitted if Chapter 4 has not been covered.

Solution We write $X(s) = \mathcal{L}\{x(t)\}$ and $Y(s) = \mathcal{L}\{y(t)\}$, exactly as in Chapter 4. Then the initial conditions in (18) imply that

$$\mathcal{L}\{x''(t)\} = s^2 X(s) \quad \text{and} \quad \mathcal{L}\{y''(t)\} = s^2 Y(s).$$

Because $\mathcal{L}\{\sin 3t\} = 3/(s^2 + 9)$, the transforms of the equations in (4) are the equations

$$2s^2 X(s) = -6X(s) + 2Y(s),$$

$$s^2 Y(s) = 2X(s) - 2Y(s) + \frac{120}{s^2 + 9}.$$

Thus the transformed system is

$$(s^2 + 3)X(s) - \qquad\qquad Y(s) = 0$$

$$-2X(s) + (s^2 + 2) Y(s) = \frac{120}{s^2 + 9}. \tag{19}$$

The determinant of this pair of linear equations in $X(s)$ and $Y(s)$ is

$$(s^2 + 3)(s^2 + 2) - 2 = (s^2 + 1)(s^2 + 4),$$

and we readily solve—using Cramer's rule, for instance—the system in (19) for

$$X(s) = \frac{120}{(s^2 + 1)(s^2 + 4)(s^2 + 9)} = \frac{5}{s^2 + 1} - \frac{8}{s^2 + 4} + \frac{3}{s^2 + 9} \tag{20}$$

and

$$Y(s) = \frac{120(s^2 + 3)}{(s^2 + 1)(s^2 + 4)(s^2 + 9)} = \frac{10}{s^2 + 1} + \frac{8}{s^2 + 4} - \frac{18}{s^2 + 9}. \tag{21}$$

The partial fraction decompositions in (20) and (21) are readily found using the methods of Section 4.4. For instance, if we write

$$\frac{120}{(s^2 + 1)(s^2 + 4)(s^2 + 9)} = \frac{As + B}{s^2 + 1} + H(s),$$

where $H(s)$ denotes the sum of the partial fractions corresponding to the two factors $s^2 + 4$ and $s^2 + 9$, then multiplication of each term by $s^2 + 1$ yields

$$\frac{120}{(s^2 + 4)(s^2 + 9)} = As + B + (s^2 + 1)H(s).$$

Substitution of $s = i$ (one of the zeros of $s^2 + 1$) in this last equation then gives

$$Ai + B = \frac{120}{(i^2 + 4)(i^2 + 9)} = \frac{120}{(3)(8)} = 5,$$

so $A = 0$ and $B = 5$. This gives the first term in (20), and the others can be found similarly.

At any rate, the inverse Laplace transforms of the expressions in (20) and (21) give the solution

$$x(t) = 5 \sin t - 4 \sin 2t + \sin 3t,$$

$$y(t) = 10 \sin t + 4 \sin 2t - 6 \sin 3t. \tag{22}$$

It is of interest to rewrite the solutions in (22) in the form of a single vector equation

$$(x, y) = 5(x_1, y_1) + 4(x_2, y_2) + (x_p, y_p),\tag{23}$$

where

$$x_1(t) = \sin t, \qquad y_1(t) = 2 \sin t;\tag{24a}$$

$$x_2(t) = -\sin 2t, \qquad y_2(t) = \sin 2t;\tag{24b}$$

$$x_p(t) = \sin 3t, \qquad y_p(t) = -6 \sin 3t.\tag{24c}$$

In Problem 25 we ask you to verify that x_p, y_p is a particular solution of the nonhomogeneous system in (4), while x_1, y_1 and x_2, y_2 are solutions of the associated homogeneous system

$$2x'' = -6x + 2y,\tag{25}$$

$$y'' = 2x - 2y.$$

Thus x_1, y_1 and x_2, y_2 represent *free* oscillations of the two masses (no external force), while x_p, y_p is a particular forced oscillation in response to the force $f(t) = 40 \sin 3t$ with the imposed frequency $\omega_p = 3$. Note that in the free oscillation x_1, y_1 the two masses move in the same direction with the same frequency $\omega_1 = 1$, but with the amplitude of motion of the second mass twice that of the first mass. In the free oscillation x_2, y_2 the two masses move in opposite directions with equal amplitudes of motion and with frequency $\omega_2 = 2$.

The homogeneous system in (25) has the two additional solutions

$$x_3(t) = \cos t, \qquad y_3(t) = 2 \cos t;\tag{26a}$$

$$x_4(t) = -\cos 2t, \qquad y_4(t) = \cos 2t;\tag{26b}$$

these correspond to (24a) and (24b) with cosines in place of sines. In Problem 25 we ask you to show that if $c_1, c_2, c_3,$ and c_4 are constants, then

$$x_h = c_1 x_1 + c_2 x_2 + c_3 x_3 + c_4 x_4,$$

$$y_h = c_1 y_1 + c_2 y_2 + c_3 y_3 + c_4 y_4\tag{27}$$

is a general solution of the homogeneous system in (25), while

$$x = x_h + x_p, \qquad y = y_h + y_p\tag{28}$$

is a general solution of the original nonhomogeneous system in (4). Thus (28) exhibits the general solution of the nonhomogeneous system as the (vector) sum of a particular solution and the general solution of the associated homogeneous system. (This should have a familiar ring!) In Section 5.3 we will see that this situation is typical of linear systems.

EXAMPLE 6 Because the system in (4) involves no first derivatives, we might well have foreseen that it would have a particular solution of the form $x_p = A \sin 3t$, $y_p = B \sin 3t$. Determine A and B by substituting this trial solution into the system.

Solution The substitution in question yields the equations

$$-18A \sin 3t = -6A \sin 3t + 2B \sin 3t$$

and

$$-9B \sin 3t = 2A \sin 3t - 2B \sin 3t + 40 \sin 3t.$$

After we remove the common factor $\sin 3t$ and collect coefficients, we see that A and B must satisfy the equations

$$12A + 2B = 0, \qquad -2A - 7B = 40.$$

The solution is $A = 1$ and $B = -6$, and thus $x_p = \sin 3t$ and $y_p = -6 \sin 3t$, exactly as in (24c).

As indicated by Example 5, the Laplace transform method often readily yields the solution of a specific system with given numerical initial conditions. Unfortunately, this method fails to provide an adequate understanding of the general situation for linear systems. The remaining sections of this chapter are devoted to alternative methods for the solution of linear systems that, in addition to yielding solutions, will exhibit more clearly the general nature of the solutions of linear systems.

5.1 Problems

In each of Problems 1–10, transform the given differential equation or system into an equivalent system of first order differential equations.

1. $x'' + 3x' + 7x = t^2$.
2. $x^{(4)} + 6x'' - 3x' + x = \cos 3t$.
3. $t^2 x'' + tx' + (t^2 - 1)x = 0$.
4. $t^3 x''' - 2t^2 x'' + 3tx' + 5x = \ln t$.
5. $x''' = (x')^2 + \cos x$.
6. $x'' - 5x + 4y = 0, y'' + 4x - 5y = 0$.
7. $x'' = \dfrac{-kx}{(x^2 + y^2)^{3/2}}, y'' = \dfrac{-ky}{(x^2 + y^2)^{3/2}}$.
8. $x'' + 3x' + 4x - 2y = 0, y'' + 2y' - 3x + y = \cos t$.
9. $x'' = 3x - y + 2z, y'' = x + y - 4z, z'' = 5x - y - z$.
10. $x'' = x(1 - y), y'' = y(1 - x)$.

In each of Problems 11–13, find a particular solution of the given system, either of the form $x = A \cos \omega t, y = B \cos \omega t$ or of the form $x = A \sin \omega t, y = B \sin \omega t$.

11. $x'' + 5x - 3y = \sin 2t, y'' - 3x + 5y = 0$.
12. $x'' + 5x - 4y = 0, y'' - 4x + 5y = 2 \cos 4t$.
13. $x'' + 3x - y = 2 \cos 3t, y'' - 2x + 2y = -\cos 3t$.

In each of Problems 14–20, use Laplace transforms to solve the given initial value problem.

14. $x' = x + 2y, y' = 2x + y; x(0) = 1, y(0) = 0$.
15. $x' = 2x + y, y' = 6x + 3y; x(0) = 1, y(0) = -2$.
16. $x' = x + 2y, y' = x + e^{-t}; x(0) = y(0) = 0$.
17. $x' + 2y' + x = 0, x' - y' + y = 0; x(0) = 0, y(0) = 1$.
18. $x'' + 2x + 4y = 0, y'' + x + 2y = 0; x(0) = y(0) = 0, x'(0) = y'(0) = -1$.

19. $x'' + x' + y' + 2x - y = 0$, $y'' + x' + y' + 4x - 2y = 0$;
$x(0) = y(0) = 1$, $x'(0) = y'(0) = 0$.

20. $x' = x + z$, $y' = x + y$, $z' = -2x - z$; $x(0) = 1$, $y(0) = 0$, $z(0) = 0$.

21. If each of the two brine tanks in Example 2 initially contains 50 lb of salt, find $x(t)$ and $y(t)$.

22. If no currents are flowing initially in the electrical network of Example 3 and the battery is connected at time $t = 0$, find $I_1(t)$ and $I_2(t)$. Show that $I_1(t)$ approaches 2 (A) as $t \to +\infty$, while $I_2(t)$ approaches zero. You may use the inverse Laplace transform

$$\mathcal{L}^{-1}\left\{\frac{b(a^2 + b^2)}{s[(s - a)^2 + b^2]}\right\} = b + e^{at}(a \sin bt - b \cos bt).$$

Problems 23 and 24 deal with the linear system

$$\begin{aligned} x' &= ax + by + f(t), \\ y' &= cx + dy + g(t). \end{aligned} \tag{29}$$

23. Suppose that c_1 and c_2 are constants and that x_1, y_1 and x_2, y_2 are two solutions of the homogeneous system associated with (29). Show that $x_h = c_1 x_1 + c_2 x_2$, $y_h = c_1 y_1 + c_2 y_2$ is also a solution of the associated homogeneous system.

24. Suppose that x_p, y_p is a particular solution of (29) and that x_h, y_h is a solution of the associated homogeneous system. Show that $x = x_h + x_p$, $y = y_h + y_p$ is a solution of (29).

25. (a) Verify that the pairs defined in Eqs. (24a), (24b), (26a), and (26b) are all solutions of the associated homogeneous system in (25). (b) Verify that the pair x_h, y_h defined in (27) is a solution of the homogeneous system in (25). (c) Verify that the pair x, y defined in (28) is a solution of the original nonhomogeneous system.

26. Derive the equations

$$\begin{aligned} m_1 x_1'' &= -(k_1 + k_2)x_1 + k_2 x_2, \\ m_2 x_2'' &= k_2 x_1 - (k_2 + k_3)x_2 \end{aligned}$$

for the displacements (from equilibrium) of the two masses shown in Fig. 5.6.

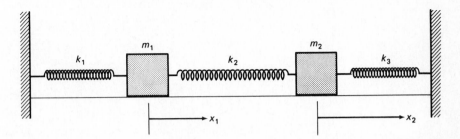

Figure 5.6 The system of Problem 26.

27. Two particles each of mass m are attached to a string under (constant) tension T as indicated in Fig. 5.7. Assume that the particles oscillate vertically (that is, parallel to the y-axis) with amplitudes so small that the sines of the angles shown are accurately approximated by their tangents. Show that the displace-

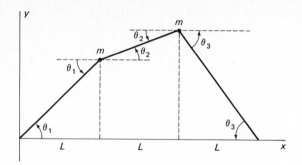

Figure 5.7 The mechanical system of Problem 27.

ments y_1 and y_2 of the two masses satisfy the equations

$$ky_1'' = -2y_1 + y_2 \qquad ky_2'' = y_1 - 2y_2$$

where $k = mL/T$.

28. Three 100-gal fermentation vats are connected as indicated in Fig 5.8, and the mixtures in each tank are kept uniform by stirring. Denote by $x_i(t)$ the amount (in pounds) of alcohol in tank T_i at time t $(i = 1, 2, 3)$. Suppose that the mixture circulates between the tanks at the rate of 10 gal/min. Derive the equations

$$10x_1' = -x_1 \qquad\quad + x_3,$$
$$10x_2' = \quad x_1 - x_2,$$
$$10x_3' = \qquad\quad x_2 - x_3.$$

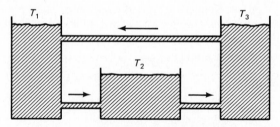

Figure 5.8 The fermentation tanks of Problem 28.

29. Set up a system of first order differential equations for the currents in the electrical circuit shown in Fig. 5.9.

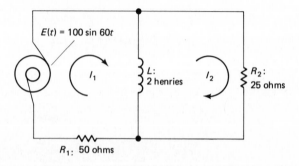

Figure 5.9 The electrical circuit of Problem 29.

30. Repeat Problem 29, except with the generator replaced with a battery supplying an emf of 100 V and with the inductor replaced by a 1-millifarad (mF) capacitor.

31. A particle of mass m moves in the plane with coordinates $(x(t), y(t))$ under the influence of a force that is directed toward the origin and has magnitude $k/(x^2 + y^2)$—an inverse-square-law central force field. Show that

$$mx'' = -\frac{kx}{r^3} \quad \text{and} \quad my'' = -\frac{ky}{r^3}$$

where $r = \sqrt{x^2 + y^2}$.

32. Suppose that a projectile of mass m moves in a vertical plane in the atmosphere near the surface of the earth under the influence of two forces: a downward gravitational force of magnitude mg, and a resistive force $\mathbf{F}_R$ that is directed opposite to the velocity vector $\mathbf{v}$ and has magnitude kv^2 (where $v = |\mathbf{v}|$ is the speed of the projectile; see Fig. 5.10). Show that the equations of motion of the projectile are

$$mx'' = -kvx', \qquad my'' = -kvy' - mg$$

where $v = \sqrt{(x')^2 + (y')^2}$.

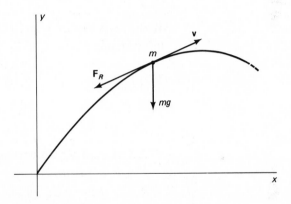

Figure 5.10   Trajectory of the projectile of Problem 32.

33. Suppose that a particle with mass m and electrical charge q moves in the xy plane under the influence of the magnetic field $\mathbf{B} = B\mathbf{k}$ (thus a uniform field parallel to the z-axis), so the force on the particle is $\mathbf{F} = q\mathbf{v} \times \mathbf{B}$ if its velocity is $\mathbf{v}$. Show that the equations of motion of the particle are

$$mx'' = +qBy', \qquad my'' = -qBx'.$$

5.2
The Method
of Elimination

The most elementary approach to linear systems of differential equations with constant coefficients involves the elimination of dependent variables by appropriately combining pairs of equations. The object of this procedure is to eliminate dependent variables in succession until there remains only a single equation containing only one dependent variable. This equation will usually be a linear equation of high order and can frequently be solved by the methods of Chapter 2. After the solution has been found, the other dependent variables can be found in turn, using either the original differential equations or those that have appeared in the elimination process.

This *method of elimination* for linear differential systems is quite similar to the solution of linear algebraic systems by elimination of variables until only one remains. It is most convenient in the case of manageably small systems: those consisting of no more than four equations. For large systems of differential equations, as well as for theoretical discussions, the matrix methods of the subsequent sections are preferable.

EXAMPLE 1 Find the solution of the system

$$x' = 4x - 3y, \qquad y' = 6x - 7y \tag{1}$$

that satisfies the initial conditions $x(0) = 2$ and $y(0) = -1$.

Solution If we solve the second equation in (1) for x, we get

$$x = \tfrac{1}{6}y' + \tfrac{7}{6}y, \tag{2}$$

so that

$$x' = \tfrac{1}{6}y'' + \tfrac{7}{6}y'. \tag{3}$$

We then substitute these expressions for x and x' in the first equation of the system in (1); this yields

$$\tfrac{1}{6}y'' + \tfrac{7}{6}y' = 4(\tfrac{1}{6}y' + \tfrac{7}{6}y) - 3y,$$

which can be simplified to

$$y'' + 3y' - 10y = 0.$$

This second order equation has characteristic equation

$$r^2 + 3r - 10 = (r - 2)(r + 5) = 0,$$

so its general solution is

$$y = c_1 e^{2t} + c_2 e^{-5t}. \tag{4}$$

Next, substitution of (4) in (2) gives

$$x = \tfrac{1}{6}(2c_1 e^{2t} - 5c_2 e^{-5t}) + \tfrac{7}{6}(c_1 e^{2t} + c_2 e^{-5t});$$

that is,

$$x = \tfrac{3}{2}c_1 e^{2t} + \tfrac{1}{3}c_2 e^{-5t}. \tag{5}$$

Thus (4) and (5) describe the general solution of the system in (1).

The given initial conditions imply that

$$x(0) = \tfrac{3}{2}c_1 + \tfrac{1}{3}c_2 = 2$$

and that

$$y(0) = c_1 + c_2 = -1;$$

these equations are readily solved for $c_1 = 2$ and $c_2 = -3$. Hence the desired solution is

$$x = 3e^{2t} - e^{-5t}, \qquad y = 2e^{2t} - 3e^{-5t}.$$

In Example 1 we used an *ad hoc* procedure to eliminate one of the dependent variables by expressing it in terms of the other. We now describe a systematic elimination procedure. Operator notation is most convenient for these purposes. A **linear differential operator** of order n with *constant* coefficients is one of the form

$$L = a_n D^n + a_{n-1} D^{n-1} + \cdots + a_1 D + a_0, \tag{6}$$

where D denotes differentiation with respect to the independent variable t. If L_1 and L_2 are two such operators, then their product $L_1 L_2$ is defined by means of the equation

$$L_1 L_2[x] = L_1[L_2 x]. \tag{7}$$

For instance, if $L_1 = D + a$ and $L_2 = D + b$, then

$$L_1 L_2[x] = (D + a)[(D + b)x]$$
$$= D[Dx + bx] + a(Dx + bx) = [D^2 + (a + b)D + ab]x.$$

This illustrates the fact that two linear operators L_1 and L_2 with constant coefficients can be multiplied as if they were ordinary polynomials in the "variable" D. Because the multiplication of such polynomials is commutative, it follows that

$$L_1 L_2 x = L_2 L_1 x \tag{8}$$

if the necessary derivatives of $x(t)$ exist. By contrast, this property of commutativity generally fails for linear operators with variable coefficients; see Problems 21 and 22.

Any system of two linear differential equations with constant coefficients can be written in the form

$$L_1 x + L_2 y = f_1(t),$$
$$L_3 x + L_4 y = f_2(t), \tag{9}$$

where $L_1, L_2, L_3,$ and L_4 are linear differential operators (perhaps of different orders) as in (6), and $f_1(t)$ and $f_2(t)$ are given functions. For instance, the system in (1) of Example 1 can be written in the form

$$(D - 4)x + \qquad 3y = 0,$$
$$-6x + (D + 7)y = 0, \tag{10}$$

with $L_1 = D - 4, L_2 = 3, L_3 = -6,$ and $L_4 = D + 7$.

To eliminate the dependent variable x from the system in (9), we operate with L_3 on the first equation and with L_1 on the second; thus we obtain the system

$$L_3 L_1 x + L_3 L_2 y = L_3 f_1(t),$$
$$L_1 L_3 x + L_1 L_4 y = L_1 f_2(t). \tag{11}$$

Subtraction of the first from the second of these equations yields the equation

$$(L_1 L_4 - L_2 L_3)y = L_1 f_2(t) - L_3 f_1(t) \tag{12}$$

in the single dependent variable y. After solving for y we can substitute the result into either of the original equations in (9) and then solve for x.

Alternatively, we could eliminate similarly the dependent variable y from the original system in (9) and get the equation

$$(L_1L_4 - L_2L_3)x = L_4f_1(t) - L_2f_2(t), \tag{13}$$

which can now be solved for x.

Note that the same operator $L_1L_4 - L_2L_3$ appears on the left-hand side in both Eq. (12) and Eq. (13). This is the **operational determinant**

$$\begin{vmatrix} L_1 & L_2 \\ L_3 & L_4 \end{vmatrix} = L_1L_4 - L_2L_3 \tag{14}$$

of the system in (9). In determinant notation Eqs. (12) and (13) can be rewritten as

$$\begin{vmatrix} L_1 & L_2 \\ L_3 & L_4 \end{vmatrix} x = \begin{vmatrix} f_1(t) & L_2 \\ f_2(t) & L_4 \end{vmatrix},$$

$$\begin{vmatrix} L_1 & L_2 \\ L_3 & L_4 \end{vmatrix} y = \begin{vmatrix} L_1 & f_1(t) \\ L_3 & f_2(t) \end{vmatrix}. \tag{15}$$

It is important to note that the determinants on the right-hand side in (15) are evaluated by means of the operators operating on the functions. The equations in (15) are strongly reminiscent of Cramer's rule for the solution of two linear equations in two (algebraic) variables and are thereby easily remembered. Indeed, you can solve a system of two linear differential equations either by carrying out the systematic elimination procedure described above or by directly employing the determinant equations in (15). Either process is especially simple if the system is homogeneous ($f_1(t) = f_2(t) = 0$), because in this case the right-hand sides of the equations in (12), (13), and (15) are zero.

EXAMPLE 2 Find the general solution of the system

$$(D - 4)x \qquad + 3y = 0,$$
$$-6x + (D + 7)y = 0. \tag{10}$$

Solution The operational determinant of this system is

$$(D - 4)(D + 7) - (3)(-6) = D^2 + 3D - 10. \tag{16}$$

Hence Eqs. (13) and (12) are

$$x'' + 3x' - 10x = 0,$$
$$y'' + 3y' - 10y = 0.$$

The characteristic equation of each is $r^2 + 3r - 10 = (r - 2)(r + 5) = 0$, so their (separate) general solutions are

$$x = a_1e^{2t} + a_2e^{-5t},$$
$$y = b_1e^{2t} + b_2e^{-5t}. \tag{17}$$

At this point we appear to have *four* arbitrary constants a_1, a_2, b_1, and b_2. But it follows from the theorem in Section 5.1 that the general solution of a system of two first order equations involves only two arbitrary constants. This apparent difficulty demands a resolution.

The explanation is simple: There must be some relation between

our four constants. We can discover it by substituting the solutions in
(17) into either of the original equations in (10). On substitution in the
first equation, we get

$$0 = x' - 4x + 3y$$
$$= (2a_1 e^{2t} - 5a_2 e^{-5t}) - 4(a_1 e^{2t} + a_2 e^{-5t}) + 3(b_1 e^{2t} + b_2 e^{-5t});$$

that is,

$$0 = (-2a_1 + 3b_1)e^{2t} + (-9a_2 + 3b_2)e^{-5t}.$$

But e^{2t} and e^{-5t} are linearly independent functions, and so it follows that
$a_1 = \frac{3}{2}b_1$ and $a_2 = \frac{1}{3}b_2$. Therefore the desired general solution is given by

$$x = \tfrac{3}{2}b_1 e^{2t} + \tfrac{1}{3}b_2 e^{-5t}, \qquad y = b_1 e^{2t} + b_2 e^{-5t}.$$

Note that this result is in accord with the general solution (Eqs. (4) and
(5)) that we obtained by a different method in Example 1.

As illustrated by Example 2, the elimination procedure ordinarily will
introduce a number of interdependent—thus unnecessary and undesirable—
arbitrary constants. These must be eliminated by substitution of the proposed
general solution into one or more of the original differential equations. If the
operational determinant defined in (15) is not identically zero, then the number
of independent arbitrary constants in a general solution of the system in (9) is
equal to the order of the operational determinant—that is, its degree as a
polynomial in D. (For a proof of this fact, see pages 144–150 of E. L. Ince's
Ordinary Differential Equations (New York: Dover, 1956).) Thus the general
solution of the system in (10) of Example 2 involves two arbitrary constants
because its operational determinant $D^2 + 3D - 10$ is of order 2.

If the operational determinant *is* identically zero, then the system is said
to be **degenerate**. A degenerate system may have either no solution or infinitely
many independent solutions. For instance, the equations

$$Dx - \quad Dy = 0,$$
$$2Dx - 2Dy = 1$$

with operational determinant zero are obviously inconsistent and thus have no
solution. On the other hand, the equations

$$Dx + \quad Dy = t,$$
$$2Dx + 2Dy = 2t$$

with operational determinant zero are obviously redundant; we can substitute
any (continuously differentiable) function for x and then integrate to obtain y.
Roughly speaking, every degenerate system is equivalent to either an inconsistent system or a redundant system.

Although the procedures and results above are described for the case of
a system of two equations, they can be generalized readily to systems of three
or more equations. For the system

$$\begin{aligned}
L_{11}x + L_{12}y + L_{13}z &= f_1(t), \\
L_{21}x + L_{22}y + L_{23}z &= f_2(t), \\
L_{31}x + L_{32}y + L_{33}z &= f_3(t)
\end{aligned} \qquad (18)$$

of three linear equations, the dependent variable x satisfies the single linear equation

$$\begin{vmatrix} L_{11} & L_{12} & L_{13} \\ L_{21} & L_{22} & L_{23} \\ L_{31} & L_{32} & L_{33} \end{vmatrix} x = \begin{vmatrix} f_1(t) & L_{12} & L_{13} \\ f_2(t) & L_{22} & L_{23} \\ f_3(t) & L_{23} & L_{33} \end{vmatrix}, \tag{19}$$

with analogous equations for y and z. For most systems of more than three equations, however, the evaluation of operational determinants is too tedious for the method to be practical.

EXAMPLE 3 In Example 1 of Section 5.1 we derived the equations

$$\begin{aligned} (D^2 + 3)x - \quad\quad\quad y &= 0, \\ -2x + (D^2 + 2)y &= 0 \end{aligned} \tag{20}$$

for the displacements of the two masses in Fig. 5.11; here $f(t) = 0$ because we assume no external force. Find the general solution of the system in (20).

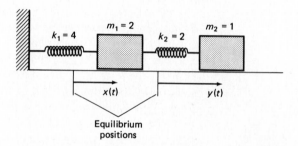

Figure 5.11   The mass-and-spring system of Example 3.

Solution   The operational determinant of the system in (20) is

$$(D^2 + 3)(D^2 + 2) - (-1)(-2) = D^4 + 5D^2 + 4$$
$$= (D^2 + 1)(D^2 + 4).$$

Hence the equations for x and y are

$$\begin{aligned} (D^2 + 1)(D^2 + 4)x &= 0, \\ (D^2 + 1)(D^2 + 4)y &= 0. \end{aligned} \tag{21}$$

The characteristic equation $(r^2 + 1)(r^2 + 4) = 0$ has roots i, $-i$, $2i$, and $-2i$. So the general solutions of the equations in (21) are

$$\begin{aligned} x &= a_1 \cos t + a_2 \sin t + b_1 \cos 2t + b_2 \sin 2t, \\ y &= c_1 \cos t + c_2 \sin t + d_1 \cos 2t + d_2 \sin 2t. \end{aligned} \tag{22}$$

Because the operational determinant is of order 4, the general solution should involve four (rather than eight) arbitrary constants. When we substitute x and y from (22) in the first equation in (20), we get

$$0 = x'' + 3x - y$$
$$= (-a_1 \cos t - a_2 \sin t - 4b_1 \cos 2t - 4b_2 \sin 2t)$$
$$+ 3(a_1 \cos t + a_2 \sin t + b_1 \cos 2t + b_2 \sin 2t)$$
$$- (c_1 \cos t + c_2 \sin t + d_1 \cos 2t + d_2 \sin 2t);$$

thus

$$0 = (2a_1 - c_1) \cos t + (2a_2 - c_2) \sin t$$
$$+ (-b_1 - d_1) \cos 2t + (-b_2 - d_2) \sin 2t.$$

It follows that

$$c_1 = 2a_1, \quad c_2 = 2a_2, \quad d_1 = -b_1, \quad \text{and} \quad d_2 = -b_2.$$

Hence the desired general solution may be written in the form

$$x = a_1 \cos t + a_2 \sin t + b_1 \cos 2t + b_2 \sin 2t,$$
$$y = 2a_1 \cos t + 2a_2 \sin t - b_1 \cos 2t - b_2 \sin 2t. \tag{23}$$

The equations in (23) describe **free oscillations** of the mass-and-spring system of Fig. 5.11—motion subject to *no* external forces. Four initial conditions (typically, initial displacements and velocities) would be needed to determine the values of $a_1, a_2, b_1,$ and b_2. Granted that these computations have been carried out, we can (by the usual trigonometric machinations) write

$$x_1 = a_1 \cos t + a_2 \sin t = A \cos (t - \alpha),$$
$$y_1 = 2a_1 \cos t + 2a_2 \sin t = 2A \cos (t - \alpha) \tag{24}$$

and

$$x_2 = b_1 \cos 2t + b_2 \sin 2t = B \cos (2t - \beta),$$
$$y_2 = -b_1 \cos 2t - b_2 \sin 2t = -B \cos (2t - \beta). \tag{25}$$

The particular solutions (x_1, y_1) and (x_2, y_2) of the system in (20) represent the two **natural modes of oscillation** of the mass-and-spring system, and they exhibit its two (circular) **natural frequencies** $\omega_1 = 1$ and $\omega_2 = 2$. The equations in (23) can be written as a single vector equation

$$(x, y) = (x_1, y_1) + (x_2, y_2). \tag{26}$$

This oscillation represents an arbitrary free oscillation of the mass-and-spring system as a superposition of its two natural modes of oscillation, with the constants $A, \alpha, B,$ and β in Eqs. (24) and (25) determined by the initial conditions. In the natural mode (x_1, y_1), the two masses move in synchrony in the same direction and with the same frequency $\omega_1 = 1$, but with the amplitude of motion of m_2 twice that of m_1. In the natural mode (x_2, y_2), the two masses move in synchrony in opposite directions with the same frequency $\omega_2 = 2$ and with equal amplitudes of oscillation. These remarks generalize the discussion following Example 5 in Section 5.1.

The systematic elimination procedure employed in Examples 2 and 3 is not always optimal. The lesson of the following example is that in solving differential equations, as in doing anything else, it pays to keep your eyes open.

EXAMPLE 4 Find the general solution of the system

$$x' = 3x - 2y + 30t - 1,$$
$$y' = -x + 3y - 2z + 2e^{3t}, \tag{27}$$
$$z' = -y + 3z.$$

Solution In operator form, the system in (27) is

$$(D - 3)x + 2y = 30t - 1,$$
$$x + (D - 3)y + 2z = 2e^{3t}, \tag{28}$$
$$y + (D - 3)z = 0.$$

The crucial observation is this: If z is found first, then the last equation will give y immediately, and then the second equation will give x. This will involve much less labor than solving separately for x, y, and z and then substituting the results into (two of) the equations in (28) to find six relations between the nine arbitrary constants that would be introduced.

The equation for z analogous to Eq. (19) is

$$\begin{vmatrix} D-3 & 2 & 0 \\ 1 & D-3 & 2 \\ 0 & 1 & D-3 \end{vmatrix} z = \begin{vmatrix} D-3 & 2 & 30t-1 \\ 1 & D-3 & 2e^{3t} \\ 0 & 1 & 0 \end{vmatrix}. \tag{29}$$

We expand the operational determinant along the first row to obtain

$$(D-3)\begin{vmatrix} D-3 & 2 \\ 1 & D-3 \end{vmatrix} - 2\begin{vmatrix} 1 & 2 \\ 0 & D-3 \end{vmatrix}$$
$$= (D-3)[(D-3)^2 - 2] - 2(D-3)$$
$$= (D-3)[(D-3)^2 - 4]$$
$$= (D-3)(D-1)(D-5) \tag{30a}$$
$$= D^3 - 9D^2 + 23D - 15. \tag{30b}$$

Upon expansion of the right-hand determinant in (29) along the third row, we get

$$(-1)\begin{vmatrix} D-3 & 30t-1 \\ 1 & 2e^{3t} \end{vmatrix} = 30t - 1$$

because $(D-3)e^{3t} = 3e^{3t} - 3e^{3t} = 0$. Thus the equation for z is

$$(D-1)(D-3)(D-5)z = 30t - 1;$$

that is,

$$z''' - 9z'' + 23z' - 15z = 30t - 1. \tag{31}$$

The complementary function of (31) is $z_c = ae^t + be^{3t} + ce^{5t}$ where a, b, and c are arbitrary constants. To find a particular solution, we substitute $z_p = At + B$ in (31) and find routinely that $A = -2$ and $B = -3$. Thus the general solution of (31) is

$$z = ae^t + be^{3t} + ce^{5t} - 2t - 3.$$

Substituting this solution for z in the last equation in (27), we get $y = 3z - z'$; that is,

$$y = 2ae^t - 2ce^{5t} - 6t - 7. \tag{33}$$

We substitute these results in the second equation in (27) to find that

$$x = 3y - 2z - y' + 2e^{3t},$$

and consequently that

$$x = 2ae^t - 2(b-1)e^{3t} + 2ce^{5t} - 14t - 9. \tag{34}$$

Thus we have directly obtained a general solution—Eqs. (32), (33), and (34)—involving the proper number (three) of constants.

5.2 Problems

Find general solutions of the linear systems in Problems 1–20. If initial conditions are given, find the solution that satisfies them.

1. $x' = -x + 3y, y' = 2y.$
2. $x' = x - 2y, y' = 2x - 3y.$
3. $x' = -3x + 2y, y' = -3x + 4y; x(0) = 0, y(0) = 2.$
4. $x' = 3x - y, y' = 5x - 3y; x(0) = 1, y(0) = -1.$
5. $x' = -3x - 4y, y' = 2x + y.$
6. $x' = x + 9y, y' = -2x - 5y; x(0) = 3, y(0) = 2.$
7. $x' = 4x + y + 2t, y' = -2x + y.$
8. $x' = 2x + y, y' = x + 2y - e^{2t}.$
9. $x' = 2x - 3y + 2\sin 2t, y' = x - 2y - \cos 2t.$
10. $x' + 2y' = 4x + 5y, 2x' - y' = 3x; x(0) = 1, y(0) = -1.$
11. $2y' - x' = x + 3y + e^t, 3x' - 4y' = x - 15y + e^{-t}.$
12. $x'' = 6x + 2y, y'' = 3x + 7y.$
13. $x'' = -5x + 2y, y'' = 2x - 8y.$
14. $x'' = -4x + \sin t, y'' = 4x - 8y.$
15. $x'' - 3y' - 2x = 0, y'' + 3x' - 2y = 0.$
16. $x'' + 13y' - 4x = 6\sin t, y'' - 2x' - 9y = 0.$
17. $x'' + y'' - 3x' - y' - 2x + 2y = 0,$
    $2x'' + 3y'' - 9x' - 2y' - 4x + 6y = 0.$
18. $x' = x + 2y + z, y' = 6x - y, z' = -x - 2y - z.$
19. $x' = 4x - 2y, y' = -4x + 4y - 2z, z' = -4y + 4z.$
20. $x' = y + z + e^{-t}, y' = x + z, z' = x + y.$
    (*Suggestion:* Solve the characteristic equation by inspection.)
21. Suppose that $L_1 = a_1 D^2 + b_1 D + c_1$ and $L_2 = a_2 D^2 + b_2 D + c_2$, where the coefficients are all constants, and that $x(t)$ is a twice differentiable function. Verify that $L_1 L_2 x = L_2 L_1 x.$
22. Suppose that $L_1 x = t\, Dx + x$ and that $L_2 x = Dx + tx$. Show that $L_1 L_2 \neq L_2 L_1$. Thus linear operators with *variable* coefficients generally do not commute.

Show that each of the systems in Problems 23–25 is degenerate. In each problem determine—by attempting to solve the system—whether it has infinitely many solutions or no solutions.

23. $(D + 2)x + (D + 2)y = e^{-3t}$, $(D + 3)x + (D + 3)y = e^{-2t}$.

24. $(D + 2)x + (D + 2)y = t$, $(D + 3)x + (D + 3)y = t^2$.

25. $(D^2 + 5D + 6)x + D(D + 2)y = 0$, $(D + 3)x + Dy = 0$.

26. Suppose that the salt concentration in each of the two brine tanks of Example 2 in Section 5.1 initially $(t = 0)$ is 0.5 lb/gal. Then solve the system in (5) there to find the amounts $x(t)$ and $y(t)$ of salt in the two tanks at time t.

27. Suppose that the electrical network of Example 3 in Section 5.1 is initially open —no currents are flowing. Assume that it is closed at time $t = 0$; solve the system in (9) there to find $I_1(t)$ and $I_2(t)$.

28. Repeat Problem 27, except use the electrical network of Problem 29 in Section 5.1.

29. Repeat Problem 28, except use the electrical network of Problem 30 in Section 5.1 and assume that $I_1(0) = 2$, $Q(0) = 0$, and that at time $t = 0$ there is no charge on the capacitor.

30. Three 100-gal brine tanks are connected as indicated in Fig. 5.8 of Section 5.1. Assume that the first tank initially contains 100 lb of salt, while the other two are filled with fresh water. Find the amounts of salt in each of the three tanks at time t. (*Suggestion:* Examine the equations to be derived in Problem 28 of Section 5.1.)

31. From Problem 33 in Section 5.1, recall the equations of motion

$$mx'' = qBy', \qquad my'' = -qBx'$$

for a particle of mass m and electrical charge q under the influence of the uniform magnetic field $\mathbf{B} = B\mathbf{k}$. Suppose that the initial conditions are $x(0) = r_0$, $y(0) = 0$, $x'(0) = 0$, and $y'(0) = \omega r_0$ where $\omega = qB/m$. Show that the trajectory of the particle is a circle of radius r_0.

32. If, in addition to the magnetic field $\mathbf{B} = B\mathbf{k}$, the charged particle of Problem 31 moves with velocity $\mathbf{v}$ under the influence of a uniform electric field $\mathbf{E} = E\mathbf{i}$, then the force acting on it is $\mathbf{F} = q(\mathbf{E} + \mathbf{v} \times \mathbf{B})$. Assume that the particle starts from rest at the origin. Show that its trajectory is the cycloid

$$x = a(1 - \cos \omega t), \qquad y = -a(\omega t - \sin \omega t)$$

where $a = E/\omega B$ and $\omega = qB/m$. The graph of such a cycloid is shown in Fig. 5.12.

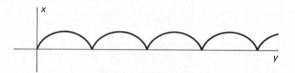

Figure 5.12   Cycloidal path of the particle of Problem 32.

33. In the mass-and-spring system of Example 3, suppose instead that $m_1 = 2$, $m_2 = 0.5$, $k_1 = 75$, and $k_2 = 25$. (a) Find the general solution of the equations of motion of the system. In particular, show that its natural frequencies are

$\omega_1 = 5$ and $\omega_2 = 5\sqrt{3}$. (b) Describe the natural modes of oscillation of the system.

34. Consider the system of two masses and three springs shown in Fig. 5.13. Derive the equations of motion

$$m_1 x'' = -(k_1 + k_2)x + k_2 y,$$
$$m_2 y'' = k_2 x - (k_2 + k_3)y.$$

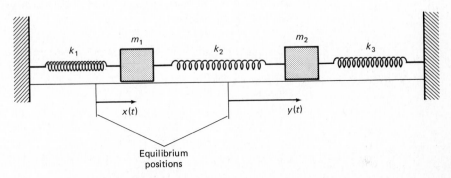

Figure 5.13 The mechanical system of Problem 34.

In each of Problems 35–39, find the general solution of the system in Problem 34 with the given masses and spring constants. Find the natural frequencies of the mass-and-spring system, and describe its natural modes of oscillation.

35. $m_1 = 1, m_2 = 1$; $k_1 = 1, k_2 = 4, k_3 = 1$.
36. $m_1 = 1, m_2 = 2$; $k_1 = 1, k_2 = 2, k_3 = 2$.
37. $m_1 = 1, m_2 = 1$; $k_1 = 1, k_2 = 2, k_3 = 1$.
38. $m_1 = 1, m_2 = 1$; $k_1 = 2, k_2 = 1, k_3 = 2$.
39. $m_1 = 1, m_2 = 2$; $k_1 = 2, k_2 = 4, k_3 = 4$.
40. (a) For the system shown in Fig. 5.14, derive the equations of motion

$$mx'' = -2kx + ky,$$
$$my'' = kx - 2ky + kz,$$
$$mz'' = ky - 2kz.$$

(b) Assume that $m = k = 1$. Show that the natural frequencies of oscillation of the system are $\omega_1 = \sqrt{2}$, $\omega_2 = \sqrt{2 - \sqrt{2}}$, and $\omega_3 = \sqrt{2 + \sqrt{2}}$.

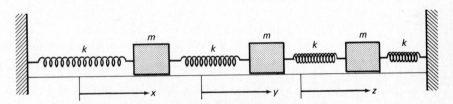

Figure 5.14 The mechanical system of Problem 40.

5.3
Linear Systems and Matrices

Although the simple elimination techniques of Section 5.2 suffice for the solution of small linear systems containing only two or three equations with constant coefficients, the general properties of linear systems—as well as solution methods suitable for larger systems—are most easily and concisely described in matrix notation. It is best that the reader of this section has had some prior exposure to matrices and determinants. If not, we have included—for ready reference—a complete and self-contained account of the notation and terminology that is needed.

REVIEW OF MATRIX NOTATION AND TERMINOLOGY

An $m \times n$ **matrix A** is a rectangular array of mn numbers (or **elements**) arranged in m (horizontal) **rows** and n (vertical) **columns**:

$$\mathbf{A} = \begin{pmatrix} a_{11} & a_{12} & a_{13} & \cdots & a_{1j} & \cdots & a_{1n} \\ a_{21} & a_{22} & a_{23} & \cdots & a_{2j} & \cdots & a_{2n} \\ a_{31} & a_{32} & a_{33} & \cdots & a_{3j} & \cdots & a_{3n} \\ \vdots & \vdots & \vdots & & \vdots & & \vdots \\ a_{i1} & a_{i2} & a_{i3} & \cdots & a_{ij} & \cdots & a_{in} \\ \vdots & \vdots & \vdots & & \vdots & & \vdots \\ a_{m1} & a_{m2} & a_{m3} & \cdots & a_{mj} & \cdots & a_{mn} \end{pmatrix}. \tag{1}$$

We will ordinarily denote matrices by bold-faced capital letters. Sometimes we use the abbreviation $\mathbf{A} = (a_{ij})$ for the matrix with the element a_{ij} in the ith row and jth column, as above. We denote the **zero matrix**, each entry of which is zero, by

$$\mathbf{0} = \begin{pmatrix} 0 & 0 & \cdots & 0 \\ 0 & 0 & \cdots & 0 \\ \vdots & \vdots & & \vdots \\ 0 & 0 & \cdots & 0 \end{pmatrix} \tag{2}$$

Actually, for each $m > 0$ and $n > 0$, there is an $m \times n$ zero matrix, but the single symbol $\mathbf{0}$ will suffice for all these zero matrices.

Two $m \times n$ matrices $\mathbf{A} = (a_{ij})$ and $\mathbf{B} = (b_{ij})$ are said to be **equal** if corresponding elements are equal; that is, if $a_{ij} = b_{ij}$ for $1 \leq i \leq m$ and $1 \leq j \leq n$. We **add** $\mathbf{A}$ and $\mathbf{B}$ by adding corresponding entries:

$$\mathbf{A} + \mathbf{B} = (a_{ij}) + (b_{ij}) = (a_{ij} + b_{ij}). \tag{3}$$

Thus the element in row i and column j of $\mathbf{C} = \mathbf{A} + \mathbf{B}$ is $c_{ij} = a_{ij} + b_{ij}$. To multiply the matrix $\mathbf{A} = (a_{ij})$ by the number c, we simply multiply each of its entries by c:

$$c\mathbf{A} = \mathbf{A}c = (ca_{ij}). \tag{4}$$

Thus

$$\begin{pmatrix} 2 & -3 \\ 4 & 7 \end{pmatrix} + \begin{pmatrix} -13 & 10 \\ 7 & -5 \end{pmatrix} = \begin{pmatrix} -11 & 7 \\ 11 & 2 \end{pmatrix}$$

and

$$6\begin{pmatrix} 3 & 0 \\ 5 & -7 \end{pmatrix} = \begin{pmatrix} 18 & 0 \\ 30 & -42 \end{pmatrix}.$$

We denote $(-1)\mathbf{A}$ by $-\mathbf{A}$ and define **subtraction** of matrices as follows:

$$\mathbf{A} - \mathbf{B} = \mathbf{A} + (-\mathbf{B}). \tag{5}$$

The matrix operations just defined have the following properties, each of which is analogous to a familiar algebraic property of the real number system:

$$\mathbf{A} + \mathbf{0} = \mathbf{0} + \mathbf{A} = \mathbf{A}, \quad \mathbf{A} - \mathbf{A} = \mathbf{0}; \tag{6}$$

$$\mathbf{A} + \mathbf{B} = \mathbf{B} + \mathbf{A} \qquad \text{(commutativity)}; \tag{7}$$

$$\mathbf{A} + (\mathbf{B} + \mathbf{C}) = (\mathbf{A} + \mathbf{B}) + \mathbf{C} \quad \text{(associativity)}; \tag{8}$$

$$\left.\begin{aligned} c(\mathbf{A} + \mathbf{B}) &= c\mathbf{A} + c\mathbf{B}, \\ (c + d)\mathbf{A} &= c\mathbf{A} + d\mathbf{A}. \end{aligned}\right\} \quad \text{(distributivity)} \tag{9}$$

Each of these properties is readily verified by elementwise application of a corresponding property of the real numbers. For example, $a_{ij} + b_{ij} = b_{ij} + a_{ij}$ for all i and j because addition of real numbers is commutative. Consequently

$$\mathbf{A} + \mathbf{B} = (a_{ij} + b_{ij}) = (b_{ij} + a_{ij}) = \mathbf{B} + \mathbf{A}.$$

The **transpose** $\mathbf{A}^T$ of the $m \times n$ matrix $\mathbf{A} = (a_{ij})$ is the $n \times m$ (note!) matrix whose jth column is the ith row of $\mathbf{A}$ (and, consequently, whose ith row is the jth column of $\mathbf{A}$). Thus $\mathbf{A}^T = (a_{ji})$, although this is not notationally perfect; it must be remembered that $\mathbf{A}^T$ will not have the same shape as $\mathbf{A}$ unless $\mathbf{A}$ is a **square** matrix—that is, unless $m = n$.

An $m \times 1$ matrix—one having only a single column—is called a **column vector**, or simply a **vector**. We often denote column vectors by boldface lowercase letters, as in

$$\mathbf{b} = \begin{pmatrix} 3 \\ -7 \\ 0 \end{pmatrix} \quad \text{or} \quad \mathbf{x} = \begin{pmatrix} x_1 \\ x_2 \\ \vdots \\ x_m \end{pmatrix}.$$

Similarly, a **row vector** is a $1 \times n$ matrix—one having only a single row, such as $\mathbf{c} = (5 \quad 17 \quad 0 \quad -3)$. For esthetic and typographical reasons, it is customary to write a column vector as the transpose of a row vector; for example, the two column vectors above may be written in the forms

$$\mathbf{b} = (3 \quad -7 \quad 0)^T \quad \text{and} \quad \mathbf{x} = (x_1 \quad x_2 \quad \cdots \quad x_n)^T.$$

Sometimes it is convenient to describe an $m \times n$ matrix in terms of either its m row vectors or its n column vectors. Thus if we write

$$\mathbf{A} = \begin{pmatrix} \mathbf{a}_1 \\ \mathbf{a}_2 \\ \cdot \\ \cdot \\ \cdot \\ \mathbf{a}_m \end{pmatrix} \quad \text{and} \quad \mathbf{B} = (\mathbf{b}_1 \quad \mathbf{b}_2 \quad \cdots \quad \mathbf{b}_n),$$

it is understood that $\mathbf{a}_1, \mathbf{a}_2, \ldots, \mathbf{a}_m$ are the row vectors of the matrix $\mathbf{A}$, while $\mathbf{b}_1, \mathbf{b}_2, \ldots, \mathbf{b}_n$ are the column vectors of the matrix $\mathbf{B}$.

MATRIX MULTIPLICATION

The properties listed in Eqs. (6)–(9) are quite natural and expected. The first surprises in the realm of matrix arithmetic come with multiplication. We define first the **scalar product** $\mathbf{a} \cdot \mathbf{b}$ of a row vector $\mathbf{a}$ and a column vector $\mathbf{b}$, each having the same number p of entries. If

$$\mathbf{a} = (a_1 \quad a_2 \quad \cdots \quad a_p) \quad \text{and} \quad \mathbf{b} = (b_1 \quad b_2 \quad \cdots \quad b_p)^T,$$

then $\mathbf{a} \cdot \mathbf{b}$ is defined by means of the equation

$$\mathbf{a} \cdot \mathbf{b} = \sum_{k=1}^{p} a_k b_k = a_1 b_1 + a_2 b_2 + \cdots + a_p b_p, \tag{10}$$

exactly as in the scalar or *dot* product of two vectors—a familiar topic from elementary calculus.

The product $\mathbf{AB}$ of two matrices is defined only if the number of columns of $\mathbf{A}$ is equal to the number of rows of $\mathbf{B}$. If $\mathbf{A}$ is an $m \times p$ matrix and $\mathbf{B}$ is a $p \times n$ matrix, then their product $\mathbf{AB}$ is the $m \times n$ matrix $\mathbf{C} = (c_{ij})$, where c_{ij} is the scalar product of the ith row vector $\mathbf{a}_i$ of $\mathbf{A}$ and the jth column vector $\mathbf{b}_j$ of $\mathbf{B}$. Thus

$$\mathbf{AB} = (\mathbf{a}_i \cdot \mathbf{b}_j). \tag{11}$$

In terms of the individual entries of $\mathbf{A} = (a_{ij})$ and $\mathbf{B} = (b_{ij})$, Eq. (11) can be recast in the form

$$c_{ij} = \sum_{k=1}^{p} a_{ik} b_{kj}. \tag{12}$$

For purposes of hand computation, the definition in (11) and (12) is easy to remember by visualizing the picture

$$\begin{pmatrix} a_{11} & a_{12} & \cdots & a_{1p} \\ a_{21} & a_{21} & \cdots & a_{2p} \\ \cdot & \cdot & & \cdot \\ \cdot & \cdot & & \cdot \\ \boxed{a_{i1} \quad a_{i2} \quad \cdots \quad a_{ip}} \\ \cdot & \cdot & & \cdot \\ \cdot & \cdot & & \cdot \\ a_{m1} & a_{m2} & \cdots & a_{mp} \end{pmatrix} \begin{pmatrix} b_{11} & b_{12} & \cdots & b_{1j} & \cdots & b_{1n} \\ b_{21} & b_{22} & \cdots & b_{2j} & \cdots & b_{2n} \\ \cdot & \cdot & & \cdot & & \cdot \\ \cdot & \cdot & & \cdot & & \cdot \\ \cdot & \cdot & & \cdot & & \cdot \\ b_{p1} & b_{p2} & \cdots & b_{pj} & \cdots & b_{pn} \end{pmatrix},$$

in which the row vector $\mathbf{a}_i$ and the column vector $\mathbf{b}_j$ are spotlighted; c_{ij} is simply their scalar product. It may help to think of "pouring the rows of $\mathbf{A}$ down the columns of $\mathbf{B}$." This also reminds us that the number of columns of $\mathbf{A}$ must be equal to the number of rows of $\mathbf{B}$.

You should pause to check your understanding of the definition of matrix multiplication by verifying that

$$\begin{pmatrix} 2 & -3 \\ -1 & 5 \end{pmatrix}\begin{pmatrix} 13 & 9 \\ 4 & 0 \end{pmatrix} = \begin{pmatrix} 14 & 18 \\ 7 & -9 \end{pmatrix}$$

and that

$$\begin{pmatrix} 2 & -3 & 1 \\ 4 & 5 & -2 \\ 6 & -7 & 0 \end{pmatrix}\begin{pmatrix} x \\ y \\ z \end{pmatrix} = \begin{pmatrix} 2x - 3y + z \\ 4x + 5y - 2z \\ 6x - 7y \end{pmatrix}.$$

The BASIC program shown in Fig. 5.15 can be used to compute the product $\mathbf{C} = \mathbf{AB}$ of an $m \times p$ matrix $\mathbf{A}$ and a $p \times n$ matrix $\mathbf{B}$. For instance, if we enter the values $M = 4$, $P = 2$, and $N = 3$ at line 110, insert the data 1, 2, 3, 4, 5, 6, 7, 8 in line 130 and the data 2, 1, 3, -1, 3, -2 in line 140, when we execute this program we find that

$$\begin{pmatrix} 1 & 2 \\ 3 & 4 \\ 5 & 6 \\ 7 & 8 \end{pmatrix}\begin{pmatrix} 2 & 1 & 3 \\ -1 & 3 & -2 \end{pmatrix} = \begin{pmatrix} 0 & 7 & -1 \\ 2 & 15 & 1 \\ 4 & 23 & 3 \\ 6 & 31 & 5 \end{pmatrix}.$$

```
100 REM        A IS M BY P, B IS P BY N
110            INPUT M, P, N
120            DIM A(M,P), B(P,N), C(M,N)
130            DATA [elements of A go here]
140            DATA [elements of B go here]
150 REM          LOAD MATRIX A
160            FOR I = 1 TO M : FOR J = 1 TO P
170            READ A(I,J)
180            NEXT J : NEXT I
190 REM          LOAD MATRIX B
200            FOR I = 1 TO P : FOR J = 1 TO N
210            READ B(I,J)
220            NEXT J : NEXT I
230 REM          CALCULATE MATRIX C = AB
240            FOR I = 1 TO M : FOR J = 1 TO N
250            X = 0
260                FOR K = 1 TO P
270                X = X + A(I,K)*B(K,J)
280                NEXT K
290            C(I,J) = X
300            NEXT I : NEXT J
310 REM          PRINT MATRIX C
320            FOR I = 1 TO M : FOR J = 1 TO N
330            PRINT C(I,J)
340            NEXT J : PRINT : NEXT I
350            END
```

Figure 5.15 BASIC program for finding the product of two matrices $\mathbf{A}$ and $\mathbf{B}$.

It can be shown by direct (though lengthy) computation based on its definition that matrix multiplication is associative and is also distributive with respect to matrix addition:

$$\mathbf{A}(\mathbf{BC}) = (\mathbf{AB})\mathbf{C} \tag{13}$$

and

$$\mathbf{A}(\mathbf{B} + \mathbf{C}) = \mathbf{AB} + \mathbf{AC} \tag{14}$$

provided that the matrices are of such sizes that the indicated multiplications and additions are possible.

But matrix multiplication is *not* commutative. That is, if $\mathbf{A}$ and $\mathbf{B}$ are both $n \times n$ matrices (so that both the products $\mathbf{AB}$ and $\mathbf{BA}$ are defined and have the same dimensions—$n \times n$), then, in general,

$$\mathbf{AB} \neq \mathbf{BA}. \tag{15}$$

Moreover, it can happen that

$$\mathbf{AB} = \mathbf{0} \quad \text{even though} \quad \mathbf{A} \neq \mathbf{0} \quad \text{and} \quad \mathbf{B} \neq \mathbf{0}. \tag{16}$$

Examples illustrating the phenomena in (15) and (16) may be found in the problems, though you can easily construct your own examples using 2×2 matrices with integral entries.

A square $n \times n$ matrix is said to have **order** n. The **identity** matrix of order n is the square matrix

$$\mathbf{I} = \begin{pmatrix} 1 & 0 & 0 & 0 & \cdots & 0 \\ 0 & 1 & 0 & 0 & \cdots & 0 \\ 0 & 0 & 1 & 0 & \cdots & 0 \\ 0 & 0 & 0 & 1 & \cdots & 0 \\ \cdot & \cdot & \cdot & \cdot & & \cdot \\ \cdot & \cdot & \cdot & \cdot & & \cdot \\ \cdot & \cdot & \cdot & \cdot & & \cdot \\ 0 & 0 & 0 & 0 & \cdots & 1 \end{pmatrix} \tag{17}$$

for which each entry on the **principal diagonal** is 1 and all off-diagonal entries are zero. It is quite easy to verify that

$$\mathbf{AI} = \mathbf{IA} = \mathbf{A} \tag{18}$$

for every square matrix $\mathbf{A}$ of the same order as $\mathbf{I}$.

If $\mathbf{A}$ is a square matrix, then an **inverse** of $\mathbf{A}$ is a square matrix $\mathbf{B}$ of the same order as $\mathbf{A}$ such that *both*

$$\mathbf{AB} = \mathbf{I} \quad \text{and} \quad \mathbf{BA} = \mathbf{I}.$$

It is not difficult to show that if the matrix $\mathbf{A}$ has an inverse, then this inverse is unique. Consequently we may speak of *the* inverse of $\mathbf{A}$ and denote it by $\mathbf{A}^{-1}$. Thus

$$\mathbf{AA}^{-1} = \mathbf{I} = \mathbf{A}^{-1}\mathbf{A}, \tag{19}$$

given the existence of $\mathbf{A}^{-1}$. It is clear that some square matrices do not have inverses; consider any square zero matrix.

In linear algebra it is proved that $\mathbf{A}^{-1}$ exists if and only if the determinant of the square matrix $\mathbf{A}$ (denoted by det $(\mathbf{A})$) is nonzero, in which case the matrix $\mathbf{A}$ is said to be **nonsingular**; if det $(\mathbf{A}) = 0$, then $\mathbf{A}$ is called a **singular** matrix.

We assume that the student has computed 2×2 and 3×3 determinants in earlier courses. If $\mathbf{A} = (a_{ij})$ is a 2×2 matrix, then its **determinant** det $(\mathbf{A})$ $= |\mathbf{A}|$ is defined as

$$|\mathbf{A}| = \begin{vmatrix} a_{11} & a_{12} \\ a_{21} & a_{22} \end{vmatrix} = a_{11}a_{22} - a_{12}a_{21}. \tag{20}$$

Determinants of higher order may be defined by induction, as follows. If $\mathbf{A} = (a_{ij})$ is an $n \times n$ matrix, let $\mathbf{A}_{ij}$ denote the $(n-1) \times (n-1)$ matrix obtained from $\mathbf{A}$ by deleting its ith row and its jth column. The *expansion* of the determinant $|\mathbf{A}|$ along its ith row is given by

$$|\mathbf{A}| = \sum_{j=1}^{n} (-1)^{i+j} a_{ij} |\mathbf{A}_{ij}|, \tag{21a}$$

while the expansion along its jth column is given by

$$|\mathbf{A}| = \sum_{i=1}^{n} (-1)^{i+j} a_{ij} |\mathbf{A}_{ij}|. \tag{21b}$$

It is shown in linear algebra that whichever row we use in (21a) and whichever column we use in (21b), the results are the same; hence $|\mathbf{A}|$ is well-defined by these formulas. For example, if

$$\mathbf{A} = \begin{pmatrix} 3 & 1 & -2 \\ 4 & 2 & 1 \\ -2 & 3 & 5 \end{pmatrix},$$

then the expansion of $|\mathbf{A}|$ along its second row is

$$|\mathbf{A}| = -(4)\begin{vmatrix} 1 & -2 \\ 3 & 5 \end{vmatrix} + (2)\begin{vmatrix} 3 & -2 \\ -2 & 5 \end{vmatrix} - (1)\begin{vmatrix} 3 & 1 \\ -2 & 3 \end{vmatrix}$$
$$= -(4)(11) + (2)(11) - (1)(11) = -33,$$

while the expansion of $|\mathbf{A}|$ along its third column is

$$|\mathbf{A}| = (-2)\begin{vmatrix} 4 & 2 \\ -2 & 3 \end{vmatrix} - (1)\begin{vmatrix} 3 & 1 \\ -2 & 3 \end{vmatrix} + (5)\begin{vmatrix} 3 & 1 \\ 4 & 2 \end{vmatrix}$$
$$= (-2)(16) - (1)(11) + (5)(2) = -33.$$

MATRIX-VALUED FUNCTIONS

A **matrix-valued function**, or simply matrix function, is a matrix such as

$$\mathbf{x}(t) = \begin{pmatrix} x_1(t) \\ \cdot \\ \cdot \\ \cdot \\ x_n(t) \end{pmatrix} \tag{22a}$$

or

$$\mathbf{A}(t) = \begin{pmatrix} a_{11}(t) & a_{12}(t) & \cdots & a_{1n}(t) \\ a_{21}(t) & a_{22}(t) & \cdots & a_{2n}(t) \\ \cdot & \cdot & & \cdot \\ \cdot & \cdot & & \cdot \\ \cdot & \cdot & & \cdot \\ a_{m1}(t) & a_{m2}(t) & \cdots & a_{mn}(t) \end{pmatrix} \tag{22b}$$

in which each entry is a function of t. We say that the matrix function $\mathbf{A}(t)$ is continuous (or differentiable) at a point (or on an interval) if each of its entries has the same property. The **derivative** of a differentiable matrix function is defined by elementwise differentiation; that is,

$$\mathbf{A}'(t) = \frac{d\mathbf{A}}{dt} = \left(\frac{da_{ij}}{dt} \right). \tag{23}$$

Thus if

$$\mathbf{x}(t) = \begin{pmatrix} t \\ t^2 \\ e^{-t} \end{pmatrix} \quad \text{and} \quad \mathbf{A}(t) = \begin{pmatrix} \sin t & 1 \\ t & \cos t \end{pmatrix},$$

then

$$\frac{d\mathbf{x}}{dt} = \begin{pmatrix} 1 \\ 2t \\ -e^{-t} \end{pmatrix} \quad \text{and} \quad \mathbf{A}'(t) = \begin{pmatrix} \cos t & 0 \\ 1 & -\sin t \end{pmatrix}.$$

The differentiation rules

$$\frac{d}{dt}(\mathbf{A} + \mathbf{B}) = \frac{d\mathbf{A}}{dt} + \frac{d\mathbf{B}}{dt} \tag{24}$$

and

$$\frac{d}{dt}(\mathbf{AB}) = \mathbf{A}\frac{d\mathbf{B}}{dt} + \frac{d\mathbf{A}}{dt}\mathbf{B} \tag{25}$$

follow readily by elementwise application of the analogous differentiation rules of elementary calculus for real-valued functions. If c is a (constant) real number and $\mathbf{C}$ is a constant matrix, then

$$\frac{d}{dt}(c\mathbf{A}) = c\frac{d\mathbf{A}}{dt}, \qquad \frac{d}{dt}(\mathbf{CA}) = \mathbf{C}\frac{d\mathbf{A}}{dt}, \qquad \text{and} \qquad \frac{d}{dt}(\mathbf{AC}) = \frac{d\mathbf{A}}{dt}\mathbf{C}. \tag{26}$$

Because of the noncommutativity of matrix multiplication, it is important not to reverse the order of the factors in (25) and (26).

FIRST ORDER LINEAR SYSTEMS

At first reading it may seem that we have erected an imposing edifice of matrix theory. On the contrary, it is merely a plethora of notation and terminology that one assimilates readily with practice. Our main use for matrix notation will be the simplification of computations with systems of differential equations, especially those computations that would be burdensome in scalar notation.

We discuss here the general system of n first order linear equations

$$x'_1 = p_{11}(t)x_1 + p_{12}(t)x_2 + \cdots + p_{1n}(t)x_n + f_1(t),$$
$$x'_2 = p_{21}(t)x_1 + p_{22}(t)x_2 + \cdots + p_{2n}(t)x_n + f_2(t),$$
$$\vdots$$
$$x'_n = p_{n1}(t)x_1 + p_{n2}(t)x_2 + \cdots + p_{nn}(t)x_n + f_n(t). \tag{27}$$

If we introduce the *coefficient matrix* $\mathbf{P}(t) = (p_{ij}(t))$ and the column vectors $\mathbf{x} = (x_i)$ and $\mathbf{f}(t) = (f_i(t))$, then the system in (27) takes the form of a single matrix equation:

$$\mathbf{x}' = \mathbf{P}(t)\mathbf{x} + \mathbf{f}(t). \tag{28}$$

We will see that the general theory of the linear system in (27) closely parallels that of a single nth order linear equation. The matrix notation used in Eq. (28) not only emphasizes this analogy, but also saves a great deal of space.

A **solution** of Eq. (28) on the open interval I is a column vector function $\mathbf{x}(t) = (x_i(t))$ such that the component functions of $\mathbf{x}$ satisfy the system in (27) identically on I. If the functions $p_{ij}(t)$ and $f_i(t)$ are all continuous on I, then the theorem in Section 5.1 guarantees the existence on I of a unique solution $\mathbf{x}(t)$ satisfying preassigned initial conditions $\mathbf{x}(a) = \mathbf{b}$.

To investigate the general nature of the solutions of Eq. (28), we consider first the **associated homogeneous equation**

$$\mathbf{x}' = \mathbf{P}(t)\mathbf{x}, \tag{29}$$

which has the form in Eq. (28), though with $\mathbf{f}(t) \equiv \mathbf{0}$. We expect it to have n solutions $\mathbf{x}_1, \mathbf{x}_2, \ldots, \mathbf{x}_n$ that are independent in some appropriate sense, and such that every solution is a linear combination of these n particular solutions. Given n solutions $\mathbf{x}_1, \mathbf{x}_2, \ldots, \mathbf{x}_n$ of (29), let us write

$$\mathbf{x}_j(t) = \begin{pmatrix} x_{1j}(t) \\ \vdots \\ x_{ij}(t) \\ \vdots \\ x_{nj}(t) \end{pmatrix}. \tag{30}$$

Thus $x_{ij}(t)$ denotes the ith component of the vector $\mathbf{x}_j(t)$, so the second subscript refers to the vector function $\mathbf{x}_j(t)$, while the first subscript refers to a component of this function. The following theorem is analogous to Theorem 1 in Section 2.2.

THEOREM 1: PRINCIPLE OF SUPERPOSITION

Let $\mathbf{x}_1, \mathbf{x}_2, \ldots, \mathbf{x}_j$ be n solutions of the homogeneous linear equation in (29) on the open interval I. If $c_1, c_2, \ldots, c_n$ are constants, then the linear combination

$$\mathbf{x} = c_1\mathbf{x}_1 + c_2\mathbf{x}_2 + \cdots + c_n\mathbf{x}_n \tag{31}$$

is also a solution of (29) on I.

Proof We know that $\mathbf{x}'_i = \mathbf{P}(t)\mathbf{x}_i$ for each i $(1 \le i \le n)$, so it follows immediately that

$$\mathbf{x}' = c_1\mathbf{x}'_1 + c_2\mathbf{x}'_2 + \cdots + c_n\mathbf{x}'_n$$
$$= c_1\mathbf{P}(t)\mathbf{x}_1 + c_2\mathbf{P}(t)\mathbf{x}_2 + \cdots + c_n\mathbf{P}(t)\mathbf{x}_n$$
$$= \mathbf{P}(t)(c_1\mathbf{x}_1 + c_2\mathbf{x}_2 + \cdots + c_n\mathbf{x}_n).$$

That is, $\mathbf{x}' = \mathbf{P}(t)\mathbf{x}$, as desired. The remarkable simplicity of this proof demonstrates clearly the advantages of matrix notation.

EXAMPLE 1 It is readily verified directly that

$$\mathbf{x}_1 = \begin{pmatrix} \frac{3}{2}e^{2t} \\ e^{2t} \end{pmatrix} \quad \text{and} \quad \mathbf{x}_2 = \begin{pmatrix} \frac{1}{3}e^{-5t} \\ e^{-5t} \end{pmatrix}$$

are each solutions of the equation

$$\mathbf{x}' = \begin{pmatrix} 4 & -3 \\ 6 & -7 \end{pmatrix}\mathbf{x}.$$

Hence Theorem 1 guarantees that the linear combination

$$\mathbf{x} = c_1\mathbf{x}_1 + c_2\mathbf{x}_2 = \begin{pmatrix} \frac{3}{2}c_1e^{2t} + \frac{1}{3}c_2e^{-5t} \\ c_1e^{2t} + c_2e^{-5t} \end{pmatrix}$$

is also a solution; it is, in fact, the general solution that we found by the method of elimination in Example 2 of Section 5.2.

Linear independence is defined in the same way for vector-valued functions as for real-valued functions (Section 2.2). The vector-valued functions $\mathbf{x}_1, \mathbf{x}_2, \ldots, \mathbf{x}_n$ are **linearly dependent** on the interval I provided that there exist constants $c_1, c_2, \ldots, c_n$ *not all zero* such that

$$c_1\mathbf{x}_1(t) + c_2\mathbf{x}_2(t) + \cdots + c_n\mathbf{x}_n(t) = \mathbf{0} \tag{32}$$

for all t in I. Otherwise they are **linearly independent**. Equivalently, they are linearly independent provided that no one of them is a linear combination of the others.

Just as in the case of a single nth order equation, there is a Wronskian determinant that tells us whether or not n given solutions of the homogeneous equation in (29) are linearly dependent. If $\mathbf{x}_1, \mathbf{x}_2, \ldots, \mathbf{x}_n$ are such solutions, then their *Wronskian* is the $n \times n$ determinant

$$W = \begin{vmatrix} x_{11}(t) & x_{12}(t) & \cdots & x_{1n}(t) \\ x_{21}(t) & x_{22}(t) & \cdots & x_{2n}(t) \\ \cdot & \cdot & & \cdot \\ \cdot & \cdot & & \cdot \\ \cdot & \cdot & & \cdot \\ x_{n1}(t) & x_{n2}(t) & \cdots & x_{nn}(t) \end{vmatrix}, \tag{33}$$

using the notation in (30) for the components of the solutions. We may write either $W(t)$ or $W(\mathbf{x}_1, \mathbf{x}_2, \ldots, \mathbf{x}_n)$. Note that W is the determinant of the matrix which has as its *column* vectors the solutions $\mathbf{x}_1, \mathbf{x}_2, \ldots, \mathbf{x}_n$. The following theorem is analogous to Theorem 3 in Section 2.2. Moreover, its proof is essentially the same, with the above definition of $W(\mathbf{x}_1, \mathbf{x}_2, \ldots, \mathbf{x}_n)$ substituted for the definition of the Wronskian of n solutions of a single nth order equation; see Problems 24–26.

THEOREM 2: WRONSKIANS OF SOLUTIONS

Suppose that $\mathbf{x}_1, \mathbf{x}_2, \ldots, \mathbf{x}_n$ are n solutions of the homogeneous linear equation $\mathbf{x}' = \mathbf{P}(t)\mathbf{x}$ on an open interval I. Suppose also that $\mathbf{P}(t) = (p_{ij}(t))$ and that each function $p_{ij}(t)$ is continuous on I. Let $W = W(\mathbf{x}_1, \mathbf{x}_2, \ldots, \mathbf{x}_n)$. Then:

(a) *If $\mathbf{x}_1, \mathbf{x}_2, \ldots, \mathbf{x}_n$ are linearly dependent on I, then $W \equiv 0$ on I.*
(b) *If $\mathbf{x}_1, \mathbf{x}_2, \ldots, \mathbf{x}_n$ are linearly independent on I, then $W \neq 0$ at each point of I.*

Thus there are only two possibilities for solutions of homogeneous systems: Either $W = 0$ everywhere on I, or $W = 0$ for no point of I.

EXAMPLE 2 It is readily verified directly that

$$\mathbf{x}_1 = (2e^t \quad 2e^t \quad e^t)^T,$$
$$\mathbf{x}_2 = (2e^{3t} \quad 0 \quad -e^{3t})^T,$$

and

$$\mathbf{x}_3 = (2e^{5t} \quad -2e^{5t} \quad e^{5t})^T$$

are solutions of the equation

$$\mathbf{x}' = \begin{pmatrix} 3 & -2 & 0 \\ -1 & 3 & -2 \\ 0 & -1 & 3 \end{pmatrix} \mathbf{x}. \tag{34}$$

The Wronskian of these solutions is

$$W = \begin{vmatrix} 2e^t & 2e^{3t} & 2e^{5t} \\ 2e^t & 0 & -2e^{5t} \\ e^t & -e^{3t} & e^{5t} \end{vmatrix} = e^{9t} \begin{vmatrix} 2 & 2 & 2 \\ 2 & 0 & -2 \\ 1 & -1 & 1 \end{vmatrix}$$

$$= e^{9t} \left(-(2) \begin{vmatrix} 2 & 2 \\ -1 & 1 \end{vmatrix} - (-2) \begin{vmatrix} 2 & 2 \\ 1 & -1 \end{vmatrix} \right),$$

and consequently $W = -16e^{9t}$, which is never zero. Hence Theorem 2 implies that the solutions $\mathbf{x}_1$, $\mathbf{x}_2$, and $\mathbf{x}_3$ are linearly independent (on any open interval).

The following theorem is analogous to Theorem 4 in Section 2.2. It says that the *general solution* of $\mathbf{x}' = \mathbf{P}(t)\mathbf{x}$ is a linear combination

$$\mathbf{x} = c_1\mathbf{x}_1 + c_2\mathbf{x}_2 + \cdots + c_n\mathbf{x}_n \tag{35}$$

of any n given linearly independent solutions. For instance, the general solution of Eq. (34) in Example 2 is

$$\mathbf{x} = c_1(2e^t \quad 2e^t \quad e^t)^T + c_2(2e^{3t} \quad 0 \quad -e^{3t})^T + c_3(2e^{5t} \quad -2e^{5t} \quad e^{5t})^T$$

$$= \begin{pmatrix} 2c_1e^t + 2c_2e^{3t} + 2c_3e^{5t} \\ 2c_1e^t \qquad\quad - 2c_3e^{5t} \\ c_1e^t - c_2e^{3t} + c_3e^{5t} \end{pmatrix}.$$

THEOREM 3: GENERAL SOLUTIONS OF HOMOGENEOUS SYSTEMS

Let $\mathbf{x}_1, \mathbf{x}_2, \ldots, \mathbf{x}_n$ be n linearly independent solutions of the homogeneous linear equation $\mathbf{x}' = \mathbf{P}(t)\mathbf{x}$ on an open interval I where the entries $p_{ij}(t)$ of $\mathbf{P}(t)$ are continuous. If $\mathbf{x}(t)$ is any solution whatsoever of the equation $\mathbf{x}' = \mathbf{P}(t)\mathbf{x}$ on I, then there exist numbers $c_1, c_2, \ldots, c_n$ such that

$$\mathbf{x}(t) = c_1\mathbf{x}_1(t) + c_2\mathbf{x}_2(t) + \cdots + c_n\mathbf{x}_n(t) \tag{35}$$

for all t in I.

Proof Let a be a fixed point of I. We show first that there exist numbers $c_1, c_2, \ldots, c_n$ such that the solution

$$\mathbf{y}(t) = c_1\mathbf{x}_1(t) + c_2\mathbf{x}_2(t) + \cdots + c_n\mathbf{x}_n(t) \tag{36}$$

has the same initial values at $t = a$ as does the given solution $\mathbf{x}(t)$; that is, such that

$$c_1\mathbf{x}_1(a) + c_2\mathbf{x}_2(a) + \cdots + c_n\mathbf{x}_n(a) = \mathbf{x}(a). \tag{37}$$

Let $\mathbf{X}(t)$ be the $n \times n$ matrix with column vectors $\mathbf{x}_1(t)$, $\mathbf{x}_2(t)$, $\ldots$, $\mathbf{x}_n(t)$, and let $\mathbf{c}$ be the column vector with components $c_1, c_2, \ldots, c_n$. Then Eq. (37) may be written in the form

$$\mathbf{X}(a)\mathbf{c} = \mathbf{x}(a). \tag{38}$$

The Wronskian determinant $W(a) = |\mathbf{X}(a)|$ is nonzero because the solutions $\mathbf{x}_1, \mathbf{x}_2, \ldots, \mathbf{x}_n$ are linearly independent. Hence the matrix $\mathbf{X}(a)$ has an inverse matrix $\mathbf{X}^{-1}(a)$. Therefore the vector $\mathbf{c} = \mathbf{X}^{-1}(a)\mathbf{x}(a)$ satisfies Eq. (38), as desired.

Finally, note that the given solution $\mathbf{x}(t)$ and the solution $\mathbf{y}(t)$ of Eq. (36)—with the values of c_i determined by the equation $\mathbf{c} = \mathbf{X}^{-1}(a)\mathbf{x}(a)$—have the same initial values (at $t = a$). It follows from the existence-uniqueness theorem of Section 5.1 that $\mathbf{x}(t) = \mathbf{y}(t)$ for all t in I. This establishes Eq. (35). ▲

Remark: Every equation $\mathbf{x}' = \mathbf{P}(t)\mathbf{x}$ with continuous coefficient matrix does have a set of n linearly independent solutions $\mathbf{x}_1, \mathbf{x}_2, \ldots, \mathbf{x}_n$, as in the hypotheses of Theorem 3. It suffices to choose for $\mathbf{x}_j(t)$ the unique solution such that

$$\mathbf{x}_j(a) = (0 \quad 0 \quad \cdots \quad 0 \quad 1 \quad 0 \quad \cdots \quad 0)^T,$$

position j

the jth column of the identity matrix. Then

$$W(\mathbf{x}_1, \mathbf{x}_2, \ldots, \mathbf{x}_n)\big|_{t=a} = |\mathbf{I}| = 1 \neq 0,$$

so the solutions $\mathbf{x}_1, \mathbf{x}_2, \ldots, \mathbf{x}_n$ are linearly independent by Theorem 2. How actually to find these solutions explicitly is another matter—one which we address in Section 5.5 (for the case of constant coefficient matrices).

We now return to the *nonhomogeneous* linear equation

$$\mathbf{x}' = \mathbf{P}(t)\mathbf{x} + \mathbf{f}(t). \tag{28}$$

The following theorem is analogous to Theorem 5 in Section 2.2 and is proved in precisely the same way, substituting the preceding theorems in this section for the analogous theorems of Section 2.2. In brief, Theorem 4 means that the general solution of (28) has the form

$$\mathbf{x} = \mathbf{x}_c + \mathbf{x}_p, \tag{39}$$

where $\mathbf{x}_p$ is a single particular solution of (28) and the **complementary function** $\mathbf{x}_c$ is the general solution of the associated homogeneous equation $\mathbf{x}' = \mathbf{P}(t)\mathbf{x}$.

THEOREM 4: SOLUTIONS OF NONHOMOGENEOUS SYSTEMS

Let $\mathbf{x}_p$ be a particular solution of the nonhomogeneous linear equation in (28) on an open interval I on which the function entries p_{ij} of $\mathbf{P}(t)$ and f_i of $\mathbf{f}(t)$ are continuous. Let $\mathbf{x}_1, \mathbf{x}_2, \ldots, \mathbf{x}_n$ be linearly independent solutions of the associated homogeneous equation on I. If $\mathbf{x}(t)$ is any solution whatsoever of Eq. (28) on I, then there exist numbers $c_1, c_2, \ldots, c_n$ such that

$$\mathbf{x}(t) = c_1\mathbf{x}_1(t) + c_2\mathbf{x}_2(t) + \cdots + c_n\mathbf{x}_n(t) + \mathbf{x}_p(t) \tag{40}$$

for all t in I.

Thus finding the general solution of a nonhomogeneous linear system involves two separate steps:

1. Finding the general solution $\mathbf{x}_c$ of the associated homogeneous system;
2. Finding a single particular solution $\mathbf{x}_p$ of the nonhomogeneous system.

The sum $\mathbf{x} = \mathbf{x}_c + \mathbf{x}_p$ will then be the general solution of the nonhomogeneous system. In contrast with the systematic elimination procedure of Section 5.2, *ad hoc* techniques for carrying out steps 1 and 2 are sometimes more efficient. The following example illustrates this.

EXAMPLE 3 Find the displacements $x_1(t)$ and $x_2(t)$ of the two masses in the system shown in Fig. 5.16, given the initial conditions $x_1(0) = x_2(0) = x_1'(0) = x_2'(0) = 0$.

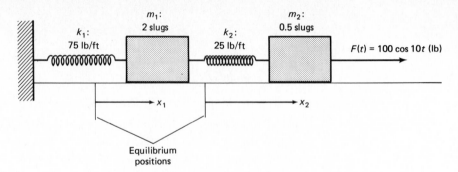

Figure 5.16 The mechanical system of Example 3.

Solution With the numerical values indicated in Fig. 5.16, Example 1 in Section 5.1—in particular, the equations in (3) there—give us the nonhomogeneous system

$$2x_1'' + 100x_1 - 25x_2 = 0,$$
$$\tfrac{1}{2}x_2'' - 25x_1 + 25x_2 = 100 \cos 10t. \tag{41}$$

The associated homogeneous system is

$$2x_1'' + 100x_1 - 25x_2 = 0,$$
$$\tfrac{1}{2}x_2'' - 25x_1 + 25x_2 = 0. \tag{42}$$

As in Example 4 of Section 5.1, this system is equivalent to a system of four first order linear equations. Hence its general solution $\mathbf{x}_c$ will be a linear combination of four linearly independent solutions.

To attempt to find these solutions, we note that neither equation in (42) contains a first derivative. Hence it seems likely that there will be solutions of the form

$$x_1 = a_1 \cos \omega t, \qquad x_2 = a_2 \cos \omega t. \tag{43}$$

For if we substituted the equations in (43) in (42), every term would be a multiple of $\cos \omega t$, so it would merely be a question of whether the constant coefficients sum correctly. When we carry out this substitution, the result is

$$-2a_1\omega^2 \cos \omega t + 100a_1 \cos \omega t - 25a_2 \cos \omega t = 0,$$
$$-\tfrac{1}{2}a_2\omega^2 \cos \omega t - 25a_1 \cos \omega t + 25a_2 \cos \omega t = 0.$$

We cancel $\cos \omega t$ throughout and collect coefficients to obtain

$$(100 - 2\omega^2)a_1 - \qquad\qquad 25a_2 = 0,$$
$$-25a_1 + (25 - \tfrac{1}{2}\omega^2)a_2 = 0. \tag{44}$$

But these equations can have a nontrivial solution for a_1 and a_2 only if the determinant of coefficients vanishes:

$$(100 - 2\omega^2)(25 - \tfrac{1}{2}\omega^2) - (25)^2 = 0;$$
$$(50 - \omega^2)^2 = (25)^2;$$
$$\omega^2 = 50 \pm 25.$$

Thus ω must have either the value $\omega_1 = 5$ rad/s or $\omega_2 = 5\sqrt{3}$ rad/s; these are the two natural frequencies of the system.

With $\omega_1 = 5$, each equation in (44) reduces to $50a_1 - 25a_2 = 0$, so $a_2 = 2a_1$. With $a_1 = 1$, we get the solution

$$x_1 = \cos 5t, \qquad x_2 = 2 \cos 5t \tag{45}$$

of the homogeneous system in (42). With $\omega_2 = 5\sqrt{3}$, each equation in (44) reduces to $-50a_1 - 25a_2 = 0$, so that $a_2 = -2a_1$. With $a_1 = 1$ here, we get the second solution

$$x_1 = \cos 5\sqrt{3}\, t, \qquad x_2 = -2 \cos 5\sqrt{3}\, t. \tag{46}$$

To obtain two more solutions, we observe that precisely the same computations would ensue if we had begun with sines rather than with cosines in our trial solution in (43). Hence two additional solutions are

$$x_1 = \sin 5t, \qquad x_2 = 2 \sin 5t \tag{47}$$

and

$$x_1 = \sin 5\sqrt{3}\, t, \qquad x_2 = -2 \sin 5\sqrt{3}\, t. \tag{48}$$

It can be verified (Problem 27) that these four solutions

$$\begin{aligned}
\mathbf{x}_1 &= (\cos 5t \quad 2 \cos 5t)^T, \\
\mathbf{x}_2 &= (\sin 5t \quad 2 \sin 5t)^T, \\
\mathbf{x}_3 &= (\cos 5\sqrt{3}\, t \quad -2 \cos 5\sqrt{3}\, t)^T, \\
\mathbf{x}_4 &= (\sin 5\sqrt{3}\, t \quad -2 \sin 5\sqrt{3}\, t)^T
\end{aligned} \tag{49}$$

are linearly independent. Hence the general solution of the system in (42) is

$$\mathbf{x}_c = c_1 \mathbf{x}_1 + c_2 \mathbf{x}_2 + c_3 \mathbf{x}_3 + c_4 \mathbf{x}_4 \tag{50}$$

where c_1, c_2, c_3, and c_4 are arbitrary constants.

Now we turn our attention to the particular solution of the orginal nonhomogeneous system

$$\begin{aligned}
2x_1'' + 100x_1 - 25x_2 &= 0, \\
\tfrac{1}{2}x_2'' - 25x_1 + 25x_2 &= 100 \cos 10t.
\end{aligned} \tag{41}$$

Because of the presence of the term $100 \cos 10t$, as well as the absence of any first derivatives, it is reasonable to try

$$x_1 = A_1 \cos 10t, \qquad x_2 = A_2 \cos 10t. \tag{51}$$

After substitution of (51) in (41) and cancellation of the common term $\cos 10t$, we get the equations

$$\begin{aligned}
-200A_1 + 100A_1 - 25A_2 &= 0, \\
-50A_2 - 25A_1 + 25A_2 &= 100,
\end{aligned}$$

which are readily solved for $A_1 = \tfrac{4}{3}$ and $A_2 = -\tfrac{16}{3}$. Thus the particular solution is

$$x_p = \tfrac{4}{3}(\cos 10t \quad -4 \cos 10t)^T. \tag{52}$$

We now assemble the information in Eqs. (49)–(52): The components of the general solution $\mathbf{x} = \mathbf{x}_c + \mathbf{x}_p$ of the nonhomogeneous system in (41) are

$$x_1 = c_1 \cos 5t + c_2 \sin 5t$$
$$+ c_3 \cos 5\sqrt{3}\,t + c_4 \sin 5\sqrt{3}\,t + \tfrac{4}{3} \cos 10t, \qquad (53)$$
$$x_2 = 2c_1 \cos 5t + 2c_2 \sin 5t$$
$$- 2c_3 \cos 5\sqrt{3}\,t - 2c_4 \sin 5\sqrt{3}\,t - \tfrac{16}{3} \cos 10t.$$

Finally imposing the initial conditions, we get the equations

$$x_1(0) = c_1 + c_3 + \tfrac{4}{3} = 0,$$
$$x_2(0) = 2c_1 - 2c_3 - \tfrac{16}{3} = 0,$$
$$x_1'(0) = 5c_2 + 5\sqrt{3}\,c_4 = 0,$$
$$x_2'(0) = 10c_2 - 10\sqrt{3}\,c_4 = 0.$$

The simultaneous solution of these equations is $c_1 = \tfrac{2}{3}$, $c_2 = 0$, $c_3 = -2$, and $c_4 = 0$. Therefore the motion of the two masses is described by

$$x_1(t) = \tfrac{2}{3} \cos 5t - 2 \cos 5\sqrt{3}\,t + \tfrac{4}{3} \cos 10t,$$
$$x_2(t) = \tfrac{4}{3} \cos 5t + 4 \cos 5\sqrt{3}\,t - \tfrac{16}{3} \cos 10t.$$

We have a superposition of two oscillations with the natural frequencies $\omega_1 = 5$ and $\omega_2 = 5\sqrt{3}$ and a forced oscillation with frequency $\omega = 10$. In each of the two natural oscillations the amplitude of motion of m_2 is twice that of m_1, while in the forced oscillation the amplitude of motion of m_2 is four times that of m_1.

5.3 Problems

1. Let

$$\mathbf{A} = \begin{pmatrix} 2 & -3 \\ 4 & 7 \end{pmatrix} \quad \text{and} \quad \mathbf{B} = \begin{pmatrix} 3 & -4 \\ 5 & 1 \end{pmatrix}.$$

Find (a) $2\mathbf{A} + 3\mathbf{B}$; (b) $3\mathbf{A} - 2\mathbf{B}$; (c) $\mathbf{AB}$; (d) $\mathbf{BA}$.

2. Verify that (a) $\mathbf{A(BC)} = \mathbf{(AB)C}$ and that (b) $\mathbf{A(B + C)} = \mathbf{AB} + \mathbf{AC}$, where $\mathbf{A}$ and $\mathbf{B}$ are the matrices given in Problem 1 and

$$\mathbf{C} = \begin{pmatrix} 0 & 2 \\ 3 & -1 \end{pmatrix}.$$

3. Find $\mathbf{AB}$ and $\mathbf{BA}$ given

$$\mathbf{A} = \begin{pmatrix} 2 & 0 & -1 \\ 3 & -4 & 5 \end{pmatrix} \quad \text{and} \quad \mathbf{B} = \begin{pmatrix} 1 & 3 \\ -7 & 0 \\ 3 & -2 \end{pmatrix}.$$

4. Let $\mathbf{A}$ and $\mathbf{B}$ be the matrices given in Problem 3, and let

$$\mathbf{x} = \begin{pmatrix} 2t \\ e^{-t} \end{pmatrix} \quad \text{and} \quad \mathbf{y} = \begin{pmatrix} t^2 \\ \sin t \\ \cos t \end{pmatrix}.$$

Find $\mathbf{Ay}$ and $\mathbf{Bx}$. Are the products $\mathbf{Ax}$ and $\mathbf{By}$ defined? Explain your answer.

5. Let

$$A = \begin{pmatrix} 3 & 2 & -1 \\ 0 & 4 & 3 \\ -5 & 2 & 7 \end{pmatrix} \quad \text{and} \quad B = \begin{pmatrix} 0 & -3 & 2 \\ 1 & 4 & -3 \\ 2 & 5 & -1 \end{pmatrix}.$$

Find (a) $7A + 4B$; (b) $3A - 5B$; (c) AB; (d) BA; (e) $A - tI$.

6. Let

$$A_1 = \begin{pmatrix} 2 & 1 \\ -3 & 2 \end{pmatrix}, \quad A_2 = \begin{pmatrix} 1 & 3 \\ -1 & -2 \end{pmatrix}, \quad B = \begin{pmatrix} 2 & 4 \\ 1 & 2 \end{pmatrix}.$$

(a) Show that $A_1 B = A_2 B$ and note that $A_1 \neq A_2$. Thus the cancellation law does not hold for matrices; that is, if $A_1 B = A_2 B$ and $B \neq 0$, it does not follow that $A_1 = A_2$. (b) Let $A = A_1 - A_2$, and show that $AB = 0$. Thus the product of two nonzero matrices may be the zero matrix.

7. Compute the determinants of the matrices A and B in Problem 6. Are your results consistent with the theorem to the effect that $\det(AB) = [\det(A)][\det(B)]$ for any two square matrices A and B of the same order?

8. Suppose that A and B are the matrices of Problem 5. Verify that $\det(AB) = \det(BA)$.

In each of Problems 9 and 10, verify the product law for differentiation, $(AB)' = A'B + AB'$.

9.

$$A(t) = \begin{pmatrix} t & 2t - 1 \\ t^3 & \dfrac{1}{t} \end{pmatrix} \quad \text{and} \quad B(t) = \begin{pmatrix} 1 - t & 1 + t \\ 3t^2 & 4t^3 \end{pmatrix}.$$

10.

$$A(t) = \begin{pmatrix} e^t & t & t^2 \\ -t & 0 & 2 \\ 8t & -1 & t^3 \end{pmatrix} \quad \text{and} \quad B(t) = \begin{pmatrix} 3 \\ 2e^{-t} \\ 3t \end{pmatrix}.$$

In each of Problems 11–15, write the given system in the form $x' = P(t)x + f(t)$; identify x, $P(t)$, and $f(t)$.

11. $x' = 3x - 2y,$
$\quad\;\; y' = 2x + y.$

12. $x' = 2x + 4y + 3e^t,$
$\quad\;\; y' = 5x - y - t^2.$

13. $x' = \quad tx - e^t y + \cos t,$
$\quad\;\; y' = e^{-t}x + t^2 y - \sin t.$

14. $x' = 3x - 4y + z + t,$
$\quad\;\; y' = x \qquad\quad - 3z + t^2,$
$\quad\;\; z' = \qquad 6y - 7z + t^3.$

15. $x' = \quad tx - \quad y + e^t z,$
$\quad\;\; y' = \quad 2x + t^2 y - \quad z,$
$\quad\;\; z' = e^{-t}x + 3ty + t^3 z.$

In each of Problems 16–22, first verify that the given vectors are solutions of the given system, and then use the Wronskian to show that they are linearly independent. Finally write the general solution of the system.

16. $x' = \begin{pmatrix} -3 & 2 \\ -3 & 4 \end{pmatrix} x; \quad x_1 = \begin{pmatrix} e^{3t} \\ 3e^{3t} \end{pmatrix}, x_2 = \begin{pmatrix} 2e^{-2t} \\ e^{-2t} \end{pmatrix}.$

17. $x' = \begin{pmatrix} 3 & -1 \\ 5 & -3 \end{pmatrix} x; \quad x_1 = e^{2t} \begin{pmatrix} 1 \\ 1 \end{pmatrix}, x_2 = e^{-2t} \begin{pmatrix} 1 \\ 5 \end{pmatrix}.$

18. $x' = \begin{pmatrix} 4 & 1 \\ -2 & 1 \end{pmatrix} x; \quad x_1 = e^{3t} \begin{pmatrix} 1 \\ -1 \end{pmatrix}, x_2 = e^{2t} \begin{pmatrix} 2 \\ -1 \end{pmatrix}.$

19.
$$\mathbf{x}' = \begin{pmatrix} 4 & -3 \\ 6 & -7 \end{pmatrix} \mathbf{x}; \quad \mathbf{x}_1 = \begin{pmatrix} 3e^{2t} \\ 2e^{2t} \end{pmatrix}, \mathbf{x}_2 = \begin{pmatrix} e^{-5t} \\ 3e^{-5t} \end{pmatrix}.$$

20.
$$\mathbf{x}' = \begin{pmatrix} 3 & -2 & 0 \\ -1 & 3 & -2 \\ 0 & -1 & 3 \end{pmatrix} \mathbf{x}; \quad \mathbf{x}_1 = e^t \begin{pmatrix} 2 \\ 2 \\ 1 \end{pmatrix}, \mathbf{x}_2 = e^{3t} \begin{pmatrix} -2 \\ 0 \\ 1 \end{pmatrix}, \mathbf{x}_3 = e^{5t} \begin{pmatrix} 2 \\ -2 \\ 1 \end{pmatrix}.$$

21.
$$\mathbf{x}' = \begin{pmatrix} 0 & 1 & 1 \\ 1 & 0 & 1 \\ 1 & 1 & 0 \end{pmatrix} \mathbf{x}; \quad \mathbf{x}_1 = e^{2t} \begin{pmatrix} 1 \\ 1 \\ 1 \end{pmatrix}, \mathbf{x}_2 = e^{-t} \begin{pmatrix} 1 \\ 0 \\ -1 \end{pmatrix}, \mathbf{x}_3 = e^{-t} \begin{pmatrix} 0 \\ 1 \\ -1 \end{pmatrix}.$$

22.
$$\mathbf{x}' = \begin{pmatrix} 1 & 2 & 1 \\ 6 & -1 & 0 \\ -1 & -2 & -1 \end{pmatrix} \mathbf{x}; \quad \mathbf{x}_1 = \begin{pmatrix} 1 \\ 6 \\ -13 \end{pmatrix}, \mathbf{x}_2 = e^{3t} \begin{pmatrix} 2 \\ 3 \\ -2 \end{pmatrix}, \mathbf{x}_3 = e^{-4t} \begin{pmatrix} -1 \\ 2 \\ 1 \end{pmatrix}.$$

23. (a) Show that the vector functions

$$\mathbf{x}_1 = \begin{pmatrix} t \\ t^2 \end{pmatrix} \quad \text{and} \quad \mathbf{x}_2 = \begin{pmatrix} t^2 \\ t^3 \end{pmatrix}$$

are linearly independent on the real line. (b) Why does it follow from Theorem 2 that there is *no* continuous matrix $\mathbf{P}(t)$ such that $\mathbf{x}_1$ and $\mathbf{x}_2$ are both solutions of $\mathbf{x}' = \mathbf{P}(t)\mathbf{x}$?

24. Suppose that one of the vector functions

$$\mathbf{x}_1(t) = \begin{pmatrix} x_{11}(t) \\ x_{21}(t) \end{pmatrix} \quad \text{and} \quad \mathbf{x}_2(t) = \begin{pmatrix} x_{12}(t) \\ x_{22}(t) \end{pmatrix}$$

is a constant multiple of the other on the open interval I. Show that their Wronskian $W(t) = |(x_{ij}(t))|$ must vanish identically on I. This proves part (a) of Theorem 2 in the case $n = 2$.

25. Suppose that the vectors $\mathbf{x}_1(t)$ and $\mathbf{x}_2(t)$ of Problem 24 are solutions of the equation $\mathbf{x}' = \mathbf{P}(t)\mathbf{x}$, where the 2×2 matrix $\mathbf{P}(t)$ is continuous on the open interval I. Show that, if there exists a point a of I at which their Wronskian $W(a)$ is zero, then there exist numbers c_1 and c_2 not both zero such that $c_1\mathbf{x}_1(a) + c_2\mathbf{x}_2(a) = \mathbf{0}$. Then conclude from the uniqueness of solutions of the equation $\mathbf{x}' = \mathbf{P}(t)\mathbf{x}$ that $c_1\mathbf{x}_1(t) + c_2\mathbf{x}_2(t) = \mathbf{0}$ for all t in I; that is, that $\mathbf{x}_1$ and $\mathbf{x}_2$ are linearly dependent. This proves part (b) of Theorem 2 in the case $n = 2$.

26. Generalize Problems 24 and 25 to prove Theorem 2 for n an arbitrary positive integer.

27. Let $\mathbf{x}_1(t), \mathbf{x}_2(t), \ldots, \mathbf{x}_n(t)$ be vector functions whose ith components (for some fixed i) $x_{i1}(t), x_{i2}(t), \ldots, x_{in}(t)$ are linearly independent real-valued functions. Conclude that the vector functions are themselves linearly independent. Does this imply that the solution vectors $\mathbf{x}_1, \mathbf{x}_2, \mathbf{x}_3$, and $\mathbf{x}_4$ in (49) are linearly independent?

Apply the method of Example 3 to find the general solution of each of the systems in Problems 28–32.

28. $x'' = -5x + 4y, \ y'' = 4x - 5y.$

29. $x_1'' = -3x_1 + 2x_2 + \cos 3t, \ 2x_2'' = 2x_1 - 4x_2.$

30. $x_1'' = -3x_1 + 2x_2 + 2 \sin 2t, \ x_2'' = 2x_1 - 3x_2.$

31. $x'' = -3x + y + \cos t, \ y'' = x - 3y + 2 \cos t.$

32. $x_1'' = -6x_1 + 4x_2, \ 2x_2'' = 4x_1 - 8x_2.$

In this section we apply the theory developed in Section 5.3 to investigate the oscillations of typical mass-and-spring systems having two or more degrees of freedom. Our examples are chosen to illustrate phenomena that are generally characteristic of complex mechanical systems.

Example 1 illustrates the effect of frictional resistance in a mechanical system subjected to a periodic external force that provides a *nonhomogeneous term* in the corresponding system of linear differential equations. On the basis of Theorem 4 of Section 5.3, we know that the solution satisfying given initial conditions will be of the form

$$\mathbf{x}(t) = \mathbf{x}_c(t) + \mathbf{x}_p(t) \tag{1}$$

where $\mathbf{x}_p(t)$ is a particular solution of the nonhomogeneous system and $\mathbf{x}_c(t)$ is a solution of the corresponding homogeneous system. It is typical for the effect of frictional resistance in mechanical systems to damp out the complementary function solution $\mathbf{x}_c(t)$, so that

$$\mathbf{x}_c(t) \longrightarrow \mathbf{0} \quad \text{as} \quad t \longrightarrow +\infty. \tag{2}$$

Hence $\mathbf{x}_c(t)$ is a **transient solution** that depends only upon the initial conditions; it dies out with time, leaving the **steady periodic solution** $\mathbf{x}_p(t)$ resulting from the external driving force:

$$\mathbf{x}(t) \longrightarrow \mathbf{x}_p(t) \quad \text{as} \quad t \longrightarrow +\infty. \tag{3}$$

As a practical matter, every physical system includes frictional resistance (however small) that damps out transient solutions in this manner.

EXAMPLE 1 We begin with the same system of two masses and two springs that we considered in Example 3 of Section 5.3. But we now include damping or frictional resistance; as indicated in Fig. 5.17, we incorporate dashpots connected to the two masses, each providing resistance proportional to the velocity of the mass to which it is connected. As the figure shows, the constants of proportionality are $c_1 = 4$ lb/ft/s and $c_2 = 1$ lb/ft/s. Also as before, a periodic external force $F(t) = 100 \cos 10t$

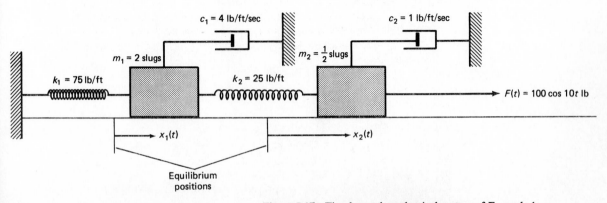

Figure 5.17 The damped mechanical system of Example 1.

(pounds) acts on the second mass. We want to determine the resulting transient and steady periodic motions of the two masses.

Solution Without the two dashpots—see the equations in (41) of Section 5.3—we derived from $F = ma$ the equations

$$2x_1'' = -100x_1 + 25x_2,$$
$$\tfrac{1}{2}x_2'' = 25x_1 - 25x_2 + 100 \cos 10t$$

for the displacements x_1 and x_2 of the two masses from their equilibrium positions at time t. The resistive force of the dashpots is $-c_1x_1' = -4x_1'$ on the first mass and $-c_2x_2' = -x_2'$ on the second mass. When we include these additional forces, we obtain the nonhomogeneous system

$$2x_1'' + 4x_1' + 100x_1 - 25x_2 = 0,$$
$$\tfrac{1}{2}x_2'' + x_2' - 25x_1 + 25x_2 = 100 \cos 10t. \tag{4}$$

THE TRANSIENT SOLUTION

First we need to find the general solution of the associated homogeneous system

$$2x_1'' + 4x_1' + 100x_1 - 25x_2 = 0,$$
$$\tfrac{1}{2}x_2'' + x_2' - 25x_1 + 25x_2 = 0; \tag{5}$$

we expect it to give the transient motions of the two masses. Because of the first derivative terms in (5), we can no longer expect the simple sine-cosine solutions we found in Example 3 of Section 5.3. Instead we apply the method of elimination of Section 5.2, in which we first write the system in the operational form

$$(2D^2 + 4D + 100)x_1 - 25x_2 = 0,$$
$$-25x_1 + (\tfrac{1}{2}D^2 + D + 25)x_2 = 0. \tag{6}$$

The operational determinant of this system is

$$(2D^2 + 4D + 100)(\tfrac{1}{2}D^2 + D + 25) - (25)^2 = (D^2 + 2D + 50)^2 - (25)^2.$$

Hence the characteristic equation—for either x_1 or x_2—is

$$(r^2 + 2r + 50)^2 - (25)^2 = 0, \tag{7}$$

which we solve as follows:

$$r^2 + 2r + 50 = \pm 25;$$
$$r^2 + 2r + 50 \pm 25 = 0;$$
$$r = \frac{-2 \pm \sqrt{4 - 4(50 \pm 25)}}{2}.$$

Thus we have two pairs of complex conjugates,

$$r = -1 \pm i\sqrt{24} \quad \text{and} \quad r = -1 \pm i\sqrt{74}. \tag{8}$$

Observe that had the characteristic equation in (7) not assumed so fortuitous a form, we might have needed a computer program for solving a fourth degree equation that has only complex roots.

The characteristic roots in (8) yield the general solutions

$$x_1 = e^{-t}(a_1 \cos \sqrt{24}t + a_2 \sin \sqrt{24}t)$$
$$+ e^{-t}(b_1 \cos \sqrt{74}t + b_2 \sin \sqrt{74}t),$$
$$x_2 = e^{-t}(c_1 \cos \sqrt{24}t + c_2 \sin \sqrt{24}t)$$
$$+ e^{-t}(d_1 \cos \sqrt{74}t + d_2 \sin \sqrt{74}t). \tag{9}$$

Because we have eight arbitrary constants rather than the four that ought to appear, we substitute x_1 and x_2 in the first equation in (6) to find the relations between these constants. A routine (but long) computation yields

$$(2D^2 + 4D + 100)x_1 = 50e^{-t}(a_1 \cos \sqrt{24}t + a_2 \sin \sqrt{24}t)$$
$$- 50e^{-t}(b_1 \cos \sqrt{74}t + b_2 \sin \sqrt{74}t).$$

In order that this expression be equal to $25x_2$, it is necessary that

$$c_1 = 2a_1, \quad c_2 = 2a_2, \quad d_1 = -2b_1, \quad \text{and} \quad d_2 = -2b_2. \tag{10}$$

Thus the general solution of the homogeneous system in (5) is given by

$$x_1 = e^{-t}(a_1 \cos \sqrt{24}t + a_2 \sin \sqrt{24}t)$$
$$+ e^{-t}(b_1 \cos \sqrt{74}t + b_2 \sin \sqrt{74}t),$$
$$x_2 = 2e^{-t}(a_1 \cos \sqrt{24}t + a_2 \sin \sqrt{24}t)$$
$$- 2e^{-t}(b_1 \cos \sqrt{74}t + b_2 \sin \sqrt{74}t). \tag{11}$$

Because of the presence of the factor e^{-t}, we do indeed have a transient solution. It may be written in the form

$$\mathbf{x}_{tr} = a_1 \begin{pmatrix} 1 \\ 2 \end{pmatrix} e^{-t} \cos \sqrt{24}t + a_2 \begin{pmatrix} 1 \\ 2 \end{pmatrix} e^{-t} \sin \sqrt{24}t$$
$$+ b_1 \begin{pmatrix} 1 \\ -2 \end{pmatrix} e^{-t} \cos \sqrt{74}t + a_2 \begin{pmatrix} 1 \\ -2 \end{pmatrix} e^{-t} \sin \sqrt{74}t. \tag{12}$$

Note that the two circular frequencies exhibited in (12) are $\omega_1 = \sqrt{24} \approx 4.8990$ rad/s and $\omega_2 = \sqrt{74} \approx 8.6023$ rad/s, as compared with the undamped natural frequencies $\omega_1 = 5$ rad/s and $\omega_2 = 5\sqrt{3} \approx 8.6603$ rad/s that we found in Example 3 of Section 5.3. Thus the effect of the frictional resistance in this example is both exponential damping of the amplitude of the oscillations and a decrease in their frequencies (just as we observed in the one-dimensional systems in Section 2.6). Equations (11) and (12) describe the "free" damped oscillations that would take place if there were no external force.

THE STEADY PERIODIC SOLUTION

Now we want to find a particular solution of the original nonhomogeneous system

$$2x_1'' + 4x_1' + 100x_1 - 25x_2 = 0,$$
$$\tfrac{1}{2}x_2'' + x_2' - 25x_1 + 25x_2 = 100 \cos 10t. \tag{4}$$

Again because of the presence of the first derivative terms, we cannot expect a solution of the simple form $x_1 = A \cos 10t$, $x_2 = B \cos 10t$. We will need both sine and cosine terms, so we could substitute

$$x_1 = A_1 \cos 10t + A_2 \sin 10t, \qquad x_2 = B_1 \cos 10t + B_2 \sin 10t$$

in each equation in (4) and attempt to determine the four constants A_1, A_2, B_1, and B_2.

Here it is more efficient to employ an alternative method that involves complex exponentials. We note first that

$$100 \cos 10t = \Re e(100 e^{10ti}). \tag{13}$$

This observation motivates us to introduce the new nonhomogeneous system

$$
\begin{aligned}
2\tilde{x}_1'' + 4\tilde{x}_1' + 100\tilde{x}_1 - 25\tilde{x}_2 &= 0, \\
\tfrac{1}{2}\tilde{x}_2'' + \tilde{x}_2' - 25\tilde{x}_1 + 25\tilde{x}_2 &= 100 e^{10ti}
\end{aligned}
\tag{14}
$$

in the (new) unknown functions $\tilde{x}_1(t)$ and $\tilde{x}_2(t)$ that we expect to be complex-valued. Because of (13), the *real parts* $x_1(t) = \Re e(\tilde{x}_1(t))$, $x_2(t) = \Re e(\tilde{x}_2(t))$ of any solution $\tilde{x}_1(t)$, $\tilde{x}_2(t)$ of the system in (14) will constitute a solution $x_1(t)$, $x_2(t)$ of the system in (4).

If we substitute in (14) the trial solution

$$\tilde{x}_1 = A e^{10ti}, \qquad \tilde{x}_2 = B e^{10ti}, \tag{15}$$

then every term in each equation will be a constant multiple of e^{10ti}. We can therefore hope to determine the (complex) constants A and B, and finally take the real parts of $\tilde{x}_1$ and $\tilde{x}_2$ to obtain x_1 and x_2.

When we substitute the expressions in (15) in the equations in (14), we get

$$-200 A e^{10ti} + 40i A e^{10ti} + 100 A e^{10ti} - 25 B e^{10ti} = 0,$$

$$-50 B e^{10ti} + 10i B e^{10ti} - 25 A e^{10ti} + 25 B e^{10ti} = 100 e^{10ti}.$$

We cancel the common factor e^{10ti} and collect coefficients to obtain the equations

$$
\begin{aligned}
(-100 + 40i)A - 25B &= 0, \\
-25A + (-25 + 10i)B &= 100.
\end{aligned}
$$

The determinant of this system of equations is

$$\Delta = (-100 + 40i)(-25 + 10i) - 625 = 1475 - 2000i.$$

Hence Cramer's rule yields

$$
A = \frac{1}{\Delta}\begin{vmatrix} 0 & -25 \\ 100 & -25 + 10i \end{vmatrix} = \frac{2500}{1475 - 2000i}
$$

$$
\approx \frac{2500}{(2485.08)e^{5.3478i}} \approx (1.0060)e^{-5.3478i}
$$

and

$$
B = \frac{1}{\Delta}\begin{vmatrix} -100 + 40i & 0 \\ -25 & 100 \end{vmatrix} = \frac{-10000 + 4000i}{1475 - 2000i}
$$

$$
\approx \frac{(-10770.33)e^{-0.3805i}}{(2485.08)e^{5.3478i}} \approx (-4.3340)e^{-5.7283i}.
$$

Using the above values for A and B in (15), we get

$$\tilde{x}_1 \approx (1.0060)e^{i(10t - 5.3478)}$$

and

$$\tilde{x}_2 \approx (-4.3340)e^{i(10t - 5.7283)}.$$

Taking real parts, we finally obtain the components

$$\begin{aligned} x_1 &= \Re e(x_1(t)) \approx (1.0060) \cos (10t - 5.3478), \\ x_2 &= \Re e(x_2(t)) \approx (-4.3340) \cos (10t - 5.7283) \end{aligned} \tag{16}$$

of the damped steady periodic solution $\mathbf{x}_p(t)$. It is of interest to compare (16) with the undamped steady periodic solution

$$x_1(t) = \tfrac{4}{3} \cos 10t, \qquad x_2(t) = -\tfrac{16}{3} \cos 10t \tag{17}$$

that we found in Example 3 of Section 5.3. Note that the equations in (16) describe approximately the same situation—with the two masses oscillating approximately 180° out of phase and with the amplitude of motion of m_2 about four times that of m_1—but with both amplitudes reduced by the frictional resistance.

Finally, let us combine the particular solution in (16) of the nonhomogeneous system in (4) with the complementary solution in (11) of the associated homogeneous system in (5). We then get the general solution

$$\begin{aligned} x_1(t) &= e^{-t}(a_1 \cos \sqrt{24}t + a_2 \sin \sqrt{24}t) \\ &\quad + e^{-t}(b_1 \cos \sqrt{74}t + b_2 \sin \sqrt{74}t) + 1.0060 \cos (10t - 5.3478), \\ x_2(t) &= 2e^{-t}(a_1 \cos \sqrt{24}t + a_2 \sin \sqrt{24}t) \\ &\quad - 2e^{-t}(b_1 \cos \sqrt{74}t + b_2 \sin \sqrt{74}t) - 4.3340 \cos (10t - 5.7283) \end{aligned} \tag{18}$$

of the nonhomogeneous system in (4). If initial values $x_1(0)$, $x_2(0)$, $x_1'(0)$, and $x_2'(0)$ were given, we could then determine the values of the constants $a_1, a_2, b_1,$ and b_2 in (18).

THE TWO-AXLE AUTOMOBILE

In Example 2 of Section 2.8 we investigated the vertical oscillations of a one-axle car—actually a unicycle. Now we can analyze a more realistic model: a car with two axles and with separate front and rear suspension systems. Figure 5.18 represents the suspension system of such a car. We assume that the car body acts as would a solid bar of mass m and length $L = L_1 + L_2$. It has moment of inertia I about its center of mass C, which is at distance L_1 from the front end of the car. The car has front and back suspension springs with Hooke's constants k_1 and k_2, respectively.

When the car is in motion, let x_1 and x_2 denote the elevations of the front and rear ends of the car body, respectively, above their equilibrium positions. Because these are the distances the two suspension springs are stretched or compressed, the two vertical forces on the two ends of the car body are

$$F_1 = -k_1 x_1 \quad \text{(front)} \quad \text{and} \quad F_2 = -k_2 x_2 \quad \text{(rear).} \tag{19}$$

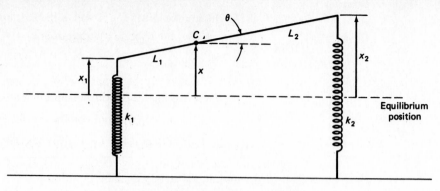

Figure 5.18 Model of the two-axle automobile.

If θ denotes the counterclockwise angular displacement (in radians) of the car body from the horizontal, then

$$x_1 = x - L_1\theta, \qquad x_2 = x + L_2\theta \tag{20}$$

where x is the elevation of the center of mass C. We assume that θ is so small that θ itself is an adequate approximation to $\sin \theta$. If we substitute (20) in (19), then Newton's law of motion yields

$$mx'' = F_1 + F_2 = -k_1(x - L_1\theta) - k_2(x + L_2\theta),$$

so that

$$mx'' = -(k_1 + k_2)x + (k_1L_1 - k_2L_2)\theta. \tag{21}$$

The counterclockwise torques of the forces F_1 and F_2 about C are

$$T_1 = -L_1F_1 \quad \text{and} \quad T_2 = +L_2F_2 \tag{22}$$

because their moment arms are L_1 and L_2. Therefore Newton's law of motion for *angular* acceleration yields

$$I\theta'' = T_1 + T_2 = k_1L_1(x - L_1\theta) - k_2L_2(x + L_2\theta);$$

thus

$$I\theta'' = (k_1L_1 - k_2L_2)x - (k_1L_1{}^2 + k_2L_2{}^2)\theta. \tag{23}$$

The equations in (21) and (23) constitute a linear system in the car's vertical and angular displacements $x(t)$ and $\theta(t)$, respectively.

EXAMPLE 2 Suppose that $m = 75$ slugs (the car weighs 2400 lb), $L_1 = 7$ ft, $L_2 = 3$ ft (it's a rear-engine car), $k_1 = k_2 = 2000$ lb/ft, and $I = 1000$ ft-lb-s^2. Then Eqs. (21) and (23) become

$$75x'' + 4000x - 8000\theta = 0,$$
$$1000\theta'' - 8000x + 116{,}000\theta = 0; \tag{24}$$

that is,

$$(3D^2 + 160)x - 320\theta = 0,$$
$$-8x + (D^2 + 116)\theta = 0. \tag{25}$$

The operational determinant of this system is

$$(3D^2 + 160)(D^2 + 116) - 2560 = 3D^4 + 508D^2 + 16{,}000. \tag{26}$$

Upon solving the characteristic equation

$$3r^4 + 508r^2 + 16{,}000 = 0 \qquad (27)$$

for r^2 and then taking square roots, we get $r \approx \pm(6.4675)i$ and $r \approx \pm(11.2918)i$. Hence the general solution for x is

$$x(t) = a_1 \cos \omega_1 t + b_1 \sin \omega_1 t + a_2 \cos \omega_2 t + b_2 \sin \omega_2 t$$

(and similarly for θ), where the natural frequencies of the system are

$$\omega_1 \approx 6.4675 \text{ rad/s} \approx 1.0293 \text{ Hz,}$$
$$\omega_2 \approx 11.2918 \text{ rad/s} \approx 1.7971 \text{ Hz.} \qquad (28)$$

The natural frequencies in (28) are the *two* frequencies at which we would expect the car to be subject to resonance vibrations. For instance, suppose that the car is driven at a speed of v feet per second along a washboard surface shaped like a sine curve with a wavelength of 40 ft. The result is a periodic force on the car with frequency $\omega = 2\pi v/40 = \pi v/20$. Resonance occurs when either $\omega = \omega_1$ or $\omega = \omega_2$; that is, at either of the *two* critical speeds

$$v_1 = \frac{20\omega_1}{\pi} \approx 41 \text{ ft/s} \approx 28 \text{ mi/h}$$

or

$$v_2 = \frac{20\omega_2}{\pi} \approx 72 \text{ ft/s} \approx 49 \text{ mi/h.}$$

At a speed close to either of these two critical speeds, the amplitudes of the car's oscillations (both vertical and angular) would be quite large—see the table in Fig. 5.19—and the ride most uncomfortable for the passengers.

Car speed v (ft/sec)	x-Amplitude (in.)	θ-Amplitude (deg)
10	1.06	0.02
20	1.28	0.10
30	2.01	0.41
40	15.60	7.29
50	-1.56	-1.80
60	-0.33	-1.87
70	2.14	-8.85
80	-0.88	1.71
90	-0.47	0.67
100	-0.32	0.39
110	-0.24	0.26
120	-0.19	0.19

Figure 5.19 This table shows the amplitudes of the vertical and angular oscillations that result when the car of Example 2 is tested (at various simulated speeds) on a road test simulator.

5.4 Problems

Problems 1–5 deal with the mass-and-spring system of Example 1, but with given damping constants c_1 and c_2 and with given external forces $F_1(t)$ and $F_2(t)$ acting on the two masses. The differential equations for $x_1(t)$ and $x_2(t)$ are then

$$2x_1'' + c_1 x_1' + 100x_1 - 25x_2 = F_1(t),$$
$$\tfrac{1}{2}x_2'' + c_2 x_2' - 25x_1 + 25x_2 = F_2(t).$$

Use the complex method to find a steady periodic solution in each problem. For $i = 1$ and 2, first write $F_i(t)$ in the form $F_i(t) = \mathcal{Re}(C_i e^{10ti})$. Then substitute the trial solution $\tilde{x}_1 = Ae^{10ti}$, $\tilde{x}_2 = Be^{10ti}$. Note that $\sin 10t = \mathcal{Re}(-ie^{10ti})$.

1. $c_1 = c_2 = 0$; $F_1(t) = 50 \cos 10t$, $F_2(t) = 100 \cos 10t$.
2. $c_1 = c_2 = 0$; $F_1(t) = 50 \sin 10t$, $F_2(t) = 100 \cos 10t$.
3. $c_1 = c_2 = 0$; $F_1(t) = 30 \cos 10t + 40 \sin 10t$, $F_2(t) = 0$.
4. $c_1 = c_2 = 2$; $F_1(t) = 100 \sin 10t$, $F_2(t) = 0$.
5. $c_1 = 4, c_2 = 1$; $F_1(t) = 50 \cos 10t$, $F_2(t) = 100 \sin 10t$.

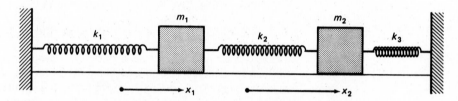

Figure 5.20 Mass-and-spring system for Problems 6–10.

Each of Problems 6–10 deals with the undamped system shown in Fig. 5.20, with given fps values for the masses and spring constants. Find the two natural frequencies of the system and describe its two natural modes of free oscillation.

6. $m_1 = m_2 = 1$; $k_1 = 1, k_2 = 4, k_3 = 1$.
7. $m_1 = 1, m_2 = 2$; $k_1 = 1, k_2 = k_3 = 2$.
8. $m_1 = m_2 = 1$; $k_1 = 1, k_2 = 2, k_3 = 1$.
9. $m_1 = m_2 = 1$; $k_1 = 2, k_2 = 1, k_3 = 2$.
10. $m_1 = 1, m_2 = 2$; $k_1 = 2, k_2 = k_3 = 4$.

Problems 11 and 12 deal with the mass-and-spring system of Fig. 5.20, except that now the masses m_1 and m_2 are also connected to dashpots with constants c_1 and c_2, respectively. With the given values of the mass-spring-dashpot parameters, find $x_1(t)$ and $x_2(t)$ given that the masses are released from rest with $x_1(0) = x_2(0) = 1$.

11. $m_1 = m_2 = 1$; $c_1 = c_2 = 2$; $k_2 = 6, k_1 = k_3 = 5$.
12. $m_1 = m_2 = 1$; $c_1 = c_2 = 2$; $k_2 = 10, k_1 = k_3 = 17$.
13. In the system of Fig. 5.21, assume that $m_1 = 1$ slug, $k_1 = 50$ lb/ft, $k_2 = 10$ lb/ft, $F_0 = 5$ lb, and $\omega = 10$ rad/s. Then find m_2 so that in the resulting steady periodic oscillations, the mass m_1 will remain at rest (!). Thus the effect of the second mass-and-spring pair will be to neutralize the effect of the force on the first mass. This is an example of a *dynamic damper*.

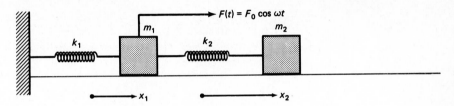

Figure 5.21 The mechanical system of Problem 13.

14. In the system of Fig. 5.22, suppose that $m_1 = m_2 = 1$, $c = 1$, $k_2 = 4$, and $k_1 = k_3 = 9$. Set up the system of differential equations for x_1 and x_2 and find its general solution. Describe the two natural modes of oscillation (one is periodic and the other is exponentially damped).

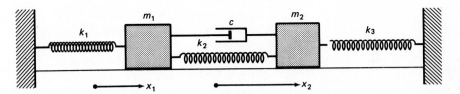

Figure 5.22 The mechanical system of Problem 14.

15. Consider a four-story building that is an elastic frame structure with its mass concentrated at the levels of each of its four *floors*. The first floor is anchored to the ground; the second floor weighs 16 tons; the third and fourth floors weigh 8 tons each. The structure, like a spring, resists horizontal shear between any two adjacent floors—the restoring force is 5 tons per foot of (horizontal) relative displacement. Find the natural frequencies and natural modes of horizontal oscillation of the building. Give the ratios of the amplitudes A, B, and C of oscillations of the second, third, and fourth floors in the form $A : B : C$ with $A = 1$.

16. Suppose that the building of Problem 15 is subjected to an earthquake in which the ground undergoes horizontal sinusoidal oscillations with a period of 3 s and an amplitude of 3 in. Find the amplitudes of the resulting steady periodic oscillations of the upper three floors. Assume the fact that a motion $E \sin \omega t$ of the ground, with acceleration $a = -E\omega^2 \sin \omega t$, produces an opposite inertial force $F = -ma = mE\omega^2 \sin \omega t$ on a floor of mass m.

17. Suppose that $k_1 = k_2 = k$ and $L_1 = L_2 = \frac{1}{2}L$ in Fig. 5.18 (the symmetric situation). Then show that every free oscillation is a combination of a vertical oscillation with frequency $\omega_1 = \sqrt{2k/m}$ and an angular oscillation with frequency $\omega_2 = \sqrt{kL^2/2I}$.

In each of Problems 18–20, the system of Fig. 5.18 is taken as a model for an undamped car with the given parameters in fps units. (a) Find the two natural frequencies of oscillation (in hertz). (b) Assume that this car is driven along a sinusoidal washboard surface with a wavelength of 40 ft. Find the two critical speeds.

18. $m = 100$, $I = 800$, $L_1 = L_2 = 5$, $k_1 = k_2 = 2000$.
19. $m = 100$, $I = 1000$, $L_1 = 6$, $L_2 = 4$, $k_1 = k_2 = 2000$.
20. $m = 100$, $I = 800$, $L_1 = L_2 = 5$, $k_1 = 1000$, $k_2 = 2000$.

5.5
The Eigenvalue Method for Homogeneous Linear Systems

We now introduce a powerful alternative to the method of elimination for constructing the general solution of a *homogeneous* first order linear system with *constant* coefficients,

$$
\begin{aligned}
x_1' &= a_{11}x_1 + a_{12}x_2 + \cdots + a_{1n}x_n, \\
x_2' &= a_{21}x_1 + a_{22}x_2 + \cdots + a_{2n}x_n, \\
&\ \ \vdots \\
x_n' &= a_{n1}x_1 + a_{n2}x_2 + \cdots + a_{nn}x_n.
\end{aligned}
\tag{1}
$$

By Theorem 3 in Section 5.3, we know that it suffices to find n linearly independent solution vectors $x_1, x_2, \ldots, x_n$; the linear combination

$$
x = c_1 x_1 + c_2 x_2 + \cdots + c_n x_n
\tag{2}
$$

with arbitrary coefficients will then be a general solution of the system in (1).

To search for the n needed linearly independent solution vectors, we proceed by analogy with the characteristic root method for solving a single homogeneous linear equation with constant coefficients (Section 2.3). It is reasonable to anticipate solution vectors of the form

$$
x = \begin{pmatrix} x_1 \\ x_2 \\ x_3 \\ \vdots \\ x_n \end{pmatrix} = \begin{pmatrix} v_1 e^{\lambda t} \\ v_2 e^{\lambda t} \\ v_3 e^{\lambda t} \\ \vdots \\ v_n e^{\lambda t} \end{pmatrix} = \begin{pmatrix} v_1 \\ v_2 \\ v_3 \\ \vdots \\ v_n \end{pmatrix} e^{\lambda t} = v e^{\lambda t}
\tag{3}
$$

where $\lambda, v_1, v_2, \ldots, v_n$ are constants. For if we substitute $x_i = v_i e^{\lambda t}$ $(i = 1, 2, \ldots, n)$ in (1), then the factor $e^{\lambda t}$ will cancel throughout, leaving us with n linear equations, which—for appropriate values of λ—we can hope to solve for the coefficients $v_1, v_2, \ldots, v_n$.

To investigate this possibility, it is more efficient to write the system in (1) in the matrix form

$$
x' = Ax
\tag{4}
$$

where $A = (a_{ij})$. When we substitute the trial solution $x = v e^{\lambda t}$ with derivative $x' = \lambda v e^{\lambda t}$ in Eq. (4), the result is

$$
\lambda v e^{\lambda t} = A v e^{\lambda t}.
$$

We cancel the nonzero scalar factor $e^{\lambda t}$ to get

$$
A v = \lambda v.
\tag{5}
$$

This means that $x = v e^{\lambda t}$ will be a nontrivial solution of (4) provided that v is a nonzero vector and λ is a constant such that (5) holds; that is, the matrix product Av is a scalar multiple of the vector v. The question now is this: How do we find v and λ?

To answer this, we rewrite (5) in the form

$$(\mathbf{A} - \lambda\mathbf{I})\mathbf{v} = \mathbf{0}. \tag{6}$$

Given λ, this is a system of n homogeneous linear equations in the unknowns $v_1, v_2, \ldots, v_n$. By a standard theorem of linear algebra, it has a nontrivial solution if and only if the determinant of its coefficient matrix vanishes; that is, if and only if

$$|\mathbf{A} - \lambda\mathbf{I}| = \det(\mathbf{A} - \lambda\mathbf{I}) = 0. \tag{7}$$

In its simplest formulation, the **eigenvalue method** for solving the system $\mathbf{x}' = \mathbf{A}\mathbf{x}$ consists of finding λ so that Eq. (7) holds and next solving (6) with this value of λ to obtain $v_1, v_2, \ldots, v_n$. Then $\mathbf{x} = \mathbf{v}e^{\lambda t}$ will be a solution vector. The name of the method comes from the following definition.

DEFINITION: EIGENVALUES AND EIGENVECTORS

*The number λ (either zero or nonzero) is called an **eigenvalue** of the $n \times n$ matrix $\mathbf{A}$ provided that*

$$|\mathbf{A} - \lambda\mathbf{I}| = 0. \tag{7}$$

*An **eigenvector** associated with the eigenvalue λ is a nonzero vector $\mathbf{v}$ such that $\mathbf{A}\mathbf{v} = \lambda\mathbf{v}$, so that*

$$(\mathbf{A} - \lambda\mathbf{I})\mathbf{v} = \mathbf{0}. \tag{6}$$

Note that if $\mathbf{v}$ is an eigenvector associated with the eigenvalue λ, then so is any nonzero constant scalar multiple $c\mathbf{v}$ of $\mathbf{v}$—this follows upon multiplication of each side in Eq. (6) by $c \neq 0$.

The prefix *eigen* is a German word with the approximate translation *characteristic* in this context; the terms *characteristic value* and *characteristic vector* are in common use. For this reason, the equation

$$|\mathbf{A} - \lambda\mathbf{I}| = \begin{vmatrix} a_{11} - \lambda & a_{12} & \cdots & a_{1n} \\ a_{21} & a_{22} - \lambda & \cdots & a_{2n} \\ \cdot & \cdot & & \cdot \\ \cdot & \cdot & & \cdot \\ \cdot & \cdot & & \cdot \\ a_{n1} & a_{n2} & \cdots & a_{nn} - \lambda \end{vmatrix} = 0 \tag{8}$$

is called the **characteristic equation** of the matrix $\mathbf{A}$; its roots are the eigenvalues of $\mathbf{A}$. Upon expanding the determinant in (8), we evidently get an nth degree polynomial of the form

$$(-1)^n\lambda^n + b_{n-1}\lambda^{n-1} + \cdots + b_1\lambda + b_0 = 0. \tag{9}$$

By the fundamental theorem of algebra, this equation has n roots—possibly some are complex, possibly some are repeated—and thus an $n \times n$ matrix has n eigenvalues (counting repetitions, if any). Although we assume that the elements of $\mathbf{A}$ are real numbers, we allow the possibility of complex eigenvalues and complex-valued eigenvectors.

In outline, the eigenvalue method for solving the system $\mathbf{x}' = \mathbf{A}\mathbf{x}$ goes as follows. We first solve the characteristic equation in (8) for the eigenvalues

$\lambda_1, \lambda_2, \ldots, \lambda_n$ of the matrix $\mathbf{A}$. Next we attempt to find n linearly independent eigenvectors $\mathbf{v}_1, \mathbf{v}_2, \ldots, \mathbf{v}_n$ associated with these eigenvalues. This will not always be possible, but when it is, we get n linearly independent solutions

$$\mathbf{x}_1(t) = \mathbf{v}_1 e^{\lambda_1 t}, \ \mathbf{x}_2 = \mathbf{v}_2 e^{\lambda_2 t}, \ldots, \mathbf{x}_n = \mathbf{v}_n e^{\lambda_n t}, \tag{10}$$

and the general solution is then a linear combination of these n solutions. We will discuss the various cases that can occur separately, depending upon whether the eigenvalues are distinct or repeated, real or complex.

DISTINCT REAL EIGENVALUES

If the eigenvalues $\lambda_1, \lambda_2, \ldots, \lambda_n$ are real and distinct, then we substitute each of them in turn in (6) and solve for the associated eigenvectors $\mathbf{v}_1, \mathbf{v}_2, \ldots, \mathbf{v}_n$. In this case it can be proved that the solution vectors given in (10) are always linearly independent. In any particular example such linear independence can always be verified by using the Wronskian determinant of Section 5.3. The following example illustrates the procedure.

EXAMPLE 1 Find a general solution of the system

$$\begin{aligned} x_1' &= 4x_1 + 2x_2, \\ x_2' &= 3x_1 - x_2. \end{aligned} \tag{11}$$

Solution The matrix form of the system in (11) is

$$\mathbf{x}' = \begin{pmatrix} 4 & 2 \\ 3 & -1 \end{pmatrix} \mathbf{x}. \tag{12}$$

The characteristic equation of the coefficient matrix is

$$\begin{aligned} \begin{vmatrix} 4 - \lambda & 2 \\ 3 & -1 - \lambda \end{vmatrix} &= (4 - \lambda)(-1 - \lambda) - 6 \\ &= \lambda^2 - 3\lambda - 10 \\ &= (\lambda + 2)(\lambda - 5) = 0, \end{aligned}$$

so we have the distinct real eigenvalues $\lambda_1 = -2$ and $\lambda_2 = 5$.

For the coefficient matrix $\mathbf{A}$ in (12) the eigenvector equation $(\mathbf{A} - \lambda\mathbf{I})\mathbf{v} = \mathbf{0}$ takes the form

$$\begin{pmatrix} 4 - \lambda & 2 \\ 3 & -1 - \lambda \end{pmatrix} \begin{pmatrix} v_1 \\ v_2 \end{pmatrix} = \begin{pmatrix} 0 \\ 0 \end{pmatrix}. \tag{13}$$

When we substitute the first eigenvalue $\lambda_1 = -2$, we find that the two scalar equations that follow from (13) are

$$6v_1 + 2v_2 = 0 \quad \text{and} \quad 3v_1 + v_2 = 0.$$

These equations are equivalent; we can choose v_1 arbitrary (but nonzero) and solve for v_2. The simplest choice is $v_1 = 1$, which yields $v_2 = -3$, and thus

$$\mathbf{v}_1 = \begin{pmatrix} 1 \\ -3 \end{pmatrix}$$

is an eigenvector associated with $\lambda_1 = -2$ (as is any nonzero constant multiple of v_1).

Remark: If, instead of the "simplest" choice $v_1 = 1$, $v_2 = -3$, we had made another choice, $v_1 = c$, $v_2 = -3c$, we would have obtained the eigenvector

$$\mathbf{v}_1 = \begin{pmatrix} c \\ -3c \end{pmatrix} = c\begin{pmatrix} 1 \\ -3 \end{pmatrix}.$$

Because this is a constant multiple of our previous result, any choice we make leads to the same solution

$$\mathbf{x}_1(t) = \begin{pmatrix} 1 \\ -3 \end{pmatrix}e^{-2t}.$$

Next we substitute in (13) the second eigenvalue $\lambda_2 = 5$ and get the equivalent scalar equations

$$-v_1 + 2v_2 = 0 \quad \text{and} \quad 3v_1 - 6v_2 = 0.$$

With $v_2 = 1$ we obtain $v_1 = 2$, so

$$\mathbf{v}_2 = \begin{pmatrix} 2 \\ 1 \end{pmatrix}$$

is an eigenvector associated with $\lambda_2 = 5$. A different choice $v_1 = 2c$, $v_2 = c$ would merely give a (constant) multiple of $\mathbf{v}_2$.

These two eigenvalues and associated eigenvectors yield the two solutions

$$\mathbf{x}_1 = \begin{pmatrix} 1 \\ -3 \end{pmatrix}e^{-2t} \quad \text{and} \quad \mathbf{x}_2 = \begin{pmatrix} 2 \\ 1 \end{pmatrix}e^{5t}.$$

They are linearly independent because their Wronskian

$$\begin{vmatrix} e^{-2t} & 2e^{5t} \\ -3e^{-2t} & e^{5t} \end{vmatrix} = 7e^{3t}$$

is nonzero. Hence a general solution of the system in (11) is

$$\mathbf{x} = c_1\mathbf{x}_1 + c_2\mathbf{x}_2 = c_1\begin{pmatrix} 1 \\ -3 \end{pmatrix}e^{-2t} + c_2\begin{pmatrix} 2 \\ 1 \end{pmatrix}e^{5t};$$

in scalar form,

$$x_1 = c_1e^{-2t} + 2c_2e^{5t}, \qquad x_2 = -3c_1e^{-2t} + c_2e^{5t}.$$

COMPLEX EIGENVALUES

Even if some of the eigenvalues are complex, so long as they are distinct the method described above still yields n linearly independent solutions. The only complication is that the eigenvectors associated with complex eigenvalues are ordinarily complex-valued, so we will have complex-valued solutions.

To obtain real-valued solutions, we note that—because we are assuming that the matrix $\mathbf{A}$ has only real entries—the coefficients in the characteristic equation in (18) all will be real. Consequently any complex eigenvalues must appear in complex conjugate pairs. Suppose then that $\lambda = p + qi$ and $\bar{\lambda} =$

$p - qi$ are such a pair of eigenvalues. If $\mathbf{v}$ is an eigenvector associated with λ, so that

$$(\mathbf{A} - \lambda\mathbf{I})\mathbf{v} = \mathbf{0},$$

then taking complex conjugates in this equation yields

$$(\mathbf{A} - \bar{\lambda}\mathbf{I})\bar{\mathbf{v}} = \mathbf{0}.$$

Thus the conjugate $\bar{\mathbf{v}}$ of $\mathbf{v}$ is an eigenvector associated with $\bar{\lambda}$. Of course the conjugate of a vector is defined componentwise; if

$$\mathbf{v} = \begin{pmatrix} a_1 + b_1 i \\ a_2 + b_2 i \\ \cdot \\ \cdot \\ \cdot \\ a_n + b_n i \end{pmatrix} = \begin{pmatrix} a_1 \\ a_2 \\ \cdot \\ \cdot \\ \cdot \\ a_n \end{pmatrix} + \begin{pmatrix} b_1 \\ b_2 \\ \cdot \\ \cdot \\ \cdot \\ b_n \end{pmatrix} i = \mathbf{a} + \mathbf{b}i \tag{14}$$

then $\bar{\mathbf{v}} = \mathbf{a} - \mathbf{b}i$. The complex-valued solution associated with λ and $\mathbf{v}$ is then

$$\mathbf{x}(t) = \mathbf{v}e^{\lambda t} = \mathbf{v}e^{(p+qi)t}$$

$$= (\mathbf{a} + \mathbf{b}i)e^{pt}(\cos qt + i \sin qt);$$

that is,

$$\mathbf{x}(t) = e^{pt}(\mathbf{a} \cos qt - \mathbf{b} \sin qt) + ie^{pt}(\mathbf{b} \cos qt + \mathbf{a} \sin qt). \tag{15}$$

Because the real and imaginary parts of a complex-valued solution are also solutions, we thus get the two *real-valued* solutions

$$\begin{aligned} \mathbf{x}_1(t) &= \mathcal{R}e(\mathbf{x}(t)) = e^{pt}(\mathbf{a} \cos qt - \mathbf{b} \sin qt), \\ \mathbf{x}_2(t) &= \mathcal{I}m(\mathbf{x}(t)) = e^{pt}(\mathbf{b} \cos qt + \mathbf{a} \sin qt) \end{aligned} \tag{16}$$

associated with the complex conjugate eigenvalues $p \pm qi$; it is easy to check that the same two real-valued solutions result from taking real and imaginary parts of $\bar{\mathbf{v}}e^{\bar{\lambda}t}$. Rather than memorizing the formulas in (16), it is preferable in a specific example to find the complex-valued solution explicitly, and then compute its real and imaginary parts.

EXAMPLE 2 Find a general solution of the system

$$\begin{aligned} x_1' &= 3x_1 &&+ x_3, \\ x_2' &= 9x_1 - x_2 + 2x_3, \\ x_3' &= -9x_1 + 4x_2 - x_3. \end{aligned} \tag{17}$$

Solution The characteristic equation of the coefficient matrix in (17) is

$$\begin{vmatrix} 3 - \lambda & 0 & 1 \\ 9 & -1 - \lambda & 2 \\ -9 & 4 & -1 - \lambda \end{vmatrix}$$

$$= (3 - \lambda)[(-1 - \lambda)^2 - 8] + (1)[36 - (-9)(-1 - \lambda)]$$

$$= (3 - \lambda)(\lambda + 1)^2 - 24 + 8\lambda + 36 - 9 - 9\lambda$$

$$= (3 - \lambda)(\lambda + 1)^2 + 3 - \lambda$$

$$= (3 - \lambda)[(\lambda + 1)^2 + 1] = 0.$$

Thus we have the real eigenvalue $\lambda_1 = 3$ and the conjugate pair $\lambda =$

$-1 + i, \bar{\lambda} = -1 - i$ of complex eigenvalues. If we had expanded the determinant in a more slapdash manner, we might have obtained the cubic equation in a form somewhat more difficult to solve.

The eigenvectors will be solutions of

$$
\begin{pmatrix} 3 - \lambda & 0 & 1 \\ 9 & -1 - \lambda & 2 \\ -9 & 4 & -1 - \lambda \end{pmatrix} \begin{pmatrix} v_1 \\ v_2 \\ v_3 \end{pmatrix} = \begin{pmatrix} 0 \\ 0 \\ 0 \end{pmatrix}. \tag{18}
$$

With the real eigenvalue $\lambda_1 = 3$, (18) reduces to the three equations

$$v_3 = 0,$$
$$9v_1 - 4v_2 + 2v_3 = 0,$$
$$-9v_1 + 4v_2 - 4v_3 = 0.$$

Because $v_3 = 0$, we may choose $v_1 = 4$ and $v_2 = 9$ to satisfy the last two equations. This gives the eigenvector and corresponding solution

$$
\mathbf{v}_1 = \begin{pmatrix} 4 \\ 9 \\ 0 \end{pmatrix}, \qquad \mathbf{x}_1 = \mathbf{v}_1 e^{3t} = \begin{pmatrix} 4 \\ 9 \\ 0 \end{pmatrix} e^{3t}. \tag{19}
$$

As in Example 1, a different choice $v_1 = 4c$, $v_2 = 9c$ (such as $v_1 = \frac{1}{9}$, $v_2 = \frac{1}{4}$ with $c = \frac{1}{36}$) would merely yield a constant multiple of $\mathbf{v}_1$ and hence would lead to the same solution $\mathbf{x}_1(t)$.

With the complex eigenvalue $\lambda = -1 + i$, Eq. (18) is equivalent to the three simultaneous equations

$$(4 - i)v_1 \qquad\quad + v_3 = 0,$$
$$9v_1 - iv_2 + 2v_3 = 0,$$
$$-9v_1 - 4v_2 - iv_3 = 0.$$

The choice $v_1 = 1$ in the first equation gives $v_3 = -4 + i$. Then the third equation gives $v_2 = \frac{1}{4}(9v_1 + iv_3) = 2 - i$. The corresponding complex-valued solution is

$$
\mathbf{x}(t) = \begin{pmatrix} 1 \\ 2 - i \\ -4 + i \end{pmatrix} e^{(-1+i)t}
$$

$$
= \left[\begin{pmatrix} 1 \\ 2 \\ -4 \end{pmatrix} + i \begin{pmatrix} 0 \\ -1 \\ 1 \end{pmatrix} \right] e^{-t}(\cos t + i \sin t),
$$

and therefore

$$
\mathbf{x}(t) = e^{-t}\left[\begin{pmatrix} 1 \\ 2 \\ -4 \end{pmatrix} \cos t - \begin{pmatrix} 0 \\ -1 \\ 1 \end{pmatrix} \sin t \right]
$$

$$
+ ie^{-t}\left[\begin{pmatrix} 0 \\ -1 \\ 1 \end{pmatrix} \cos t + \begin{pmatrix} 1 \\ 2 \\ -4 \end{pmatrix} \sin t \right].
$$

The real and imaginary parts of $\mathbf{x}$ are the real-valued solutions

$$\mathbf{x}_2 = \mathcal{Re}(\mathbf{x}) = e^{-t} \begin{pmatrix} \cos t \\ 2\cos t + \sin t \\ -4\cos t - \sin t \end{pmatrix},$$

$$\mathbf{x}_3 = \mathcal{Im}(\mathbf{x}) = e^{-t} \begin{pmatrix} \sin t \\ -\cos t + 2\sin t \\ \cos t - 4\sin t \end{pmatrix}.$$

(20)

By computing the Wronskian, it can be verified that the three solution vectors $\mathbf{x}_1$, $\mathbf{x}_2$, and $\mathbf{x}_3$ in (19) and (20) are linearly independent. Hence, putting it all together, we finally obtain the desired general solution of the system in (17):

$$\begin{aligned} \mathbf{x} &= c_1\mathbf{x}_1 + c_2\mathbf{x}_2 + c_3\mathbf{x}_3 \\ &= \begin{pmatrix} 4c_1e^{3t} + e^{-t}[c_2\cos t + c_3\sin t] \\ 9c_1e^{3t} + e^{-t}[(2c_2 - c_3)\cos t + (c_2 + 2c_3)\sin t] \\ e^{-t}[(-4c_2 + c_3)\cos t + (-c_2 - 4c_3)\sin t] \end{pmatrix}. \end{aligned}$$

REPEATED EIGENVALUES

Suppose that the $n \times n$ matrix $\mathbf{A}$ has $m < n$ distinct eigenvalues $\lambda_1, \lambda_2, \ldots, \lambda_n$. Then at least one of these eigenvalues is a repeated root of the characteristic equation

$$|\mathbf{A} - \lambda\mathbf{I}| = 0. \tag{7}$$

An eigenvalue is of **multiplicity** k if it is a k-fold root of Eq. (7). For each eigenvalue λ_i, the eigenvector equation

$$(\mathbf{A} - \lambda_i\mathbf{I})\mathbf{v} = \mathbf{0} \tag{21}$$

has at least one nontrivial solution, so there is at least one eigenvector associated with λ_i. But if some eigenvalue λ_i is of multiplicity $k > 1$, and Eq. (21) has *fewer* than k linearly independent solutions—so there are fewer than k linearly independent eigenvectors associated with λ_i—then the eigenvalue method as described thusfar will produce *fewer* than the needed n linearly independent solutions of the system $\mathbf{x}' = \mathbf{A}\mathbf{x}$.

We call an eigenvalue of multiplicity k **complete** if it has k linearly independent associated eigenvectors and **incomplete** if there are fewer than k. There are then two cases that we must consider:

1. Every eigenvalue of $\mathbf{A}$ is complete;
2. Some eigenvalue of $\mathbf{A}$ is incomplete.

In case 1, the matrix $\mathbf{A}$ has n linearly independent eigenvectors $\mathbf{v}_1, \mathbf{v}_2, \ldots, \mathbf{v}_n$ associated with the eigenvalues $\lambda_1, \lambda_2, \ldots, \lambda_n$ (each repeated with its multiplicity), and the general solution of $\mathbf{x}' = \mathbf{A}\mathbf{x}$ is

$$\mathbf{x} = c_1\mathbf{v}_1e^{\lambda_1 t} + c_2\mathbf{v}_2e^{\lambda_2 t} + \cdots + c_n\mathbf{v}_ne^{\lambda_n t}. \tag{22}$$

This case always occurs when the matrix $\mathbf{A}$ is **symmetric**—$\mathbf{A} = \mathbf{A}^T$. Indeed, it is proved in most linear algebra texts that each eigenvalue of a symmetric (real) matrix is both real and complete. The following example illustrates the approach in case 1.

EXAMPLE 3 Find a general solution of the system

$$
\begin{aligned}
x_1' &= && 2x_2 + 2x_3, \\
x_2' &= 2x_1 && + 2x_3, \\
x_3' &= 2x_1 + 2x_2.
\end{aligned}
\tag{23}
$$

Solution The characteristic equation is

$$
0 = \begin{vmatrix} -\lambda & 2 & 2 \\ 2 & -\lambda & 2 \\ 2 & 2 & -\lambda \end{vmatrix}
$$

$$
\begin{aligned}
&= -\lambda(\lambda^2 - 4) - 2(-2\lambda - 4) + 2(4 + 2\lambda) \\
&= -\lambda(\lambda + 2)(\lambda - 2) + 8\lambda + 16 \\
&= (\lambda + 2)(-\lambda^2 + 2\lambda + 8);
\end{aligned}
$$

that is,

$$
-(\lambda + 2)(\lambda + 2)(\lambda - 4) = 0.
$$

Thus we have the unrepeated eigenvalue $\lambda_1 = 4$ and the eigenvalue $\lambda_2 = -2$ of multiplicity 2.

When we substitute $\lambda_1 = 4$ in the eigenvector equation

$$
\begin{pmatrix} -\lambda & 2 & 2 \\ 2 & -\lambda & 2 \\ 2 & 2 & -\lambda \end{pmatrix} \begin{pmatrix} v_1 \\ v_2 \\ v_3 \end{pmatrix} = \begin{pmatrix} 0 \\ 0 \\ 0 \end{pmatrix},
\tag{24}
$$

we obtain three scalar equations that are readily solved (to within a constant multiple) for the eigenvector

$$
\mathbf{v}_1 = \begin{pmatrix} 1 \\ 1 \\ 1 \end{pmatrix}
$$

associated with $\lambda_1 = 4$.

When we substitute the repeated eigenvalue $\lambda_2 = -2$ in (24), the three scalar equations reduce to the single equation

$$
2v_1 + 2v_2 + 2v_3 = 0.
$$

We can choose v_1 and v_2 arbitrarily and solve for v_3. With $v_1 = 1$ and $v_2 = 0$, we get $v_3 = -1$, which we also get with $v_1 = 0$ and $v_2 = 1$. Thus we find the *two* linearly independent eigenvectors

$$
\mathbf{v}_2 = \begin{pmatrix} 1 \\ 0 \\ -1 \end{pmatrix} \quad \text{and} \quad \mathbf{v}_3 = \begin{pmatrix} 0 \\ 1 \\ -1 \end{pmatrix}
$$

associated with $\lambda_2 = -2$.

The three eigenvectors we have found yield the general solution

$$\mathbf{x}(t) = c_1 \begin{pmatrix} 1 \\ 1 \\ 1 \end{pmatrix} e^{4t} + c_2 \begin{pmatrix} 1 \\ 0 \\ -1 \end{pmatrix} e^{-2t} + c_3 \begin{pmatrix} 0 \\ 1 \\ -1 \end{pmatrix} e^{-2t}, \tag{25}$$

because the Wronskian can be used to verify that the three solutions of (23) displayed in (25) are indeed linearly independent.

Remark: The "arbitrariness" in Example 3 is a bit more interesting than that in Examples 1 and 2. Suppose that, after finding the first two eigenvectors $\mathbf{v}_1$ and $\mathbf{v}_2$ as above, we had chosen $v_1 = a$ and $v_2 = b$ (not both zero) in the equation $2v_1 + 2v_2 + 2v_3 = 0$, so that $v_3 = -a - b$. Then our third eigenvector would have been

$$\mathbf{v} = \begin{pmatrix} a \\ b \\ -a - b \end{pmatrix} = a \begin{pmatrix} 1 \\ 0 \\ -1 \end{pmatrix} + b \begin{pmatrix} 0 \\ 1 \\ -1 \end{pmatrix}.$$

It would follow that $\mathbf{v} = a\mathbf{v}_2 + b\mathbf{v}_3$. Because $\mathbf{v}$ is a linear combination of $\mathbf{v}_2$ and $\mathbf{v}_3$, our new general solution

$$\mathbf{x}(t) = c_1\mathbf{v}_1 e^{4t} + c_2\mathbf{v}_2 e^{-2t} + c\mathbf{v}e^{-2t}$$

would be equivalent to the one in (25). Thus we need not worry about making the "right" choice; any choice will do.

In case 2, we need to tell how to find the missing solutions corresponding to an incomplete eigenvalue λ of multiplicity $k > 1$. We will discuss in detail only the case $k = 2$. Suppose we have found that there is only one eigenvector $\mathbf{v}$ associated with λ, so we have the single solution $\mathbf{v}e^{\lambda t}$. By analogy with the case of a repeated characteristic root for a single differential equation (Section 2.3), we might expect to find a second solution of the form

$$\mathbf{x} = \mathbf{w}te^{\lambda t}. \tag{26}$$

But when we substitute (26) in $\mathbf{x}' = \mathbf{A}\mathbf{x}$, we get the equation

$$\mathbf{w}e^{\lambda t} + \mathbf{w}te^{\lambda t} = \mathbf{A}\mathbf{w}te^{\lambda t}. \tag{27}$$

Because the coefficients of both $e^{\lambda t}$ and $te^{\lambda t}$ must balance, it follows that $\mathbf{w} = \mathbf{0}$. Thus there is *no* nontrivial solution of the form in (26).

Still trying to make this idea work, we turn to the next most plausible possibility: a second solution of the form

$$\mathbf{x} = \mathbf{v}te^{\lambda t} + \mathbf{w}e^{\lambda t}. \tag{28}$$

When we substitute (28) in $\mathbf{x}' = \mathbf{A}\mathbf{x}$, we get the equation

$$\mathbf{v}e^{\lambda t} + \lambda\mathbf{v}te^{\lambda t} + \lambda\mathbf{w}e^{\lambda t} = \mathbf{A}\mathbf{v}te^{\lambda t} + \mathbf{A}\mathbf{w}e^{\lambda t}. \tag{29}$$

We equate coefficients of $te^{\lambda t}$ and of $e^{\lambda t}$ in (29); thus we obtain the two equations

$$(\mathbf{A} - \lambda\mathbf{I})\mathbf{v} = \mathbf{0} \tag{30}$$

and

$$(\mathbf{A} - \lambda\mathbf{I})\mathbf{w} = \mathbf{v} \tag{31}$$

that the vectors **v** and **w** must satisfy in order for (28) to be a solution. (Note that Eq. (30) merely means that **v** is an eigenvector associated with λ.) Having found **v**, we must then attempt to solve (31) for **w**. Despite the initially discouraging fact that $|\mathbf{A} - \lambda\mathbf{I}| = 0$, it turns out that a nontrivial solution **w** of Eq. (31) can always be found, as in the following example.

EXAMPLE 4 Find a general solution of the system

$$
\begin{aligned}
x_1' &= x_1 - 3x_2, \\
x_2' &= 3x_1 + 7x_2.
\end{aligned}
\tag{32}
$$

Solution The characteristic equation is

$$
0 = \begin{vmatrix} 1 - \lambda & -3 \\ 3 & 7 - \lambda \end{vmatrix} = (1 - \lambda)(7 - \lambda) + 9,
$$

so that $0 = \lambda^2 - 8\lambda + 16 = (\lambda - 4)^2$. So we have the single eigenvalue $\lambda_1 = 4$ of multiplicity 2. When we substitute $\lambda_1 = 4$ in the eigenvector equation

$$
\begin{pmatrix} 1 - \lambda & -3 \\ 3 & 7 - \lambda \end{pmatrix}\begin{pmatrix} v_1 \\ v_2 \end{pmatrix} = \begin{pmatrix} 0 \\ 0 \end{pmatrix},
$$

we obtain the equivalent scalar equations

$$
-3v_1 - 3v_2 = 0 = 3v_1 + 3v_2.
$$

We choose $v_1 = 1$ and $v_2 = -1$; this yields the eigenvector and solution

$$
\mathbf{v} = \begin{pmatrix} 1 \\ -1 \end{pmatrix} \quad \text{and} \quad \mathbf{x}_1 = \begin{pmatrix} 1 \\ -1 \end{pmatrix}e^{4t}.
\tag{33}
$$

With this eigenvector **v** and eigenvalue $\lambda = 4$, Eq. (31) takes the form

$$
\begin{pmatrix} -3 & -3 \\ 3 & 3 \end{pmatrix}\begin{pmatrix} w_1 \\ w_2 \end{pmatrix} = \begin{pmatrix} 1 \\ -1 \end{pmatrix},
$$

so that $3w_1 + 3w_2 = -1$. A nontrivial solution is $w_1 = 0$, $w_2 = -\frac{1}{3}$. Then the second solution given by (28) is

$$
\begin{aligned}
\mathbf{x}_2(t) &= \mathbf{v}te^{4t} + \mathbf{w}e^{4t} \\
&= \begin{pmatrix} 1 \\ -1 \end{pmatrix}te^{4t} + \begin{pmatrix} 0 \\ -\frac{1}{3} \end{pmatrix}e^{4t} = \begin{pmatrix} t \\ -t - \frac{1}{3} \end{pmatrix}e^{4t}.
\end{aligned}
\tag{34}
$$

The Wronskian of the two solutions $\mathbf{x}_1$ and $\mathbf{x}_2$ is

$$
W = \begin{vmatrix} e^{4t} & te^{4t} \\ -e^{4t} & -(t + \frac{1}{3})e^{4t} \end{vmatrix} = -\tfrac{1}{3}e^{8t} \neq 0,
$$

so $\mathbf{x}_1$ and $\mathbf{x}_2$ are linearly independent. Therefore a general solution of (32) is

$$
\begin{aligned}
\mathbf{x} = c_1\mathbf{x}_1 + c_2\mathbf{x}_2 &= c_1\begin{pmatrix} 1 \\ -1 \end{pmatrix}e^{4t} + c_2\begin{pmatrix} t \\ -t - \frac{1}{3} \end{pmatrix}e^{4t} \\
&= \begin{pmatrix} c_1 + c_2 t \\ -c_1 - c_2(t + \frac{1}{3}) \end{pmatrix}e^{4t}.
\end{aligned}
\tag{35}
$$

The situation is more complicated in the case of an incomplete eigenvalue λ of multiplicity $k > 2$. If $k = 3$, for instance, there may be either one or two linearly independent eigenvectors associated with λ, and thus either two or one missing solutions to be found. A general discussion of systems having incomplete eigenvalues of higher multiplicity is beyond the scope of our discussion, but some of the possibilities are illustrated in the problems that follow.

5.5 Problems

In each of Problems 1–20, apply the eigenvalue method of this section to find a general solution of the given system. If initial values are given, find also the corresponding particular solution.

1. $x_1' = x_1 + 2x_2, x_2' = 2x_1 + x_2.$

2. $x_1' = 2x_1 + 3x_2, x_2' = 2x_1 + x_2.$

3. $x_1' = 3x_1 + 4x_2, x_2' = 3x_1 + 2x_2; x_1(0) = x_2(0) = 1.$

4. $x_1' = 4x_1 + x_2, x_2' = 6x_1 - x_2.$

5. $x_1' = 6x_1 - 7x_2, x_2' = x_1 - 2x_2.$

6. $x_1' = 9x_1 + 5x_2, x_2' = -6x_1 - 2x_2; x_1(0) = 1, x_2(0) = 0.$

7. $x_1' = -3x_1 + 4x_2, x_2' = 6x_1 - 5x_2.$

8. $x_1' = x_1 - 5x_2, x_2' = x_1 - x_2.$

9. $x_1' = 2x_1 - 5x_2, x_2' = 4x_1 - 2x_2; x_1(0) = 2, x_2(0) = 3.$

10. $x_1' = -3x_1 - 2x_2, x_2' = 9x_1 + 3x_2.$

11. $x_1' = x_1 - 2x_2, x_2' = 2x_1 + x_2; x_1(0) = 0, x_2(0) = 4.$

12. $x_1' = x_1 - 5x_2, x_2' = x_1 + 3x_2.$

13. $x_1' = 5x_1 - 9x_2, x_2' = 2x_1 - x_2.$

14. $x_1' = 3x_1 - 4x_2, x_2' = 4x_1 + 3x_2.$

15. $x_1' = 7x_1 - 5x_2, x_2' = 4x_1 + 3x_2.$

16. $x_1' = 3x_1 - x_2, x_2' = x_1 + x_2.$

17. $x_1' = x_1 - 2x_2, x_2' = 2x_1 + 5x_2.$

18. $x_1' = 3x_1 - x_2, x_2' = x_1 + 5x_2.$

19. $x_1' = 7x_1 + x_2, x_2' = -4x_1 + 3x_2.$

20. $x_1' = x_1 - 4x_2, x_2' = 4x_1 + 9x_2.$

In each of Problems 21–25, the coefficient matrix has three distinct eigenvalues, which can be found by inspection and factoring. Find general solutions of the systems in these problems.

21. $\begin{aligned} x_1' &= 5x_1 & - 6x_3, \\ x_2' &= 2x_1 - x_2 - 2x_3, \\ x_3' &= 4x_1 - 2x_2 - 4x_3. \end{aligned}$

22. $\begin{aligned} x_1' &= 3x_1 + 2x_2 + 2x_3, \\ x_2' &= -5x_1 - 4x_2 - 2x_3, \\ x_3' &= 5x_1 + 5x_2 + 3x_3. \end{aligned}$

23. $\begin{aligned} x_1' &= 3x_1 + x_2 + x_3, \\ x_2' &= -5x_1 - 3x_2 - x_3, \\ x_3' &= 5x_1 + 5x_2 + 3x_3. \end{aligned}$

24. $\begin{aligned} x_1' &= 2x_1 + x_2 - x_3, \\ x_2' &= -4x_1 - 3x_2 - x_3, \\ x_3' &= 4x_1 + 4x_2 + 2x_3. \end{aligned}$

25. $\begin{aligned} x_1' &= 5x_1 + 5x_2 + 2x_3, \\ x_2' &= -6x_1 - 6x_2 - 5x_3, \\ x_3' &= 6x_1 + 6x_2 + 5x_3. \end{aligned}$

26. Show that the coefficient matrix of the system

$$
\begin{aligned}
x_1' &= 2x_1 + 3x_2 + 3x_3, \\
x_2' &= \quad\;\; - x_2 - 3x_3, \\
x_3' &= \qquad\qquad\quad 2x_3
\end{aligned}
$$

has the eigenvalue $\lambda = 2$ of multiplicity 2. Find two linearly independent eigenvectors associated with $\lambda = 2$, and then find a general solution of the system.

27. Show that the coefficient matrix of the system

$$
\begin{aligned}
x_1' &= \quad 3x_1 + 2x_2 + \;\; x_3, \\
x_2' &= -x_1 \qquad\qquad - x_3, \\
x_3' &= \quad x_1 + \;\; x_2 + 2x_3
\end{aligned}
$$

has the eigenvalue $\lambda = 2$ of multiplicity 2, but with only one linearly independent eigenvector associated with it. Use the method of Example 4 to find a general solution of this system.

28. (a) Show that the coefficient matrix $\mathbf{A}$ of the system

$$
\begin{aligned}
x_1' &= \quad 3x_1 + \;\; x_2, \\
x_2' &= -x_1 \qquad\quad - x_3, \\
x_3' &= \quad x_1 + 2x_2 + 3x_3
\end{aligned}
$$

has the single eigenvalue $\lambda = 2$ of multiplicity 3, but with only one linearly independent eigenvector $\mathbf{u}$ associated with it. Then one solution is $\mathbf{x}_1(t) = \mathbf{u}e^{2t}$.
(b) Find a second solution of the form $\mathbf{x}_2(t) = \mathbf{u}te^{2t} + \mathbf{v}e^{2t}$ where $\mathbf{v}$ is a nontrivial solution of $(\mathbf{A} - 2\mathbf{I})\mathbf{v} = \mathbf{u}$. (c) Find a third solution of the form

$$
\mathbf{x}_3(t) = \tfrac{1}{2}\mathbf{u}t^2 e^{2t} + \mathbf{v}te^{2t} + \mathbf{w}e^{2t}
$$

where $\mathbf{w}$ is a nontrivial solution of $(\mathbf{A} - 2\mathbf{I})\mathbf{w} = \mathbf{v}$.

29. (a) Show that $\lambda = 2$ is a triple eigenvalue of the coefficient matrix $\mathbf{A}$ of the system

$$
\begin{aligned}
x_1' &= \quad 3x_1 + x_2, \\
x_2' &= -x_1 + x_2, \\
x_3' &= \quad x_1 + x_2 + 2x_3,
\end{aligned}
$$

and that there are only two linearly independent eigenvectors $\mathbf{v}_1 = (1 \quad -1 \quad 0)^T$ and $\mathbf{v}_2 = (1 \quad -1 \quad 1)^T$ associated with this eigenvalue. Then $\mathbf{x}_1(t) = \mathbf{v}_1 e^{2t}$ and $\mathbf{x}_2(t) = \mathbf{v}_2 e^{2t}$ are two linearly independent solutions.
(b) Show that if $\mathbf{x}_3(t) = \mathbf{v}te^{2t} + \mathbf{w}e^{2t}$ is a third solution, then $\mathbf{v}$ and $\mathbf{w}$ must satisfy the equations

$$(\mathbf{A} - 2\mathbf{I})\mathbf{v} = 0, \tag{36a}$$

$$(\mathbf{A} - 2\mathbf{I})\mathbf{w} = \mathbf{v}. \tag{36b}$$

(c) Every solution of Eq. (36a) is of the form $\mathbf{v} = c_1\mathbf{v}_1 + c_2\mathbf{v}_2$, where c_1 and c_2 are constants. According to a fundamental theorem of linear algebra, Eq. (36b) has a nontrivial solution $\mathbf{w}$ if and only if the vector $\mathbf{v}$ is orthogonal to every solution of the system $(\mathbf{A}^T - 2\mathbf{I})\mathbf{y} = 0$. Show that the general solution of this system is

$$\mathbf{y} = \alpha \begin{pmatrix} 1 \\ 0 \\ -1 \end{pmatrix} + \beta \begin{pmatrix} 0 \\ 1 \\ 1 \end{pmatrix}.$$

(d) Show that $\mathbf{v} \cdot \mathbf{y} = c_1 \alpha - c_1 \beta$, so $\mathbf{v}$ and $\mathbf{y}$ are orthogonal provided that $c_1 = 0$. Thus, taking $c_2 = 1$, we can solve Eq. (36b) with $\mathbf{v} = \mathbf{v}_2$. (e) Solve the equation $(\mathbf{A} - 2\mathbf{I})\mathbf{w} = \mathbf{v}_2$ for $\mathbf{w} = (1 \quad 0 \quad 0)^T$. Conclude that

$$\mathbf{x}_3(t) = \begin{pmatrix} t + 1 \\ -t \\ t \end{pmatrix} e^{2t}$$

is a third independent solution of the given system.

30. Show that the coefficient matrix of the system

$$
\begin{aligned}
x_1' &= & x_2 - x_3 - & x_4, \\
x_2' &= & -x_2, \\
x_3' &= -2x_1 + 2x_2 + x_3 - & 2x_4, \\
x_4' &= -x_1 - & x_2 + x_3
\end{aligned}
$$

has the repeated eigenvalue $\lambda = -1$ with two linearly independent eigenvectors and the repeated eigenvalue $\lambda = +1$ with only one linearly independent eigenvector. Use the method of Example 4 to find a general solution.

5.6
Nonhomogeneous Linear Systems

In Sections 2.5 and 2.9 we discussed two methods of finding a particular solution of a single nonhomogeneous nth order linear differential equation—undetermined coefficients and variation of parameters. Each of these methods can be generalized to nonhomogeneous linear systems.

Given the nonhomogeneous first order linear system

$$\mathbf{x}' = \mathbf{P}(t)\mathbf{x} + \mathbf{f}(t) \tag{1}$$

with $\mathbf{P}(t)$ and $\mathbf{f}(t)$ continuous, we know from Theorem 4 of Section 5.3 that a general solution of (1) has the form

$$\mathbf{x}(t) = \mathbf{x}_c(t) + \mathbf{x}_p(t), \tag{2}$$

where $\mathbf{x}_c$ denotes a general solution of the associated homogeneous system $\mathbf{x}' = \mathbf{P}(t)\mathbf{x}$, and $\mathbf{x}_p$ is a single particular solution of the equation in (1). Section 5.5 was devoted to $\mathbf{x}_c$; our task now is to find $\mathbf{x}_p$.

UNDETERMINED COEFFICIENTS

Just as in the case of a single differential equation, the method of undetermined coefficients must be restricted to systems of equations with constant coefficients. Let us then consider the system

$$\mathbf{x}' = \mathbf{A}\mathbf{x} + \mathbf{f}(t) \tag{3}$$

where $\mathbf{A}$ is a matrix of constants, and $\mathbf{f}(t)$ is a linear combination (with constant vector coefficients) of products of polynomials, exponential functions, and sines and cosines. The method is essentially the same for systems as for single differential equations—we make an intelligent guess as to the general form of a particular solution $\mathbf{x}_p$ and then attempt to determine the coefficients in $\mathbf{x}_p$ by substitution in (3). Moreover, the choice of the general form of $\mathbf{x}_p$ is essentially the same as given in the table of Fig. 2.14 in Section 2.5; we modify it only by using undetermined *vector* coefficients rather than undetermined scalars. We will therefore confine the present discussion mainly to illustrative examples.

EXAMPLE 1 Find a general solution of the system

$$
\begin{aligned}
x' &= 4x + 2y - 8t, \\
y' &= 3x - y + 2t + 3.
\end{aligned}
\tag{4}
$$

Solution In Example 1 of Section 5.5, we found the general solution

$$
\mathbf{x}_c = \begin{pmatrix} x_c \\ y_c \end{pmatrix} = c_1 \begin{pmatrix} 1 \\ -3 \end{pmatrix} e^{2t} + c_2 \begin{pmatrix} 2 \\ 1 \end{pmatrix} e^{5t}
\tag{5}
$$

of the associated homogeneous system. Because there is no duplication between the terms of $\mathbf{x}_c$ and the nonhomogeneous terms in (4), we assume a trial solution of the form

$$
\mathbf{x}_p = \begin{pmatrix} x_p \\ y_p \end{pmatrix} = \mathbf{a}t + \mathbf{b} = \begin{pmatrix} a_1 \\ a_2 \end{pmatrix} t + \begin{pmatrix} b_1 \\ b_2 \end{pmatrix}.
\tag{6}
$$

Upon subsitution of (6) in (4) we get

$$
\begin{aligned}
\begin{pmatrix} a_1 \\ a_2 \end{pmatrix} &= \begin{pmatrix} 4 & 2 \\ 3 & -1 \end{pmatrix} \begin{pmatrix} a_1 t + b_1 \\ a_2 t + b_2 \end{pmatrix} + \begin{pmatrix} -8t \\ 2t + 3 \end{pmatrix} \\
&= \begin{pmatrix} 4a_1 + 2a_2 - 8 \\ 3a_1 - a_2 + 2 \end{pmatrix} t + \begin{pmatrix} 4b_1 + 2b_2 \\ 3b_1 - b_2 + 3 \end{pmatrix}.
\end{aligned}
\tag{7}
$$

When we equate the coefficients of t and the constant terms (in both x- and y-components) in (7), we get the equations

$$
\begin{aligned}
4a_1 + 2a_2 &= 8, \\
3a_1 - a_2 &= -2, \\
4b_1 + 2b_2 &= a_1, \\
3b_1 - b_2 &= a_2 - 3.
\end{aligned}
\tag{8}
$$

We solve the first two equations in (8) for $a_1 = \frac{2}{5}$ and $a_2 = \frac{16}{5}$ and then solve the last two equations for $b_1 = \frac{2}{25}$ and $b_2 = \frac{1}{25}$. Thus our particular solution is

$$
\mathbf{x}_p = \begin{pmatrix} \frac{2}{5}t + \frac{2}{25} \\ \frac{16}{5}t + \frac{1}{25} \end{pmatrix},
$$

and the general solution of the system in (4) is

$$x = c_1 e^{-2t} + 2c_2 e^{5t} + \tfrac{2}{5}t + \tfrac{2}{25},$$
$$y = -3c_1 e^{-2t} + c_2 e^{5t} + \tfrac{16}{5}t + \tfrac{1}{25}. \tag{9}$$

In the case of duplicate expressions in the complementary function and the nonhomogeneous terms, there is one difference between the method of undetermined coefficients for systems and for single equations (Rule 2 in Section 2.5). For a system, the usual choice for a trial solution must be multiplied not only by the smallest integral power of t that will eliminate the duplication, but also by all lower (nonnegative integral) powers as well; all the resulting terms must be included in the actual trial solution. For instance, if $\lambda = 0$ had been an (unrepeated) eigenvalue of the associated homogeneous system in Example 1, our trial solution would have been not $\mathbf{a}t^2 + \mathbf{b}t$ (the "usual" choice), but instead $\mathbf{a}t^2 + \mathbf{b}t + \mathbf{c}$.

EXAMPLE 2 Find a general solution of the system

$$x' = 4x + 2y,$$
$$y' = 3x - y + e^{-2t}. \tag{10}$$

Solution Because of duplication of the term e^{-2t} in the nonhomogeneous system above and in the complementary function

$$\mathbf{x}_c = \begin{pmatrix} x_c \\ y_c \end{pmatrix} = c_1 \begin{pmatrix} 1 \\ -3 \end{pmatrix} e^{2t} + c_2 \begin{pmatrix} 2 \\ 1 \end{pmatrix} e^{5t} \tag{5}$$

previously found, we assume a trial solution of the form

$$\mathbf{x}_p = \mathbf{a}t e^{-2t} + \mathbf{b}e^{-2t} = \begin{pmatrix} a_1 t + b_1 \\ a_2 t + b_2 \end{pmatrix} e^{-2t} \tag{11}$$

(rather than $\mathbf{a}t e^{-2t}$ alone). Then

$$\mathbf{x}_p' = (\mathbf{a} - 2\mathbf{b})e^{-2t} - 2\mathbf{a}t e^{-2t}. \tag{12}$$

Substitution of (11) and (12) in (10), and cancellation of e^{-2t} throughout, yields

$$\begin{pmatrix} a_1 - 2b_1 - 2a_1 t \\ a_2 - 2b_2 - 2a_2 t \end{pmatrix} = \begin{pmatrix} 4 & 2 \\ 3 & -1 \end{pmatrix} \begin{pmatrix} a_1 t + b_1 \\ a_2 t + b_2 \end{pmatrix} + \begin{pmatrix} 0 \\ 1 \end{pmatrix}$$
$$= \begin{pmatrix} [4a_1 + 2a_2]t + 4b_1 + 2b_2 \\ [3a_1 - a_2]t + 3b_1 - b_2 + 1 \end{pmatrix}.$$

We equate the coefficients of t and the constant terms to get the equations

$$6a_1 + 2a_2 = 0,$$
$$3a_1 + a_2 = 0,$$
$$6b_1 + 2b_2 = a_1,$$
$$3b_1 + b_2 = a_2 - 1. \tag{13}$$

The first two equations in (13) imply that $a_2 = -3a_1$. In order for the last two equations to be consistent, we must have

$$a_1 = 2(a_2 - 1) = 2(-3a_1 - 1) = -6a_1 - 2.$$

It follows that $a_1 = -\frac{2}{7}$, and so $a_2 = \frac{6}{7}$. Each of the last two equations in (13) now reduces to $6b_1 + 2b_2 = -\frac{2}{7}$, so $b_2 = -3b_1 - \frac{1}{7}$. Thus our particular solution is

$$\mathbf{x}_p = \begin{pmatrix} -\frac{2}{7} \\ \frac{6}{7} \end{pmatrix} te^{-2t} + b_1 \begin{pmatrix} 1 \\ -3 \end{pmatrix} e^{-2t} + \begin{pmatrix} 0 \\ -\frac{1}{7} \end{pmatrix} e^{-2t}.$$

The middle term on the right-hand side is a solution of the associated homogeneous system and can therefore be absorbed into the complementary function in (5). Hence the general solution of the system in (10) is given by

$$\begin{aligned} x &= c_1 e^{-2t} + 2c_2 e^{5t} - \tfrac{2}{7} te^{-2t}, \\ y &= -3c_1 e^{-2t} + c_2 e^{5t} + \tfrac{6}{7} te^{-2t} - \tfrac{1}{7} e^{-2t}. \end{aligned} \tag{15}$$

FUNDAMENTAL MATRICES

In order to generalize the method of variation of parameters so it applies to the problem of finding a particular solution of a nonhomogeneous linear system, we need the concept of a fundamental matrix of a homogeneous system. Suppose that $\mathbf{x}_1(t), \mathbf{x}_2(t), \dots, \mathbf{x}_n(t)$ are n linearly independent solutions on some open interval of the homogeneous system

$$\mathbf{x}' = \mathbf{P}(t)\mathbf{x} \tag{16}$$

of n linear equations. Then the $n \times n$ matrix

$$\mathbf{\Phi}(t) = \begin{pmatrix} x_{11}(t) & x_{12}(t) & \cdots & x_{1n}(t) \\ x_{21}(t) & x_{22}(t) & \cdots & x_{2n}(t) \\ \cdot & \cdot & & \cdot \\ \cdot & \cdot & & \cdot \\ \cdot & \cdot & & \cdot \\ x_{n1}(t) & x_{n2}(t) & \cdots & x_{nn}(t) \end{pmatrix}, \tag{17}$$

having as column vectors the solution vectors $\mathbf{x}_1, \mathbf{x}_2, \dots, \mathbf{x}_n$, is called a **fundamental matrix** for the system. Because its column vectors are linearly independent, it follows that the matrix $\mathbf{\Phi}(t)$ is nonsingular and therefore has an inverse matrix $\mathbf{\Phi}^{-1}(t)$.

In terms of the fundamental matrix $\mathbf{\Phi}(t)$, the general solution

$$\mathbf{x}(t) = c_1 \mathbf{x}_1(t) + c_2 \mathbf{x}_2(t) + \cdots + c_n \mathbf{x}_n(t) \tag{18}$$

can be written in the form

$$\mathbf{x}(t) = \mathbf{\Phi}(t)\mathbf{c} \tag{19}$$

where

$$\mathbf{c} = \begin{pmatrix} c_1 \\ c_2 \\ c_3 \\ \cdot \\ \cdot \\ \cdot \\ c_n \end{pmatrix}.$$

In order that the solution $\mathbf{x}(t)$ satisfy the initial condition

$$\mathbf{x}(a) = \mathbf{b} \tag{20}$$

where $b_1, b_2, \ldots, b_n$ are given, it therefore will suffice for the coefficient vector $\mathbf{c}$ in (19) to satisfy $\mathbf{\Phi}(a)\mathbf{c} = \mathbf{b}$; that is,

$$\mathbf{c} = \mathbf{\Phi}^{-1}(a)\mathbf{b}. \tag{21}$$

When we combine (19) and (21), it becomes clear that the solution of the initial value problem

$$\mathbf{x}' = \mathbf{P}(t)\mathbf{x}, \qquad \mathbf{x}(a) = \mathbf{b} \tag{22}$$

is given by

$$\mathbf{x}(t) = \mathbf{\Phi}(t)\mathbf{\Phi}^{-1}(a)\mathbf{b}. \tag{23}$$

In order to apply the formula in (23), we must be able to compute the inverse matrix $\mathbf{\Phi}^{-1}(a)$. The inverse of the nonsingular 2×2 matrix

$$\mathbf{A} = \begin{pmatrix} a & b \\ c & d \end{pmatrix}$$

is

$$\mathbf{A}^{-1} = \frac{1}{\Delta}\begin{pmatrix} d & -b \\ -c & a \end{pmatrix} \tag{24}$$

where $\Delta = \det(\mathbf{A}) = ad - bc \neq 0$. The inverse of the nonsingular 3×3 matrix $\mathbf{A} = (a_{ij})$ is given by

$$\mathbf{A}^{-1} = \frac{1}{\Delta}\begin{pmatrix} +A_{11} & -A_{12} & +A_{13} \\ -A_{21} & +A_{22} & -A_{23} \\ +A_{31} & -A_{32} & +A_{33} \end{pmatrix}^T \tag{25}$$

where $\Delta = \det(\mathbf{A}) \neq 0$, and A_{ij} denotes the determinant of the 2×2 submatrix of $\mathbf{A}$ obtained by deleting the ith row and jth column of $\mathbf{A}$. (Do not overlook the symbol T for *transpose* in Eq. (25).) The formula in (25) is also valid upon generalization to $n \times n$ matrices, but in practice inverses of larger matrices are usually computed instead by row reduction methods (see any linear algebra text).

EXAMPLE 3 Find a fundamental matrix for the system

$$\begin{aligned} x' &= 4x + 2y, \\ y' &= 3x - \ y \end{aligned} \tag{26}$$

and use it to find the solution of (26) that satisfies the initial conditions $x(0) = 1, y(0) = -1$.

Solution The linearly independent solutions

$$\mathbf{x}_1(t) = \begin{pmatrix} e^{-2t} \\ -3e^{-2t} \end{pmatrix} \quad \text{and} \quad \mathbf{x}_2(t) = \begin{pmatrix} 2e^{5t} \\ e^{5t} \end{pmatrix}$$

displayed in (5) yield the fundamental matrix

$$\Phi(t) = \begin{pmatrix} e^{-2t} & 2e^{5t} \\ -3e^{-2t} & e^{5t} \end{pmatrix}. \tag{27}$$

Then

$$\Phi(0) = \begin{pmatrix} 1 & 2 \\ -3 & 1 \end{pmatrix}, \tag{28}$$

and the formula in (24) gives the inverse matrix

$$\Phi^{-1}(0) = \tfrac{1}{7} \begin{pmatrix} 1 & -2 \\ 3 & 1 \end{pmatrix}. \tag{29}$$

Hence the formula in (23) gives the solution

$$\mathbf{x}(t) = \begin{pmatrix} e^{-2t} & 2e^{5t} \\ -3e^{-2t} & e^{5t} \end{pmatrix} (\tfrac{1}{7}) \begin{pmatrix} 1 & -2 \\ 3 & 1 \end{pmatrix} \begin{pmatrix} 1 \\ -1 \end{pmatrix}$$

$$= \tfrac{1}{7} \begin{pmatrix} e^{-2t} + 6e^{5t} & -2e^{-2t} + 2e^{5t} \\ -3e^{-2t} + 3e^{5t} & 6e^{-2t} + e^{5t} \end{pmatrix} \begin{pmatrix} 1 \\ -1 \end{pmatrix},$$

and so

$$\mathbf{x}(t) = \tfrac{1}{7} \begin{pmatrix} 3e^{-2t} + 4e^{5t} \\ -9e^{-2t} + 2e^{5t} \end{pmatrix}. \tag{30}$$

Thus the solution of the original initial value problem is given by

$$\begin{aligned} x &= \tfrac{3}{7}e^{-2t} + \tfrac{4}{7}e^{5t}, \\ y &= -\tfrac{9}{7}e^{-2t} + \tfrac{2}{7}e^{5t}. \end{aligned} \tag{30a}$$

VARIATION OF PARAMETERS

Recall from Section 2.9 that the method of variation of parameters may be applied to a linear differential equation with variable coefficients, and is not restricted to nonhomogeneous terms involving only polynomials, exponentials, and sinusoidal functions. The method of variation of parameters for systems enjoys the same flexibility, and has a concise matrix formulation that is convenient for both practical and theoretical purposes.

We want to find a particular solution $\mathbf{x}_p$ of the nonhomogeneous linear system

$$\mathbf{x}' = \mathbf{P}(t)\mathbf{x} + \mathbf{f}(t), \tag{31}$$

given that we have found already a general solution

$$\mathbf{x}_c = c_1\mathbf{x}_1 + c_2\mathbf{x}_2 + \cdots + c_n\mathbf{x}_n \tag{32}$$

of the associated homogeneous system

$$\mathbf{x}'_c = \mathbf{P}(t)\mathbf{x}. \tag{33}$$

We first use the fundamental matrix $\mathbf{\Phi}(t)$ with column vectors $\mathbf{x}_1$, $\mathbf{x}_2$, $\ldots$, $\mathbf{x}_n$ to rewrite the complementary function in (32) as

$$\mathbf{x}_c = \mathbf{\Phi}(t)\mathbf{c}. \tag{34}$$

Our idea is to replace the vector "parameter" $\mathbf{c}$ with a variable vector $\mathbf{u}(t)$, and thus we seek a particular solution of the form

$$\mathbf{x}_p(t) = \mathbf{\Phi}(t)\mathbf{u}(t). \tag{35}$$

We must determine $\mathbf{u}(t)$ so that $\mathbf{x}_p$ does, indeed, satisfy Eq. (31).

The derivative of $\mathbf{x}_p(t)$ is (by the product rule)

$$\mathbf{x}'_p(t) = \mathbf{\Phi}'(t)\mathbf{u}(t) + \mathbf{\Phi}(t)\mathbf{u}'(t). \tag{36}$$

Hence substitution of (35) and (36) in (31) yields

$$\mathbf{\Phi}'(t)\mathbf{u}(t) + \mathbf{\Phi}(t)\mathbf{u}'(t) = \mathbf{P}(t)\mathbf{\Phi}(t)\mathbf{u}(t) + \mathbf{f}(t). \tag{37}$$

But

$$\mathbf{\Phi}'(t) = \mathbf{P}(t)\mathbf{\Phi}(t) \tag{38}$$

because each column vector of $\mathbf{\Phi}(t)$ satisfies Eq. (33). Therefore (37) reduces to

$$\mathbf{\Phi}(t)\mathbf{u}'(t) = \mathbf{f}(t). \tag{39}$$

Thus it suffices to choose $\mathbf{u}(t)$ so that

$$\mathbf{u}'(t) = \mathbf{\Phi}^{-1}(t)\mathbf{f}(t); \tag{40}$$

that is, so that

$$\mathbf{u}(t) = \int \mathbf{\Phi}^{-1}(t)\mathbf{f}(t)\,dt. \tag{41}$$

Upon substitution of (41) in (35), we finally obtain the desired particular solution

$$\mathbf{x}_p(t) = \mathbf{\Phi}(t)\int \mathbf{\Phi}^{-1}(t)\mathbf{f}(t)\,dt. \tag{42}$$

This is the variation of parameters formula for first order linear systems. If we add this particular solution and the complementary function in (34), we get the general solution

$$\mathbf{x}(t) = \mathbf{\Phi}(t)\mathbf{c} + \mathbf{\Phi}(t)\int \mathbf{\Phi}^{-1}(t)\mathbf{f}(t)\,dt \tag{43}$$

of the nonhomogeneous system in (31).

The choice of the constant of integration in (42) is immaterial, for we need only a single particular solution. In solving initial value problems it often is convenient to choose the constant of integration so that $\mathbf{x}_p(a) = \mathbf{0}$, and thus integrate from a to t:

$$\mathbf{x}_p(t) = \mathbf{\Phi}(t)\int_a^t \mathbf{\Phi}^{-1}(s)\mathbf{f}(s)\,ds. \tag{42'}$$

If we add this particular solution and the homogeneous solution in (23), we get the solution

$$x(t) = \Phi(t)\Phi^{-1}(a)b + \Phi(t) \int_a^t \Phi^{-1}(s)f(s) \, ds \qquad (43')$$

of the initial value problem $x' = P(t)x + f(t)$, $x(a) = b$.

EXAMPLE 4 Solve the initial value problem

$$x' = \begin{pmatrix} 4 & 2 \\ 3 & -1 \end{pmatrix} x - \begin{pmatrix} 15 \\ 4 \end{pmatrix} te^{-2t}, \qquad x(0) = \begin{pmatrix} 1 \\ -1 \end{pmatrix}. \qquad (44)$$

Solution If we were to use the method of undetermined coefficients, our trial solution would be of the form

$$x_p(t) = at^2e^{-2t} + bte^{-2t} + ce^{-2t},$$

and we would have six scalar coefficients to determine. Here it will be simpler to use the method of variation of parameters, for we already know from Example 3 the fundamental matrix

$$\Phi(t) = \begin{pmatrix} e^{-2t} & 2e^{5t} \\ -3e^{-2t} & e^{5t} \end{pmatrix} \qquad (45)$$

of the associated homogeneous system.

The determinant of $\Phi(t)$ is $\Delta = 7e^{3t}$, so the formula in (27) gives the inverse matrix

$$\Phi^{-1}(t) = \tfrac{1}{7}e^{-3t} \begin{pmatrix} e^{5t} & -2e^{5t} \\ 3e^{-2t} & e^{-2t} \end{pmatrix}. \qquad (46)$$

Then the formula in (42') gives the particular solution

$$x_p(t) = \Phi(t) \int_0^t \tfrac{1}{7}e^{-3s} \begin{pmatrix} e^{5s} & -2e^{5s} \\ 3e^{-2s} & e^{-2s} \end{pmatrix} \begin{pmatrix} -15se^{-2s} \\ -4se^{-2s} \end{pmatrix} ds$$

$$= \Phi(t) \int_0^t \tfrac{1}{7}e^{-3s} \begin{pmatrix} -7se^{3s} \\ -49se^{-4s} \end{pmatrix} ds$$

$$= \Phi(t) \int_0^t \begin{pmatrix} -s \\ -7se^{-7s} \end{pmatrix} ds$$

$$= \Phi(t) \left[\begin{pmatrix} -\tfrac{1}{2}s^2 \\ se^{-7s} + \tfrac{1}{7}e^{-7s} \end{pmatrix} \right]_{s=0}^{s=t}$$

$$= \begin{pmatrix} e^{-2t} & 2e^{5t} \\ -3e^{-2t} & e^{5t} \end{pmatrix} \begin{pmatrix} -\tfrac{1}{2}t^2 \\ te^{-7t} + \tfrac{1}{7}e^{-7t} - \tfrac{1}{7} \end{pmatrix}.$$

Therefore

$$x_p(t) = \tfrac{1}{14}e^{-2t} \begin{pmatrix} 14 + 28t - 7t^2 \\ 2 + 14t + 21t^2 \end{pmatrix} - \tfrac{1}{7}e^{5t} \begin{pmatrix} 2 \\ 1 \end{pmatrix}. \qquad (47)$$

This is a particular solution such that $x_p(0) = 0$. In Example 3 we found the solution

$$x_c(t) = \tfrac{1}{7} \begin{pmatrix} 3e^{-2t} + 4e^{5t} \\ -9e^{-2t} + 2e^{5t} \end{pmatrix} \qquad (48)$$

of the associated homogeneous system such that

$$x_c(0) = \begin{pmatrix} 1 \\ -1 \end{pmatrix}.$$

Upon adding the solutions in (47) and (48), we find that the solution of the initial value problem in (44) is given by

$$\begin{aligned} x &= \tfrac{1}{14}(10 + 28t - 7t^2)e^{-2t} + \tfrac{2}{7}e^{5t}, \\ y &= \tfrac{1}{14}(-16 + 14t + 21t^2)e^{-2t} + \tfrac{1}{7}e^{5t}. \end{aligned} \tag{49}$$

The variation of parameters formula in (43) is a generalization of the formula

$$x(t) = ce^{at} + e^{at}\int e^{-at}f(t)\, dt \tag{50}$$

for the general solution of the nonhomogeneous first order (scalar) equation

$$x' = ax + f(t), \tag{51}$$

because a solution of the associated homogeneous equation $x' = ax$ is $\phi(t) = e^{at}$. A more direct generalization of (50) is obtained if we define the **exponential matrix** $e^{\mathbf{A}t}$, where $\mathbf{A}$ is a constant square matrix, as follows:

$$e^{\mathbf{A}t} = \mathbf{I} + \mathbf{A}t + \mathbf{A}^2\frac{t^2}{2!} + \cdots + \mathbf{A}^n\frac{t^n}{n!} + \cdots \tag{52}$$

in analogy with the scalar exponential series. Here $\mathbf{A}^2 = \mathbf{AA}$, $\mathbf{A}^3 = \mathbf{AAA}$, and so on; inductively, $\mathbf{A}^{n+1} = \mathbf{AA}^n$ if $n \geq 1$. It is shown in advanced texts that the series in (52) converges for all t and every constant matrix $\mathbf{A}$.

In Problem 34 we ask you to show that if the $n \times n$ matrices $\mathbf{A}$ and $\mathbf{B}$ commute ($\mathbf{AB} = \mathbf{BA}$), then $e^{\mathbf{A}+\mathbf{B}} = e^{\mathbf{A}}e^{\mathbf{B}}$ (exponential law). In Problem 35 we ask you to conclude that $(e^{\mathbf{A}})^{-1} = e^{-\mathbf{A}}$. In particular, the matrix $e^{\mathbf{A}}$ is nonsingular for all $\mathbf{A}$ (reminiscent of the fact that $e^t \neq 0$ for all t).

The series in (52) can be differentiated term by term, with the result

$$\frac{d}{dt}(e^{\mathbf{A}t}) = \mathbf{A} + \mathbf{A}^2 t + \mathbf{A}^3\frac{t^2}{2!} + \cdots$$

$$= \mathbf{A}\left(\mathbf{I} + \mathbf{A}t + \mathbf{A}^2\frac{t^2}{2!} + \cdots\right);$$

that is,

$$\frac{d}{dt}(e^{\mathbf{A}t}) = \mathbf{A}e^{\mathbf{A}t}. \tag{53}$$

Because the matrix $e^{\mathbf{A}t}$ is always nonsingular, Eq. (53) implies that

$$\mathbf{\Phi}(t) = e^{\mathbf{A}t} \tag{54}$$

is a fundamental matrix for the homogeneous linear system

$$\mathbf{x}' = \mathbf{Ax} \tag{55}$$

with constant coefficient matrix $\mathbf{A}$. Substituting (54) in (43), and using the fact that $(e^{\mathbf{A}t})^{-1} = e^{-\mathbf{A}t}$, we therefore obtain the general solution

$$\mathbf{x}(t) = e^{\mathbf{A}t}\mathbf{c} + e^{\mathbf{A}t}\int e^{-\mathbf{A}t}\mathbf{f}(t)\, dt \tag{56}$$

of the nonhomogeneous linear system

$$\mathbf{x}' = \mathbf{A}\mathbf{x} + \mathbf{f}(t) \tag{57}$$

with constant coefficients.

EXAMPLE 5 Use the exponential matrix to find a general solution of

$$\mathbf{x}' = \begin{pmatrix} 1 & 1 \\ 0 & 1 \end{pmatrix}\mathbf{x}. \tag{58}$$

Solution The coefficient matrix in (58) is $\mathbf{A} = \mathbf{I} + \mathbf{B}$ where

$$\mathbf{B} = \begin{pmatrix} 0 & 1 \\ 0 & 0 \end{pmatrix}.$$

Hence a fundamental matrix for the system in (58) is

$$\mathbf{\Phi}(t) = e^{\mathbf{A}t} = e^{\mathbf{I}t}e^{\mathbf{B}t}$$

by Problem 34. Now

$$e^{\mathbf{I}t} = \begin{pmatrix} e^t & 0 \\ 0 & e^t \end{pmatrix}$$

by Problem 33, and $\mathbf{B}^2 = \mathbf{0}$ by direct multiplication, so it follows from (52) that

$$e^{\mathbf{B}t} = \mathbf{I} + \mathbf{B}t = \begin{pmatrix} 1 & t \\ 0 & 1 \end{pmatrix}.$$

Hence

$$\mathbf{\Phi}(t) = \begin{pmatrix} e^t & 0 \\ 0 & e^t \end{pmatrix}\begin{pmatrix} 1 & t \\ 0 & 1 \end{pmatrix} = \begin{pmatrix} e^t & te^t \\ 0 & e^t \end{pmatrix}.$$

Thus a general solution of (58) is

$$\mathbf{x}(t) = c_1\begin{pmatrix} e^t \\ 0 \end{pmatrix} + c_2\begin{pmatrix} te^t \\ e^t \end{pmatrix}.$$

Remark: The simplicity of the computations in Example 5 may be a bit misleading. Only because of the special (so-called *upper triangular*) form of the matrix $\mathbf{A}$ was the computation of $e^{\mathbf{A}t}$ relatively easy. For a general matrix $\mathbf{A}$ the computation of the exponential matrix $e^{\mathbf{A}t}$ can be a formidable task, one that involves the *canonical forms* of matrices that are studied in advanced linear algebra.

5.6 Problems

Apply the method of undetermined coefficients to find a particular solution of each of the systems in Problems 1–14. If initial conditions are given, find the particular solution that satisfies these conditions.

1. $x' = x + 2y + 3$, $y' = 2x + y - 2$.
2. $x' = 2x + 3y + 5$, $y' = 2x + y - 2t$.
3. $x' = 3x + 4y$, $y' = 3x + 2y + t^2$; $x(0) = y(0) = 0$.

4. $x' = 4x + y + e^t$, $y' = 6x - y - e^t$; $x(0) = y(0) = 1$.

5. $x' = 6x - 7y + 10$, $y' = x - 2y - 2e^{-t}$.

6. $x' = 9x + y + 2e^t$, $y' = -8x - 2y + te^t$.

7. $x' = -3x + 4y + \sin t$, $y' = 6x - 5y$; $x(0) = 1$, $y(0) = 0$.

8. $x' = x - 5y + 2\sin t$, $y' = x - y - 3\cos t$.

9. $x' = x - 5y + \cos 2t$, $y' = x - y$.

10. $x' = x - 2y$, $y' = 2x - y + e^t \sin t$.

11. $x' = 2x + 4y + 2$, $y' = x + 2y + 3$; $x(0) = 1$, $y(0) = -1$.

12. $x' = x + y + 2t$, $y' = x + y - 2t$.

13. $x' = 2x + y + 2e^t$, $y' = x + 2y - 3e^t$.

14. $x' = 2x + y + 1$, $y' = 4x + 2y + e^{4t}$.

Find a fundamental matrix of each of the systems in Problems 15–22 and then apply the formula in Eq. (23) to find a solution satisfying the given initial conditions.

15. $\mathbf{x}' = \begin{pmatrix} 2 & 1 \\ 1 & 2 \end{pmatrix} \mathbf{x}$, $\mathbf{x}(0) = \begin{pmatrix} 3 \\ -2 \end{pmatrix}$.

16. $\mathbf{x}' = \begin{pmatrix} 2 & -1 \\ -4 & 2 \end{pmatrix} \mathbf{x}$, $\mathbf{x}(0) = \begin{pmatrix} 2 \\ -1 \end{pmatrix}$.

17. $\mathbf{x}' = \begin{pmatrix} 2 & -5 \\ 4 & -2 \end{pmatrix} \mathbf{x}$, $\mathbf{x}(0) = \begin{pmatrix} 0 \\ 1 \end{pmatrix}$.

18. $\mathbf{x}' = \begin{pmatrix} 3 & -1 \\ 1 & 1 \end{pmatrix} \mathbf{x}$, $\mathbf{x}(0) = \begin{pmatrix} 1 \\ 0 \end{pmatrix}$.

19. $\mathbf{x}' = \begin{pmatrix} -3 & -2 \\ 9 & 3 \end{pmatrix} \mathbf{x}$, $\mathbf{x}(0) = \begin{pmatrix} 1 \\ -1 \end{pmatrix}$.

20. $\mathbf{x}' = \begin{pmatrix} 7 & -5 \\ 4 & 3 \end{pmatrix} \mathbf{x}$, $\mathbf{x}(0) = \begin{pmatrix} 2 \\ 0 \end{pmatrix}$.

21. $\mathbf{x}' = \begin{pmatrix} 5 & 0 & -6 \\ 2 & -1 & -2 \\ 4 & -2 & -4 \end{pmatrix} \mathbf{x}$, $\mathbf{x}(0) = \begin{pmatrix} 2 \\ 1 \\ 0 \end{pmatrix}$.

22. $\mathbf{x}' = \begin{pmatrix} 3 & 2 & 2 \\ -5 & -4 & -2 \\ -5 & 5 & 3 \end{pmatrix} \mathbf{x}$, $\mathbf{x}(0) = \begin{pmatrix} 1 \\ 0 \\ -1 \end{pmatrix}$.

In each of Problems 23–32, apply the method of variation of parameters to find a particular solution of the given system.

23. $\mathbf{x}' = \begin{pmatrix} 6 & -7 \\ 1 & -2 \end{pmatrix} \mathbf{x} + \begin{pmatrix} 2t \\ 3 \end{pmatrix}$.

24. $\mathbf{x}' = \begin{pmatrix} 1 & 2 \\ 2 & -2 \end{pmatrix} \mathbf{x} + \begin{pmatrix} 3e^{2t} \\ 5t \end{pmatrix}$.

25. $\mathbf{x}' = \begin{pmatrix} 4 & -1 \\ 5 & -2 \end{pmatrix} \mathbf{x} + \begin{pmatrix} e^{-t} \\ 2e^{-t} \end{pmatrix}$.

26. $\mathbf{x}' = \begin{pmatrix} 3 & -1 \\ 9 & -3 \end{pmatrix} \mathbf{x} + \begin{pmatrix} \frac{1}{t^2} \\ \frac{1}{t^3} \end{pmatrix}$.

27. $\mathbf{x}' = \begin{pmatrix} 2 & -4 \\ 5 & -2 \end{pmatrix} \mathbf{x} + \begin{pmatrix} \cos 3t \\ \sin 3t \end{pmatrix}$.

28. $\mathbf{x}' = \begin{pmatrix} 3 & -5 \\ 5 & -3 \end{pmatrix} \mathbf{x} + \begin{pmatrix} \cos 4t \\ \sin 4t \end{pmatrix}$.

29. $\mathbf{x}' = \begin{pmatrix} 2 & -1 \\ 5 & -2 \end{pmatrix} \mathbf{x} + \begin{pmatrix} \sec t \\ \tan t \end{pmatrix}.$

30. $\mathbf{x}' = \begin{pmatrix} 2 & -1 \\ 1 & 2 \end{pmatrix} \mathbf{x} + \begin{pmatrix} e^{2t} \tan t \\ 0 \end{pmatrix}.$

31. $\mathbf{x}' = \begin{pmatrix} 3 & -2 \\ 2 & 3 \end{pmatrix} \mathbf{x} + \begin{pmatrix} 0 \\ e^{3t} \cos 2t \end{pmatrix}.$

32. $\mathbf{x}' = \begin{pmatrix} 2 & -4 \\ 1 & -2 \end{pmatrix} \mathbf{x} + \begin{pmatrix} \ln t \\ t \end{pmatrix}.$

Find the particular solution such that $x_1(1) = x_2(1) = 0$.

33. Apply the definition in Eq. (52) to show that $e^{\mathbf{I}t} = e^t \mathbf{I}$.

34. Suppose that the matrices $\mathbf{A}$ and $\mathbf{B}$ commute; that is, that $\mathbf{AB} = \mathbf{BA}$. Prove that $e^{\mathbf{A}+\mathbf{B}} = e^{\mathbf{A}} e^{\mathbf{B}}$. (*Suggestion:* Group the terms in the product of the two series on the right-hand side to obtain the series on the left.)

35. Deduce from the result of Problem 34 that, for every square matrix $\mathbf{A}$, the matrix $e^{\mathbf{A}}$ is nonsingular with $(e^{\mathbf{A}})^{-1} = e^{-\mathbf{A}}$.

36. Suppose that

$$\mathbf{A} = \begin{pmatrix} 0 & 1 \\ 1 & 0 \end{pmatrix}.$$

Show that $\mathbf{A}^{2n} = \mathbf{I}$ while $\mathbf{A}^{2n+1} = \mathbf{A}$. Conclude that $e^{\mathbf{A}t} = \mathbf{I} \cosh t + \mathbf{A} \sinh t$, and apply this fact to find a general solution of $\mathbf{x}' = \mathbf{Ax}$. Verify that it is equivalent to the general solution found by the eigenvalue method.

37. Suppose that

$$\mathbf{A} = \begin{pmatrix} 0 & 2 \\ -2 & 0 \end{pmatrix}.$$

Show that $e^{\mathbf{A}t} = \mathbf{I} \cos 2t + \frac{1}{2}\mathbf{A} \sin 2t$. Apply this fact to find a general solution of $\mathbf{x}' = \mathbf{Ax}$, and verify that it is equivalent to the one found by the eigenvalue method.

38. Apply the technique of Example 5 to find a general solution of

$$\mathbf{x}' = \begin{pmatrix} 1 & 2 & 0 \\ 0 & 1 & 2 \\ 0 & 0 & 1 \end{pmatrix} \mathbf{x}.$$

Numerical Methods

6

6.1

Introduction:
Euler's Method

The fundamental theorem of calculus implies that every continuous function f on an open interval I has an antiderivative G defined on I. Then $y = G(x)$ is a solution of the differential equation $y' = f(x)$. But even if $f(x)$ is a simple and familiar function, the function $G(x)$ may not be. For instance, it is known that any antiderivative of $f(x) = e^{-x^2}$ is a **nonelementary** function—one that cannot be expressed as a finite combination of the familiar functions of elementary calculus. Thus no particular solution of the differential equation

$$\frac{dy}{dx} = e^{-x^2} \tag{1}$$

is finitely expressible in terms of elementary functions.

Suppose that we need the solution of (1) satisfying an initial condition $y(0) = y_0$, in a form that will enable us to compute numerical values of the solution. One approach would be to apply the methods of Chapter 3 to derive an infinite series solution. This might well suffice in the case of Eq. (1), but recall that in Chapter 3 we discussed only *linear* equations (such as Bessel's equation). Infinite series methods are poorly adapted to nonlinear equations. For example, you can easily solve the differential equation $dy/dx = 4x^3 y^3$ by separating the variables. But try the method of infinite series, and note the difficulties that result when you must cube an infinite series.

Finding a particular solution of a differential equation is, in most applications, merely a means to an end. The goal is to be able to predict the shape of a column, the trajectory of a particle, or the modes of vibration of a mechanical system, with a certain degree of accuracy. Numerical results are frequently just as useful to such ends. To see why, consider an initial value problem of the form

$$y' = f(x, y), \qquad y(a) = y_a. \tag{2}$$

For purposes such as those mentioned above, it might be quite adequate to have a table of values of the unknown particular solution $y = g(x)$ at selected points of the interval $[a, b]$. *Euler's method* provides one way to obtain such a table, and it depends only on the mean value theorem.

To describe Euler's method, we first subdivide the interval $[a, b]$ into k subintervals all of the same length $h = (b - a)/k$. The **step size** h is the principal parameter in the application of Euler's method. The points of subdivision will be denoted by $a = x_0, x_1, x_2, \ldots, x_k = b$; the nth such point will then be $x_n = a + nh$. We denote by

$$g_n = g(x_n) = g(a + nh) \tag{3}$$

the *true* value at x_n of the (unknown) particular solution $y = g(x)$ of the initial value problem $y' = f(x, y)$, $y(a) = y_a$. Our goal is to find suitable *approximations* $y_1, y_2, \ldots, y_k$ to the true values $g_1, g_2, \ldots, g_k$, so

$$y_n \approx g_n = g(x_n) = g(a + nh) \tag{4}$$

for $n = 0, 1, 2, \ldots, k$. We begin with

$$y_0 = g_0 = y(a) = y_a \tag{5}$$

in accord with the initial condition. The first question is: How do we "step" from y_0 to the approximate value y_1 of $g_1 = g(x_1)$?

Recall that if the function g is continuous and differentiable on the interval $[x, x + h]$, then g satisfies the hypotheses of the mean value theorem and thus its conclusion as well:

$$g(x + h) - g(x) = hg'(z) \tag{6}$$

for some number z between x and $x + h$. Consequently $g(x + h) = g(x) + hg'(z)$. If h is a very small positive number and g' is continuous, then $g'(x)$ will be a good approximation to $g'(z)$ (because z will be close to x), and thus we obtain the linear approximation formula of elementary calculus:

$$g(x + h) \approx g(x) + hg'(x). \tag{7}$$

With $x = a$, the formula in (7) yields

$$g_1 = g(x_1) = g(a + h),$$

and thus

$$g_1 \approx g(a) + hg'(a);$$

that is,

$$g_1 \approx y_0 + hf(a, y_0), \tag{8}$$

since $g(a) = y_0$ and

$$g'(a) = f(a, g(a)) = f(a, y_0)$$

because $g(x)$ is the (exact) solution of the initial value problem in (2). As a consequence of (8), we take

$$y_1 = y_0 + hf(a, y_0) \tag{9}$$

as our approximation to $g_1 = g(x_1)$.

Next, with $x = x_1 = a + h$, the formula in (7) yields

$$g_2 = g(x_2) = g(x_1 + h) \approx g(x_1) + hg'(x_1);$$

that is,

$$g_2 \approx g_1 + f(x_1, g_1). \tag{10}$$

Because we have only the estimate y_1 of g_1 obtained in (9), we use it in the last approximation out of necessity. This yields

$$y_2 = y_1 + hf(x_1, y_1) \tag{11}$$

as our approximation to $g_2 = g(x_2)$. Iteration of this procedure to obtain estimates y_3 of g_3, y_4 of g_4, and so on, is **Euler's method** for obtaining numerical approximations to a particular solution of the initial value problem in (2).

ALGORITHM: THE EULER METHOD

Given the initial value problem

$$y' = f(x, y), \qquad y(a) = y_a, \tag{2}$$

Euler's method with step size h consists in applying the iterative formula

$$y_{n+1} = y_n + hf(x_n, y_n) \qquad (n \geq 0) \tag{12}$$

to compute successively approximations $y_1, y_2, y_3, \ldots$ to the (true) values $g_1, g_2, g_3, \ldots$ of the (exact) solution $y = g(x)$ at the points $x_1, x_2, x_3, \ldots$, respectively.

Note that the formulas in (9) and (11) are the first two cases, $n = 0$ and $n = 1$, of the general iterative formula in (12).

In practice the iteration of Euler's method is usually carried out on a programmable calculator or a computer. Thus there are three sources of error that may make the approximation y_n to g_n unreliable for large values of n. First, there is the error in the linear approximation formula in (7) itself. Second, there is the (cumulative) error involved in replacing $f(x_n, g_n)$ by $f(x_n, y_n)$ at each stage. Finally, the computing machine itself will contribute roundoff errors at each stage because only finitely many significant digits can be used in each computation. To see how to implement Euler's method, we will now try it on an initial value problem that we can solve exactly by other methods, so that we can compare the estimated solution with the actual solution.

EXAMPLE 1 Use Euler's method with step size $h = 0.05$ to obtain a table of approximate values of the solution of the initial value problem

$$y' = -y, \qquad y(0) = 1 \tag{13}$$

on the interval [0, 1].

Solution Of course, $y = g(x) = e^{-x}$ is the exact solution. In the notation used in (2), we have $f(x, y) = -y$, $a = 0$, and $y_0 = y_a = 1$. The iterative formula in (12) of Euler's method takes the form

$$y_{n+1} = y_n - hy_n = (1 - h)y_n \tag{14}$$

with $h = 0.05$. The results obtained using (14) are shown in the table in Fig. 6.1. The error in y_{20} is a little over 2.5%, and this suggests that we should use a smaller value of h. With $h = 0.0001$, we obtain the results shown in Fig. 6.2. Now the error in $y_{10,000}$ is only about 0.005%.

n	x_n	y_n	e^{-x_n}	$e^{-x_n} - y_n$
1	0.05	0.9500000	0.9512294	0.0012294
2	0.10	0.9025000	0.9048374	0.0023374
3	0.15	0.8573750	0.8607080	0.0033330
4	0.20	0.8145063	0.8187308	0.0042245
5	0.25	0.7737809	0.7788008	0.0050198
6	0.30	0.7350919	0.7408182	0.0057263
7	0.35	0.6983373	0.7046881	0.0063508
8	0.40	0.6634204	0.6703200	0.0068996
9	0.45	0.6302494	0.6376282	0.0073787
10	0.50	0.5987369	0.6065307	0.0077937
11	0.55	0.5688001	0.5769498	0.0081497
12	0.60	0.5403601	0.5488116	0.0084515
13	0.65	0.5133421	0.5220458	0.0087037
14	0.70	0.4876750	0.4965853	0.0089103
15	0.75	0.4632912	0.4723666	0.0090753
16	0.80	0.4401267	0.4493290	0.0092023
17	0.85	0.4181203	0.4274149	0.0092946
18	0.90	0.3972143	0.4065697	0.0093553
19	0.95	0.3773536	0.3867410	0.0093874
20	1.00	0.3584859	0.3678794	0.0093935

Figure 6.1 (Data rounded).

The final column in each of Figures 6.1 and 6.2 shows the error $e^{-x_n} - y_n$ at each stage of the computation. Two observations are informative: First, in each case the first few values of y_n are quite accurate, but—as n increases and x_n gets further from the starting point x_0—the error in y_n increases. Second, a comparison of the error columns in Fig. 6.1 and 6.2 suggests that the smaller the step size used, the smaller the resulting errors in our approximations.

Why, then, do we not simply choose for the step size h the smallest value of h that the machine can use effectively? In Example 1, this would be the smallest value of h such that the machine in use does not round $1 - h$ to precisely 1. There are two good reasons for not doing so. The first is obvious— the computations in Fig. 6.2 require about 500 times as much time as those in Fig. 6.1. (For example, when carried out on a TRS-80 PC-2 pocket computer, the results in Fig. 6.1 were essentially instantaneous, while those in Fig. 6.2

n	x_n	y_n	e^{-x_n}	$e^{-x_n} - y_n$
1	0.0001	0.9999000	0.9999000	0.0000000
2	0.0002	0.9998000	0.9998000	0.0000000
3	0.0003	0.9997000	0.9997000	0.0000000
4	0.0004	0.9996001	0.9996001	0.0000000
5	0.0005	0.9995001	0.9995001	0.0000000
⋮	⋮	⋮	⋮	⋮
9995	0.9995	0.3680450	0.3680634	0.0000184
9996	0.9996	0.3680082	0.3680266	0.0000184
9997	0.9997	0.3679714	0.3679898	0.0000184
9998	0.9998	0.3679346	0.3679530	0.0000184
9999	0.9999	0.3678978	0.3679162	0.0000184
10000	1.0000	0.3678610	0.3678974	0.0000184

Figure 6.2 (Data rounded).

required over 15 min.) The second reason is more subtle but more important: The computations with $h = 0.0001$ introduced approximation and roundoff errors 500 times as often as those with $h = 0.05$. With certain differential equations, $h = 0.05$ could actually produce more accurate results than those with $h = 0.0001$.

The "best" choice of h is difficult to determine in practice as well as in theory. It depends on the nature of the function $f(x, y)$ in the initial value problem in (2), on the exact code in which the program is written, and on the computer or calculator used. With a step size that is too large, the approximations inherent in Euler's method may not be sufficiently accurate, while if h is too small, then roundoff errors may accumulate to an unacceptable degree or the program may require too much time to be practical. The subject of *error propagation* in numerical algorithms is treated in numerical analysis courses and textbooks.

PROGRAMMING

One's understanding of a numerical algorithm is enhanced by considering how it can be carried out by means of a computer program. The programs displayed in this chapter (except where specifically noted) are written in a portable version of BASIC that should run (perhaps with trivial modifications, such as replacement of commas by semicolons) on any computer (from pocket to mainframe) that accepts BASIC. Our version of BASIC is also intended to make these programs intelligible to the reader who has little or no programming experience.

Each program in this chapter was executed on either a pocket computer (a TRS-80 PC-2) or a microcomputer (an IBM PC or a TRS-80 Model II). Some programs were run on two or on all three. When essentially equivalent programs are executed on two different computers, it is normal for the numerical results to differ slightly—perhaps in the last digit or two—because of differing internal methods of performing simple arithmetic.

Figure 6.3 shows the program that was used to compute the data for Example 1 shown in Fig. 6.1. Lines 110–130 set the initial values of x, y, and h. Line 160 replaces x_n with x_{n+1}, and line 170 computes y_{n+1} from y_n using the iterative formula in (14). If the computer is connected to a printer, the PRINT statement in lines 190–220 should be replaced with LPRINT to cause results to be physically printed rather than merely displayed on a screen. The remark (REM) statements in lines 100, 140, and 180 are for explanation only; they are ignored by the computer. Some inexpensive pocket computers do not accept REM statements, in which case they should be deleted before running the program.

```
100 REM        INITIAL VALUES:
110        X = 0
120        Y = 1
130        H = 0.05
140 REM        EULER ITERATION
150        FOR N = 1 TO 20
160        X = X + H
170        Y = (1 − H)*Y
180 REM        PRINT RESULTS:
190        PRINT N, X, Y,
200        PRINT EXP(−X),
210        PRINT EXP(−X) − Y
220        PRINT
230        NEXT N
240        END
```
Figure 6.3

To obtain the results shown in Fig. 6.2, the step size was replaced with $h = 0.0001$ in line 130. We also added a line to the program to suppress the printing of results for $5 < n < 9995$ (and replace those results by a few blank lines).

EXAMPLE 2 Use the Euler method to approximate the solution of the initial value problem

$$y' = \sqrt{1 - y^2}, \qquad y(0) = 0 \tag{15}$$

with step size $h = 0.01$ on the interval $[0, 3]$.

Solution The iterative formula of Euler's method is

$$y_{n+1} = y_n + h(1 - y_n^2)^{1/2} \tag{16}$$

with $y_0 = 0$. The program used is shown in Fig. 6.4. Note in line 180 the precaution taken to avoid a negative number under the radical.

The results, shown in Fig. 6.5, deserve an explanation. It is one that illustrates some pitfalls of numerical methods. The natural solution of the initial value problem in (15) is $y = \sin x$. Euler's method produces very accurate results at first. But at some point in the program—when x_n was about equal to $\pi/2$—y_n became a number either equal to 1 or slightly above. From that point on, y_n could never change—examine

```
100 REM        INITIAL VALUES:
110        X = 0
120        Y = 0
130        H = 0.01
140 REM        EULER ITERATION:
150        FOR N = 1 TO 300
160        X = X + H
170        Z = 1 − Y*Y
180        IF Z < 0 THEN LET Z = 0
190        Y = Y + H*SQR(Z)
200 REM        PRINT RESULTS:
210        PRINT N, X, Y,
220        PRINT SIN(X), SIN(X) − Y
230        PRINT
240        NEXT N
250        END
```

Figure 6.4

n	x_n	y_n	$\sin x_n$	$(\sin x_n) - y_n$
1	0.0100	0.0100	0.0100	−0.0000
2	0.0200	0.0200	0.0200	−0.0000
3	0.0300	0.0300	0.0300	−0.0000
4	0.0400	0.0400	0.0400	−0.0000
5	0.0500	0.0500	0.0500	−0.0000
⋮	⋮	⋮	⋮	⋮
295	2.9500	1.0000	0.1904	−0.8096
296	2.9600	1.0000	0.1806	−0.8194
297	2.9700	1.0000	0.1708	−0.8293
298	2.9800	1.0000	0.1609	−0.8392
299	2.9900	1.0000	0.1510	−0.8490
300	3.0000	1.0000	0.1411	−0.8589

Figure 6.5   (Data rounded).

the effect of lines 170–190 in the program shown in Fig. 6.4. See also Fig. 6.6, in which the graph of $y = \sin x$ is shown as a solid line, while the approximate solution is plotted as a dotted line (obtained in this case with $h = 0.001$).

Above we referred to $y = \sin x$ as the "natural" solution of the initial value problem in (15). Actually, the function

$$y(x) = \begin{cases} \sin x & \text{if } 0 \le x \le \dfrac{\pi}{2}, \\ 1 & \text{if } \dfrac{\pi}{2} \le x \le 3 \end{cases}$$

is equally a solution of the same initial value problem. And this explains our apparent anomaly—the solution produced by Euler's method is a correct one, after all! The differential equation $y' = \sqrt{1 - y^2}$ does not have a unique solution beginning at the point $(\pi/2, 1)$; because the partial derivative $\partial f/\partial y$ of $f(x, y) = \sqrt{1 - y^2}$ is not continuous at this point, the existence-uniqueness theorem of Section 1.3 does not apply

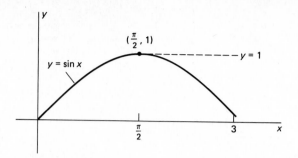

Figure 6.6 Graphs of two solutions of the initial value problem in Example 2.

(as far as uniqueness is concerned). Before applying Euler's method, it is a good idea to determine whether a unique solution exists. If not, there is no practical and generally applicable way to predict which way the numerical approximation will branch when it encounters a point of nonuniqueness.

There is one final difficulty to which most numerical methods (especially Euler's method) are particularly susceptible. In order to detect the possibility of the difficulty, some careful analysis of the initial value problem may be necessary. This difficulty is illustrated in our final example and usually crops up when the step size h is far too large.

EXAMPLE 3 Use Euler's method with step size $h = 0.075$ to approximate the solution of the initial value problem

$$7xy' + y = 0, \qquad y(-1) = 1 \tag{17}$$

on the interval $[-1, 1]$.

Solution First it's necessary to write the initial value problem in the form shown in (2); that is, $y' = -y/(7x)$, $y(-1) = 1$. The iteration to be used is

$$y_{n+1} = y_n - (0.075)\frac{y_n}{7x_n}$$

with $y_0 = 1$. The results are shown in Fig. 6.7. But these results are completely inconsistent with the graph of the actual solution, which is shown in Fig. 6.8. The reason is that the true solution increases without bound as x approaches zero from the left. The values predicted by Euler's method "jump across the discontinuity" at $x = 0$ and then travel down the graph of another solution of $7xy' + y = 0$. If h is large, the discontinuity may pass undetected; even with small values of h, the existence of the discontinuity may not be obvious.

The explanation of this situation is that the general solution of the differential equation $7xy' + y = 0$, as found easily by separation of variables, is

$$y = \frac{C}{x^{1/7}}. \tag{18}$$

n	x_n	y_n (Rounded)	True value (Rounded)
0	−1.000	1.000000	1.000000
2	−0.850	1.022421	1.023489
4	−0.700	1.049622	1.052274
6	−0.550	1.083957	1.089158
8	−0.400	1.129999	1.139852
10	−0.250	1.198518	1.219014
12	−0.100	1.326406	1.389495
14	0.050	2.097887	Values are not
16	0.200	1.507054	determined below
18	0.350	1.370748	this point by
20	0.500	1.295287	the given data.
22	0.650	1.243912	
24	0.800	1.205329	
26	0.950	1.174624	

Figure 6.7 (Data rounded).

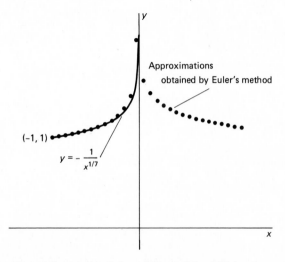

Figure 6.8 Part of the graph of the solution of the initial value problem in Example 3.

The initial condition $y(-1) = 1$ requires that $C = -1$ for $x < 0$. Because the actual solution has a discontinuity at $x = 0$, a different value of C may be chosen for $x > 0$. Thus the initial value problem in (17) does not have a unique solution on the entire interval $[-1, 1]$. Indeed, it has no (continuous) solution at all on this interval. Here our anomalous results are actually caused by nonexistence (near the infinite discontinuity at $x = 0$) rather than by nonuniqueness.

One way of avoiding a failure to detect (or jumping across) an infinite discontinuity is to run the program several times with successively smaller values of the step size h. The presence of the discontinuity may then manifest itself in approximate values that change rapidly (rather than level off) as h is decreased. See also Problem 21.

A geometric interpretation of Euler's method is based on our discussion of slope fields in Section 1.3. There we presented a number of exercises in which short line segments centered at (x, y) and having slope $y' = f(x, y)$ were given in diagrams; you were asked to drawn various solution curves fitting the given slopes. This is almost exactly what Euler's method amounts to from an analytical point of view. Figure 6.9 shows a comparison between the values $y_1, y_2, y_3, \ldots$ predicted by Euler's method and the true values on a typical solution curve.

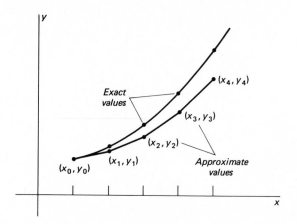

Figure 6.9 Why numerical methods don't give exact results.

6.1 Problems

In each of Problems 1–6, find the exact solution $y = g(x)$ of the given initial value problem. Then use Euler's method to find estimates $y_1, y_2, \ldots, y_{10}$ of the values of the actual solution at $x_1, x_2, \ldots, x_{10}$ over the given interval with the given step size h. Make tables similar to those in this section to show comparisons between predicted and actual values of the solution. (Note: These problems should be done with a calculator; if you have more powerful machinery available, go on to Problems 7–12.)

1. $y' = x + y, y(0) = 0$; $[0, 1]$; $h = 0.1$.
2. $y' = x - y, y(0) = 1$; $[0, 1]$; $h = 0.1$.
3. $y' = xy, y(0) = 1$; $[0, 2]$; $h = 0.2$.
4. $y' = y + 1, y(0) = -1$; $[0, 2]$; $h = 0.2$.
5. $y' = -2y, y(0) = 1$; $[0, 1]$; $h = 0.1$.
6. $y' = y, y(0) = 1$; $[-2, 0]$; $h = -0.2$.

Problems 7–12 require the use of a programmable calculator or a microcomputer; one with a printing attachment would be especially useful. Find the exact solution of the given initial value problem by earlier methods. Then use Euler's method with step size $h = 0.001$ to find approximations $y_1, y_2, \ldots, y_{100}$ to the values of the actual solution at $x_1, x_2, \ldots, x_{100}$ on the given interval. Make a table similar to those in this section to show comparisons, including percentage errors, between the approximate and true values in each case.

7. $y' = y - x - 1$, $y(0) = 0$;　[0, 1].　　**8.** $y' = 1 + y^2$, $y(0) = 0$;　[0, 1].

9. $y' = 3x^2y$, $y(0) = 1$;　[0, 1].　　**10.** $xy' = 1$, $y(1) = 0$;　[1, 2].

11. $y' = e^y$, $y(1) = 0$;　[1, 2].　　**12.** $y' = 4x^3$, $y(-1) = 1$;　[-1, 0].

Problems 13–20 require the the use of a microcomputer or a minicomputer, together with a printer. In each of these initial value problems, use Euler's method with step sizes h = 0.1, 0.02, 0.004, and 0.0008 to approximate the values of the solution at 10 equally spaced points of the given interval. Print the results in tabular form with appropriate headings to make it easy to gauge the effect of varying the step size h.

13. $y' = x^2 + y^2$, $y(0) = 0$;　[0, 1].

14. $y' = x^2 - y^2$, $y(0) = 1$;　[0, 2].

15. $y' = x + \sqrt{y}$, $y(0) = 1$;　[0, 2]

16. $y' = x + y^{1/3}$, $y(0) = -1$;　[0, 2].

17. $y' = \ln y$, $y(1) = 1$;　[1, 2].

18. $y' = x^{2/3} + y^{2/3}$, $y(0) = 1$;　[0, 2].

19. $y' = \sin x + \cos y$, $y(0) = 0$;　[0, 1].

20. $y' = \dfrac{x}{1 + y^2}$, $y(-1) = 1$;　[-1, 1].

21. More accurate results can be obtained near a discontinuity—as in Example 3— if the step size h is systematically decreased (though not *too* rapidly) as x_i approaches the discontinuity. Write a program to implement this minor variation of Euler's method and use it to obtain more accurate results, for the initial value problem of Example 3, than those shown in Fig. 6.7.

6.2
A Closer Look at the Euler Method

The Euler method as presented in Section 6.1 is not often used in practice, mainly because more accurate methods are available. But Euler's method has the advantage of simplicity, and a careful study of this method yields insights into the workings of the more accurate methods, because many of the latter are extensions or refinements of the Euler method. In order to compare two different numerical methods of approximation, we need some way to measure the accuracy of each. The following theorem tells what degree of accuracy we can expect when we use Euler's method.

THEOREM:　THE ERROR IN THE EULER METHOD

Given the initial value problem

$$y' = f(x, y), \qquad y(a) = y_a \tag{1}$$

on [a, b], suppose that the solution $y = g(x)$ has a continuous second derivative on [a, b]. (This would follow from the assumption that f, f_x, and f_y are all continuous for $a \leq x \leq b$ and $c \leq y \leq d$, where $c \leq g(x) \leq d$ for all x in [a, b].) Then there exists a constant C such that the following is true: If the approximations $y_1, y_2, y_3, \ldots, y_k$ to $g(x_1), g(x_2), g(x_3),$

..., $g(x_k)$ *(respectively) are computed using Euler's method with step size h, then*

$$|y_n - g(x_n)| \leq Ch \tag{2}$$

for each n = 1, 2, 3, ..., k.

Remark: The *error* $g(x_n) - y_n$ in (2) denotes the (cumulative) error in Euler's method after n steps in the approximation, *exclusive* of roundoff error (as though we were using a perfect machine that made no roundoff errors). The theorem can be summarized by saying that *the error in Euler's method is of order h*; that is, the error is bounded by a (predetermined) constant C multiplied by the step size h. Consequently we can (in principle) get any degree of accuracy we want by choosing h sufficiently small.

We will omit the proof of this theorem, but one can be found in Chapter 7 of *Ordinary Differential Equations* (Reading, Mass.: Addison-Wesley, 2nd ed., 1969) by G. Birkhoff and G.-C. Rota. The constant C deserves some comment. Because C tends to increase as the maximum value of $|g''(x)|$ on $[a, b]$ increases, it follows that C must depend in a fairly complicated way upon g, and actual computation of C is usually impractical. In practice, numerical analysts follow this procedure:

1. Apply Euler's method to the initial value problem in (1) with a reasonable value of h.
2. Repeat with $h/2$.
3. Repeat with $h/4$.
4. Repeat with $h/8$.
5. Continue until there is negligible change in the significant digits in the computed approximation; at this stage, the numerical approximation is considered to be almost certainly quite accurate.

EXAMPLE 1 Try the above procedure with the initial value problem

$$y' = -\frac{2xy}{x^2 + 1}, \qquad y(0) = 1. \tag{3}$$

Solution In order to illustrate the method, we begin with a value of $h = 0.01$ that is somewhat large. We will use the interval $[0, 1]$ and give only the computed values of $y(1)$. The results, computed with successively smaller values of h, are shown in the table in Fig. 6.10. The data suggest very strongly that the true value of the solution at $x = 1$ is exactly 0.5. Indeed, the exact solution of the initial value problem in (3) is $g(x) = 1/(1 + x^2)$, so $g(1) = 0.5$.

The final column of the table in Fig. 6.10 displays the ratio of the magnitude of the error to h; that is, $|(g(1) - y(1)|/h$. Observe how the data in this column substantiate the theorem above—in this computation, the error bound in (2) appears to hold with a value of C slightly larger than $\frac{1}{10}$.

h	$y(1)$ (Approximate value)	$g(1)$ (True value)	$\lvert$Error$\rvert/h$
0.01	0.50109	0.50000	0.11
0.005	0.50054	0.50000	0.11
0.0025	0.50027	0.50000	0.11
0.00125	0.50013	0.50000	0.10
0.000625	0.50007	0.50000	0.11
0.0003125	0.50003	0.50000	0.10

Figure 6.10

AN IMPROVEMENT IN EULER'S METHOD

As Fig. 6.11 shows, Euler's method is rather unsymmetrical. It uses the predicted slope of the graph of the solution at the left-hand end point of the interval $[x, x + h]$ as if it were the actual slope of the solution over that entire interval. We now turn our attention to a way in which increased accuracy can easily be obtained; it is known as the *improved Euler method.*

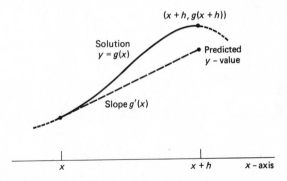

Figure 6.11 True and predicted value in Euler's method.

Given the initial value problem

$$y' = f(x, y) \qquad y(a) = y_a, \tag{1}$$

we begin with $x_0 = a$ and use $y_0 = y_a$. We then use the Euler method to obtain a first estimate—which we now call u_1 rather than y_1—of the value of the solution at $x_1 = x_0 + h$. Now that x_1 and u_1 have been computed, $m_1 = f(x_1, u_1)$ is a prediction of the slope of the solution at $x = x_1$. (It would be perfectly accurate if u_1 were actually equal to the value $g(x_1)$ of the solution, but this is unlikely.)

Of course the slope $m_0 = f(x_0, y_0)$ has already been computed at the beginning. Why not *average* these two numbers to obtain a more reliable estimate of the average slope of the solution over the entire interval $[x_0, x_1]$? This is the essence of the improved Euler method. Figure 6.12 shows the

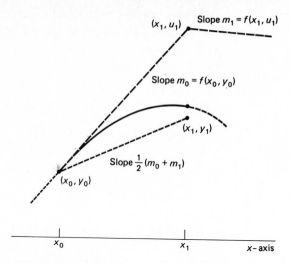

Figure 6.12 The improved Euler method: Average the slope at x_0 and the Euler approximation to the slope at x_1.

geometry behind the method; here it is in more symbolic terms:

$$\text{Let } m_0 = f(x_0, y_0).$$
$$\text{Let } u_1 = y_0 + hm_0.$$
$$\text{Let } m_1 = f(x_1, u_1). \tag{4}$$
$$\text{Let } \bar{m}_1 = \tfrac{1}{2}(m_0 + m_1).$$
$$\text{Let } y_1 = y_0 + h\bar{m}_1.$$

In general, if x_n and y_n have been computed:

$$\text{Let } m_n = f(x_n, y_n).$$
$$\text{Let } u_{n+1} = y_n + hm_n.$$
$$\text{Let } m_{n+1} = f(x_{n+1}, u_{n+1}). \tag{5}$$
$$\text{Let } \bar{m}_{n+1} = \tfrac{1}{2}(m_n + m_{n+1}).$$
$$\text{Let } y_{n+1} = y_n + h\bar{m}_{n+1}.$$

The improved Euler method is one of a class of numerical techniques known as predictor-corrector methods. First a **predictor** u_{n+1} of the next y-value is computed; then it is used to **correct** itself. Thus the **improved Euler method** with step size h consists in using the iterative formula

$$y_{n+1} = y_n + \frac{h}{2}[f(x_n, y_n) + f(x_{n+1}, u_{n+1})] \tag{6}$$

where

$$u_{n+1} = y_n + hf(x_n, y_n), \tag{7}$$

to compute successive approximations to the values of the exact solution $y = g(x)$ of the initial value problem in (1).

Under the assumption that the exact solution $y = g(x)$ of the initial value problem in (1) has a continuous third derivative, it can be proved—see

Chapter 7 of Birkhoff and Rota—that the error in the improved Euler method is of order h^2. That is, on a given bounded interval $[a, b]$, each approximate value y_n satisfies the inequality

$$|g(x_n) - y_n| \leq Ch^2 \tag{8}$$

where the constant C does not depend on h. Because h^2 is much smaller than h if h is small, this means that the improved Euler method is more accurate that Euler's method itself. This advantage is offset by the fact that about twice as many computations are required.

The program listed in Fig. 6.13 was used to compute approximations $y(1)$ to the true value $g(1) = 0.5$ of the solution $g(x) = 1/(1 + x^2)$ of the initial value problem

$$y' = -\frac{2xy}{x^2 + 1}, \qquad y(0) = 1 \tag{3}$$

```
100 REM        INITIAL VALUES:
110        H = 0.01 : K = 100
120 REM        IMPROVED EULER ITERATION:
130        X = 0 : Y = 1
140        FOR N = 1 TO K
150        M0 = -2*X*Y/(1 + X*X)
160        U = Y + H*M0
170        W = X + H
180        M1 = -2*W*U/(1+ + W*W)
190        Y = Y + (H/2)*(M0 + M1)
200        X = W
210        NEXT N
220 REM        PRINT INTERMEDIATE RESULTS:
230        PRINT H, Y, 0.5 - Y
240        PRINT
250 REM        MODIFY H AND K FOR ANOTHER
260 REM           SERIES OF COMPUTATIONS:
270        H = H/2 : K = 2*K
280        IF H > 0.0005 THEN GOTO 130
290        END
```

Figure 6.13

of Example 1, with the same values of h as in that example, but using the improved Euler method. The results appear in the table in Fig. 6.14. Note that with $h = 0.01$, *more* accuracy is attained than with $h = 0.0003125$ in the "unimproved" Euler method. Note also that the final column in Fig. 6.14 is consistent with the prediction of errors of order h^2.

h	$y(1)$ (Approximation)	Error	$|\text{Error}|/h^2$
0.01	0.500012448	-0.000012448	0.12
0.005	0.500003128	-0.000003128	0.13
0.0025	0.500000791	-0.000000791	0.13
0.00125	0.500000206	-0.000000206	0.13
0.000625	0.500000060	-0.000000060	0.15

Figure 6.14

Here is a final analysis of the initial value problem of Example 1. If we use step size $h = 0.02$ in the improved Euler method and $h = 0.01$ in the unimproved method, and examine approximate values and true values for $x = 0.1, 0.2, \ldots, 1.0$, we find that the improved method is clearly superior in terms of accuracy; see Fig. 6.15. The fact that twice as many computations are required is hardly a shortcoming, considering the speed of modern computers. (The reason for doubling the step size in the improved method is to roughly equalize the number of computations required to obtain the predicted values.)

x	Euler method ($h = 0.01$)	Improved Euler method ($h = 0.02$)	True value
0.0	1.00000	1.00000	1.00000
0.1	0.99107	0.99010	0.99010
0.2	0.96335	0.96155	0.96154
0.3	0.91982	0.91745	0.91743
0.4	0.86475	0.86209	0.86207
0.5	0.80271	0.80003	0.80000
0.6	0.73783	0.73533	0.73529
0.7	0.67337	0.67118	0.67114
0.8	0.61161	0.60980	0.60976
0.9	0.55395	0.55254	0.55249
1.0	0.50109	0.50005	0.50000

Figure 6.15

THE THREE-TERM TAYLOR SERIES METHOD

The iterative formula of Euler's method can be regarded as a consequence of the mean value theorem, and Taylor's formula is a generalization of that theorem. Indeed, using only the first two terms of the Taylor series of $g(x)$ centered at $x = a$ yields the linear approximation

$$g(a + h) \approx g(a) + hg'(a), \tag{9}$$

that we obtained in Section 6.1 as a consequence of the mean value theorem. Perhaps we could improve Euler's method by using more terms of the Taylor series of the actual solution $g(x)$ of the initial value problem

$$y' = f(x, y), \qquad y(a) = y_a. \tag{1}$$

If $g(x)$ is the actual solution, then $g'(x) = f(x, y) = f(x, g(x))$, and therefore the chain rule yields

$$g''(x) = f_x(x, g(x)) + f_y(x, g(x))g'(x)$$
$$= f_x(x, y) + f_y(x, y)f(x, y). \tag{10}$$

Because using the first *three* terms of the Taylor series of $g(x)$,

$$g(a + h) \approx g(a) + hg'(a) + \frac{h^2}{2}g''(a), \tag{11}$$

should provide a better approximation than that given in (9), the formula in (11) perhaps will provide us with a still more accurate method for approximating the solution of the problem in (1).

To carry out this idea, suppose that the approximation y_n to $g(x_n)$ has been computed. If we substitute the formula in (10) in the approximation in (11) with $a = x_n$, we get

$$g(x_{n+1}) = g(x_n + h)$$

$$\approx g(x_n) + hg'(x_n) + \frac{h^2}{2}g''(x_n)$$

$$\approx g(x_n) + hf(x_n, g(x_n)) + \frac{h^2}{2}[f_x(x_n, g(x_n)) + f_y(x_n, g(x_n))f(x_n, g(x_n))].$$

Replacing $g(x_n)$ with the approximation y_n, we finally get the approximation

$$y_{n+1} = y_n + hf(x_n, y_n) + \frac{h^2}{2}[f_x(x_n, y_n) + f_y(x_n, y_n)f(x_n, y_n)] \qquad (12)$$

to the true value $g(x_{n+1})$. The use of the iterative formula in (12), to compute successive approximations to the values of the solution $g(x)$ of the initial value problem in (1), constitutes the **three-term Taylor series method** with step size h. Like the improved Euler method on a given bounded interval $[a, b]$, this method has a (cumulative) error of order h^2.

The implementation of the three-term Taylor series method has the following algorithmic description. Assume that x_n and y_n have been calculated for $n \geq 0$ and that the step size is h. Then:

$$\text{Let } r_n = f(x_n, y_n).$$

$$\text{Let } s_n = f_x(x_n, y_n) + r_n f_y(x_n, y_n).$$

$$\text{Let } y_{n+1} = y_n + h\left(r_n + \frac{h}{2}s_n\right). \qquad (13)$$

$$\text{Let } x_{n+1} = x_n + h.$$

There are several reasons why this technique might not work as well (in a particular problem) as the improved Euler method. First, the form of f_x or f_y might be complicated enough to increase computational time greatly. More important, if the values of either f_x or f_y are sensitive to small errors in y_n, this may cause large errors in y_{n+1}. Finally, the entire effort is pointless if the third term on the right in (11) is negligible—as would be the case for very small values of h. Nevertheless, we can experiment.

EXAMPLE 2 Compare the improved Euler method with the three-term Taylor series method when each is applied to the initial value problem

$$y' = 2xy, \qquad y(0) = 1 \qquad (14)$$

on the interval $[0, 1]$.

Solution The program used is listed in Fig. 6.16, and the results are shown in Fig. 6.17. In both methods the value $h = 0.001$ was used. The final column shows values of the exact solution $g(x) = e^{x^2}$. The results of these computations suggest that the improved Euler method is somewhat more

```
100 REM        INITIAL VALUES:
110        X = 0 : YE = 1 : YT = 1
120        N = 0 : H = 0.001
130 REM        LOOP BEGINNING:
140        N = N + 1
150 REM        IMPROVED EULER METHOD:
160        M0 = 2*X*YE
170        U = YE + H*M0
180        W = X + H
190        M1 = 2*W*U
200        YE = YE + (H/2)*(M0 + M1)
210 REM        TAYLOR SERIES METHOD:
220        R = M0
230        S = 2*YT + 2*R*X
240        YT = YT + H*(R + 0.5*H*S)
250        X = X + H
260 REM        PRINT EACH 100TH VALUE:
270        K = INT(N/100)
280        IF 100*K < N THEN GOTO 140
290        PRINT X, YE, YT, EXP(X*X)
300        IF X < 1 THEN GOTO 140
310        END
```

Figure 6.16

x	Improved Euler	Three-term Taylor series	True value
0.0	1.00000000	1.00000000	1.00000000
0.1	1.01005017	1.01005016	1.01005017
0.2	1.04081077	1.04081073	1.04081077
0.3	1.09417428	1.09417418	1.09417428
0.4	1.17351085	1.17351068	1.17351087
0.5	1.28402538	1.28402509	1.28402542
0.6	1.43332933	1.43332891	1.43332942
0.7	1.63231607	1.63231545	1.63231622
0.8	1.89648060	1.89647971	1.89648088
0.9	2.24790747	2.24790623	2.24790799
1.0	2.71828089	2.71827919	2.71828183

Figure 6.17

accurate than the three-term Taylor series method for the particular initial value problem in (14).

There is no reason why the Taylor series method cannot be used with more terms of the series—this is practical if the differential equation involved is particularly simple. The following example illustrates also the use of a *negative* step size to step backward rather than forward.

EXAMPLE 3 Consider the initial value problem

$$y' = y^2, \qquad y(1) = 1 \tag{15}$$

on the interval $[-0.5, 1]$. Choose step size $h = -0.01$, and compare the results of using the Taylor polynomials of degrees 2, 4, and 6 to estimate the solution at $x = 0.9, 0.8, \ldots, -0.4, -0.5$; also compare these results with the true values at those points.

Solution To obtain the Taylor polynomial of degree four, we first compute the following derivatives of the solution $y = g(x)$:

$$y' = y^2;$$

$$y'' = 2yy' = 2y^3;$$

$$y^{(3)} = 6y^2 y' = 6y^4;$$

$$y^{(4)} = 24y^3 y' = 24y^5.$$

We now write y for $g(x)$; the Taylor polynomial approximation of degree four is then

$$g(x + h) \approx y + hy^2 + h^2 y^3 + h^3 y^4 + h^4 y^5;$$

that is,

$$g(x + h) \approx y(1 + hy\{1 + hy[1 + hy(1 + hy)]\}). \tag{16}$$

Once we have computed the approximation y_n to $g(x_n)$, we therefore take

$$y_{n+1} = y_n(1 + hy_n\{1 + hy_n[1 + hy_n(1 + hy_n)]\}) \tag{17}$$

as our approximation to $g(x_{n+1}) = g(x_n + h)$. (The polynomial has been "stacked" in (17) in order to avoid roundoff errors caused by exponentiation of numbers close to zero. This is a practice highly recommended for the numerical evaluation of any polynomial.) The numerical algorithm then takes the following form:

> Initialize $x = 1, y = 1, h = -0.01$.
> Let $u = hy$; let $v = 0$.
> Do the following four times: Let $v = (1 + v)u$.
> Then let $y = (1 + v)y$.
> Replace x by $x + h$.
> If x is not one of the values $0.9, 0.8, \ldots$, return to the second step.
> Else compute the true value $z = 1/(2 - x)$, and print x, y, and z.
> If $x > -0.5$, return to the second step; otherwise stop.

In the third line of the algorithm, replacement of the word *four* by the word *two* in one case and *six* in another will satisfy the conditions required in Example 3.

Figure 6.18 shows a program to implement the algorithm above, and the results (augmented by the data for the polynomials of degrees two and six) appear in Fig. 6.19. As you would expect, the results improve as the degree of the Taylor polynomial increases. Improvements also result if the size of h is decreased. The final column in Fig. 6.19 shows values of the exact solution $g(x) = 1/(2 - x)$ of the initial value problem in (14).

```
100 REM       INITIAL VALUES:
110     X = 0 : Y = 1
120     N = 0 : H = −0.01
130 REM       BEGINNING OF LOOP:
140     N = N + 1
150 REM       COMPUTE STACKED POLYNOMIAL:
160     U = H*Y : V = 0
170         FOR I = 1 TO 4
180         V = U*(1 + V)
190         NEXT I
200     Y = Y*(1 + V)
210     X = X + H
220 REM       PRINT EACH TENTH VALUE:
230     K = INT(N/10)
240     IF 10*K < N THEN GOTO 140
250     PRINT N, X, Y, 1/(2 − X)
260     IF X > −0.5 THEN GOTO 140
270     END
```

Figure 6.18

x	Taylor polynomial degree 2	Taylor polynomial degree 4	Taylor polynomial degree 6	True value
1.0	1.000000000	1.000000000	1.000000000	1.0
0.9	0.909098567	0.909090910	0.909090909	0.909090909
0.8	0.833345122	0.833333335	0.833333333	0.833333333
0.7	0.769244669	0.769230771	0.769230762	0.769230762
0.6	0.714300514	0.714285716	0.714285715	0.714285714
0.5	0.666681731	0.666666668	0.666666667	0.666666667
0.4	0.625014889	0.625000002	0.625000000	0.625
0.3	0.588249771	0.588235296	0.588235294	0.588235294
0.2	0.555569489	0.555555559	0.555555556	0.555555556
0.1	0.526329114	0.526315791	0.526315790	0.526315790
0.0	0.500012690	0.500000001	0.500000001	0.5
−0.1	0.476202532	0.476190477	0.476190477	0.476190476
−0.2	0.454556890	0.454545456	0.454545455	0.454545455
−0.3	0.434793449	0.434782610	0.434782609	0.434782609
−0.4	0.416676939	0.416666668	0.416666667	0.416666667
−0.5	0.400009736	0.400000001	0.400000001	0.4

Figure 6.19

6.2 Problems

In each of Problems 1–6, find the exact solution $y = g(x)$ of the given initial value problem. Then use both Euler's method and the improved Euler method to find estimates $y_1, y_2, \ldots, y_{10}$ of the values of the solution at $x_1, x_2, \ldots, x_{10}$ over the given interval with the given step size h. Make tables similar to those in this section to show comparisons between predicted and actual values of the solution.

1. $y' = x + y, y(0) = 0$;   [0, 1]; $h = 0.1$.
2. $y' = x - y, y(0) = 1$;   [0, 1]; $h = 0.1$.
3. $y' = -y^2, y(1) = 1$;   [1, 3]; $h = 0.2$.

4. $y' = y + 1$, $y(0) = -1$; [0, 2]; $h = 0.2$.

5. $y' = -2y$, $y(0) = 1$; [0, 1]; $h = 0.1$.

6. $y' = y$, $y(0) = 1$; [-2, 0]; $h = -0.2$.

Problems 7–12 require the use of a programmable calculator or a microcomputer with printer. First find the exact solution of the given initial value problem by the methods of earlier chapters. Then use Euler's method with step size $h = 0.001$ and the improved Euler method with step size $h = 0.01$ to find approximations to the true value of the solution at 10 equally spaced points (incremented by 0.1) of the given interval.

7. $y' = y - x - 1$, $y(0) = 0$; [0, 1].

8. $y' = 1 + y^2$, $y(0) = 0$; [0, 1].

9. $y' = 3x^2y$, $y(0) = 1$; [0, 1].

10. $xy' = 1$, $y(1) = 0$; [1, 2].

11. $y' = e^y$, $y(1) = 0$; [1, $\frac{3}{2}$].

12. $y' = 4x^3$, $y(-1) = 1$; [-1, 0].

In Problems 13–16, use the three-term Taylor series method with step size $h = 0.01$ to find approximations to the true value of the solution at 10 equally spaced points of the given interval. In each case make a table giving comparisons of these approximations to the true values (as in the figures of this section).

13. $y' = y - x - 1$, $y(0) = 0$; [0, 1].

14. $y' = 1 + y^2$, $y(0) = 0$; [0, 1].

15. $y' = 3x^2y$, $y(0) = 1$; [0, 1].

16. $y' = -y^2$, $y(1) = 1$; [1, 3].

Problems 17–22 require the use of a microcomputer or a minicomputer, together with a printer. In each of these initial value problems, use the improved Euler method with initial step size $h = 0.1$ to approximate the values of the solution at 10 equally spaced points of the given interval. Repeat with the step size halved, then halved again, until you feel you can trust the answers to five significant digits. Present your results in easy-to-read tabular form.

17. $y' = x^2 + y^2$, $y(0) = 0$; [0, 1].

18. $y' = x^2 - y^2$, $y(0) = 1$; [0, 2].

19. $y' = x^2 - \sqrt{y}$, $y(0) = 1$; [0, 1].

20. $y' = \ln y$, $y(1) = 1$; [1, 2].

21. $y' = \sin xy$, $y(0) = 1$; [0, 0.5].

22. $y' = xy^2 - y$, $y(0) = -1$; [-1, 0].

23. Here is one idea that might yield an improvement in both the improved Euler method and the Taylor series method, and it depends upon combining the two. Suppose that x_n and y_n have both been computed for some value of n (and all previous values, of course) in the initial value problem

$$y' = f(x, y), \qquad y(a) = y_a. \qquad (1)$$

Use the differential equation in (1) to compute estimates of $y'(x_n)$ and $y''(x_n)$ then use the Taylor series method to obtain a first estimate u_{n+1} of the solution at x_{n+1}. Next use the differential equation in (1) to compute estimates of $y'(x_{n+1})$ and $y''(x_{n+1})$. Then average the first derivatives at the points x_n and x_{n+1}, and also average the second derivatives at those two points, to obtain (possibly better) estimates of y' and y'' over the interval $[x_n, x_{n+1}]$. Finally use these averages in the three-term Taylor series algorithm to obtain a (possibly better) estimate y_{n+1} of the solution at x_{n+1}. Write a flowchart for this new algorithm, write a program to implement it, and apply it to the initial value problem

$$y' = 2xy, \qquad y(0) = 1 \qquad (14)$$

on [0, 1] discussed in Example 2 of this section. Comment on the results. (*Note:* Data for the improved Euler method and the three-term Taylor series method are shown in Fig. 6.17.)

6.3
The Runge-Kutta Method

Recall the initial value problem

$$y' = f(x, y), \qquad y(x_0) = y_0. \tag{1}$$

We now discuss a method for approximating the solution $y = g(x)$ of (1) that has a higher order of accuracy than any of the numerical methods discussed in Sections 6.1 and 6.2. It is called the *Runge-Kutta method* and is named for the German mathematicians Carl Runge (1856–1927) and Wilhelm Kutta (1867–1944).

With the usual notation, suppose that we have computed the approximations $y_1, y_2, \ldots, y_n$ to the (true) values $g(x_1), g(x_2), \ldots, g(x_n)$, and now want to compute $y_{n+1} \approx g(x_{n+1})$. Then

$$g(x_{n+1}) - g(x_n) = \int_{x_n}^{x_{n+1}} g'(x)\, dx = \int_{x_n}^{x_n + h} g'(x)\, dx \tag{2}$$

by the fundamental theorem of calculus, and Simpson's rule for numerical integration therefore yields

$$g(x_{n+1}) - g(x_n) \approx \frac{h}{6} \left[g'(x_n) + 4g'\left(x_n + \frac{h}{2}\right) + g'(x_{n+1}) \right]. \tag{3}$$

Hence we want to define y_{n+1} so that

$$y_{n+1} \approx y_n + \frac{h}{6} \left[g'(x_n) + 2g'\left(x_n + \frac{h}{2}\right) \right.$$

$$\left. + 2g'\left(x_n + \frac{h}{2}\right) + g'(x_{n+1}) \right]; \tag{4}$$

we have split $4g'(x_n + h/2)$ into a sum of two terms because we intend to approximate the slope $g'(x + h/2)$ of $y = g(x)$ at the midpoint $x_n + h/2$ of $[x_n, x_{n+1}]$ in two different ways.

On the right-hand side in (4), we replace the (true) slopes $g'(x_n)$, $g'(x_n + h/2)$, $g'(x_n + h/2)$, and $g'(x_{n+1})$ (respectively) by the estimates

$$k_{n1} = f(x_n, y_n) \qquad \text{(the slope at } x_n\text{)}, \tag{5a}$$

$$k_{n2} = f\left(x_n + \frac{h}{2}, y_n + \frac{h}{2} k_{n1}\right) \qquad \begin{array}{l}\text{(the slope at the midpoint of } [x_n, x_{n+1}],\\ \text{using the Euler method to predict the}\\ \text{ordinate there),} \end{array} \tag{5b}$$

$$k_{n3} = f\left(x_n + \frac{h}{2}, y_n + \frac{h}{2} k_{n2}\right) \qquad \begin{array}{l}\text{(an improved Euler value for the slope}\\ \text{at the midpoint),} \end{array} \tag{5c}$$

and

$$k_{n4} = f(x_{n+1}, y_n + h k_{n3}) \qquad \begin{array}{l}\text{(the Euler method slope at } x_{n+1}\text{, using}\\ \text{the improved slope } k_{n3} \text{ at the midpoint}\\ \text{to step to } x_{n+1}\text{).} \end{array} \tag{5d}$$

When these substitutions are made in (4), the result is the iterative formula

$$y_{n+1} = y_n + \frac{h}{6}(k_{n1} + 2k_{n2} + 2k_{n3} + k_{n4}). \tag{6}$$

The use of this formula to compute the approximations $y_1, y_2, y_3, \ldots$ successively constitutes the **Runge-Kutta method**.

It happens that this method is equivalent to the five-term (fourth degree) Taylor formula method that uses in place of (6) the iterative formula

$$y_{n+1} = y_n + y_n'h + \frac{y_n''}{2!}h^2 + \frac{y_n^{(3)}}{3!}h^3 + \frac{y_n^{(4)}}{4!}h^4. \qquad (7)$$

The Runge-Kutta method is, however, simpler to apply in practice because the computation of the numbers $k_{n1}, k_{n2}, k_{n3},$ and k_{n4} in (5a)–(5d) involves only evaluation of the original function $f(x, y)$ in (1), whereas use of the Taylor formula in (7) requires high-order partial derivatives of f.

The Runge-Kutta method is a *fourth order* method—it can be proved that cumulative error on a bounded interval $[a, b]$ with $a = x_0$ is of order h^4. (Thus the iteration in (6) is sometimes called the *fourth order* Runge-Kutta method because it is possible to develop Runge-Kutta methods of other orders.) That is,

$$|y_n - g(x_n)| \le Ch^4, \qquad (8)$$

where the constant C depends on the function $f(x, y)$ and the interval $[a, b]$, but does not depend on the step size h. The following example illustrates this high accuracy in comparison with the lower-order accuracy of our previous numerical methods.

EXAMPLE 1 Compare the results obtained using the Runge-Kutta method with the Euler method, the improved Euler method, and the exact solution of the initial value problem

$$y' = x + y + 1, \qquad y(0) = 1 \qquad (9)$$

on $[0, 1]$ with step size $h = 0.1$.

Solution First we decide how to implement the Runge-Kutta method algorithmically. Suppose that x_n and y_n have been computed. Then:

Let $k_1 = f(x_n, y_n) = x_n + y_n + 1.$

Let $k_2 = f\left(x_n + \frac{h}{2}, y_n + \frac{h}{2}k_1\right) = x_n + y_n + \frac{h}{2}(1 + k_1).$

Let $k_3 = f\left(x_n + \frac{h}{2}, y_n + \frac{h}{2}k_2\right) = x_n + y_n + \frac{h}{2}(1 + k_2).$

Let $x_{n+1} = x_n + h.$

Let $k_4 = f(x_{n+1}, y_n + hk_3) = x_n + y_n + h(1 + k_3).$

Let $y_{n+1} = y_n + \frac{h}{6}(k_1 + 2k_2 + 2k_3 + k_4).$

It is not difficult to write a simple program that follows this outline, together with additional steps for computing the Euler method values, the improved Euler method values, and the true values of the solution. The results appear in Fig. 6.20.

Because several calculations are carried out at each step in the

x	Euler method	Improved Euler	Runge-Kutta fourth order	Exact value
0.0	1.000000000	1.000000000	1.000000000	1.000000000
0.1	1.2	1.215	1.2155125	1.215512754
0.2	1.43	1.463075	1.464207713	1.464208274
0.3	1.693	1.747697875	1.749575492	1.749576423
0.4	1.9923	2.072706152	2.075472721	2.075474093
0.5	2.33153	2.442340298	2.446161917	2.446163812
0.6	2.714683	2.861286029	2.866353888	2.866356401
0.7	3.1461513	3.334721062	3.341254882	3.341258122
0.8	3.63076643	3.868366774	3.876618692	3.876622785
0.9	4.173843073	4.468545285	4.478804244	4.478809333
1.0	4.781227380	5.142242540	5.154839235	5.154845485

Figure 6.20

Runge-Kutta method, relatively large step sizes should be used to avoid accumulation of roundoff error. With $h = 0.1$, the Runge-Kutta method is in error at $x = 1$ by 6.25×10^{-6}, and with $h = 0.01$, the error there decreases to 2.0×10^{-9}. But with $h = 0.001$, the error *increases* to 3.6×10^{-8}. (The exact results depend, of course, on the computer, the the programming language, and the program logic itself.)

EXAMPLE 2 Consider the initial value problem of Example 1. Use the Runge-Kutta method with initial step size $h = 0.2$ to estimate the value of the solution at $x = 1$. Compare the result with the true value. Halve the step size and repeat. Continue this process as far as practical.

Solution   We omit the computational details. The results are shown in Fig. 6.21. It is also worth noting that the results do not improve with h

h	Runge-Kutta fourth order approximation	$\lvert \text{Error} \rvert / h^4$
0.2	5.154753409	0.0575
0.1	5.154839235	0.0625
0.05	5.154845077	0.0653
0.025	5.154845459	0.0666
0.0125	5.154845488	0.1229
0.00625	5.154845489	2.6214
0.003125	5.154845488	31.457
0.0015625	5.154845477	1342.2
0.00078125	5.154845475	26844
0.000390625	5.154845464	901940
0.0001953125	5.154845503	12370×10^3
0.00009765625	5.154845793	33865×10^4

Figure 6.21   True value: $y(1) \approx 5.154845845$.

< 0.0125, and in fact become worse—because of accumulated roundoff error—for $h < 0.003125$. Indeed, the last entry in the approximation column depends on the accumulated errors in over 60,000 arithmetical computations. The final column (the ratio of the magnitude of the error to h^4) exhibits dramatically the cumulative effect of roundoff error. Observe that the first four entries in this column are consistent with the error bound in (8), C apparently being less than 0.07. The size of the remaining entries indicates that roundoff error is rapidly taking over.

EXAMPLE 3   A skydiver jumps from an aircraft at initial altitude 10,000 ft. Assume that the skydiver falls vertically with initial velocity zero, weighs 128 lb, and that air resistance exerts an upward force numerically equal to $-(0.1)v + (0.001)v^2 - (0.0001)v^3$ (pounds), where v is the velocity of the skydiver in feet per second. If she does not open her parachute, what will be her terminal velocity? How fast will she be falling after 10 s have elapsed? After 15 s? After 20 s?

Solution   The skydiver's mass is 4 slugs, so Newton's law $F = ma$ yields the equation of motion

$$4\frac{dv}{dt} = -128 - (0.01)v + (0.001)v^2 - (0.0001)v^3 \tag{10}$$

(we have taken the upward direction as the positive direction). The initial condition is $v(0) = 0$, and the above equation can be written in the form

$$\frac{dv}{dt} = f(v) \tag{11}$$

where

$$f(v) = -32 - (0.0025)v + (0.00025)v^2 - (0.000025)v^3. \tag{12}$$

We can calculate her terminal velocity immediately by the simple expedient of setting the acceleration $f(v)$ equal to zero and then using Newton's method to solve the resulting algebraic equation

$$-32 - (0.0025)v + (0.00025)v^2 - (0.000025)v^3 = 0. \tag{13}$$

The result is that terminal velocity is approximately -105.046 ft/s. But this answers only the first question posed.

A Runge-Kutta approximation to the solution of the initial value problem $v'(t) = f(v)$, $v(0) = 0$ with step size $h = 0.01$ yields the velocity v as a function of time t with values shown in the table in Fig. 6.22.

Note that after 16 s the terminal velocity has been effectively attained. After 10 s, $v(10) = 104.984$ ft/s is 99.94% of terminal velocity.

The program used to generate the data is shown in Fig. 6.23. Note that in line 130 it makes use of a feature not available in all versions of BASIC, which enables us to define and use thereafter the function $f(v)$, denoted by FNF(V). When this feature is available, it enables us to write programs that are almost as transparent as flowcharts for numerical algorithms.

t (sec)	v (ft/sec)
1	-31.675
2	-60.264
3	-81.423
4	-93.933
5	-100.163
6	-102.972
7	-104.178
8	-104.685
9	-104.897
10	-104.984
11	-105.020
12	-105.035
13	-105.042
14	-105.044
15	-105.045
16	-105.046
17	-105.046
18	-105.046
19	-105.046
20	-105.046

Figure 6.22

```
100 REM      DEFINITION OF F(V):
110          A = − 32 : B = − 0.0025
120          C = − B/10 : D = − C/10
130          DEF FNF(V) = A + V*(B + V*(C + D*V))
140 REM      INITIAL VALUES:
150          T = 0 : V = 0 : N = 0 : H = 0.01
160 REM      BEGINNING OF ITERATION:
170          N = N + 1 : T = T + H
180 REM      COMPUTE SLOPES:
190          K1 = FNF(V)
200          K2 = FNF(V + (H/2)*K1)
210          K3 = FNF(V + (H/2)*K2)
220          K4 = FNF(V + H*K3)
230 REM      RUNGE-KUTTA FORMULA:
240          V = V + (H/6)*(K1 + 2*K2 + 2*K3 + K4)
250 REM      PRINT ONLY AFTER EACH WHOLE SECOND:
260          M = INT(N/100)
270          IF 100*M < N THEN GOTO 170
280          PRINT T, V
290          IF T < 20 THEN GOTO 170
300          END
```

Figure 6.23

6.3 Problems

In each of Problems 1–10, first find the exact solution $g(x)$ of the given initial value problem. Then use the Runge-Kutta method to find estimates $y_1, y_2, \ldots, y_{10}$ of the solution at $x_1, x_2, \ldots, x_{10}$ over the given interval with step size $h = 0.1$. Make a table showing a comparison between the approximate and actual solution in each case.

1. $y' = x + y, y(0) = 0;$ $[0, 1]$.
2. $y' = x - y, y(0) = 1;$ $[0, 1]$.
3. $y' = -y^2, y(1) = 1;$ $[1, 2]$.
4. $y' = y + 1, y(0) = -1;$ $[0, 1]$.
5. $y' = -2y, y(0) = 1;$ $[0, 1]$.
6. $y' = y, y(0) = 1;$ $[-1, 0]$.
7. $y' = y - x - 1, y(0) = 0;$ $[0, 1]$.
8. $y' = 1 + y^2, y(0) = 0;$ $[0, 1]$.
9. $y' = 3x^2y, y(0) = 1;$ $[0, 1]$.
10. $xy' = 1, y(1) = 0;$ $[1, 2]$.

In each of Problems 11–20, find the exact solution of the given initial value problem. Then use the Runge-Kutta method to approximate the solution at 10 equally spaced points of the given interval, using step size $h = 0.01$ in each case. Make a table showing comparisons between the true values and the approximate values. (Note: Each of these problems should be worked using a programmable calculator or a microcomputer.)

11. $y' = e^y, y(1) = 0;$ $[1, \frac{3}{2}]$.
12. $y' = 4x^3, y(-1) = 1;$ $[-1, 0]$.
13. $y' = 5x^4, y(-1) = -1;$ $[-1, 0]$.
14. $y' = (1 - y^2)^{1/2}, y(0) = 0;$ $[0, 3]$.
15. $3y^2y' = 2x + 1, y(0) = 1;$ $[0, 2]$.
16. $x^2y' = y^2 + 1, y(0.5) = 0;$ $[0.5, 1.5]$.
17. $x\,dy = (x + 1)y^2\,dx, y(1) = 1;$ $[1.0, 1.5]$.
18. $y' = e^x - 2y, y(0) = 1;$ $[0, 1]$.

19. $x^2y' + 2xy = 1, y(1) = 1;$ [1, 2].

20. $y' + 2xy = x^3, y(0) = 0;$ [−2, 0].

The next four problems require the use of a programmable calculator or a microcomputer, as well as a printer. In Problems 21–24, use the Runge-Kutta method with appropriate step size to approximate the solution of the given initial value problem at 10 equally spaced points on the given interval. Halve the step size and repeat. Continue the process until you are reasonably sure that you have five digits correct to the right of the decimal point.

21. $y' = x^2 + y^2, y(0) = 0;$ [0, 1].

22. $y' = x^2 - y^2, y(0) = 0;$ [0, 1].

23. $y' = x^2 - \sqrt{y}, y(0) = 1;$ [0, 2].

24. $y' = xy^2 - y, y(0) = -1;$ [−1, 0].

25. Given the initial value problem $y' = f(x, y), y(x_0) = y_0$, consider the following solution technique. Suppose that x_n and y_n have been computed. Let

$$y_{n+1} = y_n + hf\left(x_n + \frac{h}{2}, y_n + \frac{h}{2}f(x_n, y_n)\right).$$

The use of this algorithm for approximating the solution of the initial value problem above is called the modified Euler method, although it is rather different from the improved Euler method and more closely related to the Runge-Kutta method. Apply the modified Euler method to the initial value problem $y' = 2xy, y(0) = 1$ on the interval [0, 2] with step size $h = 0.05$. Compare the results with the improved Euler method and the Runge-Kutta method with the same step size, and compare all as well with the actual solution.

In Problems 26–30, apply the modified Euler method (Problem 25) to find numerical solutions of the given initial value problems at 10 equally spaced points of the given interval; use step size $h = 0.01$ in each case. Compare the numerical results with the actual solutions in each case.

26. $y' = x + y + 1, y(0) = 0;$ [0, 1].   **27.** $y' = 3x^2 - y, y(0) = 1;$ [0, 1].

28. $y' = -2xy, y(0) = 1;$ [0, 2].   **29.** $y' = 4x^3, y(-1) = 1;$ [−1, 0].

30. $4y^3y' = 2x + 1, y(1) = 1;$ [0, 2].

31. The simple pendulum with bob of mass m on a rod of negligible weight of length L obeys the differential equation $mLy'' = -mg \sin y$ where y is the angle the rod makes with the vertical and the derivative is with respect to time t. Assume that the rod is initially at rest and hanging vertically. At time $t = 0$ the bob is given an initial (angular) velocity of $y'(0) = v$ radians per second. Show that the subsequent motion of the pendulum is governed by the initial value problem

$$(y')^2 = \frac{1}{L}(C + 2g \cos y), \qquad y(0) = 0,$$

where $C = v^2L - 2g$. With $L = 1$ (m), $v = 0.5$, and $g = 9.80$ (m/s²), use the Runge-Kutta method to find the period T of the pendulum.

32. Repeat Problem 31 if the initial angular velocity of the bob is $v = 1.0$ rad/s.

33. A projectile of mass m is launched from the surface of the earth with an initial velocity of v_0 miles per second. Assume that air resistance can be ignored. Then the subsequent motion of the projectile is determined by Newton's law of universal gravitation, so that if $y = y(t)$ denotes the distance between the projectile and the *center* of the earth at time t, then

$$my'' = -\frac{GMm}{y^2},$$

where M is the mass of the earth and G is a constant. Because it is easy to show that $GM = R^2g$, where R is the radius of the earth and g is its surface gravity, the motion of the projectile is thus determined by the initial value problem

$$y'' = -\frac{R^2g}{y^2}, \qquad y(0) = R, \qquad y'(0) = v_0.$$

Suppose that $R = 4000$ (mi), $g = 32$ (ft/s²), and $v_0 = 6$ (mi/s). First multiply each term in the last differential equation by $2y'$ to make it easy to integrate; you will obtain a first order initial value problem. Then use the Runge-Kutta method to compute the height and velocity of the projectile at increments of 10 s for the first 5 min of its flight.

34. A 2-lb projectile is fired straight upward with an initial velocity of 2800 ft/s. Assume that air resistance is numerically equal to $(0.000001)|v|^{3/2}$ pounds where v is the velocity of the projectile in feet per second. Because the projectile will remain close to the surface of the earth, you may assume that the acceleration $g = 32$ ft/s² of gravity is constant. How high will the projectile go? How long will be required for it to reach that altitude? Use the Runge-Kutta method.

35. Repeat Problem 34 in the case that the air resistance is instead numerically equal to $(0.02)|v|^{3/2}$ pounds where v is the velocity of the projectile in feet per second.

36. A motorboat is traveling at 30 ft/s when its engine is shut off. Assume that the complex interaction between the shape of its hull and the turbulent water causes the equation of its subsequent motion to be $mv' = -kv - cv^3$ where $m = 100$ (slugs) is the mass of the motorboat, v is its velocity in feet per second, $k = 0.1$, and $c = 0.01$. Use the Runge-Kutta method to find how far the motorboat will coast in the first 10 s after its engine is shut off.

6.4
A Multistep Method

The numerical methods of the previous three sections are **one-step** methods for computing approximations $y_1, y_2, y_3, \ldots$ to the true values $g(x_1), g(x_2), g(x_3), \ldots$ of the solution of the initial value problem

$$y' = f(x, y), \qquad y(x_0) = y_0. \tag{1}$$

That is, in the Euler, Runge-Kutta, and Taylor series methods, the computation of the next approximation y_{n+1} uses only the value of the previous approximation y_n (and perhaps various values of x as well). In a **multistep** method, the computation of y_{n+1} uses two or more previous approximations (for instance, both y_n and y_{n-1}).

There are many multistep methods; here we will examine in detail only one, the *Milne method*, which is a two-step method. Assume that $y_1, y_2, \ldots, y_n$ have already been computed and that $n \geq 2$. We use Simpson's rule (much as in the derivation of the Runge-Kutta formula in Section 6.3) to write

$$g(x_{n+1}) - g(x_{n-1}) = \int_{x_{n-1}}^{x_{n+1}} g'(x)\,dx \approx \frac{(2h)}{6}[g'(x_{n-1}) + 4g'(x_n) + g'(x_{n+1})]. \tag{2}$$

Because $g'(x) = f(x, g(x))$, the replacement of $g(x_i)$ by y_i in (2) yields the iterative formula

$$y_{n+1} = y_{n-1} + \frac{h}{3}[f(x_{n-1}, y_{n-1}) + 4f(x_n, y_n) + f(x_{n+1}, y_{n+1})] \qquad (3)$$

The successive application of this formula constitutes the **Milne method**.

Observe that Eq. (3) is an *implicit* formula for y_{n+1}. It appears that in order to obtain y_{n+1} on the left-hand side, we must know in advance the value of y_{n+1} on the right-hand side. To skirt this problem, we simply solve (3) for y_{n+1} to express it in terms of y_n, y_{n-1}, and various values of x. If solving for y_{n+1} is difficult, the Milne method may not be a good choice; for certain simple differential equations, such as the linear equation of the following example, solving for y_{n+1} is simple.

EXAMPLE 1 Derive an explicit Milne formula for approximating the solution of the initial value problem

$$y' = x - y + 1, \qquad y(0) = 2. \qquad (4)$$

Solution From (3) we find that

$$y_{n+1} = y_{n-1} + \frac{h}{3}[(x_{n-1} - y_{n-1} + 1) + 4(x_n - y_n + 1)$$
$$+ (x_{n+1} - y_{n+1} + 1)].$$

Next, we employ a little elementary algebra to solve for y_{n+1}:

$$y_{n+1} = \frac{3}{3+h}\left[\frac{3-h}{3}y_{n-1} + \frac{h}{3}(x_{n-1} + 4x_n + x_{n+1} - 4y_n + 6)\right]. \qquad (5)$$

Notice that we can use the formula in (5) to compute successively the approximations $y_2, y_3, y_4, \ldots$ only if we know not only the initial value y_0 but also the first approximate value y_1. This situation is typical of multistep methods. Before proceeding to use the iterative formula of the method, we first must compute y_1 separately by a one-step method. The normal procedure is to use the Runge-Kutta method or a Taylor series method for this purpose. It is advisable to compute y_1 with a good deal of accuracy, because all subsequent values $y_2, y_3, y_4, \ldots$ depend on it. Thus, even if the step size to be used in the Milne method is $h = 0.01$, for instance, it might be well to apply the Runge-Kutta iteration ten times with step size $h = 0.001$ to obtain y_1 rather than using the step size planned for the Milne method.

EXAMPLE 2 Find y_1 for Example 1 by the fourth degree Taylor series method. Then, on the interval $[0, 2]$, approximate the solution of the initial value problem in (4) by the Milne method, using step size $h = 0.01$ throughout.

Solution Here we have

$$y' = x - y - 1,$$
$$y'' = 1 - y' = -x + y,$$
$$y^{(3)} = -1 + y' = x - y,$$

and

$$y^{(4)} = 1 - y' = -x + y.$$

The initial condition $y(0) = 2$ therefore yields $y'(0) = -1$, $y''(0) = 2$, $y^{(3)}(0) = -2$, and $y^{(4)}(0) = 2$. Hence the fourth degree Taylor polynomial for $y_1 = y(h)$ is

$$y_1 = 2 - h + h^2 - \tfrac{1}{3}h^3 + \tfrac{1}{12}h^4 \approx 1.990099668,$$

using $h = 0.01$. The table in Fig. 6.24 shows typical approximate values of the solution using the formula in (5).

x	Approximate value	True value
0.0	2.000000000	2.000000000
0.1	1.909674837	1.909674836
0.2	1.837461506	1.837461506
0.3	1.781636442	1.781636441
0.4	1.740640093	1.740640092
0.5	1.713061322	1.713061319
⋮	⋮	⋮
1.6	2.003793044	2.003793036
1.7	2.065367056	2.065367048
1.8	2.130597784	2.130587776
1.9	2.199137245	2.199137238
2.0	2.270670573	2.270670566

Figure 6.24

A comparison of the approximate and true values in Fig. 6.24 suggests that the Milne method is very accurate. Under the assumption that the first approximation y_1 is computed with an error of order h^5, it can be proved that the cumulative error (on a fixed bounded interval) in the Milne method is of order h^4, the same as in the Runge-Kutta method. But the Milne method enjoys the advantage that each iteration requires only *one* new evaluation of the function $f(x, y)$, whereas each iteration of the Runge-Kutta method requires *four* new evaluations of $f(x, y)$. Therefore Milne's method often runs faster on a computer when, as in Example 1, the implicit Eq. (3) is easily solved for y_{n+1}.

In general, some labor may be involved in solving the implicit Eq. (3) for y_{n+1} in terms of previously computed values. For example, if $f(x, y)$ is a polynomial of degree three in y, then (3) will be a cubic equation in y_{n+1}. This cubic equation could be solved using Newton's method, starting with y_n itself as a (probably very good) initial guess for the value of y_{n+1}.

It is usually more efficient, though, to employ a predictor-corrector technique to obtain y_{n+1}. Any one-step method can be used as a predictor to compute an estimate u_{n+1} of the value of y_{n+1}. If y_{n+1} is then replaced by u_{n+1} on the right-hand side in (3), then the Milne formula becomes a corrector formula. In Example 3, we use an improved Euler predictor,

$$u_{n+1} = y_n + \frac{h}{2}[f(x_n, y_n) + f(x_{n+1}, y_n + hf(x_n, y_n))], \qquad (6)$$

and a Milne corrector,

$$y_{n+1} = y_{n-1} + \frac{h}{3}[f(x_{n-1}, y_{n-1}) + 4f(x_n, y_n) + f(x_{n+1}, u_{n+1})] \qquad (7)$$

following a Runge-Kutta starter to compute y_1.

EXAMPLE 3 In Example 2 of Section 1.9 we considered a lunar lander that initially is falling freely toward the surface of the moon at a speed of 1000 mi/h. Its retrorockets, when fired in free space, provide a deceleration of 33,000 mi/h^2; in addition, the lander is subject to the gravitational attraction of the moon. We found that a soft touchdown ($v = 0$ at impact) is achieved by firing the rockets beginning at time $t = 0$ at a height of 25 mi above the lunar surface.

Here we want to compute the *descent time* of the lunar lander. In Section 1.9 we essentially derived the differential equation

$$\frac{dy}{dt} = -\left(66y + \frac{30.32}{y} - 99.36\right)^{1/2} \qquad (8)$$

where t is measured in hours. The distance y from the lunar lander to the center of the moon is measured in kilomiles. Because the radius of the moon is 1.080 kilomiles, the initial condition associated with the lunar lander problem is $y(0) = 1.105$.

The program listed in Fig. 6.25 implements the predictor-corrector method of the formulas in (6) and (7). In lines 170–210, the value of y_1 is computed using the Runge-Kutta method. Line 150 is a "stopping rule." The radicand in (8) is positive as long as the lander is above the lunar surface, is zero at impact, and would be negative were the lander to continue beneath the surface. Hence we stop the iteration as soon as this quantity becomes negative. This is the effect of the ON ERROR GOTO instruction in line 150. Whenever the square root of a negative number is encountered, an ERROR is registered, and this instruction (which is not available in all versions of BASIC) transfers program control to line 320, which prints "STOP." (One improvement: If you are using a version of BASIC that distinguishes between various sorts of errors, and attempting to take the square root of a negative number is identified as ERROR 5, modify line 150 to read, "ON ERROR 5 GOTO 320.")

Figure 6.26 shows the results obtained by running this program with step size $h = 0.004$ (14.4 s). The last iteration before stopping gave the values T = 0.0440 and Y = 1.080462. The program was then executed a second time with these as new initial values and with the reduced step size $h = 0.0004$. The results are shown in Fig. 6.27. They indicate that the lander stops just after time $t = 0.0492$ (h; that is, about 2 min, 57 s). Thus the lander's time of descent to the lunar surface is just under 3 mins.

```
100 REM          ***     LUNAR LANDER     ***
105 REM
110 REM          DEFINITION OF F(Y):
115 REM
120                  DEF FNF(Y) = − SQR(66*Y + (30.32)/Y − 99.36)
125 REM
130 REM          INITIAL VALUES:
135 REM
140                  T = 0 : Y0 = 1.105 : H = 0.004
150                  ON ERROR GOTO 320
155 REM
160 REM          COMPUTE Y1 BY RUNGE-KUTTA METHOD:
165 REM
170                  K1 = FNF(Y0)
180                  K2 = FNF(Y0 + (H/2)*K1)
190                  K3 = FNF(Y0 + (H/2)*K2)
200                  K4 = FNF(Y0 + H*K3)
210                  Y1 = Y0 + (H/6)*(K1 + 2*K2 + 2*K3 + K4)
215 REM
220 REM          BEGINNING OF ITERATION:
225 REM
230                  T = T + H : V = Y1
240                  PRINT T, Y1
245 REM
250 REM          IMPROVED EULER PREDICTOR:
255 REM
260                  M = FNF(V)
270                  U = V + (H/2)*(M + FNF(V + H*M) )
275 REM
280 REM          MILNE CORRECTOR:
285 REM
290                  Y1 = Y0 + (H/3)*(FNF(Y0) + 4*FNF(V) + FNF(U) )
300                  Y0 = V
310                  GOTO 230
314 REM
315 REM          END OF LOOP.
316 REM
320                  PRINT "STOP."
330                  END
```

Figure 6.25

T (hours)	Y (kilomiles)
0.0040	1.101147
0.0080	1.097621
0.0120	1.094422
0.0160	1.091549
0.0200	1.089000
0.0240	1.086774
0.0280	1.084871
0.0320	1.083290
0.0360	1.082030
0.0400	1.081088
0.0440	1.080462
STOP	

Figure 6.26

T (hours)	Y (kilomiles)
0.0444	1.080419
0.0448	1.080379
0.0452	1.080342
0.0456	1.080308
0.0460	1.080278
0.0464	1.080251
0.0468	1.080227
0.0472	1.080206
0.0476	1.080188
0.0480	1.080174
0.0484	1.080162
0.0488	1.080155
0.0492	1.080149
STOP	

Figure 6.27

*THE ADAMS-MOULTON METHOD

To indicate the great variety of multistep techniques, we close this section with a brief description of the four-step **Adams-Moulton** predictor-corrector method. A detailed treatment can be found in *Elementary Numerical Analysis* (New York: McGraw-Hill, 2nd, ed., 1972) by S. D. Conte and C. deBoor. Beginning with the integral

$$g(x_{n+1}) - g(x_n) = \int_{x_n}^{x_{n+1}} g'(x)\,dx, \tag{9}$$

the unknown function $g'(x)$ is approximated with a cubic polynomial whose graph passes through the four points (x_{n-3}, f_{n-3}), (x_{n-2}, f_{n-2}), (x_{n-1}, f_{n-1}), and (x_n, f_n), where $f_i = f(x_i, y_i)$. The result of this approximation and the integration in (9) is then the predictor formula

$$u_{n+1} = y_n + \frac{h}{24}(55f_n - 59f_{n-1} + 37f_{n-2} - 9f_{n-3}) \tag{10}$$

for $u_{n+1} \approx g(x_{n+1})$. Next, $g'(x)$ is approximated by a cubic polynomial that passes through the four points (x_{n-2}, f_{n-2}), (x_{n-1}, f_{n-1}), (x_n, f_n), and $(x_{n+1}, f(x_{n+1}, u_{n+1}))$. The result of the integration in (9) is then the corrector formula

$$y_{n+1} = y_n + \frac{h}{24}[9f(x_{n+1}, u_{n+1}) + 19f_n - 5f_{n-1} + f_{n-2}]. \tag{11}$$

The use of the formulas in (10) and (11) requires that the values of y_1, y_2, and y_3 be computed first. The Runge-Kutta method is natural for this purpose. Once the values of y_0, y_1, y_2, and y_3 are available, the formulas in (10) and (11) can be employed alternately to compute u_4, y_4, u_5, y_5, ... As a final elegant touch, the corrector formula may be used two or three times at each iteration to improve the approximation y_i. That is, to obtain y_{n+1}, obtain u_{n+1} with the predictor formula in (10). Improve it to v_{n+1} with the corrector formula in (11), improve it again to w_{n+1} using the corrector, and last improve it to y_{n+1} by means of a final application of the corrector.

6.4 Problems

Use the Milne method in Problems 1–12.

1. Solve the initial value problem $y' = -y$, $y(0) = 1$ on the interval $[0, 1]$ with step size $h = 0.05$. Obtain *all* computed values of y_n. Compare your results with those of Example 1 of Section 6.1 (see Fig. 6.1).

2. Solve the initial value problem $y' = -8y$, $y(0) = 1$ on the interval $[0, 2]$ with step size $h = 0.01$. Print the true values of y together with the computed values for $x_1 = 0.1$, $x_2 = 0.2$, ..., $x_{20} = 2.0$, and (important!) include the *percentage error* at each computed value.

3. Solve the initial value problem $7xy' + y = 0$, $y(-1) = 1$ with step size $h = 0.075$ on the interval $[-1, 1]$. Compare your results with those of Example 3 of Section 6.1.

4. The initial value problem $y' = -2xy/(x^2 + 1)$, $y(0) = 1$ has the exact value $y(1) = \frac{1}{2}$. What does the Milne method produce for $y(1)$ with step sizes $h = 0.1$, 0.05, 0.025, ..., and 0.003125? Compare your results with those of Example 1 of Section 6.2 (see Fig. 6.10).

5. Solve the initial value problem of Problem 4 on the interval $[0, 1]$ with $h = 0.1$; print all values of y_n. Compare your results with those shown in Fig. 6.15 (in Section 6.2).

6. Solve the initial value problem $y' = 3x^2y$, $y(0) = 1$ on the interval $[0, 1]$ with step size $h = 0.01$. Print the true values, the computed values, and the percentage errors for $x_1 = 0.1$, $x_2 = 0.2$, ..., $x_{10} = 1.0$.

7. Solve the initial value problem $y' = y + 1$, $y(0) = 1$ on the interval $[0, 2]$ with step size $h = 0.01$. Compare your results with the true values at $x_1 = 0.2$, $x_2 = 0.4$, ..., $x_{10} = 2.0$.

8. Solve the initial value problem $y' = x + y + 1$, $y(0) = 1$ on the interval $[0, 1]$ with step size $h = 0.1$. Compare your results with those of Example 1 of Section 6.3 (see Fig. 6.20).

9. Solve the initial value problem $y' = 2xy$, $y(0) = 1$ on the interval $[0, 1]$ with step size $h = 0.01$; print y_n for $x_n = 0, 0.1, 0.2, ..., 0.9$, and 1.0. Compare your results with those of Example 2 of Section 6.2 (see Fig. 6.17).

10. Find the exact solution of $y' = x + y$, $y(0) = 0$. Compare with the results predicted by the Milne method on $[0, 1]$ with step size $h = 0.01$ at the points $x_1 = 0.1$, $x_2 = 0.2$, ..., $x_{10} = 1.0$.

11. Find the exact solution of $y' = x - xy$, $y(0) = 2$. Compare with the results predicted by the Milne method on $[0, 1]$ with step size $h = 0.01$ at the points $x_1 = 0.1$, $x_2 = 0.2$, ..., $x_{10} = 1.0$.

12. Find the exact solution of $y' = 2x + 1$, $y(0) = 2$. Compare with the results predicted by the Milne method on $[0, 2]$ with step size $h = 0.01$ at $x_1 = 0.2$, $x_2 = 0.4$, ..., $x_{10} = 2.0$.

The Milne method is difficult to use if the formula in (3) is difficult to solve for y_{n+1}. Nevertheless, a single-step method may be used to compute a first estimate u_{n+1} of y_{n+1}, and this value may be used in place of y_{n+1} on the right-hand side in (3); then the Milne formula becomes a corrector formula for this predictor. At first glance this seems to be a plausible numerical technique. Apply it in Problems 13–16 with step size $h = 0.01$. Compare the results with the true values at equally spaced points of the interval (as in the previous twelve problems).

13. $y' = y^2, y(1) = 0.25;$ $[1, 3].$ **14.** $y' = (1 - y^2)^{1/2}, y(0) = 0;$ $[0, 3].$

15. $y' = 1 + y^2, y(0) = 0;$ $[0, 1].$ **16.** $3y^2 y' = 2x + 1, y(0) = 2;$ $[0, 2].$

Use the Adams-Moulton predictor-corrector method in Problems 17–21.

17. Solve the initial value problem $y' = x - y + 1, y(0) = 2$ on the interval $[0, 2]$ with step size $h = 0.01$. Print values of y_n for $x = 0.0, 0.1, 0.2, \ldots, 2.0$. Compare your results with those of Example 2 of this section (see Fig. 6.24).

18. Solve the initial value problem $y' = x - y, y(0) = 1$ on the interval $[0, 1]$ with step size $h = 0.01$. Compare your results with the exact values at $x = 0, 0.1, 0.2, \ldots, 1.0$.

19. Solve the initial value problem $y' = y + 1, y(0) = 1$ on the interval $[0, 1]$ with step size $h = 0.01$. Compare your results with the exact values at $x = 0, 0.1, 0.2, \ldots, 1.0$.

20. Repeat the instructions of the previous two problems for the initial value problem $y' = 3x^2 y, y(0) = 1$.

21. Repeat the instructions of Problems 18 and 19 for the initial value problem $y' = 3x^2 e^{-y}, y(0) = 0$.

In Problems 22–26, apply the Runge-Kutta method, the Milne method, and the Adams-Moulton predictor-corrector method with the given step size to approximate the solution of the given initial value problem on the given interval. Make a table comparing the results with the exact values at 10 equally spaced points of the interval.

22. $xy' = 1, y(1) = 0;$ $h = 0.01;$ $[1, 2].$

23. $y' = 4x^3, y(-1) = 1;$ $h = 0.05;$ $[-1, 0].$

24. $y' = e^x - 2y, y(0) = 1;$ $h = 0.02;$ $[0, 1].$

25. $y' + 2xy = x^3, y(0) = 0;$ $h = 0.02;$ $[0, 2].$

26. $y' = x^2 - y, y(0) = 1;$ $h = 0.01;$ $[0, 2].$

6.5
Systems of Differential Equations

We now examine the numerical solution of systems of differential equations. First let us consider the system of two differential equations

$$x' = f(t, x, y),$$
$$y' = g(t, x, y) \tag{1}$$

where the primes denote differentiation with respect to t. Suppose that we are also given the initial conditions $x_0 = x(t_0), y_0 = y(t_0)$. If f, g, and their partial derivatives are all continuous in a region containing the point (t_0, x_0, y_0), then Theorems 3 and 4 in Section 9.4 guarantee the existence and uniqueness of solutions $x = x(t)$ and $y = y(t)$ in a neighborhood of t_0 on the t-axis. With this assurance, we can proceed to discuss the numerical approximation of the solutions $x(t)$ and $y(t)$ of an initial value problem for the system in (1).

We begin with a step-size-h subdivision $t_0, t_1, t_2, \ldots, t_k$ of an interval with initial point t_0. Suppose that we already have computed both approximations $x_1, x_2, \ldots, x_n$ to the true values $x(t_1), x(t_2), \ldots, x(t_n)$, as well as approximations $y_1, y_2, \ldots, y_n$ to the true values $y(t_1), y(t_2), \ldots, y(t_n)$. Next we want to compute approximations x_{n+1} and y_{n+1} to the true values $x(t_{n+1})$ and $y(t_{n+1})$. For this purpose, any of the one-step methods of Sections 6.1 through 6.3 may be applied twice—once to obtain x_{n+1} and again to obtain y_{n+1}.

Using the original Euler method, for instance, we approximate by $f(t_n, x_n, y_n)$ the rate of change of $x(t)$ on the interval $[t_n, t_{n+1}]$ of length h; similarly, we approximate by $g(t_n, x_n, y_n)$ the rate of change of $y(t)$ on the same interval. These approximations yield the iterative formulas

$$x_{n+1} = x_n + hf(t_n, x_n, y_n),$$
$$y_{n+1} = y_n + hg(t_n, x_n, y_n). \tag{2}$$

The successive application of these formulas constitutes the Euler method for the system in (1) of two first order differential equations. As with single equations, the cumulative errors in the Euler method for systems are of order h (on a fixed bounded interval).

EXAMPLE 1 Approximate on the interval $0 \leq t \leq 1$ the solutions of the initial value problem

$$x' = x + y, \qquad x(0) = 1;$$
$$y' = 3x - y, \qquad y(0) = 0. \tag{3}$$

Solution Here we have $f(x, y) = x + y$ and $g(x, y) = 3x - y$, so the iterative equations in (2) take the form

$$x_{n+1} = x_n + h(x_n + y_n) \qquad \text{and} \qquad y_{n+1} = y_n + h(3x_n - y_n).$$

With step size $h = 0.001$, we obtain the results shown in Fig. 6.28.

t	x_n	True value	y_n	True value
0.0	1.000000	1.000000	0.000000	0.000000
0.1	1.120528	1.120735	0.302291	0.302004
0.2	1.286012	1.286449	0.616683	0.616128
0.3	1.503089	1.503792	0.955792	0.954980
0.4	1.780466	1.781488	1.333224	1.332459
0.5	2.129268	2.130681	1.764123	1.762802
0.6	2.563485	2.130681	2.265776	2.264192
0.7	3.100536	3.103049	2.858308	2.856452
0.8	3.761962	3.765248	3.565489	3.563352
0.9	4.574295	4.578560	4.415687	4.413261
1.0	5.570121	5.575626	5.443009	5.440291

Figure 6.28

Note that each formula in (2) has the form of a single Euler iteration, but with y_n inserted like a parameter in the first formula (for x_{n+1}) and with x_n inserted like a parameter in the second formula (for y_{n+1}). The generalization to systems of the other one-step methods of this chapter follows a similar pattern. For instance, when we generalize to the system in (1) the Runge-

Kutta method, we obtain the two iterative formulas

$$x_{n+1} = x_n + \frac{h}{6}(k_{n1} + 2k_{n2} + 2k_{n3} + k_{n4}),$$

$$y_{n+1} = y_n + \frac{h}{6}(j_{n1} + 2j_{n2} + 2j_{n3} + j_{n4}),$$

(4)

each of which has the same form as the original Runge-Kutta formula (Eq. (6) in Section 6.3). The four slopes k_{n1}, k_{n2}, k_{n3}, and k_{n4} used in stepping from x_n to x_{n+1} are the appropriate values of $f(x, y)$, while the four slopes j_{n1}, j_{n2}, j_{n3}, and j_{n4} are appropriate values of $g(x, y)$. Each of these two quadruples of numbers is defined by a quadruple of equations analogous to Eqs. 5(a)–5(d) in Section 6.3:

$$k_{n1} = f(t_n, x_n, y_n),$$

$$j_{n1} = g(t_n, x_n, y_n),$$

$$k_{n2} = f\left(t_n + \frac{h}{2}, x_n + \frac{h}{2}k_{n1}, y_n + \frac{h}{2}j_{n1}\right),$$

$$j_{n2} = g\left(t_n + \frac{h}{2}, x_n + \frac{h}{2}k_{n1}, y_n + \frac{h}{2}j_{n1}\right),$$

$$k_{n3} = f\left(t_n + \frac{h}{2}, x_n + \frac{h}{2}k_{n2}, y_n + \frac{h}{2}j_{n2}\right),$$

$$j_{n3} = g\left(t_n + \frac{h}{2}, x_n + \frac{h}{2}k_{n2}, y_n + \frac{h}{2}j_{n2}\right),$$

$$k_{n4} = f(t_n + h, x_n + hk_{n3}, y_n + hj_{n3}),$$

$$j_{n4} = g(t_n + h, x_n + hk_{n3}, y_n + hj_{n3}).$$

(5)

As with single equations, the Runge-Kutta method for systems leads to cumulative errors (on a fixed bounded interval) of order h^4. We defer to the problems the numerical illustrations of the Runge-Kutta method for systems.

HIGHER ORDER SYSTEMS

As we saw in Section 5.1, any system of higher order differential equations can be replaced with an equivalent system of first order differential equations. For instance, consider the system

$$x'' = F(t, x, y, x', y'),$$

$$y'' = G(t, x, y, x', y'),$$

(6)

of two second order equations. If we define the two auxiliary functions $p(t) = x'(t)$ and $q(t) = y'(t)$, we get the equivalent system

$$x' = p,$$

$$y' = q,$$

$$p' = F(t, x, y, p, q),$$

$$q' = G(t, x, y, p, q),$$

(7)

of four first order equations in the four unknown functions $x(t)$, $y(t)$, $p(t)$, and $q(t)$. As illustrated in the following example, the application of the Euler method to a system like that in (7) is a straightforward generalization of the iteration described in (2) for a pair of equations—we simply have four Euler predictors instead of two.

EXAMPLE 2   Suppose that a batted baseball starts at $x_0 = 0$, $y_0 = 0$ with initial velocity $v_0 = 160$ ft/s and with initial angle of inclination $\theta = 30°$. If air resistance is ignored, we find by the elementary methods of Section 1.2 that the baseball travels a (horizontal) distance of $400\sqrt{3}$ ft (approximately 693 ft) in 5 s before striking the ground. Now suppose that in addition to a downward gravitational acceleration ($g = 32$ ft/s²), the baseball experiences an acceleration due to air resistance of $0.0025v^2$ feet per second per second, directed opposite to its instantaneous direction of motion. Apply the Euler method to determine how far it will then travel horizontally.

Solution   According to Problem 32 in Section 5.1, the equations of motion of the baseball are

$$x'' = -cvx',$$
$$y'' = -cvy' - g \tag{8}$$

where $v = \sqrt{(x')^2 + (y')^2}$ is the speed of the ball, and where $c = 0.0025$ and $g = 32$ in fps units. We convert to a first order system as in (7), and thereby obtain

$$x' = p,$$
$$y' = q,$$
$$p' = -cp\sqrt{p^2 + q^2}, \tag{9}$$
$$q' = -cq\sqrt{p^2 + q^2} - g,$$

four first order equations with initial conditions $x_0 = y_0 = 0$, $p_0 = 80\sqrt{3}$, and $q_0 = 80$. Note that $p(t)$ and $q(t)$ are simply the x- and y-components of the velocity vector of the baseball, so that $v = \sqrt{p^2 + q^2}$.

The iterative equations for the Euler method in this situation are

$$x_{n+1} = x_n + hx'_n = x_n + hp_n,$$
$$y_{n+1} = y_n + hy'_n = y_n + hq_n,$$
$$p_{n+1} = p_n + hp'_n = p_n - chp_nv_n,$$
$$q_{n+1} = q_n + hq'_n = q_n - chq_nv_n - gh,$$

where $v_n = \sqrt{p_n^2 + q_n^2}$. Figure 6.29 shows a program using the iteration described by these formulas. To test the validity of the program with step size $h = 0.005$, it was first run with $c = 0$ (no air resistance) inserted in line 160, and the correct range of 693 ft (to the nearest foot) was indeed obtained.

```
100 REM              BASEBALL TRAJECTORY, AIR RESISTANCE
110 REM                 PROPORTIONAL TO SQUARE OF VELOCITY
120 REM       INITIAL VALUES:
130       X = 0 : Y = 0 : T = 0
140       P = 80*SQR(3) : Q = 80
150       V = SQR(P*P + Q*Q)
160       G = 32 : C = 0.0025 : H = 0.005
170 REM        BEGINNING OF ITERATION:
180       X = X + H*P : Y = Y + H*Q
190       P = P − C*H*P*V
200       Q = Q − C*H*Q*V − G*H
210       V = SQR(P*P + Q*Q) : T = T + H
220       PRINT T, X, Y, P, Q, V
230 REM        STOP WHEN BALL HITS THE GROUND:
240       IF Y > −5 THEN GOTO 180
250       END
```

Figure 6.29

The program was then executed with $c = 0.0025$ for air resistance
proportional to v^2 (this is a fairly accurate model of air resistance for a
batted baseball). Figure 6.30 shows the relevant portion of the output.
It indicates that with air resistance, the baseball travels a distance of
only 342 ft in 4.02 s. Thus air resistance converts what otherwise would
be a massive home run into a simple fly ball (if hit straightway to center
field). Note also that when the ball strikes the ground, it has slightly
under *half* its initial velocity and is falling at a steeper angle—approximately $\tan^{-1}\left(\frac{57}{54}\right)$; that is, about 47°. The trajectory of the baseball is
shown in Fig. 6.31; the data for this graph were obtained by means of
the program listed in Fig. 6.29.

T	X	Y	V_X	V_Y	V
4.00	339.70	1.05	54.75	−56.50	78.67
4.01	340.25	0.49	54.64	−56.71	78.75
4.02	340.80	−0.08	54.53	−56.92	78.82
4.03	341.34	−0.65	54.42	−57.13	78.90
4.04	341.88	−1.22	54.32	−57.33	78.98
4.05	342.43	−1.80	54.21	−57.54	79.05

Figure 6.30

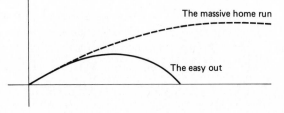

Figure 6.31 An "easy out" or
a home run?

Had we used the Runge-Kutta formulas in (4) and (5) instead of the simple Euler method in Example 2, a much larger step size (and hence fewer iterations) would have sufficed. But perhaps the most common application of the Runge-Kutta method for systems is to the solution of second order initial value problems of the form

$$\frac{d^2x}{dt^2} = g(t, x),$$

$$x(0) = x_0, \qquad x'(0) = y_0. \tag{11}$$

The substitution $x' = y$ yields the system

$$x' = y, \qquad y' = g(t, x) \tag{12}$$

for which the formulas for the parameters in (5) become shorter and simpler; they appear in the table in Fig. 6.32. The successive approximations x_n and y_n are then computed as before, using the iterative formulas in (4).

i	k_{ni}	j_{ni}
1	hy_n	$hg(t_n, x_n)$
2	$h(y_n + \frac{h}{2} j_{n1})$	$hg(t_n + \frac{h}{2}, x_n + \frac{h}{2} k_{n1})$
3	$h(y_n + \frac{h}{2} j_{n2})$	$hg(t_n + \frac{h}{2}, x_n + \frac{h}{2} k_{n2})$
4	$h(y_n + h j_{n3})$	$hg(t_n + h, x_n + h k_{n3})$

Figure 6.32

6.5 Problems

Use the Euler method to solve the systems in Problems 1–4 numerically on $[0, 1]$ with step size $h = 0.001$. Compare the results with the true values at $x = 0.1, 0.2, \ldots, 1.0$. Primes denote derivatives with respect to x.

1. $y' = 2y - z$, $z' = 2y$; $\quad y(0) = 1$, $z(0) = 1$.
2. $y' = 2y - z$, $z' = y + 2z$; $\quad y(0) = 1$, $z(0) = 0$.
3. $y' = y + 2z$, $z' = 2y + z$; $\quad y(0) = 1$, $z(0) = 0$.
4. $w' = w + z$, $y' = w + y$, $z' = -2w - z$; $\quad w(0) = 1$, $y(0) = 0$, $z(0) = 0$.

Use the Runge-Kutta method to solve the systems in Problems 5–12 numerically on the interval $[0, 1]$ with step size $h = 0.01$. Compare the results with the true values at $0.1, 0.2, \ldots, 1.0$. Primes denote derivatives with respect to t.

5. $x' = 2x + y$, $y' = 6x + 3y$; $\quad x(0) = 1$, $y(0) = 3$.
6. $x' = x + 2y$, $y' = x + e^{-t}$; $\quad x(0) = y(0) = 0$.
7. $x' + 2y' + x = 0$, $x' - y' + y = 0$; $\quad x(0) = 0$, $y(0) = 1$.
8. $x'' + 2x + 4y = 0$, $y'' + x + 2y = 0$; $\quad x(0) = y(0) = 0$, $x'(0) = y'(0) = -1$.

9. $x' = 2x - 3y + 2 \sin 2t$, $y' = x - 2y - \cos 2t$; $x(0) = 0 = y(0)$.

10. $x'' + x = \sin t$; $x(0) = x'(0) = 0$.

11. Adapt the Milne method to the approximation of solutions of systems of two first order equations.

Use the Milne predictor-corrector method (Problem 11) to solve the systems in Problems 12–16 numerically on the interval [0, 1] *with step size* $h = 0.01$. *Compare the results with the true values at* $0.1, 0.2, \ldots, 1.0$. *Primes denote derivatives with respect to t.*

12. The system of Problem 5.

13. The system of Problem 6.

14. The system of Problem 7.

15. The system of Problem 9.

16. The system of Problem 10.

17. Verify that

$$x = e^{-t} \sin t - te^{-t}, \qquad y = e^{-t} \cos t + t^3 e^{-t}$$

is the solution of the initial value problem

$$x' = -x + y - (t^3 + 1)e^{-t}, \qquad x(0) = 0;$$
$$y' = -x - y + (3t^2 - t)e^{-t}; \qquad y(0) = 1.$$

Then solve this system numerically by a method of your choice on the interval [0, 1]. Compare the results with the true values at $x = 0.1, 0.2, \ldots, 1.0$; take care that the maximum error at these points does not exceed 0.0001.

18. Write the van der Pol equation $y'' - \mu(1 - y^2)y' + y = 0$ as a system of two first order equations. Use the fourth order Runge-Kutta method to approximate its solution at $x = 0.1, 0.2, \ldots, 1.0$ given $\mu = 0.1$, $y(0) = 2$, and $y'(0) = 0$. Begin with a large step size; then continue reducing it until you are reasonably sure you have three digits correct to the right of the decimal point.

19. A weight is located at the point (0, 10) in the xy-plane. A rope of length 10 connects the weight to a horse at the origin. At time $t = 0$ the horse begins to walk out the positive x-axis at unit speed. Find the coordinates of the weight at the times $t = 1, 2, 3, \ldots, 10$.

20. A nonlinear spring is attached to a mass that slides with friction proportional to its velocity, so that the system satisfies the differential equation

$$3x'' + x' + 2x^3 = 0,$$

where $x = x(t)$ is the displacement of the mass from its equilibrium position. Take $x(0) = 1$, $x'(0) = 0$, and describe the subsequent motion of the mass.

21. An external force is applied to the system of Problem 20, so that the motion of the mass satisfies the initial value problem

$$3x'' + x' + 2x^3 = 2 \sin 5t,$$
$$x(0) = x'(0) = 0.$$

Describe the motion of the mass for $t \geq 0$.

22. A lake is initially stocked with 100 bass and 600 redear. There is ample food for the redear. Because bass prey on redear, the population of bass will increase at a rate proportional to the number of encounters between the species; bass will also die at a rate proportional to the bass population. The redear multiply at

a rate proportional to their population and die off at a rate proportional to the number of encounters between the two species. This implies that the populations $B(t)$ of bass and $R(t)$ of redear satisfy a system of differential equations of the following form:

$$B'(t) = pBR - qB,$$
$$R'(t) = uR - vBR.$$

Suppose it is known that $p = 0.00004$, $q = 0.02$, $u = 0.05$, and $v = 0.0004$. In Chapter 9 we will see that the populations of the two species go through periodic cycles with the same period. Find the period in this case.

23. In a certain mass-and-spring system, the mass slides with nonlinear friction; its displacement x at time t satisfies the initial value problem

$$x'' + 2|x'|x' + x = 0;$$
$$x(0) = 1, \ x'(0) = 0.$$

Approximate $x(t)$ for $t = 0.5, 1.0, 1.5, 2.0, \ldots, 9.5, 10.0$.

24. As a skydiver falls and before opening his parachute, his air resistance is proportional to $|v|^{3/2}$, where v is his velocity at time t. He observes that his terminal velocity is 120 mi/h. If he jumps from 10,000 ft and does not open his parachute, how long will it take him to reach the ground?

25. A crossbow bolt that weighs 8 oz is fired from ground level at a 45° angle with initial speed 288 ft/s. Under the assumption that air resistance produces a force opposite the direction of motion of the bolt and of magnitude $(0.000003)|v|^2$ lb, find how far the bolt will travel (horizontally) and how fast it will be moving when it strikes the ground.

Fourier Series and Separation of Variables

7

7.1
Periodic Functions and Trigonometric Series

As motivation for the subject of Fourier series, we consider the differential equation

$$x'' + \omega_0^2 x = f(t) \tag{1}$$

that models the behavior of a mass-and-spring system with natural (circular) frequency ω_0, moving under the influence of an external force of magnitude $f(t)$ per unit mass. As we saw in Section 2.5, a particular solution of Eq. (1) can easily be found by the method of undetermined coefficients if $f(t)$ is a simple harmonic function—a sine or cosine function. For instance, the equation

$$x'' + \omega_0^2 x = A \cos \omega t \tag{2}$$

with $\omega^2 \neq \omega_0^2$ has the particular solution

$$x_p(t) = \frac{A}{\omega_0^2 - \omega^2} \cos \omega t, \tag{3}$$

which is readily found by beginning with the trial solution $x_p(t) = a \cos \omega t$.

Now suppose, more generally, that the force function $f(t)$ in (1) is a linear combination of simple harmonic functions. Then, on the basis of (3) and the analogous formula with sine in place of cosine, we can apply the principle of superposition to construct a particular solution of Eq. (1). For example, consider the equation

$$x'' + \omega_0^2 x = \sum_{n=1}^{N} A_n \cos \omega_n t, \tag{4}$$

in which ω_0^2 is equal to none of the ω_n^2. Equation (4) has the particular solution

$$x_p(t) = \sum_{n=1}^{N} \frac{A_n}{\omega_0{}^2 - \omega_n{}^2} \cos \omega_n t \tag{5}$$

obtained by adding the solutions given in (3) corresponding to the n terms on the right-hand side in Eq. (4).

Mechanical (and electrical) systems often involve periodic forcing functions that are not (simply) finite linear combinations of sines and cosines. Nevertheless, as we will soon see, any reasonably nice periodic function $f(t)$ has a representation as an *infinite series* of trigonometric terms. This fact opens the way toward solving Eq. (1) by superposition of trigonometric "building blocks," with the finite sum in (5) replaced by an infinite series.

The function $f(t)$ defined for all t is called **periodic** provided that there exists a positive number p such that

$$f(t + p) = f(t) \tag{6}$$

for all t. The number p is then called a **period** of the function f. Note that the period of a periodic function is not unique; for example, if p is a period of $f(t)$, then so is np for each positive integer n. Indeed, every positive number is a period of any constant function.

If there exists a smallest positive number P such that $f(t)$ is periodic with period P, then we will call P *the* period of f. For instance, the period of the functions $\cos nt$ and $\sin nt$ (where n is a positive integer) is $2\pi/n$ because

$$\cos n\left(t + \frac{2\pi}{n}\right) = \cos(nt + 2\pi) = \cos nt$$

and (7)

$$\sin n\left(t + \frac{2\pi}{n}\right) = \sin(nt + 2\pi) = \sin nt.$$

Moreover, 2π itself is *a* period of the functions $\cos nt$ and $\sin nt$. Ordinarily we will have no need to refer to the smallest period of a function $f(t)$ and will simply say that $f(t)$ has period p if p is a period of $f(t)$; that is, if $f(t + p) = f(t)$.

In Section 4.5 we saw several examples of piecewise continuous periodic functions. For instance, the square wave function having the graph shown in Fig. 7.1 has period 2π.

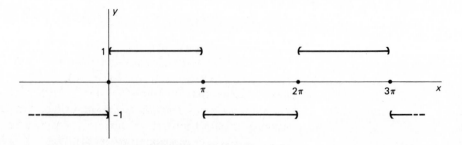

Figure 7.1 A square wave function.

Because cos nt and sin nt each have period 2π, any linear combination of sines and cosines of integral multiples of t, such as

$$f(t) = 3 + \cos t - \sin t + 5 \cos 2t + 17 \sin 3t,$$

has period 2π. But every such linear combination is continuous, so the square wave function cannot be expressed in such manner. In his celebrated treatise *The Analytical Theory of Heat* (1822), the French scientist Joseph Fourier (1768–1830) made the remarkable assertion that every function $f(t)$ with period 2π can be represented by a trigonometric infinite series of the form

$$\frac{a_0}{2} + \sum_{n=1}^{\infty} (a_n \cos nt + b_n \sin nt). \tag{8}$$

We will see in Section 7.2 that under rather mild restrictions on the function $f(t)$, this is so! An infinite series of the form in (8) is called a *Fourier series*, and the representation of functions by Fourier series is one of the most widely used techniques in applied mathematics, especially for the solution of partial differential equations (see Sections 7.5–7.7).

In this section we will confine our attention to functions of period 2π. We want to determine what the coefficients in the Fourier series in (8) must be if it is to converge to a given function $f(t)$ of period 2π. For this purpose we need the following integrals, in which m and n denote positive integers (see Problems 27–29):

$$\int_{-\pi}^{\pi} \cos mt \cos nt \, dt = \begin{cases} 0 & \text{if } m \neq n, \\ \pi & \text{if } m = n. \end{cases} \tag{9}$$

$$\int_{-\pi}^{\pi} \sin mt \sin nt \, dt = \begin{cases} 0 & \text{if } m \neq n, \\ \pi & \text{if } m = n. \end{cases} \tag{10}$$

$$\int_{-\pi}^{\pi} \cos mt \sin nt \, dt = 0 \quad \text{for all } m \text{ and } n. \tag{11}$$

These formulas imply that the functions cos nt and sin nt for $n = 1, 2, 3, \ldots$ constitute a *mutually orthogonal* set of functions on the interval $-\pi \leq t \leq \pi$. Two real-valued functions $u(t)$ and $v(t)$ are said to be **orthogonal** on the interval $[a, b]$ provided that

$$\int_{a}^{b} u(t)v(t) \, dt = 0. \tag{12}$$

Suppose now that the piecewise continuous function $f(t)$ of period 2π has a Fourier series representation

$$f(t) = \frac{a_0}{2} + \sum_{m=1}^{\infty} (a_m \cos mt + b_m \sin mt), \tag{13}$$

in the sense that the infinite series on the right converges to the value $f(t)$ for every t. We assume in addition that, when the infinite series in (13) is multiplied by any continuous function, the resulting series can be integrated term by term. Then the result of termwise integration of Eq. (13) itself from $t = -\pi$ to $t = \pi$ is

$$\int_{-\pi}^{\pi} f(t)\, dt = \frac{a_0}{2} \int_{-\pi}^{\pi} 1 \, dt + \sum_{m=1}^{\infty} a_m \int_{-\pi}^{\pi} \cos mt \, dt$$

$$+ \sum_{m=1}^{\infty} b_m \int_{-\pi}^{\pi} \sin mt \, dt = \pi a_0$$

because all the trigonometric integrals vanish. Hence

$$a_0 = \frac{1}{\pi} \int_{-\pi}^{\pi} f(t) \, dt. \tag{14}$$

If we first multiply each side in Eq. (13) by $\cos nt$ and then integrate termwise, the result is

$$\int_{-\pi}^{\pi} f(t) \cos nt \, dt = \frac{a_0}{2} \int_{-\pi}^{\pi} \cos nt \, dt$$

$$+ \sum_{m=1}^{\infty} a_m \int_{-\pi}^{\pi} \cos mt \cos nt \, dt$$

$$+ \sum_{m=1}^{\infty} b_m \int_{-\pi}^{\pi} \cos mt \sin nt \, dt;$$

it then follows from (11) that

$$\int_{-\pi}^{\pi} f(t) \cos nt \, dt = \sum_{m=1}^{\infty} a_m \int_{-\pi}^{\pi} \cos mt \cos nt \, dt. \tag{15}$$

But, of all the integrals (for $m = 1, 2, 3, \ldots$) on the right-hand side in (15), only the one for which $m = n$ is nonzero. Thus (9) yields

$$\int_{-\pi}^{\pi} f(t) \cos nt \, dt = a_n \int_{-\pi}^{\pi} \cos^2 nt \, dt = \pi a_n,$$

so the value of the coefficient a_n is

$$a_n = \frac{1}{\pi} \int_{-\pi}^{\pi} f(t) \cos nt \, dt. \tag{16}$$

Note that with $n = 0$, the formula in (16) reduces to (14); this explains why we denoted the constant term in the original Fourier series by $\frac{1}{2}a_0$ (rather than simply a_0). If we multiply each side in Eq. (13) by $\sin nt$ and then integrate termwise, we find in a similar way that

$$b_n = \frac{1}{\pi} \int_{-\pi}^{\pi} f(t) \sin nt \, dt. \tag{17}$$

In short, we have found that if the series in (13) converges to $f(t)$ and if the termwise integrations carried out above are valid, then the coefficients in the series must have the values given in (16) and (17). This motivates us to *define* the Fourier series of a periodic function by means of these formulas, whether or not the resulting series converges to the function (or even converges at all).

DEFINITION: FOURIER SERIES AND FOURIER COEFFICIENTS

*Let $f(t)$ be a piecewise continuous function of period 2π that is defined for all t. Then the **Fourier series** of $f(t)$ is the series*

$$\frac{a_0}{2} + \sum_{n=1}^{\infty} (a_n \cos nt + b_n \sin nt) \tag{18}$$

where the **Fourier coefficients** a_n and b_n are defined by means of the formulas

$$a_n = \frac{1}{\pi} \int_{-\pi}^{\pi} f(t) \cos nt \, dt \qquad (16)$$

for $n = 0, 1, 2, 3, \ldots$ and

$$b_n = \frac{1}{\pi} \int_{-\pi}^{\pi} f(t) \sin nt \, dt \qquad (17)$$

for $n = 1, 2, 3, \ldots$.

It can (and often does) happen that the Fourier series of a function fails to converge to the function at a few points in the domain of the function. We will therefore write

$$f(t) \sim \frac{a_0}{2} + \sum_{n=1}^{\infty} (a_n \cos nt + b_n \sin nt), \qquad (19)$$

not using an equal sign between the function and its Fourier series until we have discussed convergence of Fourier series in Section 7.2.

Suppose that the piecewise continuous function $f(t)$ as given initially is defined only on the interval $-\pi \leqq t \leqq \pi$, and assume that $f(-\pi) = f(\pi)$. Then we can extend f so that its domain includes all real numbers by means of the periodicity condition $f(t + 2\pi) = f(t)$ for all t. We continue to denote this extension of the original function by f, and note that it automatically has period 2π. Its graph looks the same on every interval of the form $(2n - 1)\pi \leqq t \leqq (2n + 1)\pi$; each is a copy of its graph on the original domain $-\pi \leqq t \leqq \pi$. See Fig. 7.2. For instance, the square wave function of Fig. 7.1 can be described as the period 2π function such that

$$f(t) = \begin{cases} -1 & \text{if } -\pi < t < 0; \\ +1 & \text{if } 0 < t < \pi; \\ 0 & \text{if } t = -\pi, 0, \text{ or } \pi. \end{cases} \qquad (20)$$

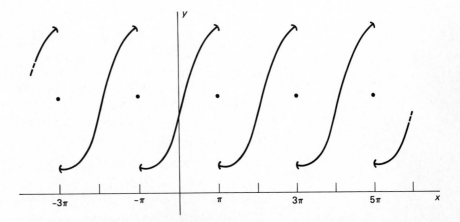

Figure 7.2 Extending a function to produce a periodic function.

Thus the square wave function is the period 2π function defined on one full period by means of (20).

We need to consider Fourier series of piecewise continuous functions because many functions that appear in applications are only piecewise continuous, not continuous. Note that the integrals in (16) and (17) exist if $f(t)$ is piecewise continuous, so every piecewise continuous function has a Fourier series.

EXAMPLE 1 Find the Fourier series of the square wave function under discussion above.

Solution It is always a good idea to calculate a_0 separately using Eq. (14). Thus

$$a_0 = \frac{1}{\pi} \int_{-\pi}^{\pi} f(t)\, dt$$

$$= \frac{1}{\pi} \int_{-\pi}^{0} (-1)\, dt + \frac{1}{\pi} \int_{0}^{\pi} (+1)\, dt$$

$$= \frac{1}{\pi}(-\pi) + \frac{1}{\pi}(\pi) = 0.$$

We split the integral in the first line into two integrals in the second because $f(t)$ is defined by different formulas on the intervals $(-\pi, 0)$ and $(0, \pi)$; the values of $f(t)$ at the endpoints of these intervals do not affect the values of the integrals.

Formula (16) yields (for $n > 0$)

$$a_n = \frac{1}{\pi} \int_{-\pi}^{\pi} f(t) \cos nt\, dt$$

$$= \frac{1}{\pi} \int_{-\pi}^{0} (-\cos nt)\, dt + \frac{1}{\pi} \int_{0}^{\pi} \cos nt\, dt$$

$$= \frac{1}{\pi}\left[-\frac{1}{n} \sin nt \right]_{-\pi}^{0} + \frac{1}{\pi}\left[\frac{1}{n} \sin nt \right]_{0}^{\pi} = 0.$$

Formula (17) yields

$$b_n = \frac{1}{\pi} \int_{-\pi}^{0} (-\sin nt)\, dt + \frac{1}{\pi} \int_{0}^{\pi} \sin nt\, dt$$

$$= \frac{1}{\pi}\left[\frac{1}{n} \cos nt \right]_{-\pi}^{0} + \frac{1}{\pi}\left[-\frac{1}{n} \cos nt \right]_{0}^{\pi}$$

$$= \frac{2}{n\pi}(1 - \cos n\pi) = \frac{2}{n\pi}[1 - (-1)^n].$$

Thus $a_n = 0$ for all $n \geq 0$, and

$$b_n = \begin{cases} \dfrac{4}{n\pi} & \text{for } n \text{ odd}; \\[2mm] 0 & \text{for } n \text{ even}. \end{cases}$$

The last result follows because $\cos(-n\pi) = \cos(n\pi) = (-1)^n$. With these values of the Fourier coefficients, we obtain the Fourier series

$$f(t) \sim \frac{4}{\pi} \sum_{n \text{ odd}} \frac{\sin nt}{n}$$

$$= \frac{4}{\pi} \left(\sin t + \frac{1}{3} \sin 3t + \frac{1}{5} \sin 5t + \cdots \right). \qquad (21)$$

Here we have introduced the useful abbreviation

$$\sum_{n \text{ odd}} \quad \text{for} \quad \sum_{\substack{n=1 \\ n \text{ odd}}}^{\infty}.$$

For example,

$$\sum_{n \text{ odd}} \frac{1}{n} = 1 + \frac{1}{3} + \frac{1}{5} + \cdots.$$

Figure 7.3 shows the graphs of several of the partial sums

$$S_N(t) = \frac{4}{\pi} \sum_{n=1}^{N} \frac{\sin (2n-1)t}{2n-1}$$

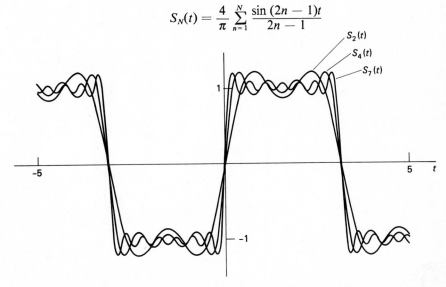

Figure 7.3a Some partial sums of the Fourier series of $f(t)$.

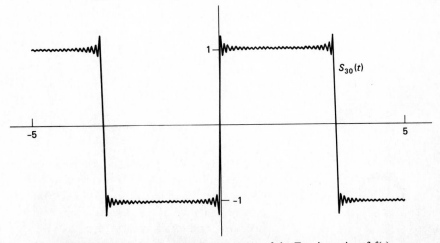

Figure 7.3b The sum of the first 30 terms of the Fourier series of $f(t)$.

of the Fourier series in (21). Note that as t approaches a discontinuity of $f(t)$ from either side, the value of $S_N(t)$ tends to overshoot the limiting value of $f(t)$—either $+1$ or -1 in this case. This behavior of a Fourier series near a point of discontinuity of its function is typical and is known as *Gibbs' phenomenon*.

The following integral formulas, easily derived by integration by parts, are useful in computing Fourier series of polynomial functions:

$$\int u \cos u \, du = \cos u + u \sin u + C; \tag{22}$$

$$\int u \sin u \, du = \sin u - u \cos u + C; \tag{23}$$

$$\int u^n \cos u \, du = u^n \sin u - n \int u^{n-1} \sin u \, du; \tag{24}$$

$$\int u^n \sin u \, du = -u^n \cos u + n \int u^{n-1} \cos u \, du. \tag{25}$$

EXAMPLE 2 Find the Fourier series of the period 2π function that is defined in one period to be

$$f(t) = \begin{cases} 0 & \text{if } -\pi < t \leq 0; \\ t & \text{if } 0 \leq t < \pi; \\ \dfrac{\pi}{2} & \text{if } t = \pm\pi. \end{cases} \tag{26}$$

The graph of f is shown in Fig. 7.4.

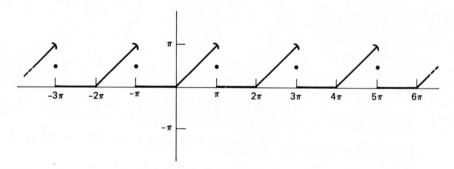

Figure 7.4 The periodic function of Example 2.

Solution The values of $f(\pm\pi)$ are irrelevant because they have no effect on the values of the integrals that yield the Fourier coefficients. Because $f(t) \equiv 0$ on the interval $(-\pi, 0)$, each integral from $t = -\pi$ to $t = \pi$ may be replaced by an integral from $t = 0$ to $t = \pi$. The formulas in (14), (16), and (17) therefore give

$$a_0 = \frac{1}{\pi} \int_0^\pi t \, dt = \frac{1}{\pi} \left[\frac{1}{2} t^2 \right]_0^\pi = \frac{\pi}{2};$$

$$a_n = \frac{1}{\pi} \int_0^\pi t \cos nt \, dt$$

$$= \frac{1}{n^2 \pi} \int_0^{n\pi} u \cos u \, du \qquad \left(u = nt, \quad t = \frac{u}{n} \right)$$

$$= \frac{1}{n^2 \pi} \Big[\cos u + u \sin u \Big]_0^{n\pi} \qquad \text{(by Eq. (22))}$$

$$= \frac{1}{n^2 \pi} [(-1)^n - 1].$$

Consequently, $a_n = 0$ if n is even and $n \geqq 2$;

$$a_n = -\frac{2}{n^2 \pi} \quad \text{if } n \text{ is odd.}$$

$$b_n = \frac{1}{\pi} \int_0^\pi t \sin nt \, dt$$

$$= \frac{1}{n^2 \pi} \int_0^{n\pi} u \sin u \, du$$

$$= \frac{1}{n^2 \pi} \Big[\sin u - u \cos u \Big]_0^{n\pi} \qquad \text{(by Eq. (23))}$$

$$= -\frac{1}{n} \cos n\pi.$$

Thus

$$b_n = \frac{(-1)^{n+1}}{n} \quad \text{for all } n \geqq 1.$$

Therefore the Fourier series of $f(t)$ is

$$f(t) \sim \frac{\pi}{4} - \frac{2}{\pi} \sum_{n \text{ odd}} \frac{\cos nt}{n^2} + \sum_{n=1}^\infty \frac{(-1)^{n+1} \sin nt}{n}.$$

If $f(t)$ is a function of period 2π, it is readily verified (Problem 30) that

$$\int_{-\pi}^\pi f(t) \, dt = \int_a^{a+2\pi} f(t) \, dt \qquad (28)$$

for all a. That is, the integral of $f(t)$ over one interval of length 2π is equal to its integral over any other such interval. In case $f(t)$ is given explicitly on the interval $[0, 2\pi]$ rather than on $[-\pi, \pi]$, it may be more convenient to compute its Fourier coefficients as

$$a_n = \frac{1}{\pi} \int_0^{2\pi} f(t) \cos nt \, dt$$

and $\qquad\qquad\qquad\qquad\qquad\qquad\qquad\qquad\qquad\qquad\qquad\qquad (29)$

$$b_n = \frac{1}{\pi} \int_0^{2\pi} f(t) \sin nt \, dt.$$

7.1 Problems

In each of Problems 1–10, determine whether or not the given function is periodic. If so, find its smallest period.

1. $f(t) = \sin 3t$.

2. $f(t) = \cos 2\pi t$.

3. $f(t) = \cos \dfrac{3t}{2}$.

4. $f(t) = \sin \pi \dfrac{t}{3}$.

5. $f(t) = \tan t$.

6. $f(t) = \cot 2\pi t$.

7. $f(t) = \cosh 3t$.

8. $f(t) = \sinh \pi t$.

9. $f(t) = |\sin t|$.

10. $f(t) = \cos^2 3t$.

In each of Problems 11–16, the values of a period 2π function $f(t)$ in one full period are given. Sketch several periods of its graph and find its Fourier series.

11. $f(t) = 1,\ -\pi \leqq t \leqq \pi$.

12. $f(t) = \begin{cases} -1, & -\pi < t \leqq 0; \\ +1, & 0 < t \leqq \pi. \end{cases}$

13. $f(t) = \begin{cases} 0, & -\pi < t \leqq 0; \\ 1, & 0 < t \leqq \pi. \end{cases}$

14. $f(t) = \begin{cases} 3, & -\pi < t \leqq 0; \\ -2, & 0 < t \leqq \pi. \end{cases}$

15. $f(t) = t,\ -\pi < t \leqq \pi$.

16. $f(t) = t,\ 0 < t < 2\pi$.

17. $f(t) = |t|,\ -\pi \leqq t \leqq \pi$.

18. $f(t) = \begin{cases} \pi + t, & -\pi \leqq t \leqq 0; \\ \pi - t, & 0 \leqq t \leqq \pi. \end{cases}$

19. $f(t) = \begin{cases} \pi + t, & -\pi \leqq t < 0; \\ 0, & 0 \leqq t \leqq \pi. \end{cases}$

20. $f(t) = \begin{cases} 0, & -\pi \leqq t < -\dfrac{\pi}{2}; \\ 1, & -\dfrac{\pi}{2} \leqq t \leqq \dfrac{\pi}{2}; \\ 0, & \dfrac{\pi}{2} < t \leqq \pi. \end{cases}$

21. $f(t) = t^2,\ -\pi \leqq t \leqq \pi$.

22. $f(t) = t^2,\ 0 \leqq t < 2\pi$.

23. $f(t) = \begin{cases} 0, & -\pi \leqq t \leqq 0; \\ t^2, & 0 \leqq t < \pi. \end{cases}$

24. $f(t) = |\sin t|,\ -\pi \leqq t \leqq \pi$.

25. $f(t) = \cos^2 2t,\ -\pi \leqq t \leqq \pi$.

26. $f(t) = \begin{cases} 0, & -\pi \leqq t \leqq 0; \\ \sin t, & 0 \leqq t \leqq \pi. \end{cases}$

27. Verify the formula in Eq. (9). (*Suggestion:* Use the trigonometric identity

$$\cos A \cos B = \tfrac{1}{2}[\cos (A + B) + \cos (A - B)].)$$

28. Verify the formula in Eq. (10).

29. Verify the formula in Eq. (11).

30. Let $f(t)$ be a piecewise continuous function with period P. (a) Suppose that $0 \leqq a < P$. Substitute $u = t - P$ to show that

$$\int_P^{a+P} f(t)\, dt = \int_0^a f(t)\, dt.$$

Conclude that

$$\int_a^{a+P} f(t)\, dt = \int_0^P f(t)\, dt.$$

(b) Given A, choose n so that $A = nP + a$ with $0 \leq a < P$. Then substitute $v = t - nP$ to show that

$$\int_A^{A+P} f(t)\, dt = \int_a^{a+P} f(t)\, dt = \int_0^P f(t)\, dt.$$

31. Multiply each side in Eq. (13) by $\sin nt$ and then integrate term by term to derive the formula in Eq. (17).

7.2
General Fourier Series and Convergence

In Section 7.1 we defined the Fourier series of a periodic function of period 2π. Now let $f(t)$ be a function that is piecewise continuous for all t and has arbitrary period $P > 0$. We write

$$P = 2L, \tag{1}$$

so L is the **half-period** of the periodic function f. Let us define the function g as follows:

$$g(u) = f\left(\frac{Lu}{\pi}\right) \tag{2}$$

for all u. Then

$$g(u + 2\pi) = f\left(\frac{Lu}{\pi} + 2L\right) = f\left(\frac{Lu}{\pi}\right) = g(u),$$

and hence $g(u)$ is also periodic and has period 2π. Consequently g has the Fourier series

$$g(u) \sim \frac{a_0}{2} + \sum_{n=1}^{\infty} (a_n \cos nu + b_n \sin nu) \tag{3}$$

with Fourier coefficients

$$a_n = \frac{1}{\pi} \int_{-\pi}^{\pi} g(u) \cos nu\, du \tag{4a}$$

and

$$b_n = \frac{1}{\pi} \int_{-\pi}^{\pi} g(u) \sin nu\, du. \tag{4b}$$

If we now write

$$t = \frac{Lu}{\pi}, \qquad u = \frac{\pi t}{L}, \qquad f(t) = g(u), \tag{5}$$

then

$$f(t) = g\left(\frac{\pi t}{L}\right) \sim \frac{a_0}{2} + \sum_{n=1}^{\infty} \left(a_n \cos \frac{n\pi t}{L} + b_n \sin \frac{n\pi t}{L}\right),$$

and the substitution of (5) in (4) yields

$$a_n = \frac{1}{\pi} \int_{-\pi}^{\pi} g(u) \cos nu \, du \qquad \left(u = \frac{\pi t}{L}, \; du = \frac{\pi \, dt}{L} \right)$$

$$= \frac{1}{L} \int_{-L}^{L} g\left(\frac{\pi t}{L}\right) \cos \frac{n\pi t}{L} \, dt.$$

Therefore

$$a_n = \frac{1}{L} \int_{-L}^{L} f(t) \cos \frac{n\pi t}{L} \, dt; \qquad (7)$$

similarly,

$$b_n = \frac{1}{L} \int_{-L}^{L} f(t) \sin \frac{n\pi t}{L} \, dt. \qquad (8)$$

This computation motivates the following definition of the Fourier series of a periodic function of period $2L$.

DEFINITION: FOURIER SERIES AND FOURIER COEFFICIENTS

*Let $f(t)$ be a piecewise continuous function of period $2L$ that is defined for all t. Then the **Fourier series** of $f(t)$ is the series*

$$f(t) \sim \frac{a_0}{2} + \sum_{n=1}^{\infty} \left(a_n \cos \frac{n\pi t}{L} + b_n \sin \frac{n\pi t}{L} \right), \qquad (6)$$

*where the **Fourier coefficients** $\{a_n\}_0^{\infty}$ and $\{b_n\}_1^{\infty}$ are defined to be*

$$a_n = \frac{1}{L} \int_{-L}^{L} f(t) \cos \frac{n\pi t}{L} \, dt \qquad (7)$$

and

$$b_n = \frac{1}{L} \int_{-L}^{L} f(t) \sin \frac{n\pi t}{L} \, dt. \qquad (8)$$

With $n = 0$ the formula in (7) takes the simple form

$$a_0 = \frac{1}{L} \int_{-L}^{L} f(t) \, dt, \qquad (9)$$

which demonstrates that the constant term $\frac{1}{2}a_0$ in the Fourier series of f is simply the average value of $f(t)$ on the interval $[-L, L]$.

As a consequence of Problem 30 in Section 7.1, we may evaluate the integrals in (7) and (8) over any other interval of length $2L$. For instance, if $f(t)$ is given by a single formula for $0 < t < 2L$, it may be more convenient to compute the integrals

$$a_n = \frac{1}{L} \int_{0}^{2L} f(t) \cos \frac{n\pi t}{L} \, dt$$

and　(10)

$$b_n = \frac{1}{L} \int_0^{2L} f(t) \sin \frac{n\pi t}{L}\, dt.$$

EXAMPLE 1　Figure 7.5 shows the graph of a square wave function with period 4. Find its Fourier series.

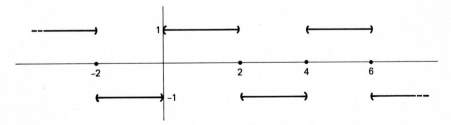

Figure 7.5　The square wave function of Example 1.

Solution　Here $L = 2$; also $f(t) = -1$ if $-2 < t < 0$, while $f(t) = 1$ if $0 < t < 2$. Hence the formulas in (7)–(9) yield

$$a_0 = \frac{1}{2} \int_{-2}^{2} f(t)\, dt$$

$$= \frac{1}{2} \int_{-2}^{0} (-1)\, dt + \frac{1}{2} \int_{0}^{2} (+1)\, dt = 0,$$

$$a_n = \frac{1}{2} \int_{-2}^{0} (-1) \cos \frac{n\pi t}{2}\, dt + \frac{1}{2} \int_{0}^{2} (+1) \cos \frac{n\pi t}{2}\, dt$$

$$= \frac{1}{2}\left[-\frac{2}{n\pi} \sin \frac{n\pi t}{2}\right]_{-2}^{0} + \frac{1}{2}\left[\frac{2}{n\pi} \sin \frac{n\pi t}{2}\right]_{0}^{2} = 0,$$

and

$$b_n = \frac{1}{2} \int_{-2}^{0} (-1) \sin \frac{n\pi t}{2}\, dt + \frac{1}{2} \int_{0}^{2} (+1) \sin \frac{n\pi t}{2}\, dt$$

$$= \frac{1}{2}\left[\frac{2}{n\pi} \cos \frac{n\pi t}{2}\right]_{-2}^{0} + \frac{1}{2}\left[-\frac{2}{n\pi} \cos \frac{n\pi t}{2}\right]_{0}^{2}$$

$$= \frac{2}{n\pi}[1 - (-1)^n] = \begin{cases} \dfrac{4}{n\pi} & \text{if } n \text{ is odd,} \\ 0 & \text{if } n \text{ is even.} \end{cases}$$

Thus the Fourier series is

$$f(t) \sim \frac{4}{\pi} \sum_{n \text{ odd}} \frac{1}{n} \sin \frac{n\pi t}{2} \tag{11a}$$

$$= \frac{4}{\pi}\left(\sin \frac{\pi t}{2} + \frac{1}{3} \sin \frac{3\pi t}{2} + \frac{1}{5} \sin \frac{5\pi t}{2} + \cdots\right). \tag{11b}$$

THE CONVERGENCE THEOREM

We want to impose conditions on the periodic function f that are enough to guarantee that its Fourier series actually converges to $f(t)$ at least at those values of t at which f is continuous. Recall that the function f is said to be *piecewise continuous* on the interval $a \leq t \leq b$ provided that there is a finite partition of $[a, b]$ with endpoints $a = t_0 < t_1 < t_2 < \cdots < t_{n-1} < t_n = b$ such that:

1. f is continuous on each open subinterval $t_{i-1} < t < t_i$, and
2. at each endpoint t_i of such a subinterval the limit of $f(t)$, as t approaches t_i from within the subinterval, exists and is finite.

The function f is called *piecewise continuous* for all t if it is piecewise continuous on every bounded interval. It follows that a piecewise continuous function is continuous except possibly at isolated points, and that at each such point of discontinuity, the one-sided limits

$$f(t+) = \lim_{u \to t^+} f(u) \quad \text{and} \quad f(t-) = \lim_{u \to t^-} f(u) \tag{12}$$

both exist and are finite. Thus a piecewise continuous function has only isolated "finite jump" discontinuities like the one shown in Fig. 7.6.

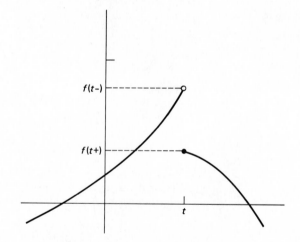

Figure 7.6 A finite jump discontinuity.

The square wave and sawtooth functions that we saw in earlier chapters are typical examples of periodic piecewise continuous functions. The function $f(t) = \tan t$ is a periodic function (of period π) that is not piecewise continuous because it has infinite discontinuities. The function $g(t) = \sin(1/t)$ is not piecewise continuous on $[-1, 1]$ because its one-sided limits at $t = 0$ do not exist. The function

$$h(t) = \begin{cases} t & \text{if } t = \dfrac{1}{n} \quad (n \text{ an integer}), \\ 0 & \text{otherwise} \end{cases}$$

on $[-1, 1]$ has one-sided limits everywhere, but is not piecewise continuous because its discontinuities are not isolated—it has the infinite sequence $\{1/n\}_1^\infty$ of discontinuities; a piecewise continuous function can have only finitely many discontinuities in any bounded interval.

Note that a piecewise continuous function need not be defined at its isolated points of discontinuity. Alternatively, it can be defined arbitrarily at such points. For instance, the square wave function f of Fig. 7.5 is piecewise continuous no matter what its values might be at the points $\{2n\}_1^\infty$ at which it is discontinuous. Its derivative f' is also piecewise continuous; $f'(t) = 0$ unless t is an even integer, in which case $f'(t)$ is undefined.

The piecewise continuous function f is said to be **piecewise smooth** provided that its derivative f' is piecewise continuous. The theorem below tells us that the Fourier series of a piecewise smooth function converges everywhere. More general Fourier convergence theorems—with weaker hypotheses on the periodic function f—are known. But the hypothesis that f is piecewise smooth is easy to check and is satisfied by most functions encountered in practical applications. A proof of the following theorem may be found in G. P. Tolstov, *Fourier Series* (New York: Dover, 1976).

THEOREM: CONVERGENCE OF FOURIER SERIES

Suppose that the periodic function f is piecewise smooth. Then its Fourier series in (6) converges:

(a) to the value $f(t)$ at each point where f is continuous, and

(b) to the value $\frac{1}{2}[f(t+) + f(t-)]$ at each point where f is discontinuous.

Note that $\frac{1}{2}[f(t+) + f(t-)]$ is the *average* of the right-hand and left-hand limits of f at the point t. If f is continuous at t, then $f(t) = f(t+) = f(t-)$, so

$$f(t) = \frac{f(t+) + f(t-)}{2}. \tag{13}$$

Hence the theorem above could be rephrased as follows: The Fourier series of a piecewise smooth function f converges for *every* t to the average value in (13). For this reason it is customary to write

$$f(t) = \frac{a_0}{2} + \sum_{n=1}^{\infty} \left(a_n \cos \frac{n\pi t}{L} + b_n \sin \frac{n\pi t}{L} \right), \tag{14}$$

with the understanding that the piecewise smooth function f has been redefined (if necessary) at each of its points of discontinuity in order to satisfy the average value condition in (13).

EXAMPLE 1 (CONTINUED)   Figure 7.5 shows us at a glance that if t_0 is an even integer, then

$$\lim_{t \to t_0^-} f(t) = -1 \quad \text{and} \quad \lim_{t \to t_0^+} f(t) = +1.$$

Hence

$$\frac{f(t_0+) + f(t_0-)}{2} = 0.$$

Note that, in accord with the theorem above, the Fourier series of $f(t)$ in (11) clearly converges to zero if t is an even integer (because $\sin n\pi = 0$).

EXAMPLE 2 Let $f(t)$ be a function of period 2 with $f(t) = t^2$ if $0 < t < 2$. We define $f(t)$ for t an even integer by the average value condition in (13); consequently $f(t) = 2$ if t is an even integer. The graph of the function f appears in Fig. 7.7. Find its Fourier series.

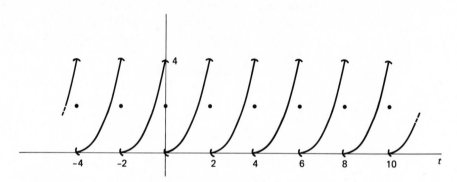

Figure 7.7 The period 2 function of Example 2.

Solution Here $L = 1$, and it is most convenient to integrate from $t = 0$ to $t = 2$. Then

$$a_0 = \frac{1}{1} \int_0^2 t^2 \, dt = \left[\frac{1}{3} t^3 \right]_0^2 = \frac{8}{3}.$$

With the aid of the integral formulas in (22)–(25) of Section 7.1, we obtain

$$a_n = \int_0^2 t^2 \cos n\pi t \, dt$$

$$= \frac{1}{n^3 \pi^3} \int_0^{2n\pi} u^2 \cos u \, du \qquad \left(u = n\pi t, \ t = \frac{u}{n\pi} \right)$$

$$= \frac{1}{n^3 \pi^3} \left[u^2 \sin u - 2 \sin u + 2u \cos u \right]_0^{2n\pi} = \frac{4}{n^2 \pi^2};$$

$$b_n = \int_0^2 t^2 \sin n\pi t \, dt$$

$$= \frac{1}{n^3 \pi^3} \int_0^{2n\pi} u^2 \sin u \, du$$

$$= \frac{1}{n^3 \pi^3} \left[-u^2 \cos u + 2 \cos u + 2u \sin u \right]_0^{2n\pi} = -\frac{4}{n\pi}.$$

Hence the Fourier series of f is

$$f(t) = \frac{4}{3} + \frac{4}{\pi^2} \sum_{n=1}^{\infty} \frac{\cos n\pi t}{n^2} - \frac{4}{\pi} \sum_{n=1}^{\infty} \frac{\sin n\pi t}{n}, \qquad (15)$$

and the convergence theorem assures us that this series converges to $f(t)$ for all t.

We can draw some interesting consequences from the Fourier series in (15). If we substitute $t = 0$ on each side, we find that

$$f(0) = 2 = \frac{4}{3} + \frac{4}{\pi^2} \sum_{n=1}^{\infty} \frac{1}{n^2}.$$

On solving for the series, we obtain the lovely summation

$$\sum_{n=1}^{\infty} \frac{1}{n^2} = 1 + \frac{1}{2^2} + \frac{1}{3^2} + \frac{1}{4^2} + \cdots = \frac{\pi^2}{6} \tag{16}$$

that was discovered by Euler. If we substitute $t = 1$ in (15), we get

$$f(1) = 1 = \frac{4}{3} + \frac{4}{\pi^2} \sum_{n=1}^{\infty} \frac{(-1)^n}{n^2},$$

which yields

$$\sum_{n=1}^{\infty} \frac{(-1)^{n+1}}{n^2} = 1 - \frac{1}{2^2} + \frac{1}{3^2} - \frac{1}{4^2} + \cdots = \frac{\pi^2}{12}. \tag{17}$$

If we add the series in (16) and (17) and then divide by 2, the "even" terms cancel and the result is

$$\sum_{n \text{ odd}} \frac{1}{n^2} = 1 + \frac{1}{3^2} + \frac{1}{5^2} + \frac{1}{7^2} + \cdots = \frac{\pi^2}{8}. \tag{18}$$

7.2 Problems

In each of Problems 1–14, the values of a periodic function $f(t)$ in one full period are given; at each discontinuity the value of $f(t)$ is that given by the average value condition in (13). Sketch the graph of f and find its Fourier series.

1. $f(t) = \begin{cases} -2, & -3 < t < 0; \\ 2, & 0 < t < 3. \end{cases}$

2. $f(t) = \begin{cases} 0, & -5 < t < 0; \\ 1, & 0 < t < 5. \end{cases}$

3. $f(t) = \begin{cases} 2, & -2\pi < t < 0; \\ -1, & 0 < t < 2\pi. \end{cases}$

4. $f(t) = t, -2 < t < 2.$

5. $f(t) = t, -2\pi < t < 2\pi.$

6. $f(t) = t, 0 < t < 3.$

7. $f(t) = |t|, -1 < t < 1.$

8. $f(t) = \begin{cases} 0, & 0 < t < 1; \\ 1, & 1 < t < 2; \\ 0, & 2 < t < 3. \end{cases}$

9. $f(t) = t^2, -1 < t < 1.$

10. $f(t) = \begin{cases} 0, & -2 < t < 0; \\ t^2, & 0 < t < 2. \end{cases}$

11. $f(t) = \cos \frac{\pi t}{2}, -1 < t < 1.$

12. $f(t) = \sin \pi t, 0 < t < 1.$

13. $f(t) = \begin{cases} 0, & -1 < t < 0; \\ \sin \pi t, & 0 < t < 1. \end{cases}$

14. $f(t) = \begin{cases} 0, & -2\pi < t < 0; \\ \sin t, & 0 < t < 2\pi. \end{cases}$

15. (a) Suppose that f is a function of period 2π with $f(t) = t^2$ for $0 < t < 2\pi$. Show that

$$f(t) = \frac{4\pi^2}{3} + 4 \sum_{n=1}^{\infty} \frac{\cos nt}{n^2} - 4\pi \sum_{n=1}^{\infty} \frac{\sin nt}{n}.$$

(b) Deduce the series summations in (16) and (17) from the Fourier series in part (a).

16. (a) Suppose that f is a function of period 2 such that $f(t) = 0$ if $-1 < t < 0$ and $f(t) = t$ if $0 < t < 1$. Show that

$$f(t) = \frac{1}{4} - \frac{2}{\pi^2} \sum_{n \text{ odd}} \frac{\cos n\pi t}{n^2} + \frac{1}{\pi} \sum_{n=1}^{\infty} (-1)^{n+1} \frac{\sin n\pi t}{n}.$$

(b) Deduce the series summation in (18) from the Fourier series in part (a).

17. (a) Suppose that f is a function of period 2 with $f(t) = t$ for $0 < t < 2$. Show that

$$f(t) = 1 - \frac{2}{\pi} \sum_{n=1}^{\infty} \frac{\sin n\pi t}{n}.$$

(b) Substitute an appropriate value of t to deduce **Leibniz's series**

$$1 - \frac{1}{3} + \frac{1}{5} - \frac{1}{7} + \cdots = \frac{\pi}{4}.$$

Derive the Fourier series listed in Problems 18–21.

18. $\displaystyle\sum_{n=1}^{\infty} \frac{\sin nt}{n} = \frac{\pi - t}{2}$ $(0 < t < 2\pi)$.

19. $\displaystyle\sum_{n=1}^{\infty} (-1)^{n+1} \frac{\sin nt}{n} = \frac{t}{2}$ $(-\pi < t < \pi)$.

20. $\displaystyle\sum_{n=1}^{\infty} \frac{\cos nt}{n^2} = \frac{3t^2 - 6\pi t + 2\pi^2}{12}$ $(0 < t < 2\pi)$.

21. $\displaystyle\sum_{n=1}^{\infty} (-1)^{n+1} \frac{\cos nt}{n^2} = \frac{\pi^2 - 3t^2}{12}$ $(-\pi < t < \pi)$.

22. Suppose that $p(t)$ is a polynomial of degree n. Show by repeated integration by parts that

$$\int p(t)g(t)\, dt = p(t)G_1(t) - p'(t)G_2(t)$$

$$+ p''(t)G_3(t) - \cdots + (-1)^n p^{(n)}(t)G_{n+1}(t)$$

where $G_k(t)$ denotes the kth iterated antiderivative $G_k(t) = (D^{-1})^k g(t)$. This formula is useful in computing Fourier coefficients of polynomials.

23. Apply the integral formula of Problem 22 to show that

$$\int t^4 \cos t\, dt = t^4 \sin t + 4t^3 \cos t$$

$$- 12t^2 \sin t - 24t \cos t + 24 \sin t + C$$

and

$$\int t^4 \sin t\, dt = -t^4 \cos t + 4t^3 \sin t$$

$$+ 12t^2 \cos t - 24t \sin t - 24 \cos t + C.$$

24. (a) Show that, for $0 < t < 2\pi$,

$$t^4 = \frac{16\pi^4}{5} + 16 \sum_{n=1}^{\infty} \left(\frac{2\pi^2}{n^2} - \frac{3}{n^4} \right) \cos nt + 16\pi \sum_{n=1}^{\infty} \left(\frac{3}{n^3} - \frac{\pi^2}{n} \right) \sin nt.$$

(b) From the Fourier series in part (a), deduce the summations

$$\sum_{n=1}^{\infty} \frac{1}{n^4} = \frac{\pi^4}{90}, \qquad \sum_{n=1}^{\infty} \frac{(-1)^{n+1}}{n^4} = \frac{7\pi^4}{720},$$

and

$$1 + \frac{1}{3^4} + \frac{1}{5^4} + \frac{1}{7^4} + \cdots = \frac{\pi^4}{96}.$$

25. Evaluate

$$1 - \frac{1}{3^3} + \frac{1}{5^3} - \frac{1}{7^3} + \frac{1}{9^3} - \cdots$$

by methods similar to those in this section.

26. Use the methods of this section to *attempt* to evaluate

$$1 + \frac{1}{2^3} + \frac{1}{3^3} + \frac{1}{4^3} + \frac{1}{5^3} + \cdots;$$

describe clearly the difficulties that arise.

7.3
Even-Odd Functions and Termwise Differentiation

Certain properties of functions are reflected prominently in their Fourier series. The function f defined for all t is called **even** if

$$f(-t) = f(t) \tag{1}$$

for all t; f is said to be **odd** if

$$f(-t) = -f(t) \tag{2}$$

for all t. The first condition implies that the graph of $y = f(t)$ is symmetric with respect to the y-axis, while the condition in (2) implies that the graph of an odd function is symmetric with respect to the origin; see Fig. 7.8. The functions $f(t) = t^{2n}$ and $g(t) = \cos t$ are even functions, while the functions $f(t) = t^{2n+1}$ and $g(t) = \sin t$ are odd. We will see that the Fourier series of an even periodic function has only cosine terms, while the Fourier series of an odd periodic function has only sine terms.

Addition and cancellation of areas as indicated in Fig. 7.9 reminds us of the following basic facts about integrals of even and odd functions over an interval $[-a, a]$ that is symmetric about the origin.

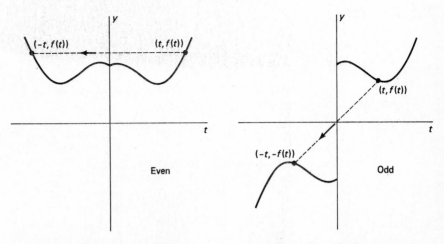

Figure 7.8a An even function.   **Figure 7.8b** An odd function.

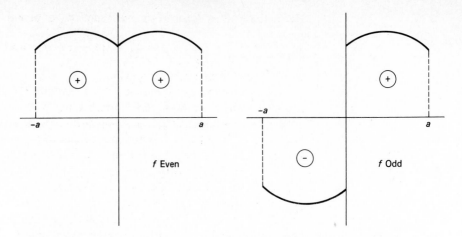

Figure 7.9a Area under the graph of an even function.

Figure 7.9b Area under the graph of an odd function.

$$\text{If } f \text{ is even:} \quad \int_{-a}^{a} f(t)\, dt = 2 \int_{0}^{a} f(t)\, dt. \tag{3}$$

$$\text{If } f \text{ is odd:} \quad \int_{-a}^{a} f(t)\, dt = 0. \tag{4}$$

These facts are easy to verify analytically (see Problem 17).

It follows immediately from (1) and (2) that the product of two even functions is even, as is the product of two odd functions; the product of an even function and an odd function is odd. In particular, if $f(t)$ is an *even* periodic function of period $2L$, then $f(t) \cos{(n\pi t/L)}$ is even, while $f(t) \sin{(n\pi t/L)}$ is odd, because the cosine function is even while the sine function is odd. When we compute the Fourier coefficients of f, we therefore get

$$a_n = \frac{1}{L} \int_{-L}^{L} f(t) \cos \frac{n\pi t}{L}\, dt = \frac{2}{L} \int_{0}^{L} f(t) \cos \frac{n\pi t}{L}\, dt \tag{5a}$$

and

$$b_n = \frac{1}{L} \int_{-L}^{L} f(t) \sin \frac{n\pi t}{L}\, dt = 0 \tag{5b}$$

because of (3) and (4). Hence the Fourier series of the *even* function f of period $2L$ has only *cosine* terms:

$$f(t) = \frac{a_0}{2} + \sum_{n=1}^{\infty} a_n \cos \frac{n\pi t}{L} \qquad (f \text{ even}) \tag{6}$$

with a_n given by the formula in (5). If $f(t)$ is *odd*, then $f(t) \cos{(n\pi t/L)}$ is odd, while $f(t) \sin{(n\pi t/L)}$ is even, so

$$a_n = \frac{1}{L} \int_{-L}^{L} f(t) \cos \frac{n\pi t}{L}\, dt = 0 \tag{7a}$$

while

$$b_n = \frac{1}{L} \int_{-L}^{L} f(t) \sin \frac{n\pi t}{L}\, dt = \frac{2}{L} \int_{0}^{L} f(t) \sin \frac{n\pi t}{L}\, dt. \tag{7b}$$

Hence the Fourier series of the *odd* function f of period $2L$ has only *sine* terms:

$$f(t) = \sum_{n=1}^{\infty} b_n \sin \frac{n\pi t}{L} \qquad (f \text{ odd}) \tag{8}$$

with the coefficients b_n given in (7).

In all our earlier discussions and examples, we began with a periodic function defined *for all* t; the Fourier series of such a function is uniquely determined by the Fourier coefficient formulas. In many practical situations, however, we begin with a function f defined only on an interval of the form $0 < t < L$, and we want to represent its values on this interval by a Fourier series of period $2L$. The first step is the necessary extension of f to the interval $-L < t < 0$. Granted this, we may extend f to the entire real line by the periodicity condition $f(t + 2L) = f(t)$ (and use the average value property should any discontinuities arise). But *how* we define f for $-L < t < 0$ is our choice, and the Fourier series representation for $f(t)$ on $(0, L)$ that we obtain will depend upon that choice. Specifically, different choices of the extension of f to the interval $(-L, 0)$ will yield different Fourier series that converge to the same function $f(t)$ in the original interval $0 < t < L$, but which converge to the different extensions of f on the interval $-L < t < 0$.

In practice, given $f(t)$ defined for $0 < t < L$, we generally make one of two natural choices—we extend f to obtain either an even function or an odd function. The **even period $2L$ extension** of f is the function f_E defined as

$$f_E(t) = \begin{cases} f(t) & \text{if } 0 < t < L; \\ f(-t) & \text{if } -L < t < 0 \end{cases} \tag{9}$$

and by $f_E(t + 2L) = f_E(t)$ for all t. The values of $f_E(t)$ for t an integral multiple of L are not important because they cannot affect the Fourier series of f_E. The Fourier series of the even extension f_E, given by the formulas in (5) and (6), will contain only cosine terms and is called the *Fourier cosine series* of the original function f.

The **odd period $2L$ extension** of f is the function f_O defined as

$$f_O(t) = \begin{cases} f(t) & \text{if } 0 < t < L; \\ -f(-t) & \text{if } -L < t < 0 \end{cases} \tag{10}$$

and by $f_O(t + 2L) = f_O(t)$ for all t. The Fourier series of the odd extension f_O, given by the formulas in (7) and (8), will contain only sine terms and is called the *Fourier sine series* of the original function f.

DEFINITION: FOURIER COSINE AND SINE SERIES

*Suppose that the function $f(t)$ is piecewise continuous on the interval $[0, L]$. Then the **Fourier cosine series** of f is the series*

$$f(t) = \frac{a_0}{2} + \sum_{n=1}^{\infty} a_n \cos \frac{n\pi t}{L} \tag{11}$$

with

$$a_n = \frac{2}{L} \int_0^L f(t) \cos \frac{n\pi t}{L} \, dt. \tag{12}$$

The **Fourier sine series** of f is the series

$$f(t) = \sum_{n=1}^{\infty} b_n \sin \frac{n\pi t}{L} \tag{13}$$

with

$$b_n = \frac{2}{L} \int_0^L f(t) \sin \frac{n\pi t}{L} \, dt. \tag{14}$$

Assuming that f is piecewise smooth and satisfies the average value condition $f(t) = \frac{1}{2}[f(t+) + f(t-)]$ at each of its isolated discontinuities, the convergence theorem in Section 7.2 implies that each of the two series in (11) and (12) converges to $f(t)$ for all t in the interval $0 < t < L$. Outside this interval, the cosine series in (11) converges to the even period $2L$ extension of f, while the sine series in (13) converges to the odd period $2L$ extension of f. In many cases of interest we have no concern with the values of f outside the original interval $(0, L)$, and therefore the choice between (11) and (12) or (13) and (14) is determined by whether we prefer to represent $f(t)$ in the interval $(0, L)$ by a cosine series or a sine series. See Example 2 below for a situation that dictates our choice between a Fourier cosine series and a Fourier sine series to represent a given function.

EXAMPLE 1 Suppose that $f(t) = t$ for $0 < t < L$. Find both the Fourier cosine series and the Fourier sine series for f.

Solution The formula in (12) gives

$$a_0 = \frac{2}{L} \int_0^L t \, dt = \frac{2}{L} \left[\frac{1}{2} t^2 \right]_0^L = L,$$

and

$$a_n = \frac{2}{L} \int_0^L t \cos \frac{n\pi t}{L} \, dt = \frac{2L}{n^2 \pi^2} \int_0^{n\pi} u \cos u \, du$$

$$= \frac{2L}{n^2 \pi^2} \left[u \sin u + \cos u \right]_0^{n\pi} = \begin{cases} -\dfrac{4L}{n^2 \pi^2} & \text{for } n \text{ odd}; \\ 0 & \text{for } n \text{ even.} \end{cases}$$

Thus the Fourier cosine series of f is

$$t = \frac{L}{2} - \frac{4L}{\pi^2} \left(\cos \frac{\pi t}{L} + \frac{1}{3^2} \cos \frac{3\pi t}{L} + \frac{1}{5^2} \cos \frac{5\pi t}{L} + \cdots \right) \tag{15}$$

for $0 < t < L$. Next, the formula in (14) gives

$$b_n = \frac{2}{L} \int_0^L t \sin \frac{n\pi t}{L} \, dt = \frac{2L}{n^2 \pi^2} \int_0^{n\pi} u \sin u \, du$$

$$= \frac{2L}{n^2 \pi^2} \left[-u \cos u + \sin u \right]_0^{n\pi} = \frac{2L}{n\pi} (-1)^{n+1}.$$

Thus the Fourier sine series of f is

$$t = \frac{2L}{\pi}\left(\sin\frac{\pi t}{L} - \frac{1}{2}\sin\frac{2\pi t}{L} + \frac{1}{3}\sin\frac{3\pi t}{L} - \cdots\right) \qquad (16)$$

for $0 < t < L$. The series in (15) converges to the even period $2L$ extension of f shown in Fig. 7.10, while the series in (16) converges to the odd period $2L$ extension shown in Fig. 7.11.

Figure 7.10 The even period $2L$ extension of f.

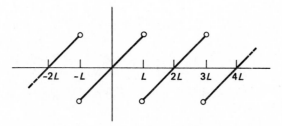

Figure 7.11 The odd period $2L$ extension of f.

TERMWISE DIFFERENTIATION OF FOURIER SERIES

In this and subsequent sections, we want to consider Fourier series as possible solutions of differential equations. In order to substitute a Fourier series for the unknown dependent variable in a differential equation to check whether it is a solution, we first need to differentiate the series in order to compute the derivatives that appear in the equation. Care is required here; term-by-term differentiation of an infinite series of variable terms is not always valid. The following theorem gives sufficient conditions for the validity of termwise differentiation of a Fourier series.

THEOREM 1: TERMWISE DIFFERENTIATION OF FOURIER SERIES

Suppose that the function f is continuous for all t, periodic with period $2L$, and that its derivative f' is piecewise smooth for all t. Then the Fourier series of f' is the series

$$f'(t) = \sum_{n=1}^{\infty}\left(-\frac{n\pi}{L}a_n\sin\frac{n\pi t}{L} + \frac{n\pi}{L}b_n\cos\frac{n\pi t}{L}\right) \qquad (17)$$

obtained by termwise differentiation of the Fourier series

$$f(t) = \frac{a_0}{2} + \sum_{n=1}^{\infty}\left(a_n\cos\frac{n\pi t}{L} + b_n\sin\frac{n\pi t}{L}\right). \qquad (18)$$

Proof The point of the theorem is that the differentiated series in (17) actually converges to $f'(t)$ (with the usual proviso about average values). But, because f' is periodic and piecewise smooth, we know from the convergence theorem in Section 7.2 that the Fourier series of f' converges to $f'(t)$:

$$f'(t) = \frac{\alpha_0}{2} + \sum_{n=1}^{\infty} \left(\alpha_n \cos \frac{n\pi t}{L} + \beta_n \sin \frac{n\pi t}{L} \right). \tag{19}$$

In order to prove Theorem 1, it therefore suffices to show that the series in (17) and (19) are identical. We will do so under the additional hypothesis that f' is continuous everywhere. Then

$$\alpha_0 = \frac{1}{L} \int_{-L}^{L} f'(t)\, dt = \frac{1}{L} \Big[f(t) \Big]_{-L}^{L} = 0$$

because $f(L) = f(-L)$ by periodicity, and

$$\alpha_n = \frac{1}{L} \int_{-L}^{L} f'(t) \cos \frac{n\pi t}{L}\, dt$$

$$= \frac{1}{L} \Big[f(t) \cos \frac{n\pi t}{L} \Big]_{-L}^{L} + \frac{n\pi}{L} \cdot \frac{1}{L} \int_{-L}^{L} f(t) \sin \frac{n\pi t}{L}\, dt$$

—integration by parts. It follows that

$$\alpha_n = \frac{n\pi}{L} b_n.$$

Similarly we find that

$$\beta_n = -\frac{n\pi}{L} a_n,$$

and therefore the series in (17) and (19) are, indeed, identical.

Whereas the assumption that the derivative f' is continuous is merely a convenience—the proof above can be strengthened to allow isolated discontinuities in f'—it is important to note that the conclusion of Theorem 1 generally fails when f itself is discontinuous. For example, consider the Fourier series

$$t = \frac{2L}{\pi} \left(\sin \frac{\pi t}{L} - \frac{1}{2} \sin \frac{2\pi t}{L} + \frac{1}{3} \sin \frac{3\pi t}{L} - \cdots \right), \tag{16}$$

$-L < t < L$, of the discontinuous sawtooth function having the graph shown in Fig. 7.11. All the hypotheses of Theorem 1 are satisfied apart from the continuity of f, and f has only isolated jump discontinuities. But the series

$$\frac{2\pi^2}{L^2} \left(\cos \frac{\pi t}{L} - \cos \frac{2\pi t}{L} + \cos \frac{3\pi t}{L} - \cdots \right) \tag{20}$$

obtained by differentiating the series in (16) term by term diverges (for instance, when $t = 0$ and when $t = L$), and therefore termwise differentiation of the series in (16) is not valid.

By contrast, consider the (continuous) triangular wave function $f(t)$ having the graph shown in Fig. 7.10, with $f(t) = |t|$ for $-L < t < L$. This function satisfies all the hypotheses of Theorem 1, so its Fourier series

$$f(t) = \frac{L}{2} - \frac{4L}{\pi^2}\left(\cos\frac{\pi t}{L} + \frac{1}{3^2}\cos\frac{3\pi t}{L} + \frac{1}{5^2}\cos\frac{5\pi t}{L} + \cdots\right) \qquad (15)$$

can be differentiated termwise. The result is

$$f'(t) = \frac{4}{\pi}\left(\sin\frac{\pi t}{L} + \frac{1}{3}\sin\frac{3\pi t}{L} + \frac{1}{5}\sin\frac{5\pi t}{L} + \cdots\right), \qquad (21)$$

which is the Fourier series of the period $2L$ square wave function that takes the value -1 for $-L < t < 0$ and $+1$ for $0 < t < L$.

FOURIER SERIES SOLUTIONS OF DIFFERENTIAL EQUATIONS

In the remainder of this chapter and in Chapter 8, we will frequently need to solve endpoint value problems of the general form

$$ax'' + bx' + cx = f(t) \qquad (0 < t < L); \qquad (22)$$
$$x(0) = x(L) = 0, \qquad (23)$$

where the function $f(t)$ is given. We might consider applying the techniques of Chapter 2, solving the problem by:

1. first finding the general solution $x_c = c_1 x_1 + c_2 x_2$ of the associated homogeneous differential equation;
2. then finding a single particular solution x_p of the nonhomogeneous equation in (22);
3. and finally, determining the constants c_1 and c_2 so that $x = x_c + x_p$ satisfies the endpoint conditions in (23).

   In many problems, the following Fourier series method is more convenient and more useful. We first extend the definition of the function $f(t)$ to the interval $-L < t < 0$ in an appropriate way, and then to the entire real line by the periodicity condition $f(t + 2L) = f(t)$. Then the function f, if piecewise smooth, has a Fourier series

$$f(t) = \frac{A_0}{2} + \sum_{n=1}^{\infty}\left(A_n\cos\frac{n\pi t}{L} + B_n\sin\frac{n\pi t}{L}\right), \qquad (24)$$

which has coefficients A_n and B_n that we can and do compute. We then assume that the differential equation in (22) has a solution $x(t)$ with a Fourier series

$$x(t) = \frac{a_0}{2} + \sum_{n=1}^{\infty}\left(a_n\cos\frac{n\pi t}{L} + b_n\sin\frac{n\pi t}{L}\right) \qquad (25)$$

that may legally be differentiated twice termwise. We attempt to determine the coefficients in (25) by first substituting the series in (24) and (25) into the differential equation in (22) and then equating coefficients of like terms. If this process is carried out in such a way that the resulting series in (25) also satisfies the endpoint conditions in (23), then we have a "formal Fourier series solution" of the original endpoint value problem; that is, a solution subject to verification of the assumed termwise differentiability. The following example illustrates this process.

EXAMPLE 2 Find a formal Fourier series solution of the endpoint value problem

$$x'' + 4x = 4t, \tag{26}$$

$$x(0) = x(1) = 0. \tag{27}$$

Solution Here $f(t) = 4t$ for $0 < t < 1$. A crucial first step—which we did not make explicit in the exposition above—is to choose a periodic extension of $f(t)$ so that each term in its Fourier series satisfies the endpoint conditions in (27). For this purpose we choose the odd period 2 extension, because each term of the form $\sin n\pi t$ satisfies (27). Then from the series in (16) with $L = 1$, we get the Fourier series

$$4t = \frac{8}{\pi} \sum_{n=1}^{\infty} \frac{(-1)^{n+1}}{n} \sin n\pi t \tag{28}$$

for $0 < t < 1$. We therefore anticipate a sine series solution

$$x(t) = \sum_{n=1}^{\infty} b_n \sin n\pi t, \tag{29}$$

noting that any such series will satisfy the endpoint conditions in (27). When we substitute the series in (28) and (29) in (26), the result is

$$\sum_{n=1}^{\infty} (-n^2\pi^2 + 4)b_n \sin n\pi t = \frac{8}{\pi} \sum_{n=1}^{\infty} \frac{(-1)^{n+1}}{n} \sin n\pi t. \tag{30}$$

We next equate coefficients of like terms in (30). This yields

$$b_n = \frac{(8)(-1)^{n+1}}{\pi n(4 - n^2\pi^2)},$$

so our formal Fourier series solution is

$$x(t) = \frac{8}{\pi} \sum_{n=1}^{\infty} \frac{(-1)^{n+1} \sin n\pi t}{n(4 - n^2\pi^2)}. \tag{31}$$

In Problem 16 we ask you to derive the exact solution

$$x(t) = t - \frac{\sin 2t}{\sin 2} \quad (0 \leq t \leq 1), \tag{32}$$

and to verify that (31) is the Fourier series of the odd period 2 extension of this solution.

TERMWISE INTEGRATION OF FOURIER SERIES

The following theorem guarantees that the Fourier series of a piecewise continuous periodic function can always be integrated term by term, whether or not it converges! A proof is outlined in Problem 25.

THEOREM 2: TERMWISE INTEGRATION OF FOURIER SERIES

Suppose that f is a piecewise continuous periodic function with period $2L$ and Fourier series

$$f(t) \sim \frac{a_0}{2} + \sum_{n=1}^{\infty} \left(a_n \cos \frac{n\pi t}{L} + b_n \sin \frac{n\pi t}{L} \right), \qquad (33)$$

which may not converge. Then, for all t,

$$\int_0^t f(s)\, ds = \frac{a_0 t}{2} + \sum_{n=1}^{\infty} \frac{L}{n\pi} \left[a_n \sin \frac{n\pi t}{L} - b_n \left(\cos \frac{n\pi t}{L} - 1 \right) \right], \qquad (34)$$

with the series on the right-hand side convergent for all t. Note that the series in (34) is the result of term-by-term integration of the series in (33).

EXAMPLE 3 Let us attempt to verify the conclusion of Theorem 2 in the case that $f(t)$ is the period 2π function such that

$$f(t) = \begin{cases} -1, & -\pi < t < 0; \\ +1, & 0 < t < \pi. \end{cases} \qquad (35)$$

By Example 1 of Section 7.1, the Fourier series of f is

$$f(t) = \frac{4}{\pi} \left(\sin t + \frac{1}{3} \sin 3t + \frac{1}{5} \sin 5t + \cdots \right). \qquad (36)$$

Theorem 2 then implies that

$$F(t) = \int_0^t f(s)\, ds$$

$$= \int_0^t \frac{4}{\pi} \left(\sin s + \frac{1}{3} \sin 3s + \frac{1}{5} \sin 5s + \cdots \right) ds$$

$$= \frac{4}{\pi} \left[(1 - \cos t) + \frac{1}{3^2}(1 - \cos 3t) + \frac{1}{5^2}(1 - \cos 5t) + \cdots \right].$$

Thus

$$F(t) = \frac{4}{\pi} \left(1 + \frac{1}{3^2} + \frac{1}{5^2} + \cdots \right)$$

$$- \frac{4}{\pi} \left(\cos t + \frac{1}{3^2} \cos 3t + \frac{1}{5^2} \cos 5t + \cdots \right). \qquad (37)$$

On the other hand, direct integration of (35) yields

$$F(t) = \int_0^t f(s)\, ds = |t| = \begin{cases} -t, & -\pi < t < 0, \\ t, & 0 < t < \pi. \end{cases}$$

We know from Example 1 in this section (with $L = \pi$) that

$$|t| = \frac{\pi}{2} - \frac{4}{\pi} \left(\cos t + \frac{1}{3^2} \cos 3t + \frac{1}{5^2} \cos 5t + \cdots \right). \qquad (38)$$

We also know from the formula in (18) of Section 7.2 that

$$1 + \frac{1}{3^2} + \frac{1}{5^2} + \cdots = \frac{\pi^2}{8},$$

so it follows that the two series in (37) and (38) are indeed identical.

7.3 Problems

In each of Problems 1–10, a function $f(t)$ is defined on an interval $0 < t < L$. Find the Fourier cosine and sine series of f, and sketch the graphs of the two extensions of f to which these two series converge.

1. $f(t) = 1, 0 < t < L$.

2. $f(t) = 1 - t, 0 < t < 1$.

3. $f(t) = 1 - t, 0 < t < 2$.

4. $f(t) = \begin{cases} t, & 0 < t \le 1; \\ 2 - t, & 1 \le t < 2. \end{cases}$

5. $f(t) = \begin{cases} 0, & 0 < t < 1; \\ 1, & 1 < t < 2; \\ 0, & 2 < t < 3. \end{cases}$

6. $f(t) = t^2, 0 < t < L$.

7. $f(t) = t(\pi - t), 0 < t < \pi$.

8. $f(t) = t - t^2, 0 < t < 1$.

9. $f(t) = \sin t, 0 < t < \pi$.

10. $f(t) = \begin{cases} \sin t, & 0 < t \le \pi; \\ 0, & \pi \le t < 2\pi. \end{cases}$

Find a formal Fourier series solution of each of the endpoint value problems in Problems 11–14.

11. $x'' + 2x = 1, x(0) = x(\pi) = 0$.

12. $x'' - 4x = 1, x(0) = x(\pi) = 0$.

13. $x'' + x = t, x(0) = x(1) = 0$.

14. $x'' + 2x = t, x(0) = x(2) = 0$.

15. Find a formal Fourier series solution of the endpoint value problem

$$x'' + 2x = t, \qquad x'(0) = x'(\pi) = 0.$$

(*Suggestion:* Use a Fourier cosine series in which each term satisfies the endpoint conditions.)

16. (a) Derive the solution $x(t) = t - (\sin 2t)/(\sin 2)$ of the endpoint value problem

$$x'' + 4x = 4t, \qquad x(0) = x(1) = 0.$$

(b) Show that the series in Eq. (31) is the Fourier sine series of the solution in part (a).

17. (a) Suppose that f is an even function. Show that

$$\int_{-a}^{0} f(t)\, dt = \int_{0}^{a} f(t)\, dt.$$

(b) Suppose that f is an odd function. Show that

$$\int_{-a}^{0} f(t)\, dt = -\int_{0}^{a} f(t)\, dt.$$

18. By Example 2 in Section 7.2, the Fourier series of the period 2 function f with $f(t) = t^2$ for $0 < t < 2$ is

$$f(t) = \frac{4}{3} + \frac{4}{\pi^2} \sum_{n=1}^{\infty} \frac{\cos n\pi t}{n^2} - \frac{4}{\pi} \sum_{n=1}^{\infty} \frac{\sin n\pi t}{n}.$$

Show that the termwise derivative of this series does not converge to $f'(t)$.

19. Begin with the Fourier series

$$t = 2 \sum_{n=1}^{\infty} \frac{(-1)^{n+1}}{n} \sin nt, \qquad -\pi < t < \pi,$$

and integrate termwise three times in succession to obtain the series

$$\frac{1}{24}t^4 = \frac{\pi^2 t^2}{12} - 2\sum_{n=1}^{\infty} \frac{(-1)^n}{n^4} \cos nt + 2\sum_{n=1}^{\infty} \frac{(-1)^n}{n^4}.$$

20. Substitute $t = \pi/2$ and $t = \pi$ in the series of Problem 19 to obtain the summations

$$\sum_{n=1}^{\infty} \frac{1}{n^4} = \frac{\pi^4}{90},$$

$$\sum_{n=1}^{\infty} \frac{(-1)^{n+1}}{n^4} = \frac{7\pi^4}{720},$$

and

$$1 + \frac{1}{3^4} + \frac{1}{5^4} + \frac{1}{7^4} + \cdots = \frac{\pi^4}{96}.$$

21. (Odd half-multiple sine series)   Let $f(t)$ be given for $0 < t < L$, and define $F(t)$ for $0 < t < 2L$ as follows:

$$F(t) = \begin{cases} f(t), & 0 < t < L; \\ f(2L - t), & L < t < 2L. \end{cases}$$

Thus the graph of $F(t)$ is symmetric about the line $t = L$ (see Fig. 7.12). Then the period $4L$ Fourier sine series of F is

$$F(t) = \sum_{n=1}^{\infty} b_n \sin \frac{n\pi t}{2L},$$

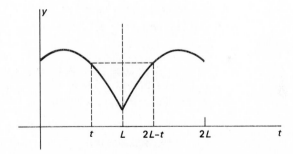

Figure 7.12   Construction of F from f in Problem 21.

where

$$b_n = \frac{1}{L} \int_0^L f(t) \sin \frac{n\pi t}{2L} \, dt + \frac{1}{L} \int_L^{2L} f(2L - t) \sin \frac{n\pi t}{2L} \, dt.$$

Substitute $s = 2L - t$ in the second integral to derive the series (for $0 < t < L$)

$$f(t) = \sum_{n \text{ odd}} b_n \sin \frac{n\pi t}{2L},$$

where

$$b_n = \frac{2}{L} \int_0^L f(t) \sin \frac{n\pi t}{2L} \, dt \qquad (n \text{ odd}).$$

22. (Odd half-multiple cosine series)   Let $f(t)$ be given for $0 < t < L$, and define $G(t)$ for $0 < t < 2L$ as follows:

$$G(t) = \begin{cases} f(t), & 0 < t < L; \\ -f(2L - t), & L < t < 2L. \end{cases}$$

Use the period $4L$ Fourier cosine series of $G(t)$ to derive the series (for $0 < t < L$)

$$f(t) = \sum_{n \text{ odd}} a_n \cos \frac{n\pi t}{2L} \, dt,$$

where

$$a_n = \frac{2}{L} \int_0^L f(t) \cos \frac{n\pi t}{2L} \, dt \qquad (n \text{ odd}).$$

23. Given: $f(t) = t$, $0 < t < 2\pi$. Derive the odd half-multiple sine series (Problem 21)

$$f(t) = \frac{8}{\pi} \sum_{n \text{ odd}} \frac{(-1)^{(n-1)/2}}{n^2} \sin \frac{nt}{2}.$$

24. Given the endpoint value problem

$$x'' - x = t, \qquad x(0) = 0, \qquad x'(\pi) = 0,$$

note that any constant multiple of $\sin (nt/2)$ with n odd satisfies the endpoint conditions. Hence use the odd half-multiple sine series of Problem 23 to derive the formal Fourier series solution

$$x(t) = \frac{32}{\pi} \sum_{n \text{ odd}} \frac{(-1)^{(n+1)/2}}{n^2(n^2 + 4)} \sin \frac{nt}{2}.$$

25. In this problem we outline the proof of Theorem 2. Suppose that $f(t)$ is a piecewise continuous period $2L$ function. Define

$$F(t) = \int_0^t [f(s) - \tfrac{1}{2}a_0] \, ds$$

where a_n and b_n denote the Fourier coefficients of $f(t)$.

(a) Show directly that $F(t + 2L) = F(t)$, so that F is a continuous period $2L$ function, and therefore has a convergent Fourier series

$$F(t) = \frac{1}{2} A_0 + \sum_{n=1}^{\infty} \left(A_n \cos \frac{n\pi t}{L} + B_n \sin \frac{n\pi t}{L} \right).$$

(b) Suppose that $n \geq 1$. Show by direct computation that

$$A_n = -\frac{L}{n\pi} b_n \quad \text{and} \quad B_n = \frac{L}{n\pi} a_n.$$

(c) Thus

$$\int_0^t f(s) \, ds = \frac{t}{2} a_0 + \frac{1}{2} A_0 + \sum_{n=1}^{\infty} \frac{L}{n\pi} \left(a_n \sin \frac{n\pi t}{L} - b_n \cos \frac{n\pi t}{L} \right).$$

Finally substitute $t = 0$ to see that

$$\frac{1}{2} A_0 = \sum_{n=1}^{\infty} \frac{L}{n\pi} b_n.$$

---

*7.4
Applications
of Fourier Series

We consider first the motion of a mass m on a spring with Hooke's constant k under the influence of a *periodic* external force $F(t)$, as indicated in Fig. 7.13. Its displacement from equilibrium, $x(t)$, satisfies the familiar equation

$$mx'' + kx = F(t). \tag{1}$$

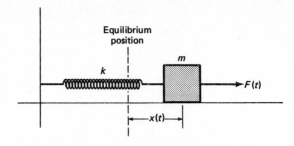

Figure 7.13 A mass-and-spring system with external force.

The general solution of Eq. (1) is of the form

$$x(t) = c_1 \cos \omega_0 t + c_2 \sin \omega_0 t + x_p(t), \tag{2}$$

where $\omega_0 = \sqrt{k/m}$ is the natural frequency of the system and $x_p(t)$ is a particular solution of Eq. (1). The values c_1 and c_2 would be determined by the initial conditions. Here we want to use Fourier series to find a *periodic* particular solution of (1). We will denote it by $x_{sp}(t)$ and call it a **steady periodic solution**.

We assume for simplicity that $F(t)$ is an odd function with period $2L$, so its Fourier series has the form

$$F(t) = \sum_{n=1}^{\infty} B_n \sin \frac{n\pi t}{L}. \tag{3}$$

If $n\pi/L$ is not equal to ω_0 for any positive integer n, we can determine a steady periodic solution of the form

$$x_{sp}(t) = \sum_{n=1}^{\infty} b_n \sin \frac{n\pi t}{L} \tag{4}$$

by substituting the series in (3) and (4) in Eq. (1) to find the coefficients in (4). The following example illustrates this procedure.

EXAMPLE 1 Suppose that $m = 2$ slugs, $k = 32$ lb/ft, and that $F(t)$ is an odd periodic force with a period of 2 s given in one period by

$$F(t) = \begin{cases} +10 \text{ lb} & \text{if } 0 < t < 1; \\ -10 \text{ lb} & \text{if } 1 < t < 2. \end{cases} \tag{5}$$

Find the steady periodic motion $x_{sp}(t)$.

Solution The graph of the periodic forcing function $F(t)$ is shown in Fig. 7.14. By essentially the same computation as in Example 1 of Section 7.1, the Fourier series of $F(t)$ is

$$F(t) = \frac{40}{\pi} \sum_{n \text{ odd}} \frac{\sin n\pi t}{n}; \tag{6}$$

note that it contains terms corresponding only to n odd. When we substitute this series and

$$x_{sp}(t) = \sum_{n \text{ odd}} b_n \sin n\pi t, \tag{7}$$

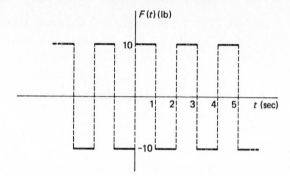

Figure 7.14 Graph of the forcing function of Example 1.

a trial solution likewise containing only odd terms, into Eq. (1) with $m = 2$ and $k = 32$, we get

$$\sum_{n \text{ odd}} b_n(-2n^2\pi^2 + 32) \sin n\pi t = \frac{40}{\pi} \sum_{n \text{ odd}} \frac{\sin n\pi t}{n}.$$

We equate coefficients of like terms; the result is that

$$b_n = \frac{20}{n\pi(16 - n^2\pi^2)} \quad \text{for } n \text{ odd.}$$

Hence

$$x_{sp}(t) = \frac{20}{\pi} \sum_{n \text{ odd}} \frac{\sin n\pi t}{n(16 - n^2\pi^2)}. \tag{8}$$

The fact that each term in (8) is symmetric about $t = \frac{1}{2}$ suggests that $x_{sp}(t)$ is maximal when $t = \frac{1}{2}$. Assuming this, we find the amplitude of the steady periodic motion to be

$$x_{sp}\left(\frac{1}{2}\right) = \frac{20}{\pi} \sum_{n \text{ odd}} \frac{\sin \frac{n\pi}{2}}{n(16 - n^2\pi^2)}.$$

The first five terms yield $x(\frac{1}{2}) \approx 1.064$ ft.

If there is a nonzero term $B_N \sin (N\pi t/L)$ in the Fourier series of $F(t)$ for which $N\pi/L = \omega_0$, then this term causes **pure resonance**. The reason is that the equation

$$mx'' + kx = B_N \sin \omega_0 t$$

has the resonance solution

$$x(t) = -\frac{B_N}{2m\omega_0} t \cos \omega_0 t.$$

The solution corresponding to (4) in this case is then

$$x(t) = -\frac{B_N}{2m\omega_0} t \cos \omega_0 t + \sum_{n \neq N} \frac{B_n \sin \frac{n\pi t}{L}}{m\left(\omega_0^2 - \frac{n^2\pi^2}{L^2}\right)}. \tag{9}$$

EXAMPLE 2   Suppose that $m = 2$ slugs and $k = 32$ lb/ft as in Example 1. Determine whether pure resonance will occur if $F(t)$ is an odd periodic function defined in one period to be

(a) $F(t) = \begin{cases} +10, & 0 < t < \pi; \\ -10, & \pi < t < 2\pi. \end{cases}$

(b) $F(t) = 10t, \; -\pi < t < \pi.$

Solution   (a) The natural frequency is $\omega_0 = 4$, and the Fourier series of $F(t)$ is

$$F(t) = \frac{40}{\pi}\left(\sin t + \frac{1}{3}\sin 3t + \frac{1}{5}\sin 5t + \cdots\right).$$

Because this series contains no $\sin 4t$ term, no resonance occurs.
(b). In this case the Fourier series is

$$F(t) = \sum_{n=1}^{\infty} \frac{(-1)^{n+1}}{n}\sin nt.$$

Pure resonance occurs because of the presence of the $\sin 4t$ term.

The following example illustrates the *near resonance* that can occur when a single term in the solution series is magnified because its frequency is close to the natural frequency ω_0.

EXAMPLE 3   Find a steady periodic solution of

$$x'' + 10x = F(t) \tag{10}$$

where $F(t)$ is the function of period 4 with $F(t) = 5t$ for $-2 < t < 2$ and Fourier series

$$F(t) = \frac{20}{\pi}\sum_{n=1}^{\infty} \frac{(-1)^{n+1}}{n}\sin\frac{n\pi t}{2}. \tag{11}$$

Solution   When we substitute (11) and

$$x_{sp} = \sum_{n=1}^{\infty} b_n \sin\frac{n\pi t}{2}$$

in (10), we obtain

$$\sum_{n=1}^{\infty} b_n\left(-\frac{n^2\pi^2}{4} + 10\right)\sin\frac{n\pi t}{2} = \frac{20}{\pi}\sum_{n=1}^{\infty} \frac{(-1)^{n+1}}{n}\sin\frac{n\pi t}{2}.$$

We equate coefficients of like terms and then solve for b_n to get the steady periodic solution

$$x_{sp}(t) = \frac{80}{\pi}\sum_{n=1}^{\infty} \frac{(-1)^{n+1}}{n(40 - n^2\pi^2)}\sin\frac{n\pi t}{2}$$

$$\approx (0.8452)\sin\frac{\pi t}{2} - (24.4111)\sin\frac{2\pi t}{2} - (0.1739)\sin\frac{3\pi t}{2} + \cdots.$$

The large magnitude of the second term results from the fact that $\omega_0 = \sqrt{10} \approx \pi = 2\pi/2$. Thus the dominant motion of a spring with the differential equation in (10) would be an oscillation with frequency π radians per second and amplitude about 24.

DAMPED FORCED OSCILLATIONS

Now we consider the motion of a mass m attached both to a spring with Hooke's constant k and to a dashpot with damping constant c, under the influence of a *periodic* external force $F(t)$; see Fig. 7.15. The displacement $x(t)$ of the mass from equilibrium satisfies the equation

$$mx'' + cx' + kx = F(t). \tag{12}$$

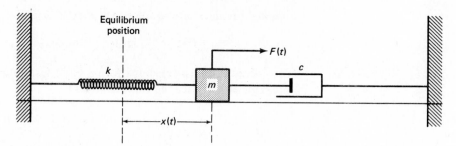

Figure 7.15   A damped mass-and-spring system with external force.

We recall from Problem 21 in Section 2.8 that the steady periodic solution of Eq. (12) with $F(t) = F_0 \sin \omega t$ is

$$x(t) = \frac{F_0}{\sqrt{(k - m\omega^2)^2 + (c\omega)^2}} \sin (\omega t - \alpha), \tag{13}$$

where

$$\alpha = \tan^{-1} \frac{c\omega}{k - m\omega^2}, \qquad 0 \leq \alpha \leq \pi. \tag{14}$$

If $F(t)$ is an odd period $2L$ function with Fourier series

$$F(t) = \sum_{n=1}^{\infty} B_n \sin \frac{n\pi t}{L}, \tag{15}$$

then the formulas above yield, by superposition, the steady periodic solution

$$x_{sp}(t) = \sum_{n=1}^{\infty} \frac{B_n \sin (\omega_n t - \alpha_n)}{\sqrt{(k - m\omega_n^2)^2 + (c\omega_n)^2}} \tag{16}$$

where $\omega_n = n\pi/L$ and α_n is the angle determined by Eq. (14) with this value of ω. The following example illustrates the interesting fact that the dominant frequency of the steady periodic solution can be an integral multiple of the frequency of the force $F(t)$

EXAMPLE 4   Suppose that $m = 3$ slugs, $c = 0.02$ lb/ft/s, $k = 27$ lb/ft, and $F(t)$ is the odd period 2π function with $F(t) = \pi t - t^2$ if $0 < t < \pi$. Find the steady periodic motion $x_{sp}(t)$.

Solution   We find that the Fourier series of $F(t)$ is

$$F(t) = \frac{8}{\pi} \left(\sin t + \frac{1}{3^3} \sin 3t + \frac{1}{5^3} \sin 5t + \cdots \right). \tag{17}$$

Thus $B_n = 0$ for n even, $B_n = 8/\pi n^3$ for n odd, and $\omega_n = n$. The formula in (16) gives

$$x_{sp}(t) = \frac{8}{\pi} \sum_{n \text{ odd}} \frac{\sin(nt - \alpha_n)}{n^3 \sqrt{(27 - 3n^2)^2 + (0.02n)^2}} \tag{18}$$

with

$$\alpha_n = \tan^{-1} \frac{(0.02)n}{27 - 3n^2}, \qquad 0 \le \alpha_n \le \pi. \tag{19}$$

With the aid of a programmable calculator, we find that

$$x_{sp} \approx (0.1061) \sin(t - 0.0008) + (1.5719) \sin\left(3t - \frac{\pi}{2}\right)$$

$$+ (0.0004) \sin(5t - 3.1395) + (0.0001) \sin(7t - 3.1404) + \cdots. \tag{20}$$

Because the coefficient corresponding to $n = 3$ is much larger than the others, the response of this system is approximately a sinusoidal motion with frequency three times that of the input force. Figure 7.16 shows $x_{sp}(t)$ in comparison with the reduced force $F(t)/k$ that has the appropriate dimension of distance.

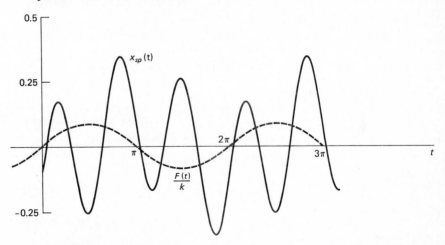

Figure 7.16 The imposed force and the resulting steady periodic motion in Example 4.

What is happening here is this: The mass $m = 3$ on a spring with $k = 27$ has (if we ignore the small effect of the dashpot) a natural frequency $\omega_0 = \sqrt{k/m} = 3$ rad/s. The imposed external force $F(t)$ has a (smallest) period of 2π s and hence a fundamental frequency of 1 rad/s. Consequently, the term corresponding to $n = 3$ in the Fourier series of $F(t)$ (in Eq. (17)) has the same frequency as the natural frequency of the system. Thus near resonance vibrations occur, with the mass completing essentially *three* oscillations for every single oscillation of the external force. This is the physical effect of the dominant $n = 3$ term on the right-hand side in (20). For instance, you can push a friend in a swing quite high even if you push the swing only every *third* time it returns to you.

This is a general phenomenon that must be taken into account in the design of mechanical systems. To avoid the occurrence of abnormally large and potentially destructive near resonance vibrations, the system must be so designed that it is not subject to any external periodic force, some *multiple* of whose fundamental frequency is close to any natural frequency of vibration.

7.4 Problems

Find a steady periodic solution of each of the differential equations in Problems 1–6.

1. $x'' + 5x = F(t)$, where $F(t)$ is the function of period 2π such that $F(t) = 3$ if $0 < t < \pi$, $F(t) = -3$ if $\pi < t < 2\pi$.

2. $x'' + 10x = F(t)$, where $F(t)$ is the even function of period 4 such that $F(t) = 3$ if $0 < t < 1$, $F(t) = -3$ if $1 < t < 2$.

3. $x'' + 3x = F(t)$, where $F(t)$ is the odd function of period 2π with $F(t) = 2t$ if $0 < t < \pi$.

4. $x'' + 4x = F(t)$, where $F(t)$ is the even function of period 4 with $F(t) = 2t$ if $0 < t < 2$.

5. $x'' + 10x = F(t)$, where $F(t)$ is the odd function of period 2 with $F(t) = t - t^2$ if $0 < t < 1$.

6. $x'' + 2x = F(t)$, where $F(t)$ is the even function of period 2π with $F(t) = \sin t$ if $0 < t < \pi$.

In each of Problems 7–12, the mass m and Hooke's constant k for a mass-and-spring system are given. Determine whether or not pure resonance will occur under the influence of the given external periodic force F(t).

7. $m = 1, k = 9$; $F(t)$ is the odd function of period 2π with $F(t) = 1$ for $0 < t < \pi$.

8. $m = 2, k = 10$; $F(t)$ is the odd function of period 2 with $F(t) = 1$ for $0 < t < 1$.

9. $m = 3, k = 12$; $F(t)$ is the odd function of period 2π with $F(t) = 3$ for $0 < t < \pi$.

10. $m = 1, k = 4\pi^2$; $F(t)$ is the odd function of period 2 with $F(t) = 2t$ if $0 < t < 1$.

11. $m = 3, k = 48$; $F(t)$ is the even function of period 2π with $F(t) = t$ if $0 < t < \pi$.

12. $m = 2, k = 50$; $F(t)$ is the odd function of period 2π with $F(t) = \pi t - t^2$ if $0 < t < \pi$.

In each of Problems 13–16, the values of m, c, and k for a damped mass-and-spring system are given. Find the steady periodic motion—in the form of Eq. (16)—of the mass under the influence of the given external force F(t). Compute the coefficients and phase angles for the first three nonzero terms.

13. $m = 1, c = 0.1, k = 4$; $F(t)$ is the force of Problem 1.

14. $m = 2, c = 0.1, k = 18$; $F(t)$ is the force of Problem 3.

15. $m = 3$, $c = 1$, $k = 30$; $F(t)$ is the force of Problem 5.

16. $m = 1$, $c = 0.01$, $k = 4$; $F(t)$ is the force of Problem 4.

17. Consider a forced, damped mass-and-spring system with $m = \frac{1}{4}$ slug, $c = 0.6$ lb/ft/s, $k = 36$ lb/ft. The force $F(t)$ is the period 2 (s) function with $F(t) = 15$ if $0 < t < 1$, $F(t) = -15$ if $1 < t < 2$. (a) Find the steady periodic solution in the form $x = \sum b_n \sin(n\pi t - \alpha_n)$. (b) Find the location—to the nearest tenth of an inch—of the mass when $t = 5$ s.

18. Consider a forced, damped mass-and-spring system with $m = 1$, $c = 0.01$, $k = 25$. The force $F(t)$ is the odd function of period 2π with $F(t) = t$ if $0 < t < \pi/2$, $F(t) = \pi - t$ if $\pi/2 < t < \pi$. Find the steady periodic motion; compute enough terms to see that the dominant frequency of the motion is five times that of the force.

7.5
Heat Conduction and Separation of Variables

The most important applications of Fourier series are to the solution of partial differential equations by means of the method of separation of variables that we introduce in this section. Recall that a *partial differential equation* is one containing one or more *partial* derivatives of a dependent variable that is a function of at least two independent variables. An example is the **one-dimensional heat equation**

$$\frac{\partial u}{\partial t} = k\frac{\partial^2 u}{\partial x^2}, \tag{1}$$

in which the dependent variable u is an unknown function of x and t, and k is a given constant.

THE HEATED ROD

Equation (1) models the variation of temperature u with position x and time t in a heated rod that extends along the x-axis. We assume that the rod has uniform cross section with area A perpendicular to the axis, and that it is made of a homogeneous material. We assume further that the cross section of the rod is so small that u is constant on each cross section, and that the lateral surface of the rod is insulated so that no heat can pass through it. Then u will, indeed, be a function of x and t, and heat will flow along the rod in only the x-direction. In general, we envision heat as flowing like a fluid from the warmer to the cooler parts of a body.

The **heat flux** $\phi(x, t)$ in the rod is the rate of flow of heat at time t across a unit area of the cross section at x. Typical units for ϕ are calories per second per square centimeter. The derivation of Eq. (1) is based upon the empirical principle that

$$\phi = -K\frac{\partial u}{\partial x}, \tag{2}$$

where the proportionality constant K is called the **thermal conductivity** of the material of the rod. The minus sign in (2) corresponds to the fact that heat flows in the negative x-direction when $u_x > 0$, while ϕ is positive for heat flow in the positive x-direction; in short, heat flows from a hot place to a cold place, not vice versa.

Now consider a small segment of the rod corresponding to the interval $[x, x + \Delta x]$, as shown in Fig. 7.17. The rate of flow R (in calories per second) of heat *into* this segment through its two ends is

$$R = A\phi(x, t) - A\phi(x + \Delta x, t) = KA[u_x(x + \Delta x, t) - u_x(x, t)]. \qquad (3)$$

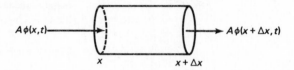

$$A\phi(x,t) \longrightarrow \qquad \longrightarrow A\phi(x + \Delta x, t)$$

$$x \qquad\qquad x + \Delta x$$

Figure 7.17   Net flow of heat into a short segment of the rod.

The resulting time rate of change u_t of the temperature in the segment depends upon its density δ (grams per cubic centimeter) and specific heat c (both assumed constant). The **specific heat** c is the amount of heat (in calories) required to raise by $1°$ the temperature of 1 g of material. Consequently $c\delta u$ calories of heat are required to raise 1 cm^3 of the material from temperature zero to temperature u. A short slice of the rod of length dx has volume $A\,dx$, so $c\delta uA\,dx$ calories of heat are required to raise the temperature of this slice from 0 to u. The **heat content**

$$Q(t) = \int_x^{x+\Delta x} c\delta Au(x, t)\,dx \qquad (4)$$

of the segment $[x, x + \Delta x]$ of the rod is the amount of heat needed to raise it from zero to the given temperature $u(x, t)$. Because heat enters and leaves the segment only through its ends, we see from (3) that

$$Q'(t) = KA[u_x(x + \Delta x, t) - u_x(x, t)]. \qquad (5)$$

On the other hand, note that $R = dQ/dt$. Thus by differentiating (4) within the integral and applying the mean value theorem for integrals, we see that

$$Q'(t) = \int_x^{x+\Delta x} c\delta Au_t(x, t)\,dx = c\delta Au_t(\bar{x}, t)\,\Delta x \qquad (6)$$

for some $\bar{x}$ in $(x, x + \Delta x)$. Upon equating the values in (5) and (6), we get

$$c\delta Au_t(\bar{x}, t)\,\Delta x = KA[u_x(x + \Delta x, t) - u_x(x, t)], \qquad (7)$$

so

$$u_t(\bar{x}, t) = k\frac{u_x(x + \Delta x, t) - u_x(x, t)}{\Delta x}, \qquad (8)$$

where

$$k = \frac{K}{c\delta} \qquad (9)$$

is the **thermal diffusivity** of the material. Finally, by taking the limit in (8) as $\Delta x \to 0$ and $\bar{x} \to x$, we obtain the one-dimensional heat equation

$$\frac{\partial u}{\partial t} = k\frac{\partial^2 u}{\partial x^2}. \qquad (1)$$

Thus the temperature $u(x, t)$ in a thin rod with insulated sides must satisfy this partial differential equation.

BOUNDARY CONDITIONS

Now suppose that the rod has finite length L, extending from $x = 0$ to $x = L$. Its temperature function $u(x, t)$ will be determined among all possible solutions of (1) by appropriate subsidiary conditions. In fact, whereas a solution of an ordinary differential equation involves arbitrary constants, a solution of a partial differential equation generally involves arbitrary *functions*. In the case of our rod, we can specify its temperature function $f(x)$ at time $t = 0$. This gives the *initial condition*

$$u(x, 0) = f(x). \tag{10}$$

We may also specify fixed temperatures at the two ends of the rod. For instance, if each end were clamped against a large block of ice at temperature zero, we would have the *endpoint conditions*

$$u(0, t) = u(L, t) = 0 \quad \text{(for all } t). \tag{11}$$

Combining all this, we get the **boundary value problem**

$$\frac{\partial u}{\partial t} = k\frac{\partial^2 u}{\partial x^2} \quad (0 < x < L, \quad t > 0); \tag{12a}$$

$$u(0, t) = u(L, t) = 0 \tag{12b}$$

$$u(x, 0) = f(x). \tag{12c}$$

Figure 7.18 gives a geometric interpretation of the boundary value problem above: We are to find a function $u(x, t)$ that is continuous on the unbounded strip (and its boundary) shaded in the xt-plane. This function must satisfy the differential equation in (12a) at each interior point of the strip, and on the boundary of the strip must have the values prescribed by the boundary conditions in (12b) and (12c). Physical intuition suggests that if $f(x)$ is a reasonable function, then there will exist one and only one such function $u(x, t)$.

Instead of having fixed temperatures, the two ends of the rod might be

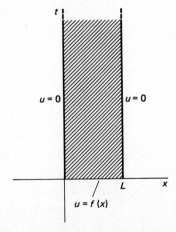

$u = 0$ $u = 0$

$u = f(x)$

Figure 7.18 A geometric interpretation of the boundary value problem in Equations [12(a)(b)(c)].

insulated. In this case no heat would flow through either end, so we see from Eq. (2) that the conditions in (12b) would be replaced in the boundary value problem by the endpoint conditions

$$u_x(0, t) = u_x(L, t) = 0 \tag{13}$$

(for all t). Alternatively, the rod could be insulated at one end and have a fixed temperature at the other. This and other endpoint possibilities are discussed in the problems.

SUPERPOSITION OF SOLUTIONS

Note that the heat equation in (12a) is *linear*. That is, any linear combination $u = c_1 u_1 + c_2 u_2$ of two solutions of (12a) is also a solution of (12a); this follows immediately from the linearity of partial differentiation. It is also true that if u_1 and u_2 each satisfy the conditions in (12b), then so does any linear combination $u = c_1 u_1 + c_2 u_2$. The conditions in (12b) are therefore called **homogeneous** boundary conditions (though a more descriptive term might be *linear* boundary conditions). By contrast, the final boundary condition in (12c) is not homogeneous; it is a **nonhomogeneous** boundary condition.

Our overall strategy for solving the boundary value problem in (12a)–(12c) will be to find functions $u_1, u_2, u_3, \ldots$ that satisfy both the partial differential equation in (12a) and the homogeneous boundary conditions in (12b), and then attempt to combine these functions by superposition, as if they were building blocks, in the hope of obtaining a solution $u = c_1 u_1 + c_2 u_2 + \cdots$ that satisfies the nonhomogeneous condition in (12c) as well. The following example illustrates this approach.

EXAMPLE 1 It is easy to verify by direct substitution that each of the functions $u_1 = e^{-t} \sin x$, $u_2 = e^{-4t} \sin 2x$, and $u_3 = e^{-9t} \sin 3x$ satisfies the equation $u_t = u_{xx}$. Use these functions to construct a solution of the boundary value problem

$$\frac{\partial u}{\partial t} = \frac{\partial^2 u}{\partial x^2} \quad (0 < x < \pi, \quad t > 0); \tag{14a}$$

$$u(0, t) = u(\pi, t) = 0, \tag{14b}$$

$$u(x, 0) = 80 \sin^3 x = 60 \sin x - 20 \sin 3x. \tag{14c}$$

Solution Any linear combination of the form

$$u(x, t) = c_1 u_1(x, t) + c_2 u_2(x, t) + c_3 u_3(x, t)$$
$$= c_1 e^{-t} \sin x + c_2 e^{-4t} \sin 2x + c_3 e^{-9t} \sin 3x$$

satisfies both the differential equation in (14a) and the homogeneous conditions in (14b). Because

$$u(x, 0) = c_1 \sin x + c_2 \sin 2x + c_3 \sin 3x,$$

we see that we can also satisfy the nonhomogeneous condition in (14c) simply by choosing $c_1 = 60$, $c_2 = 0$, and $c_3 = -20$. Thus a solution of the given boundary value problem is

$$u(x, t) = 60e^{-t} \sin x - 20e^{-9t} \sin 3x.$$

The boundary value problem in Example 1 is exceptionally simple in that only a finite number of homogeneous solutions are needed to satisfy by superposition the nonhomogeneous boundary condition. It is more usual that an infinite sequence $u_1, u_2, u_3, \ldots$ of functions satisfying (12a) and (12b) is required. If so, we write the the infinite series

$$u(x, t) = \sum_{n=1}^{\infty} c_n u_n(x, t), \tag{15}$$

and then attempt to determine the coefficients $c_1, c_2, c_3, \ldots$ in order to satisfy (12c) as well. The following theorem (stated without proof) summarizes the properties of this infinite series that must be verified in order to insure that we have a solution of the boundary value problem in (12a)–(12c).

THEOREM: SUPERPOSITION OF SOLUTIONS

Suppose that each of the functions $u_1, u_2, u_3, \ldots$ satisfies both the differential equation in (12a) (for $0 < x < L$ and $t > 0$) and the homogeneous conditions in (12b). Suppose also that the coefficients in Eq. (15) are chosen to meet the following three criteria:

(a) *For $0 < x < L$ and $t > 0$, the function determined by the series in (15) is continuous and termwise differentiable (once with respect to t and twice with respect to x).*

(b) $\sum_{n=1}^{\infty} c_n u_n(x, 0) = f(x).$

(c) *The function $u(x, t)$ determined by Eq. (15) interior to the strip $0 \leqq x \leqq L$ and $t \geqq 0$, and by the boundary conditions in (12b) and (12c) on its boundary, is continuous.*

Then $u(x, t)$ is a solution of the boundary value problem in (12a)–(12c).

In the method of separation of variables described below, we concentrate on finding the solutions $u_1, u_2, u_3, \ldots$ satisfying the homogeneous conditions and on determining the coefficients so that the series in (15) satisfies the nonhomogeneous condition upon substitution of $t = 0$. At this point we have only a *formal* series solution of the boundary value problem—one that is subject to verification of the continuity and differentiability conditions given in Part (a) above. If the function $f(x)$ in (12c) is piecewise smooth, it can be proved that a formal series solution always satisfies these restrictions and, moreover, is the unique solution of the boundary value problem. For this proof, see Chapter 6 of R. V. Churchill and J. W. Brown, *Fourier Series and Boundary Value Problems*, (New York: McGraw-Hill, 3rd ed., 1978).

SEPARATION OF VARIABLES

This method of solving the boundary value problem of (12a)–(12c) for the heated rod was introduced by Fourier in his study of heat cited in Section 7.1. We first search for the *building block* functions $u_1, u_2, u_3, \ldots$ that satisfy the differential equation $u_t = ku_{xx}$ and the homogeneous conditions $u(0, t) =$

$u(L, t) = 0$, with each of these functions being of the special form

$$u(x, t) = X(x)T(t) \tag{16}$$

in which the variables x and t are "separated." Substitution of (16) in $u_t = ku_{xx}$ yields $XT' = kX''T$, where for brevity we write T' for $T'(t)$ and X'' for $X''(x)$. Division of both sides by kXT then gives

$$\frac{X''}{X} = \frac{T'}{kT}. \tag{17}$$

The left side in (17) is a function of x alone, while the right side is a function of t alone. If t is held constant on the right-hand side, then the left-hand side X''/X must remain constant as x varies. Similarly, if x is held constant on the left-hand side, then the right-hand side T'/kT must remain constant as t varies. Consequently, equality can hold only if each of these two expressions is the same constant, which for convenience we denote by $-\lambda$. Thus (17) becomes

$$\frac{X''}{X} = \frac{T'}{kT} = -\lambda, \tag{18}$$

which consists of the two equations

$$X''(x) + \lambda X(x) = 0 \tag{19}$$

and

$$T'(t) + \lambda kT(t) = 0. \tag{20}$$

It follows that the product function $u(x, t) = X(x)T(t)$ satisfies the partial differential equation $u_t = ku_{xx}$ if $X(x)$ and $T(t)$ separately satisfy the ordinary differential equations in (19) and (20) for some value of the constant λ.

We focus first on $X(x)$. The homogeneous endpoint conditions give

$$u(0, t) = X(0)T(t) = 0, \qquad u(L, t) = X(L)T(t) = 0. \tag{21}$$

If $T(t)$ is to be a nontrivial function of t, then (21) can hold only if $X(0) = X(L) = 0$. Thus $X(x)$ must satisfy the endpoint value problem

$$X'' + \lambda X = 0,$$
$$X(0) = 0, \qquad X(L) = 0. \tag{22}$$

This is actually an eigenvalue problem of the type we discussed in Section 2.10. Indeed, we saw in Example 3 of that section that (22) has a nontrivial solution if and only if λ is one of the eigenvalues

$$\lambda_n = \frac{n^2\pi^2}{L^2}, \qquad n = 1, 2, 3, \dots, \tag{23}$$

and that an eigenfunction associated with λ_n is

$$X_n(x) = \sin\frac{n\pi x}{L}, \qquad n = 1, 2, 3, \dots. \tag{24}$$

Recall that the reasoning behind (23) and (24) is as follows. If $\lambda = 0$, then (22) obviously implies $X(x) \equiv 0$. If $\lambda = -\alpha^2 < 0$, then $X(x) = A \cosh \alpha x + B \sinh \alpha x$, and then the conditions $X(0) = 0 = X(L)$ imply that $A = B = 0$. Hence the only possibility for a nontrivial eigenfunction is that $\lambda = \alpha^2 > 0$.

Then $X(x) = A \cos \alpha x + B \sin \alpha x$, and the conditions $X(0) = 0 = X(L)$ then imply that $A = 0$ and that $\alpha = n\pi/L$ for some positive integer n. Whenever separation of variables leads to an unfamiliar eigenvalue problem, we generally must consider separately the cases $\lambda = 0$, $\lambda = -\alpha^2 < 0$, and $\lambda = \alpha^2 > 0$.

Now we turn our attention to Eq. (20), knowing that the constant λ must be one of the eigenvalues listed in (23). For the nth of these possibilities we rewrite Eq. (20) as

$$T_n' + \frac{n^2 \pi^2 k}{L^2} T_n = 0, \tag{25}$$

in anticipation of a different solution $T_n(t)$ for each different integer n. A nontrivial solution of this equation is

$$T_n(t) = e^{-n^2 \pi^2 kt/L^2}. \tag{26}$$

We omit the arbitrary constant because it will (in effect) be inserted later.

To summarize our progress, we have discovered the two associated sequences $\{X_n\}_1^\infty$ and $\{T_n\}_1^\infty$ of functions given in (24) and (26). Together they yield the sequence of product functions

$$u_n(x, t) = X_n(x)T_n(t) = e^{-n^2 \pi^2 kt/L^2} \sin \frac{n\pi x}{L}, \tag{27}$$

$n = 1, 2, 3, \ldots$. Each of these functions satisfies both the heat equation $u_t = ku_{xx}$ and the homogeneous conditions $u(0, t) = u(L, t) = 0$. Now we combine these functions (*superposition*) to attempt to satisfy the nonhomogeneous condition $u(x, 0) = f(x)$ as well. We therefore form the infinite series

$$u(x, t) = \sum_{n=1}^\infty c_n u_n(x, t) = \sum_{n=1}^\infty c_n e^{-n^2 \pi^2 kt/L^2} \sin \frac{n\pi x}{L}. \tag{28}$$

It remains only to determine the constant coefficients $\{c_n\}_1^\infty$ so that

$$u(x, 0) = \sum_{n=1}^\infty c_n \sin \frac{n\pi x}{L} = f(x). \tag{29}$$

But this will be the Fourier sine series of $f(x)$ on $[0, L]$ provided we choose

$$c_n = \frac{2}{L} \int_0^L f(x) \sin \frac{n\pi x}{L} \, dx \tag{30}$$

for each $n = 1, 2, 3, \ldots$. Therefore, we finally see that if the coefficients $\{c_n\}_1^\infty$ are computed by means of the formula in (30), then

$$u(x, t) = \sum_{n=1}^\infty c_n e^{-n^2 \pi^2 kt/L^2} \sin \frac{n\pi x}{L} \tag{31}$$

will be a formal series solution of the boundary value problem in (12a)–(12c). Note that by taking the limit in (31) termwise as $t \to +\infty$, we get $u(x, \infty) \equiv 0$, as we expect because the two ends of the rod are held at temperature zero.

The series solution in (31) usually converges quite rapidly, unless t is quite small, because of the presence of the negative exponential factors. Therefore it is practical for numerical computations. For use in problems and examples, values of the thermal diffusivity constant for some common materials are listed in the table in Fig. 7.19.

Material	k (cm^2/s)
Silver	1.70
Copper	1.15
Aluminum	0.85
Iron	0.15
Concrete	0.005

Figure 7.19 Some thermal diffusivity constants.

EXAMPLE 2 Suppose that a rod 50 cm long is immersed in steam until its temperature is 100°C throughout. At time $t = 0$, its lateral surface is insulated and its two ends embedded in ice at 0°C. Calculate the temperature at its midpoint after a half hour if the rod is made of (a) iron; (b) concrete.

Solution The boundary value problem for this rod is

$$\begin{cases} u_t = k u_{xx}; \\ u(0, t) = u(50, t) = 0, \\ u(x, 0) = 100. \end{cases}$$

The Fourier sine series of $f(x) = 100$ is

$$f(x) = 100 = \frac{400}{\pi} \sum_{n \text{ odd}} \frac{1}{n} \sin \frac{n\pi x}{50}$$

for $0 < x < 50$. The coefficients $\{c_n\}$ in (31) are simply the Fourier sine coefficients of $f(x)$, so the solution of the boundary value problem is

$$u(x, t) = \frac{400}{\pi} \sum_{n \text{ odd}} \frac{1}{n} e^{-n^2 \pi^2 kt/2500} \sin \frac{n\pi x}{50}.$$

Hence the temperature at the midpoint $x = 25$ cm at time $t = 1800$ s is

$$u(25, 1800) = \frac{400}{\pi} \sum_{n \text{ odd}} \frac{(-1)^{n+1}}{n} e^{-18n^2\pi^2 k/25}.$$

(a) With $k = 0.15$ for iron, this series gives

$$u \approx 43.8519 - 0.0029 + 0.0000 - \cdots \approx 43.85°C.$$

(b) With $k = 0.005$ for concrete, it gives

$$u \approx 122.8795 - 30.8257 + 10.4754 - 3.1894 + 0.7958 - 0.1572$$
$$+ 0.0242 - 0.0029 + 0.0003 - 0.0000 + \cdots \approx 100.00°C.$$

Thus concrete is a very effective insulator.

We now consider the boundary value problem

$$\frac{\partial u}{\partial t} = k \frac{\partial^2 u}{\partial x^2} \qquad (0 < x < L, \qquad t > 0); \tag{32a}$$

$$u_x(0, t) = u_x(L, t) = 0, \tag{32b}$$

$$u(x, 0) = f(x), \tag{32c}$$

which corresponds to a rod of length L with initial temperature $f(x)$, but with its two ends insulated. The separation of variables $u(x, t) = X(x)T(t)$ proceeds as in Eqs. (16)–(20) without change. But the homogeneous endpoint conditions in (32b) yield $X'(0) = X'(L) = 0$. Thus $X(t)$ must satisfy the endpoint value problem

$$X'' + \lambda X = 0,$$
$$X'(0) = 0, \qquad X'(L) = 0. \tag{33}$$

We must consider separately the possibilities $\lambda = 0$, $\lambda = -\alpha^2 < 0$, and $\lambda = +\alpha^2 > 0$ for the eigenvalues.

With $\lambda = 0$, the general solution of $X'' = 0$ is $X(x) = Ax + B$, so $X'(x) = A$. Hence the endpoint conditions in (33) require $A = 0$, but B may be nonzero. Because a constant multiple of an eigenfunction is an eigenfunction, we can choose any constant value we wish for B. Thus, with $B = 1$, we have the zero eigenvalue and associated eigenfunction

$$\lambda_0 = 0, \qquad X_0(x) = 1. \tag{34}$$

With $\lambda = 0$ in Eq. (20), we get $T'(t) = 0$, so we may take $T_0(t) = 1$ as well.

With $\lambda = -\alpha^2 < 0$, the general solution of the equation $X' - \alpha^2 X = 0$ is $X(x) = A \cosh \alpha x + B \sinh \alpha x$, and we readily verify that $X'(0) = X'(L) = 0$ only if $A = B = 0$. Thus there are no negative eigenvalues.

With $\lambda = +\alpha^2 > 0$, the general solution of $X'' + \alpha^2 X = 0$ is

$$X(x) = A \cos \alpha x + B \sin \alpha x, \qquad X'(x) = -A\alpha \sin \alpha x + B\alpha \cos \alpha x.$$

Hence $X'(0) = 0$ implies that $B = 0$, and then

$$X'(L) = -A\alpha \sin \alpha L = 0$$

requires that αL be an integral multiple of π, because $\alpha \neq 0$ and $A \neq 0$ if we are to have a nontrivial solution. Thus we have the infinite sequence of eigenvalues and associated eigenfunctions

$$\lambda_n = \alpha_n^2 = \frac{n^2 \pi^2}{L^2}, \qquad X_n(x) = \cos \frac{n\pi x}{L} \tag{35}$$

for $n = 1, 2, 3, \ldots$. Just as before, the solution of (20) with $\lambda = n^2\pi^2/L^2$ is $T_n(t) = e^{-n^2\pi^2 kt/L^2}$.

Therefore our product functions satisfying the nonhomogeneous conditions are

$$u_0(x, t) = 1; \qquad u_n(x, t) = e^{-n^2\pi^2 kt/L^2} \cos \frac{n\pi x}{L} \tag{36}$$

for $n = 1, 2, 3, \ldots$. Hence our trial solution is

$$u(x, t) = c_0 + \sum_{n=1}^{\infty} c_n e^{-n^2\pi^2 kt/L^2} \cos \frac{n\pi x}{L}. \tag{37}$$

In order to satisfy the nonhomogeneous condition $u(x, 0) = f(x)$, we obviously want (37) to reduce when $t = 0$ to the Fourier cosine series

$$f(x) = \frac{a_0}{2} + \sum_{n=1}^{\infty} a_n \cos \frac{n\pi x}{L}, \tag{38}$$

where

$$a_n = \frac{2}{L} \int_0^L f(x) \cos \frac{n\pi x}{L} \, dx \tag{39}$$

for $n = 0, 1, 2, \ldots$. Our formal series solution of the boundary value problem in (32a)–(32c) is therefore

$$u(x, t) = \frac{a_0}{2} + \sum_{n=1}^{\infty} a_n e^{-n^2 \pi^2 kt/L^2} \cos \frac{n\pi x}{L}, \tag{40}$$

with the coefficients $\{a_n\}$ given by the formula in (39). Note that

$$\lim_{t \to \infty} u(x, t) = \frac{1}{2} a_0 = \frac{1}{L} \int_0^L f(x)\, dx, \tag{41}$$

the average value of the initial temperature. With both the lateral surface and the ends of the rod insulated, its original heat content ultimately distributes itself uniformly throughout the rod.

Finally, we point out that, while we set up the boundary value problems (12a)–(12c) and (32a)–(32c) for a rod of length L, they also model the temperature $u(x, t)$ within the infinite slab $0 \leq x \leq L$ in three-dimensional space, if its initial temperature $f(x)$ depends only upon x, and its two faces $x = 0$ and $x = L$ are either both insulated or both held at temperature zero.

7.5 Problems

Solve the boundary value problems in Problems 1–12.

1. $u_t = 3u_{xx}, 0 < x < \pi, t > 0$; $u(0, t) = u(\pi, t) = 0, u(x, 0) = 4 \sin 2x$.
2. $u_t = 10u_{xx}, 0 < x < 5, t > 0$; $u_x(0, t) = u_x(5, t) = 0, u(x, 0) = 7$.
3. $u_t = 2u_{xx}, 0 < x < 1, t > 0$; $u(0, t) = u(1, t) = 0$,
   $u(x, 0) = 5 \sin x\pi - \frac{1}{3} \sin 3\pi x$.
4. $u_t = u_{xx}, 0 < x < \pi, t > 0$; $u(0, t) = u(\pi, t) = 0, u(x, 0) = 4 \sin 4x \cos 2x$.
5. $u_t = 2u_{xx}, 0 < x < 3, t > 0$; $u_x(0, t) = u_x(3, t) = 0$,
   $u(x, 0) = 4 \cos \frac{2\pi x}{3} - 2 \cos \frac{4\pi x}{3}$.
6. $2u_t = u_{xx}, 0 < x < 1, t > 0$; $u(0, t) = u(1, t) = 0$,
   $u(x, 0) = 4 \sin \pi x \cos^3 \pi x$.
7. $3u_t = u_{xx}, 0 < x < 2, t > 0$; $u_x(0, t) = u_x(2, t) = 0, u(x, 0) = \cos^2 2\pi x$.
8. $u_t = u_{xx}, 0 < x < 2, t > 0$; $u_x(0, t) = u_x(2, t) = 0$,
   $u(x, 0) = 10 \cos \pi x \cos 3\pi x$.
9. $10u_t = u_{xx}, 0 < x < 5, t > 0$; $u(0, t) = u(5, t) = 0, u(x, 0) = 25$.
10. $5u_t = u_{xx}, 0 < x < 10, t > 0$; $u(0, t) = u(10, t) = 0, u(x, 0) = 4x$.
11. $5u_t = u_{xx}, 0 < x < 10, t > 0$; $u_x(0, t) = u_x(10, t) = 0, u(x, 0) = 4x$.
12. $u_t = u_{xx}, 0 < x < 100, t > 0$; $u(0, t) = u(100, t) = 0$,
    $u(x, 0) = x(100 - x)$.
13. Suppose that a rod 40 cm long with insulated lateral surface is heated to a uniform temperature of 100°C, and that at time $t = 0$ its two ends are embedded in ice at 0°C. (a) Find the formal series solution for its temperature $u(x, t)$. (b) In the case the rod is made of copper, show that after 5 min the temperature at its midpoint is about 15°C. (c) In the case the rod is made of concrete, use the first term of the series to find the time required for its midpoint to cool to 15°C.

14. A copper rod 50 cm long with insulated lateral surface has initial temperature $u(x, 0) = 2x$, and at time $t = 0$ its two ends are insulated. (a) Find $u(x, t)$. (b) What will its temperature be at $x = 10$ after 1 min? (c) After approximately how long will its temperature at $x = 10$ be $45°C$?

15. The two faces of the slab $0 \leq x \leq L$ are kept at temperature zero, and the slab's initial temperature is given by $u(x, 0) = A$ (constant) for $0 < x < L/2$, $u(x, 0) = 0$ for $L/2 < x < L$. Derive the formal series solution

$$u(x, t) = \frac{4A}{\pi} \sum_{n=1}^{\infty} \frac{\sin^2 (n\pi/4)}{n} e^{-n^2\pi^2 kt/L^2} \sin \frac{n\pi x}{L}.$$

16. Two iron slabs are each 25 cm thick. Initially one has temperature $100°C$ and the other has temperature $0°C$ throughout. At time $t = 0$ they are placed face to face, and their outer faces are kept at $0°C$. (a) Use the result of Problem 15 to find that after a half hour the temperature at their common face is approximately $22°C$. (b) Suppose that the two slabs are instead made of concrete. How long will it take for their common face to reach a temperature of $22°C$?

17. (Steady-state and transient temperatures) Let a laterally insulated rod with initial temperature $u(x, 0) = f(x)$ have fixed end temperatures $u(0, t) = A$ and $u(L, t) = B$. (a) It is observed empirically that as $t \rightarrow +\infty$, $u(x, t)$ approaches a **steady-state temperature** $u_{ss}(x)$ that corresponds to setting $u_t = 0$ in the boundary value problem. Thus $u_{ss}(x)$ is the solution of the endpoint value problem

$$\frac{\partial^2 u_{ss}}{\partial x^2} = 0, \qquad u_{ss}(0) = A, \qquad u_{ss}(L) = B.$$

Find $u_{ss}(x)$. (b) The **transient temperature** $u_{tr}(x, t)$ is defined to be

$$u_{tr}(x, t) = u(x, t) - u_{ss}(x).$$

Show that u_{tr} satisfies the boundary value problem

$$\begin{cases} \dfrac{\partial u_{tr}}{\partial t} = k\dfrac{\partial^2 u_{tr}}{\partial x^2}; \\ u_{tr}(0, t) = u_{tr}(L, t) = 0, \\ u_{tr}(0, t) = g(x) = f(x) - u_{ss}(x). \end{cases}$$

(c) Conclude from the formulas in (30) and (31) that

$$u(x, t) = u_{ss}(x) + u_{tr}(x, t) = u_{ss}(x) + \sum_{n=1}^{\infty} c_n e^{-n^2\pi^2 kt/L^2} \sin \frac{n\pi x}{L},$$

where

$$c_n = \frac{2}{L} \int_0^L [f(x) - u_{ss}(x)] \sin \frac{n\pi x}{L} \, dx.$$

18. Suppose that a laterally insulated rod with length $L = 50$ and thermal diffusivity $k = 1$ has initial temperature $u(x, 0) = 0$ and end temperatures $u(0, t) = 0$, $u(50, t) = 100$. Apply the result of Problem 17 to show that

$$u(x, t) = 2x - \frac{200}{\pi} \sum_{n=1}^{\infty} \frac{(-1)^{n+1}}{n} e^{-n^2\pi^2 kt/2500} \sin \frac{n\pi x}{50}.$$

19. Suppose that heat is generated within a laterally insulated rod at the rate of $q(x, t)$ calories per second per cubic centimeter. Extend the derivation of the heat equation in this section to derive the equation

$$\frac{\partial u}{\partial t} = k\frac{\partial^2 u}{\partial x^2} + \frac{q(x, t)}{c\delta}.$$

20. Suppose that current flowing through a laterally insulated rod generates heat at a constant rate; then Problem 19 yields the equation

$$\frac{\partial u}{\partial t} = k\frac{\partial^2 u}{\partial x^2} + C.$$

Assume the boundary conditions $u(0, t) = u(L, t) = 0$ and $u(x, 0) = f(x)$. (a) Find the steady-state temperature $u_{ss}(x)$ determined by

$$0 = k\frac{d^2 u_{ss}}{dx^2} + C, \qquad u_{ss}(0) = u_{ss}(L) = 0.$$

(b) Show that the transient temperature $u_{tr}(x, t) = u(x, t) - u_{ss}(x)$ satisfies the boundary value problem

$$\begin{cases} \dfrac{\partial u_{tr}}{\partial t} = k\dfrac{\partial^2 u_{tr}}{\partial x^2}; \\ u_{tr}(0, t) = u_{tr}(L, t) = 0, \\ u_{tr}(x, 0) = g(x) = f(x) - u_{ss}(x). \end{cases}$$

Hence conclude from the formulas in (34) and (35) that

$$u(x, t) = u_{ss}(x) + \sum_{n=1}^{\infty} c_n e^{-n^2\pi^2 kt/L^2} \sin\frac{n\pi x}{L},$$

where

$$c_n = \frac{2}{L}\int_0^L [f(x) - u_{ss}(x)] \sin\frac{n\pi x}{L}\, dx.$$

21. The answer to Problem 20(a) is $u_{ss}(x) = Cx(L - x)/2k$. If $f(x) \equiv 0$ in Problem 20, so the rod being heated is initially at temperature zero, deduce from the result of part (b) that

$$u(x, t) = \frac{Cx}{2k}(L - x) - \frac{4CL^2}{k\pi^3}\sum_{n\ odd}\frac{1}{n^3} e^{-n^2\pi^2 kt/L^2} \sin\frac{n\pi x}{L}.$$

22. Consider the temperature $u(x, t)$ in a bare slender wire with $u(0, t) = u(L, t) = 0$ and $u(x, 0) = f(x)$. Instead of being laterally insulated, the wire loses heat to a surrounding medium (at temperature zero) at a rate proportional to $u(x, t)$. (a) Conclude from Problem 19 that

$$\frac{\partial u}{\partial t} = k\frac{\partial^2 u}{\partial x^2} - hu$$

where h is a positive constant. (b) Substitute $u(x, t) = e^{-ht}v(x, t)$ to show that $v(x, t)$ satisfies the boundary value problem having the solution given in (30) and (31). Hence conclude that

$$u(x, t) = e^{-ht}\sum_{n=1}^{\infty} c_n e^{-n^2\pi^2 kt/L^2} \sin\frac{n\pi x}{L},$$

where

$$c_n = \frac{2}{L}\int_0^L f(x) \sin\frac{n\pi x}{L}\, dx.$$

23. Consider a slab with thermal conductivity K occupying the region $0 \leqq x \leqq L$. Suppose that, in accord with Newton's law of cooling, each face of the slab

loses heat to the surrounding medium (at temperature zero) at the rate of H calories per second per square centimeter. Deduce from Eq. (2) that the temperature $u(x, t)$ in the slab satisfies the boundary conditions

$$hu(0, t) - u_x(0, t) = 0 = hu(L, t) + u_x(L, t)$$

where $h = H/K$.

24. Suppose that a laterally insulated rod with length L, thermal diffusivity k, and initial temperature $u(x, 0) = f(x)$ is insulated at the end $x = L$ and held at temperature zero at $x = 0$. (a) Separate the variables to show that the eigenfunctions are $X_n(x) = \sin(n\pi x/2L)$ for n odd. (b) Use the odd half-multiple sine series of Problem 21 in Section 7.3 to derive the solution

$$u(x, t) = \sum_{n \text{ odd}} c_n e^{-n^2\pi^2 kt/4L^2} \sin \frac{n\pi x}{2L}$$

where

$$c_n = \frac{2}{L} \int_0^L f(x) \sin \frac{n\pi x}{2L} \, dx.$$

7.6
Vibrating Strings and the One-Dimensional Wave Equation

Although Fourier systematized the method of separation of variables, trigonometric series solutions of partial differential equations had appeared earlier in eighteenth century investigations of vibrating strings by Euler, d'Alembert, and Daniel Bernoulli. To derive the partial differential equation that models the vibrations of a string, we begin with a flexible uniform string with linear density ρ (in grams per centimeters or slugs per foot) stretched under a tension of T (dynes or pounds) between the fixed points $x = 0$ and $x = L$. Suppose that, as the string vibrates in the xy-plane about its equilibrium position, each point moves parallel to the y-axis, so we can denote by $y(x, t)$ the displacement at time t of the point x of the string. Then, for any fixed value of t, the shape of the string at time t is the curve $y = y(x, t)$. We assume also that the deflection of the string remains so slight that the approximation $\sin \theta \approx \tan \theta = y_x(x, t)$ is quite accurate (see Fig. 7.20). Finally we assume that, in addition to the

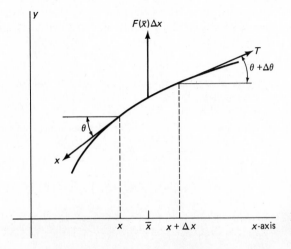

Figure 7.20 Forces on a short segment of the vibrating string.

internal forces of tension acting tangentially to the string, it is acted on by an external vertical force with linear density $F(x)$ in such units as dynes per centimeters or pounds per foot.

We want to apply Newton's law $F = ma$ to the short segment of string of mass $\rho \Delta x$ corresponding to the interval $[x, x + \Delta x]$, with a being the vertical acceleration $y_{tt}(\bar{x}, t)$ of its midpoint. Reading the vertical components of the forces shown in Fig. 7.20, we get

$$(\rho \Delta x) y_{tt}(\bar{x}, t) \approx T \sin (\theta + \Delta\theta) - T \sin \theta + F(\bar{x}) \, \Delta x$$
$$\approx T y_x(x + \Delta x, t) - T y_x(x, t) + F(\bar{x}) \, \Delta x,$$

so division by Δx yields

$$\rho y_{tt}(\bar{x}, t) \approx T \frac{y_x(x + \Delta x, t) - y_x(x, t)}{\Delta x} + F(\bar{x}).$$

When we take the limit as $\Delta x \longrightarrow 0$ and $\bar{x} \longrightarrow x$, we obtain the partial differential equation

$$\rho \frac{\partial^2 y}{\partial t^2} = T \frac{\partial^2 y}{\partial x^2} + F(x) \tag{1}$$

that describes the vertical vibrations of a flexible string with constant linear density ρ and tension T under the influence of an external vertical force with linear density $F(x)$.

If we set

$$a^2 = \frac{T}{\rho} \tag{2}$$

and $F(x) \equiv 0$ in (1), we get the **one-dimensional wave equation**

$$\frac{\partial^2 y}{\partial t^2} = a^2 \frac{\partial^2 y}{\partial x^2} \tag{3}$$

that models the *free* vibrations of a uniform flexible string.

The fixed ends of the string at the points $x = 0$ and $x = L$ on the x-axis correspond to the *endpoint conditions*

$$y(0, t) = y(L, t) = 0. \tag{4}$$

Our intuition about the physics of the situation suggests that the motion of the string will be determined if we specify both its **initial position function**

$$y(x, 0) = f(x) \qquad (0 < x < L) \tag{5}$$

and its **initial velocity function**

$$y_t(x, 0) = g(x) \qquad (0 < x < L). \tag{6}$$

On combining Eqs. (3)–(6), we get the boundary value problem

$$\frac{\partial^2 y}{\partial t^2} = a^2 \frac{\partial^2 y}{\partial x^2} \qquad (0 < x < L, \qquad t > 0); \tag{7a}$$

$$y(0, t) = y(L, t) = 0, \tag{7b}$$

$$y(x, 0) = f(x) \qquad (0 < x < L), \tag{7c}$$

$$y_t(x, 0) = g(x) \qquad (0 < x < L) \tag{7d}$$

for the displacement function $y(x, t)$ of a freely vibrating string with fixed ends, initial position $f(x)$, and initial velocity $g(x)$.

Like the heat equation, the wave equation in (7a) is linear: Any linear combination of two solutions is again a solution. Another similarity is that the endpoint conditions in (7b) are homogeneous. Unfortunately, the conditions in both (7c) and (7d) are nonhomogeneous; we must deal with *two* nonhomogeneous boundary conditions.

As we described it in Section 7.5, the method of separation of variables involves superposition of solutions satisfying the homogeneous conditions to obtain a solution that also satisfies a single nonhomogeneous boundary condition. To adapt the technique to the present situation, we adopt the "divide and conquer" strategy of splitting the problem in (7a)–(7d) into the following two separate boundary value problems, each involving only a single nonhomogeneous boundary condition:

$$\text{Problem A} \qquad\qquad\qquad \text{Problem B}$$

$$
\begin{cases}
y_{tt} = a^2 y_{xx}; \\
y(0, t) = y(L, t) = 0, \\
y(x, 0) = f(x), \\
y_t(x, 0) = 0.
\end{cases}
\qquad
\begin{cases}
y_{tt} = a^2 y_{xx}; \\
y(0, t) = y(L, t) = 0, \\
y(x, 0) = 0, \\
y_t(x, 0) = g(x).
\end{cases}
$$

If we can separately find a solution $y_A(x, t)$ of Problem A and a solution $y_B(x, t)$ of Problem B, then their sum $y(x, t) = y_A(x, t) + y_B(x, t)$ will be a solution of the original problem in (7a)–(7d) because

$$y(x, 0) = y_A(x, 0) + y_B(x, 0) = f(x) + 0 = f(x)$$

and

$$y_t(x, 0) = y_{At}(x, 0) + y_{Bt}(x, 0) = 0 + g(x) = g(x).$$

So let us attack Problem A by separation of variables. Substitution of

$$y(x, t) = X(x)T(t) \tag{8}$$

in $y_{tt} = a^2 y_{xx}$ yields $XT'' = a^2 X''T$; that is,

$$\frac{X''}{X} = \frac{T''}{a^2 T}. \tag{9}$$

The functions X''/X of x and $T''/a^2 T$ of t can agree for all x and t only if each is equal to the same constant. Consequently we may conclude that

$$\frac{X''}{X} = \frac{T''}{a^2 T} = -\lambda \tag{10}$$

for some constant λ; the minus sign is inserted here merely to facilitate recognition of the eigenvalue problem in (13), given below. Thus our partial differential equation separates into the two ordinary differential equations

$$X'' + \lambda X = 0, \tag{11}$$

$$T'' + \lambda a^2 T = 0. \tag{12}$$

The endpoint conditions

$$y(0, t) = X(0)T(t) = 0, \qquad y(L, t) = X(L)T(t) = 0$$

require that $X(0) = X(L) = 0$ if $T(t)$ is a nontrivial function. Hence $X(x)$ must satisfy the now familiar eigenvalue problem

$$X'' + \lambda X = 0, \qquad X(0) = X(L) = 0. \tag{13}$$

As in Eqs. (23) and (24) of Section 7.5, the eigenvalues of this problem are the numbers

$$\lambda_n = \frac{n^2 \pi^2}{L^2}, \qquad n = 1, 2, 3, \ldots, \tag{14}$$

and the eigenfunction associated with λ_n is

$$X_n(x) = \sin \frac{n\pi x}{L}, \qquad n = 1, 2, 3, \ldots. \tag{15}$$

Now we turn to Eq. (12). The homogeneous initial condition

$$y_t(x, 0) = X(x)T'(0) = 0$$

implies that $T'(0) = 0$. Therefore the solution $T_n(t)$ associated with the eigenvalue $\lambda_n = n^2\pi^2/L^2$ must satisfy the conditions

$$T_n'' + \frac{n^2\pi^2 a^2}{L^2} T_n = 0, \qquad T_n'(0) = 0. \tag{16}$$

The general solution of the differential equation in (16) is

$$T_n(t) = A_n \cos \frac{n\pi a t}{L} + B_n \sin \frac{n\pi a t}{L}. \tag{17}$$

Its derivative

$$T_n'(t) = \frac{n\pi a}{L} \left(-A_n \sin \frac{n\pi a t}{L} + B_n \cos \frac{n\pi a t}{L} \right)$$

satisfies the condition $T_n'(0) = 0$ if $B_n = 0$. Thus a nontrivial solution of (16) is

$$T_n(t) = \cos \frac{n\pi a t}{L}. \tag{18}$$

We combine the results in (15) and (18) to obtain the infinite sequence of product functions,

$$y_n(x, t) = X_n(x)T_n(t) = \cos \frac{n\pi a t}{L} \sin \frac{n\pi x}{L}, \tag{19}$$

$n = 1, 2, 3, \ldots$. Each of these *building block* functions satisfies both the wave equation $y_{tt} = a^2 y_{xx}$ and the homogeneous boundary conditions in Problem A, above. By superposition we get the infinite series

$$y(x, t) = \sum_{n=1}^{\infty} A_n X_n(x) T_n(t) = \sum_{n=1}^{\infty} A_n \cos \frac{n\pi a t}{L} \sin \frac{n\pi x}{L}. \tag{20}$$

It remains only to choose the coefficients $\{A_n\}$ to satisfy the nonhomogeneous boundary condition

$$y(x, 0) = \sum_{n=1}^{\infty} A_n \sin \frac{n\pi x}{L} = f(x) \tag{21}$$

for $0 \leq x \leq L$. But this will be the Fourier sine series of $f(x)$ on $[0, L]$ provided we choose

$$A_n = \frac{2}{L} \int_0^L f(x) \sin \frac{n\pi x}{L} \, dx. \tag{22}$$

Thus we see finally that a formal series solution of Problem A is

$$y(x, t) = \sum_{n=1}^{\infty} A_n \cos \frac{n\pi a t}{L} \sin \frac{n\pi x}{L}, \tag{23}$$

with the coefficients $\{A_n\}_1^{\infty}$ computed using the formula in (22). Note that the series in (23) is obtained from the Fourier sine series of $f(x)$ simply by inserting the factor $\cos(n\pi a t/L)$ in the nth term. Note also that this term has (circular) frequency $\omega_n = n\pi a/L$.

EXAMPLE 1 It follows immediately that the solution of the boundary value problem

$$\begin{cases} \dfrac{\partial^2 y}{\partial t^2} = 4\dfrac{\partial^2 y}{\partial x^2} & (0 < x < \pi, \quad t > 0); \\[2mm] y(0, t) = y(\pi, t) = 0, \\[2mm] y(x, 0) = \dfrac{1}{10} \sin^3 x = \dfrac{3}{40} \sin x - \dfrac{1}{40} \sin 3x, \\[2mm] y_t(x, 0) = 0, \end{cases}$$

for which $L = \pi$ and $a = 2$, is

$$y(x, t) = \frac{3}{40} \cos 2t \sin x - \frac{1}{40} \cos 6t \sin 3x.$$

Solution The reason is that we are given explicitly the Fourier sine series of $f(x)$ with $A_1 = \frac{3}{40}$, $A_3 = -\frac{1}{40}$, and $A_n = 0$ otherwise.

EXAMPLE 2 (A plucked string) Figure 7.21 shows the initial position function $f(x)$ for a stretched string (length L) that is set in motion by moving its midpoint $x = L/2$ aside the distance $\frac{1}{2}bL$ and then releasing it from rest at time $t = 0$. The corresponding boundary value problem is

$$\begin{cases} y_{tt} = a^2 y_{xx} & (0 < x < L, \quad t > 0); \\[2mm] y(0, t) = y(L, t) = 0, \\[2mm] y(x, 0) = f(x), \\[2mm] y_t(x, 0) = 0, \end{cases}$$

where $f(x) = bx$ for $0 \leq x \leq L/2$ and $f(x) = b(L - x)$ for $L/2 \leq x \leq L$. Find $y(x, t)$.

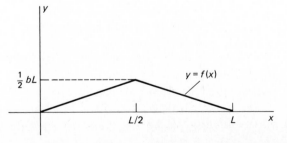

Figure 7.21 Initial position of the plucked string of Example 2.

Solution The nth Fourier sine coefficient of $f(x)$ is

$$A_n = \frac{2}{L} \int_0^L f(x) \sin \frac{n\pi x}{L} \, dx$$

$$= \frac{2}{L} \int_0^{L/2} bx \sin \frac{n\pi x}{L} \, dx + \frac{2}{L} \int_{L/2}^L b(L - x) \sin \frac{n\pi x}{L} \, dx;$$

it follows that

$$A_n = \frac{4bL}{n^2\pi^2} \sin \frac{n\pi}{2}.$$

Hence (23) yields the formal series solution

$$y(x, t) = \frac{4bL}{\pi^2} \sum_{n=1}^{\infty} \frac{1}{n^2} \sin \frac{n\pi}{2} \cos \frac{n\pi a t}{L} \sin \frac{n\pi x}{L} \qquad (24)$$

$$= \frac{4bL}{\pi^2} \left(\cos \frac{\pi a t}{L} \sin \frac{\pi x}{L} - \frac{1}{3^2} \cos \frac{3\pi a t}{L} \sin \frac{3\pi x}{L} + \cdots \right).$$

MUSIC

Numerous familiar musical instruments employ vibrating strings to generate the sounds they produce. When a string vibrates with a given frequency, vibrations at this frequency are transmitted through the air—in the form of periodic variations in air density called **sound waves**—to the ear of the listener. For instance, middle C is a *tone* with a frequency of 256 Hz. When several tones are heard simultaneously, the combination is perceived as harmonious if the ratios of their frequencies are small whole numbers; otherwise many perceive the combination as dissonance.

The series in (23) represents the motion of a string as a superposition of infinitely many vibrations with different frequencies. The nth term $A_n \cos (n\pi a t/L) \sin (n\pi x/L)$ represents a vibration with frequency

$$\nu_n = \frac{\omega_n}{2\pi} = \frac{n\pi a/L}{2\pi} = \frac{n}{2L} \sqrt{\frac{T}{\rho}} \quad \text{(Hz)}. \qquad (25)$$

The lowest of these frequencies,

$$\nu_1 = \frac{1}{2L} \sqrt{\frac{T}{\rho}} \quad \text{(Hz)}, \qquad (26)$$

is called the **fundamental frequency**, and it is ordinarily predominant in the sound we hear. The frequency $\nu_n = n\nu_1$ of the nth **overtone** or **harmonic** is an integral multiple of ν_1, and this is why the sound of a single vibrating string is harmonious rather than dissonant.

Note in (26) that the fundamental frequency ν_1 is proportional to $\sqrt{T}$ and inversely proportional to L and to $\sqrt{\rho}$. Thus we can double this frequency —and hence get a fundamental tone one octave higher—either by halving the length L or by quadrupling the tension T. The initial conditions do *not* affect ν_1; instead, they determine the coefficients in (23) and hence the extent to which the higher harmonics contribute to the sound produced. Therefore the initial conditions affect the **timbre**, or overall frequency mixture, rather than the fundamental frequency. (Technically this is true only for relatively small displacements of the string; if you strike a piano key rather forcefully you may be able to detect a very slight and very brief initial deviation from the usual frequency of the note.)

According to one (rather simplistic) theory of hearing, the loudness of the sound produced by a vibrating string is proportional to its total (kinetic plus potential) energy, which is given by

$$E = \frac{1}{2} \int_0^L \left[\rho \left(\frac{\partial y}{\partial t} \right)^2 + T \left(\frac{\partial y}{\partial x} \right)^2 \right] dx. \tag{27}$$

In Problem 17 we ask you to show that substitution of the series in (23) in the formula above yields

$$E = \frac{\pi^2 T}{4L} \sum_{n=1}^{\infty} n^2 A_n{}^2. \tag{28}$$

The ratio of the nth term $n^2 A_n{}^2$ to the sum $\sum n^2 A_n{}^2$ is then the portion of the whole sound attributable to the nth harmonic.

We illustrate this theory with the series in (24), which describes the motion of the plucked string of Example 2. Note that the even harmonics are missing, and that $A_n = 4bL/\pi^2 n^2$ for n odd. Hence (28) gives

$$E = \frac{\pi^2 T}{4L} \sum_{n \text{ odd}} n^2 \frac{16 b^2 L^2}{\pi^4 n^4} = \frac{4 b^2 LT}{\pi^2} \sum_{n \text{ odd}} \frac{1}{n^2},$$

and so the proportion of the sound associated with the nth harmonic (for n odd) is

$$\frac{1/n^2}{\sum\limits_{n \text{ odd}} 1/n^2} = \frac{1/n^2}{\pi^2/8} = \frac{8}{\pi^2 n^2}.$$

Substituting $n = 1$ and $n = 3$, we find that 81.06% of the sound of the string of Example 2 is associated with the fundamental tone and 9.01% with the harmonic corresponding to $n = 3$.

THE d'ALEMBERT SOLUTION

In contrast to series solutions of the heat equation, formal series solutions of the wave equation ordinarily do *not* possess sufficient termwise differentiability to permit verification of the solution by application of a superposition theorem analogous to the one stated in Section 7.5. For instance, termwise differentiation of the series in (24) would yield the series

$$\frac{\partial^2 y}{\partial x^2} = -\frac{4b}{L} \sum_{n=1}^{\infty} \sin \frac{n\pi}{2} \cos \frac{n\pi a t}{L} \sin \frac{n\pi x}{L},$$

which generally fails to converge, because the "convergence factor" $1/n^2$ has disappeared after the second differentiation.

There is an alternative approach, however, that both verifies the solution in (23) and yields valuable additional information about it. If we apply the trigonometric identity $2 \sin A \cos B = \sin (A + B) + \sin (A - B)$ with A

$= n\pi x/L$ and $B = n\pi at/L$, then (23) gives

$$y(x, t) = \sum_{n=1}^{\infty} A_n \sin \frac{n\pi x}{L} \cos \frac{n\pi at}{L}$$

$$= \frac{1}{2} \sum_{n=1}^{\infty} A_n \sin \frac{n\pi}{L}(x + at) + \frac{1}{2} \sum_{n=1}^{\infty} A_n \sin \frac{n\pi}{L}(x - at). \quad (29)$$

But, by definition of the coefficients,

$$\sum_{n=1}^{\infty} A_n \sin \frac{n\pi x}{L} = F(x)$$

for all x, where $F(x)$ is the odd extension of period $2L$ of the initial position function $f(x)$. Hence (29) means that

$$y(x, t) = \tfrac{1}{2}[F(x + at) + F(x - at)]. \quad (30)$$

Therefore the series in (23) converges to the expression on the right-hand side in (30), which is known as the **d'Alembert form** of the solution of Problem A for the vibrating string. Moreover, using the chain rule it is easy to verify (see Problems 13 and 14) that the function $y(x, t)$ defined in (30) does, indeed, satisfy the equation $y_{tt} = a^2 y_{xx}$ (under the assumption that F is twice differentiable), as well as the boundary conditions $y(0, t) = y(L, t) = 0$ and $y(x, 0) = f(x)$.

For any function $F(x)$, the functions $F(x + at)$ and $F(x - at)$ represent "waves" moving to the left and right, respectively, along the x-axis with speed a. This fact is illustrated by Fig. 7.22, which shows the graphs of these three functions for t fixed, in the case $F(x)$ is a pulse function centered at $x = x_0$. Thus the d'Alembert solution in (30) expresses $y(x, t)$ as a superposition of two waves moving in opposite directions with speed a. This is why the equation $y_{tt} = a^2 y_{xx}$ is called the *wave* equation.

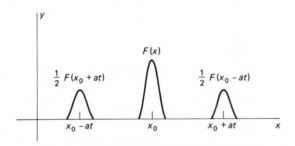

Figure 7.22 The pulse $F(x)$ produces two waves—one moving to the left, one to the right.

STRINGS WITH INITIAL VELOCITY

The separation of variables solution of Problem B, with initial conditions $y(x, 0) = 0$ and $y_t(x, t) = g(x)$, proceeds precisely as did that of Problem A until Eq. (16) is reached. But then $y(x, 0) = X(x)T(0) = 0$ implies that $T(0) = 0$, so instead of (16) we have

$$\frac{d^2 T_n}{dt^2} + \frac{n^2 \pi^2 a^2}{L^2} T_n = 0, \qquad T_n(0) = 0, \quad (31)$$

From Eq. (17) we see that a nontrivial solution of (31) is

$$T_n(t) = \sin\frac{n\pi at}{L}. \tag{32}$$

Our formal power series solution is therefore of the form

$$y(x, t) = \sum_{n=1}^{\infty} B_n \sin\frac{n\pi at}{L} \sin\frac{n\pi x}{L}, \tag{33}$$

and so we want to choose the coefficients $\{B_n\}$ so that

$$y_t(x, 0) = \sum_{n=1}^{\infty} B_n \frac{n\pi a}{L} \sin\frac{n\pi x}{L} = g(x). \tag{34}$$

Thus we want $B_n \cdot n\pi a/L$ to be the Fourier sine coefficient b_n of $g(x)$ on $[0, L]$:

$$B_n \frac{n\pi a}{L} = b_n = \frac{2}{L}\int_0^L g(x) \sin\frac{n\pi x}{L}\,dx.$$

Hence we choose

$$B_n = \frac{2}{n\pi a}\int_0^L g(x) \sin\frac{n\pi x}{L}\,dx \tag{35}$$

in order for $y(x, t)$ in (33) to be a formal series solution of Problem B.

EXAMPLE 3 Consider a string on a guitar lying crosswise in the back of a pickup truck that at time $t = 0$ slams into a brick wall with speed v_0. Then $g(x) \equiv v_0$, so

$$B_n = \frac{2}{n\pi a}\int_0^L v_0 \sin\frac{n\pi x}{L}\,dx = \frac{2v_0 L}{n^2\pi^2 a}[1 - (-1)^n].$$

Hence the series in (33) gives

$$y(x, t) = \frac{4v_0 L}{\pi^2 a} \sum_{n\ \text{odd}} \frac{1}{n^2} \sin\frac{n\pi at}{L} \sin\frac{n\pi x}{L}.$$

If we differentiate the series in (33) termwise with respect to t, we get

$$y_t(x, t) = \sum_{n=1}^{\infty} b_n \sin\frac{n\pi x}{L} \cos\frac{n\pi at}{L} = \frac{1}{2}[G(x + at) + G(x - at)], \tag{36}$$

where G is the odd period $2L$ extension of the initial velocity function $g(x)$, using the same device as in the derivation of Eq. (30). In Problem 15 we ask you to deduce that

$$y(x, t) = \frac{1}{2a}[H(x + at) + H(x - at)], \tag{37}$$

where the function $H(x)$ is defined to be

$$H(x) = \int_0^x G(s)\,ds. \tag{38}$$

If, finally, a string has both an initial position $y(x, 0) = f(x)$ and an initial velocity $y_t(x, 0) = g(x)$, then we can obtain its displacement function $y(x, t)$ by superposing the d'Alembert solutions of Problems A and B given in Eqs. (30) and (37), respectively. Hence the vibrations of this string with general initial conditions are described by

$$y(x, t) = \frac{1}{2}[F(x + at) + F(x - at)] + \frac{1}{2a}[H(x + at) + H(x - at)], \quad (39)$$

a superposition of four waves moving along the x-axis with speed a, two moving to the left and two to the right.

7.6 Problems

Solve the boundary value problems in Problems 1–10.

1. $y_{tt} = 4y_{xx}, 0 < x < \pi, t > 0; \quad y(0, t) = y(\pi, t) = 0, y(x, 0) = \frac{1}{10} \sin 2x,$
   $y_t(x, 0) = 0.$

2. $y_{tt} = y_{xx}, 0 < x < 1, t > 0; \quad y(0, t) = y(1, t) = 0,$
   $y(x, 0) = \frac{1}{10} \sin \pi x - \frac{1}{20} \sin 3\pi x, y_t(x, 0) = 0.$

3. $4y_{tt} = y_{xx}, 0 < x < \pi, t > 0; \quad y(0, t) = y(\pi, t) = 0,$
   $y(x, 0) = y_t(x, 0) = \frac{1}{10} \sin x.$

4. $4y_{tt} = y_{xx}, 0 < x < 2, t > 0; \quad y(0, t) = y(2, t) = 0,$
   $y(x, 0) = \frac{1}{5} \sin \pi x \cos \pi x, y_t(x, 0) = 0.$

5. $y_{tt} = 25y_{xx}, 0 < x < 3, t > 0; \quad y(0, t) = y(3, t) = 0, y(x, 0) = \frac{1}{4} \sin \pi x,$
   $y_t(x, 0) = 10 \sin 2\pi x.$

6. $y_{tt} = 100y_{xx}, 0 < x < \pi, t > 0; \quad y(0, t) = y(\pi, t) = 0,$
   $y(x, 0) = x(\pi - x), y_t(x, 0) = 0.$

7. $y_{tt} = 100y_{xx}, 0 < x < 1, t > 0; \quad y(0, t) = y(1, t) = 0, y(x, 0) = 0,$
   $y_t(x, 0) = x.$

8. $y_{tt} = 4y_{xx}, 0 < x < \pi, t > 0; \quad y(0, t) = y(\pi, t) = 0, y(x, 0) = \sin x,$
   $y_t(x, 0) = 1.$

9. $y_{tt} = 4y_{xx}, 0 < x < 1, t > 0; \quad y(0, t) = y(1, t) = 0, y(x, 0) = 0,$
   $y_t(x, 0) = x(1 - x).$

10. $y_{tt} = 25y_{xx}, 0 < x < \pi, t > 0; \quad y(0, t) = y(\pi, t) = 0,$
    $y(x, 0) = y_t(x, 0) = \sin^2 x.$

11. Suppose that a string 2 ft long weighs $\frac{1}{32}$ oz and is subjected to a tension of 32 lb. Find the fundamental frequency with which it vibrates and the velocity with which the vibration waves travel along it.

12. Show that the amplitude of the oscillations of the midpoint of the string of Example 3 is

$$y\left(\frac{L}{2}, \frac{L}{2a}\right) = \frac{4v_0 L}{\pi^2 a} \sum_{n \text{ odd}} \frac{1}{n^2} = \frac{v_0 L}{2a}.$$

If the string is the string of Problem 11 and the impact speed of the pickup truck is 60 mph, show that this amplitude is approximately 1 in.

13. Suppose that the function $F(x)$ is twice differentiable for all x. Use the chain rule to verify that the functions $y = F(x + at)$ and $y = F(x - at)$ satisfy the equation $y_{tt} = a^2 y_{xx}$.

14. Given the differentiable odd period $2L$ function $F(x)$, show that the function $y = \frac{1}{2}[(F(x + at) + F(x - at)]$ satisfies the conditions $y(0, t) = y(L, t) = 0$, $y(x, 0) = F(x)$, and $y_t(x, 0) = 0$.

15. If $y(x, 0) = 0$, then Eq. (36) implies (why?) that

$$y(x, t) = \frac{1}{2} \int_0^t G(x + a\tau) \, d\tau + \frac{1}{2} \int_0^t G(x - a\tau) \, d\tau.$$

Make appropriate substitutions in these integrals to derive Eqs. (37) and (38).

16. (a) Show that the substitutions $u = x + at$ and $v = x - at$ transform the equation $y_{tt} = a^2 y_{xx}$ into the equation $y_{uv} = 0$. (b) Conclude that every solution of $y_{tt} = a^2 y_{xx}$ is of the form $y(x, t) = F(x + at) + G(x - at)$, which represents two waves traveling in opposite directions, each with speed a.

17. Suppose that $y = \sum A_n \cos (n\pi at/L) \sin (n\pi x/L)$. Square the derivatives y_t and y_x and then integrate termwise—applying the orthogonality of the sine and cosine functions—to verify that

$$E = \frac{1}{2} \int_0^L (\rho y_t{}^2 + T y_x{}^2) \, dx = \frac{\pi^2 T}{4L} \sum_{n=1}^{\infty} n^2 A_n{}^2.$$

18. Consider a stretched string, initially at rest; its end at $x = 0$ is fixed, but its end at $x = L$ is free—it is allowed to slide without friction along the vertical line $x = L$. The corresponding boundary value problem is

$$\begin{cases} y_{tt} = a^2 y_{xx} & (0 < x < L, \quad t > 0); \\ y(0, t) = y_x(L, t) = 0, \\ y(x, 0) = f(x), \quad y_t(x, 0) = 0. \end{cases}$$

Separate the variables and use the odd half-multiple sine series of $f(x)$, as in Problem 24 of Section 7.5, to derive the solution

$$y(x, t) = \sum_{n \text{ odd}} A_n \cos \frac{n\pi at}{2L} \sin \frac{n\pi x}{2L},$$

where

$$A_n = \frac{2}{L} \int_0^L f(x) \sin \frac{n\pi x}{2L} \, dx.$$

Problems 19 and 20 deal with the vibrations of a string under the influence of the downward force $F(x) = -\rho g$ of gravity. According to Eq. (1), its displacement function satisfies the partial differential equation

$$\frac{\partial^2 y}{\partial t^2} = a^2 \frac{\partial^2 y}{\partial x^2} - g \tag{40}$$

with endpoint conditions $y(0, t) = y(L, t) = 0$.

19. Suppose first that the string hangs in a stationary position, so that $y = y(x)$ and $y_{tt} = 0$, and hence its differential equation of motion takes the simple form $a^2 y'' = g$. Derive the stationary solution

$$y = \phi(x) = \frac{gx}{2a^2}(x - L).$$

20. Now suppose that the string is released from rest in equilibrium; consequently the initial conditions are $y(x, 0) = 0$ and $y_t(x, 0) = 0$. Define $v(x, t) = y(x, t) - \phi(x)$, where $\phi(x)$ is the stationary solution of Problem 19. Deduce from Eq. (40) that $v(x, t)$ satisfies the boundary value problem

$$\begin{cases} v_{tt} = a^2 v_{xx}; \\ v(0, t) = v(x, t) = 0, \\ v(x, 0) = -\phi(x), \quad v_t(x, 0) = 0. \end{cases}$$

Conclude from Eqs. (22) and (23) that

$$y(x, t) = \phi(x) + \sum_{n=1}^{\infty} A_n \cos \frac{n\pi at}{L} \sin \frac{n\pi x}{L}.$$

where the coefficients $\{A_n\}$ are the Fourier sine coefficients of $f(x) = -\phi(x)$. Finally, explain why it follows that the string oscillates between the positions $y = 0$ and $y = 2\phi(x)$.

21. For a string vibrating in air with resistance proportional to velocity, the boundary value problem is

$$\begin{cases} y_{tt} = a^2 y_{xx} - 2h y_t; \\ y(0, t) = y(L, t) = 0, \\ y(x, 0) = f(x), \qquad y_t(x, 0) = 0. \end{cases} \tag{41}$$

Assume that $0 < h < \pi a/L$. (a) Substitute $y = X(x)T(t)$ in (41) to obtain the equations

$$X'' + \lambda X = 0, \qquad X(0) = X(L) = 0 \tag{42}$$

and

$$T'' + 2hT' + a^2 \lambda T = 0, \qquad T'(0) = 0. \tag{43}$$

(b) The eigenvalues and eigenfunctions of (42) are $\lambda_n = n^2\pi^2/L^2$ and $X_n(x) = \sin(n\pi x/L)$ (as usual). Show that the general solution with $\lambda = n^2\pi^2/L^2$ of the differential equation in (43) is

$$T_n(t) = e^{-ht}(A_n \cos \omega_n t + B_n \sin \omega_n t)$$

where $\omega_n = \sqrt{(n^2\pi^2 a^2/L^2) - h^2} < n\pi a/L$. (c) Show that $T_n'(0) = 0$ implies that $B_n = hA_n/\omega_n$, and hence that to within a constant multiplicative coefficient,

$$T_n(t) = e^{-ht} \cos(\omega_n t - \alpha_n)$$

where $\alpha_n = \tan^{-1}(h/\omega_n)$. (d) Finally conclude that

$$y(x, t) = e^{-ht} \sum_{n=1}^{\infty} c_n \cos(\omega_n t - \alpha_n) \sin \frac{n\pi x}{L},$$

where

$$c_n = \frac{2}{L \cos \alpha_n} \int_0^L f(x) \sin \frac{n\pi x}{L} \, dx.$$

From this formula we see that the air resistance has three main effects: exponential damping of amplitudes, decreased frequencies $\omega_n < n\pi a/L$, and the production of the phase delay angles α_n.

22. Rework Problem 21 as follows: First substitute $y(x, t) = e^{-ht} v(x, t)$ in Eq. (41) and then show that the boundary value problem for $v(x, t)$ is

$$\begin{cases} v_{tt} = a^2 v_{xx} + h^2 v; \\ v(0, t) = v(L, t) = 0, \\ v(x, 0) = f(x), \qquad v_t(x, 0) = hf(x). \end{cases}$$

Next show that the substitution $v(x, t) = X(x)T(t)$ leads to the equations

$$X'' + \lambda X = 0, \qquad X(0) = X(L) = 0,$$
$$T'' + (\lambda a^2 - h^2)T = 0.$$

Proceed in this manner to derive the solution $y(x, t)$ given in part (d) of Problem 21.

We now consider the temperature in a two-dimensional plate, or lamina, that occupies a region R in the xy-plane bounded by a piecewise smooth curve C, as shown in Fig. 7.23. We assume that the faces of the plate are insulated, and that it is so thin that the temperature within it does not vary in the direction perpendicular to the xy-plane. We want to determine, under various conditions, the temperature $u(x, y, t)$ at the point (x, y) at time t.

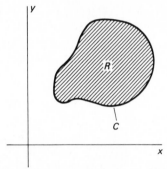

Figure 7.23   A plane region R and its bounding curve C.

Let the plate consist of material with density δ (mass per unit volume), specific heat c (per unit mass), and thermal conductivity K, all assumed constant throughout the plate. Under these assumptions, it can be shown (by a generalization of the derivation of the one-dimensional heat equation in Section 7.5) that the temperature function $u(x, y, t)$ satisfies the **two-dimensional heat equation**

$$\frac{\partial u}{\partial t} = k\left(\frac{\partial^2 u}{\partial x^2} + \frac{\partial^2 u}{\partial y^2}\right) \tag{1}$$

As in Section 7.5, k denotes the thermal diffusivity,

$$k = \frac{K}{c\delta}, \tag{2}$$

of the material of the plate. The sum of second derivatives on the right-hand side in (1) is the **Laplacian** of the function u, commonly denoted by

$$\nabla^2 u = \frac{\partial^2 u}{\partial x^2} + \frac{\partial^2 u}{\partial y^2}, \tag{3}$$

so the two-dimensional heat equation may be written as

$$\frac{\partial u}{\partial t} = k\nabla^2 u. \tag{1'}$$

In comparing (1') with the one-dimensional heat equation $u_t = ku_{xx}$, we see that in passing from one to two dimensions, the second order space derivative u_{xx} is replaced by the Laplacian $\nabla^2 u$. This is an instance of a general phenomenon. For instance, if a flexible stretched membrane occupies in equilibrium a region in the xy-plane and vibrates in the (perpendicular) z-direction, then its displacement function $z = z(x, y, t)$ satisfies the **two-dimensional wave**

equation

$$\frac{\partial^2 z}{\partial t^2} = a^2 \left(\frac{\partial^2 z}{\partial x^2} + \frac{\partial^2 z}{\partial y^2} \right) = a^2 \nabla^2 z. \tag{4}$$

This equation has the same relation to the one-dimensional wave equation $z_{tt} = a^2 z_{xx}$ (here writing $z(x, t)$ for the displacement of a string) as Eq. (1') has to the one-dimensional heat equation.

In this section we confine our attention to the steady-state situation in which the temperature u does not vary with time, and so is a function only of x and y. Thus we are interested in the *steady-state temperature* of a plate. In this case $u_t = 0$, so Eq. (1) becomes the **two-dimensional Laplace equation**

$$\nabla^2 u = \frac{\partial^2 u}{\partial x^2} + \frac{\partial^2 u}{\partial y^2} = 0. \tag{5}$$

This important partial differential equation is also known as the **potential equation**. The three-dimensional Laplace equation $u_{xx} + u_{yy} + u_{zz} = 0$ is satisfied (in empty space) by electric and gravitational potential functions. It is also satisfied by the velocity potential function for the steady irrotational flow of an incompressible and inviscid liquid.

DIRICHLET PROBLEMS

A particular solution of Laplace's equation in a bounded plane region R is determined by appropriate boundary conditions. For example, it is plausible on physical grounds that the steady-state temperature $u(x, y)$ in a plate is determined if we know that $u(x, y)$ agrees with a given function $f(x, y)$ at each point of the boundary curve C of the plate. To find the steady-state temperature in a plate with assigned boundary values, we need to solve the boundary value problem

$$\left. \begin{array}{ll} \dfrac{\partial^2 u}{\partial x^2} + \dfrac{\partial^2 u}{\partial y^2} = 0 & \text{(within } R\text{);} \\[2mm] u(x, y) = f(x, y) & \text{(if } (x, y) \text{ is on } C\text{).} \end{array} \right\} \tag{6}$$

Such a problem, finding a solution of Laplace's equation in a region R with given boundary values, is called a **Dirichlet problem**. It is known that, if the boundary curve C and the boundary value function are reasonably well behaved, then there exists a unique solution of the Dirichlet problem in (6).

Figure 7.24 shows a rectangular plate with indicated boundary values along its four edges. The corresponding Dirichlet problem is

$$\left. \begin{array}{ll} u_{xx} + u_{yy} = 0 & \text{(in } R\text{);} \\[2mm] u(x, 0) = f_1(x),\ u(x, b) = f_2(x) & (0 < x < a), \\[2mm] u(0, y) = g_1(y),\ u(a, y) = g_2(y) & (0 < y < b). \end{array} \right\} \tag{7}$$

Because there are four nonhomogeneous conditions (rather than one), this boundary value problem is not directly susceptible to the method of separation of variables. In Section 7.6, when confronted with this difficulty, we split the problem of the vibrating string with both initial position and initial velocity

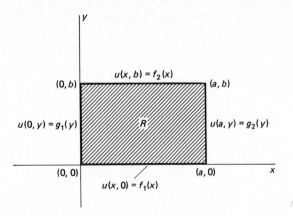

Figure 7.24 A rectangular plate with given boundary values.

into two problems with a single nonhomogeneous condition each. In a similar way, the boundary value problem in (7) can be split into *four* problems, each with a single nonhomogeneous boundary condition. For each of these four problems, $u(x, y)$ will be zero along three edges of the rectangle and will have assigned values on the fourth. Each of these four boundary value problems can be solved by separation of variables, and the sum of the four solutions is the solution of the original problem in (7). In the following example we solve one of these four problems—the one illustrated in Fig. 7.25.

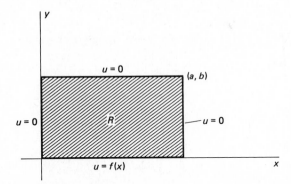

Figure 7.25 The boundary value problem of Example 1.

EXAMPLE 1   Solve the boundary value problem

$$\begin{cases} u_{xx} + u_{yy} = 0; \\ \quad u(0, y) = u(a, y) = u(x, b) = 0, \\ \quad u(x, 0) = f(x) \end{cases} \tag{8}$$

for the rectangle in Fig. 7.25.

Solution   Substitution of $u(x, y) = X(x)Y(y)$ gives $X'' Y + X Y'' = 0$, so

$$\frac{X''}{X} = -\frac{Y''}{Y} = -\lambda \tag{9}$$

for some constant λ. Thus $X(x)$ must satisfy the familiar eigenvalue problem

$$X'' + \lambda X = 0,$$
$$X(0) = X(a) = 0.$$

The eigenvalues and associated eigenfunctions are

$$\lambda_n = \frac{n^2\pi^2}{a^2}, \qquad X_n(x) = \sin\frac{n\pi x}{a} \tag{10}$$

for $n = 1, 2, 3, \ldots$. From (9), with $\lambda = n^2\pi^2/a^2$ and the remaining homogeneous boundary condition in (8), we get

$$Y_n'' - \frac{n^2\pi^2}{a^2}Y_n = 0, \qquad Y_n(b) = 0. \tag{11}$$

The general solution of the differential equation in (11) is

$$Y_n(y) = A_n\cosh\frac{n\pi y}{a} + B_n\sinh\frac{n\pi y}{a},$$

and the condition

$$Y_n(b) = A_n\cosh\frac{n\pi b}{a} + B_n\sinh\frac{n\pi b}{a} = 0$$

implies that $B_n = -[A_n\cosh(n\pi b/a)]/\sinh(n\pi b/a)$, so

$$Y_n(y) = A_n\cosh\frac{n\pi y}{a} - \frac{A_n\cosh(n\pi b/a)}{\sinh(n\pi b/a)}\sinh\frac{n\pi y}{a}$$

$$= \frac{A_n}{\sinh(n\pi b/a)}\left(\sinh\frac{n\pi b}{a}\cosh\frac{n\pi y}{a} - \cosh\frac{n\pi b}{a}\sinh\frac{n\pi y}{a}\right).$$

Therefore

$$Y_n(y) = C_n\sinh\frac{n\pi}{a}(b - y) \tag{12}$$

where $C_n = A_n/\sinh(n\pi b/a)$. From (10) and (12) we obtain the formal series solution

$$u(x, y) = \sum_{n=1}^{\infty} c_n X_n(x)Y_n(y) = \sum_{n=1}^{\infty} c_n\sin\frac{n\pi x}{a}\sinh\frac{n\pi}{a}(b - y). \tag{13}$$

It remains only to choose the coefficients $\{c_n\}$ to satisfy the nonhomogeneous condition

$$u(x, 0) = \sum_{n=1}^{\infty}\left(c_n\sinh\frac{n\pi b}{a}\right)\sin\frac{n\pi x}{a} = f(x).$$

For this purpose we want

$$c_n\sinh\frac{n\pi b}{a} = b_n = \frac{2}{a}\int_0^a f(x)\sin\frac{n\pi x}{a}\,dx,$$

so

$$c_n = \frac{2}{a\sinh(n\pi b/a)}\int_0^a f(x)\sin\frac{n\pi x}{a}\,dx. \tag{14}$$

With this choice of coefficients, the series in (13) is a formal series solution of the Dirichlet problem in (8).

For instance, suppose that

$$f(x) = T_0 = \frac{4T_0}{\pi} \sum_{n \text{ odd}}^{\infty} \frac{1}{n} \sin \frac{n\pi x}{a}$$

if $0 < x < a$, so that $b_n = 4T_0/\pi n$ for n odd and $b_n = 0$ for n even. Then the formulas in (13) and (14) yield

$$u(x, y) = \frac{4T_0}{\pi} \sum_{n \text{ odd}} \frac{\sin(n\pi x/a)\sinh(n\pi[b - y]/a)}{n \sinh(n\pi b/a)} \tag{15}$$

for the steady-state temperature in a rectangular plate with its base held at temperature T_0 and its other three edges at temperature zero. For instance, suppose that $T_0 = 100$ and that $a = b = 10$. Then from (15) we find that the temperature at the center of the plate is

$$u(5, 5) = \frac{400}{\pi} \sum_{n \text{ odd}} \frac{\sin(n\pi/2)\sinh(n\pi/2)}{n \sinh n\pi}$$

$$\approx 25.3716 - 0.3812 + 0.0099 - 0.0003;$$

thus $u(5, 5)$ is approximately 25.00.

Question: Can you think of an argument—perhaps using symmetry—that proves (without using the series solution) that $u(5, 5)$ is exactly 25?

EXAMPLE 2 Let R be the "semi-infinite" strip shown in Fig. 7.26. Solve the boundary value problem

$$\left. \begin{aligned} u_{xx} + u_{yy} &= 0 && \text{(in } R\text{)}; \\ u(x, 0) = u(x, b) &= 0 && (0 < x < \infty), \\ u(x, y) \text{ is bounded as } & x \longrightarrow +\infty, \\ u(0, y) &= g(y). \end{aligned} \right\} \tag{16}$$

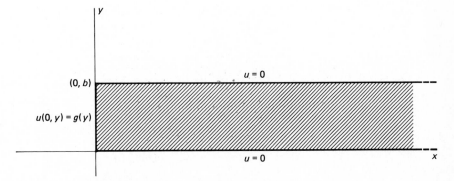

Figure 7.26 The "semi-infinite" strip of Example 2.

Solution The condition that $u(x, y)$ is bounded as $x \rightarrow +\infty$ plays the role of a homogeneous boundary condition associated with the "missing" right edge of the "rectangle"; this is typical of Dirichlet problems for unbounded regions. With $u(x, y) = X(x)Y(y)$, the actual homogeneous conditions give $Y(0) = Y(b) = 0$. Hence it is $Y(y)$ that will satisfy an eigenvalue problem. We therefore write the separation in (9) as

$$\frac{Y''}{Y} = -\frac{X''}{X} = -\lambda, \tag{17}$$

changing the sign of λ to get the familiar eigenvalue problem

$$Y'' + \lambda Y = 0, \qquad Y(0) = Y(b) = 0$$

with eigenvalues and associated eigenfunctions

$$\lambda_n = \frac{n^2\pi^2}{b^2}, \qquad Y_n(x) = \sin\frac{n\pi y}{b}. \tag{18}$$

The general solution of

$$X_n'' - \frac{n^2\pi^2}{b^2}X_n = 0$$

is $X_n(x) = A_n e^{n\pi x/b} + B_n e^{-n\pi x/b}$. We write the solution in exponential form because then the condition that $u(x, y)$, and hence $X(x)$, is bounded as $x \longrightarrow +\infty$ immediately implies that $A_n = 0$. Suppressing the constant, this leaves

$$X_n(x) = e^{-n\pi x/b}. \tag{19}$$

From (18) and (19) we get the formal series solution

$$u(x, y) = \sum_{n=1}^{\infty} b_n X_n(x)Y_n(y) = \sum_{n=1}^{\infty} b_n e^{-n\pi x/b} \sin\frac{n\pi y}{b} \tag{20}$$

that satisfies the homogeneous boundary conditions and the boundedness condition. In order also to satisfy the nonhomogeneous condition

$$u(0, y) = \sum_{n=1}^{\infty} b_n \sin\frac{n\pi y}{b} = g(y),$$

we choose b_n to be the Fourier sine coefficient

$$b_n = \frac{2}{b}\int_0^b g(y) \sin\frac{n\pi y}{b}\, dy \tag{21}$$

of the function $g(y)$ on $[0, b]$.

For instance, if

$$g(y) = T_0 = \frac{4T_0}{\pi}\sum_{n\,\text{odd}} \frac{1}{n}\sin\frac{n\pi y}{b}$$

for $0 < y < b$, then the formulas in (20) and (21) yield

$$u(x, y) = \frac{4T_0}{\pi}\sum_{n\,\text{odd}} \frac{1}{n} e^{-n\pi x/b} \sin\frac{n\pi y}{b}.$$

THE DIRICHLET PROBLEM FOR A CIRCULAR DISK

We now investigate the steady state temperature in a circular disk of radius a with insulated faces and with given boundary temperatures. Obviously we should accommodate the geometry of the circle by expressing $u = u(r, \theta)$ in terms of polar coordinates r and θ, with $x = r \cos \theta$ and $y = r \sin \theta$. When these equations are used to transform the Laplacian, the result is **Laplace's equation in polar coordinates**:

$$\nabla^2 u = \frac{\partial^2 u}{\partial r^2} + \frac{1}{r} \frac{\partial u}{\partial r} + \frac{1}{r^2} \frac{\partial^2 u}{\partial \theta^2} = 0. \tag{22}$$

We prescribe assigned boundary temperatures (as indicated in Fig. 7.27) by requiring that

$$u(a, \theta) = f(\theta) \tag{23}$$

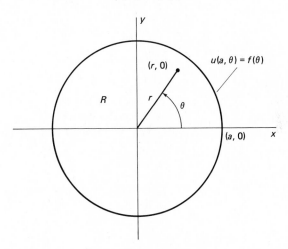

Figure 7.27 The Dirichlet problem for a circular disk.

where the function $f(\theta)$ of period 2π is given. In addition, because (r, θ) and $(r, \theta + 2\pi)$ are the polar coordinates of the same point, we impose upon u the periodicity condition

$$u(r, \theta) = u(r, \theta + 2\pi) \tag{24}$$

for $r < a$ and all θ; this will play the role of a homogeneous boundary condition.

To solve the boundary value problem in (22)–(24), we substitute $u(r, t) = R(t)\Theta(\theta)$ in Eq. (22); the result is

$$R''\Theta + \frac{1}{r} R'\Theta + \frac{1}{r^2} R\Theta'' = 0.$$

Division by $R\Theta/r^2$ gives

$$\frac{r^2 R'' + r R'}{R} + \frac{\Theta''}{\Theta} = 0,$$

so it follows that

$$\frac{r^2 R'' + r R'}{R} = -\frac{\Theta''}{\Theta} = \lambda$$

for some constant λ. This yields the two ordinary differential equations

$$r^2 R'' + r R' - \lambda R = 0 \tag{25}$$

and

$$\Theta'' + \lambda\Theta = 0. \tag{26}$$

The general solution of Eq. (26) is

$$\Theta(\theta) = A \cos \alpha\theta + B \sin \alpha\theta \quad \text{if } \lambda = \alpha^2 > 0, \tag{27a}$$

$$\Theta(\theta) = A + B\theta \quad\quad\quad\quad \text{if } \lambda = 0, \tag{27b}$$

$$\Theta(\theta) = Ae^{\alpha\theta} + Be^{-\alpha\theta} \quad\quad \text{if } \lambda = -\alpha^2 < 0. \tag{27c}$$

Now we apply the condition in (24), which implies that $\Theta(\theta) = \Theta(\theta + 2\pi)$; that is, that $\Theta(\theta)$ has period 2π. In Problems 22 and 23 we ask you to show that this is so only if $\lambda = n^2$ (n an integer) in (27a) or if $\lambda = 0$ and $B = 0$ in (27b). Thus we have the eigenvalues and associated eigenfunctions

$$\lambda_0 = 0, \quad\quad \Theta_0(\theta) = 1; \tag{28a}$$

$$\lambda_n = n^2, \quad\quad \Theta_n(\theta) = A_n \cos n\theta + B_n \sin n\theta \tag{28b}$$

for $n = 1, 2, 3, \ldots$.

Next we turn our attention to Eq. (25). With $\lambda_0 = 0$ it reduces to the equation $r^2 R_0'' + r R_0' = 0$ with general solution

$$R_0(r) = C_0 + D_0 \ln r. \tag{29}$$

Of course we want $u(r, \theta)$, and hence $R(r)$, to be continuous at $r = 0$; this requires that $D_0 = 0$ in (29), so $R_0(r) = C_0$. With $\lambda = \lambda_n = n^2$, Eq. (25) is the Euler equation

$$r^2 R_n'' + r R_n' - n^2 R_n = 0.$$

By the method of Section 2.6, we easily find that its general solution is

$$R_n(r) = C_n r^n + \frac{D_n}{r^n}; \tag{30}$$

continuity at $r = 0$ requires that $D_n = 0$, so $R_n(r) = C_n r^n$.

Finally we combine the results in Eqs. (28)–(30) with $D_n \equiv 0$ to obtain a formal series solution of the form

$$u(r, \theta) = \sum_{n=0}^{\infty} R_n(r)\Theta_n(\theta);$$

that is,

$$u(r, \theta) = \frac{a_0}{2} + \sum_{n=1}^{\infty} (a_n \cos n\theta + b_n \sin n\theta)r^n. \tag{31}$$

In order that the boundary condition $u(a, \theta) = f(\theta)$ be satisfied, we want

$$u(a, \theta) = \frac{a_0}{2} + \sum_{n=1}^{\infty} (a_n a^n \cos n\theta + b_n a^n \sin n\theta)$$

to be the Fourier series of $f(\theta)$ on $[0, 2\pi]$. Hence we choose

$$a_n = \frac{1}{\pi a^n} \int_0^{2\pi} f(\theta) \cos n\theta \, d\theta \quad (n = 0, 1, 2, \ldots) \tag{32a}$$

and

$$b_n = \frac{1}{\pi a^n} \int_0^{2\pi} f(\theta) \sin n\theta \, d\theta \quad (n = 1, 2, 3, \ldots). \tag{32b}$$

For instance, if $F(\theta) = +T_0$ for $0 < \theta < \pi$ and $f(\theta) = -T_0$ for $\pi < \theta < 2\pi$, so that

$$f(\theta) = \frac{4T_0}{\pi} \sum_{n \text{ odd}} \frac{1}{n} \sin n\theta,$$

then $a_n = 0$ for all $n \geq 0$, $b_n = 0$ for n even, and $b_n = 4T_0/\pi n a^n$ for n odd. So Eqs. (31) and (32) yield

$$u(r, \theta) = \frac{4T_0}{\pi} \sum_{n \text{ odd}} \frac{r^n}{na^n} \sin n\theta.$$

Note: In the Dirichlet problems of this section, we have concentrated on temperature functions with preassigned boundary values. In the case of a plate with an *insulated* edge—across which no heat can flow—the boundary condition is that the derivative of $u(x, y)$ in the direction normal to this edge vanishes. For instance, consider the rectangular plate in Fig 7.24. If the edge $x = 0$ is insulated, then $u_x(0, y) \equiv 0$, while if the edge $y = b$ is insulated, then $u_y(x, b) \equiv 0$.

7.7 Problems

In each of Problems 1–3, solve the Dirichlet problem for the rectangle $0 < x < a$, $0 < y < b$ consisting of Laplace's equation $u_{xx} + u_{yy} = 0$ and the given boundary value conditions.

1. $u(x, 0) = u(x, b) = u(0, y) = 0$, $u(a, y) = g(y)$.
2. $u(x, 0) = u(x, b) = u(a, y) = 0$, $u(0, y) = g(y)$.
3. $u(x, 0) = u(0, y) = u(a, y) = 0$, $u(x, b) = f(x)$.
4. Consider the boundary value problem

$$\begin{cases} u_{xx} + u_{yy} = 0; \\ u_x(0, y) = u_x(a, y) = u(x, 0) = 0, \\ u(x, b) = f(x) \end{cases}$$

corresponding to a rectangular plate $0 < x < a, 0 < y < b$ with the edges $x = a$ and $x = b$ insulated. Derive the solution

$$u(x, y) = \frac{a_0 y}{2b} + \sum_{n=1}^{\infty} a_n \left(\cos \frac{n\pi x}{a} \right) \left[\frac{\sinh (n\pi y/a)}{\sinh (n\pi b/a)} \right],$$

where

$$a_n = \frac{2}{a} \int_0^a f(x) \cos \frac{n\pi x}{a} \, dx \qquad (n = 0, 1, 2, \ldots).$$

(*Suggestion:* Show first that $\lambda_0 = 0$ is an eigenvalue with $X_0(x) = 1$ and $Y_0(y) = y$.)

In each of Problems 5 and 6, find a solution of Laplace's equation $u_{xx} + u_{yy} = 0$ in the rectangle $0 < x < a, 0 < y < b$ that satisfies the given boundary conditions. See Problem 4.

5. $u_y(x, 0) = u_y(x, b) = u(a, y) = 0$, $u(0, y) = g(y)$.
6. $u_x(0, y) = u_x(a, y) = u_y(x, 0) = 0$, $u(x, b) = f(x)$.

In each of Problems 7 and 8, find a solution of Laplace's equation in the semi-infinite strip $0 < x < a, y > 0$ that satisfies the given boundary conditions and the additional condition that $u(x, y)$ is bounded as $y \longrightarrow +\infty$.

7. $u(0, y) = u(a, y) = 0, u(x, 0) = f(x)$.

8. $u_x(0, y) = u_x(a, y) = 0, u(x, 0) = f(x)$.

9. Suppose that $a = 10$ and $f(x) = 10x$ in Problem 8. Show that

$$u(x, y) = 50 - \frac{400}{\pi^2} \sum_{n \text{ odd}} \frac{1}{n^2} e^{-n\pi y/10} \cos \frac{n\pi x}{10}.$$

Then compute (with two-decimal-place accuracy) the values $u(0, 10)$, $u(5, 10)$, and $u(10, 10)$.

10. The edge $x = a$ of the rectangular plate $0 < x < a, 0 < y < b$ is insulated, the edges $x = 0$ and $y = 0$ are held at temperature zero, and $u(x, b) = f(x)$. Use the odd half-multiple sine series of Problem 21 in Section 7.3 to derive a solution of the form

$$u(x, t) = \sum_{n \text{ odd}} c_n \sin \frac{n\pi x}{2a} \sinh \frac{n\pi y}{2a}.$$

Give a formula for c_n.

11. The edge $y = 0$ of the rectangular plate $0 < x < a, 0 < y < b$ is insulated, the edges $x = a$ and $y = b$ are held at temperature zero, and $u(0, y) = g(y)$. Use the odd half-multiple cosine series of Problem 22 in Section 7.3 to find $u(x, y)$.

12. A vertical cross section of a long, high wall 30 cm thick has the shape of the semi-infinite strip $0 < x < 30, y > 0$. The face $x = 0$ is held at temperature zero, while the face $x = 30$ is insulated. Given $u(x, 0) = 25$, derive the formula

$$u(x, y) = \frac{100}{\pi} \sum_{n \text{ odd}} \frac{1}{n} e^{-n\pi y/60} \sin \frac{n\pi x}{60}$$

for steady-state temperature within the wall.

Problems 13–15 deal with the semicircular plate of radius a shown in Fig. 7.28. The circular edge has a given temperature $u(a, \theta) = f(\theta)$. In each problem, derive the given series for the steady state temperature $u(r, \theta)$ satisfying the given boundary conditions along $\theta = 0$ and $\theta = \pi$, and give the formula for the coefficients c_n.

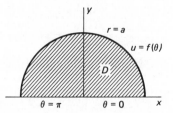

Figure 7.28 The semicircular plate of Problems 13–15.

13. $u(r, 0) = u(r, \pi) = 0$;

$$u(r, \theta) = \sum_{n=1}^{\infty} c_n r^n \sin n\theta.$$

14. $u_\theta(r, 0) = u_\theta(r, \pi) = 0$;

$$u(r, \theta) = \frac{c_0}{2} + \sum_{n=1}^{\infty} c_n r^n \cos n\theta.$$

15. $u(r, 0) = u_\theta(r, \pi) = 0$;

$$u(r, \theta) = \sum_{n \text{ odd}} c_n r^{n/2} \sin \frac{n\theta}{2}.$$

16. Consider Dirichlet's problem for the region *exterior* to the circle $r = a$. You want to find a solution of

$$r^2 u_{rr} + r u_r + u_{\theta\theta} = 0$$

such that $u(a, \theta) = f(\theta)$ and $u(r, \theta)$ is bounded as $r \to +\infty$. Derive the series

$$u(r, \theta) = \frac{a_0}{2} + \sum_{n=1}^{\infty} \frac{1}{r^n}(a_n \cos n\theta + b_n \sin n\theta),$$

and give formulas for the coefficients $\{a_n\}$ and $\{b_n\}$.

17. The velocity potential function $u(r, \theta)$ for steady flow of an ideal fluid around a cylinder of radius $r = a$ satisfies the boundary value problem

$$\begin{cases} r^2 u_{rr} + r u_r + u_{\theta\theta} = 0 & (r > a); \\ u_r(a, \theta) = 0, \quad u(r, \theta) = u(r, -\theta), \\ \lim_{r \to \infty} [u(r, \theta) - U_0 r \cos \theta] = 0. \end{cases}$$

(a) By separation of variables, derive the solution

$$u(r, \theta) = \frac{U_0}{r}(r^2 + a^2) \cos \theta.$$

(b) Hence show that the velocity components of the flow are

$$v_x = \frac{\partial u}{\partial x} = \frac{U_0}{r^2}(r^2 - a^2 \cos 2\theta)$$

and

$$v_y = \frac{\partial u}{\partial y} = -\frac{U_0}{r^2} a^2 \sin 2\theta.$$

The streamlines for this fluid flow around the cylinder are shown in Fig. 7.29.

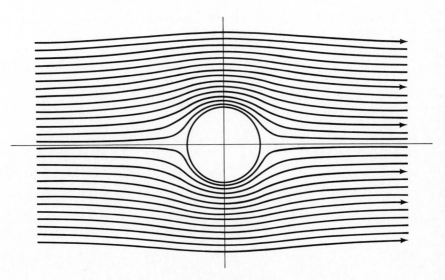

Figure 7.29 Streamlines for ideal fluid flow around a cylinder.

Comment: The streamlines in Fig. 7.29 are the level curves of the function $\psi(x, y)$ given in Problem 18(b). It is troublesome to write a computer program to plot such level curves. It is far easier to plot some solutions of the differential equation $dy/dx = v_y/v_x$. We did so with initial conditions $x_0 = -5$, $y_0 = -2.7$, -2.5, -2.3, ..., 2.5, 2.7. The solutions were obtained numerically using the improved Euler method (see Section 6.2) with step size 0.02.

18. (a) Show that the velocity potential u in Problem 17(a) can be written in rectangular coordinates as

$$u(x, y) = U_0 x \left(1 + \frac{a^2}{x^2 + y^2}\right).$$

(b) The stream function for the flow is

$$\psi(x, y) = U_0 y \left(\frac{a^2}{x^2 + y^2} - 1\right).$$

Show that $\nabla u \cdot \nabla \psi \equiv 0$. Because $\mathbf{v} = \nabla u$ is the velocity vector, this shows that the streamlines of the flow are the level curves of $\psi(x, y)$.

Problems 19–21 deal with a solid spherical ball of radius $r = a$ with initial temperature T_0 throughout. At time $t = 0$ it is packed in ice, so its temperature at the surface $r = a$ is zero thereafter. Its temperature then depends only upon time t and the distance r from the center of the ball, so we write $u = u(r, t)$.

19. (a) The heat content of the spherical shell with inner and outer radii r and $r + \Delta r$ is

$$Q(t) = \int_r^{r+\Delta r} c\delta u(s, t) \cdot 4\pi s^2 \, ds.$$

Show that $Q'(t) = 4\pi c\delta r^2 u_t(\bar{r}, t)$ for some $\bar{r}$ in $(r, r + \Delta r)$. (b) The radial heat flux is $\phi = -Ku_r$ across the bounding spherical surfaces of the shell in part (a). Conclude that

$$Q'(t) = 4\pi K[(r + \Delta r)^2 u_r(r + \Delta r, t) - r^2 u_r(r, t)].$$

(c) Equate the values of $Q'(t)$ in parts (a) and (b); then take the limit as $\Delta r \to 0$ to get the equation

$$\frac{\partial u}{\partial t} = \frac{k}{r^2} \frac{\partial}{\partial r}\left(r^2 \frac{\partial u}{\partial r}\right).$$

(d) Finally show that this last equation is equivalent to

$$\frac{\partial}{\partial t}(ru) = k \frac{\partial^2}{\partial r^2}(ru).$$

20. It follows from Problem 19(d) that $u(r, t)$ satisfies the boundary value problem

$$\begin{cases} \dfrac{\partial}{\partial t}(ru) = k \dfrac{\partial^2}{\partial r^2}(ru) & (r < a); \\[2mm] u(a, t) = 0, \quad u(r, 0) = T_0. \end{cases}$$

Show that the new function $v(r, t) = ru(r, t)$ satisfies a familiar boundary value problem (in Section 7.5), and thus derive the solution

$$u(r, t) = \frac{2aT_0}{\pi} \sum_{n=1}^{\infty} \frac{(-1)^{n+1}}{nr} e^{-n^2\pi^2 kt/a^2} \sin \frac{n\pi r}{a}.$$

21. (a) Deduce from the solution to Problem 20 that the temperature at the center

of the ball at time t is

$$u(0, t) = 2T_0 \sum_{n=1}^{\infty} (-1)^{n+1} e^{-n^2\pi^2 kt/a^2}.$$

(b) Let $a = 30$ cm and $T_0 = 100°C$. Compute $u(0, t)$ after 15 min if the ball is made of iron with $k = 0.15$ cgs. (The answer is approximately 45°C.) (c) If you have access to a programmable calculator, repeat part (b) for a ball made of concrete with $k = 0.005$ in cgs units. About 15 terms are required to show that $u = 100.00°C$ accurate to two decimal places.

22. In the discussion of the Dirichlet problem for a circular disk in this section, we obtained the ordinary differential equation $\Theta'' + \lambda\Theta = 0$ with the periodicity condition $\Theta(\theta) = \Theta(\theta + 2\pi)$. (a) Suppose that $\lambda = \alpha^2 > 0$. Show that the general solution $\Theta = A \cos \alpha\theta + B \sin \alpha\theta$ has period 2π only if $\lambda = n$, an integer. (b) In the case $\lambda = 0$, show that the general solution $\Theta = A\theta + B$ is periodic only if $A = 0$.

23. If $\lambda = -\alpha^2 < 0$ in Problem 22, then the general solution of the differential equation is $\Theta = Ae^{\alpha\theta} + Be^{-\alpha\theta}$. Show that this function is not periodic unless $A = B = 0$.

Eigenvalues and Boundary Value Problems

8

In the last three sections of Chapter 7, we saw that numerous different boundary value problems all lead—by separation of variables—to the same ordinary differential equation

$$X''(x) + \lambda X(x) = 0 \qquad (0 < x < L), \tag{1}$$

containing an eigenvalue λ, but with different endpoint conditions, such as

$$X(0) = X(L) = 0 \tag{2}$$

or

$$X'(0) = X'(L) = 0 \tag{3}$$

or

$$X(0) = X'(L) = 0, \tag{4}$$

depending on the original boundary conditions. For example, recall the problem (in Section 7.5) of finding the temperature $u(x, t)$ in a rod $0 \leq x \leq L$ with given initial temperature $u(x, 0) = f(x)$. As a boundary value problem, the problem of the rod is the same as the problem of finding the temperature within a large slab that occupies the region $0 \leq x \leq L$ in xyz-space. If its initial temperature depends only on x and is independent of y and z—that is, if $u(x, 0) = f(x)$—then the same will be true of its temperature $u = u(x, t)$ at time t. When we substitute $u(x, t) = X(x)T(t)$ in the heat equation $u_t = ku_{xx}$, we find that $X(x)$ satisfies the endpoint conditions in (2) if the faces $x = 0$ and $x = L$ of the slab are held at temperature zero, those in (3) if both faces are insulated, and those in (4) if one face is insulated and the other is held at temperature zero. But if each face loses heat to a surrounding medium (at temperature zero) in accordance with Newton's law of cooling, then (according to Problem 23 in Section 7.5) the endpoint conditions take the form

$$hX(0) - X'(0) = 0 = hX(L) + X'(L) \tag{5}$$

where h is a nonnegative heat transfer coefficient.

The point is that when we impose different endpoint conditions on the solutions of Eq. (1), we get different eigenvalue problems, and hence different eigenvalues $\{\lambda_n\}$ and different eigenfunctions $\{X_n(x)\}$ to use in constructing a formal power series solution

$$u(x, t) = \sum c_n X_n(x) T_n(t) \tag{6}$$

of the boundary value problem. The final step in this construction is to choose the coefficients in (6) so that

$$u(x, 0) = \sum c_n T_n(0) X_n(x) = f(x). \tag{7}$$

Thus we need finally an *eigenfunction expansion* of the given function $f(x)$ in terms of the eigenfunctions of the pertinent endpoint value problem.

In order to unify and generalize the method of separation of variables, it is useful to formulate a general type of eigenvalue problem that includes as special cases each of those mentioned above. Equation (1), with y instead of X as the dependent variable, can be written in the form

$$\frac{d}{dx}\left[p(x)\frac{dy}{dx}\right] - q(x)y + \lambda r(x)y = 0, \tag{8}$$

where $p(x) = r(x) \equiv 1$ and $q(x) \equiv 0$. Indeed, we indicate in Problem 16 that essentially any linear second order differential equation of the form

$$A(x)y'' + B(x)y' + C(x)y + \lambda D(x)y = 0$$

takes the form in (8) after multiplication by a suitable factor. For instance, if we multiply the parametric Bessel equation

$$x^2 y'' + xy' + (\lambda x^2 - n^2)y = 0 \qquad (x > 0)$$

of order n by $1/x$, the result may be written as

$$[xy']' - \frac{n^2}{x}y + \lambda xy = 0,$$

which is of the form in (8) with $p(x) = r(x) = x$ and $q(x) = n^2/x$.

Now, on solutions of Eq. (8) in a bounded open interval (a, b), let us impose homogeneous (linear) endpoint conditions of the form

$$\alpha_1 y(a) - \alpha_2 y'(a) = 0, \qquad \beta_1 y(b) + \beta_2 y'(b) = 0, \tag{9}$$

where the coefficients α_1, α_2, β_1, and β_2 are constants. In addition to being homogeneous, these two conditions are *separated*, in that one involves the values of $y(x)$ and $y'(x)$ at one endpoint $x = a$, while the other involves values at the other endpoint $x = b$. For instance, note that the conditions $y(a) = y'(b) = 0$ are of the form in (9) with $\alpha_1 = \beta_2 = 1$ and $\alpha_2 = \beta_1 = 0$.

An eigenvalue problem of the form

$$[p(x)y']' - q(x)y + \lambda r(x)y = 0 \qquad (a < x < b);$$ (8)

$$\alpha_1 y(a) - \alpha_2 y'(a) = 0, \qquad \beta_1 y(b) + \beta_2 y'(b) = 0,$$ (9)

(with neither α_1 and α_2 nor β_1 and β_2 both zero) is called a **Sturm-Liouville problem,** named for the French mathematicians Charles Sturm (1803–1855) and Joseph Liouville (1809–1882), who investigated this type of problem in the 1830s. Note that the problem in (8)–(9) always has the trivial solution $y(x) \equiv 0$. We seek the values of λ—the **eigenvalues**—for which this problem has a non-trivial real-valued solution (an **eigenfunction**), and for each eigenvalue its associated eigenfunction (or eigenfunctions). Note that any constant (nonzero) multiple of an eigenfunction will also be an eigenfunction. The following theorem provides sufficient conditions under which the problem in (8)–(9) has an infinite sequence $\{\lambda_n\}_1^\infty$ of nonnegative eigenvalues, with each eigenvalue λ_n having (to within a constant multiple) exactly one associated eigenfunction $y_n(x)$. A proof of this theorem is outlined in Chapter 9 of G. P. Tolstov, *Fourier Series* (New York: Dover, 1976).

THEOREM 1: STURM-LIOUVILLE EIGENVALUES

Suppose that the functions $p(x), p'(x), q(x),$ and $r(x)$ in Eq. (8) are continuous on the interval $[a, b]$ and that $p(x) > 0$ and $r(x) > 0$ at each point of $[a, b]$. Then the eigenvalues of the Sturm-Liouville problem in (8)–(9) constitute an increasing infinite sequence

$$\lambda_1 < \lambda_2 < \lambda_3 < \cdots < \lambda_{n-1} < \lambda_n < \cdots$$ (10)

of real numbers with

$$\lim_{n \to \infty} \lambda_n = +\infty.$$ (11)

To within a constant factor, only a single eigenfunction $y_n(x)$ is associated with each eigenvalue λ_n. Moreover, if $q(x) \geqq 0$ on $[a, b]$ and the coefficients $\alpha_1, \alpha_2, \beta_1,$ and β_2 in (9) are all nonnegative, then the eigenvalues are all nonnegative.

Note: It is important to observe the signs in (8) and (9) when verifying the hypotheses of Theorem 1. Sometimes the Sturm-Liouville problem in (8)–(9) is called **regular** if it satisfies the hypotheses of Theorem 1; otherwise it is **singular.** In this section we will confine our attention to regular Sturm-Liouville problems. Singular Sturm-Liouville problems associated with Bessel's equation will appear in Section 8.4.

EXAMPLE 1 In Example 3 of Section 2.10, we saw that the Sturm-Liouville problem

$$y'' + \lambda y = 0 \qquad (0 < x < L);$$
$$y(0) = 0, \qquad y(L) = 0$$ (12)

has eigenvalues $\lambda_n = n^2\pi^2/L^2$ and associated eigenfunctions $y_n(x) = \sin(n\pi x/L)$ $(n = 1, 2, 3, \ldots)$. There we had to consider the cases $\lambda =$

$-\alpha^2 < 0$, $\lambda = 0$, and $\lambda = \alpha^2 > 0$ separately. Here $p(x) = r(x) \equiv 1$, $q(x) \equiv 0$, $\alpha_1 = \beta_1 = 1$, and $\alpha_2 = \beta_2 = 0$, so we know from Theorem 1 that the problem in (12) has only nonnegative eigenvalues. Hence only the two cases $\lambda = 0$ and $\lambda = \alpha^2$ would need to be considered if we were starting afresh.

We are also familiar (from the problems in Section 7.5) with the Sturm-Liouville problem

$$y'' + \lambda y = 0 \qquad (0 < x < L);$$
$$y'(0) = 0, \qquad y'(L) = 0, \tag{13}$$

which has eigenvalues $\lambda_0 = 0$, $\lambda_n = n^2\pi^2/L^2$ $(n = 1, 2, 3, \ldots)$ and associated eigenfunctions $y_0(x) \equiv 1$, $y_n(x) = \cos(n\pi x/L)$. We will customarily write $\lambda_0 = 0$ if 0 is an eigenvalue and otherwise write λ_1 for the smallest eigenvalue; thus λ_1 always denotes the smallest positive eigenvalue.

EXAMPLE 2 Find the eigenvalues and associated eigenfunctions of the Sturm-Liouville problem

$$y'' + \lambda y = 0 \qquad (0 < x < L);$$
$$y'(0) = 0, \qquad y(L) = 0. \tag{14}$$

Solution This is a Sturm-Liouville problem satisfying the hypotheses of Theorem 1 with $\alpha_1 = \beta_2 = 0$ and $\alpha_2 = \beta_1 = 1$, so there are no negative eigenvalues. If $\lambda = 0$, then $y(x) = Ax + B$, and thus $y'(0) = A = 0$ and $y(L) = B = 0$. Therefore 0 is not an eigenvalue. If $\lambda = \alpha^2$, then $y(x) = A \cos \alpha x + B \sin \alpha x$ and $y'(x) = -A\alpha \sin \alpha x + B\alpha \cos \alpha x$. Hence $y'(0) = 0$ implies that $B = 0$, and then $y(L) = A \cos \alpha L = 0$, so it follows that αL is an *odd* multiple of $\pi/2$: $\alpha L = (2n - 1)\pi/2$. Thus we have eigenvalues and associated eigenfunctions

$$\lambda_n = {\alpha_n}^2 = \frac{(2n - 1)^2\pi^2}{4L^2}$$

and

$$y_n(x) = \cos\frac{(2n - 1)\pi}{2L}$$

for $n = 1, 2, 3, \ldots$.

EXAMPLE 3 Find the eigenvalues and associated eigenfunctions of the Sturm-Liouville problem

$$y'' + \lambda y = 0 \qquad (0 < x < L);$$
$$y(0) = 0, \qquad hy(L) + y'(L) = 0 \qquad (h > 0). \tag{15}$$

Solution This problem satisfies the hypotheses of Theorem 1 with $\alpha_1 = \beta_2 = 1$, $\alpha_2 = 0$, and $\beta_1 = h > 0$, so there are no negative eigenvalues. If $\lambda = 0$, then $y(x) = Ax + B$, thus $y(0) = B = 0$. Then

$$hy(L) + y'(L) = h(AL) + A = A(hL + 1) = 0,$$

and it follows that $A = 0$ as well. Thus 0 is not an eigenvalue.

If $\lambda = \alpha^2$, then $y(x) = A \cos \alpha x + B \sin \alpha x$. Hence $y(0) = A = 0$, so $y(x) = B \sin \alpha x$ and $y'(x) = B\alpha \cos \alpha x$. Therefore

$$hy(L) + y'(L) = hB \sin \alpha L + B\alpha \cos \alpha L = 0.$$

If $B \neq 0$, it follows that

$$\tan \alpha L = -\frac{\alpha}{h} = -\frac{\alpha L}{hL}. \tag{16}$$

Thus $\beta = \alpha L$ is a (positive) solution of the equation

$$\tan \beta = -\frac{\beta}{hL}. \tag{17}$$

The solutions of (17) are the points of intersection of the graphs of $y = -\tan x$ and $y = x/hL$, as indicated in Fig. 8.1. It is apparent from the figure that there is an infinite sequence of positive roots $\beta_1, \beta_2, \beta_3, \ldots$, and that when n is large, β_n is only slightly larger than $(2n - 1)\pi/2$. See the table in Fig. 8.2, in which the first eight solutions of Eq. (17) are listed for the case $hL = 1$. At any rate, the eigenvalues and associated

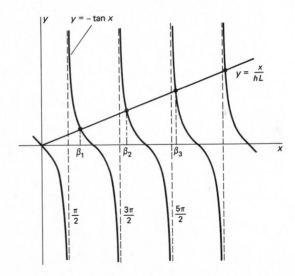

Figure 8.1 The solutions of Equation (17) in geometric form.

n	x_n	$(2n-1)\pi/2$
1	2.0288	1.5708
2	4.9132	4.7124
3	7.9787	7.8540
4	11.0855	10.9956
5	14.2074	14.1372
6	17.3364	17.2788
7	20.4692	20.4204
8	23.6043	23.5619

Figure 8.2 Approximate values of the first eight solutions of $\tan x = -x$.

eigenfunctions of the problem in (15) are given by

$$\lambda_n = \alpha_n^2 = \frac{\beta_n^2}{L^2}, \qquad y_n(x) = \sin \alpha_n x = \sin \frac{\beta_n x}{L} \qquad (18)$$

for $n = 1, 2, 3, \ldots$. Equation (17) appears frequently in certain applications (mechanical vibrations and heat conduction are only two of many examples), and its solutions for various values of hL are tabulated in Table 4.19 of Abramowitz and Stegun, *Handbook of Mathematical Functions* (New York: Dover, 1965).

EIGENVALUE EXPANSIONS

Recall from Section 7.1 that the familiar formulas for the coefficients in a Fourier series follow from the orthogonality of the sine and cosine functions. Similarly, the expansion of a given function in terms of the eigenfunctions of a Sturm-Liouville problem depends upon a crucial orthogonality property of these eigenfunctions. The functions $\phi(x)$ and $\psi(x)$ are said to be **orthogonal** on the interval $[a, b]$ with respect to the **weight function** $r(x)$ provided that

$$\int_a^b \phi(x)\psi(x)r(x)\, dx = 0. \qquad (19)$$

The following theorem implies that any two eigenfunctions of a regular Sturm-Liouville problem that are associated with distinct eigenvalues are orthogonal with respect to the weight function $r(x)$ in Eq. (8).

THEOREM 2: ORTHOGONALITY OF EIGENFUNCTIONS

Suppose that the functions p, q, and r in the Sturm-Liouville problem in (8)–(9) satisfy the hypotheses of Theorem 1, and let $y_i(x)$ and $y_j(x)$ be eigenfunctions associated with the distinct eigenvalues λ_i and λ_j. Then

$$\int_a^b y_i(x)y_j(x)r(x)\, dx = 0. \qquad (20)$$

Proof We begin with the equations

$$
\begin{aligned}
[p(x)y_i']' - q(x)y_i + \lambda_i r(x)y_i = 0, \\
[p(x)y_j']' - q(x)y_j + \lambda_j r(x)y_j = 0.
\end{aligned}
\qquad (21)
$$

These equations imply that λ_i, y_i and λ_j, y_j are eigenvalue-eigenfunction pairs. If we multiply the first by y_j and the second by y_i, then subtract the results, we get

$$y_j[p(x)y_i']' - y_i[p(x)y_j']' + (\lambda_i - \lambda_j)r(x)y_iy_j = 0.$$

Hence

$$(\lambda_i - \lambda_j)y_iy_j r(x) = y_i[p(x)y_j']' - y_j[p(x)y_i']' = \frac{d}{dx}[p(x)(y_iy_j' - y_jy_i')],$$

the latter of the equalities being verifiable by direct differentiation. Therefore integration from $x = a$ to $x = b$ yields

$$(\lambda_i - \lambda_j)\int_a^b y_i(x)y_j(x)r(x)\, dx = \left[p(x)(y_i(x)y_j'(x) - y_j(x)y_i'(x)) \right]_a^b. \qquad (22)$$

From the endpoint condition in (9), we have

$$\alpha_1 y_i(a) - \alpha_2 y_i'(a) = 0$$

and

$$\alpha_1 y_j(a) - \alpha_2 y_j'(a) = 0.$$

Because α_1 and α_2 are not both zero, it follows that the determinant of coefficients must be zero:

$$y_i(a)y_j'(a) - y_j(a)y_i'(a) = 0.$$

Similarly, the second endpoint condition in (9) implies that

$$y_i(b)y_j'(b) - y_j(b)y_i'(b) = 0.$$

Thus the right-hand side in Eq. (22) vanishes. Inasmuch as $\lambda_i \neq \lambda_j$, the result in (20) follows, and the proof is complete.

Now suppose that the function $f(x)$ can be represented in the interval $[a, b]$ by an **eigenfunction series**

$$f(x) = \sum_{m=1}^{\infty} c_m y_m(x), \tag{23}$$

where the functions $y_1, y_2, y_3, \ldots$ are the eigenfunctions of the regular Sturm-Liouville problem in Eqs. (8)–(9). To determine the coefficients $c_1, c_2, c_3, \ldots$, we generalize the technique by which we determined the ordinary Fourier coefficients in Section 7.1. First we multiply each side in (23) by $y_n(x)r(x)$, then integrate from $x = a$ to $x = b$. Under the assumption that termwise integration is valid, we obtain

$$\int_a^b f(x)y_n(x)r(x)\,dx = \sum_{m=1}^{\infty} c_m \int_a^b y_m(x)y_n(x)r(x)\,dx. \tag{24}$$

But because of the orthogonality in (20), the only nonzero term on the right-hand side in (24) is the one for which $m = n$. Thus (24) takes the form

$$\int_a^b y(x)y_n(x)r(x)\,dx = c_n \int_b^b [y_n(x)]^2 r(x)\,dx,$$

and therefore

$$c_n = \frac{\displaystyle\int_a^b f(x)y_n(x)r(x)\,dx}{\displaystyle\int_a^b [y_n(x)]^2 r(x)\,dx}. \tag{25}$$

We therefore *define* the eigenfunction series in (23)—representing $f(x)$ in terms of the eigenfunctions of the given Sturm-Liouville problem—by means of the choice of the coefficients specified by the formula in (25).

For instance, suppose that the Sturm-Liouville problem is the familiar

$$y'' + \lambda y = 0 \qquad (0 < x < \pi);$$
$$y(0) = y(\pi) = 0, \tag{26}$$

for which $r(x) \equiv 1$ and the eigenfunctions are $y_n(x) = \sin nx$ for $n = 1, 2,$

3, ... Then the formula in (25) yields

$$c_n = \frac{\int_0^\pi f(x) \sin nx \, dx}{\int_0^\pi \sin^2 nx \, dx} = \frac{2}{\pi} \int_0^\pi f(x) \sin nx \, dx,$$

because $\int_0^\pi \sin^2 nx \, dx = \pi/2$. This is the familiar formula for the Fourier sine coefficients, and so the eigenfunction series in (23) is simply the familiar Fourier sine series $f(x) = \sum c_n \sin nx$ of $f(x)$ on $[0, \pi]$.

The following theorem, stated without proof, generalizes the Fourier convergence theorem of Section 7.2.

THEOREM 3 : CONVERGENCE OF EIGENFUNCTION SERIES

Let $y_1, y_2, y_3, \ldots$ be the eigenfunctions of a regular Sturm-Liouville problem on $[a, b]$. If the function $f(x)$ is piecewise smooth on $[a, b]$, then the eigenfunction series in (23) converges for $a < x < b$ to the value $f(x)$ wherever f is continuous, and to the average value $\frac{1}{2}[f(x+) + f(x-)]$ at each point of discontinuity.

EXAMPLE 4 For the Sturm-Liouville problem $y'' + \lambda y = 0$ $(0 < x < L)$, $y'(0) = y(L) = 0$ of Example 2 we found the eigenfunctions $y_n(x) = \cos (2n - 1)\pi x/2L$, $n = 1, 2, 3, \ldots$. The corresponding eigenfunction series for a function $f(x)$ is

$$f(x) = \sum_{n=1}^\infty c_n \cos \frac{(2n - 1)\pi x}{2L} \tag{27}$$

with

$$c_n = \frac{\int_0^L f(x) \cos ([2n - 1]\pi x/2L) \, dx}{\int_0^L \cos^2 ([2n - 1]\pi x/2L) \, dx} = \frac{2}{L} \int_0^L f(x) \cos \frac{(2n - 1)\pi x}{2L} \, dx, \tag{28}$$

because $\int_0^L \cos^2 (2n - 1)\pi x/2L \, dx = L/2$. Thus (27) is the odd half-multiple cosine series of Problem 22 in Section 7.3. Similarly, the Sturm-Liouville problem $y'' + \lambda y = 0$, $y(0) = y'(L) = 0$ leads to the odd half-multiple sine series

$$f(x) = \sum_{n=1}^\infty c_n \sin \frac{(2n - 1)\pi x}{2L},$$

$$c_n = \frac{2}{L} \int_0^L f(x) \sin \frac{(2n - 1)\pi x}{2L} \, dx \tag{29}$$

of Problem 21 in Section 7.3.

EXAMPLE 5 Represent the function $f(x) = A$ (constant) for $0 < x < 1$ as a series of eigenfunctions of the Sturm-Liouville problem

$$y'' + \lambda y = 0 \qquad (0 < x < 1);$$
$$y(0) = 0, \qquad y(1) + 2y'(1) = 0. \tag{30}$$

Solution Comparing (30) with (15), we see that this is the Sturm-Liouville problem of Example 3 with $L = 1$ and $h = \frac{1}{2}$. From (17) and (18) we see that the eigenfunctions of this problem are therefore $y_n(x) = \sin \beta_n x$, where $\beta_1, \beta_2, \beta_3, \ldots$ are the positive roots of the equation $\tan x = -2x$. Hence the coefficients in the desired series are given by

$$c_n = \frac{\int_0^1 A \sin \beta_n x \, dx}{\int_0^1 \sin^2 \beta_n x \, dx}. \tag{31}$$

Now $\int_0^1 A \sin \beta_n x \, dx = A(1 - \cos \beta_n)/\beta_n$ and

$$\int_0^1 \sin^2 \beta_n x \, dx = \int_0^1 \frac{1}{2}(1 - \cos 2\beta_n x) \, dx$$

$$= \frac{1}{2}\left[x - \frac{1}{2\beta_n} \sin 2\beta_n x \right]_0^1 = \frac{1}{2}\left(1 - \frac{\sin \beta_n \cos \beta_n}{\beta_n} \right).$$

Consequently,

$$\int_0^1 \sin^2 \beta_n x \, dx = \frac{1}{2}(1 + 2\cos^2 \beta_n).$$

In the last step we used the fact from Eq. (17) that $(\sin \beta_n)/\beta_n = -2\cos \beta_n$. Substituting these values for the integrals in (31), we get the eigenfunction series

$$f(x) = \sum_{n=1}^{\infty} \frac{2A(1 - \cos \beta_n)}{\beta_n(1 + 2\cos^2 \beta_n)} \sin \beta_n x. \tag{32}$$

In summary, every regular Sturm-Liouville problem has the following three properties: It has an infinite sequence of eigenvalues diverging to infinity (Theorem 1); the eigenfunctions are orthogonal with appropriate weight function (Theorem 2); and any piecewise smooth function can be represented by an eigenfunction series (Theorem 3). There are other types of eigenvalue problems in applied mathematics that also enjoy these three important properties. Some isolated examples will appear in subsequent sections, though we will confine our attention largely to applications of Sturm-Liouville problems.

8.1 Problems

The problems for Section 2.10 deal with eigenvalues and eigenfunctions and may be used here also. In Problems 1–5 below, verify that the eigenvalues and eigenfunctions for the indicated Sturm-Liouville problem are those listed.

1. $y'' + \lambda y = 0$, $y'(0) = y'(L) = 0$; $\lambda_0 = 0$, $y_0(x) = 1$ and $\lambda_n = n^2\pi^2/L^2$, $y_n(x) = \cos n\pi x/L$.

2. $y'' + \lambda y = 0$, $y(0) = y'(L) = 0$; $\lambda_n = (2n - 1)^2\pi^2/4L^2$, $y_n(x) = \sin (2n - 1)\pi x/2L$, $n \geq 1$.

3. $y'' + \lambda y = 0$, $y'(0) = hy(L) + y'(L) = 0$ $(h > 0)$; $\lambda_n = \beta_n^2/L^2$, $y_n(x) = \cos \beta_n x/L$ $(n \geq 1)$ where β_n is the nth positive root of $\tan x = hL/x$. Sketch $y = \tan x$ and $y = hL/x$ to estimate the value of β_n for n large.

4. $y'' + \lambda y = 0$, $hy(0) - y'(0) = y(L) = 0$ $(h > 0)$; $\quad \lambda_n = \beta_n^2/L^2$, $y_n(x) = \beta_n \cos \beta_n x/L + hL \sin \beta_n x/L$ $(n \geq 1)$ where β_n is the nth positive root of $\tan x = -x/hL$.

5. $y'' + \lambda y = 0$, $hy(0) - y'(0) = hy(L) + y'(L) = 0$ $(h > 0)$; $\quad \lambda_n = \beta_n^2/L^2$, $y_n(x) = \beta_n \cos \beta_n x/L + hL \sin \beta_n x/L$ $(n \geq 1)$ where β_n is the nth positive root of $\tan x = 2hLx/(x^2 - h^2L^2)$. Estimate β_n for n large by sketching the graphs of $y = 2hL \cot x$ and the hyperbola $y = (x^2 - h^2L^2)/x$.

6. Show that the Sturm-Liouville problem $y'' + \lambda y = 0$, $y(0) = y'(L) = 0$ leads to the odd half-multiple sine series in (29). See Problem 2.

In each of Problems 7–10, represent the given function $f(x)$ as a series of eigenfunctions of the indicated Sturm-Liouville problem.

7. $f(x) = 1$; the Sturm-Liouville problem of Example 3.
8. $f(x) = 1$; the Sturm-Liouville problem of Problem 3.
9. $f(x) = x$; the Sturm-Liouville problem of Example 3 with $L = 1$.
10. $f(x) = x$; the Sturm-Liouville problem of Problem 3 with $L = 1$.

Problems 11–14 deal with the regular Sturm-Liouville problem

$$y'' + \lambda y = 0 \quad (0 < x < L);$$
$$y(0) = 0, \quad hy(L) - y'(L) = 0 \tag{33}$$

where $h > 0$. Note that Theorem 1 does not exclude the possibility of negative eigenvalues.

11. Show that $\lambda_0 = 0$ is an eigenvalue if and only if $hL = 1$, in which case the associated eigenfunction is $y_0(x) = x$.

12. Show that the problem in (33) has a single negative eigenvalue λ_0 if and only if $hL > 1$, in which case $\lambda_0 = -\beta_0^2/L^2$ and $y_0(x) = \sinh \beta_0 x/L$, where β_0 is the positive root of the equation $\tanh x = x/hL$. (*Suggestion:* Sketch the graphs $y = \tanh x$ and $y = x/hL$.)

13. Show that the positive eigenvalues and associated eigenfunctions of the problem in (33) are $\lambda_n = \beta_n^2/L^2$ and $y_n(x) = \sin \beta_n x/L$ $(n \geq 1)$, where β_n is the nth positive root of $\tan x = x/hL$.

14. Suppose that $hL = 1$ in (33) and that $f(x)$ is piecewise smooth. Then show that

$$f(x) = c_0 x + \sum_{n=1}^{\infty} c_n \sin \frac{\beta_n x}{L},$$

where $\{\beta_n\}_1^{\infty}$ are the positive roots of $\tan x = x$, and

$$c_0 = \frac{3}{L^3} \int_0^L xf(x)\, dx,$$

$$c_n = \frac{2}{L \sin^2 \beta_n} \int_0^L f(x) \sin \frac{\beta_n x}{L}\, dx.$$

15. Show that the eigenvalues and eigenfunctions of the Sturm-Liouville problem

$$y'' + \lambda y = 0 \quad (0 < x < 1);$$
$$y(0) + y'(0) = 0, \quad y(1) = 0$$

are given by $\lambda_0 = 0$, $y_0(x) = x - 1$ and $\lambda_n = \beta_n^2$, $y_n(x) = \beta_n \cos \beta_n x -$ $\sin \beta_n x$ for $n \geq 1$, where $\{\beta_n\}_1^\infty$ are the positive roots of $\tan x = x$.

16. Starting with the equation

$$A(x)y'' + B(x)y' + C(x)y + \lambda D(x)y = 0,$$

first divide by $A(x)$ and then multiply by

$$p(x) = \exp\left(\int \frac{B(x)}{A(x)}\,dx\right).$$

Show that the resulting equation can be written in the Sturm-Liouville form $[p(x)y']' - q(x)y + \lambda r(x)y = 0$ with $q(x) = -p(x)C(x)/A(x)$ and $r(x) = p(x)D(x)/A(x)$.

Loaded Uniform Beam

Consider a uniform beam with (downward) load $w(x)$, whose deflection function $y(x)$ satisfies the fourth order equation $EIy^{(4)} = w(x)$ for $0 < x < L$ and the endpoint conditions $y = y'' = 0$ at a hinged (or simply supported) end, $y = y' = 0$ at a fixed (built-in) end, $y'' = y''' = 0$ at a free end. In the hinged/hinged case (both ends hinged), $y(x)$ can be found by the Fourier series method discussed in Section 7.4—that is, by substituting the Fourier series $y(x) = \sum b_n \sin n\pi x/L$ in the differential equation

$$EIy^{(4)} = w(x) = \sum_{n=1}^{\infty} c_n \sin \frac{n\pi x}{L} \tag{34}$$

(where c_n is the nth Fourier sine coefficient of $w(x)$) to determine the coefficients $\{b_n\}$.

17. Suppose that w is constant in Eq. (34). Apply the method described above to obtain the deflection function

$$y(x) = \frac{4wL^4}{EI\pi^5} \sum_{n \text{ odd}} \frac{1}{n^5} \sin \frac{n\pi x}{L}.$$

18. Suppose that $w = bx$ in (34). Derive the deflection function

$$y(x) = \frac{2bL^5}{EI\pi^5} \sum_{n=1}^{\infty} \frac{(-1)^{n+1}}{n^5} \sin \frac{n\pi x}{L}.$$

The method used in Problems 17 and 18 succeeds because the functions $\sin n\pi x/L$ satisfy the hinged/hinged conditions $y(0) = y''(0) = y(L) = y''(L) = 0$, so that $y(x)$ does also.

If, instead, both ends of the beam are fixed, in place of the sine functions we can use the eigenfunctions of the problem

$$\begin{aligned} y^{(4)} - \lambda y = 0 \quad &(0 < x < L); \\ y(0) = y'(0) = 0, \quad y(L) = y'(L) = 0, \end{aligned} \tag{35}$$

because these eigenfunctions satisfy the fixed/fixed endpoint conditions. The eigenvalues of this problem are all positive, and by Problem 22 below the associated eigenfunctions are orthogonal with weight function $r(x) \equiv 1$. Hence we can write

$$w(x) = \sum_{n=1}^{\infty} c_n y_n(x), \qquad c_n = \frac{\int_0^L w(x) y_n(x)\, dx}{\int_0^L [y_n(x)]^2\, dx}, \tag{36}$$

according to the analogue of Theorem 3 that holds for the problem in (35). If we write $\lambda = \alpha^4$, then $y_n(x)$ is of the form

$$y(x) = A \cosh \alpha x + B \sinh \alpha x + C \cos \alpha x + D \sin \alpha x, \tag{37}$$

with $\alpha = \alpha_n$, so it follows that $y_n^{(4)}(x) = \alpha_n^4 y_n(x)$. When we substitute the series $y(x) = \sum b_n y_n(x)$—which evidently satisfies the fixed/fixed endpoint conditions —in (35), we obtain

$$EI \sum_{n=1}^{\infty} b_n \alpha_n^4 y_n(x) = \sum_{n=1}^{\infty} c_n y_n(x).$$

Hence $EIb_n \alpha_n^4 = c_n$, so the deflection function of the beam is

$$y(x) = \sum_{n=1}^{\infty} \frac{c_n}{EI\alpha_n^4} y_n(x). \tag{38}$$

The following problems deal with the eigenvalues and eigenfunctions of the problem in (35) and similar problems.

19. Begin with the general solution in (37) of $y^{(4)} - \alpha^4 y = 0$. First note that $y(0) = y'(0) = 0$ implies that $C = -A$ and $D = -B$. Then impose the conditions $y(L) = y'(L) = 0$ to get two homogeneous linear equations in A and B. Hence the determinant of coefficients of A and B must vanish; deduce from this that $\cosh \alpha L \cos \alpha L = 1$. Conclude that the nth eigenvalue is $\lambda_n = \beta_n^4/L^4$ where $\{\beta_n\}_1^\infty$ are the positive roots of $\cosh x \cos x = 1$ (see Fig. 8.3). Finally, show that an associated eigenfunction is

$$y_n = (\sinh \beta_n - \sin \beta_n)\left(\cosh \frac{\beta_n x}{L} - \cos \frac{\beta_n x}{L}\right)$$

$$- (\cosh \beta_n - \cos \beta_n)\left(\sinh \frac{\beta_n x}{L} - \sin \frac{\beta_n x}{L}\right).$$

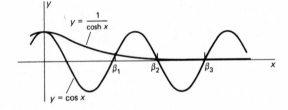

Figure 8.3 The solutions of $\cosh x \cos x = 1$ (the graph is not drawn to scale).

20. For the case of a cantilever (fixed/free) beam, we need to solve the eigenvalue problem

$$y^{(4)} - \lambda y = 0 \qquad (0 < x < L);$$

$$y(0) = y'(0) = 0, \qquad y''(L) = y'''(L) = 0.$$

Proceeding as in Problem 19, show that the nth eigenvalue is $\lambda_n = \beta_n^4/L^4$, where $\{\beta_n\}_1^\infty$ are the positive roots of $\cosh x \cos x = -1$. Then find the associated eigenfunction.

21. For the eigenvalue problem $y^{(4)} - \lambda y = 0$, $y(0) = y'(0) = 0 = y(L) = y''(L)$

corresponding to a fixed/hinged beam, show that the nth eigenvalue is $\lambda_n = \beta_n^4/L^4$, where $\{\beta_n\}_1^\infty$ are the positive roots of the equation $\tanh x = \tan x$.

22. Suppose that $y_m^{(4)} = \lambda_m y_m$ and $y_n^{(4)} = \lambda_n y_n$. Apply the method of proof of Theorem 2 and integrate by parts twice to show that

$$(\lambda_m - \lambda_n)\int_0^L y_m y_n\,dx = \left[y_n y_m''' - y_m y_n''' - y_n' y_m'' + y_m' y_n'' \right]_0^L.$$

Conclude that if each endpoint condition is either $y = y' = 0$, $y = y'' = 0$, or $y'' = y''' = 0$, then y_m and y_n are orthogonal if $\lambda_m \neq \lambda_n$.

8.2
Applications of Eigenfunction Series

This section is devoted to three examples that illustrate the application of the eigenfunction series of Section 8.1 to boundary value problems. In each, the method of separation of variables leads to a Sturm-Liouville problem in which the eigenfunctions are used as building blocks to construct a solution satisfying the nonhomogeneous boundary condition in the original problem.

EXAMPLE 1 A uniform slab of material with thermal diffusivity k occupies the region $0 \leq x \leq L$ and initially has temperature U_0 throughout. Beginning at time $t = 0$, the face $x = 0$ is held at temperature zero; at the face $x = L$, heat exchange takes place with a surrounding medium at temperature zero, so $hu + u_x = 0$ there (by Problem 23 in Section 7.5; the constant h is an appropriate heat transfer coefficient). We want to find the temperature $u(x, t)$ of the slab at position x at time t; $u(x, t)$ satisfies the boundary value problem

$$u_t = ku_{xx} \qquad (0 < x < L, \qquad t > 0); \qquad (1)$$
$$u(0, t) = 0, \qquad (2)$$
$$hu(L, t) + u_x(L, t) = 0, \qquad (3)$$
$$u(x, 0) = U_0. \qquad (4)$$

Solution As in Section 7.5, we begin by substituting $u(x, t) = X(x)T(t)$ in (1) and thereby obtain

$$\frac{X''}{X} = \frac{T'}{kT} = -\lambda$$

as usual, so

$$X'' + \lambda X = T' + \lambda kT = 0. \qquad (5)$$

Now (2) gives $X(0) = 0$, and (3) yields $hX(L)T(t) + X'(L)T(t) = 0$. We naturally assume that $T(t)$ is not identically zero (because we are not looking for a trivial solution); it follows that $X(x)$ and λ satisfy the Sturm-Liouville problem

$$X'' + \lambda X = 0 \qquad (0 < x < L);$$
$$X(0) = 0, \qquad hX(L) + X'(L) = 0. \qquad (6)$$

In Example 3 of Section 8.1, we found that the eigenvalues and associated eigenfunctions of this problem are

$$\lambda_n = \frac{\beta_n^2}{L^2}, \qquad X_n(x) = \sin\frac{\beta_n x}{L} \tag{7}$$

for $n = 1, 2, 3, \ldots$, where β_n denotes the nth positive root of the equation

$$\tan x = -\frac{x}{hL}. \tag{8}$$

When we substitute $\lambda = \beta_n^2/L^2$ in the right-hand equation in (5), we get the first order equation

$$T_n' = -\frac{\beta_n^2 k}{L^2} T_n$$

with solution (to within a multiplicative constant)

$$T_n(t) = e^{-\beta_n^2 kt/L^2}. \tag{9}$$

Consequently, each of the functions

$$u_n(x, t) = X_n(x)T_n(t) = e^{-\beta_n^2 kt/L^2} \sin\frac{\beta_n x}{L}$$

satisfies the homogeneous conditions in (1)–(3) of our problem. It remains only for us to choose the coefficients so that the formal series

$$u(x, t) = \sum_{n=1}^{\infty} c_n e^{-\beta_n^2 kt/L^2} \sin\frac{\beta_n x}{L} \tag{10}$$

also satisfies the nonhomogeneous condition

$$u(x, 0) = \sum_{n=1}^{\infty} c_n \sin\frac{\beta_n x}{L} = U_0. \tag{11}$$

Now $r(x) \equiv 1$ in the Sturm-Liouville problem in (6), so by Theorem 3 and Eq. (25) in Section 8.1, we can satisfy (11) by choosing

$$c_n = \frac{\displaystyle\int_0^L U_0 \sin(\beta_n x/L)\, dx}{\displaystyle\int_0^L \sin^2(\beta_n x/L)\, dx}.$$

Now

$$\int_0^L U_0 \sin\frac{\beta_n x}{L}\, dx = \frac{U_0 L}{\beta_n}(1 - \cos\beta_n),$$

and by essentially the same computation as in Example 5 of Section 8.1 we find that

$$\int_0^L \sin^2\frac{\beta_n x}{L}\, dx = \frac{1}{2h}(hL + \cos^2\beta_n).$$

Hence

$$c_n = \frac{2U_0 hL(1 - \cos\beta_n)}{\beta_n(hL + \cos^2\beta_n)},$$

and substitution of this value in (10) yields the formal series solution

$$u(x, t) = 2U_0 hL \sum_{n=1}^{\infty} \frac{1 - \cos\beta_n}{\beta_n(hL + \cos^2\beta_n)} e^{-\beta_n^2 kt/L^2} \sin\frac{\beta_n x}{L}. \tag{12}$$

LONGITUDINAL VIBRATIONS OF BARS

Suppose that a uniform elastic bar has length L, cross-sectional area A, and density δ (mass per unit volume) and occupies the interval $0 \leq x \leq L$ when it is unstretched. We consider longitudinal vibrations in which each cross section (normal to the x-axis) moves only in the x-direction. We can then describe the motion of the bar in terms of the displacement $u(x, t)$ at time t of the cross section whose unstretched position is x. It follows (see Problem 13) from Hooke's law and the definition of the Young's modulus E of the material of the bar that the force $F(x, t)$ exerted on the cross section x by the part of the bar to the *left* of this section is

$$F(x, t) = -AEu_x(x, t) \tag{13}$$

the minus sign signifying that F acts to the left when $u_x > 0$ (so the bar is then stretched near x rather than compressed). See Fig. 8.4.

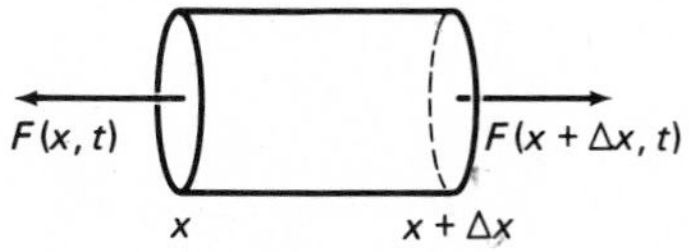

Figure 8.4 A small segment of the bar.

We take (13) as the starting point for our derivation of the partial differential equation that the displacement function $u(x, t)$ satisfies when the displacements are sufficiently small that Hooke's law may be applied. If we apply Newton's second law of motion to the segment of the bar between cross section x and cross section $x + \Delta x$, we get

$$\delta A \, \Delta x \, u_{tt}(\bar{x}, t) \approx -F(x + \Delta x, t) + F(x, t)$$
$$= AE[u_x(x + \Delta x, t) - u_x(x, t)], \tag{14}$$

where $\bar{x}$ denotes the midpoint of $[x, x + \Delta x]$, because this segment has mass $\delta A \, \Delta x$ and approximate acceleration $u_{tt}(\bar{x}, t)$. When we divide the expressions in (14) by $\delta A \, \Delta x$ and then take the limit as $\Delta x \to 0$, the result is the one-dimensional wave equation

$$\frac{\partial^2 u}{\partial t^2} = a^2 \frac{\partial^2 u}{\partial x^2} \tag{15}$$

where

$$a^2 = \frac{E}{\delta}. \tag{16}$$

Because (15) is identical to the equation of the vibrating string, it follows from our discussion of the d'Alembert solution in Section 7.6 that the (free) longitudinal vibrations of a bar with fixed ends are represented by waves of the form $u = F(x \pm at)$. The velocity $a = \sqrt{E/\delta}$ with which these waves travel is the velocity of sound in the material of the bar. Indeed, the wave equation in (15) also describes ordinary one-dimensional sound waves in a gas in a pipe. In this case, E in (16) denotes the bulk modulus (fractional increase in density per unit increase in pressure) of the gas and δ is its equilibrium density.

EXAMPLE 2 A bar has length L, density δ, cross-sectional area A, Young's modulus E, and mass $M = \delta AL$. Its end $x = 0$ is fixed, while a mass m is attached to its free end (see Fig. 8.5). The bar initially is stretched linearly by moving m a distance $d = bL$ to the right; at time $t = 0$ the system is released from rest. To determine the subsequent vibrations of the bar, we must solve the boundary value problem

$$u_{tt} = a^2 u_{xx} \qquad (0 < x < L, \qquad t > 0); \tag{17a}$$

$$u(0, t) = 0, \tag{17b}$$

$$m u_{tt}(L, t) = -AE u_x(L, t), \tag{17c}$$

$$u(x, 0) = bx, \qquad u_t(x, 0) = 0. \tag{17d}$$

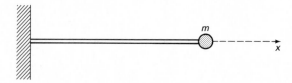

Figure 8.5   The bar-and-mass system of Example 2.

Solution   The endpoint condition at $x = L$ in (17c) comes from equating $ma = m u_{tt}$ for the mass with the force $F = -AE u_x$ given in (13)—the mass being acted on only by the bar. Substitution of $u(x, t) = X(x)T(t)$ in $u_{tt} = a^2 u_{xx}$ leads, as usual, to the equations

$$X'' + \lambda X = 0, \qquad T'' + \lambda a^2 T = 0. \tag{18}$$

The boundary condition in (17b) yields $u(0, t) = X(0)T(t) = 0$, so one endpoint condition is $X(0) = 0$. Because $u_{tt} = XT''$ and $u_x = X'T$, (17c) yields

$$m X(L) T''(t) = -AE X'(L) T(t)$$

as the other endpoint condition. Substitution of

$$T''(t) = -\lambda a^2 T(t) = -\frac{\lambda E}{\delta} T(t),$$

followed by division by $-ET(t)/\delta$, gives $m\lambda X(L) = A\delta X'(L)$. Thus the eigenvalue problem for $X(x)$ is

$$X'' + \lambda X = 0;$$
$$X(0) = 0, \qquad m\lambda X(L) = A\delta X'(L). \tag{19}$$

It is important to note that—because of the presence of λ in the right-endpoint condition—this is *not* a Sturm-Liouville problem, so Theorems 1–3 in Section 8.2 do not apply. Nevertheless, all eigenvalues of (19) are positive (see Problem 9). We therefore write $\lambda = \alpha^2$, and note that $X(x) = \sin \alpha x$ satisfies $X(0) = 0$. The right-endpoint condition in (19) then yields

$$m\alpha^2 \sin \alpha L = A\delta\alpha \cos \alpha L,$$

and thus

$$\tan \alpha L = \frac{A\delta}{m\alpha} = \frac{M/m}{\alpha L} \tag{20}$$

where $M = A\delta L$. We put $\beta = \alpha L$; it follows that the eigenvalues of (19) and associated eigenfunctions are

$$\lambda_n = \frac{\beta_n^2}{L^2}, \qquad X_n(x) = \sin \frac{\beta_n x}{L} \tag{21}$$

for $n = 1, 2, 3, \ldots$, where $\{\beta_n\}_1^\infty$ are the positive roots of the equation

$$\tan x = \frac{M/m}{x}, \tag{22}$$

indicated in Fig. 8.6.

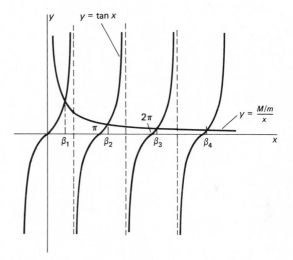

Figure 8.6

Next, we deduce in the usual way from

$$T_n'' + \frac{\beta_n^2 a^2}{L^2} T_n = 0, \qquad T_n'(0) = 0$$

that $T_n(t) = \cos \beta_n at/L$ to within a multiplicative constant. Thus it remains only to find coefficients $\{c_n\}_1^\infty$ so that the series

$$u(x, t) = \sum_{n=1}^\infty c_n \cos \frac{\beta_n at}{L} \sin \frac{\beta_n x}{L} \tag{23}$$

satisfies the nonhomogeneous condition

$$u(x, 0) = \sum_{n=1}^\infty c_n \sin \frac{\beta_n x}{L} = f(x) = bx. \tag{24}$$

Caution is required because (19) is not a Sturm-Liouville problem. Indeed, we ask you to show in Problems 14 and 15 that the eigenfunctions $\{\sin \beta_n x/L\}_1^\infty$ are *not* orthogonal on $[0, L]$, so the coefficient formula of (25) in Section 8.1 is not applicable here.

But the following sly trick succeeds. Differentiation of the expressions in (24) yields

$$\sum_{n=1}^{\infty} \frac{c_n \beta_n}{L} \cos \frac{\beta_n x}{L} = f'(x) = b. \tag{25}$$

The extreme members in (24) are zero when $x = 0$, so it suffices (assuming the validity of termwise integration of (25)) to find $\{c_n\}_1^{\infty}$ so that (25) holds. It happens that the terms $\{\cos \beta_n x/L\}_1^{\infty}$ *are* orthogonal on $[0, L]$. In fact, we ask you in Problem 8 to show that they are precisely the eigenfunctions of the regular Sturm-Liouville problem

$$y''(x) + \lambda y(x) = 0 \qquad (0 < x < L);$$
$$y'(0) = 0, \qquad My(L) + mLy'(L) = 0. \tag{26}$$

Consequently we *can* use Eq. (25) in Section 8.1 to find the coefficients in (25):

$$\frac{c_n \beta_n}{L} = \frac{\int_0^L b \cos (\beta_n x/L)\, dx}{\int_0^L \cos^2 (\beta_n x/L)\, dx}.$$

Routine computations give

$$\int_0^L b \cos \frac{\beta_n x}{L}\, dx = \frac{bL}{\beta_n} \sin \beta_n = \frac{bLM}{m\beta_n^2} \cos \beta_n$$

and

$$\int_0^L \cos^2 \frac{\beta_n x}{L}\, dx = \frac{L}{4\beta_n}(2\beta_n + \sin 2\beta_n),$$

so

$$c_n = \frac{4bLM}{m} \cdot \frac{\cos \beta_n}{\beta_n^2(2\beta_n + \sin 2\beta_n)}.$$

On substituting these values in (23), we finally get the formal series solution

$$u(x, t) = \frac{4bLM}{m} \sum_{n=1}^{\infty} \frac{\cos \beta_n \sin (\beta_n x/L) \cos (\beta_n at/L)}{\beta_n^2(2\beta_n + \sin 2\beta_n)}. \tag{27}$$

It is of interest to note that this series converges rapidly because $\beta_n \approx (n-1)\pi$ for n large, and that the nth term frequency (in Hz) is

$$v_n = \frac{\beta_n a}{2\pi L} = \frac{\beta_n}{2\pi L}\sqrt{\frac{E}{\delta}}. \tag{28}$$

TRANSVERSE VIBRATIONS OF BARS

We now discuss vibrations of a uniform elastic bar in which the motion of each point is not longitudinal, but instead perpendicular to the x-axis (the axis of the bar in its equilibrium position). Let $y(x, t)$ denote the transverse displacement of the cross section at x at time t, as indicated in Fig. 8.7. We merely want to outline a derivation of the partial differential equation that the deflection function $f(x, t)$ satisfies. Recall first that in Section 1.2 we introduced the

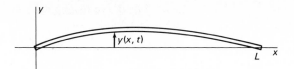

Figure 8.7 A bar undergoing transverse vibrations.

equation $EIy^{(4)} = F$ for the static deflection of a bar or beam under the influence of a transverse external force F (per unit length). According to a general dynamical principle, we can transform the static equation $EIy^{(4)} = F$ to a dynamical equation (with no external force) by replacing F with the reversed inertial force $F = -\rho y_{tt}$, where ρ is the linear density (mass/length) of the bar, and also by replacing $y^{(4)}$ with y_{xxxx}. This gives

$$EI\frac{\partial^4 y}{\partial x^4} = -\frac{\partial^2 y}{\partial t^2},$$

which may be written in the form

$$\frac{\partial^2 y}{\partial t^2} + a^4 \frac{\partial^4 y}{\partial x^4} = 0 \tag{29}$$

where

$$a^4 = \frac{EI}{\rho}. \tag{30}$$

The following example illustrates the solution of this fourth order partial differential equation by the method of separation of variables.

EXAMPLE 3 A uniform bar with linear density ρ, Young's modulus E, and cross-sectional moment of inertia I is simply supported (or hinged) at its two ends $x = 0$ and $x = L$. If the bar is set in motion from rest with given initial position $f(x)$, then its displacement function $y(x, t)$ satisfies the boundary value problem

$$\begin{cases} y_{tt} + a^4 y_{xxxx} = 0 \quad (0 < x < L, \quad t > 0); & \text{(31a)} \\ y(0, t) = y_{xx}(0, t) = y(L, t) = y_{xx}(L, t) = 0, & \text{(31b)} \\ y(x, 0) = f(x), \quad y_t(x, 0) = 0. & \text{(31c)} \end{cases}$$

Find $y(x, t)$. The boundary conditions in (31b) are the hinged end conditions that we accept without proof, and those in (31c) are the initial conditions.

Solution Substitution of $y(x, t) = X(x)T(t)$ in the differential equation yields $XT'' + a^4 X^{(4)}T = 0$, so

$$\frac{X^{(4)}}{X} = -\frac{T''}{a^4 T} = -\lambda. \tag{32}$$

In order to determine the sign of λ, we reason that the equation

$$T'' + \lambda a^4 T = 0 \tag{33}$$

must have trigonometric rather than exponential solutions. The reason is that on the basis of practical experience—with tuning forks or xylophone bars, for instance—one expects periodic vibrations to take place. This could not occur if λ were negative and Eq. (33) had exponential solutions. Hence λ must be positive, and it is convenient to write $\lambda = \alpha^4 > 0$. Then $X(x)$ must satisfy the equation

$$X^{(4)}(x) - \alpha^4 X(x) = 0, \tag{34}$$

which has the general solution

$$X(x) = A \cos \alpha x + B \sin \alpha x + C \cosh \alpha x + D \sinh \alpha x,$$

with

$$X''(x) = \alpha^2(-A \cos \alpha x - B \sin \alpha x + C \cosh \alpha x + D \sinh \alpha x).$$

The endpoint conditions in (31) yield

$$X(0) = X''(0) = X(L) = X''(L) = 0. \tag{35}$$

Hence

$$X(0) = A + C = 0 \quad \text{and} \quad X''(0) = -A + C = 0,$$

and these two equations imply that $A = C = 0$. Therefore

$$X(L) = B \sin \alpha L + D \sinh \alpha L = 0$$

and

$$X''(L) = \alpha^2(-B \sin \alpha L + D \sinh \alpha L) = 0.$$

It follows that

$$B \sin \alpha L = 0 \quad \text{and} \quad D \sinh \alpha L = 0.$$

But $\sinh \alpha L \neq 0$ because $\alpha \neq 0$; consequently, $D = 0$. Hence $B \neq 0$ if we are to have a nontrivial solution, so $\sin \alpha L = 0$. Thus α must be an integral multiple of π/L. Therefore the eigenvalues and associated eigenfunctions of the problem determined by Eqs. (34) and (35) are

$$\lambda_n = \alpha_n^4 = \frac{n^4 \pi^4}{L^4}, \qquad X_n(x) = \sin \frac{n\pi x}{L} \tag{36}$$

for $n = 1, 2, 3, \ldots$.

With $\lambda = n^4 \pi^4 / L^4$ in (33) we get the equation

$$T_n'' + \frac{n^4 \pi^4 a^4}{L^4} T_n = 0. \tag{37}$$

Because the initial condition $y_t(x, 0) = 0$ yields $T_n'(0) = 0$, we take

$$T_n(t) = \cos \frac{n^2 \pi^2 a^2 t}{L^2}. \tag{38}$$

Combining the results in (36) and (38), we construct the series

$$y(x, t) = \sum_{n=1}^{\infty} c_n \cos \frac{n^2 \pi^2 a^2 t}{L^2} \sin \frac{n\pi x}{L} \tag{39}$$

that formally satisfies the partial differential equation in (31) and the homogeneous boundary conditions. The nonhomogeneous condition is

$$y(x, 0) = \sum_{n=1}^{\infty} c_n \sin \frac{n\pi x}{L} = f(x),$$

so we choose the Fourier sine coefficients, given by

$$c_n = \frac{2}{L} \int_0^L f(x) \sin \frac{n\pi x}{L} \, dx, \tag{40}$$

in order for (39) to provide a formal series solution.

Note that the frequency (in Hz) of the nth term in (39) is

$$\nu_n = \frac{n^2 \pi^2 a^2}{2\pi L^2} = n^2 \nu_1 \tag{41}$$

where the fundamental frequency (in Hz) of the bar is

$$\nu_1 = \frac{\pi a^2}{2L^2} = \frac{\pi}{2L^2} \sqrt{\frac{EI}{\rho}}. \tag{42}$$

Since the frequency of each of the higher harmonics is an integral multiple of ν_1, the sound of a vibrating bar with simply supported ends is musical. Because the higher frequencies $\{n^2\nu_1\}_2^\infty$ are sparser than the higher frequencies $\{n\nu_1\}_2^\infty$ of a vibrating string, the tone of a vibrating bar is purer than that of a vibrating string. This partly accounts for the silkiness of the sound of the vibraphone as played by Milt Jackson, former member of the Modern Jazz Quartet.

8.2 Problems

Find a formal series solution of each of the boundary value problems in Problems 1–6.
Express each answer in the form given in Problem 1.

1. $u_t = k u_{xx}$ $(0 < x < L, t > 0)$; $\quad u_x(0, t) = h u(L, t) + u_x(L, t) = 0$, $\quad u(x, 0)$
   $= f(x)$. (*Answer:*

   $$u(x, t) = \sum_{n=1}^\infty c_n e^{-\beta_n^2 kt/L^2} \cos \frac{\beta_n x}{L}$$

   where $\{\beta_n\}_1^\infty$ are the positive roots of the equation $\tan x = hL/x$ and

   $$c_n = \frac{2h}{hL + \sin^2 \beta_n} \int_0^L f(x) \cos \frac{\beta_n x}{L} \, dx.)$$

2. $u_{xx} + u_{yy} = 0$ $(0 < x < L, \, 0 < y < L)$; $\quad u(0, y) = h u(L, y) + u_x(L, y) = 0$, $u(x, L) = 0, u(x, 0) = f(x)$.

3. $u_{xx} + u_{yy} = 0$ $(0 < x < L, \, 0 < y < L)$; $\quad u_y(x, 0) = h u(x, L) + u_y(x, L) = 0$, $u(L, y) = 0, u(0, y) = g(y)$.

4. $u_{xx} + u_{yy} = 0$ $(0 < x < L, y > 0)$; $\quad u(0, y) = h u(L, y) + u_x(L, y) = 0$, $u(x, y)$ bounded as $y \longrightarrow +\infty, u(x, 0) = f(x)$.

5. $u_t = k u_{xx}$ $(0 < x < L, \; t > 0)$; $\quad h u(0, t) - u_x(0, t) = u(L, t) = 0$, $\quad u(x, 0)$
   $= f(x)$.

6. $u_t = k u_{xx}$ $(0 < x < L, t > 0)$; $\quad h u(0, t) - u_x(0, t) = h u(L, t) + u_x(L, t) = 0$, $u(x, 0) = f(x)$.

7. Let $u(x, y)$ denote the bounded steady-state temperature in an infinitely high wall with base $y = 0$ and faces $x = 0$ and $x = 1$. The face $x = 0$ is insulated, the base $y = 0$ is kept at temperature 100°C, and heat transfer with $h = 1$

takes place at the face $x = 1$. Derive the solution

$$u(x, y) = 200 \sum_{n=1}^{\infty} \frac{e^{-\alpha_n y} \sin \alpha_n \cos \alpha_n x}{\alpha_n + \sin \alpha_n \cos \alpha_n},$$

where $\{\alpha_n\}_1^{\infty}$ are the positive roots of the equation $\cot x = x$. Given $\alpha_1 = 0.860$, $\alpha_2 = 3.426$, $\alpha_3 = 6.437$, and $\alpha_4 = 9.529$, calculate the temperature $u(1, 1)$ accurate to $0.1°$.

8. Verify that the eigenvalues and eigenfunctions of the regular Sturm-Liouville problem in (26) are

$$\lambda_n = \frac{\beta_n^2}{L^2} \quad \text{and} \quad y_n(x) = \cos \frac{\beta_n x}{L}$$

for $n = 1, 2, 3, \ldots,$ where $\{\beta_n\}_1^{\infty}$ are the positive roots of the equation $\tan x = M/mx$.

9. (a) Show that $\lambda = 0$ is not an eigenvalue of the problem in (19). (b) Show that this problem has no negative eigenvalues. (*Suggestion:* Sketch the graphs $y = \tanh x$ and $y = -k/x$ with $k > 0$.)

10. Calculate the speed (in miles per hour) of longitudinal sound waves in each case.
   (a) Steel, with $\delta = 7.75$ g/cm³ and $E = 2 \times 10^{12}$ in cgs units.
   (b) Water, with $\delta = 1$ g/cm³ and bulk modulus $K = 2.25 \times 10^{10}$ in cgs units.

11. Consider a mass $m = nm_0$ of an ideal gas of molecular weight m_0 whose pressure P and volume V satisfy the law $PV = nRT_K$, where n is the number of moles of the gas, $R = 8314$ in mks units, and $T_K = T_C + 273$ where T_C is the Celsius temperature. The bulk modulus of the gas is $K = \gamma P$, where the value of the dimensionless constant γ is 1.4 for air having molecular weight $m_0 = 29$.
   (a) Show that the velocity of sound in this gas is

$$a = \sqrt{\frac{K}{\delta}} = \sqrt{\frac{\gamma R T_K}{m_0}}.$$

   (b) Use this formula to show that the speed of sound in air at Celsius temperature T_C is approximately $740 + (1.36)T_C$ miles per hour.

12. Suppose that the free end of the bar of Example 2 is attached to a spring (rather than to a mass), as shown in Fig. 8.8. The endpoint condition then becomes $ku(L, t) + AEu_x(L, t) = 0$. (Why?) Assume that $u(x, 0) = f(x)$ and that $u_t(x, 0) = 0$. Derive a solution of the form

$$u(x, t) = \sum_{n=1}^{\infty} c_n \cos \frac{\beta_n at}{L} \sin \frac{\beta_n x}{L},$$

where $\{\beta_n\}_1^{\infty}$ are the positive roots of the equation $\tan x = -AEx/kL$.

Figure 8.8   The bar of Problem 12.

13. If a bar has natural length L, cross-sectional area A, and Young's modulus E, then (as a consequence of Hooke's law) the axial force at each end required to stretch it the small amount ΔL is $F = AE\,\Delta L/L$. Apply this result to a segment of the bar of natural length $L = \Delta x$ between the cross sections at x and $x + \Delta x$ that is stretched by the amount $\Delta L = u(x + \Delta x, t) - u(x, t)$. Then let $\Delta x \to 0$ to derive Eq. (13).

14. Show that the eigenfunctions $\{X_n(x)\}_1^\infty$ of the problem in (19) are not orthogonal. (*Suggestion:* Apply Eq. (22) of Section 8.1 to show that if $m \neq n$, then

$$\int_0^L X_m(x) X_n(x)\, dx = -\frac{m}{A\delta} X_m(L) X_n(L).)$$

15. Show that the eigenfunctions $\{\sin \beta_n x/L\}_1^\infty$ of the problem in (19) are not orthogonal; do so by obtaining the explicit value of the integral

$$\int_0^L \sin \frac{\beta_m x}{L} \sin \frac{\beta_n x}{L}\, dx.$$

(*Suggestion:* Use the fact that $\{\beta_n\}_1^\infty$ are the roots of the equation $x \tan x = M/m$.)

16. According to Problem 19 in Section 7.7, the temperature $u(r, t)$ in a uniform solid spherical ball of radius a satisfies the partial differential equation $(ru)_t = k(ru)_{rr}$. Suppose that the ball has initial temperature $u(r, 0) = f(r)$ and that its surface $r = a$ is insulated, so that $u_r(a, t) = 0$. Substitute $v(r, t) = ru(r, t)$ to derive the solution

$$u(r, t) = c_0 + \sum_{n=1}^\infty \frac{c_n}{r} e^{-\beta_n^2 kt/a^2} \sin \frac{\beta_n r}{a},$$

where $\{\beta_n\}_1^\infty$ are the positive roots of the equation $\tan x = x$ and

$$c_0 = \frac{3}{a^3} \int_0^a r^2 f(r)\, dr,$$

$$c_n = \frac{2}{a \sin^2 \beta_n} \int_0^a rf(r) \sin \frac{\beta_n r}{a}\, dr.$$

See Problem 14 in Section 8.1.

17. A problem concerning the diffusion of gas through a membrane leads to the boundary value problem

$$\begin{cases} u_t = k u_{xx} & (0 < x < L, \quad t > 0); \\ u(0, t) = u_t(L, t) + hk u_x(L, t) = 0, \\ u(x, 0) = 1. \end{cases}$$

Derive the solution

$$u(x, t) = 4 \sum_{n=1}^\infty \frac{1 - \cos \beta_n}{2\beta_n - \sin 2\beta_n} e^{-\beta_n^2 kt/L^2} \sin \frac{\beta_n x}{L},$$

where $\{\beta_n\}_1^\infty$ are the positive roots of the equation $x \tan x = hL$.

18. Suppose that the simply supported uniform bar of Example 3 has, instead, initial position $y(x, 0) = 0$ and initial velocity $y_t(x, 0) = g(x)$. Then derive the solution

$$y(x, t) = \sum_{n=1}^\infty c_n \sin \frac{n^2 \pi^2 a^2 t}{L^2} \sin \frac{n\pi x}{L},$$

where

$$c_n = \frac{2L}{n^2\pi^2 a^2} \int_0^L g(x) \sin \frac{n\pi x}{L}\, dx.$$

19. To approximate the effect of an initial momentum impulse P applied at the mid-point $x = L/2$ of a simply supported bar, substitute

$$g(x) = \begin{cases} \dfrac{P}{2\rho\epsilon} & \text{if } \dfrac{L}{2} - \epsilon < x < \dfrac{L}{2} + \epsilon, \\ 0 & \text{otherwise} \end{cases}$$

in the result of Problem 18. Then let $\epsilon \rightarrow 0$ to obtain the solution

$$y(x, t) = C \sum_{n=1}^{\infty} \frac{1}{n^2} \sin \frac{n\pi}{2} \sin \frac{n^2\pi^2 a^2 t}{L^2} \sin \frac{n\pi x}{L}$$

where

$$C = \frac{2PL}{\pi^2 \sqrt{EI\rho}}.$$

20. (a) If $g(x) = v_0$ (constant) in Problem 18, show that

$$y(x, t) = \frac{4v_0 L^2}{\pi^3 a^2} \sum_{n \text{ odd}} \frac{1}{n^3} \sin \frac{n^2\pi^2 a^2 t}{L^2} \sin \frac{n\pi x}{L}.$$

This describes the vibrations of a hinged bar lying crosswise in the back of a pickup truck that hits a brick wall with speed v_0 at time t. (b) Now suppose that the bar is made of steel ($E = 2 \times 10^{12}$ dyn/cm^2, $\delta = 7.75$ g/cm^3), has a square cross section with edge $a = 1$ in. (so that $I = \frac{1}{12}a^4$), and length $L = 19$ in. What is its fundamental frequency (in hertz)?

*8.3
Steady Periodic Solutions and Natural Frequencies

In Section 7.6 we derived the solution

$$y(x, t) = \sum_{n=1}^{\infty} \left(A_n \cos \frac{n\pi a t}{L} + B_n \sin \frac{n\pi a t}{L} \right) \sin \frac{n\pi x}{L}$$

$$= \sum_{n=1}^{\infty} C_n \cos \left(\frac{n\pi a t}{L} - \gamma_n \right) \sin \frac{n\pi x}{L} \tag{1}$$

of the vibrating string problem

$$\begin{cases} \dfrac{\partial^2 y}{\partial t^2} = a^2 \dfrac{\partial^2 y}{\partial x^2} & \left(a = \sqrt{\dfrac{T}{\rho}} \right); \tag{2} \\[2mm] y(0, t) = y(L, t) = 0, \tag{3} \\[2mm] y(x, 0) = f(x), \qquad y_t(x, 0) = g(x). \tag{4} \end{cases}$$

The solution in (1) describes the *free* vibrations of a string with length L and linear density ρ under tension T; the constant coefficients in (1) are determined by the initial conditions in (4).

In particular, we see from the terms in (1) that the natural (circular) frequencies of vibration (in radians per second) of the string are given by

$$\omega_n = \frac{n\pi a}{L}, \qquad n = 1, 2, 3, \ldots. \tag{5}$$

These are the only values of ω for which Eq. (2) has a steady periodic solution of the form

$$y(x, t) = X(x) \cos (\omega t - \gamma) \tag{6}$$

that satisfies the endpoint conditions in (3). For if we substitute (6) in (2) and cancel the factor $\cos (\omega t - \gamma)$, we find that $X(x)$ must satisfy the equation

$$a^2 X''(x) - \omega^2 X(x) = 0,$$

whose general solution

$$X(x) = A \cos \frac{\omega x}{a} + B \sin \frac{\omega x}{a}$$

satisfies the conditions in (3) only if $A = 0$ and $\omega = n\pi a/L$ for some positive integer n.

FORCED VIBRATIONS AND RESONANCE

Now suppose that the string is influenced by a periodic external force $F(t) = F_0 \cos \omega t$ (force per unit mass) that acts uniformly on the string along its length. Then, according to Eq. (1) in Section 7.6, the displacement $y(x, t)$ of the string will satisfy the nonhomogeneous partial differential equation

$$\frac{\partial^2 y}{\partial t^2} = a^2 \frac{\partial^2 y}{\partial x^2} + F_0 \cos \omega t \tag{7}$$

together with boundary conditions such as those in (3) and (4). For instance, if the string is initially at rest in equilibrium when the external force begins to act, we want to find a solution of (7) that satisfies the conditions

$$y(0, t) = y(L, t) = y(x, 0) = y_t(x, 0) = 0. \tag{8}$$

To do this, it suffices first to find a particular solution $y_p(x, t)$ of (7) that satisfies the fixed endpoint conditions in (3), and second a solution $y_c(x, t)$ like (1) of the familiar problem in (2)–(4) with $f(x) = -y_p(x, 0)$ and $g(x) = -D_t y_p(x, 0)$. Evidently $y(x, t) = y_c(x, t) + y_p(x, t)$ will then satisfy (7) and (8).

So our new task is to find $y_p(x, t)$. Examination of the individual terms in (7) suggests that we try

$$y_p(x, t) = X(x) \cos \omega t. \tag{9}$$

Substitution of this in (7) and cancellation of the common factor $\cos \omega t$ yields the ordinary differential equation

$$a^2 X'' + \omega^2 X = -F_0$$

with general solution

$$X(x) = A \cos \frac{\omega x}{a} + B \sin \frac{\omega x}{a} - \frac{F_0}{\omega^2}. \tag{10}$$

The condition $X(0) = 0$ requires that $A = F_0/\omega^2$, and then $X(L) = 0$ requires that

$$X(L) = \frac{F_0}{\omega^2}\left(\cos \frac{\omega L}{a} - 1\right) + B \sin \frac{\omega L}{a} = 0. \tag{11}$$

Now suppose that the frequency ω of the periodic external force is *not* equal to any one of the natural frequencies $\omega_n = n\pi a/L$ of the string. Then $\sin \omega L/a \neq 0$, so we can solve Eq. (11) for B and then substitute the result in (10) to obtain

$$X(x) = \frac{F_0}{\omega^2}\left(\cos\frac{\omega x}{a} - 1\right) + \frac{F_0[\cos(\omega L/a) - 1]}{\omega^2 \sin(\omega L/a)} \sin\frac{\omega x}{a}. \tag{12}$$

Then (9), with this choice of $X(x)$, gives the desired particular solution $y_p(x, t)$.

Note, however, that as the value of ω approaches $\omega_n = n\pi a/L$ with n odd, the coefficient of $\sin \omega x/a$ in (12) approaches $+\infty$; thus *resonance* occurs. This explains the fact that when (only) one of two nearby identical strings is plucked, the other will begin to vibrate as well, due to its being acted on (through the medium of the air) by an external periodic force at its fundamental frequency. Observe also that if $\omega = \omega_n = n\pi a/L$ with n even, then we can choose $B = 0$ in (11), so resonance does not occur in this case. Problem 20 explains why some of the resonance possibilities are absent.

The vibrating string is typical of continuous systems that have an infinite sequence of natural frequencies of vibration. When a periodic external force acts on such a system, potentially destructive resonance vibrations may occur if the imposed frequency is close to one of the natural frequencies of the system. Hence an important aspect of proper structural design is the avoidance of such resonance vibrations.

NATURAL FREQUENCIES OF BEAMS

Figure 8.9 shows a uniform beam of length L, linear density ρ, and Young's modulus E, clamped at each end. For $0 < x < L$ and $t > 0$, its deflection function $y(x, t)$ satisfies the fourth order equation

$$\frac{\partial^2 y}{\partial t^2} + a^4 \frac{\partial^4 y}{\partial x^4} = 0 \qquad \left(a^4 = \frac{EI}{\rho}\right) \tag{13}$$

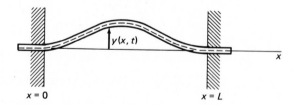

Figure 8.9   A bar clamped at each end.

that we discussed in Section 8.2; I denotes the moment of inertia of the cross section of the beam about its horizontal axis of symmetry. Because both the displacement and the slope are zero at each fixed end, the endpoint conditions are

$$y(0, t) = y_x(0, t) = 0 \tag{14}$$

and

$$y(L, t) = y_x(L, t) = 0. \tag{15}$$

Here we want only to find the natural frequencies of vibration of the beam, so we will not be concerned with initial conditions. The natural frequencies are the values of ω for which Eq. (13) has a nontrivial solution of the form

$$y(x, t) = X(x) \cos(\omega t - \gamma) \tag{16}$$

that satisfies the conditions in (14) and (15). When we substitute $y(x, t)$ from (16) in (13) and then cancel the common factor $\cos(\omega t - \gamma)$, we obtain the fourth order ordinary differential equation $-\omega^2 X + a^4 X^{(4)} = 0$; that is,

$$X^{(4)} - \frac{\omega^2}{a^4} X = 0. \tag{17}$$

If we write $\alpha^4 = \omega^2/a^4$, we can express the general solution of (17) as

$$X(x) = A \cosh \alpha x + B \sinh \alpha x + C \cos \alpha x + D \sin \alpha x,$$

with

$$X'(x) = \alpha(A \sinh \alpha x + B \cosh \alpha x - C \sin \alpha x + D \cos \alpha x).$$

The conditions in (14) give

$$X(0) = A + C = 0 \quad \text{and} \quad X'(0) = \alpha(B + D) = 0,$$

so $C = -A$ and $D = -B$. Hence the conditions in (15) give

$$X(L) = A(\cosh \alpha L - \cos \alpha L) + B(\sinh \alpha L - \sin \alpha L) = 0$$

and

$$\frac{1}{\alpha} X'(L) = A(\sinh \alpha L + \sin \alpha L) + B(\cosh \alpha L - \cos \alpha L) = 0.$$

In order for these two linear homogeneous equations in A and B to have a nontrivial solution, the determinant of coefficients must be zero:

$$(\cosh \alpha L - \cos \alpha L)^2 - (\sinh^2 \alpha L - \sin^2 \alpha L) = 0;$$

$$(\cosh^2 \alpha L - \sinh^2 \alpha L) + (\cos^2 \alpha L + \sin^2 \alpha L) - 2 \cosh \alpha L \cos \alpha L = 0;$$

$$2 - 2 \cosh \alpha L \cos \alpha L = 0.$$

Thus $\beta = \alpha L$ must be a nonzero root of the equation

$$\cosh x \cos x = 1. \tag{18}$$

From Fig. 8.10 we see that this equation has an infinite increasing sequence of positive roots $\{\beta_n\}_1^\infty$. Now $\omega = \alpha^2 a^2 = \beta^2 a^2/L^2$ and $a^2 = \sqrt{EI/\rho}$, so it follows that the natural (circular) frequencies of vibration of the beam with clamped ends are given by

$$\omega_n = \frac{\beta_n^2}{L^2} \sqrt{\frac{EI}{\rho}} \quad \text{(rad/s)} \tag{19}$$

for $n = 1, 2, 3, \ldots$. The roots of Eq (18) are $\beta_1 \approx 4.73004$, $\beta_2 \approx 7.85320$, $\beta_3 \approx 10.99561$, and $\beta_n \approx (2n + 1)\pi/2$ for $n \geq 4$ (as indicated in Fig. 8.10). For example, suppose the basic structural element of a bridge is a 120-ft-

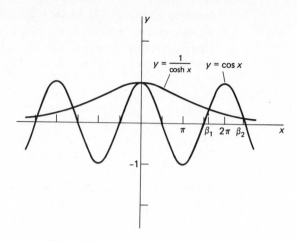

Figure 8.10 Solutions of $\cosh x \cos x = 1$.

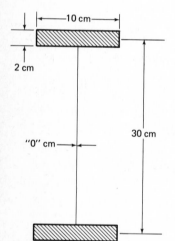

Figure 8.11 An idealized I beam.

long steel I beam with the cross section indicated in Fig. 8.11, having moment of inertia $I = 9000$ cm⁴. If we substitute the values

$$L = (120 \text{ ft})\left(30.48 \frac{\text{cm}}{\text{ft}}\right),$$

$$\rho = \left(7.75 \frac{\text{g}}{\text{cm}^3}\right)(40 \text{ cm}^2),$$

$$E = 2 \times 10^{12} \frac{\text{dyne}}{\text{cm}^2},$$

and the values of β_1 and β_2 in Eq. (19), we find that the lowest two natural frequencies of the beam are

$$\omega_1 \approx 12.74 \ \frac{\text{rad}}{\text{s}} \qquad \left(122 \frac{\text{cycles}}{\text{min}}\right)$$

and

$$\omega_2 \approx 35.13 \ \frac{\text{rad}}{\text{s}} \qquad \left(335 \frac{\text{cycles}}{\text{min}}\right).$$

If a company of soldiers marching at about 120 steps per minute approaches this bridge, they therefore would be well-advised to break cadence before crossing. From time to time bridges collapse because of resonance vibrations. Recall the Kansas City hotel disaster of July 17, 1981, in which a skywalk filled with dancers collapsed; newspapers quoted investigators who speculated that the rhythmic movement of the dancers had set up destructive resonance vibrations in the steel I beams that supported the skywalk.

UNDERGROUND TEMPERATURE OSCILLATIONS

Let us assume that the underground temperature at a particular location is a function $u(x, t)$ of time t and the depth x beneath the surface. Then u satisfies the heat equation $u_t = k u_{xx}$, with k being the thermal diffusivity of the soil. We may regard the temperature $u(0, t)$ at the surface $x = 0$ as being known from weather records. In fact, the periodic seasonal variation of monthly average surface temperatures, with a maximum in midsummer (July) and a minimum in midwinter (January) in the northern hemisphere, is very close to a sine or cosine oscillation. We shall therefore assume that

$$u(0, t) = T_0 + A_0 \cos \omega t, \tag{20}$$

where we take $t = 0$ at midsummer. Here T_0 is the annual average temperature, A_0 the amplitude of seasonal temperature variation, and ω is chosen to make the period of the above function 1 year. (In cgs units, for instance, ω would be 2π divided by the number of seconds in a year.)

It is reasonable to assume that the temperature at a fixed depth also varies periodically with t. If we introduce $U(x, t) = u(x, t) - T_0$ for convenience, then we are interested in periodic solutions of the form

$$U(x, t) = A(x) \cos (\omega t - \gamma) = V(x) \cos \omega t + W(x) \sin \omega t \tag{21}$$

of the problem

$$\frac{\partial U}{\partial t} = k \frac{\partial^2 U}{\partial x^2} \qquad (x > 0, \qquad t > 0), \tag{22}$$

$$U(0, t) = A_0 \cos \omega t. \tag{23}$$

To solve this problem, let us regard $U(x, t)$ in (21) as the real part of the complex-valued function

$$\tilde{U}(x, t) = X(x)e^{i\omega t}. \tag{24}$$

Then we want $\tilde{U}(x, t)$ to satisfy the conditions

$$\tilde{U}_t = k\tilde{U}_{xx}, \tag{22'}$$

$$\tilde{U}(0, t) = A_0 e^{i\omega t} \tag{23'}$$

If we substitute (24) in (22'), we get $i\omega X = kX''$; that is,

$$X'' - \alpha^2 X = 0 \tag{25}$$

where

$$\alpha = \pm\sqrt{\frac{i\omega}{k}} = \pm(1 + i)\sqrt{\frac{\omega}{2k}} \tag{26}$$

because $\sqrt{i} = \pm(1 + i)/\sqrt{2}$. Hence the general solution of (25) is

$$X(x) = Ae^{-(1+i)x\sqrt{\omega/2k}} + Be^{+(1+i)x\sqrt{\omega/2k}}. \tag{27}$$

In order that $U(x, t)$, and hence $X(x)$, be bounded as $x \to +\infty$, it is necessary that $B = 0$. Also we see from (23') and (24) that $A = X(0) = A_0$, so

$$X(x) = A_0 e^{-(1+i)x\sqrt{\omega/2k}}. \tag{28}$$

Finally, the solution of our original problem in (22)–(23) is

$$U(x, t) = \mathcal{R}e\tilde{U}(x, t) = \mathcal{R}eX(x)e^{i\omega t}$$

$$= \mathcal{R}e(A_0 e^{i\omega t}e^{-(1+i)x\sqrt{\omega/2k}})$$

$$= \mathcal{R}e(A_0 e^{-x\sqrt{\omega/2k}}e^{i(\omega t - x\sqrt{\omega/2k})});$$

hence

$$U(x, t) = A_0 e^{-x\sqrt{\omega/2k}} \cos\left(\omega t - x\sqrt{\frac{\omega}{2k}}\right). \tag{29}$$

Thus the amplitude $A(x)$ of the annual temperature is exponentially damped as a function of the depth x:

$$A(x) = A_0 e^{-x\sqrt{\omega/2k}}. \tag{30}$$

In addition, there is a phase delay $\gamma(x) = x\sqrt{\omega/2k}$ at the depth x. With $k = 0.005$ (a typical value for soil in cgs units) and the value of ω previously mentioned, we find that $\sqrt{\omega/2k} \approx 0.00446$ cm^{-1}. For instance, we then see from (30) that the amplitude is one-half the surface amplitude, $A(x) = \frac{1}{2}A_0$, when $(0.00446)x = \ln 2$; that is, when $x \approx 155.30$ cm ≈ 5.10 ft. If $A_0 = 16°$, it follows that at a depth of about 20 ft, the amplitude of annual temperature variation is only $1°$. Another interesting consequence of (29) is the "reversal of seasons" that occurs when $\gamma(x) = (0.00446)x = \pi$; that is, at a depth of $x \approx 704.39$ cm ≈ 23.11 ft.

8.3 Problems

A uniform bar of length L is made of material with density δ and Young's modulus E. In each of Problems 1–6, substitute $u(x, t) = X(x) \cos \omega t$ in $\delta u_{tt} = E u_{xx}$ to find the natural frequencies of longitudinal vibration of the bar with the two given conditions at its ends $x = 0$ and $x = L$.

1. Both ends are fixed.

2. Both ends are free.

3. The end at $x = 0$ is fixed; the end at $x = L$ is free.

4. The end at $x = 0$ is fixed; the free end at $x = L$ is attached to a mass m as in Example 2 of Section 8.2.

5. Each end is free, but the end at $x = L$ is attached to a spring with Hooke's constant k as in Problem 12 in Section 8.2.

6. The free ends are attached to masses m_0 and m_1.

7. Suppose that the mass on the free end at $x = L$ in Problem 4 is attached also to the spring of Problem 5. Show that the natural frequencies are given by $\omega_n = (\beta_n/L)\sqrt{E/\delta}$, where $\{\beta_n\}_1^\infty$ are the positive roots of the equation

$$(mEx^2 - k\delta L^2) \sin x = MEx \cos x.$$

(*Note:* The condition at $x = L$ is $mu_{tt} = -AEu_x - ku$.)

Problems 8–14 deal with transverse vibrations of the uniform beam of this section, but with various end conditions. In each case show that the natural frequencies are given by the formula in (19), with $\{\beta_n\}_1^\infty$ being the positive roots of the given frequency equation. Recall that $y = y' = 0$ at a fixed end, $y = y'' = 0$ at a hinged end, and $y'' = y''' = 0$ at a free end (primes denote differentiation with respect to x).

8. The ends at $x = 0$ and $x = L$ are both hinged; the frequency equation is $\sin x = 0$, so that $\beta_n = n\pi$.

9. The end at $x = 0$ is fixed and the end at $x = L$ is hinged; the frequency equation is $\tanh x = \tan x$.

10. The beam is a cantilever with the end at $x = 0$ fixed and the end at $x = L$ free; the frequency equation is $\cosh x \cos x = -1$.

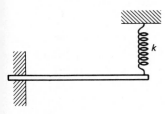

11. The end at $x = 0$ is fixed and the end at $x = L$ is attached to a vertically sliding clamp, so $y' = y''' = 0$ there; the frequency equation is $\tanh x + \tan x = 0$.

12. The cantilever of Problem 10 has total mass $M = \rho L$ and has a mass m attached to its free end; the frequency equation is

$$M(1 + \cosh x \cos x) = mx(\cosh x \sin x - \sinh x \cos x).$$

The conditions at $x = L$ are $y_{xx} = 0$ and $my_{tt} = EIy_{xxx}$.

13. The free end at $x = L$ of the cantilever of Problem 10 is attached (as in Fig. 8.12) to a spring with Hooke's constant k; the frequency equation is

$$EIx^3(1 + \cosh x \cos x) = kL^3(\sinh x \cos x - \cosh x \sin x).$$

The conditions at $x = L$ are $y_{xx} = 0$ and $ky = EIy_{xxx}$.

14. Suppose that the mass m on the free end of the cantilever of Problem 12 is attached to the spring of Problem 13. The conditions at $x = L$ are $y_{xx} = 0$ and $my_{tt} = EIy_{xxx} - ky$. Derive the frequency equation

$$MEIx^3(1 + \cosh x \cos x) = (kML^3 - mEIx^4)(\sinh x \cos x - \cosh x \sin x).$$

Note that the frequency equations in Problem 12 and 13 are the special cases $k = 0$ and $m = 0$, respectively.

15. If a uniform bar hinged at each end is subjected to an axial force of compression P, then its transverse vibrations satisfy the equation

$$\rho \frac{\partial^2 y}{\partial t^2} + P \frac{\partial^2 y}{\partial x^2} + EI \frac{\partial^4 y}{\partial x^4} = 0.$$

Show that its natural frequencies are given by

$$\omega_n = \frac{n^2 \pi^2}{L^2} \left(1 + \frac{PL^2}{n^2 \pi^2 EI}\right)^{1/2} \sqrt{\frac{EI}{\rho}}$$

for $n = 1, 2, 3, \ldots$. Note that with $P = 0$, this reduces to the result in Example 3 of Section 8.2 and that the effect of $P > 0$ is to increase the value of each of the ω_n.

16. A beam hinged at each end is sufficiently thick that its kinetic energy of rotation must be taken into account. Then the differential equation for its transverse vibrations is

$$\rho \frac{\partial^2 y}{\partial t^2} - \frac{I}{A} \frac{\partial^4 y}{\partial x^2 \partial t^2} + EI \frac{\partial^4 y}{\partial x^4} = 0.$$

Show that its natural frequencies are given by

$$\omega_n = \frac{n^2 \pi^2}{L^2} \left(1 + \frac{n^2 \pi^2 I}{AL^2}\right)^{-1/2} \sqrt{\frac{EI}{\rho}}$$

for $n = 1, 2, 3, \ldots$.

17. Suppose that the end $x = 0$ of a uniform bar with cross-sectional area A and Young's modulus E is fixed, while the longitudinal force $F(t) = F_0 \sin \omega t$ acts on its end $x = L$, so $AEu_x(L, t) = F_0 \sin \omega t$. Derive the steady periodic solution

$$u(x, t) = \frac{F_0 a \sin (\omega x/a) \sin \omega t}{AE\omega \cos (\omega L/a)}.$$

18. Repeat Problem 17, except here a transverse force $F(t) = F_0 \sin \omega t$ acts at the free end at $x = L$, so

$$y_{xx}(L, t) = EIy_{xxx}(L, t) + F_0 \sin \omega t = 0.$$

Figure 8.12 The cantilever of Problem 13.

Determine the steady periodic transverse oscillations of the cantilever in the form $y(x, t) = X(x) \sin \omega t$.

19. A uniform bar with simply supported ends at $x = 0$ and $x = L$ is acted on by a uniformly distributed force with density $F(t) = F_0 \sin \omega t$ per unit mass, so

$$y_{tt} + a^4 y_{xxxx} = F_0 \sin \omega t.$$

Derive the steady periodic solution (with $\alpha^2 = \omega/a^2$) $y(x, t) = G(t)H(x)$, where

$$G(t) = \frac{a^4 F_0 \sin \omega t}{2\omega^2}$$

and

$$H(x) = \frac{\sinh \alpha(L - x) + \sinh \alpha x}{\sinh \alpha L} + \frac{\sin \alpha(L - x) + \sin \alpha x}{\sin \alpha L} - 2.$$

20. A string with fixed ends is acted on by a periodic force $F(x, t) = F(x) \sin \omega t$ per unit mass, so

$$y_{tt} = a^2 y_{xx} + F(x) \sin \omega t.$$

Substitute

$$y(x, t) = \sum_{n=1}^{\infty} c_n \sin \frac{n\pi x}{L} \sin \omega t$$

and

$$F(x) = \sum_{n=1}^{\infty} F_n \sin \frac{n\pi x}{L}$$

to derive the steady periodic solution

$$y(x, t) = \sum_{n=1}^{\infty} \frac{F_n \sin (n\pi x/a) \sin \omega t}{\omega_n^2 - \omega^2}$$

where $\omega_n = n\pi a/L$. Hence resonance does not result if $\omega = \omega_n$ but $F_n = 0$.

21. Suppose that the string of Problem 20 is also subject to air resistance proportional to its velocity, so

$$y_{tt} = a^2 y_{xx} - cy_t + F(x) \sin \omega t.$$

Generalize the method of Problem 20 to derive the steady periodic solution

$$y(x, t) = \sum_{n=1}^{\infty} \rho_n F_n \sin \frac{n\pi x}{L} \sin (\omega t - \alpha_n),$$

where $\rho_n = [(\omega_n^2 - \omega^2)^2 + \omega^2 c^2]^{-1/2}$ and $\alpha = \tan^{-1}[\omega c/(\omega_n^2 - \omega^2)]$. Note the similarity to damped forced motion of a mass on a spring.

22. The **telephone equation** for the voltage $e(x, t)$ in a long transmission line at the point $x \geq 0$ at time t is

$$\frac{\partial^2 e}{\partial x^2} = LC \frac{\partial^2 e}{\partial t^2} + (RC + LG)\frac{\partial e}{\partial t} + RGe,$$

where $R, L, G,$ and C denote resistance, inductance, conductance, and capacitance (all per unit length of line), respectively. The condition $e(0, t) = E_0 \cos \omega t$ represents a periodic signal voltage at the origin of transmission at $x = 0$. Assume that $e(x, t)$ is bounded as $x \longrightarrow +\infty$. Substitute $\tilde{e}(x, t) = E(x)e^{i\omega t}$ to derive the steady periodic solution

$$e(x, t) = E_0 e^{-\alpha x} \cos (\omega t - \beta x),$$

where α and β are the real and imaginary parts, respectively, of the complex number $[(RG - LC\omega^2) + i\omega(RC + LG)]^{1/2}$.

23. The temperature $u(x, t)$ at the point $x \geq 0$ at time t of water moving with velocity $\gamma \geq 0$ in a long pipe satisfies the equation

$$\frac{\partial u}{\partial t} = k\frac{\partial^2 u}{\partial x^2} - \gamma\frac{\partial u}{\partial x}.$$

Suppose that $u(0, t) = A_0 \cos \omega t$ and that $u(x, t)$ is bounded as $x \to +\infty$. Substitute $\tilde{u}(x, t) = X(x)e^{i\omega t}$ to derive the steady periodic solution

$$u(x, t) = A_0 e^{-\alpha x} \cos(\omega t - \beta x),$$

where

$$\alpha = \frac{1}{2k}(\gamma^4 + 16k^2\omega^2)^{1/4} \cos\frac{\phi}{2} - \frac{\gamma}{2k},$$

$$\beta = \frac{1}{2k}(\gamma^4 + 16k^2\omega^2)^{1/4} \sin\frac{\phi}{2},$$

$$\phi = \tan^{-1}\frac{4k}{\gamma^2}.$$

Show also that when $\gamma = 0$, this solution reduces to that in Eq. (29).

8.4
Applications of Bessel Functions

When the Laplacian $\nabla^2 u = u_{xx} + u_{yy} + u_{zz}$ of a function $u = u(x, y, z)$ is transformed by means of the substitution $x = r\cos\theta$, $y = r\sin\theta$, the result is the **Laplacian in cylindrical coordinates**,

$$\nabla^2 u = \frac{\partial^2 u}{\partial r^2} + \frac{1}{r}\frac{\partial u}{\partial r} + \frac{1}{r^2}\frac{\partial^2 u}{\partial \theta^2} + \frac{\partial^2 u}{\partial z^2}. \tag{1}$$

For an application of this, consider a very long uniform solid cylinder of radius c that lies along the z-axis. Suppose that it is heated to an initial temperature depending only on the distance r from the z-axis, and that at time $t = 0$ the condition

$$\beta_1 u + \beta_2 u_r|_{r=c} = 0 \tag{2}$$

is imposed on its surface $r = c$. Note that (2) reduces to the condition $u = 0$ if $\beta_1 = 1$ and $\beta_2 = 0$, to the insulation condition $u_r = 0$ if $\beta_1 = 0$ and $\beta_2 = 1$, and to the heat transfer condition $hu + u_r = 0$ if $\beta_1 = h$ and $\beta_2 = 1$. It is reasonable to expect the temperature u within the cylinder at time t to depend only on r, so we write $u = u(r, t)$. Then $u_{\theta\theta} = u_{zz} = 0$, so substitution of (1) in the heat equation $u_t = k\nabla^2 u$ yields the boundary value problem

$$\frac{\partial u}{\partial t} = k\left(\frac{\partial^2 u}{\partial r^2} + \frac{1}{r}\frac{\partial u}{\partial r}\right) \qquad (r < c, \qquad t > 0); \tag{3}$$

$$\beta_1 u(c, t) + \beta_2 u_r(c, t) = 0, \tag{4}$$

$$u(r, 0) = f(r) \qquad \text{(initial temperature)}. \tag{5}$$

To solve this problem by separation of variables, we substitute $u(r, t) = R(r)T(t)$ in Eq. (3); thus we obtain

$$RT' = k\left(R''T + \frac{1}{r}R'T\right). \tag{6}$$

Division by kRT yields

$$\frac{R'' + R'/r}{R} = \frac{T'}{kT} = -\lambda. \tag{7}$$

Hence $R(r)$ must satisfy the equation

$$R'' + \frac{1}{r}R' + \lambda R = 0, \tag{8}$$

as well as the condition $\beta_1 R(r) + \beta_2 R'(r) = 0$ that follows from (4). Moreover, the equation $T' = -\lambda kT$ implies that $T(t) = e^{-\lambda kt}$ to within a constant multiple. Because the diffusivity k is positive, it follows that λ must be nonnegative if $T(t)$ is to remain bounded as $t \longrightarrow \infty$, as is required by the physical problem that Eqs. (3)–(5) model. We therefore write $\lambda = \alpha^2$.

Then Eq. (8) is, in different notation, the parametric Bessel equation

$$x^2 y'' + xy' + \alpha^2 x^2 y = 0 \tag{9}$$

of order zero that we discussed in Section 3.5. More generally, recall that the Bessel equation

$$x^2 y'' + xy' + (\alpha^2 x^2 - n^2)y = 0 \tag{10}$$

has general solution

$$y(x) = AJ_n(\alpha x) + BY_n(\alpha x) \tag{11}$$

if $\alpha > 0$. Upon division by x, the Bessel equation in (10) takes the Sturm-Liouville form

$$\frac{d}{dx}[xy'] - \frac{n^2}{x}y + \lambda xy = 0 \tag{12}$$

with $p(x) = x$, $q(x) = n^2/x$, $r(x) = x$, and $\lambda = \alpha^2$. We want to determine the nonnegative values of λ for which there is a solution of (12) in $(0, c)$ that is *continuous* (along with its derivative y') on the closed interval $[0, c]$ and satisfies the endpoint condition

$$\beta_1 y(c) + \beta_2 y'(c) = 0, \tag{13}$$

where β_1 and β_2 are not both zero.

The Sturm-Liouville problem associated with Eqs. (12) and (13) is *singular* because $p(0) = r(0) = 0$ and $q(x) \longrightarrow +\infty$ as $x \longrightarrow 0^+$, whereas we assumed in Theorem 1 of Section 8.1 that $p(x)$ and $r(x)$ were positive and that $q(x)$ was continuous on the whole interval. This problem also fails to fit the pattern of Section 8.1 in that no condition like (13) is imposed at the left endpoint $x = 0$. Nevertheless, the requirement that $y(x)$ be continuous on $[0, c]$ plays the role of such a condition. Because $Y_n(x) \longrightarrow -\infty$ as $x \longrightarrow 0$, the solution in (11) for $\alpha > 0$ can be continuous at $x = 0$ only if $B = 0$, so $y(x) = J_n(\alpha x)$ to within a constant multiple. It remains only to impose the condition in (13) at $x = c$.

It is convenient to distinguish the cases $\beta_2 = 0$ and $\beta_2 \neq 0$. If $\beta_2 = 0$, then (13) takes the simple form

$$y(c) = 0. \tag{13a}$$

If $\beta_2 \neq 0$, we multiply each term in (13) by c/β_2 and then write $h = c\beta_1/\beta_2$

to obtain the equivalent condition

$$hy(c) + cy'(c) = 0. \tag{13b}$$

We assume hereafter that $h \geq 0$.

First we consider the possibility of a zero eigenvalue $\lambda = 0$. If both $\lambda = 0$ and $n = 0$, then (12) reduces to the equation $[xy']' = 0$ with general solution $y = A \ln x + B$, and continuity on $[0, c]$ requires that $A = 0$. But then (13a) implies that $B = 0$ as well, as does (13b) unless $h = 0$, in which case $\lambda = 0$ is an eigenvalue with associated eigenfunction $y(x) = 1$.

If $\lambda = 0$ but $n > 0$, then (12) is simply the Euler-Cauchy equation

$$x^2 y'' + xy' - n^2 y = 0$$

with general solution $y = Ax^n + Bx^{-n}$, and continuity on $[0, c]$ requires that $B = 0$. But it is easy to check that $y = Ax^n$ satisfies neither (13a) nor (13b) unless $A = 0$. Thus $\lambda = 0$ is not an eigenvalue if $n > 0$. We have therefore shown that $\lambda = 0$ *is an eigenvalue of the problem in* (12)–(13) *if and only if* $n = h = 0$ *and the endpoint condition at* $x = c$ *is* $y'(c) = 0$, *in which case the associated eigenfunction is* $y(x) \equiv 1$. In this case we write $\lambda_0 = 0$ and $y_0(x) = 1$.

Now suppose that $\lambda = \alpha^2 > 0$, in which case the only solution of Eq. (12) that is continuous on $[0, c]$ is, to within a constant multiple, $y = J_n(\alpha x)$. Then the condition in (13a) requires that $J_n(\alpha c) = 0$—that is, that αc be a positive root of the equation

$$J_n(x) = 0. \tag{14a}$$

Recall from Section 3.5 that graphs of $J_0(x)$ and $J_1(x)$ look as indicated in Fig. 8.13. The graph of $J_n(x)$ for $n > 1$ resembles that of $J_1(x)$, with $J_n(0) = 0$. In particular, for each $n = 1, 2, 3, \ldots$, Eq. (14a) has an increasing infinite sequence $\{\gamma_{nk}\}_{k=1}^{\infty}$ of positive roots with $\lim_{k \to \infty} \gamma_{nk} = \infty$. These roots for $n \leq 8$ and $k \leq 20$ are given in Table 9.5 of Abramowitz and Stegun's *Handbook of Mathematical Functions*.

If $y = J_n(\alpha x)$, so that $dy/dx = \alpha J_n'(\alpha x)$, then the condition in (13b) requires that $h J_n(\alpha c) + \alpha c J_n'(\alpha c) = 0$—that is, that αc be a positive root of

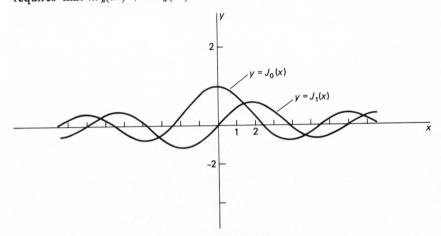

Figure 8.13 The graphs of $J_0(x)$ and $J_1(x)$.

the equation

$$hJ_n(x) + xJ'_n(x) = 0. \tag{14b}$$

It is known that this equation also has an increasing infinite sequence $\{\gamma_{nk}\}_{k=1}^{\infty}$ of positive roots diverging to $+\infty$. If $h = 0$, (14b) reduces to the equation $J'_n(x) = 0$; the roots of this equation appear in Table 9.5 of Abramowitz and Stegun. In the important case $n = 0$, the first five roots of (14b) for various values of h can be found in Table 9.7 of the same reference.

If either of the boundary conditions in (13a) and (13b) holds, then the kth positive eigenvalue is $\lambda_k = \gamma_k^2/c^2$, where we write γ_k for the kth positive root of the appropriate one of the Eqs. (14a) and (14b); the associated eigenfunction is $y_k(x) = J_n(\gamma_k x/c)$. The table in Fig. 8.14 summarizes this situation for ready reference. The exceptional case $n = h = 0$ corresponding to the endpoint condition $y'(c) = 0$ is listed separately. We have discussed only nonnegative eigenvalues, but it can be proved that the problem in (12)–(13) has no negative eigenvalues; see Section 78 of R. V. Churchill and J. W. Brown, *Fourier Series and Boundary Value Problems* (New York: McGraw-Hill, 3rd ed., 1978).

	Endpoint condition	Eigenvalues	Associated eigenfunctions
CASE 1:	$y(c) = 0$	$\lambda_k = \gamma_k^2/c^2$; $\{\gamma_k\}_1^{\infty}$ the positive roots of $J_n(x) = 0$	$y_k(x) = J_n\left(\dfrac{\gamma_k x}{c}\right)$
CASE 2:	$hy(c) + cy'(c) = 0$; h and n not both zero	$\lambda_k = \gamma_k^2/c^2$; $\{\gamma_k\}_1^{\infty}$ the positive roots of $hJ_n(x) + xJ_n'(x) = 0$	$y_k(x) = J_n\left(\dfrac{\gamma_k x}{c}\right)$
CASE 3:	$y'(c) = 0$, $n = 0$	$\gamma_0 = 0, \ \gamma_k = \lambda_k^2/c^2$; $\{\gamma_k\}_1^{\infty}$ the positive roots of $J_0'(x) = 0$	$y_0(x) = 1,$ $y_k(x) = J_0\left(\dfrac{\gamma_k x}{c}\right)$

Figure 8.14

Now that we know that the singular Sturm-Liouville problem in (12)–(13) has an infinite sequence of eigenvalues and associated eigenfunctions similar to those of a regular Sturm-Liouville problem, we can discuss eigenfunction expansions. In either Case 1 or Case 2 of Fig. 8.14, we expect a piecewise smooth function $f(x)$ on $[0, c]$ to have an eigenfunction series of the form

$$f(x) = \sum_{k=1}^{\infty} c_k y_k(x) = \sum_{k=1}^{\infty} c_k J_n\left(\frac{\gamma_k x}{c}\right), \tag{15}$$

while in the exceptional Case 3, the series will also contain a constant term c_0

corresponding to $\lambda_0 = 0$, $y_0(x) = 1$. If the conclusion of Theorem 2 in Section 8.1 is to hold (despite the fact that its hypotheses are not satisfied), then the eigenfunctions

$$J_n\left(\frac{\lambda_k x}{c}\right), \qquad k = 1, 2, \ldots$$

must be orthogonal on $[0, c]$ with weight function $r(x) = x$. Indeed, if we substitute $p(x) = r(x) = x$ and

$$y_k(x) = J_n\left(\frac{\gamma_k x}{c}\right), \qquad y_k'(x) = \frac{\gamma_k}{c}J_n'\left(\frac{\gamma_k x}{c}\right)$$

in Eq. (22) of Section 8.1, the result is

$$(\lambda_i - \lambda_j)\int_0^c xJ_n\left(\frac{\gamma_i x}{c}\right)J_n\left(\frac{\gamma_j x}{c}\right)dx$$

$$= \left[x\left(\frac{\gamma_j}{c}J_n\left(\frac{\gamma_i x}{c}\right)J_n'\left(\frac{\gamma_j x}{c}\right) - \frac{\gamma_i}{c}J_n\left(\frac{\gamma_j x}{c}\right)J_n'\left(\frac{\gamma_i x}{c}\right)\right)\right]_0^c$$

$$= \gamma_j J_n(\gamma_i)J_n'(\gamma_j) - \gamma_i J_n(\gamma_j)J_n'(\gamma_i). \tag{16}$$

It is clear that the quantity in (16) is zero if γ_i and γ_j are both roots of Eq. (14a), $J_n(x) = 0$, while if both are roots of Eq. (14b) it reduces to

$$J_n(\gamma_i)[-hJ_n(\gamma_j)] - J_n(\gamma_j)[-hJ_n(\gamma_i)] = 0.$$

In either event we thus see that if $i \neq j$, then

$$\int_0^c xJ_n\left(\frac{\gamma_i x}{c}\right)J_n\left(\frac{\gamma_j x}{c}\right)dx = 0. \tag{17}$$

This orthogonality with weight function x is what we need to determine the coefficients in the eigenfunction series in (15). If we multiply each term in (15) by $xJ_n(\gamma_k x/c)$ and then integrate termwise, we get

$$\int_0^c xf(x)J_n\left(\frac{\gamma_k x}{c}\right)dx = \sum_{j=1}^\infty c_j\int_0^c xJ_n\left(\frac{\gamma_j x}{c}\right)J_n\left(\frac{\gamma_k x}{c}\right)dx$$

$$= c_k\int_0^c x\left[J_n\left(\frac{\gamma_k x}{c}\right)\right]^2 dx$$

by (17). Hence

$$c_k = \frac{\int_0^c xf(x)J_n(\gamma_k x/c)\,dx}{\int_0^c x[J_n(\gamma_k x/c)]^2\,dx}. \tag{18}$$

With these coefficients, series of the form in (15) are often called **Fourier-Bessel series**. It is known that a Fourier-Bessel series for a piecewise smooth function $f(x)$ satisfies the convergence conclusion of Theorem 3 in Section 8.1. That is, it converges to the average value $\frac{1}{2}[f(x+) + f(x-)]$ at each point of $(0, c)$, and hence to the value $f(x)$ at each interior point of continuity.

In spite of their appearance, the denominator integrals in (18) are not difficult to evaluate. Suppose that $y(x) = J_n(\alpha x)$, so that y satisfies the parametric Bessel equation of order n,

$$\frac{d}{dx}[xy'] + \left(\alpha^2 x - \frac{n^2}{x}\right)y = 0. \tag{19}$$

By multiplying this equation by $2xy'$ and integrating by parts—see Problem 10—one can easily derive the formula

$$2\alpha^2 \int_0^c x[J_n(\alpha x)]^2\, dx = \alpha^2 c^2 [J_n'(\alpha c)]^2 + (\alpha^2 c^2 - n^2)[J_n(\alpha c)]^2. \tag{20}$$

Now suppose that $\alpha = \gamma_k/c$, where γ_k is a root of the equation $J_n(x) = 0$. We apply Eq. (20) as well as the recurrence formula

$$xJ_n'(x) = nJ_n(x) - xJ_{n+1}(x)$$

of Section 3.5, which implies that $J_n'(\gamma_k) = -J_{n+1}(\gamma_k)$. The result is

$$\int_0^c x\left[J_n\!\left(\frac{\gamma_k x}{c}\right)\right]^2 dx = \frac{c^2}{2}[J_n'(\gamma_k)]^2 = \frac{c^2}{2}[J_{n+1}(\gamma_k)]^2. \tag{21}$$

The other entries in the table of Fig. 8.15 follow similarly from Eq. (20). Fourier-Bessel series with $n = 0$ are the most common (see Problem 9). The forms they take in our three cases are listed below (we have used m in place of k for the index of summation).

$\{\gamma_k\}_1^\infty$ the positive roots of the equation	Value of $\int_0^c x\left[J_n\!\left(\frac{\gamma_k x}{c}\right)\right]^2 dx$
Case 1: $J_n(x) = 0$	$\dfrac{c^2}{2}[J_{n+1}(\gamma_k)]^2$
Case 2: $hJ_n(x) + xJ_n'(x) = 0$ (n and h not both zero)	$\dfrac{c^2(\gamma_k^2 - n^2 - h^2)}{2\gamma_k^2}[J_n(\gamma_k)]^2$
Case 3: $J_0'(x) = 0$	$\dfrac{c^2}{2}[J_0(\gamma_k)]^2$

Figure 8.15

Case 1 with n = 0

If $\{\gamma_m\}_{m=1}^\infty$ are the positive roots of the equation $J_0(x) = 0$, then

$$f(x) = \sum_{m=1}^\infty c_m J_0\!\left(\frac{\gamma_m x}{c}\right);$$

$$c_m = \frac{2}{c^2[J_1(\gamma_m)]^2} \int_0^c xf(x)J_0\!\left(\frac{\gamma_m x}{c}\right) dx. \tag{22}$$

Case 2 with n = 0

If $\{\gamma_m\}_1^\infty$ are the positive roots of $hJ_0(x) + xJ_0'(x) = 0$ with $h > 0$, then

$$f(x) = \sum_{m=1}^\infty c_m J_0\!\left(\frac{\gamma_m x}{c}\right);$$

$$c_m = \frac{2\gamma_m^2}{c^2(\gamma_m^2 + h^2)[J_0(\gamma_m)]^2} \int_0^c xf(x)J_0\!\left(\frac{\gamma_m x}{c}\right) dx. \tag{23}$$

Case 3

If $\{\gamma_m\}_1^\infty$ are the positive roots of $J_0'(x) = 0$, then (see Problem 10)

$$f(x) = c_0 + \sum_{m=1}^\infty c_m J_0\left(\frac{\gamma_m x}{c}\right); \tag{24a}$$

$$c_0 = \frac{2}{c^2} \int_0^c x f(x)\, dx, \tag{24b}$$

$$c_m = \frac{2}{c^2[J_0(\gamma_m)]^2} \int_0^c x f(x) J_0\left(\frac{\gamma_m x}{c}\right) dx. \tag{24c}$$

EXAMPLE 1 Suppose that a long solid circular cylinder of radius c has initial temperature $u(r, 0) = u_0$ (constant). Find $u(r, t)$ if (a) $u(c, t) = 0$; (b) the boundary of the cylinder is insulated, so that $u_r(c, t) = 0$; (c) heat transfer takes place across its boundary, so that $Hu(c, t) + Ku_r(c, t) = 0$ with H and K positive.

Solution   (a) With $u(r, t) = R(r)T(t)$, we saw previously that $\lambda = \alpha^2 > 0$ in Eq. (7), so

$$r^2 R'' + rR' + \alpha^2 r^2 R = 0 \tag{25}$$

and

$$T' = -\alpha^2 kT. \tag{26}$$

Equation (25) is Bessel's parametric equation of order zero, and its only continuous nontrivial solutions on $[0, c]$ are of the form $R(r) = AJ_0(\alpha r)$. Now $u(c, t) = 0$ yields $R(c) = AJ_0(\alpha c) = 0$, so c must be one of the roots $\{\gamma_n\}_1^\infty$ of the equation $J_0(x) = 0$. Thus the eigenvalues and eigenfunctions are

$$\lambda_n = \frac{\gamma_n^2}{c^2}, \qquad R_n(r) = J_0\left(\frac{\gamma_n r}{c}\right) \tag{27}$$

for $n = 1, 2, 3, \ldots$. Then the equation $T_n' = -(\gamma_n^2/c^2)kT_n$ yields $T_n(t) = \exp(-\gamma_n^2 kt/c^2)$ to within a constant coefficient. Hence the series

$$u(r, t) = \sum_{n=1}^\infty c_n \exp(-\gamma_n^2 kt/c^2) J_0\left(\frac{\gamma_n r}{c}\right) \tag{28}$$

satisfies formally the heat equation and the boundary condition $u(c, t) = 0$. It remains to choose the coefficients so that

$$u(r, 0) = \sum_{n=1}^\infty c_n J_0\left(\frac{\gamma_n r}{c}\right) = u_0.$$

Because $J_0(\gamma_n) = 0$, the formula in (22) yields

$$c_n = \frac{2u_0}{c^2[J_1(\gamma_n)]^2} \int_0^c r J_0\left(\frac{\gamma_n r}{c}\right) dr$$

$$= \frac{2u_0}{\gamma_n^2[J_1(\gamma_n)]^2} \int_0^{\gamma_n} x J_0(x)\, dx$$

$$= \frac{2u_0}{\gamma_n^2[J_1(\gamma_n)]^2} \Big[x J_1(x) \Big]_0^{\gamma_n} = \frac{2u_0}{\gamma_n J_1(\gamma_n)}.$$

Here we have used the integral

$$\int x J_0(x)\, dx = x J_1(x) + C.$$

On substitution of the coefficients above in (28), we finally obtain

$$u(r, t) = 2u_0 \sum_{n=1}^{\infty} \frac{1}{\gamma_n J_1(\gamma_n)} \exp\left(-\gamma_n^2 kt/c^2\right) J_0\left(\frac{\gamma_n r}{c}\right). \qquad (29)$$

Because of the presence of the exponential factors, only a few terms are ordinarily needed for numerical computations. The data required for the first five terms are listed in Fig. 8.16. For instance, suppose that the cylinder has radius $c = 10$ cm, is made of iron with thermal diffusivity $k = 0.15$, and has initial temperature $u_0 = 100°C$ throughout. Then, because $J_0(0) = 1$, we find from (29) that the temperature at its axis ($r = 0$) after two minutes ($t = 120$) will be

$$u(0, 120) = 200 \sum_{n=1}^{\infty} \frac{1}{\gamma_n J_1(\gamma_n)} \exp\left(-[0.18]\gamma_n^2\right)$$

$$\approx 200(0.28283 - 0.00221 + 0.00000 - \cdots);$$

thus $u(0, 120)$ will be approximately $56.13°C$.

n	γ_n	$J_1(\gamma_n)$
1	2.40483	+0.51915
2	5.52008	−0.34026
3	8.65373	+0.27145
4	11.79153	−0.23246
5	14.93092	+0.20655

Figure 8.16

(b) If the boundary is insulated, we certainly should find that $u(r, t) \equiv u_0$. Inasmuch as $R'(c) = 0$, we have the case $n = h = 0$ in (25), so $\lambda_0 = 0$ is an eigenvalue with eigenfunction $R_0(r) = 1$. A corresponding solution of (26) is $T_0(t) = 1$. The positive eigenvalues are again given by (27), except that now the numbers $\{\gamma_n\}_1^{\infty}$ are the positive roots of $J_0'(x) = 0$. The solution is therefore of the form

$$u(r, t) = c_0 + \sum_{n=1}^{\infty} c_n \exp\left(-\gamma_n^2 kt/c^2\right) J_0\left(\frac{\gamma_n r}{c}\right),$$

and we want

$$u(r, 0) = c_0 + \sum_{n=1}^{\infty} c_n J_0\left(\frac{\gamma_n r}{c}\right) = u_0.$$

Thus we have Case 3, and the formulas in (24) yield

$$c_0 = \frac{2}{c^2} \int_0^c r u_0 \, dr = \frac{2u_0}{c^2}\left[\frac{1}{2} r^2\right]_0^c = u_0,$$

$$c_n = \frac{2u_0}{c^2[J_0(\gamma_n)]^2} \int_0^c r J_0\left(\frac{\gamma_n r}{c}\right) dr = \frac{2u_0}{\gamma_n^2[J_0(\gamma_n)]^2} \int_0^{\gamma_n} x J_0(x)\, dx$$

$$= \frac{2u_0}{\gamma_n^2[J_0(\gamma_n)]^2}\left[x J_1(x)\right]_0^{\gamma_n} = \frac{2u_0 J_1(\gamma_n)}{\gamma_n[J_0(\gamma_n)]^2} = 0$$

because $J_1(\gamma_n) = -J_0'(\gamma_n) = 0$. Thus we find that $u(r, t) \equiv u_0$, as expected.

(c) The substitution $u(r, t) = R(r)T(t)$ now leads to

$$r^2R'' + rR' + \lambda r^2 R = 0, \tag{30}$$

$$HR(c) + KR'(c) = 0.$$

The endpoint condition at $r = c$ can be written in the form

$$hR(c) + cR'(c) = 0 \tag{31}$$

where $h = cH/K > 0$. In comparing (31) with (14b), we see that we have Case 2 with $n = 0$. Hence the eigenvalues and associated eigenfunctions of the problem in (30) are

$$\lambda_n = \frac{\gamma_n^2}{c^2}, \qquad R_n(r) = J_0\left(\frac{\gamma_n r}{c}\right), \tag{32}$$

where $\{\gamma_n\}_1^\infty$ are the positive roots of the equation $hJ_0(x) + xJ_0'(x) = 0$. Now $T' = -\lambda kt$ as before; consequently, the solution is of the form

$$u(r, t) = \sum_{n=1}^\infty c_n \exp\left(-\gamma_n^2 kt/c^2\right)J_0\left(\frac{\gamma_n r}{c}\right), \tag{33}$$

and the formula in (23) yields

$$c_n = \frac{2u_0\gamma_n^2}{c^2(\gamma_n^2 + h^2)[J_0(\gamma_n)]^2} \int_0^c rJ_0\left(\frac{\gamma_n r}{c}\right) dr$$

$$= \frac{2u_0}{(\gamma_n^2 + h^2)[J_0(\gamma_n)]^2} \int_0^{\gamma_n} xJ_0(x)\, dx = \frac{2u_0\gamma_n J_1(\gamma_n)}{(\gamma_n^2 + h^2)[J_0(\gamma_n)]^2}.$$

Hence

$$u(r, t) = 2u_0 \sum_{n=1}^\infty \frac{\gamma_n J_1(\gamma_n)}{(\gamma_n^2 + h^2)[J_0(\gamma_n)]^2} \exp\left(-\gamma_n^2 kt/c^2\right)J_0\left(\frac{\gamma_n r}{c}\right). \tag{34}$$

EXAMPLE 2 Suppose that a flexible circular membrane of radius c vibrates under tension T in such a way that its (normal) displacement u depends only upon time t and the distance r from its center; this is the case of *radial vibrations*. Then $u = u(r, t)$ satisfies the partial differential equation

$$\frac{\partial^2 u}{\partial t^2} = a^2\nabla^2 u = a^2\left(\frac{\partial^2 u}{\partial r^2} + \frac{1}{r}\frac{\partial u}{\partial r}\right) \tag{35}$$

where $a^2 = T/\rho$, and $u(c, t) = 0$ if the boundary of the membrane is fixed. Find the natural frequencies and normal modes of radial vibration of the membrane.

Solution We apply the method of Section 8.3: Substitution of $u(r, t) = R(r)\sin \omega t$ in (35) yields

$$-\omega^2 R \sin \omega t = a^2(R'' \sin \omega t + \frac{1}{r}R' \sin \omega t).$$

Thus ω and $R(r)$ must satisfy the equation

$$r^2 R'' + rR' + \frac{\omega^2}{a^2}r^2 R = 0, \tag{36}$$

as well as the condition $R(c) = 0$ that follows from $u(c, t) = 0$. Equation

(36) is the Bessel equation of order zero with parameter $\alpha = \omega/a$, and its only nontrivial continuous solution is (to within a constant multiple) $R(r) = J_0(\omega r/a)$. Hence $R(c) = J_0(\omega c/a) = 0$, so $\omega c/a$ must be one of the positive roots $\{\gamma_n\}_1^\infty$ of the equation $J_0(x) = 0$. Thus the nth natural (circular) frequency and a corresponding natural mode of vibration are

$$\omega_n = \frac{\gamma_n a}{c}, \qquad u_n(r, t) = J_0\left(\frac{\gamma_n r}{c}\right) \sin \frac{\gamma_n a t}{c}. \qquad (37)$$

On inspection of the table of values of $\{\gamma_n\}$ in Fig. 8.16, we see that the higher natural frequencies ω_n for $n > 1$ are *not* integral multiples of $\omega_1 = \gamma_1 a/c$; this is why the sound of a vibrating circular drumhead is not perceived as being harmonious. The vibration of the membrane in the nth normal mode in (37) is illustrated in Fig. 8.17, which shows a vertical cross section through its center. In addition to the boundary $r = c$, there are $n - 1$ fixed circles, called *nodal circles*, with radii $r_i = \gamma_i c/\gamma_n$, $i = 1, 2, \ldots, n - 1$. The annular regions of the membrane between consecutive pairs of nodal circles move alternately up and down between the surfaces $u = \pm J_0(\gamma_n r/c)$.

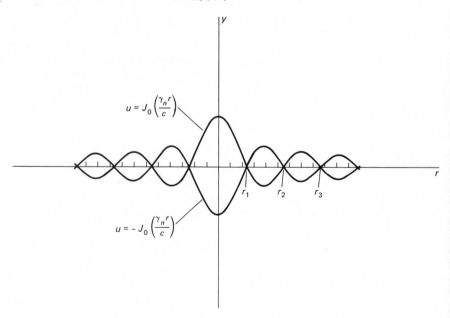

Figure 8.17 Cross section of the vibrating circular membrane.

8.4 Problems

1. Suppose that the circular membrane of Example 2 has initial position $u(r, 0) = f(r)$ and initial velocity $u_t(r, t) = 0$. Derive by separation of variables the solution

$$u(r, t) = \sum_{n=1}^\infty c_n J_0\left(\frac{\gamma_n r}{c}\right) \cos \frac{\gamma_n a t}{c}$$

where $\{\gamma_n\}_1^\infty$ are the positive roots of $J_0(x) = 0$ and

$$c_n = \frac{2}{c^2[J_1(\gamma_n)]^2} \int_0^c rf(r)J_0\left(\frac{\gamma_n r}{c}\right) dr.$$

2. Assume that the circular membrane of Example 2 has initial position $u(r, 0) = 0$ and initial velocity $u_t(r, 0) = v_0$ (constant). Derive the solution

$$u(r, t) = \frac{2cv_0}{a} \sum_{n=1}^\infty \frac{J_0(\gamma_n r/c) \sin (\gamma_n at/c)}{\gamma_n^2 J_1(\gamma_n)}$$

where $\{\gamma_n\}_1^\infty$ are the positive roots of $J_0(x) = 0$.

3. (a) Find $u(r, t)$ in the case the circular membrane of Example 2 has initial position $u(r, 0) = 0$ and initial velocity

$$u_t(r, t) = \begin{cases} \dfrac{P_0}{\rho\pi\epsilon^2} & \text{if } 0 \leq r < \epsilon, \\ 0 & \text{if } \epsilon < r \leq c. \end{cases}$$

(b) Use the fact that $(1/x)J_1(x) \to 1$ as $x \to 0$ to find the limiting value of the result in part (a) as $\epsilon \to 0$. You should obtain

$$u(r, t) = \frac{2aP_0}{\pi cT} \sum_{n=1}^\infty \frac{J_0(\gamma_n r/c) \sin (\gamma_n at/c)}{\gamma_n[J_1(\gamma_n)]^2},$$

where $\{\gamma_n\}$ are the roots of $J_0(x) = 0$. This function describes the motion of a drumhead resulting from an initial momentum impulse P_0 at its center.

4. (a) A circular plate of radius c has insulated faces and heat capacity s calories per degree per square centimeter. Find $u(r, t)$ given $u(c, t) = 0$ and

$$u(r, 0) = \begin{cases} \dfrac{q_0}{s\pi\epsilon^2} & \text{if } 0 \leq r < \epsilon, \\ 0 & \text{if } \epsilon < r \leq c. \end{cases}$$

(b) Take the limit as $\epsilon \to 0$ of the result in part (a) to obtain

$$u(r, t) = \frac{2q_0}{s\pi c^2} \sum_{n=1}^\infty \frac{1}{[J_1(\gamma_n)]^2} \exp\left(-\gamma_n^2 kt/c^2\right) J_0\left(\frac{\gamma_n r}{c}\right)$$

(where $\{\gamma_n\}$ are the roots of $J_0(x) = 0$) for the temperature resulting from an initial injection of q_0 calories of heat at the center.

Problems 5–7 deal with the steady-state temperature $u = u(r, z)$ in a solid cylinder of radius $r = c$ with bottom $z = 0$ and top $z = L$, given that u satisfies Laplace's equation

$$\frac{\partial^2 u}{\partial r^2} + \frac{1}{r}\frac{\partial u}{\partial r} + \frac{\partial^2 u}{\partial z^2} = 0.$$

5. If $u(r, L) = u_0$ and the rest of the surface of the cylinder is held at temperature zero, use separation of variables to derive the solution

$$u(r, z) = 2u_0 \sum_{n=1}^\infty \left(\frac{J_0(\gamma_n r/c)}{\gamma_n J_1(\gamma_n)}\right)\left(\frac{\sinh \gamma_n z/c}{\sinh \gamma_n L/c}\right),$$

where $\{\gamma_n\}$ are the roots of $J_0(x) = 0$.

6. (a) If $u(r, L) = f(r)$, $u(r, 0) = 0$, and the cylindrical surface $r = c$ is insulated, derive a solution of the form

$$u(r, z) = c_0 z + \sum_{n=1}^\infty c_n J_0\left(\frac{\gamma_n r}{c}\right) \sinh \frac{\gamma_n z}{c}$$

where $\{\gamma_n\}$ are the positive roots of $J_0'(x) = 0$.

(b) Suppose that $f(r) = u_0$ (constant). Deduce from the result in part (a) that $u(r, z) = u_0 z/L$.

7. Let $c = 1$ and $L = +\infty$, so that the cylinder is semi-infinite. If $u(r, 0) = u_0$, $hu(1, z) + u_r(1, z) = 0$ (heat transfer on the cylindrical surface), and $u(r, z)$ is bounded as $z \longrightarrow +\infty$, derive the solution

$$u(r, z) = 2hu_0 \sum_{n=1}^{\infty} \frac{\exp(-\gamma_n z)J_0(\gamma_n r)}{(\gamma_n^2 + h^2)J_0(\gamma_n)},$$

where $\{\gamma_n\}$ are the positive roots of the equation $hJ_0(x) = xJ_1(x)$.

8. Begin with the parametric Bessel equation of order n,

$$[xy']' + \left(\alpha^2 x - \frac{n^2}{x}\right)y = 0. \tag{38}$$

Multiply each term by $2xy'$, then write the result as

$$[(xy')^2]' + (\alpha^2 x^2 - n^2)[y^2]' = 0.$$

Integrate each term, using integration by parts for the second term, to obtain

$$\left[(xy')^2 + (\alpha^2 x^2 - n^2)y^2\right]_0^c - 2\alpha^2 \int_0^c xy^2 \, dx = 0.$$

Finally substitute $y = J_n(\alpha x)$, a solution of Eq. (38), to get the formula in Eq. (20) in the text.

9. This problem provides the coefficient integrals for Fourier-Bessel series with $n = 0$.
   (a) Substitute $n = 0$ in the result of Problem 8 to obtain the integral formula

   $$\int_0^c x[J_0(\alpha x)]^2 \, dx = \frac{c^2}{2}\{[J_0(\alpha c)]^2 + [J_1(\alpha c)]^2\}. \tag{39}$$

   (b) Suppose that $\alpha = \gamma_k/c$ where γ_k is a root of the equation $J_0(x) = 0$. Deduce that

   $$\int_0^c x\left[J_0\left(\frac{\gamma_k x}{c}\right)\right]^2 \, dx = \frac{c^2}{2}[J_1(\gamma_k)]^2.$$

   (c) Suppose that $\alpha = \gamma_k/c$, where γ_k is a root of the equation $J_0'(x) = 0$. Deduce that

   $$\int_0^c x\left[J_0\left(\frac{\gamma_k x}{c}\right)\right]^2 \, dx = \frac{c^2}{2}[J_0(\gamma_k)]^2.$$

   (d) Suppose that $\alpha = \gamma_k/c$, where γ_k is a root of the equation $hJ_0(x) + xJ_0'(x) = 0$. Deduce from Eq. (39) that

   $$\int_0^c x\left[J_0\left(\frac{\gamma_k x}{c}\right)\right]^2 \, dx = \frac{c^2(\gamma_k^2 + h^2)}{2\gamma_k^2}[J_0(\gamma_k)]^2.$$

10. Suppose that $\{\gamma_m\}_1^{\infty}$ are the positive roots of the equation $J_0'(x) = 0$ and that

    $$f(x) = c_0 + \sum_{m=1}^{\infty} c_m J_0\left(\frac{\gamma_m x}{c}\right). \tag{40}$$

    (a) Multiply each side in (40) by x and then integrate termwise from $x = 0$ to $x = c$ to show that

    $$c_0 = \frac{2}{c^2} \int_0^c xf(x) \, dx.$$

(b) Multiply each side in Eq. (40) by $xJ_0(\gamma_n x/c)$ and then integrate termwise to show that

$$c_m = \frac{2}{c^2[J_0(\gamma_m)]^2} \int_0^c xf(x)J_0\left(\frac{\gamma_m x}{c}\right) dx.$$

11. If a circular membrane with fixed boundary is subjected to a periodic force $F_0 \sin \omega t$ per unit mass uniformly distributed over the membrane, then its displacement function $u(r, t)$ satisfies the equation

$$\frac{\partial^2 u}{\partial t^2} = a^2\left(\frac{\partial^2 u}{\partial r^2} + \frac{1}{r}\frac{\partial u}{\partial r}\right) + F_0 \sin \omega t.$$

Substitute $u(r, t) = R(r) \sin \omega t$ to find a steady periodic solution.

12. Consider a vertically hanging cable of length L and weight w per unit length, with fixed upper end at $x = L$ and free lower end at $x = 0$, as shown in Fig. 8.18. When the cable vibrates transversely, its displacement function $y(x, t)$ satisfies the equation

$$\frac{w}{g}\frac{\partial^2 y}{\partial t^2} = \frac{\partial}{\partial x}\left(wx\frac{\partial y}{\partial x}\right)$$

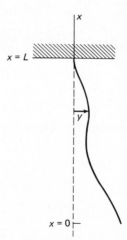

$x = L$

y

$x = 0$

Figure 8.18 The vertical hanging cable of Problem 12.

because the tension is $T(x) = wx$. Substitute $y(x, t) = X(x) \sin \omega t$; then apply the theorem of Section 3.6 to solve the ordinary differential equation that results. Deduce from the solution that the natural frequencies of vibration of the hanging cable are given by

$$\omega_n = \frac{\gamma_n}{2}\sqrt{\frac{g}{L}} \quad \left(\frac{\text{rad}}{\text{s}}\right)$$

where $\{\gamma_n\}_1^\infty$ are the roots of $J_0(x) = 0$. Historically, this problem was the first in which Bessel functions appeared.

Problems 13–15 deal with a canal of length L off the ocean (see Fig. 8.19 for a top view). Its vertical cross section at x is rectangular with width w(x) and depth h(x); the latter is the equilibrium depth of the water at x. Consider a periodic canal tide such that the vertical displacement of the water surface is y(x, t) = X(x) cos ωt at time t. Then

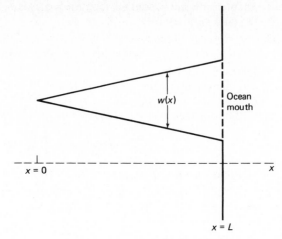

w(x)

Ocean
mouth

x = 0

x

x = L

Figure 8.19 The canal of
Problems 13–15.

y(x, t) satisfies the equation

$$\frac{w(x)}{g}\frac{\partial^2 y}{\partial t^2} = \frac{\partial}{\partial x}\left[w(x)h(x)\frac{\partial y}{\partial x}\right].$$

Let $y_0 = X(L)$ be the amplitude of the tide at the canal mouth.

13. Suppose that $w(x) = wx$ and $h(x) = h$ (constant). Show that

$$y(x, t) = y_0\frac{J_0(\omega x/\sqrt{gh})}{J_0(\omega L/\sqrt{gh})}\cos \omega t.$$

   Apply the theorem in Section 3.6.

14. Suppose that $w(x) = w$ (constant) and $h(x) = hx$. Show that

$$y(x, t) = y_0\frac{J_0(2\omega\sqrt{x/gh})}{J_0(2\omega\sqrt{L/gh})}\cos \omega t.$$

15. Suppose that $w(x) = wx$ and $h(x) = hx$ with both w and h constant. Show that

$$y(x, t) = y_0\sqrt{\frac{L}{x}}\,\frac{J_1(2\omega\sqrt{x/gh})}{J_1(2\omega\sqrt{L/gh})}\cos \omega t.$$

16. If $0 < a < b$, then the eigenvalue problem

$$\frac{d}{dx}[xy'] + \lambda xy = 0, \qquad y(a) = y(b) = 0$$

   for the parametric Bessel equation of order zero is a regular Sturm-Liouville problem. By Theorem 1 in Section 8.1, it therefore has an infinite sequence of nonnegative eigenvalues. (a) Prove that zero is not an eigenvalue. (b) Show that the nth eigenvalue is $\lambda_n = \gamma_n^2$, where $\{\gamma_n\}_1^\infty$ are the positive roots of the equation

$$J_0(ax)Y_0(bx) - J_0(bx)Y_0(ax) = 0. \tag{41}$$

   The first five roots of (41) for various values of a/b are listed in Table 9.7 of Abramowitz and Stegun, *Handbook of Mathematical Functions*. (c) Show that an associated eigenfunction is

$$R_n(x) = Y_0(\gamma_n a)J_0(\gamma_n x) - J_0(\gamma_n a)Y_0(\gamma_n x). \tag{42}$$

17. Suppose that an annular membrane with constant density ρ (per unit area) is stretched under constant tension T between the circles $r = a$ and $r = b\,(a < b)$.

Show that its nth natural (circular) frequency is $\omega_n = \gamma_n \sqrt{T/\rho}$, where $\{\gamma_n\}_1^\infty$ are the positive roots of Eq. (41).

18. Suppose that the infinite cylindrical shell $a \leq r \leq b$ has initial temperature $u(r, 0) = f(r)$, and thereafter $u(a, t) = u(b, t) = 0$. By separation of variables, derive the solution

$$u(r, t) = \sum_{n=1}^{\infty} c_n \exp\left(-\gamma_n^2 k t\right) R_n(r),$$

where $R_n(r)$ is the function in (42) and

$$c_n \int_a^b r[R_n(r)]^2 \, dr = \int_a^b rf(r)R_n(r) \, dr.$$

19. Consider the semi-infinite cylindrical shell $a \leq r \leq b$, $z \geq 0$. If $u(a, z) = u(b, z) = 0$ and $u(r, 0) = f(r)$, derive by separation of variables the steady-state temperature

$$u(r, z) = \sum_{n=1}^{\infty} c_n e^{-\gamma_n z} R_n(r),$$

where $\{c_n\}$ and $\{R_n\}$ are as given in Problems 18 and 16.

8.5
Nuclear Reactors and Other Applications

This section consists largely of four annotated problem sets, each of which will provide the student with an opportunity to work through a fairly substantial application of the method of separation of variables and eigenfunction series. As preparation, Example 1 and the problems that immediately follow it introduce the technique of representing a function $f(x, y)$ by a Fourier series in two variables.

EXAMPLE 1 Suppose that the thin rectangular plate $0 \leq x \leq a$, $0 \leq y \leq b$ has insulated faces and that its four edges are held at temperature zero. If it has initial temperature $u(x, y, 0) = f(x, y)$, then u satisfies the boundary value problem

$$u_t = k(u_{xx} + u_{yy}); \tag{1}$$

$$u(0, y, t) = u(a, y, t) = u(x, 0, t) = u(x, b, t) = 0, \tag{2}$$

$$u(x, y, 0) = f(x, y). \tag{3}$$

Find $u(x, y, t)$.

Solution We substitute $u(x, y, t) = X(x)Y(y)T(t)$ in (1). After division by $kXYT$, we obtain

$$\frac{T'}{kT} = \frac{X''}{X} + \frac{Y''}{Y}.$$

This relation can hold for all x, y, t only if each term is a constant, so we write

$$\frac{X''}{X} = -\lambda, \qquad \frac{Y''}{Y} = -\mu, \qquad \frac{T'}{kT} = -(\lambda + \mu). \tag{4}$$

Taking into account the boundary conditions in (2), we see that $X(x)$

and $Y(y)$ satisfy the separate Sturm-Liouville problems

$$X'' + \lambda X = 0, \qquad X(0) = X(a) = 0; \qquad (5)$$

$$Y'' + \mu Y = 0, \qquad Y(0) = Y(b) = 0. \qquad (6)$$

The eigenvalues and eigenfunctions of the familiar problem in (5) are

$$\lambda_m = \frac{m^2\pi^2}{a^2}, \qquad X_m(x) = \sin \frac{m\pi x}{a} \qquad (7)$$

for $m = 1, 2, 3, \ldots$. Similarly, the eigenvalues and eigenfunctions of the problem in (6) are

$$\mu_n = \frac{n^2\pi^2}{b^2}, \qquad Y_n(y) = \sin \frac{n\pi y}{b} \qquad (8)$$

for $n = 1, 2, 3, \ldots$. We use distinct indices m and n in (7) and (8) because the two problems in (5) and (6) are independent of each other.

For each pair m, n of positive integers, we must solve the third equation in (4),

$$T'_{mn} = -(\lambda_m + \mu_n)kT_{mn} = -\left(\frac{m^2}{a^2} + \frac{n^2}{b^2}\right)\pi^2 k T_{mn}. \qquad (9)$$

To within a multiplicative constant, the solution of Eq. (9) is

$$T_{mn}(t) = \exp\left(-\omega_{mn}^2 kt\right), \qquad (10)$$

where

$$\omega_{mn} = \pi\sqrt{\frac{m^2}{a^2} + \frac{n^2}{b^2}}. \qquad (11)$$

Thus we have found a "doubly infinite" collection of building blocks, and it follows that the series

$$u(x, y, t) = \sum_{m=1}^{\infty} \sum_{n=1}^{\infty} c_{mn} \exp\left(-\omega_{mn}^2 kt\right) \sin \frac{m\pi x}{a} \sin \frac{n\pi y}{b} \qquad (12)$$

formally satisfies the heat equation in (1) and the homogeneous boundary conditions in (2). It remains only to determine the coefficients $\{c_{mn}\}$ so that the series satisfies also the nonhomogeneous condition

$$u(x, y, 0) = \sum_{m=1}^{\infty} \sum_{n=1}^{\infty} c_{mn} \sin \frac{m\pi x}{a} \sin \frac{n\pi y}{b} = f(x, y). \qquad (13)$$

To do so, we first group the terms in this double series to display the total coefficient of $\sin(n\pi y/b)$ and write

$$f(x, y) = \sum_{n=1}^{\infty} \left(\sum_{m=1}^{\infty} c_{mn} \sin \frac{m\pi x}{a}\right) \sin \frac{n\pi y}{b}. \qquad (14)$$

For each fixed x, we want (14) to be the Fourier sine series of $f(x, y)$ on $0 \le y \le b$. This will be so provided that

$$\sum_{m=1}^{\infty} c_{mn} \sin \frac{m\pi x}{a} = \frac{2}{b} \int_0^b f(x, y) \sin \frac{n\pi y}{b} \, dy. \qquad (15)$$

The right-hand side in (15) is, for each fixed n, a function $F_n(x)$; viz,

$$F_n(x) = \sum_{m=1}^{\infty} c_{mn} \sin \frac{m\pi x}{a}. \qquad (16)$$

This requires that c_{mn} be the mth Fourier sine coefficient of $F_n(x)$ on $0 \leq x \leq a$; that is, that

$$c_{mn} = \frac{2}{a} \int_0^a F_n(x) \sin \frac{m\pi x}{a} \, dx. \tag{17}$$

Substituting the right-hand side in (15) for $F_n(x)$ in (17), we finally get

$$c_{mn} = \frac{4}{ab} \int_0^a \int_0^b f(x, y) \sin \frac{m\pi x}{a} \sin \frac{n\pi y}{b} \, dy \, dx \tag{18}$$

for $m = 1, 2, 3, \ldots$. With these coefficients, the series in (13) is the double Fourier sine series of $f(x, y)$ on the rectangle $0 \leq x \leq a, 0 \leq y \leq b$, and the series in (12) formally satisfies the boundary value problem in (1)–(3).

PROBLEM 1   Suppose that $f(x, y) = u_0$, a constant. Compute the coefficients in (18) to obtain the solution

$$u(x, y, t) = \frac{16u_0}{\pi^2} \sum_{m \text{ odd}} \sum_{n \text{ odd}} \frac{\exp(-\omega_{mn}^2 kt)}{mn} \sin \frac{m\pi x}{a} \sin \frac{n\pi y}{b}.$$

PROBLEM 2   Replace the boundary conditions (2) in Example 1 by

$$u(0, y, t) = u(x, 0, t) = 0,$$

$$hu(a, y, t) + u_x(a, y, t) = hu(x, b, t) + u_y(x, b, t) = 0.$$

Thus the edges $x = 0$ and $y = 0$ are still held at temperature zero, but now heat transfer takes place along the edges $x = a$ and $x = b$. Then derive the solution

$$u(x, y, t) = \sum_{m=1}^{\infty} \sum_{n=1}^{\infty} c_{mn} \exp(-\omega_{mn}^2 kt) \sin \frac{\alpha_m x}{a} \sin \frac{\beta_n y}{b}$$

where $\omega_{mn}^2 = (\alpha_m/a)^2 + (\beta_n/b)^2$, $\{\alpha_m\}$ are the positive roots of $x \tan x = -ha$, $\{\beta_n\}$ are the positive roots of $x \tan x = -hb$, and

$$c_{mn} = \frac{4}{ab} \int_0^a \int_0^b f(x, y) \sin \frac{\alpha_m x}{a} \sin \frac{\beta_n y}{b} \, dy \, dx.$$

PROBLEM 3   The displacement function $u(x, y, t)$ for a vibrating rectangular membrane satisfies the boundary value problem

$$u_{tt} = c^2(u_{xx} + u_{yy}),$$

$$u(0, y, t) = u(a, y, t) = u(x, 0, t) = u(x, b, t) = 0,$$

$$u(x, y, 0) = f(x, y), \qquad u_t(x, y, 0) = 0.$$

Derive the solution

$$u(x, y, t) = \sum_{m=1}^{\infty} \sum_{n=1}^{\infty} c_{mn} \cos \omega_{mn} ct \sin \frac{m\pi x}{a} \sin \frac{n\pi y}{b},$$

where the numbers $\{\omega_{mn}\}$ and the coefficients $\{c_{mn}\}$ are given by the formulas in (11) and (18), respectively. Because the numbers $\{\omega_{mn}c\}$ are not integral multiples of a fundamental frequency, the sound of a vibrating rectangular membrane is not harmonious.

PROBLEM 4   For the vibrating rectangular membrane of Problem 3, suppose instead that the initial conditions are

$$u(x, y, 0) = 0, \qquad u_t(x, y, 0) = v_0 \quad \text{(a constant)}.$$

Derive the solution

$$u(x, y, t) = \frac{16v_0}{\pi^2 c} \sum_{m \text{ odd}} \sum_{n \text{ odd}} \frac{\sin \omega_{mn} ct}{mn\omega_{mn}} \sin \frac{m\pi x}{a} \sin \frac{n\pi y}{b}.$$

NUCLEAR REACTORS

Suppose that a nuclear reactor core consists of a homogeneous mixture of a fissionable material (for example, $^{235}U_{92}$) and a moderator (for example, graphite $^{12}C_6$), and occupies the region R bounded by the surface S. We make the simplifying assumption that the free neutrons in the core are all traveling at the same speed (about 2.2×10^5 cm/s for the *thermal neutrons* that produce fissions); the main purpose of moderating material in a reactor core is to slow the neutrons to the velocity at which collisions with fissionable nuclei are most likely to occur. Under this assumption, the behavior of the reactor core is determined by the density $u(x, y, z, t)$ of neutrons (per unit volume) at the point (x, y, z) at time t.

It is found empirically that the transport of neutrons in a reactor core is a diffusion process much like the flow of heat. In particular, the neutron flux vector $\mathbf{\Phi}(x, y, z, t)$ satisfies the equation

$$\mathbf{\Phi} = -K\nabla u \quad (K \text{ constant})$$

from which, in Section 7.7, we derived the partial differential equation

$$\frac{1}{k} \frac{\partial u}{\partial t} = \nabla^2 u.$$

In a reactor, though, new neutrons are produced by fissions that occur at a rate proportional to the neutron density u. We therefore include a source term and get the equation

$$\frac{1}{k} \frac{\partial u}{\partial t} = \nabla^2 u + B^2 u \tag{1}$$

for the neutron density $u(x, y, z, t)$ in a homogeneous reactor core. The constant B is the so-called buckling constant that must be determined empirically for each specific mixture of fissionable and moderating material. We want to discuss the solution of Eq. (1) with an initial condition $u(x, y, z, 0) = f(x, y, z)$ and on the surface S the boundary condition $u = 0$ that corresponds to a "bare" core.

PROBLEM 1   Show that the substitution

$$v(x, y, z, t) = e^{-B^2 kt} u(x, y, z, t) \tag{2}$$

transforms (1) into the source-free diffusion equation

$$\frac{1}{k} \frac{\partial v}{\partial t} = \nabla^2 v. \tag{3}$$

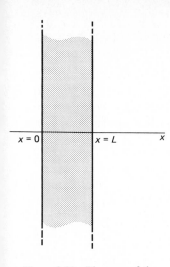

Figure 8.20 The core of the slab reactor of Problem 2.

PROBLEM 2 (Slab Reactor) Suppose that the reactor core is a very large slab of thickness L, as indicated in Fig. 8.20. Then (3) reduces to the one-dimensional diffusion equation

$$\frac{\partial v}{\partial t} = k \frac{\partial^2 v}{\partial x^2} \qquad (0 < x < L)$$

with boundary conditions

$$v(0, t) = v(L, t) = 0, \qquad v(x, 0) = f(x).$$

Solve this boundary value problem to find the neutron density

$$u(x, t) = \sum_{n=1}^{\infty} b_n \exp(P_n kt) \sin \frac{n\pi x}{L} \tag{4}$$

where $P_n = B^2 - (n^2\pi^2/L^2)$.

PROBLEM 3 Suppose that the buckling constant B satisfies the critical condition $B = \pi/L$. Deduce from Eq. (4) that

$$\lim_{t \to \infty} u(x, t) = b_1 \sin \frac{\pi x}{L} = \phi(x).$$

Thus, in this case, we have a **stable** (steady) solution. The **critical thickness** of the slab is $L = \pi/B$. Show that

$$\lim_{t \to \infty} u(x, t) = 0$$

if $L < \pi/B$. Show that

$$\lim_{t \to \infty} u(x, t) = \infty$$

if $L > \pi/B$, so in this case the slab is a bomb rather than a reactor.

PROBLEM 4 (Spherical reactor) Now suppose that the reactor core is a solid sphere of radius $r = a$, as indicated in Fig. 8.21, and that the initial neutron density is spherically symmetric; that is, that $u(r, 0) = f(r)$. Then $u(r, t)$ and $v(r, t) = \exp(-B^2 kt)u(r, t)$ are spherically symmetric, so

$$\nabla^2 v = \frac{1}{r} \frac{\partial^2}{\partial r^2}(rv),$$

by the computation of the spherical Laplacian in Section 3.7. Show that the substitution

$$w(r, t) = rv(r, t) \tag{5}$$

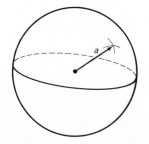

Figure 8.21 The core of the spherical reactor of Problem 4.

transforms (3) into the one-dimensional diffusion equation $w_t = k w_{rr}$. Solve this equation with the transformed boundary conditions to find the neutron density

$$u(r, t) = \sum_{n=1}^{\infty} \frac{b_n}{r} \left[\exp\left(B^2 - \frac{n^2\pi^2}{a^2}\right) kt \right] \sin \frac{n\pi r}{a}. \tag{6}$$

PROBLEM 5 Suppose that the buckling constant B satisfies the critical condition

$$B = \frac{\pi}{a}. \tag{7}$$

Deduce from (6) that

$$\lim_{t\to\infty} u(r, t) = \frac{b_1}{r} \sin \frac{\pi r}{a} = \phi(r),$$

a stable solution, while $u \to 0$ if $B < \pi/a$, and $u \to +\infty$ if $B > \pi/a$. Hence (7) determines the **critical radius** $a = \pi/B$ of a homogeneous spherical reactor core; if $a > \pi/B$, then it is a bomb rather than a reactor.

PROBLEM 6 For a homogeneous mixture of slightly enriched uranium (3% $^{235}U_{92}$, 97% $^{238}U_{92}$) and graphite moderator, with 500 $^{12}C_6$ atoms per uranium atom, the value of the buckling constant is $B = 0.0191$ cm^{-1}. Compute the critical mass (in feet) and the weight (in pounds) of the uranium that a critical spherical reactor core contains. Use the densities 1.6 gm/cm^3 for graphite and 18.9 gm/cm^3 for uranium. (*Answer:* $a \approx 5.40$ ft with about 2600 lb of uranium.)

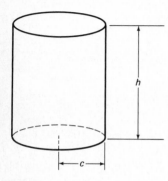

For a *cylindrical* nuclear reactor core of radius $r = c$ and height $z = h$ (as in Fig. 8.22), Equation (1) in cylindrical coordinates is

$$\frac{1}{k}\frac{\partial u}{\partial t} = \frac{\partial^2 u}{\partial r^2} + \frac{1}{r}\frac{\partial u}{\partial r} + \frac{\partial^2 u}{\partial z^2} + B^2 u \qquad (8)$$

with boundary conditions

$$u(r, 0, t) = u(r, h, t) = u(c, z, t) = 0 \qquad (9)$$

and initial condition $u(r, z, 0) = f(r, z)$, f given. According to Problem 1, the substitution

$$v(r, z, t) = \exp(-B^2 kt)u(r, z, t) \qquad (10)$$

Figure 8.22 The core of the cylindrical reactor.

transforms (8) into the source-free equation

$$\frac{1}{k}\frac{\partial v}{\partial t} = \frac{\partial^2 v}{\partial r^2} + \frac{1}{r}\frac{\partial v}{\partial r} + \frac{\partial^2 v}{\partial z^2}. \qquad (11)$$

PROBLEM 7 Show that the substitution $v(r, z, t) = R(r)Z(z)T(t)$ in (11) yields the separation of variables

$$r^2 R'' + rR' + \alpha^2 r^2 R = 0, \qquad R(c) = 0; \qquad (12)$$

$$Z'' + \beta^2 Z = 0, \qquad Z(0) = Z(h) = 0; \qquad (13)$$

$$T' = -(\alpha^2 + \beta^2)kT. \qquad (14)$$

Note that the problem in (12) has a solution in Bessel functions given in Section 8.4. Derive the formal series solution

$$u(r, z, t) = e^{B^2 kt} \sum_{m=1}^{\infty} \sum_{n=1}^{\infty} c_{mn} \exp\left(-\left[\frac{\gamma_m^2}{c^2} + \frac{n^2\pi^2}{h^2}\right]kt\right) J_0\left(\frac{\gamma_m r}{c}\right) \sin \frac{n\pi z}{h}$$

$$(15)$$

where $\{\gamma_m\}_1^{\infty}$ are the positive roots of the equation $J_0(x) = 0$. Show also that

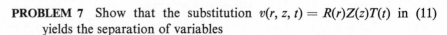

$$c_{mn} = \frac{4}{hc^2[J_1(\gamma_m)]^2} \int_0^c \int_0^h rf(r, z)J_0\left(\frac{\gamma_m r}{c}\right) \sin \frac{n\pi z}{h} \, dz \, dr. \qquad (16)$$

PROBLEM 8 Suppose that $f(r, z) = u_0$ (constant). Deduce from (15) and (16) that

$$u(r, z, t) = \frac{8u_0}{\pi} \exp(B^2 kt) \sum_{n \text{ odd}} \sum_{m=1}^{\infty} \frac{\lambda_m}{n\gamma_m J_1(\gamma_m)} J_0\left(\frac{\gamma_m r}{c}\right) \sin\frac{n\pi z}{h} \qquad (17)$$

where $\lambda_m = \exp(-[\{\gamma_m/c\}^2 + \{n\pi/h\}^2]kt)$.

PROBLEM 9 In continuation of Problem 8, show that if the radius c and the height h of the reactor core satisfy the **critical equation**

$$\frac{\gamma_1^2}{c^2} + \frac{\pi^2}{h^2} = B^2, \qquad (18)$$

then there is a stable solution

$$\lim_{t\to\infty} u(r, z, t) = \frac{8u_0}{\pi\gamma_1 J_1(\gamma_1)} J_0\left(\frac{\gamma_1 r}{c}\right) \sin\frac{\pi z}{h}.$$

PROBLEM 10 Now let us change the notation to core radius R and core height H. Then the critical equation takes the form

$$g(R, H) = \frac{\gamma_1^2}{R^2} + \frac{\pi^2}{H^2} - B^2 = 0. \qquad (19)$$

We want to minimize the critical volume

$$V = f(R, H) = \pi R^2 H \qquad (20)$$

of the cylindrical reactor core subject to the constraint that R and H satisfy Eq. (19). According to the Lagrange multiplier method, the optimal values R and H must satisfy the conditions

$$\frac{\partial f}{\partial R} = \lambda \frac{\partial g}{\partial R}, \qquad \frac{\partial f}{\partial H} = \lambda \frac{\partial g}{\partial H}$$

for some number λ. Eliminate λ to deduce that the minimum critical volume is obtained with

$$R = \frac{\gamma_1}{B}\sqrt{\frac{3}{2}}, \qquad H = \frac{\pi}{B}\sqrt{3}. \qquad (21)$$

Recall that $\gamma_1 \approx 2.40483$ is the smallest positive solution of $J_0(x) = 0$.

PROBLEM 11 For the mixture of slightly enriched uranium and graphite specified in Problem 6, compute from (21) the critical radius and height (in feet) of a cylindrical reactor core, and also the weight (in pounds) of uranium that such a core contains. (*Answer:* $R \approx 5.06$ ft and $H \approx 9.35$ ft, with about 2950 lb of uranium—more than a spherical core requires).

TORSION AND THE TWISTING OF BARS

Consider a bar of length L built in at the end $z = 0$ and free at the end $z = L$, as indicated in Fig. 8.23. A moment or torque $M = Fd$ is applied to the end $z = L$, twisting it through an angle of θ radians; the *twist* of the bar is $\alpha = \theta/L$ radians per unit length. We want to find the relationship between M and α. Let R denote the cross section of the bar in the xy-plane.

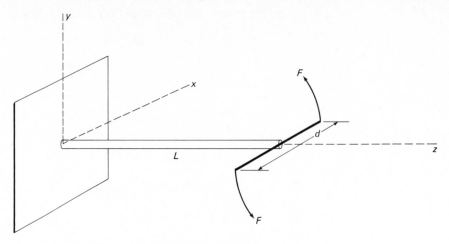

Figure 8.23 Twisting a bar.

It is known from elasticity theory that the **stress function** $\psi(x, y)$ of the bar satisfies the Poisson equation

$$\nabla^2\psi = \frac{\partial^2\psi}{\partial x^2} + \frac{\partial^2\psi}{\partial y^2} = -2 \qquad \text{in } R, \tag{1}$$

$$\psi = 0 \qquad \text{on the boundary of } R;$$

moreover,

$$M = 2G\alpha \iint_R \psi(x, y) \, dx \, dy \tag{2}$$

where G is the **shear modulus** of the material of the bar.

PROBLEM 1   Suppose first that R is a circular disk of radius a. (a) Recall that the Laplacian in polar coordinates is

$$\nabla^2\psi = \frac{\partial^2\psi}{\partial r^2} + \frac{1}{r}\frac{\partial\psi}{\partial r}.$$

Verify directly that $\psi = \frac{1}{2}(a^2 - r^2)$ satisfies the boundary value problem in (1). (b) Then evaluate (2) with this choice of ψ to obtain

$$M = G\alpha I_0 \tag{3}$$

for a circular bar, where $I_0 = \frac{1}{2}\pi a^4$ is the usual polar moment of inertia. This tells us the conceptual meaning of the shear modulus G—Eq. (3) implies that (for a circular bar) the torque M is proportional to the product of the twist α and the moment of inertia I_0, with G being the constant of proportionality. In particular, G is the torque required to impart a unit (1 rad) twist to a bar with unit moment of inertia, so the shear modulus of a given material is readily measured by empirical means. Deduce from (3) that the units of G are force over the square of distance.

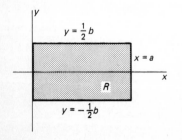

Figure 8.24 Cross section of the bar of Problem 2.

PROBLEM 2   Now suppose that the bar has a rectangular cross section of dimensions a by b, as shown in Fig. 8.24. (a) Make the substitution

$$\psi(x, y) = u(x, y) + ax - x^2$$

in (1) to show that $u(x, y)$ satisfies the boundary value problem

$$\frac{\partial^2 u}{\partial x^2} + \frac{\partial^2 u}{\partial y^2} = 0 \quad \text{in } R;$$

$$u(0, y) = u(a, y) = 0, \tag{4}$$

$$u\left(x, \pm\frac{1}{2}b\right) = x^2 - ax.$$

(b) Separate the variables to derive the solution

$$u(x, y) = -\frac{8a^2}{\pi^3} \sum_{n \text{ odd}} \frac{\cosh(n\pi y/a) \sin(n\pi x/a)}{n^3 \cosh(n\pi b/2a)}.$$

Hence the stress function is

$$\psi(x, y) = ax - x^2 - \frac{8a^2}{\pi^3} \sum_{n \text{ odd}} \frac{\cosh(n\pi y/a) \sin(n\pi x/a)}{n^3 \cosh(n\pi b/2a)}. \tag{5}$$

(*Suggestion:* Note that it suffices (why?) to find $u(x, y)$ an *even* function of y that satisfies Laplace's equation and the boundary conditions $u(0, y) = u(a, y) = 0$, $u(x, \frac{1}{2}b) = x^2 - ax$.)

(c) Substitute (5) in (2) and integrate termwise to obtain

$$M = 2G\alpha\left(\frac{1}{6}a^3 b - \frac{32a^4}{\pi^5} \sum_{n \text{ odd}} \frac{1}{n^5} \tanh\frac{n\pi b}{2a}\right). \tag{6}$$

(d) Sum enough terms in (6) to demonstrate these results:

$$M \approx (0.1406)G\alpha a^4 \quad \text{if } a = b;$$
$$M \approx (0.2287)G\alpha ab^3 \quad \text{if } a = 2b.$$

PROBLEM 3 Suppose that in Fig. 8.23 we have a steel bar (with shear modulus $G = 1.7 \times 10^9$ lb/ft^2) of length $L = 6$ ft and moment arm $d = 4$ ft. Compute the force F required to twist its end through an angle of $30°$ for each given cross section.

(a) A circle with diameter 1 in.
(b) A square with the same area as the circle in part (a).
(c) A rectangle with $a = 2b$ and the same area as before.

(*Answers:* (a) 176 lb; (b) 155 lb; (c) 126 lb.)

DEFLECTION OF A LOADED PLATE

Consider a rectangular plate with **flexural rigidity** D that occupies the region $-\frac{1}{2}a \leq x \leq \frac{1}{2}a$, $0 \leq y \leq b$ in the (horizontal) xy-plane, as indicated in Fig. 8.25. If the plate is subjected to a (vertical) load of p (in units of force per unit area), then its vertical deflection $z(x, y)$ satisfies the fourth order equation

$$\nabla^4 z = \frac{\partial^4 z}{\partial x^4} + 2\frac{\partial^4 z}{\partial^2 x \, \partial^2 y} + \frac{\partial^4 z}{\partial y^4} = \frac{p}{D}. \tag{1}$$

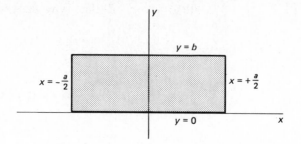

Figure 8.25 A rectangular plate.

If the plate is simply supported along its four edges, the boundary conditions are

$$z(x, 0) = z(x, b) = 0, \tag{2}$$

$$z_{yy}(x, 0) = z_{yy}(x, b) = 0, \tag{3}$$

$$z\left(\pm\frac{a}{2}, y\right) = z_{xx}\left(\pm\frac{a}{2}, y\right) = 0. \tag{4}$$

PROBLEM 1 Show that the substitution

$$z(x, y) = u(x, y) + \frac{p}{24D}(b^3 y - 2by^3 + y^4) \tag{5}$$

transforms (1) into the **biharmonic equation**

$$\nabla^4 u = 0 \tag{6}$$

with boundary conditions

$$u(x, 0) = u(x, b) = 0, \tag{7a}$$

$$u_{yy}(x, 0) = u_{yy}(x, b) = 0; \tag{7b}$$

$$u\left(\pm\frac{a}{2}, y\right) = -\frac{p}{24D}(b^3 y - 2by^3 + y^4), \tag{8a}$$

$$u_{xx}\left(\pm\frac{a}{2}, y\right) = 0. \tag{8b}$$

PROBLEM 2 Show that the separation of variables $u_n(x, y) = X_n(x)Y_n(y)$ in (6) gives

$$X_n^{(4)}Y_n + 2X_n''Y_n'' + X_nY_n^{(4)} = 0. \tag{9}$$

Verify the happy circumstance that

$$Y_n(y) = \sin\frac{n\pi y}{b} \tag{10}$$

satisfies all the conditions imposed on $Y_n(y)$ by the boundary conditions in (7). Hence substitute (10) in (9) and cancel to obtain the linear homogeneous ordinary differential equation

$$X_n^{(4)} - 2\frac{n^2\pi^2}{b^2}X_n'' + \frac{n^4\pi^4}{b^4}X_n = 0. \tag{11}$$

PROBLEM 3 Explain why the general solution of (11) can be written in the form

$$X_n(x) = a_n \cosh \frac{n\pi x}{b} + b_n x \sinh \frac{n\pi x}{b} + c_n \sinh \frac{n\pi x}{b} + d_n x \cosh \frac{n\pi x}{b}.$$

The fact that $X_n(x)$ is an even function (is this obvious?) implies that $c_n = d_n = 0$. Apply the boundary conditions in (8) with the positive sign to solve for

$$a_n = -\left(\frac{4pb^4}{\pi^5 Dn^5}\right)\left(\frac{2\cosh \alpha_n + \alpha_n \sinh \alpha_n}{2\cosh^2 \alpha_n}\right)$$

and

$$b_n = +\left(\frac{4pb^4}{\pi^5 Dn^5}\right)\left(\frac{n\pi}{b}\right)\left(\frac{\cosh \alpha_n}{2\cosh^2 \alpha_n}\right)$$

for n odd, where $\alpha_n = n\pi a/2b$; deduce also that $a_n = b_n = 0$ if n is even. You may use the Fourier series

$$\frac{p}{24D}(b^3 y - 2by^3 + y^4) = \frac{4pb^4}{\pi^5 D}\sum_{n \text{ odd}} \frac{1}{n^5}\sin \frac{n\pi y}{b}.$$

PROBLEM 4 Finally assemble all these results to obtain

$$z(x, y) = \frac{p}{24D}(b^3 y - 2by^3 + y^4) + \sum_{n=1}^{\infty} u_n(x, y)$$

$$= \frac{4pb^4}{\pi^5 D}\sum_{n \text{ odd}} C_n(x)\sin \frac{n\pi y}{b}, \tag{12}$$

where

$$C_n(x) = \frac{1}{2n^5 \cosh^2 \alpha_n}(2\cosh^2 \alpha_n - 2\cosh \alpha_n \cosh x_n$$

$$- \alpha_n \sinh \alpha_n \cosh x_n + x_n \cosh \alpha_n \sinh x_n)$$

with $x_n = n\pi x/b$. Under the assumption that $a = 2b$, perform the computations necessary to show that $C_1(0) \approx 0.77873$, $C_3(0) \approx 0.00411$, and $C_5(0) \approx 0.00032$. Conclude that the center point deflection of the plate is given by

$$z\left(0, \frac{b}{2}\right) \approx \frac{(3.1)pb^4}{\pi^5 D}.$$

PROBLEM 5 Suppose that a floor is a $\frac{3}{4}$-in. steel plate measuring 20 ft by 10 ft, simply supported along its edges and with a uniform load of $p = 150$ lb/ft². Compute the deflection (in inches) of this floor at its center point. The flexural rigidity of a plate is given by

$$D = \frac{Eh^3}{12(1 - \sigma^2)}.$$

where h is the thickness of the plate, E is Young's modulus (here 4.2×10^9 lb/ft²), and σ is the Poisson ratio 0.29 (dimensionless). (*Answer:* Approximately 1.95 in.)

DRYING A STONE COLUMN

The problems below deal with the drying of a freshly quarried (at time $t = 0$) sandstone column in the shape of a cylinder of radius 2 ft and height 10 ft. Let $c(r, z, t)$ be the moisture concentration (as a fraction of saturation) at the point (r, z) (cylindrical coordinates) at time t. Then c satisfies the diffusion equation

$$\frac{\partial c}{\partial t} = k\nabla^2 c. \tag{1}$$

Suppose that the initial concentration

$$c(r, z, 0) = c_0 \tag{2}$$

and the surface concentration

$$c(2, z, t) = c(r, 10, t) = c_s \tag{3}$$

are constant, while the base is "insulated," so that

$$c_z(r, 0, t) = 0. \tag{4}$$

To remove the excessive inhomogeneity, we define

$$u(r, z, t) = \frac{c(r, z, t) - c_s}{c_0 - c_s}. \tag{5}$$

PROBLEM 1 Show from Eqs. (1)–(5) that

$$\frac{1}{k}\frac{\partial u}{\partial t} = \frac{\partial^2 u}{\partial r^2} + \frac{1}{r}\frac{\partial u}{\partial r} + \frac{\partial^2 u}{\partial z^2} \tag{6}$$

$$u(r, 10, t) = u(2, z, t) = u_z(r, 0, t) = 0, \tag{7}$$

and

$$u(r, z, 0) = 1. \tag{8}$$

PROBLEM 2 Separate the variables in the boundary value problem in Eqs. (6)–(8), just as in the problem concerning the cylindrical nuclear reactor, to derive a solution of the form

$$u(r, z, t) = \sum_{m,n=1}^{\infty} A_{mn} \exp\left(-[\alpha_m^2 + \beta_n^2]kt\right) J_0\left(\frac{\gamma_m r}{2}\right) \cos\frac{(2n-1)\pi z}{20},$$

where $\{\gamma_m\}$ are the roots of $J_0(x) = 0$, $\alpha_m = \frac{1}{2}\gamma_m$, and $\beta_n = (2n-1)\pi/20$.

PROBLEM 3 To satisfy the condition in (8), we must find A_{mn} so that

$$1 = \sum_{n=1}^{\infty}\left[\sum_{m=1}^{\infty} A_{mn} J_0\left(\frac{\gamma_m r}{2}\right)\right] \cos\frac{(2n-1)\pi z}{20}$$

$$= \sum_{n=1}^{\infty} a_n(r) \cos\frac{(2n-1)\pi z}{20}.$$

First compute $a_n(r)$ and then A_{mn} to derive the solution

$$u(r, z, t) = \frac{8}{\pi}\sum_{m,n=1}^{\infty}\frac{(-1)^{n-1}\exp\left(-[\alpha_m^2 + \beta_n^2]kt\right)}{(2n-1)\gamma_m J_1(\gamma_m)}$$

$$\cdot J_0\left(\frac{\gamma_m r}{2}\right)\cos\frac{(2n-1)\pi z}{20}. \tag{9}$$

PROBLEM 4 Assume that $c_0 = 12$ (%) and $c_s = 3$ (%). Compute the maximum moisture concentration in the column after $T = 5 \times 10^4$ h (this is about 5 years and 8.5 months). Use the diffusion coefficient value $k = 2 \times 10^{-5}$ ft²/hr. (*Suggestions:* Take it as physically obvious that the maximum concentration will be at the center $(0, 0)$ of the base of the column (it has the greatest average distance from the surface of the column, where moisture is escaping by evaporation). Hence compute $u(0, 0, T)$ from the series in (9). Use enough terms to obtain two-place accuracy—add all those terms that exceed 0.001 after multiplication by $8/\pi$. Finally convert to $c(0, 0, T)$.) (*Answer:* About 6.4%, so after almost 6 years the maximum moisture concentration is still more than half the original concentration.)

Qualitative Properties
and Existence
of Solutions

9

9.1
Introduction
to Stability

It is often difficult, if not impossible, to solve a given differential equation explicitly, especially one that is nonlinear. Therefore, it is important to determine whether qualitative information about the solutions of a differential equation can be obtained without the necessity of obtaining an explicit solution. For example, we may be able to establish that every solution $x(t)$ grows without bound as $t \to +\infty$, or approaches a finite limit, or is a periodic function of t. In this section we introduce—by consideration of simple differential equations that *can* be solved explicitly—some of the more important qualitative questions that can sometimes be answered for less tractable equations.

The question of whether a population $x(t)$ is bounded or unbounded as $t \to +\infty$ is of evident interest. In Section 1.8 we introduced the general population equation

$$\frac{dx}{dt} = (\beta - \delta)x \tag{1}$$

where β and δ are the birth and death rates, respectively, in births or deaths per individual per unit of time. If β and δ are constants, we have the case of natural population growth, and the solution of Eq. (1) is $x(t) = x_0 e^{kt}$ where $k = \beta - \delta$ and $x_0 = x(0)$. Thus it is clear that as $t \to +\infty$, $x(t) \to 0$ if $k < 0$, while $x(t) \to +\infty$ if k and x_0 are positive.

In more interesting situations β and δ are (known) functions of x. Then Eq. (1) takes the form

$$\frac{dx}{dt} = f(x), \tag{2}$$

This is an **autonomous** first order differential equation—one in which the independent variable t does not appear explicitly. A **critical point** of Eq. (2) is a root of $f(x) = 0$. If $x = c$ is a critical point of (2), then the differential equation has the constant solution $x(t) \equiv c$. The following example illustrates the fact that the qualitative behavior (as t increases) of an autonomous first order equation can be described in terms of its critical points.

EXAMPLE 1 Suppose that the death rate $\delta = \delta_0$ in Eq. (1) is constant but that, because of cultural sophistication or a limited food supply or for some other reason, the birth rate is a linearly decreasing function $\beta = \beta_0 - \beta_1 x$ of the population x. Then Eq. (1) becomes the autonomous equation

$$\frac{dx}{dt} = ax - bx^2 \tag{3}$$

with $a = \beta_0 - \delta_0$ and $b = \beta_1$; we assume that a and b are each positive. When we rewrite (3) in the form

$$\frac{dx}{dt} = bx(M - x) \qquad \left(M = \frac{a}{b} > 0\right), \tag{4}$$

we recognize it as the logistic equation that we introduced in Section 1.8. It has two critical points, $x = 0$ and $x = M$, and thus the two constant solutions $x(t) \equiv 0$ and $x(t) \equiv M$ corresponding to the initial populations $x_0 = 0$ and $x_0 = M$, respectively. Because the right-hand side of Eq. (4) is continuously differentiable, the equation has unique solutions, and therefore no two solution curves $x = x_1(t)$ and $x = x_2(t)$ can intersect anywhere (because such an intersection would violate uniqueness). The horizontal lines $x = 0$ and $x = M$ are solution curves of Eq. (4), so it follows that no other solution curve $x = x(t)$ can intersect either of these two lines. This fact implies that if $x_0 > M$, then $x(t) > M$ for all $t > 0$; if $0 < x_0 < M$, then $0 < x(t) < M$ for all t; and if $x_0 < 0$, then $x(t) < 0$. (Although x_0 would never be negative for an actual population, we can discuss the behavior of the solutions of Eq. (4) with $x_0 < 0$.)

We can solve Eq. (4) by separating the variables:

$$\int \frac{dx}{x(M - x)} = \int b\, dt;$$

$$\frac{1}{M} \int \left(\frac{1}{x} + \frac{1}{M - x}\right) dx = \int b\, dt;$$

$$\ln\left|\frac{x}{M - x}\right| = bMt + C. \tag{5}$$

Because of the absolute value on the left, the cases $x_0 > M, 0 < x_0 < M$, and $x_0 < 0$ must be considered separately. In Problem 30, however, we ask you to show that all three cases lead to the same formula

$$x(t) = \frac{Mx_0}{x_0 + (M - x_0)e^{-bMt}} \tag{6}$$

for the solution in terms of the initial value $x_0 = x(0)$. Note that (6) yields $x(t) \equiv 0$ if $x_0 = 0$ and $x(t) \equiv M$ if $x_0 = M$.

If $x_0 > 0$, then the denominator on the right-hand side in (6) is positive for all $t > 0$ and approaches x_0 as $t \to +\infty$, so

$$\lim_{t \to \infty} x(t) = M \quad \text{if } x_0 > 0. \tag{7a}$$

On the other hand, if $x_0 < 0$, then this denominator is equal to $(M + |x_0|)e^{-kMt} - |x_0|$, which is initially positive but approaches zero as

$$t \longrightarrow \frac{1}{kM} \ln \frac{x_0 - M}{x_0} = t_0 > 0.$$

Thus

$$\lim_{t \to t_0^-} x(t) = -\infty \quad \text{if } x_0 < 0. \tag{7b}$$

From (7a) and (7b) we see that the solution curves of Eq. (3) appear as shown in Fig. 9.1.

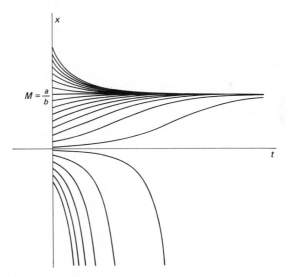

Figure 9.1   Graphs of some solutions of $\frac{dx}{dt} = bx(M - x)$.

This figure illustrates the concept of stability. A critical point $x = c$ of an autonomous first order equation is called *stable* provided that, if the initial value x_0 is sufficiently close to c, then $x(t)$ is close to c for all $t > 0$. More precisely, the critical point c is **stable** if, given $\epsilon > 0$, there exists $\delta > 0$ such that

$$|x_0 - c| < \delta \quad \text{implies} \quad |x(t) - c| < \epsilon \tag{8}$$

for all $t > 0$. The critical point c is said to be **unstable** if it is not stable. Thus the critical point $M = a/b$ of Eq. (3) is stable, while the critical point 0 is unstable. We can summarize the behavior of solutions of (3)—in terms of their initial values—by means of the *phase portrait* shown in Fig. 9.2. It indicates that $x(t) \to M$ as $t \to +\infty$ if either $x_0 > M$ or $0 < x_0 < M$, while $x(t) \to -\infty$ as t increases if $x_0 < 0$. The fact that M is a stable critical point would be important, for instance, if we wished to conduct an experiment with a population of M bacteria. It is impossible to count precisely M bacteria for M large, but (7a) shows that any initially positive population will approach M with increasing t.

Figure 9.2 The phase portrait shows a stable critical point and an unstable critical point.

Here is another aspect of stability. The coefficients a and b in Eq. (3) are unlikely to be known precisely for an actual population. But if they are replaced with close approximations a^* and b^*, then the approximate limiting population $M^* = a^*/b^*$ will be close to the actual limiting population $M = a/b$. Thus the limiting population that Eq. (3) predicts is stable with respect to small perturbations of its constant coefficients.

AUTONOMOUS SYSTEMS AND THE PHASE PLANE

If we begin with an autonomous second order differential equation

$$x''(t) = G(x, x') \tag{9}$$

and introduce the new dependent variable $y = x'$, then we obtain the equivalent system

$$\left. \begin{aligned} \frac{dx}{dt} &= y, \\ \frac{dy}{dt} &= G(x, y), \end{aligned} \right\} \tag{10}$$

—that is, two first order equations neither of which involves explicitly the independent variable t. This system is a special case of the general **autonomous** system of two first order equations:

$$\left. \begin{aligned} \frac{dx}{dt} &= F(x, y), \\ \frac{dy}{dt} &= G(x, y); \end{aligned} \right\} \tag{11}$$

again, the independent variable t does not appear explicitly. We assume that the functions F and G are continuously differentiable in some region R in the xy-plane, which is called the **phase plane** for the system in (11). Then, according to the existence-uniqueness theorem of Section 9.4, given t_0 and any point (x_0, y_0) of R, there is a *unique* solution $x = x(t)$, $y = y(t)$ of (11) that is defined on some open interval $a < t_0 < b$ and satisfies the initial conditions

$$x(t_0) = x_0, \qquad y(t_0) = y_0. \tag{12}$$

The equations $x = x(t)$, $y = y(t)$ describe a parameterized solution curve in the phase plane. Any such solution curve is called a **trajectory** of the system in (11), and precisely one trajectory passes through each point of the region R (see Problem 31). A **critical point** of the system in (11) is a point (x_0, y_0) such that $F(x_0, y_0) = G(x_0, y_0) = 0$.

Note that $x = x_0$, $y = y_0$ is a solution of the system, so the trajectory containing the critical point (x_0, y_0) consists of the single point (x_0, y_0). It turns out that any trajectory not consisting of a single point is a nondegenerate curve with no self-intersections (see Problem 32). We can demonstrate qualitatively the behavior of solutions of the system in (11) by constructing its **phase portrait**—a phase plane picture of its critical points and typical nondegenerate trajectories. The behavior of the trajectories near an isolated critical point is of particular interest.

EXAMPLE 2 Consider the autonomous linear system

$$\left.\begin{array}{l} x' = -x, \\ y' = -ky \quad (k \text{ constant}) \end{array}\right\} \tag{13}$$

that has the origin $(0, 0)$ as its only critical point. The solution with initial point (x_0, y_0) is

$$x(t) = x_0 e^{-t}, \qquad y(t) = y_0 e^{-kt}. \tag{14}$$

If $x_0 \neq 0$, we can write

$$y = y_0 e^{-kt} = \frac{y_0}{x_0^k}(x_0 e^{-t})^k = bx^k \tag{15}$$

where $b = y_0/x_0^k$.

If $k = 1$, then each curve in (15) is a straight line through the origin. Each trajectory is an open ray along which the point $(x(t), y(t)) = (x_0 e^{-t}, y_0 e^{-t})$ approaches the origin as $t \to +\infty$. This type of critical point, shown in Fig. 9.3, is called a **proper node**. Note the arrows that indicate the orientations of the trajectories; they point in the direction of increasing t.

In general, the critical point (x_0, y_0) of the autonomous system in (11) is called a **node** provided that *either* every trajectory approaches (x_0, y_0) as $t \to +\infty$ *or* every trajectory recedes from (x_0, y_0) as $t \to +\infty$, *and* every trajectory is tangent at (x_0, y_0) to some straight line through the critical point. Note that the critical point $(0, 0)$ in Fig. 9.3—where the trajectories *are* straight lines, not merely tangents to straight lines— would still be a node if all the arrows were reversed to make the trajectories recede from the critical point rather than approach it. This node is said to be *proper* because no two different pairs of "opposite" trajectories are tangent to the same straight line. In the following paragraph we consider a case in which the critical point is a node that is not proper.

If $k = 2$ and neither x_0 nor y_0 is zero in (15), then each curve is a parabola $y = bx^2$ tangent to the x-axis at the origin. The solution curve in (14) is half of the x-axis if $y_0 = 0$, and half of the y-axis if $x_0 = 0$. The trajectories are the semiaxes and the right and left halves of the parabolas

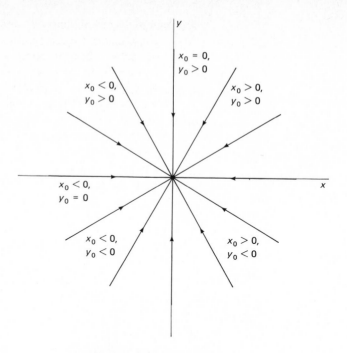

Figure 9.3 A stable proper node.

shown in Fig. 9.4. Along each trajectory the point $(x(t), y(t))$ approaches the origin as $t \rightarrow +\infty$. Thus all trajectories except for a single pair approach the origin tangent to the same line—the x-axis. This sort of critical point is called an **improper node**.

If $k = -1$, then $x(t) = x_0 e^{-t}$ and $y(t) = y_0 e^t$, so $xy = x_0 y_0 = b$. If neither x_0 nor y_0 is zero, then the trajectory is one branch of the

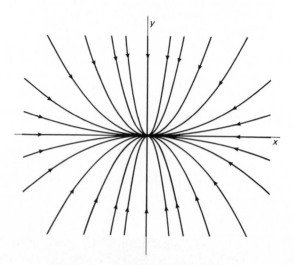

Figure 9.4 A stable improper node.

rectangular hyperbola $xy = b$, and $y \to \pm\infty$ as $t \to +\infty$. If $x_0 = 0$ or $y_0 = 0$, then the trajectory is a semiaxis of the hyperbola. The point $(x(t), y(t))$ approaches the origin along the x-axis, but recedes from it along the y-axis as $t \to +\infty$. Thus there are two trajectories that approach the critical point $(0, 0)$, and all others are unbounded as $t \to +\infty$. This type of critical point, shown in Fig. 9.5, is called a **saddle point**.

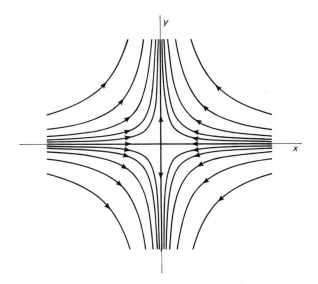

Figure 9.5 A saddle point.

A critical point (x^*, y^*) of the autonomous system in (11) is said to be *stable* provided that if the initial point (x_0, y_0) is sufficiently close to (x^*, y^*), then $(x(t), y(t))$ remains close to (x^*, y^*) for all $t > 0$. More precisely, the critical point (x^*, y^*) is **stable** if, for each $\epsilon > 0$, there exists δ such that

$$[(x_0 - x^*)^2 + (y_0 - y^*)^2]^{1/2} < \delta$$

implies

$$[(x(t) - x^*)^2 + (y(t) - y^*)^2]^{1/2} < \epsilon$$

for all $t > 0$. The critical point (x^*, y^*) is called **unstable** if it is not stable. In each of Figs. 9.3 and 9.4, the origin $(0, 0)$ is a stable critical point. The saddle point at $(0, 0)$ shown in Fig. 9.5 is an unstable critical point.

If the signs on the right-hand sides in (13) are changed to obtain the system

$$\left. \begin{array}{l} x' = x, \\ y' = ky, \end{array} \right\} \tag{16}$$

the solution is $x(t) = x_0 e^t$, $y(t) = y_0 e^{kt}$. Then with $k = 1$ and $k = 2$, the trajectories are the same as those shown in Figs. 9.3 and 9.4, respectively, but with the arrows reversed. In Fig. 9.3 with the arrows reversed, the origin is an unstable proper node; in Fig. 9.4 with arrows reversed, the origin is an unstable improper node.

It is also possible for trajectories to remain near a stable critical point without approaching it, as the following example shows.

EXAMPLE 3   Consider a mass m that oscillates without damping on a spring with Hooke's constant k, so that its position function $x(t)$ satisfies the differential equation $x'' + \omega^2 x = 0$ $(\omega^2 = k/m)$. If we introduce its velocity $y = x'$, we get the system

$$\left.\begin{aligned} x' &= y, \\ y' &= -\omega^2 x \end{aligned}\right\} \tag{17}$$

with general solution

$$x(t) = A \cos \omega t + B \sin \omega t, \tag{18a}$$

$$y(t) = -A\omega \sin \omega t + B\omega \cos \omega t. \tag{18b}$$

With $C = (A^2 + B^2)^{1/2}$ and $\alpha = \tan^{-1}(B/A)$, we can rewrite (18) as

$$x(t) = C \cos(\omega t - \alpha), \tag{19a}$$

$$y(t) = -\omega C \sin(\omega t - \alpha), \tag{19b}$$

so it becomes clear that each trajectory other than the critical point $(0, 0)$ is an ellipse with equation of the form

$$\frac{x^2}{C^2} + \frac{y^2}{\omega^2 C^2} = 1. \tag{20}$$

Each point $(x_0, y_0) \neq (0, 0)$ in the xy-plane lies on exactly one of these ellipses, and a solution $(x(t), y(t))$ with initial point (x_0, y_0) traverses the ellipse containing (x_0, y_0) in the clockwise direction with *period* $T = 2\pi/\omega$. (It is clear from (18) that $x(t + T) = x(t)$ and $y(t + T) = y(t)$ for all t.) Thus each nontrivial solution of the system in (17) is periodic and its trajectory is a simple closed curve.

It is clear in Fig. 9.6, in which $\omega < 1$, that if the distance from (x_0, y_0) to $(0, 0)$ is less than $\delta = \omega\epsilon$, then for all t the distance from $(x(t), y(t))$ is less than ϵ. Hence $(0, 0)$ is a stable critical point of the system $x' = y$, $y' = -\omega^2 x$. Unlike the situation shown in Figs. 9.3 and 9.4, though, no single trajectory

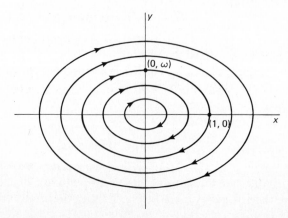

Figure 9.6   A (stable) center surrounded by closed trajectories.

approaches $(0, 0)$. A critical point surrounded by simple closed trajectories representing periodic solutions is called a **(stable) center**.

The critical point (x^*, y^*) is called **asymptotically stable** if it is stable and, moreover, every trajectory that begins sufficiently close to (x^*, y^*) also approaches (x^*, y^*) as $t \to +\infty$. That is, there exists $\delta > 0$ such that

$$[(x_0 - x^*)^2 + (y_0 - y^*)^2]^{1/2} < \delta$$

implies that both

$$\lim_{t \to \infty} x(t) = x^* \tag{21a}$$

and

$$\lim_{t \to \infty} y(t) = y^*. \tag{21b}$$

The stable nodes shown in Figs. 9.3 and 9.4 are asymptotically stable, while the center in Fig. 9.6 is stable but not asymptotically stable.

For a mechanical system as in Example 3, a critical point represents an *equilibrium state* of the system—if the velocity $y = x'$ and the acceleration $y' = x''$ vanish simultaneously, the mass remains at rest with no force acting on it. Stability of a critical point concerns the question whether, when the mass is displaced slightly from equilibrium, it (1) moves back toward the equilibrium point as $t \to +\infty$, (2) merely remains near the equilibrium point without approaching it, or (3) moves further away from equilibrium. In Case 1 the critical (equilibrium) point is asymptotically stable; in Case 2 it is stable but not asymptotically so; in Case 3 it is an unstable critical point. A marble balanced on the top of a basketball is an example of an unstable equilibrium state. A mass on a spring with damping illustrates the case of asymptotic stability of a mechanical system. The mass-and-spring without damping in Example 3 is an example of a system that is stable but not asymptotically stable.

EXAMPLE 4 Suppose that $m = 1$ and $k = 2$ for the mass and spring of Example 3 and that the mass is attached also to a dashpot with damping constant $c = 2$. Then its displacement function $x(t)$ satisfies the second order equation

$$x''(t) + 2x'(t) + 2x(t) = 0. \tag{22}$$

With $y = x'$ we obtain the equivalent first order system

$$\left.\begin{aligned} \frac{dx}{dt} &= y, \\ \frac{dy}{dt} &= -2x - 2y \end{aligned}\right\} \tag{23}$$

with critical point $(0, 0)$. The characteristic equation $r^2 + 2r + 2 = 0$ of (22) has roots $-1 + i$ and $-1 - i$, so the general solution of the system in (23) is given by

$$x(t) = e^{-t}(A \cos t + B \sin t) \tag{24a}$$

$$= Ce^{-t} \cos (t - \alpha),$$

$$y(t) = e^{-t}[(B - A) \cos t - (A + B) \sin t] \tag{24b}$$

$$= -C\sqrt{2}\, e^{-t} \sin \left(t - \alpha + \frac{\pi}{4}\right),$$

where $C = \sqrt{A^2 + B^2}$ and $\alpha = \tan^{-1}(B/A)$. We see that $x(t)$ and $y(t)$ oscillate between positive and negative values and that each approaches zero as $t \longrightarrow +\infty$. Consequently, a typical trajectory spirals toward the origin, as indicated in Fig. 9.7. It is clear that the critical point $(0, 0)$ is asymptotically stable; this type of critical point is called a (stable) **spiral point**. In the case of the mass-spring-dashpot system, a spiral point is the manifestation in the phase plane of the damped oscillations that occur because of resistance.

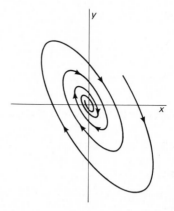

Figure 9.7 A stable spiral point and one nearby trajectory.

It can be shown that there are four possibilities for a nondegenerate trajectory of the autonomous system in (11):

$$\left. \begin{array}{l} \dfrac{dx}{dt} = F(x, y), \\[2mm] \dfrac{dy}{dt} = G(x, y). \end{array} \right\}$$

Either:

1. $x(T) = x(0)$ and $y(T) = y(0)$ for some $T > 0$, in which case the trajectory is **closed** and represents a periodic solution (see Problem 32); or
2. $(x(t), y(t))$ approaches a closed trajectory as $t \longrightarrow +\infty$, as in Fig. 9.8; or
3. $(x(t), y(t))$ is unbounded with increasing t; or
4. $(x(t), y(t))$ approaches a critical point as $t \longrightarrow +\infty$.

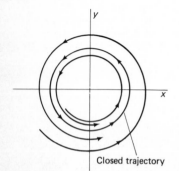

Closed trajectory

Figure 9.8 A closed trajectory and two nonclosed trajectories spiralling toward it.

As a consequence, the qualitative nature of the phase plane picture of the trajectories of an autonomous system is determined largely by the locations of its critical points and by the behavior of its trajectories near its critical points. We will see in Section 9.2 that, subject to mild restrictions on the functions F and G, each isolated critical point of the system $x' = F(x, y)$, $y' = G(x, y)$ resembles qualitatively one of the examples of this section—it is either a node (proper or improper), a saddle point, a center, or a spiral point.

In each of Problems 1–10, first solve explicitly for x(t) in terms of t and the initial value $x_0 = x(0)$. Then construct a sketch like the one in Fig. 9.1 showing the nature of the trajectories for a wide variety of possible values of x_0. Finally, determine the stability or instability of each critical point.

1. $\dfrac{dx}{dt} = -x.$

2. $\dfrac{dx}{dt} = x - 2.$

3. $\dfrac{dx}{dt} = x - x^2.$

4. $\dfrac{dx}{dt} = x^2 - 2x.$

5. $\dfrac{dx}{dt} = x + x^2.$

6. $\dfrac{dx}{dt} = (x - 1)^2.$

7. $\dfrac{dx}{dt} = 1 - x^2.$

8. $\dfrac{dx}{dt} = x^2 - 4.$

9. $\dfrac{dx}{dt} = 2 - x - x^2.$

10. $\dfrac{dx}{dt} = x - x^3.$

Find the critical points of the two-dimensional systems in Problems 11–18.

11. $\dfrac{dx}{dt} = 3x - y, \quad \dfrac{dy}{dt} = x + 3y.$

12. $\dfrac{dx}{dt} = x - 2y, \quad \dfrac{dy}{dt} = 2x - 4y.$

13. $\dfrac{dx}{dt} = 3x - 2y, \quad \dfrac{dy}{dt} = 4x - 3y + 1.$

14. $\dfrac{dx}{dt} = xy - x^2, \quad \dfrac{dy}{dt} = xy + y^2.$

15. $\dfrac{dx}{dt} = 2x - xy, \quad \dfrac{dy}{dt} = xy - 3y.$

16. $\dfrac{dx}{dt} = x - 3x^2 + xy, \quad \dfrac{dy}{dt} = 4y - y^2 - 2xy.$

17. $\dfrac{dx}{dt} = y, \quad \dfrac{dy}{dt} = -\sin x.$

18. $\dfrac{dx}{dt} = x + y - x(x^2 + y^2), \quad \dfrac{dy}{dt} = -x + y - y(x^2 + y^2).$

Solve each of the linear systems in Problems 19–26 to determine whether the critical point (0, 0) is stable, asymptotically stable, or unstable. Sketch typical trajectories, and indicate the direction of motion with increasing t. Identify the critical point as a node, a saddle point, a center, or a spiral point.

19. $x' = -2x, \quad y' = -2y.$

20. $x' = 2x, \quad y' = -2y.$

21. $x' = -2x, \quad y' = -y.$

22. $x' = x, \quad y' = 3y.$

23. $x' = y, \quad y' = -x.$

24. $x' = -y, \quad y' = 4x.$

25. $x' = 2y, \quad y' = -2x.$

26. $x' = y, \quad y' = -5x - 4y.$

27. Consider a population $x(t)$ of unsophisticated animals in which deaths occur at a natural linear rate, but births occur as a result of chance encounters, so that

$$\dfrac{dx}{dt} = bx^2 - ax \qquad (a, b > 0).$$

(a) Solve for $x(t)$ in terms of t and the initial population $x_0 = x(0)$. (b) Suppose that $0 < x_0 < a/b$. Show that $x(t) \longrightarrow 0$ as $t \longrightarrow +\infty$. (c) Suppose that $x_0 > a/b$. Show that $x(t) \longrightarrow +\infty$ as t increases. (This is a "doomsday versus extinction" situation.)

28. Suppose that the equation $x' = ax - bx^2$ of Example 1 describes the fish population $x(t)$ in a lake after t months in which no fishing occurs. Now suppose that, because of fishing, fish are removed from the lake at the rate of kx fish per month. (a) What is the new limiting population in the case $0 < k < a$? (b) Show that if $k \geq a$, then $x(t) \longrightarrow 0$ as $t \longrightarrow +\infty$.

29. In continuation of Problem 28, suppose now that k fish per month are removed from the lake, so that

$$x' = -k + ax - bx^2.$$

(a) In the case $4bk = a^2$, show that $x(t) \longrightarrow a/2b$ (half the limiting population without fishing) as $t \longrightarrow +\infty$. (b) In the case $4bk < a^2$, show that, as $t \longrightarrow +\infty$, $x(t)$ approaches $M_k = [a + (a^2 - 4bk)^{1/2}]/2b$. Note that $M_k \longrightarrow M = a/b$ as $k \longrightarrow 0$. (Suggestion: Complete the square in the differential equation.) (c) In the case $4bk > a^2$, show that $x(t) = 0$ after a finite time interval.

30. Return to Example 1; show that Eq. (5) implies Eq. (6) regardless of the initial value x_0.

Problems 31 and 32 deal with the system

$$\frac{dx}{dt} = F(x, y), \qquad \frac{dy}{dt} = G(x, y)$$

in a region where the functions F and G are continuously differentiable, so for each number a and point (x_0, y_0), there is a unique solution with $x(a) = x_0$ and $y(a) = y_0$.

31. Let $(x_1(t), y_1(t))$ and $(x_2(t), y_2(t))$ be two solutions having trajectories that meet at the point (x_0, y_0): $x_1(a) = x_2(b) = x_0$ and $y_1(a) = y_2(b) = y_0$. Then define

$$x_3(t) = x_2(t + T), \qquad y_3(t) = y_2(t + T)$$

where $T = b - a$, so $(x_2(t), y_2(t))$ and $(x_3(t), y_3(t))$ have the same trajectory. Apply the uniqueness theorem to show that $(x_1(t), y_1(t))$ and $(x_3(t), y_3(t))$ are identical solutions. Hence the original two trajectories are identical.

32. Suppose that the solution $(x_1(t), y_1(t))$ is defined for all t and that its trajectory has an apparent self-intersection:

$$x_1(a) = x_1(a + T) = x_0, \qquad y_1(a) = y_1(a + T) = y_0$$

for some $T > 0$. Introduce the solution

$$x_2(t) = x_1(t + T), \qquad y_2(t) = y_1(t + T)$$

and then apply the uniqueness theorem to show that

$$x_1(t) = x_1(t + T) \quad \text{and} \quad y_1(t) = y_1(t + T)$$

for *all* t. Thus the solution $(x_1(t), y_1(t))$ is periodic with period T and has a closed trajectory.

9.2
Linear and Almost Linear Systems

We now discuss the behavior of solutions of the autonomous system

$$\frac{dx}{dt} = F(x, y), \qquad \frac{dy}{dt} = G(x, y) \tag{1}$$

near an isolated critical point (x_0, y_0) at which $F(x_0, y_0) = G(x_0, y_0) = 0$. A critical point is called **isolated** if some neighborhood of it contains no other critical point. We assume throughout that the functions F and G are continuously differentiable in a neighborhood of (x_0, y_0).

We can assume without loss of generality that $x_0 = y_0 = 0$. Otherwise, we make the substitutions $u = x - x_0, v = y - y_0$. Then $dx/dt = du/dt$ and $dy/dt = dv/dt$, so (1) is equivalent to the system

$$\left.\begin{aligned}
\frac{du}{dt} &= F(u + x_0, v + y_0) = F_1(u, v), \\
\frac{dv}{dt} &= G(u + x_0, v + y_0) = G_1(u, v)
\end{aligned}\right\} \tag{2}$$

that has $(0, 0)$ as an isolated critical point.

EXAMPLE 1 The system

$$\left.\begin{aligned}
\frac{dx}{dt} &= 3x - x^2 - xy = x(3 - x - y), \\
\frac{dy}{dt} &= y + y^2 - 3xy = y(1 - 3x + y)
\end{aligned}\right\} \tag{3}$$

has $(1, 2)$ as one of its critical points. We substitute $u = x - 1, v = y - 2$; that is, $x = u + 1, y = v + 2$. Then

$$3 - x - y = 3 - (u + 1) - (v + 2) = -u - v$$

and

$$1 - 3x + y = 1 - 3(u + 1) + (v + 2) = -3u + v,$$

so the system in (3) takes the form

$$\left.\begin{aligned}
\frac{du}{dt} &= (u + 1)(-u - v) = -u - v - u^2 - uv, \\
\frac{dv}{dt} &= (v + 2)(-3u + v) = -6u + 2v + v^2 - 3uv
\end{aligned}\right\} \tag{4}$$

that has $(0, 0)$ as a critical point. If we can determine the trajectories of the system in (4) near $(0, 0)$, then their translates under the rigid motion that carries $(0, 0)$ to $(1, 2)$ will be the trajectories near $(1, 2)$ of the original system in (3).

We therefore assume hereafter that $(0, 0)$ is an isolated critical point of the autonomous system in (1). It then follows from Taylor's formula for functions of two variables that (1) can be written in the form

$$\left.\begin{aligned}\frac{dx}{dt} &= ax + by + f(x, y), \\[4pt] \frac{dy}{dt} &= cx + dy + g(x, y),\end{aligned}\right\} \tag{5}$$

where $a = F_x(0, 0)$, $b = F_y(0, 0)$, $c = G_x(0, 0)$, and $d = G_y(0, 0)$, and the functions $f(x, y)$ and $g(x, y)$ have the property that

$$\lim_{(x,y)\to(0,0)} \frac{f(x, y)}{\sqrt{x^2 + y^2}} = \lim_{(x,y)\to(0,0)} \frac{g(x, y)}{\sqrt{x^2 + y^2}} = 0. \tag{6}$$

That is, when (x, y) is near $(0, 0)$, the quantities $f(x, y)$ and $g(x, y)$ are small in comparison with $r = \sqrt{x^2 + y^2}$, which itself is small. Thus, when (x, y) is near $(0, 0)$, the nonlinear system in (5) is in some sense "near" the **linearized** system

$$\left.\begin{aligned}\frac{dx}{dt} &= ax + by, \\[4pt] \frac{dy}{dt} &= cx + dy.\end{aligned}\right\} \tag{7}$$

The autonomous system in (5) is therefore called **almost linear** provided that f and g satisfy the condition in (6). It turns out that in most (but not all) cases, the trajectories near $(0, 0)$ of the almost linear system in (5) strongly resemble —qualitatively—those of its "linearization" in (7). Consequently, the first step toward understanding general autonomous systems is to characterize the critical points of linear systems.

LINEAR SYSTEMS

The linear system in (7) is readily solved by the elimination method of Section 5.2 or by the eigenvalue method of Section 5.5. Using the latter method, we substitute $x = Ae^{\lambda t}$ and $y = Be^{\lambda t}$ in (7). Upon dividing the resulting equations by $e^{\lambda t}$, we get the two homogeneous linear equations

$$\begin{aligned}(a - \lambda)A + \quad\quad bB &= 0, \\[4pt] cA + (d - \lambda)B &= 0\end{aligned} \tag{8}$$

that the coefficients A and B must satisfy. In order for these equations to have a nontrivial solution, the determinant of coefficients must vanish:

$$\begin{aligned}\Delta = (a - \lambda)(d - \lambda) - bc \\[4pt] = \lambda^2 - (a + d)\lambda + (ad - bc) = 0.\end{aligned} \tag{9}$$

The nature of the solution of the system in (7) is determined by the nature of the roots of the quadratic **characteristic equation** in (9).

We assume that $(0, 0)$ is an isolated critical point of (7). It follows that $\lambda = 0$ cannot be a root of (9). The reason is that the constant term $ad - bc$

in (9) is the determinant of the system of linear equations

$$ax + by = 0, \\ cx + dy = 0. \qquad (10)$$

If $ad - bc = 0$, then the equations in (10) are the equations of the same line through (0, 0), so the system has a whole line of critical points rather than an isolated critical point at (0, 0).

Consequently the characteristic equation in (9) has either:

1. real unequal roots of the same sign;
2. real unequal roots of opposite sign;
3. real nonzero equal roots;
4. complex conjugate roots; or
5. pure imaginary roots.

These five cases are discussed separately below. In each case the critical point (0, 0) resembles one of those we saw in the examples of Section 9.1—a node (proper or improper), a saddle point, a center, or a spiral point.

Real Unequal Roots of the Same Sign

In this case the general solution of (7) is of the form

$$x = A_1 e^{\lambda_1 t} + A_2 e^{\lambda_2 t}, \\ y = B_1 e^{\lambda_1 t} + B_2 e^{\lambda_2 t} \qquad (11)$$

where only two of the four coefficients are arbitrary. If both roots are negative $(\lambda_2 < \lambda_1 < 0)$, it is clear from (11) that x and y approach 0 as $t \to +\infty$, so the critical point (0, 0) is asymptotically stable. But if both roots are positive $(0 < \lambda_2 < \lambda_1)$, then x and y are unbounded as $t \to +\infty$, so the critical point (0, 0) is unstable.

The system $x' = -x$, $y' = -2y$ considered in Example 2 of Section 9.1, and the improper node shown in Fig. 9.4 there, are typical of this case. In general, there is a linear change of coordinates—$u = a_1 x + a_2 y$, $v = b_1 x + b_2 y$—that transforms (11) into

$$u = A e^{\lambda_1 t}, \qquad v = B e^{\lambda_2 t}. \qquad (12)$$

In the (oblique) uv-coordinate system, the trajectories of the system in (12) are the straight lines $u = 0$ and $v = 0$ and curves of the form $v = Cu^k$, where $k = \lambda_2/\lambda_1 > 0$. The trajectories in the xy-coordinate system are the images of these curves under the inverse transformation from uv- to xy-coordinates. Thus (0, 0) is an *improper node* like the one shown in Fig. 9.9, where $k = 2$, so the trajectories are parabolas in the oblique uv-system. The distinguishing feature of an improper node is that one pair of trajectories lies on the line $u = 0$, while all other trajectories are tangent to the line $v = 0$ at (0, 0). If λ_1 and λ_2 are negative, then all trajectories approach (0, 0) as $t \to 0$, while if λ_1 and λ_2 are positive then the arrows in Fig. 9.9 would be reversed.

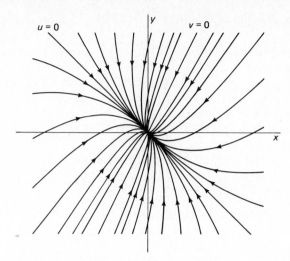

Figure 9.9 An improper node; both roots of the characteristic equation are negative.

Real Unequal Roots of Opposite Sign

In this case the situation is the same as in the previous case, except that $\lambda_2 < 0 < \lambda_1$ in (12). In the oblique uv-coordinate system, the trajectories determined by (12) are the straight lines $u = 0$ and $v = 0$ and curves of the form $uv^k = C$, where $k = -\lambda_1/\lambda_2 > 0$. The general situation resembles that shown in Fig. 9.10, where $k = 1$, so the nonlinear trajectories are hyperbolas in the uv-system. As $t \longrightarrow +\infty$, the trajectories $u = Ae^{\lambda_1 t}$, $v = 0$ go to infinity because $\lambda_1 > 0$, while the trajectories $u = 0$, $v = Be^{\lambda_2 t}$ approach the origin because $\lambda_2 < 0$. The critical point $(0, 0)$ is therefore an unstable *saddle point* in this case.

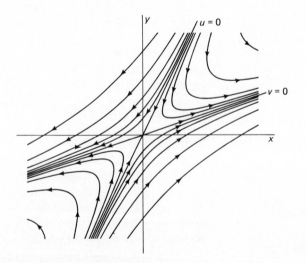

Figure 9.10 A saddle point; the roots of the characteristic equation are real and of opposite sign.

Real Equal Roots

In this case, with $\lambda_1 = \lambda_2 = \lambda$, the general solution of the system in (7) is of the form

$$x = (A_1 + A_2 t)e^{\lambda t}, \qquad y = (B_1 + B_2 t)e^{\lambda t} \tag{13}$$

where only two of the four coefficients are arbitrary. If $\lambda < 0$, then it is clear from (13) that $x \to 0$ and $y \to 0$ as $t \to +\infty$, so the critical point $(0, 0)$ is asymptotically stable. But if $\lambda > 0$, then x and y increase without bound as $t \to +\infty$, so $(0, 0)$ is an unstable critical point.

The nature of the trajectories depends on whether the term $te^{\lambda t}$ is actually present in (13). The system $x' = -x, y' = -y$ considered in Example 2 of Section 9.1 illustrates the case in which $A_2 = B_2 = 0$, so (13) reduces to

$$x = Ae^{\lambda t}, \qquad y = Be^{\lambda t}. \tag{14}$$

Then $y = (B/A)x$ if $A \neq 0$, so all trajectories lie on straight lines through the critical point $(0, 0)$, which is therefore a *proper node*; see Fig. 9.11.

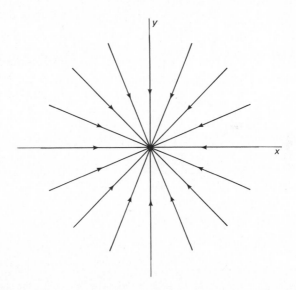

Figure 9.11 The roots of the characteristic equation are real and equal, resulting in a proper node.

The situation when the term $te^{\lambda t}$ is present is illustrated by the system $x' = -x, y' = x - y$, which has the general solution

$$x = Ae^{-t}, \qquad y = Be^{-t} + Ate^{-t}. \tag{15}$$

After we solve the first equation for $t = -(1/A) \ln x$ and substitute this in the second equation, we find that

$$y = \frac{B}{A}x - x \ln \frac{x}{A} \tag{16}$$

and

$$\frac{dy}{dx} = \frac{y'}{x'} = \frac{-Be^{-t} + Ae^{-t} - Ate^{-t}}{-Ae^{-t}}$$

$$= \frac{1}{A}(B - A + At) \tag{17}$$

if $A \neq 0$. It is clear from (17) that $dy/dx \to +\infty$ as $t \to +\infty$, so all the trajectories are tangent to the y-axis at $(0, 0)$. In this case the critical point $(0, 0)$ is an *improper node* as in Fig. 9.12. For a more general improper node, all trajectories are tangent to an arbitrary fixed line through the critical point, as in Fig. 9.13.

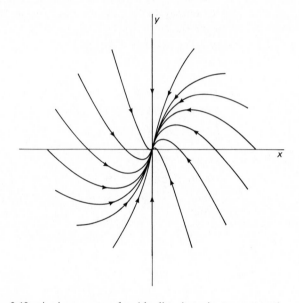

Figure 9.12 An improper node with all trajectories tangent to the y-axis.

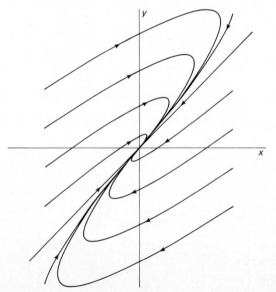

Figure 9.13 An improper node with all trajectories tangent to the graph of $y = x$.

Complex Conjugate Roots

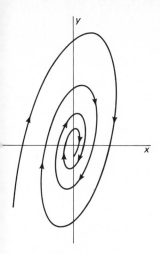

Figure 9.14 A spiral point; the roots of the characteristic equation are complex conjugates.

In this case, with $\lambda_1 = p + qi$ and $\lambda_2 = p - qi$ with p and q nonzero, the general solution has the form

$$x = e^{pt}(A_1 \cos qt + A_2 \sin qt),$$
$$y = e^{pt}(B_1 \cos qt + B_2 \sin qt). \tag{18}$$

Thus $x(t)$ and $y(t)$ oscillate between positive and negative values as t increases, and the critical point $(0, 0)$ is a *spiral point* as in Example 4 of Section 9.1. If the real part p of λ_1 and λ_2 is negative, then it is clear from (18) that (x, y) approaches $(0, 0)$ as $t \rightarrow +\infty$, so $(0, 0)$ is an asymptotically stable critical point, as shown in Fig. 9.14. But if $p > 0$ then the critical point is unstable.

Pure Imaginary Roots

If $\lambda_1 = qi$ and $\lambda_2 = -qi$ with $q \neq 0$, then the general solution is of the form

$$x = A_1 \cos qt + A_2 \sin qt, \qquad y = B_1 \cos qt + B_2 \sin qt. \tag{19}$$

As in Example 3 of Section 9.1, the trajectories are (rotated) ellipses (see Fig. 9.15), so the critical point $(0, 0)$ is a *center*: stable but not asymptotically stable.

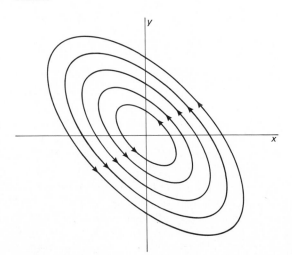

Figure 9.15 A center; the roots of the characteristic equation are pure imaginary.

For the linear system in (7) with $ad - bc \neq 0$, the table in Fig. 9.16 lists the type of critical point at $(0, 0)$ found in the five cases discussed above, according to the nature of the roots λ_1 and λ_2 of the characteristic equation in (9). Our discussion of these cases shows also that the stability of the critical point $(0, 0)$ is determined by the real parts of the characteristic roots λ_1 and λ_2, as summarized in Theorem 1. If λ_1 and λ_2 are real, then they are themselves their real parts.

Roots of characteristic equation	Type of critical point
Real, unequal, same sign	Improper node
Real, unequal, opposite sign	Saddle point
Real and equal	Proper or improper node
Complex conjugate	Spiral point
Pure imaginary	Center

Figure 9.16 Type of critical point $(0, 0)$ of the linear system in Equation (7).

THEOREM 1: STABILITY OF LINEAR SYSTEMS

Let λ_1 and λ_2 be the roots of the characteristic equation

$$\lambda^2 - (a + d)\lambda + (ad - bc) = 0 \tag{9}$$

of the linear system

$$\left.\begin{aligned}\frac{dx}{dt} &= ax + by, \\[4pt] \frac{dy}{dt} &= cx + dy\end{aligned}\right\} \tag{7}$$

with $ad - bc \neq 0$. Then the critical point $(0, 0)$ is:

(a) *asymptotically stable if the real parts of λ_1 and λ_2 are both negative;*
(b) *stable but not asymptotically stable if the real parts of λ_1 and λ_2 are both zero (so that $\lambda_1, \lambda_2 = \pm qi$);*
(c) *unstable if either λ_1 or λ_2 has a positive real part.*

It is worthwhile to consider the effect of small perturbations in the coefficients a, b, c, and d of the linear system in (7), which result in small perturbations of the characteristic roots λ_1 and λ_2. If these perturbations are sufficiently small, then positive real parts (of λ_1 and λ_2) remain positive, and negative real parts remain negative. Hence an asymptotically stable critical point remains asymptotically stable, and an unstable critical point remains unstable. Part (b) of Theorem 1 is therefore the only one in which arbitrarily small perturbations can affect the stability of the critical point $(0, 0)$. In this case pure imaginary roots $\lambda_1, \lambda_2 = \pm qi$ can be changed to nearby complex roots $\mu_1, \mu_2 = r \pm si$, with r either positive or negative; see Fig. 9.17. Consequently, a small perturbation of the coefficients of the linear system in (7) can change a stable center to a spiral point that is either unstable or asymptotically stable.

There is one other exceptional case in which the type, though not the stability, of the critical point $(0, 0)$ of a linear system can be altered by a small perturbation of its coefficients. This is the case with $\lambda_1 = \lambda_2$, equal real roots that (under a small perturbation of the coefficients) can split into two roots μ_1 and μ_2, which are either complex conjugates or unequal real roots (see

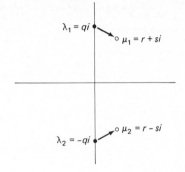

Figure 9.17 Effect of perturbation of pure imaginary roots.

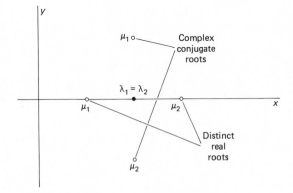

Figure 9.18 Effects of perturbation of real equal roots.

Fig. 9.18). In either case, the sign of the real parts of the roots is preserved, so the stability of the critical point is unaltered. Its nature may change, however; the table in Fig. 9.16 shows that a node with $\lambda_1 = \lambda_2$ can either remain a node (if μ_1 and μ_2 are real) or change to a spiral point (if μ_1 and μ_2 are complex conjugates).

ALMOST LINEAR SYSTEMS

We now return to the almost linear system

$$\left.\begin{aligned} \frac{dx}{dt} &= ax + by + f(x, y), \\ \frac{dy}{dt} &= cx + dy + g(x, y) \end{aligned}\right\} \tag{5}$$

having $(0, 0)$ as an isolated critical point with $ad - bc \neq 0$. Theorem 2, which we state without proof, essentially implies that—with regard to the type and stability of the critical point $(0, 0)$—the effect of the small nonlinear terms $f(x, y)$ and $g(x, y)$ is equivalent to the effect of a small perturbation in the coefficients of the associated *linear* system in (7).

THEOREM 2: STABILITY OF ALMOST LINEAR SYSTEMS

Let λ_1 and λ_2 be the characteristic roots of the linear system in (7) associated with the almost linear system in (5). Then:

(a) If $\lambda_1 = \lambda_2$ are real equal roots, then the critical point $(0, 0)$ of (5) is either a node or a spiral point, and is asymptotically stable if $\lambda_1 = \lambda_2 < 0$, unstable if $\lambda_1 = \lambda_2 > 0$.

(b) If λ_1 and λ_2 are pure imaginary, then $(0, 0)$ is either a center or a spiral point, which may be either asymptotically stable, stable, or unstable.

(c) Otherwise—that is, unless λ_1 and λ_2 are either real equal or pure imaginary—the critical point $(0, 0)$ of the almost linear system in (5) is of the same type and stability as the critical point $(0, 0)$ of the associated linear system in (7).

Thus, if $\lambda_1 \ne \lambda_2$ and $\Re e(\lambda_i) \ne 0$, then the type and stability of the critical point of the almost linear system in (5) can be determined by analysis of its associated linear system in (7), and only in the case of pure imaginary characteristic roots is the stability of $(0, 0)$ not determined by the linear system. Except in the sensitive cases $\lambda_1 = \lambda_2$ and $\Re e(\lambda_i) = 0$, the trajectories near $(0, 0)$ will resemble qualitatively those of the associated linear system—they enter or leave the critical point in the same way, but may be "deformed" in a nonlinear manner. An important consequence of the classification of cases in Theorem 2 is that *a critical point of an almost linear system is asymptotically stable if it is an asymptotically stable critical point of the linearization of the system.*

EXAMPLE 2 Determine the type and stability of the critical point $(0, 0)$ of the almost linear system

$$\begin{rcases} \dfrac{dx}{dt} = 4x + 2y + 2x^2 - 3y^2, \\[2mm] \dfrac{dy}{dt} = 4x - 3y + 7xy. \end{rcases} \tag{20}$$

Solution The characteristic equation of the associated linear system (obtained simply by deleting the quadratic terms in (20)) is

$$(4 - \lambda)(-3 - \lambda) - 8 = (\lambda - 5)(\lambda + 4) = 0,$$

so the roots $\lambda_1 = 5$ and $\lambda_2 = -4$ are real, unequal, and have opposite sign. By our discussion of this case we know that $(0, 0)$ is an unstable saddle point of the linear system, and hence by Part (c) of Theorem 2, it is also an unstable saddle point of the almost linear system in (20). The trajectories of the linear system near $(0, 0)$ are shown in Fig. 9.19, while those of the nonlinear system in (20) are shown in Fig. 9.20.

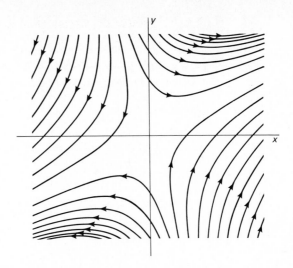

Figure 9.19 Trajectories of the linearized system of Example 2; here we have $|x| \leqq 0.5, |y| \leqq 0.4$.

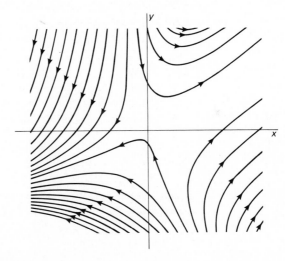

Figure 9.20 Trajectories of the nonlinear system of Example 2; $|x| \leqq 0.5$, $|y| \leqq 0.4$.

9.2 Problems

In each of Problems 1–10, determine the type of the critical point $(0, 0)$ *and whether it is asymptotically stable, stable, or unstable.*

1. $x' = -2x + y, y' = x - 2y.$

2. $x' = 4x - y, y' = 2x + y.$

3. $x' = x + 2y, y' = 2x + y.$

4. $x' = 3x + y, y' = 5x - y.$

5. $x' = x - 2y, y' = 2x - 3y.$

6. $x' = 5x - 3y, y' = 3x - y.$

7. $x' = 3x - 2y, y' = 4x - y$.

8. $x' = x - 3y, y' = 6x - 5y$.

9. $x' = 2x - 2y, y' = 4x - 2y$.

10. $x' = x - 2y, y' = 5x - y$.

Each of the systems in Problems 11–18 has a single critical point (x_0, y_0). Classify it as to type and stability. Begin by making the substitutions $u = x - x_0$, $v = y - y_0$, as in Example 1.

11. $x' = x - 2y, y' = 3x - 4y - 2$.

12. $x' = x - 2y - 8, y' = x + 4y + 10$.

13. $x' = 2x - y - 2, y' = 3x - 2y - 2$.

14. $x' = x + y - 7, y' = 3x - y - 5$.

15. $x' = x - y, y' = 5x - 3y - 2$.

16. $x' = x - 2y + 1, y' = x + 3y - 9$.

17. $x' = x - 5y - 5, y' = x - y - 3$.

18. $x' = 4x - 5y + 3, y' = 5x - 4y + 6$.

In each of Problems 19–28, investigate the type and stability of the critical point $(0, 0)$ of the given almost linear system.

19. $x' = x - 3y + 2xy, y' = 4x - 6y - xy$.

20. $x' = 6x - 5y + x^2, y' = 2x - y + y^2$.

21. $x' = x + 2y + x^2 + y^2, y' = 2x - 2y - 3xy$.

22. $x' = x + 4y - xy^2, y' = 2x - y + x^2y$.

23. $x' = 2x - 5y + x^3, y' = 4x - 6y + y^4$.

24. $x' = 5x - 5y + x(x^2 + y^2), y' = 5x - 3y + y(x^2 + y^2)$.

25. $x' = x - 2y + 3xy, y' = 2x - 3y - x^2 - y^2$.

26. $x' = 3x - 2y - x^2 - y^2, y' = 2x - y + 3xy$.

27. $x' = x - y + x^4 - y^2, y' = 2x - y + y^4 - x^2$.

28. $x' = 3x - y + x^3 + y^3, y' = 13x - 3y + 3xy$.

In each of Problems 29–32, find all critical points of the given system and investigate the type and stability of each.

29. $x' = x - y, y' = x^2 - y$.

30. $x' = y - 1, y' = x^2 - y$.

31. $x' = y^2 - 1, y' = x^3 - y$.

32. $x' = xy - 2, y' = x - 2y$.

Problems 33 and 34 illustrate the sensitive cases in which a small perturbation in the coefficients of a linear system can change the type or stability (or both) of the critical point $(0, 0)$.

33. Consider the linear system

$$\frac{dx}{dt} = hx - 4y, \qquad \frac{dy}{dt} = x + hy.$$

(a) Show that $(0, 0)$ is a center if $h = 0$. (b) Show that $(0, 0)$ is an unstable spiral point if $h > 0$. (c) Show that $(0, 0)$ is an asymptotically stable spiral point if $h < 0$. Thus small perturbations of the system $x' = -4y, y' = x$ can change both the type and the stability of the critical point $(0, 0)$.

34. Consider the linear system

$$\frac{dx}{dt} = -x + hy, \qquad \frac{dy}{dt} = x - y.$$

(a) Show that $(0, 0)$ is an asymptotically stable node if $h = 0$. (b) Show that $(0, 0)$ is an asymptotically stable spiral point if $h < 0$. (c) Show that $(0, 0)$ is an asymptotically stable node if $0 < h < 1$. Thus small perturbations of the system $x' = -x$, $y' = x - y$ can change the type of the critical point $(0, 0)$ without affecting its stability.

35. This problem deals with the almost linear system

$$\frac{dx}{dt} = y + hx(x^2 + y^2), \qquad \frac{dy}{dt} = -x + hy(x^2 + y^2),$$

in illustration of the sensitive case of Theorem 2, in which the theorem provides no information about the stability of the critical point $(0, 0)$. (a) Show that $(0, 0)$ is a center of the linear system obtained by setting $h = 0$. (b) Suppose that $h \neq 0$. Let $r^2 = x^2 + y^2$, and then apply the fact that $x(dx/dt) + y(dy/dt) = r(dr/dt)$ to show that $dr/dt = hr^3$. (c) Suppose that $h = -1$. Integrate the differential equation in (b); then show that $r \to 0$ as $t \to +\infty$. Thus $(0, 0)$ is an asymptotically stable critical point of the almost linear system in this case. (d) Suppose that $h = +1$. Show that $r \to +\infty$ as t increases, so $(0, 0)$ is an unstable critical point in this case.

9.3
Ecological Applications —Predators and Competitors

Some of the most interesting and important applications of stability theory involve the interactions between two or more biological populations occupying the same environment. We consider first a **predator-prey** situation involving two species. One species—the **predators**—feeds on the other species—the **prey**—which in turn feeds on some third food item readily available in the environment. A standard example is a population of foxes and rabbits in a woodland; the foxes (predators) eat rabbits (the prey), while the rabbits eat certain vegetation in the woodland. Other examples are sharks (predators) and food fish (prey), bass (predators) and sunfish (prey), ladybugs (predators) and aphids (prey), and beetles (predators) and scale insects (prey).

The classical mathematical model of a predator-prey situation was developed in the 1920s by the Italian mathematician Vito Volterra (1860–1940) in order to analyze the cyclic variations observed in the shark and food fish populations in the Adriatic Sea. To construct such a model, we denote the number of prey by $x(t)$ and the number of predators at time t by $y(t)$, and make the following simplifying assumptions:

1. In the absence of predators, the prey population would grow at a natural rate, with $dx/dt = at$, $a > 0$.

2. In the absence of prey, the predator population would decline at a natural rate, with $dy/dt = -cy$, $c > 0$.

3. When both predator and prey are present, there occurs, in combination with these natural rates of growth and decline, a decline in the prey population and a growth in the predator population, each at a rate proportional to the frequency of encounters between individuals of the two species. We assume further that the frequency of such encounters is proportional to the product xy, reasoning that doubling either population alone should double the frequency of encounters, while doubling both populations ought to quadruple the frequency of encounters. Consequently, the effect of predators eating prey is an *interaction rate* of decline $-bxy$ in the prey population x, and an interaction rate of growth dxy of the predator population y, with b and d positive constants.

When we add the natural and interaction rates described above, we obtain the **predator-prey equations**

$$\left.\begin{aligned}
\frac{dx}{dt} &= ax - bxy = x(a - by), \\
\frac{dy}{dt} &= -cy + dxy = y(-c + dx),
\end{aligned}\right\} \tag{1}$$

with the constants a, b, c, and d all positive. This is an almost linear system with two critical points, $(0, 0)$ and $(c/d, a/b)$. The point $(0, 0)$ is a saddle point, but the corresponding equilibrium solution $x(t) \equiv 0$, $y(t) \equiv 0$ merely describes simultaneous extinction of both species.

The critical point $(c/d, a/b)$ is of greater interest; $x(t) \equiv c/d$ and $y(t) \equiv a/b$ are the nonzero constant prey and predator populations, respectively, that can coexist in equilibrium. We would like to know, if the initial populations x_0 and y_0 are near these critical populations, whether $(x(t), y(t))$ remains near $(c/d, a/b)$ for all $t > 0$. That is, is the critical point $(c/d, a/b)$ stable? To attempt to answer this question, we substitute $u = x - c/d$, $v = y - a/b$ in (1). We thereby obtain the almost linear system

$$\left.\begin{aligned}
\frac{du}{dt} &= -\frac{bc}{d}v - buv, \\
\frac{dv}{dt} &= \frac{ad}{b}u + duv,
\end{aligned}\right\} \tag{2}$$

which has $(0, 0)$ as its critical point corresponding to the critical point $(c/d, a/b)$ of the system in (1). The corresponding linear system

$$\left.\begin{aligned}
\frac{du}{dt} &= -\frac{bc}{d}v, \\
\frac{dv}{dt} &= \frac{ad}{b}u
\end{aligned}\right\} \tag{3}$$

has characteristic equation $\lambda^2 + ac = 0$ with pure imaginary roots $\lambda_1, \lambda_2 = \pm i\sqrt{ac}$. Hence $(0, 0)$ is a stable center of (3), and the trajectories are ellipses. In fact, if we divide the second equation in (3) by the first, we obtain

$$\frac{dv}{du} = -\frac{ad/b}{bc/d}\frac{u}{v} = -\frac{ad^2u}{cb^2v},$$

$$ad^2u\,du + cb^2v\,dv = 0,$$

$$ad^2u^2 + cb^2v^2 = C,$$

$$\frac{u^2}{A^2} + \frac{v^2}{B^2} = 1, \tag{4}$$

where C is a constant of integration, $A^2 = C/ad^2$, and $B^2 = C/cb^2$. In terms of x and y, the trajectories of the linearized system are, therefore, ellipses of the form

$$\frac{1}{A^2}\left(x - \frac{c}{d}\right)^2 + \frac{1}{B^2}\left(y - \frac{a}{b}\right)^2 = 1 \tag{5}$$

centered at the critical point $(c/d, a/b)$. Some of these ellipses are shown in Fig. 9.21.

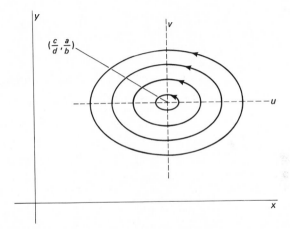

Figure 9.21 Linearized trajectories of the system in (2); the directions indicated by the arrows are determined by the signs in Equation (3).

Unfortunately, this analysis does not settle the question of the stability of the critical point $(c/d, a/b)$ of the original nonlinear system in (1), because a stable center represents the indeterminate case of Theorem 2 in Section 9.2, in which the critical point can (aside from a center) be either an unstable or an asymptotically stable spiral point. We can in this case, however, find the trajectories explicitly by dividing the second equation in (1) by the first to obtain

$$\frac{dy}{dx} = \frac{y(-c + dx)}{x(a - by)}.$$

We separate the variables to get

$$\frac{c - dx}{x}\,dx + \frac{a - by}{y}\,dy = 0,$$

and thus

$$c \ln x - dx + a \ln y - by = C, \tag{6}$$

where C is a constant of integration (and dx in Eq. (6) is not a differential, but the product of the positive constant d with x). In any case, the trajectories of the system in (1) near the critical point $(c/d, a/b)$ are the level curves of the function $f(x, y)$ that appears on the left-hand side in Eq. (6). It can be shown that these trajectories are simple closed curves enclosing $(c/d, a/b)$, which is therefore a stable center. Figure 9.22 shows a numerical plot of these trajectories for the case $a = b = c = d = 1$. It follows from Problem 32 in Section 9.1 that $x(t)$ and $y(t)$ are both periodic functions of t; this explains the fluctuations that are experimentally observed in predator-prey populations. If we follow a single trajectory in Fig. 9.22, beginning at a point where the prey population x is maximal and $y = a/b$ (Why is $y = a/b$ when x is maximal?), we see that x decreases and y increases until $x = c/d$ and the predator population y is maximal. Then both decrease until x is minimal and $y = a/b$ again. And so on around the trajectory back to the initial point. In particular, we see that if both $x_0 > 0$ and $y_0 > 0$, then both $x(t) > 0$ and $y(t) > 0$ for all t, so both populations survive in coexistence with each other. Exception: If the fluctuations are so wide that $x(t)$ is nearly zero, there is a possibility that the last few prey will be devoured, resulting in their immediate extinction and the consequent eventual extinction of the predators. This will certainly take place if ever $x(t) < 2$ in a population of mammals!

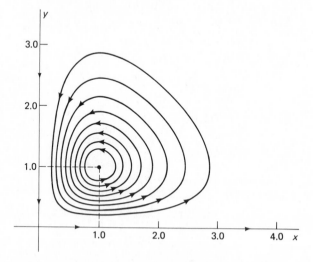

Figure 9.22 Actual trajectories of the predator-prey system $x' = x - xy$, $y' = -y + xy$.

COMPETING SPECIES

Now we consider two species (of animals, plants, or bacteria, for instance) with populations $x(t)$ and $y(t)$ and which compete with each other for the food available in their common environment. This is in marked contrast to the case in which one species preys upon the other. To construct a mathematical model that is as realistic as possible, let us assume that in the absence of either species, the other would have a bounded (logistic) population like that considered in Example 1 of Section 9.1. In the absence of any interaction or competition between the two species, their populations would then satisfy the differential equations

$$\left.\begin{aligned} \frac{dx}{dt} &= a_1 x - b_1 x^2, \\ \frac{dy}{dt} &= a_2 y - b_2 y^2, \end{aligned}\right\} \tag{7}$$

each of the form of Eq. (3) in Section 9.1. But in addition, we assume that competition has the effect of a rate of decline in each population that is proportional to their product xy. We insert such terms with *negative* proportionality constants in the equations in (7) to obtain the **competition equations**

$$\left.\begin{aligned} \frac{dx}{dt} &= a_1 x - b_1 x^2 - c_1 xy = x(a_1 - b_1 x - c_1 y), \\ \frac{dy}{dt} &= a_2 y - b_2 y^2 - c_2 xy = y(a_2 - b_2 y - c_2 x), \end{aligned}\right\} \tag{8}$$

where the coefficients a_1, a_2, b_1, b_2, c_1, and c_2 are all positive.

The almost linear system in (8) has four critical points. Upon setting the right-hand sides of the two equations equal to zero, we see that if $x = 0$, then either $y = 0$ or $y = a_2/b_2$, while if $y = 0$, then either $x = 0$ or $x = a_1/b_1$. This gives the three critical points $(0, 0)$, $(0, a_2/b_2)$, and $(a_1/b_1, 0)$. The fourth is at the intersection of the two lines

$$\left.\begin{aligned} b_1 x + c_1 y &= a_1, \\ c_2 x + b_2 y &= a_2. \end{aligned}\right\} \tag{9}$$

We assume that these two lines are not parallel and that they intersect at a point in the first quadrant. (The other cases will be explored in the exercises.) Then this point (x_E, y_E) is the fourth critical point, and it represents the possibility of peaceful coexistence of the two species, with stable populations $x(t) \equiv x_E$ and $y(t) \equiv y_E$.

We are interested in the stability of the critical point (x_E, y_E). This turns out to depend upon the relative orientation of the two lines in (9). The two possibilities are shown in Fig. 9.23, with the first line in (9) solid and the second dashed. Comparing the slopes of the two lines, we see that Fig. 9.23(a) corresponds to the condition that

$$\frac{a_2/b_2}{a_2/c_2} < \frac{a_1/c_1}{a_1/b_1}; \text{ that is, } c_1 c_2 < b_1 b_2. \tag{10}$$

Similarly, Fig. 9.23(b) corresponds to the condition that

$$\frac{a_2/b_2}{a_2/c_2} > \frac{a_1/c_1}{a_1/b_1}; \text{ that is, } c_1 c_2 > b_1 b_2. \tag{11}$$

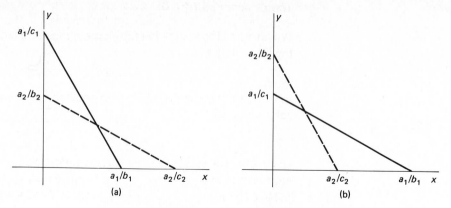

Figure 9.23 Stability of the critical point (x_E, y_E).

The conditions in (10) and (11) have a natural interpretation. In the equations in (7), we see that b_1 and b_2 represent the inhibiting effect of each population on its own growth (possibly due to limitations of food or space). On the other hand, c_1 and c_2 represent the effect of competition between the two populations. Thus $b_1 b_2$ is a measure of *inhibition*, while $c_1 c_2$ is a measure of *competition*. A general analysis of the system in (8) shows the following:

1. If $c_1 c_2 < b_1 b_2$, so that competition is small in comparison with inhibition, then (x_E, y_E) is an asymptotically stable critical point that is approached by each solution as $t \rightarrow +\infty$. Thus the two species can and do coexist in this case.

2. If $c_1 c_2 > b_1 b_2$, so that competition is large in comparison with inhibition, then (x_E, y_E) is an unstable critical point, and either $x(t)$ or $y(t)$ approaches zero as $t \rightarrow +\infty$. Thus the two species cannot coexist in this case; one survives and the other becomes locally extinct.

Rather than carrying out the general analysis that leads to the conclusions stated above, we present two examples that illustrate these two possibilities.

EXAMPLE 1 **(Survival of a single species)** Suppose that the populations x and y satisfy the equations

$$\left.\begin{array}{l} \dfrac{dx}{dt} = 60x - 4x^2 - 3xy = x(60 - 4x - 3y), \\[2mm] \dfrac{dy}{dt} = 42y - 2y^2 - 3xy = y(42 - 3x - 2y). \end{array}\right\} \quad (12)$$

These are the competition equations with $a_1 = 60$, $a_2 = 42$, $b_1 = 4$, $b_2 = 2$, and $c_1 = c_3 = 3$. Note that $c_1 c_2 = 9 > 8 = b_1 b_2$, so we should expect the results in Case 2. The four critical points are $(0, 0)$, $(0, 21)$, $(15, 0)$, and $(6, 12)$. We will analyze them individually.

The Critical Point (0, 0)

We linearize the system in (12) by dropping the quadratic terms; the result is the linear system

$$x' = 60x, \qquad y' = 42y, \tag{13}$$

in which primes denote, as usual, derivatives with respect to time t. The general solution of this system is

$$x = Ae^{60t}, \qquad y = Be^{42t}, \tag{14}$$

so (0, 0) is an unstable node for the linearized system in (13). Because the equations in (14) yield $y = Cx^{7/10}$, all the trajectories other than the x-axis are tangent to the y-axis at the origin. Figure 9.24 shows some of these trajectories near the critical point (0, 0).

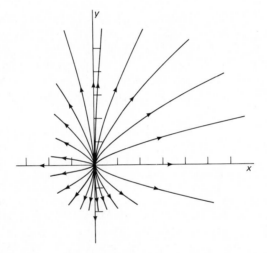

Figure 9.24 Linearized trajectories of (12) near (0, 0).

The Critical Point (0, 21)

To linearize the equations in (12) near the critical point (0, 21), we substitute $u = x$ and $v = y - 21$. The result is the almost linear system

$$\left.\begin{aligned}
\frac{du}{dt} &= -3u - 4u^2 - 3uv, \\
\frac{dv}{dt} &= -63u - 42v - 3uv - 2v^2
\end{aligned}\right\} \tag{15}$$

with critical point (0, 0). The corresponding linear system

$$u' = -3u, \qquad v' = -63u - 42v \tag{16}$$

has characteristic roots $\lambda_1 = -3$ and $\lambda_2 = -42$. So (0, 21) is an asymptotically stable node. The general solution of (16) has the form

$$u = Ae^{-3t},$$

$$v = -\frac{21}{13}Ae^{-3t} + Be^{-42t}, \qquad (17)$$

where A and B are arbitrary constants. As $t \to +\infty$, (u, v) approaches $(0, 0)$, so (x, y) approaches $(0, 21)$. With $A = 0$, (17) gives the vertical line through the critical point. The slope of any trajectory with $A \neq 0$ is

$$\frac{dy}{dx} = \frac{dy/dt}{dx/dt} = \frac{dv/dt}{du/dt} = \frac{\frac{63}{13}Ae^{-3t} - 42Be^{-42t}}{-3Ae^{-3t}} = -\frac{21}{13} + \frac{14B}{A}e^{-39t}.$$

Hence $dy/dx \to -\frac{21}{13}$ as $t \to +\infty$ and (x, y) approaches $(0, 21)$. Consequently, the trajectories near the critical point $(0, 21)$ resemble those shown in Fig. 9.25.

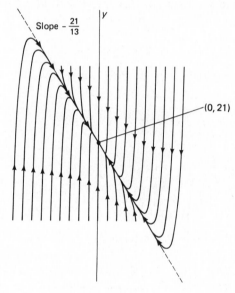

Figure 9.25 Linearized trajectories of (12) near $(0, 21)$; the asymptote has slope $-21/13$.

The Critical Point (15, 0)

The substitution $u = x - 15$, $v = y$ in the equations in (12) yields

$$\frac{du}{dt} = -60u - 45v - 4u^2 - 3uv,$$

$$\frac{dv}{dt} = -3v - 3uv - 2v^2. \qquad (18)$$

The corresponding linear system

$$u' = -60u - 45v, \qquad v' = -3v \qquad (19)$$

has characteristic roots $\lambda_1 = -3$ and $\lambda_2 = -60$, so this critical point is also an asymptotically stable node. The general solution is of the form

$$u = -\frac{15}{19}Ae^{-3t} + Be^{-60t},$$

$$v = Ae^{-3t}. \qquad (20)$$

As $t \to +\infty$, $(u, v) \to (0, 0)$ and so $(x, y) \to (15, 0)$. With $A = 0$ we get the horizontal line through the critical point. The slope of any trajectory with $A \neq 0$ is

$$\frac{dy}{dx} = \frac{dy/dt}{dx/dt} = \frac{dv/dt}{du/dt} = \frac{-3Ae^{-3t}}{\frac{45}{19}Ae^{-3t} - 60Be^{-60t}} = \frac{-A}{\frac{15}{19}A - 20Be^{-57t}}.$$

Hence $dy/dx \to -\frac{19}{15}$ as $t \to +\infty$ and (x, y) approaches $(15, 0)$. The trajectories near $(15, 0)$ therefore resemble those shown in Fig. 9.26.

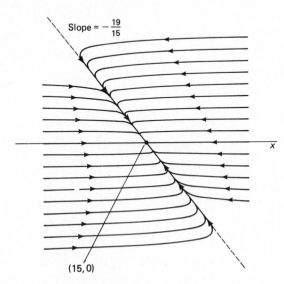

Slope $= -\frac{19}{15}$

x

$(15, 0)$

Figure 9.26 Linearized trajectories of (12) near $(15, 0)$; the asymptote has slope $-19/15$.

The Critical Point $(6, 12)$

To linearize the equations in (12) near the critical point $(6, 12)$ (the one representing the possibility of coexistence of the two species), we substitute $u = x - 6$ and $v = y - 12$. The result is the almost linear system

$$\left. \begin{array}{l} \dfrac{du}{dt} = -24u - 18v - 4u^2 - 3uv, \\[2mm] \dfrac{dv}{dt} = -36u - 24v - 3uv - 2v^2. \end{array} \right\} \tag{21}$$

The associated linear system is

$$u' = -24u - 18v, \qquad v' = -36u - 24v. \tag{22}$$

The characteristic equation associated with Eq. (22) is

$$(-24 - \lambda)^2 - (-36)(-18) = (\lambda + 24)^2 - 2(18)^2 = 0$$

with roots $\lambda_1 = -24 + 18\sqrt{2} > 0$ and $\lambda_2 = -24 - 18\sqrt{2} < 0$. Thus this critical point is an unstable saddle point. We find that the general solution of (22) is

$$\left. \begin{array}{l} u = Ae^{\lambda_1 t} + Be^{\lambda_2 t}, \\[2mm] v = -\sqrt{2}\,Ae^{\lambda_1 t} + \sqrt{2}\,Be^{\lambda_2 t}, \end{array} \right\} \tag{23}$$

where A and B are arbitrary constants. The slope of a trajectory in the linearized system is given by

$$\frac{dy}{dx} = \frac{dv/dt}{du/dt} = \sqrt{2}\left(\frac{-\lambda_1 A e^{\lambda_1 t} + \lambda_2 B e^{\lambda_2 t}}{\lambda_1 A e^{\lambda_1 t} + \lambda_2 B e^{\lambda_2 t}}\right). \tag{24}$$

If $A = 0$ then $(u, v) \rightarrow (0, 0)$, and so $(x, y) \rightarrow (6, 12)$ as $t \rightarrow +\infty$ because $\lambda_2 < 0$. From (23) and (24) we see that (x, y) then approaches the critical point $(6, 12)$ along the straight line of slope $\sqrt{2}$. If $A \neq 0$ then from (24) we find that

$$\frac{dy}{dx} = \sqrt{2}\left(\frac{-\lambda_1 A + \lambda_2 B e^{-36\sqrt{2}t}}{\lambda_1 A + \lambda_2 B c^{-36\sqrt{2}t}}\right). \tag{25}$$

With $B = 0$ we have the straight line of slope $-\sqrt{2}$ along which (x, y) leaves the saddle point as t increases, and from (25) we see that $dy/dx \rightarrow -\sqrt{2}$ along any other trajectory as $t \rightarrow +\infty$. Hence the trajectories near the critical point $(6, 12)$ resemble those shown in Fig. 9.27.

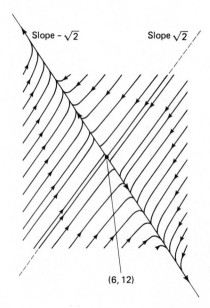

Slope $-\sqrt{2}$ Slope $\sqrt{2}$

$(6, 12)$

Figure 9.27 Linearized trajectories of (12) near $(6, 12)$; the asymptote has slope $-\sqrt{2}$.

Now that we have completed our local analysis of the four critical points of the almost linear system in (12), we want to assemble this information into a coherent whole—to construct a phase plane picture (phase portrait) of the global behavior of the trajectories in the first quadrant (where both populations are nonnegative). If we accept the facts that (1) near each critical point the trajectories look qualitatively like the linearized trajectories shown in Figs. 9.24—9.27, and (2) as $t \rightarrow +\infty$, each trajectory either approaches a critical point or diverges toward infinity somewhere in the first quadrant, then it follows that the phase plane picture should look qualitatively similar to the linearized trajectories shown in the schematic Fig. 9.28.

The two trajectories that approach the saddle point $(6, 12)$, together with that saddle point, form a curve known as the *dividing trajectory*, or **separatrix**.

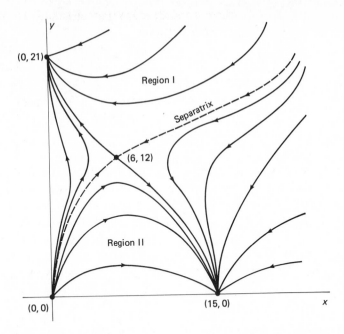

Figure 9.28 A qualitative representation of some of the trajectories of the system in (12).

It plays a crucial role in determining the long-term behavior of the two populations. If the initial point (x_0, y_0) lies precisely on the separatrix, then $(x(t), y(t))$ approaches $(6, 12)$ as $t \to +\infty$. Of course, random events make it extremely unlikely that $(x(t), y(t))$ will remain on the separatrix. If not, peaceful coexistence of the two species is impossible. If (x_0, y_0) lies in Region I above the separatrix, then $(x(t), y(t))$ approaches $(0, 21)$ as $t \to +\infty$, so the population $x(t)$ decreases to zero. Alternatively, if (x_0, y_0) lies in Region II below the separatrix, then $(x(t), y(t))$ approaches $(15, 0)$ as $t \to +\infty$, so the population $y(t)$ dies out. In short, whichever population has the initial competitive advantage survives, while the other faces extinction.

EXAMPLE 2 **(Peaceful coexistence)** Now suppose that the two populations satisfy the equations

$$\left. \begin{aligned} \frac{dx}{dt} &= 60x - 3x^2 - 4xy = x(60 - 3x - 4y), \\ \frac{dy}{dt} &= 42y - 3y^2 - 2xy = y(42 - 2x - 3y). \end{aligned} \right\} \tag{26}$$

Here $a_1 = 60$, $a_2 = 42$, $b_1 = b_2 = 3$, $c_1 = 4$, and $c_2 = 2$. So $c_1 c_2 = 8 < b_1 b_2 = 9$: Competition is smaller than inhibition. The analysis of the system in (26) follows step by step that of the system in (12) of Example 1. We will present only the results of this analysis, leaving the details to Problems 2–6. There are four critical points—$(0, 0)$, $(0, 14)$, $(20, 0)$, and $(12, 6)$.

The critical point (0, 0) is an unstable node of the linearized system, just as in Example 1.

The critical point (0, 14) is an unstable saddle point of the linearized system. The entering trajectories lie along the y-axis, while the departing trajectories lie along the line through (0, 14) with slope $-\frac{14}{23}$.

The critical point (20, 0) is also an unstable saddle point of the linearized system. The entering trajectories lie along the x-axis, while the departing trajectories lie along the straight line through (20, 0) with slope $-\frac{31}{40}$.

The critical point (12, 6) represents the possibility of peaceful coexistence because it is an asymptotically stable node. One pair of trajectories lie along the line through (12, 6) with slope $(-3 + \sqrt{73})/16 \approx 0.35$. All other trajectories of the linearized system enter the node and are tangent to the line with slope $(-3 - \sqrt{73})/16 \approx -0.72$.

Figure 9.29 shows a reasonable phase portrait that is qualitatively consistent with the results of our critical point analysis. One especially notable feature of this system is that for *any* positive initial values x_0 and y_0, $(x(t), y(t))$ approaches (12, 6) as $t \to +\infty$, so the two species both survive in stable (peaceful) coexistence.

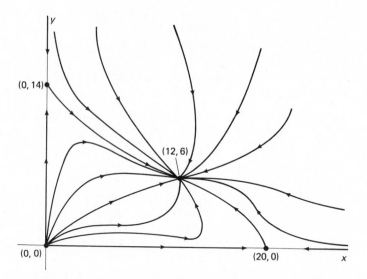

Figure 9.29 A qualitative representation of some of the trajectories of the system in (26).

Examples 1 and 2 illustrate the power of elementary critical point analysis. But we must conclude with a word of caution. Ecological systems in nature are rarely so simple as in these examples; they normally involve far more than two species, and the rates of growth of these populations and their interactions are almost always more complex than those discussed in this section.

9.3 Problems

1. Let $x(t)$ be a harmful insect population (aphids?) that under natural conditions is held somewhat in check by a benign predator insect population $y(t)$ (ladybugs?). Assume that $x(t)$ and $y(t)$ satisfy the predator-prey equations in (1), so that the stable equilibrium populations are $x_E = c/d$ and $y_E = a/b$. Now suppose that an insecticide is employed that kills (per unit time) the same fraction $f < a$ of each species of insect. Show that the harmful population x_E is increased, while the benign population y_E is decreased, so the use of the insecticide is counterproductive. This is an instance in which mathematical analysis reveals undesirable consequences of a well-intentioned interference with nature.

Problems 2–6 provide the details cited in Example 2 for the two competing populations $x(t)$ and $y(t)$ that satisfy the equations

$$\left.\begin{aligned}
\frac{dx}{dt} &= 60x - 3x^2 - 4xy = x(60 - 3x - 4y), \\
\frac{dy}{dt} &= 42y - 3y^2 - 2xy = y(42 - 2x - 3y).
\end{aligned}\right\} \tag{26}$$

2. Show that the critical points of the system in (26) are $(0, 0)$, $(0, 14)$, $(20, 0)$, and $(12, 6)$.

3. Show that the critical point $(0, 0)$ is an unstable node at which all but two trajectories are tangent to the y-axis.

4. (a) To investigate the critical point $(0, 14)$, substitute $u = x, v = y - 14$; then show that the corresponding linear system is $u' = 4u$, $v' = -28u - 42v$. (b) Show that the general solution of this linear system is $u = Ae^{4t}$, $v = -\frac{14}{23}Ae^{4t} + Be^{-42t}$. (c) Then conclude that $(0, 14)$ is an unstable saddle point at which the entering and departing trajectories lie on the y-axis and on the line through $(0, 14)$ with slope $-\frac{14}{23}$.

5. (a) To investigate the critical point $(20, 0)$, substitute $u = x - 20, v = y$; then show that the corresponding linear system is $u' = -60u - 80v$, $v' = 2v$. (b) Show that the general solution of this linear system is $u = Ae^{-60t} - \frac{40}{31}Be^{2t}$, $v = Be^{2t}$. (c) Then conclude that $(20, 0)$ is an unstable saddle point at which entering and departing trajectories lie on the x-axis and on the line through $(20, 0)$ with slope $-\frac{31}{40}$.

6. (a) To investigate the critical point $(12, 6)$, substitute $u = x - 12, v = y - 6$; then show that the corresponding linear system is $u' = -36u - 48v$, $v' = -12u - 18v$. (b) Show that the characteristic roots of this linear system are $\lambda_1 = -27 + 3\sqrt{73} < 0$ and $\lambda_2 = -27 - 3\sqrt{73} < 0$, and that its general solution is

$$x = Ae^{\lambda_1 t} + Be^{\lambda_2 t}, \qquad y = Ce^{\lambda_1 t} + De^{\lambda_2 t}$$

where $C = (-3 - \sqrt{73})A/16$ and $D = (-3 + \sqrt{73})B/16$. (c) Hence conclude that $(12, 6)$ is an asymptotically stable node, where two trajectories lie on the line through $(12, 6)$ with slope $(-3 + \sqrt{73})/16$ and the others are tangent to the line through $(12, 6)$ with slope $(-3 - \sqrt{73})/16$.

Problems 7–11 deal with the predator-prey system that is modeled by the equations

$$\left.\begin{aligned}\frac{dx}{dt} &= 5x - x^2 - xy = x(5 - x - y), \\ \frac{dy}{dt} &= xy - 2y = y(x - 2).\end{aligned}\right\} \tag{27}$$

In contrast with the system in (1) discussed in the text, the prey population $x(t)$ would—in the absence of any predators—be a bounded population described by the logistic equation $dx/dt = 5x - x^2$.

7. Show that the critical points of the system in (27) are $(0, 0)$, $(5, 0)$, and $(2, 3)$.

8. Show that the critical point $(0, 0)$ is an unstable saddle point of the linearized system, with trajectories entering along the y-axis and departing along the x-axis.

9. Show that the critical point $(5, 0)$ is an unstable saddle point of the linearized system, with trajectories entering along the x-axis and departing along the line through $(5, 0)$ with slope $-\frac{8}{5}$.

10. Show that the critical point $(2, 3)$ is an asymptotically stable spiral point of the linearized system.

11. Show that the results of Problems 7–10 are consistent with the global phase portrait for the original nonlinear system in (27) shown in Fig. 9.30. Conclude that the prey and predators coexist with stable equilibrium populations $x_E = 2$ and $y_E = 3$.

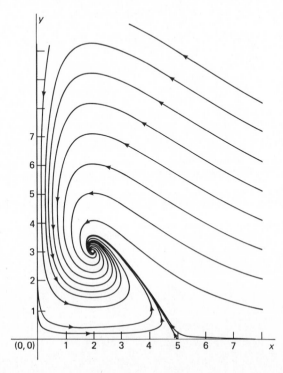

Figure 9.30 Some actual trajectories of the predator-prey system in (27).

Problems 12–17 deal with the predator-prey system modeled by the equations

$$\begin{aligned}\frac{dx}{dt} &= x^2 - 2x - xy = x(x - y - 2),\\[2mm]\frac{dy}{dt} &= y^2 - 4y + xy = y(x + y - 4).\end{aligned} \right\} \tag{28}$$

In this system, each population—the prey population $x(t)$ and the predator population $y(t)$—is an unsophisticated population (similar to the one considered in Problem 27 of Section 9.1) for each of which the only alternatives (in the absence of the other) are doomsday and extinction.

12. Show that the critical points of the system in (28) are $(0, 0)$, $(0, 4)$, $(2, 0)$, and $(3, 1)$.

13. Show that the critical point $(0, 0)$ is an asymptotically stable node of the linearized system, at which one pair of trajectories lie along the y-axis and the others are tangent to the x-axis.

14. Show that the critical point $(0, 4)$ is an unstable saddle point of the linearized system, with trajectories departing along the y-axis and entering along the line through $(0, 4)$ with slope $-\frac{2}{5}$.

15. Show that the critical point $(2, 0)$ is an unstable saddle point of the linearized system, with trajectories departing along the x-axis and entering along the line through $(2, 0)$ with slope 2.

16. Show that the critical point $(3, 1)$ is an unstable spiral point of the linearized system.

17. Show that the results of Problems 12–16 are consistent with the global phase portrait for the original nonlinear system in (28) shown in Fig. 9.31. This is a two-dimensional version of doomsday versus extinction. If the initial point (x_0, y_0) lies in Region I, then both populations increase without bound (until doomsday), while if it lies in Region II, then both populations will decrease to zero.

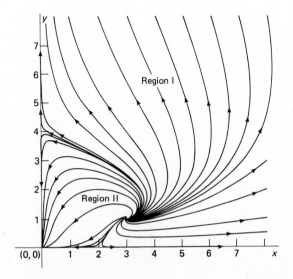

Figue 9.31 Some actual trajectories of the predator-prey system in (28).

The next two problems deal with the competition equations in (8) and the associated linearized system. The discussion in the text covered only the case in which the critical point of interest (x_E, y_E) lay in the first quadrant. You may now explore the other possibilities.

18. Determine the behavior of the linearized system associated with the system in (8) in the case that (x_E, y_E) exists but does not lie in the first quadrant. Because the populations are never negative, there are only three critical points to examine. There are two cases, depending upon whether (x_E, y_E) lies in the second quadrant or the fourth quadrant.

19. Determine the behavior of the linearized system associated with the system in (8) in the case that (x_E, y_E) does not exist because the two lines having the equations in (9) are parallel. There will be two cases, depending upon which line is above the other. (You may take for granted that the improbable—coincidence of the two lines—does not occur.)

9.4
Existence and Uniqueness of Solutions

In Chapter 1 we saw that an initial value problem of the form

$$\frac{dy}{dx} = f(x, y), \qquad y(a) = b \tag{1}$$

can fail (on a given interval containing the point $x = a$) to have a unique solution. For instance, in Example 3 of Section 1.3, we saw that the initial value problem

$$y' = -\frac{y^2}{x^2}, \qquad y(0) = 1 \tag{2}$$

has no solution at all. In Problem 37 of Section 1.3, we asked you to show that the initial value problem

$$\frac{dy}{dx} = -\sqrt{1 - y^2}, \qquad y(0) = 1 \tag{3}$$

has the two distinct solutions $y(x) \equiv 1$ and $y(x) = \cos x$ on the interval $0 \leq x \leq \pi$. In this final section, we investigate conditions on the function $f(x, y)$ that suffice to guarantee that the initial value problem in (1) has one and only one solution, and then proceed to establish appropriate versions of the existence-uniqueness theorems that were stated without proof in Sections 1.3, 2.1, 2.2, and 5.1.

EXISTENCE OF SOLUTIONS

The approach we employ is the **method of successive approximations**, which was developed by the French mathematician Emile Picard (1856–1941). This method is based on the fact that the function $f(x)$ satisfies the initial value problem in (1) on the open interval I containing $x = a$ if and only if it satisfies the integral equation

$$y(x) = b + \int_a^x f(t, y(t)) \, dt \tag{4}$$

for all x in I. In particular, if $y(x)$ satisfies (4), then clearly $y(a) = b$, and differentiation of both sides in (4)—using the fundamental theorem of calculus—yields the differential equation $y'(x) = f(x, y(x))$.

To attempt to solve Eq. (4), we begin with the initial function

$$y_0(x) = b, \tag{5}$$

and then define iteratively a sequence $y_1, y_2, y_3, \ldots$ of functions that we hope will converge to the solution. Specifically, we let

$$y_1(x) = b + \int_a^x f(t, y_0(t))\, dt$$

and $\tag{6}$

$$y_2(x) = b + \int_a^x f(t, y_1(t))\, dt.$$

In general, y_{n+1} is obtained by substitution of y_n for y in the right-hand side in (4):

$$y_{n+1}(x) = b + \int_a^x f(t, y_n(t))\, dt. \tag{7}$$

Suppose we know that each of these functions $\{y_n(x)\}_0^\infty$ is defined on some open interval (the same for each n) containing $x = a$, and that the limit

$$y(x) = \lim_{n \to \infty} y_n(x) \tag{8}$$

exists at each point of this interval. Then it will follow that

$$y(x) = \lim_{n \to \infty} y_{n+1}(x)$$

$$= \lim_{n \to \infty} \left[b + \int_a^x f(t, y_n(t))\, dt \right]$$

$$= b + \lim_{n \to \infty} \int_a^x f(t, y_n(t))\, dt \tag{9}$$

$$= b + \int_a^x f(t, \lim_{n \to \infty} y_n(t))\, dt \tag{10}$$

and hence that

$$y(x) = b + \int_a^x f(t, y(t))\, dt,$$

provided that we can validate the interchange of limit operations involved in passing from (9) to (10). It is therefore reasonable to expect that, under favorable conditions, the sequence $\{y_n(x)\}$ defined recursively in (5) and (7) will converge to a solution $y(x)$ of the integral equation in (4), and hence to a solution of the original initial value problem in (1).

EXAMPLE 1 To apply the method of successive approximations to the initial value problem

$$\frac{dy}{dx} = y, \qquad y(0) = 1, \tag{11}$$

we write Eqs. (5) and (7), thereby obtaining

$$y_0(x) \equiv 1, \qquad y_{n+1}(x) = 1 + \int_0^x y_n(t)\, dt. \tag{12}$$

The iteration formula in (12) yields

$$y_1(x) = 1 + \int_0^x 1 \, dt = 1 + x,$$

$$y_2(x) = 1 + \int_0^x (1 + t) \, dt$$

$$= 1 + x + \tfrac{1}{2}x^2,$$

$$y_3(x) = 1 + \int_0^x (1 + t + \tfrac{1}{2}t^2) \, dt$$

$$= 1 + x + \tfrac{1}{2}x^2 + \tfrac{1}{6}x^3,$$

and

$$y_4(x) = 1 + \int_0^x (1 + t + \tfrac{1}{2}t^2 + \tfrac{1}{6}t^3) \, dt$$

$$= 1 + x + \tfrac{1}{2}x^2 + \tfrac{1}{6}x^3 + \tfrac{1}{24}x^4.$$

It is clear that we are generating the sequence of partial sums of a power series solution; indeed, we immediately recognize the series as that of $y(x) = e^x$. There is no difficulty in demonstrating that the exponential function is indeed the solution of the initial value problem in (11); moreover, a diligent student can verify (using a proof by induction on n) that $y_n(x)$, obtained in the manner above, is indeed the nth partial sum for the Taylor series with center zero for $y(x) = e^x$.

EXAMPLE 2 To apply the method of successive approximations to the initial value problem

$$\frac{dy}{dx} = 4xy, \qquad y(0) = 3, \tag{13}$$

we write Eqs. (5) and (7) as in Example 1. Now we obtain

$$y_0(x) \equiv 3, \qquad y_{n+1}(x) = 3 + \int_0^x 4t y_n(t) \, dt \tag{14}$$

The iteration formula in (14) yields

$$y_1(x) = 3 + \int_0^x (4t)(3) \, dt = 3 + 6x^2,$$

$$y_2(x) = 3 + \int_0^x 4t(3 + 6t^2) \, dt$$

$$= 3 + 6x^2 + 6x^4,$$

$$y_3(x) = 3 + \int_0^x 4t(3 + 6t^2 + 6t^4) \, dt$$

$$= 3 + 6x^2 + 6x^4 + 4x^6,$$

and

$$y_4(x) = 3 + \int_0^x 4t(3 + 6t^2 + 6t^4 + 4t^6) \, dt$$

$$= 3 + 6x^2 + 6x^4 + 4x^6 + 2x^8.$$

It is again clear that we are generating partial sums of a power series solution. It is not quite so obvious what function has such a power series representation, but the initial value problem in (13) is readily solved by separation of variables:

$$y(x) = 3e^{2x^2} = 3 \sum_{n=0}^{\infty} \frac{(2x^2)^n}{n!}$$

$$= 3 + 6x^2 + 6x^4 + 4x^6 + 2x^8 + \frac{4}{5}x^{10} + \cdots.$$

In some cases it may be necessary to compute a much larger number of terms, either in order to identify the solution or to use a partial sum of its series with large subscript to approximate the solution accurately for x near its initial value. Fortunately, there are computer programs available (even for microcomputers) that will perform the symbolic integrations (as opposed to numerical integrations) of the sort in the examples above. If necessary, you could generate the first hundred terms in Example 2 in a matter of minutes.

In general, of course, we apply Picard's method because we cannot find a solution by elementary methods. Suppose that we have produced a large number of terms of what we believe to be the correct power series expansion of the solution. We *must* have conditions under which the sequence $\{y_n(x)\}$ provided by the method of successive approximation is guaranteed in advance to converge to a solution. It is just as convenient to discuss the initial value problem

$$\frac{d\mathbf{x}}{dt} = \mathbf{f}(\mathbf{x}, t), \qquad \mathbf{x}(a) = \mathbf{b} \tag{15}$$

for a system of m first order equations, where

$$\mathbf{x} = \begin{pmatrix} x_1 \\ x_2 \\ x_3 \\ \cdot \\ \cdot \\ \cdot \\ x_m \end{pmatrix}, \quad \mathbf{f} = \begin{pmatrix} f_1 \\ f_2 \\ f_3 \\ \cdot \\ \cdot \\ \cdot \\ f_m \end{pmatrix}, \quad \text{and} \quad \mathbf{b} = \begin{pmatrix} b_1 \\ b_2 \\ b_3 \\ \cdot \\ \cdot \\ \cdot \\ b_m \end{pmatrix}.$$

It turns out that with the aid of this vector notation (which we introduced in Section 5.3), most results concerning a single (scalar) equation $x' = f(x, t)$ can be generalized readily to analogous results for a system of m first order equations, as abbreviated in (15). Consequently, the effort of using vector notation is amply justified by the generality it affords.

The method of successive approximations for the system in (15) calls for us to compute the sequence $\{\mathbf{x}_n(t)\}_0^{\infty}$ of vector-valued functions of t,

$$\mathbf{x}_n(t) = \begin{pmatrix} x_{1n}(t) \\ x_{2n}(t) \\ x_{3n}(t) \\ \cdot \\ \cdot \\ \cdot \\ x_{mn}(t) \end{pmatrix},$$

defined iteratively by

$$\mathbf{x}_0(t) \equiv \mathbf{b}, \qquad \mathbf{x}_{n+1}(t) = \mathbf{b} + \int_a^t \mathbf{f}(\mathbf{x}_n(s), s) \, ds. \tag{16}$$

Recall that vector-valued functions are integrated componentwise.

EXAMPLE 3 Consider the m-dimensional initial value problem

$$\frac{d\mathbf{x}}{dt} = \mathbf{Ax}, \qquad \mathbf{x}(0) = \mathbf{b} \tag{17}$$

for a homogeneous linear system with $m \times m$ constant coefficient matrix $\mathbf{A}$. The equations in (16) take the form

$$\mathbf{x}_0(t) = \mathbf{b}, \qquad \mathbf{x}_{n+1}(t) = \mathbf{b} + \int_0^t \mathbf{Ax}_n(s) \, ds. \tag{18}$$

Thus

$$\mathbf{x}_1(t) = \mathbf{b} + \int_0^t \mathbf{Ab} \, ds$$
$$= \mathbf{b} + \mathbf{Ab}t = (\mathbf{I} + \mathbf{A}t)\mathbf{b};$$
$$\mathbf{x}_2(t) = \mathbf{b} + \int_0^t \mathbf{A}(\mathbf{b} + \mathbf{Ab}s) \, ds$$
$$= \mathbf{b} + \mathbf{Ab}t + \tfrac{1}{2}\mathbf{A}^2\mathbf{b}t^2$$
$$= (\mathbf{I} + \mathbf{A}t + \tfrac{1}{2}\mathbf{A}^2t^2)\mathbf{b},$$

and

$$\mathbf{x}_3(t) = \mathbf{b} + \int_0^t \mathbf{A}(\mathbf{b} + \mathbf{Ab}s + \tfrac{1}{2}\mathbf{A}^2\mathbf{b}s^2) \, ds$$
$$= (\mathbf{I} + \mathbf{A}t + \tfrac{1}{2}\mathbf{A}^2t^2 + \tfrac{1}{6}\mathbf{A}^3t^3)\mathbf{b}.$$

We have therefore obtained the first several partial sums of the exponential series solution

$$\mathbf{x}(t) = e^{\mathbf{A}t}\mathbf{b} = \left(\sum_{n=0}^{\infty} \frac{(\mathbf{A}t)^n}{n!} \right)\mathbf{b} \tag{19}$$

of (17), which was derived earlier in Section 5.6.

The key to establishing convergence in the method of successive approximations is an appropriate condition on the rate at which $\mathbf{f}(\mathbf{x}, t)$ changes when $\mathbf{x}$ varies but t is held fixed. If R is a region in $(m + 1)$-dimensional $(\mathbf{x}, t)$-space, then the function $\mathbf{f}(\mathbf{x}, t)$ is said to be **Lipschitz continuous** in R if there exists a constant $k > 0$ such that

$$|\mathbf{f}(\mathbf{x}_1, t) - \mathbf{f}(\mathbf{x}_2, t)| \leq k|\mathbf{x}_1 - \mathbf{x}_2| \tag{20}$$

if $(\mathbf{x}_1, t)$ and $(\mathbf{x}_2, t)$ are points of R. Recall that the norm of an m-dimensional point or vector $\mathbf{x}$ is defined to be

$$|\mathbf{x}| = (x_1^2 + x_2^2 + \cdots + x_m^2)^{1/2}. \tag{21}$$

Then $|\mathbf{x}_1 - \mathbf{x}_2|$ is simply the Euclidean distance between the points $\mathbf{x}_1$ and $\mathbf{x}_2$.

EXAMPLE 4 Let $f(x, t) = x^2 e^{-t^2} \sin t$ and let R be the strip $0 \leq x \leq 2$ in the xt-plane. If (x_1, t) and (x_2, t) are both points of R, then

$$|f(x_1, t) - f(x_2, t)| = |e^{-t^2} \sin t| |x_1 + x_2| |x_1 - x_2|$$
$$\leq 4|x_1 - x_2|,$$

because $|e^{-t^2} \sin t| \leq 1$ for all t and $|x_1 + x_2| \leq 4$ if x_1 and x_2 are both in the interval $[0, 2]$. Thus f satisfies the Lipschitz condition in (20) with $k = 4$ and is therefore Lipschitz continuous in the strip R.

EXAMPLE 5 Let $f(x, t) = t\sqrt{x}$ on the rectangle R consisting of the points (x, t) in the plane for which $0 \leq x \leq 1$ and $0 \leq t \leq 1$. Then, taking $x_1 = x$, $x_2 = 0$, and $t = 1$, we find that

$$|f(x, 1) - f(0, 1)| = \sqrt{x} = \frac{1}{\sqrt{x}}|x - 0|.$$

Because $x^{-1/2} \rightarrow +\infty$ as $x \rightarrow 0^+$, we see that the Lipschitz condition in (20) cannot be satisfied by any (finite) constant $k > 0$. Thus the function f, though obviously continuous on R, is *not* Lipschitz continuous on R.

Suppose, however, that the function $f(x, t)$ has a continuous partial derivative $f_x(x, t)$ on the closed rectangle R in the xt-plane, and denote by k the maximum value of $|f_x(x, t)|$ on R. Then the mean value theorem of differential calculus yields

$$|f(x_1, t) - f(x_2, t)| = |f_x(\bar{x}, t)(x_1 - x_2)|$$

for some $\bar{x}$ in (x_1, x_2), so it follows that

$$|f(x_1, t) - f(x_2, t)| \leq k|x_1 - x_2|$$

because $|f_x(\bar{x}, t)| \leq k$. Thus a continuously differentiable function $f(x, t)$ defined on a closed rectangle *is* Lipschitz continuous there. More generally, the multivariable mean value theorem of advanced calculus can be used similarly to prove that *a vector-valued function $\mathbf{f}(x, t)$ with continuously differentiable component functions on a closed rectangular region R in (x, t)-space is Lipschitz continuous in R.*

EXAMPLE 6 The function $f(x, t) = x^2$ is Lipschitz continuous on any closed (bounded) rectangle in the xt-plane. But consider this function on the infinite strip R consisting of the points (x, t) for which $0 \leq t \leq 1$ and x is arbitrary. Then

$$|f(x_1, t) - f(x_2, t)| = |x_1^2 - x_2^2|$$
$$= |x_1 + x_2| |x_1 - x_2|.$$

Because $|x_1 + x_2|$ can be made arbitrarily large, it follows that f is *not* Lipschitz continuous on the infinite strip R.

If I is an interval on the t-axis, then the set of all points $(\mathbf{x}, t)$ with t in I is an infinite strip or slab in $(m + 1)$-space (as indicated in Fig. 9.32). Example 6 shows that Lipschitz continuity of $\mathbf{f}(\mathbf{x}, t)$ on such an infinite slab is a very strong

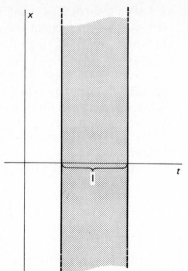

Figure 9.32 An infinite slab in $(m + 1)$-space.

condition. Nevertheless, the existence of a solution of the initial value problem

$$\frac{d\mathbf{x}}{dt} = \mathbf{f}(\mathbf{x}, t), \qquad \mathbf{x}(a) = \mathbf{b} \tag{15}$$

under the hypothesis of Lipschitz continuity of $\mathbf{f}$ in such a slab is of considerable importance.

THEOREM 1:   GLOBAL EXISTENCE OF SOLUTIONS

Let $\mathbf{f}$ be a vector-valued function (with m components) of $m + 1$ real variables, and let I be a (bounded or unbounded) open interval containing $t = a$. If $\mathbf{f}(\mathbf{x}, t)$ is continuous and satisfies the Lipschitz condition in (20) for all t in I and for all $\mathbf{x}_1$ and $\mathbf{x}_2$, then the initial value problem in (15) has a solution on the (entire) interval I.

Proof We want to show that the sequence $\{\mathbf{x}_n(t)\}_0^\infty$ of successive approximations determined iteratively by

$$\mathbf{x}_0(t) \equiv \mathbf{b}, \qquad \mathbf{x}_{n+1}(t) = \mathbf{b} + \int_a^t \mathbf{f}(\mathbf{x}_n(s), s) \, ds \tag{16}$$

converges to a solution $\mathbf{x}(t)$ of (15). We see that each of these functions in turn is continuous on I, being an (indefinite) integral of a continuous function.

We can assume that $a = 0$, because the transformation $t \to t + a$ converts (15) into an equivalent problem with initial point $t = 0$. Also, we will consider only the portion $t \geq 0$ of the interval I; the details for the case $t \leq 0$ are very similar.

The main part of the proof consists in showing that if $[0, T]$ is a closed (and bounded) interval contained in I, then the sequence $\{\mathbf{x}_n(t)\}$ converges *uniformly* on I to a limit function $\mathbf{x}(t)$. This means that, given $\epsilon > 0$, there exists an integer N such that

$$|\mathbf{x}_n(t) - \mathbf{x}(t)| < \epsilon \tag{22}$$

for all $n \geqq N$ and all t in $[0, T]$. For ordinary (perhaps nonuniform) convergence the integer N, for which (22) holds for all $n \geqq N$, may depend upon t, with no single value of N working for all t in I. Once this uniform convergence of the sequence $\{x_n(t)\}$ has been established, the following conclusions will follow from standard theorems of advanced calculus (see pages 620–622 of Taylor and Mann, *Advanced Calculus*, (New York: John Wiley, 3rd ed., 1983):

1. The limit function $\mathbf{x}(t)$ is continuous on $[0, T]$.
2. If N is chosen so that the inequality in (22) holds for $n \geq N$, then the Lipschitz continuity of $\mathbf{f}$ implies that

$$|\mathbf{f}(\mathbf{x}_n(t), t) - \mathbf{f}(\mathbf{x}(t), t)| \leqq k|\mathbf{x}_n(t) - \mathbf{x}(t)| < k\epsilon$$

for all t in $[0, T]$ and $n \geqq N$, so it follows that the sequence $\{\mathbf{f}(\mathbf{x}_n(t), t)\}_0^\infty$ converges uniformly to $\mathbf{f}(\mathbf{x}(t), t)$ on $[0, T]$.

3. But a uniformly convergent sequence or series can be integrated termwise, so it follows that, upon taking limits in the iterative formula in (16),

$$\mathbf{x}(t) = \lim_{n \to \infty} \mathbf{x}_{n+1}(t)$$

$$= \mathbf{b} + \lim_{n \to \infty} \int_0^t \mathbf{f}(\mathbf{x}_n(s), x) \, ds$$

$$= \mathbf{b} + \int_0^t \lim_{n \to \infty} \mathbf{f}(\mathbf{x}_n(s), s) \, ds;$$

thus

$$\mathbf{x}(t) = \mathbf{b} + \int_0^t \mathbf{f}(\mathbf{x}(s), s) \, ds. \tag{23}$$

4. Because the function $\mathbf{x}(t)$ is continuous on $[0, T]$, the integral equation in (23) (analogous to the one-dimensional case in (4)) implies that $\mathbf{x}'(t) = \mathbf{f}(\mathbf{x}(t), t)$ on $[0, T]$. But if this is true on every closed subinterval of the open interval I, then it is true on the entire interval I.

It therefore remains only to prove that the sequence $\{\mathbf{x}_n(t)\}_0^\infty$ converges uniformly on the closed interval $[0, T]$. Let M be the maximum value of $|\mathbf{f}(\mathbf{b}, t)|$ for t in $[0, T]$. Then

$$|\mathbf{x}_1(t) - \mathbf{x}_0(t)| = \left| \int_0^t \mathbf{f}(\mathbf{x}_0(s), s) \, ds \right| \leqq \int_0^t |\mathbf{f}(\mathbf{b}, s)| \, ds = Mt. \tag{24}$$

Next,

$$|\mathbf{x}_2(t) - \mathbf{x}_1(t)| = \left| \int_0^t [\mathbf{f}(\mathbf{x}_1(s), s) - \mathbf{f}(\mathbf{x}_0(s), s) \, ds \right|$$

$$\leqq k \int_0^t |\mathbf{x}_1(s) - \mathbf{x}_0(s)| \, ds,$$

and hence

$$|\mathbf{x}_2(t) - \mathbf{x}_1(t)| \leqq k \int_0^t Ms \, ds = \tfrac{1}{2}kMt^2. \tag{25}$$

We now proceed by induction: Assume that

$$|\mathbf{x}_n(t) - \mathbf{x}_{n-1}(t)| \leq \frac{M}{k} \cdot \frac{(kt)^n}{n!}. \tag{26}$$

It then follows that

$$|\mathbf{x}_{n+1}(t) - \mathbf{x}_n(t)| = \left| \int_0^t [\mathbf{f}(\mathbf{x}_n(s), s) - \mathbf{f}(x_{n-1}(s), s)] \, ds \right|$$

$$\leq k \int_0^t |\mathbf{x}_n(s) - \mathbf{x}_{n-1}(s)| \, ds;$$

consequently,

$$|\mathbf{x}_{n+1}(t) - \mathbf{x}_n(t)| \leq k \int_0^t \frac{M}{k} \cdot \frac{(ks)^n}{n!} \, ds.$$

It follows on evaluating the integral above that

$$|\mathbf{x}_{n+1} - \mathbf{x}_n(t)| \leq \frac{M}{k} \cdot \frac{(kt)^{n+1}}{(n+1)!}.$$

Thus (26) holds on $[0, T]$ for all $n \geq 1$.

Hence the terms of the infinite series

$$\mathbf{x}_0(t) + \sum_{n=1}^{\infty} [\mathbf{x}_n(t) - \mathbf{x}_{n-1}(t)] \tag{27}$$

are dominated (in magnitude on the interval $[0, T]$) by the terms of the convergent series

$$\sum_{n=1}^{\infty} \frac{M}{k} \cdot \frac{(kT)^{n+1}}{(n+1)!} = \frac{M}{k}(e^{kT} - 1), \tag{28}$$

which is a series of positive constants. It therefore follows (from the Weierstrass M-test on pages 618–619 of Taylor and Mann) that the series in (27) converges uniformly on $[0, T]$. But the sequence of partial sums of this series is simply our original sequence $\{\mathbf{x}_n(t)\}_0^{\infty}$ of successive approximations, so the proof of Theorem 1 is finally complete.

Linear Systems

An important application of the global existence theorem just given is to the initial value problem

$$\frac{d\mathbf{x}}{dt} = \mathbf{A}(t)\mathbf{x} + \mathbf{g}(t), \qquad \mathbf{x}(a) = \mathbf{b} \tag{29}$$

for a linear system, where the $m \times m$ matrix-valued function $\mathbf{A}(t)$ and the vector-valued function $\mathbf{g}(t)$ are continuous on a (bounded or unbounded) open interval I containing the point $t = a$. In order to apply Theorem 1 to the linear system in (29), we note first that the proof of Theorem 1 requires only that, for each closed and bounded subinterval J of I, there exists a Lipschitz constant k such that

$$|\mathbf{f}(\mathbf{x}_1, t) - \mathbf{f}(\mathbf{x}_2, t)| \leq k |\mathbf{x}_1 - \mathbf{x}_2| \tag{20}$$

for all t in J (and all $\mathbf{x}_1$ and $\mathbf{x}_2$). Thus we do not need a single Lipschitz constant for the entire open interval I.

In (29) we have $\mathbf{f}(\mathbf{x}, t) = \mathbf{Ax} + \mathbf{g}$, so

$$\mathbf{f}(\mathbf{x}_1, t) - \mathbf{f}(\mathbf{x}_2, t) = \mathbf{A}(t)(\mathbf{x}_1 - \mathbf{x}_2). \tag{30}$$

It therefore suffices to show that, if $\mathbf{A}(t)$ is continuous on the closed and bounded interval J, then there is a constant k such that

$$|\mathbf{A}(t)\mathbf{x}| \leq k|\mathbf{x}| \tag{31}$$

for all t in J. But this follows from the fact (see Problem 17) that

$$|\mathbf{Ax}| \leq \|\mathbf{A}\| \cdot |\mathbf{x}|, \tag{32}$$

where the **norm** $\|\mathbf{A}\|$ of the matrix $\mathbf{A}$ is defined to be

$$\|\mathbf{A}\| = \left(\sum_{i,j=1}^{m} a_{ij}{}^2 \right)^{1/2}. \tag{33}$$

Because $\mathbf{A}(t)$ is continuous on the closed and bounded interval J, the norm $\|\mathbf{A}(t)\|$ is bounded on J, so Eq. (31) follows, as desired. Thus we have the following global existence theorem for the linear initial value problem in (29).

THEOREM 2:   EXISTENCE FOR LINEAR SYSTEMS

Let the $m \times m$ matrix-valued function $\mathbf{A}(t)$ and the vector-valued function $\mathbf{g}(t)$ be continuous on the (bounded or unbounded) open interval I containing the point $t = a$. Then the initial value problem

$$\frac{d\mathbf{x}}{dt} = \mathbf{A}(t)\mathbf{x} + \mathbf{g}(t), \qquad \mathbf{x}(a) = \mathbf{b} \tag{29}$$

has a solution on the (entire) interval I.

As we saw in Section 5.1, the mth order initial value problem

$$\left. \begin{array}{l} x^{(m)} + a_1(t)x^{(m-1)} + \cdots + a_{m-1}(t)x' + a_m(t)x = p(t), \\ x(a) = b_0,\ x'(a) = b_1, \ldots, x^{(m-1)}(a) = b_{m-1} \end{array} \right\} \tag{34}$$

is easily transformed into an equivalent $m \times m$ system of the form in (29). It therefore follows from Theorem 2 that if the functions $a_1(t), a_2(t), \ldots, a_m(t)$, and $p(t)$ in (34) are all continuous on the (bounded or unbounded) open interval I containing $t = a$, then the initial value problem in (34) has a solution on the (entire) interval I.

Local Existence

In the case of a *nonlinear* initial value problem

$$\frac{d\mathbf{x}}{dt} = \mathbf{f}(\mathbf{x}, t), \qquad \mathbf{x}(a) = \mathbf{b}, \tag{35}$$

the hypothesis in Theorem 1 that $\mathbf{f}$ satisfies a Lipschitz condition on a slab $(\mathbf{x}, t)$ (t in I, all $\mathbf{x}$) is unrealistic and rarely satisfied. This is illustrated by the following simple example.

EXAMPLE 7 Consider the initial value problem

$$\frac{dx}{dt} = x^2, \qquad x(0) = b > 0. \tag{36}$$

As we saw in Example 6, the equation $x' = x^2$ does not satisfy a "strip Lipschitz condition." When we solve (36) by separation of variables, we get

$$x(t) = \frac{b}{1 - bt}. \tag{37}$$

Because the denominator vanishes for $t = 1/b$, (37) provides a solution of the initial value problem in (36) only for $t < 1/b$, despite the fact that the differential equation $x' = x^2$ "looks nice" on the whole real line. In particular, if b is large, then we have a solution only on a very small interval to the right of $t = 0$.

Although Theorem 2 assures us that *linear* equations have global solutions, Example 7 shows that, in general, even a "nice" nonlinear differential equation can be expected to have a solution only on a small interval about the initial point $t = a$, and that the length of this interval of existence can depend on the initial value $\mathbf{x}(a) = \mathbf{b}$, as well as on the differential equation itself. The reason is this: If $\mathbf{f}(\mathbf{x}, t)$ is continuously differentiable in a neighborhood of the point $(\mathbf{b}, a)$ in $(m + 1)$-space, then—as indicated in the discussion preceding Example 6—we can conclude that $\mathbf{f}(\mathbf{x}, t)$ satisfies a Lipschitz condition on some rectangular region R centered at $(\mathbf{b}, a)$, of the form

$$|t - a| < A, \qquad |x_i - b_i| < B_i \tag{38}$$

$(i = 1, 2, \ldots, m)$. In the proof of Theorem 1, we need to apply the Lipschitz condition on the function $\mathbf{f}$ in analyzing the iterative formula

$$\mathbf{x}_{n+1}(t) = \mathbf{b} + \int_a^t \mathbf{f}(\mathbf{x}_n(s), s) \, ds. \tag{39}$$

The potential difficulty is that unless the values of t are suitably restricted, the points $(\mathbf{x}_n(t), t)$ appearing in the integrand in (39) may not lie in the region R where $\mathbf{f}$ is known to satisfy a Lipschitz condition. On the other hand, it can be shown that—on a sufficiently small open interval J containing the point $t = a$ —the graphs of the functions $\{\mathbf{x}_n(t)\}$ given iteratively by the formula in (39) remain within the region R, so the proof of convergence can then be carried out as in the proof of Theorem 1. A proof of the following *local* existence theorem can be found in Chapter 6 of G. Birkhoff and G.-C. Rota, *Ordinary Differential Equations*, (New York: John Wiley, 2nd ed., 1969).

THEOREM 3: LOCAL EXISTENCE OF SOLUTIONS

Let $\mathbf{f}$ be a vector-valued function (with m components) of the $m + 1$ real values $x_1, x_2, \ldots, x_m$, and t. If the first order partial derivatives of $\mathbf{f}$ all exist and are continuous in some neighborhood of the point $\mathbf{x} = \mathbf{b}$, $t = a$, *then the initial value problem*

$$\frac{d\mathbf{x}}{dt} = \mathbf{f}(\mathbf{x}, t), \qquad \mathbf{x}(a) = \mathbf{b}, \tag{35}$$

has a solution on some open interval containing the point $t = a$.

UNIQUENESS OF SOLUTIONS

It is possible to establish the existence of solutions of the initial value problem in (35) under the much weaker hypothesis that $\mathbf{f}(\mathbf{x}, t)$ is merely continuous; techniques other than those used in this section are required. By contrast, the Lipschitz condition that we used in proving Theorem 1 is the key to *uniqueness* of solutions. In particular, the solution provided by Theorem 3 is unique near the point $t = a$.

THEOREM 4:   UNIQUENESS OF SOLUTIONS

Suppose that on some region R in $(m + 1)$-space, the function $\mathbf{f}$ in (35) is continuous and satisfies the Lipschitz condition

$$|\mathbf{f}(\mathbf{x}_1, t) - \mathbf{f}(\mathbf{x}_2, t)| \leqq k|\mathbf{x}_1 - \mathbf{x}_2|. \tag{20}$$

If $\mathbf{x}_1(t)$ and $\mathbf{x}_2(t)$ are two solutions of the initial value problem in (35) on some open interval I containing $t = a$, such that the solution curves $(\mathbf{x}_1(t), t)$, and $(\mathbf{x}_2(t), t)$ both lie in R for all t in I, then $\mathbf{x}_1(t) = \mathbf{x}_2(t)$ for all t in I.

We will outline the proof of Theorem 4 for the one-dimensional case in which x is a real variable. A generalization of this proof to the multivariable case can be found in Chapter 6 of Birkhoff and Rota.

Let us consider the function

$$\phi(t) = [x_1(t) - x_2(t)]^2 \tag{40}$$

for which $\phi(a) = 0$, because $x_1(a) = x_2(a) = b$. We want to show that $\phi(t) \equiv 0$, so that $x_1(t) \equiv x_2(t)$. We will consider only the case $t \geqq a$; the details are similar for the case $t \leqq a$.

If we differentiate each side in (40), we find that

$$
\begin{aligned}
|\phi'(t)| &= |2[x_1(t) - x_2(t)][x_1'(t) - x_2'(t)]| \\
&= |2[x_1(t) - x_2(t)][f(x_1(t), t) - f(x_2(t), t)]| \\
&\leqq 2k|x_1(t) - x_2(t)|^2 = 2k\phi(t),
\end{aligned}
$$

using the Lipschitz condition on f. Hence

$$\phi'(t) \leqq 2k\phi(t). \tag{41}$$

Now let us temporarily ignore the fact that $\phi(a) = 0$ and compare $\phi(t)$ with the solution of the differential equation

$$\Phi'(t) = 2k\Phi(t) \tag{42}$$

such that $\Phi(a) = \phi(a)$; clearly

$$\Phi(t) = \phi(a)e^{2k(t-a)}. \tag{43}$$

In comparing (41) with (42), it seems inevitable that

$$\phi(t) \leq \Phi(t) \quad \text{for } t \geq a, \tag{44}$$

and this is easily proved (see Problem 18). Hence

$$0 \leq [x_1(t) - x_2(t)]^2 \leq [x_1(a) - x_2(a)]^2 e^{2k(t-a)}.$$

On taking square roots, we get

$$0 \leq |x_1(t) - x_2(t)| \leq |x_1(a) - x_2(a)| e^{k(t-a)}. \tag{45}$$

But $x_1(a) - x_2(a) = 0$, so (45) implies that $x_1(t) \equiv x_2(t)$.

EXAMPLE 8 The initial value problem

$$\frac{dx}{dt} = 3x^{2/3}, \qquad x(0) = 0 \tag{46}$$

has both the obvious solution $x_1(t) \equiv 0$ and the solution $x_2(t) = t^3$ that is easily found by separation of variables. Hence the function $f(x, t) = 3x^{2/3}$ must *fail* to satisfy a Lipschitz condition near $(0, 0)$. Indeed, the mean value theorem yields

$$|f(x, 0) - f(0, 0)| = |f_x(\bar{x}, 0)||x - 0|$$

for some $\bar{x}$ between 0 and x. But $f_x(x, 0) = 2x^{-1/3}$ is unbounded as $x \to 0$, so no Lipschitz condition can be satisfied.

In addition to uniqueness, another consequence of the inequality in (45) is the fact that solutions of the differential equation

$$\frac{dx}{dt} = f(x, t) \tag{47}$$

depend *continuously* on the initial value $x(a)$; that is, if $x_1(t)$ and $x_2(t)$ are two solutions of (47) on the interval $a \leq t \leq T$ such that the initial values $x_1(a)$ and $x_2(a)$ are sufficiently close to one another, then the values $x_1(t)$ and $x_2(t)$ remain close to one another. In particular, if $|x_1(a) - x_2(a)| \leq \delta$, then (45) implies that

$$|x_1(t) - x_2(t)| \leq \delta e^{k(T-a)} = \epsilon \tag{48}$$

for all t with $a \leq t \leq T$. Obviously, we can make ϵ as small as we wish by choosing δ sufficiently close to zero.

This continuity of solutions of (47) with respect to initial values is important in practical applications where we are unlikely to know the initial value $x(a)$ with absolute precision. For example, suppose that $x(t)$ is a population, and we know that the actual initial population is within δ of the value x_0. Then (45) implies that the actual population at time t is within $\delta e^{k(t-a)}$ of the value of the particular solution of (47) having the initial value x_0. More generally, an initial value problem is usually considered "well posed" as a mathematical model for a real-world situation only if the differential equation has unique solutions that are continuous with respect to initial values. Otherwise it is unlikely that the initial value problem adequately mirrors the real-world situation.

9.4 Problems

In each of Problems 1–8, apply the successive approximations formula to compute $y_n(x)$ for $n \leq 4$. Then write the exponential series for which these approximations are partial sums (perhaps minus the first term or two; for example,

$$e^x - 1 = x + \tfrac{1}{2}x^2 + \cdots).$$

1. $y' = y$, $y(0) = 3$.
2. $y' = -2y$, $y(0) = 4$.
3. $y' = -2xy$, $y(0) = 1$.
4. $y' = 3x^2 y$, $y(0) = 2$.
5. $y' = 2y + 2$, $y(0) = 0$.
6. $y' = x + y$, $y(0) = 0$.
7. $y' = 2x(1 + y)$, $y(0) = 0$.
8. $y' = 4x(y + 2x^2)$, $y(0) = 0$.

In each of Problems 9–12, compute the successive approximations $y_n(x)$ for $n \leq 3$; then compare them with the appropriate partial sums of the Taylor series of the exact solution.

9. $y' = x + y$, $y(0) = 1$.
10. $y' = y + e^x$, $y(0) = 0$.
11. $y' = y^2$, $y(0) = 1$.
12. $y' = \tfrac{1}{2}y^3$, $y(0) = 1$.

13. Apply the iterative formula in (16) to compute the first three successive approximations to the solution of the initial value problem

$$x' = 2x - y, \qquad x(0) = 1;$$
$$y' = 3x - 2y, \qquad y(0) = -1.$$

14. Apply the matrix exponential series in (19) to solve (in closed form) the initial value problem

$$\mathbf{x}'(t) = \begin{pmatrix} 1 & 1 \\ 0 & 1 \end{pmatrix}\mathbf{x}, \ \mathbf{x}(0) = \begin{pmatrix} 1 \\ 1 \end{pmatrix}.$$

(*Suggestion:* Show first that

$$\begin{pmatrix} 1 & 1 \\ 0 & 1 \end{pmatrix}^n = \begin{pmatrix} 1 & n \\ 0 & 1 \end{pmatrix}$$

for each positive integer n.)

15. For the initial value problem $dy/dx = 1 + y^3$, $y(1) = 1$, show that the second Picard approximation is

$$y_2(x) = 1 + 2(x - 1) + 3(x - 1)^2 + 4(x - 1)^3 + 2(x - 1)^4.$$

Then compute $y_2(1.1)$ and $y_2(1.2)$. The fourth-order Runge-Kutta method with step size $h = 0.005$ yields $y(1.1) \approx 1.2391$ and $y(1.2) \approx 1.6269$.

16. For the initial value problem $dy/dx = x^2 + y^2$, $y(0) = 0$, show that the third Picard approximation is

$$y_3(x) = \frac{1}{3}x^3 + \frac{1}{63}x^7 + \frac{2}{2079}x^{11} + \frac{1}{59,535}x^{15}.$$

Compute $y_3(1)$. The fourth-order Runge-Kutta method yields $y(1) \approx 0.350232$, both with step size $h = 0.05$ and with step size $h = 0.025$.

17. Prove as follows the inequality $|\mathbf{Ax}| \leq \|\mathbf{A}\| \cdot |\mathbf{x}|$, where $\mathbf{A}$ is an $m \times m$ matrix with row vectors $\mathbf{a}_1, \mathbf{a}_2, \ldots, \mathbf{a}_m$, and $\mathbf{x}$ is an m-dimensional vector. First note

that the components of the vector $\mathbf{Ax}$ are $\mathbf{a}_1 \cdot \mathbf{x}, \mathbf{a}_2 \cdot \mathbf{x}, \ldots, \mathbf{a}_m \cdot \mathbf{x}$, so

$$|\mathbf{Ax}| = \left[\sum_{i=1}^{m} (\mathbf{a}_i \cdot \mathbf{x})^2\right]^{1/2}.$$

Then use the Cauchy-Schwarz inequality $(\mathbf{a} \cdot \mathbf{x})^2 \leq |\mathbf{a}|^2 |\mathbf{x}|^2$ for the dot product.

18. Suppose that $\phi(t)$ is a differentiable function with

$$\phi'(t) \leq k\phi(t) \qquad (k > 0)$$

for $t \geq a$. Multiply both sides by e^{-kt}; then transpose to show that

$$\frac{d}{dt}[\phi(t)e^{-kt}] \leq 0$$

for $t \geq a$. Then apply the mean value theorem to conclude that

$$\phi(t) \leq \phi(a)e^{k(t-a)}$$

for $t \geq a$.

References
for Further Study

The literature of the theory and applications of differential equations is vast. The following list includes a selection of books that might be useful to readers who wish to pursue further the topics introduced in this book.

1. M. ABRAMOWITZ and I. A. STEGUN, *Handbook of Mathematical Functions*, New York: Dover, 1965. The comprehensive collection of tables to which frequent reference is made in the text.

2. G. BIRKHOFF and G.-C. ROTA, *Ordinary Differential Equations* (2nd ed.), New York: John Wiley, 1969. An intermediate-level text that includes more complete treatment of existence and uniqueness theorems, Sturm-Liouville problems, and eigenfunction expansions.

3. F. BRAUER and J. NOHEL, *Qualitative Theory of Ordinary Differential Equations*, New York: W. A. Benjamin, 1969. A more complete treatment of linear systems and of qualitative properties of solutions.

4. M. BRAUN, *Differential Equations and Their Applications* (3rd ed.), New York: Springer-Verlag, 1983. An introductory text at a slightly higher level than this one; it includes several interesting "case study" applications.

5. R. V. CHURCHILL, *Operational Mathematics* (3rd ed.), New York: McGraw-Hill, 1972. Standard reference for theory and applications of Laplace transforms, starting at about the same level as Chapter 4 of this book.

6. R. V. CHURCHILL and J. W. BROWN, *Fourier Series and Boundary Value Problems* (3rd ed.), New York: McGraw-Hill, 1978. At about the same level as Chapters 7 and 8 of this book.

7. E. A. CODDINGTON, *An Introduction to Ordinary Differential Equations*, Englewood Cliffs, N. J.: Prentice-Hall, 1961. An intermediate-level introduction; Chapters 3

and 4 include proofs of the theorems on power series and Frobenius series solutions stated in Chapter 3 of this book.

8. E. A. CODDINGTON and N. LEVINSON, *Theory of Ordinary Differential Equations*, New York: McGraw-Hill, 1955. An advanced theoretical text; Chapter 5 discusses solution near an irregular singular point.

9. S. D. CONTE and C. DEBOOR, *Elementary Numerical Analysis* (2nd ed.), New York: McGraw-Hill, 1972. An introductory text; Chapter 6 treats numerical solution of differential equations, and includes FORTRAN programs.

10. P. HENRICI, *Discrete Variable Methods in Ordinary Differential Equations*, New York: John Wiley, 1962. A more complete and somewhat advanced treatment of numerical methods of solution.

11. M. W. HIRSH and S. SMALE, *Differential Equations, Dynamical Systems, and Linear Algebra*, New York: Academic Press, 1974. Intermediate-level treatment of linear systems and of qualitative properties of solutions.

12. E. L. INCE, *Ordinary Differential Equations*, New York: Dover, 1956. First published in 1926, this is the classic older reference work on the subject.

13. N. N. LEBEDEV, *Special Functions and Their Applications*, New York: Dover, 1972. A comprehensive account of Bessel functions and the other special functions of mathematical physics.

14. N. N. LEBEDEV, I. P. SKALSKAYA, and Y. S. UFLYAND, *Worked Problems in Applied Mathematics*, New York: Dover, 1979. A large collection of applied examples and problems similar to those discussed in Chapter 8 of this book.

15. N. W. MCLACHLAN, *Bessel Functions for Engineers* (2nd ed.), London: Oxford University Press, 1955. Includes numerous physical applications of Bessel functions.

16. N. W. MCLACHLAN, *Ordinary Non-Linear Differential Equations in Engineering and Physical Sciences*, London: Oxford University Press, 1956. A concrete introduction to the effects of nonlinearity in physical systems.

17. B. NOBLE, *Applied Linear Algebra* (2nd ed.), Englewood Cliffs, N. J.: Prentice-Hall, 1977. An introduction to linear algebra and its applications.

18. E. RAINVILLE, *Intermediate Differential Equations* (2nd ed.), New York: Macmillan, 1964. Chapters 3 and 4 include proofs of the theorems on power series and Frobenius series solutions stated in Chapter 3 of this book.

19. J. R. RICE, *Numerical Methods, Software, and Analysis: IMSL Reference Edition*, New York: McGraw-Hill, 1983. Chapter 9 contains detailed discussion of sophisticated numerical algorithms and their mainframe implementation.

20. H. SAGAN, *Boundary and Eigenvalue Problems in Mathematical Physics*, New York: John Wiley, 1961. Discusses the classical boundary value problems and the variational approach to Sturm-Liouville problems, eigenvalues, and eigenfunctions.

21. G. F. SIMMONS, *Differential Equations*, New York: McGraw-Hill, 1972. An introductory text with interesting historical notes and fascinating applications and with the most eloquent preface in any mathematics book currently in print.

22. W. G. STRANG, *Linear Algebra and Its Applications* (2nd ed.), New York: Academic Press, 1980. An introductory treatment of linear algebra with motivating applications.

23. G. P. TOLSTOV, *Fourier Series*, New York: Dover, 1976. An introductory text

including detailed discussion of both convergence and applications of Fourier series.

24. H. F. WEINBERGER, *A First Course in Partial Differential Equations*, New York: Blaisdell, 1965. Includes separation of variables, Sturm-Liouville methods, and applications of Laplace transform methods to partial differential equations.

25. R. WEINSTOCK, *Calculus of Variations*, New York: Dover, 1974. Includes variational derivations of the partial differential equations of vibrating strings and membranes, rods, and bars.

Answers

SECTION 1.1

13. $r = \frac{2}{3}$ **15.** $r = -2, 1$ **17.** $A = \frac{4}{3}, B = -\frac{1}{3}$ **19.** $A = \frac{3}{4}, B = \frac{1}{4}$ **21.** $y' = x + y$
23. $y' = x/(1 - y)$ **25.** $y' = (y - x)/(x + y)$ **27.** $dv/dt = kv^2$ **29.** $dN/dt = k(P - N)$ **31.** $y = 1$ or $y = x$
33. $y = x^2$ **35.** $y = \frac{1}{2}e^x$

SECTION 1.2

1. $y = x^2 + x + 3$ **3.** $y = \frac{2}{3}x^{3/2} - \frac{16}{3}$ **5.** $y = 2(x + 2)^{1/2} - 5$ **7.** $y = 10 \tan^{-1} x$ **9.** $y = \sin^{-1} x$
11. $x = 25t^2 + 10t + 20$ **13.** $x = \frac{1}{2}t^3 + 5t$ **15.** $x = \frac{1}{3}(t + 3)^4 - 37t - 26$ **17.** $x = 1/(2t + 2) + \frac{1}{2}t - \frac{1}{2}$
19. Approximately 38.58 m   **21.** Approximately 178.57 m   **23.** $181\frac{1}{3}$ ft/s   **25.** 150 ft/s (about 102 mi/h)
27. Hits at $20\sqrt{10}$ ft/s after $2\sqrt{10}$ s

SECTION 1.3

1.

3.

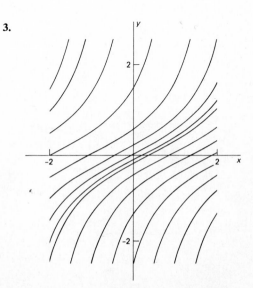

5.

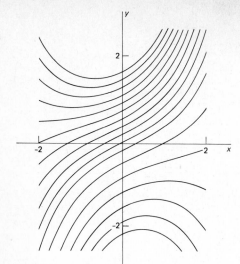

7.

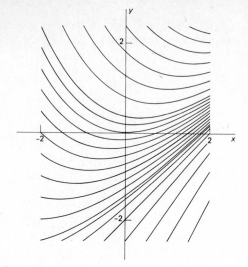

9.

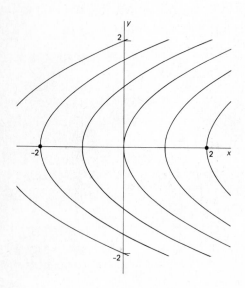

11.

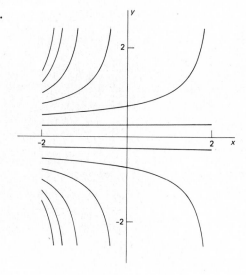

13.

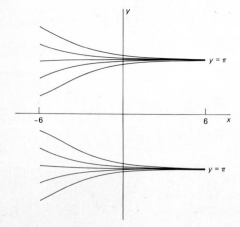

15.

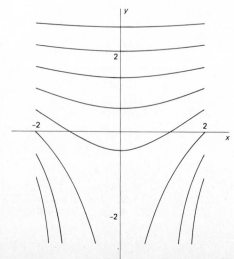

17. Solution exists and is unique. **19.** Solution exists and is unique.
21. Neither is guaranteed by the theorem. **23.** Solution exists and is unique.
25. Solution exists and is unique.

27.

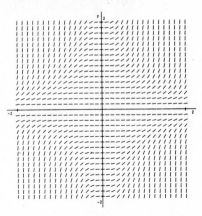

29.

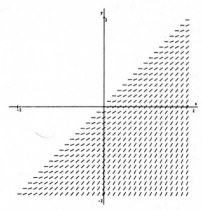

31.

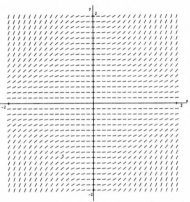

33.

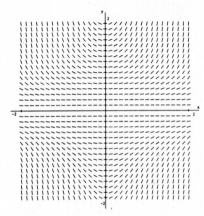

35.

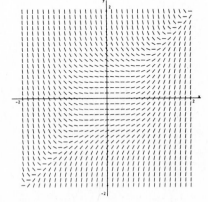

SECTION 1.4

1. $(C - x^2)y^2 = 1$ **3.** $\frac{1}{4}\ln(y^4 + 1) = \sin x + C$ **5.** $1/3y^3 - 2/y = 1/x + \ln|x| + C$
7. $\ln|1 + y| = x + \frac{1}{2}x^2 + C$ **9.** $y = 2\exp(e^x)$ **11.** $y^2 = 1 + (x^2 - 16)^{1/2}$ **13.** $\ln(2y - 1) = 2(x - 1)$
15. $\ln y = x^2 - 1 + \ln x$ **17.** About 51,840 persons **19.** About 14,735 years **21.** $21,103.48 **23.** 2585 mg

25. About 4.86 billion years ago   **27.** 1 h 3 min after putting it on the porch
29. (a) 0.495 m (b) $(8.32 \times 10^{-7})I_0$ (c) 3.29 m   **31.** (a) $A' = rA + Q$ (b) $Q = \$704.82$ per year
33. After 66 min 40 s   **35.** About 46 days after the rumor starts   **37.** 16 min 12 s   **39.** 14 min 29 s
43. 1 h $18\frac{2}{3}$ min after the plug is pulled

SECTION 1.5

1. $xe^{-3x}y = C$; $y \equiv 0$ **3.** $ye^x = \frac{1}{2}e^{2x} + C$; $y = \frac{1}{2}(e^x + e^{-x})$ **5.** $ye^{(x^2)} = \frac{1}{2}e^{(x^2)} + C$; $y = \frac{1}{2}(1 - 5e^{-x^2})$
7. $(1 + x)y = \sin x + C$; $y = (1 + \sin x)/(1 + x)$ **9.** $y \sin x = \frac{1}{2}\sin^2 x + C$
11. $x^{-3}y = \sin x + C$; $y = x^3 \sin x$ **13.** $x^{-3}e^{2x}y = 2e^{2x} + C$ **15.** $y = e^{-3x^2/2}[3(x^2 + 1)^{3/2} - 2]$
17. $xe^{-y} = \frac{1}{2}y^2 + C$ **19.** $y = e^{(x^2)}[C + (\sqrt{\pi}/2)\,\text{erf}\,(x)]$ **23.** After about 7 min 41 s
25. After about 10.85 years   **27.** 393.75 lb   **29.** (a) $x(t) = 100 \exp(-t/10)$, $y(t) = 10t \exp(-t/10)$; (b) $t = 10$(min)
31. (a) $A'(t) = (0.06)A + (3.6)e^{0.05t}$ (b) $\$1,308,283$

SECTION 1.6

1. $x^2 - 2xy - y^2 = C$ **3.** $y = x(C + \ln|x|)^2$ **5.** $\ln|xy| = C + x/y$ **7.** $y^3 = 3x^3(C + \ln|x|)$
9. $x + y(C + \ln|x|) = 0$ **11.** $y = C(x^2 + y^2)$ **13.** $y + (x^2 + y^2)^{1/2} = Cx^2$
15. $\ln(2x^4 + y^4) + 2\tan^{-1}(y/x\sqrt{2}) = C$ **17.** $\tan^{-1}((2x + y)/\sqrt{2}) = \sqrt{2}(x + C)$ **19.** $x = y^2(2 + Cx^5)$
21. $\ln(1 + y^2) = C - 2x + \ln y^2$ **23.** $y = (x + Cx^2)^{-3}$ **25.** $2x^3y^3 = 3(1 + x^4)^{1/2} + C$ **27.** $y = Ax^2 + B$
29. $y = A\cos 2x + B\sin 2x$ **31.** $y = A - \ln|x + B|$ **33.** $y = (A + Be^x)^{1/2}$ **35.** $y^3 + 3x + Ay + B = 0$
41. $x^2 - 2xy - y^2 - 2x - 6y = C$ **45.** $y = x + (C - x)^{-1}$ **53.** Approximately 8 ft

SECTION 1.7

1. $x^2 + 3xy + y^2 = C$ **3.** $x^3 + 2xy^2 + 2y^3 = C$ **5.** $\frac{1}{4}x^4 + \frac{1}{3}y^3 + y\ln x = C$ **7.** $\sin x + x\ln y + e^y = C$
9. $x^3y^3 + xy^4 + \frac{1}{5}y^5 = C$ **11.** $x^2/y + y^2/x^3 + 2y^{1/2} = C$ **13.** $2x = y^3 + Cy$ **15.** $xy - x - 2y = C$
17. $x^3y + e^y = C$ **19.** $x + \tan^{-1}(y/x) = C$ **21.** $x^4y = C$ **23.** $x^2 + y^2 = Cy$ **25.** $e^x + x\ln y + \sin y = C$
27. $x^2 \sin y = C$ **29.** $x^7y^2 - x^3y^9 = C$ **31.** $\tan^{-1}(x/y) + (x^2 + y^2)^{1/2} = C$

SECTION 1.8

1. 9.13 weeks   **3.** (b) $P(t) \rightarrow +\infty$ as $t \rightarrow 6$ (months)   **7.** (a) 271.15 million (b) 282.22 million
9. (a) 200 g (b) 0.36 s   **11.** (a) $M = 200$ million (b) In 1995
13. (a) $P = 1000$ when $t = 100 \ln \frac{9}{5} \approx 58.78$ (b) $P \rightarrow +\infty$ as $t \rightarrow 100 \ln 2 \approx 69.31$

SECTION 1.9

1. About $31\frac{1}{2}$ s   **3.** About 577 ft   **5.** $400 \ln 7 \approx 778$ ft
7. $v(t) = v_0(1 + rv_0{}^r kt)^{-1/r}$; $x(t) = x_0 + [(1 + rv_0{}^r kt)^{1-1/r} - 1]/[(r - 1)v_0{}^{r-1}k]$ **9.** 50 ft/s (about 34 mi/h)
11. $\rho = \frac{12}{55}$. The time required to fall to the ground is about 12.5 s.
15. $v(10) = 40 \tanh 1 \approx 30.46$ ft/s; the limiting velocity is 40 ft/s.
19. Descent time: 8.361 s; impact speed: 228 ft/s
21. After 30 s of free fall, he has an altitude of 4727 feet and a speed of $206\frac{1}{2}$ ft/s. It takes 228 s more to reach the ground.
23. (b) About 3915 mi/h (c) 5.26R
27. If $c > 800$ (ft/s), then liftoff is immediate. If $640/\ln(5) < c \leq 800$, the rocket will lift off sometime during the first 20 s. If $c \leq 640/\ln(5)$, then the rocket will never get off the ground.

SECTION 2.1

1. $y = \frac{3}{2}e^x - \frac{1}{2}e^{-x}$ **3.** $y = 3\cos 2x + 4\sin 2x$ **5.** $y = 2e^x - e^{2x}$ **7.** $y = 6 - 8e^{-x}$ **9.** $y = 2e^{-x} + xe^{-x}$
11. $y = 5e^x \sin x$ **13.** $y = 5x - 2x^2$ **15.** $y = 7x - 5x\ln x$ **21.** Linearly independent
23. Linearly independent   **25.** Linearly independent
29. There is no contradiction because if the given differential equation is divided by x^2 to get the form in Eq. (8), then the resulting functions $p(x) = -4/x$ and $q(x) = 6/x^2$ are not continuous at $x = 0$.

SECTION 2.2

1. $\frac{5}{2}(2x) - \frac{8}{3}(3x^2) - (5x - 8x^2) = 0$ **3.** $1(0) + 0(\sin x) + 0(e^x) = 0$ **5.** $17 - 34(\cos^2 x) + 17(\cos 2x) = 0$
13. $y = \frac{4}{3}e^x - \frac{1}{3}e^{-2x}$ **15.** $y = (2 - 2x + x^2)e^x$ **17.** $y = \frac{1}{9}(29 - 2\cos 3x - 3\sin 3x)$ **19.** $y = x + 2x^2 + 3x^3$
21. $y = 2\cos x - 5\sin x + 3x$ **23.** $y = e^{-x} + 4e^{3x} - 2$

SECTION 2.3

1. $y = c_1 e^{2x} + c_2 e^{-2x}$ **3.** $y = c_1 e^{2x} + c_2 e^{-5x}$ **5.** $y = c_1 e^{-3x} + c_2 x e^{-3x}$ **7.** $y = c_1 e^{(3x/2)} + c_2 x e^{(3x/2)}$
9. $y = e^{-4x}(c_1 \cos 3x + c_2 \sin 3x)$ **11.** $y = c_1 + c_2 x + c_3 e^{4x} + c_4 x e^{4x}$ **13.** $y = c_1 + c_2 e^{-(2x/3)} + c_3 x e^{-(2x/3)}$
15. $y = c_1 e^{2x} + c_2 x e^{2x} + c_3 e^{-2x} + c_4 x e^{-2x}$
17. $y = c_1 \cos (x/\sqrt{2}) + c_2 \sin (x/\sqrt{2}) + c_3 \cos (2x/\sqrt{3}) + c_4 \sin (2x/\sqrt{3})$
19. $y = c_1 e^x + c_2 e^{-x} + c_3 x e^{-x}$ **21.** $y = 5e^x + 2e^{3x}$ **23.** $y = e^{3x}(3\cos 4x - 2\sin 4x)$
25. $y = \frac{1}{4}(-13 + 6x + 9e^{-2x/3})$ **27.** $y = c_1 e^x + c_2 e^{-2x} + c_3 x e^{-2x}$
29. $y = c_1 e^{-3x} + e^{3x/2}(c_1 \cos \frac{3}{2}\sqrt{3}x + c_2 \sin \frac{3}{2}\sqrt{3}x)$ **31.** $y = c_1 e^{0.6527x} + c_2 e^{2.8974x} + c_3 e^{-0.5321x}$
33. $y = c_1 e^{1.2115x} + c_2 e^{-1.9631x} + e^{0.3758x}(c_1 \cos 1.6739x + c_2 \sin 1.6739x)$
35. $y = c_1 e^{-x/2} + c_2 e^{-x/3} + c_3 \cos 2x + c_4 \sin 2x$ **39.** $y = c_1 e^{-ix} + c_2 e^{3ix}$
41. $y = c_1 \exp ([1 + i\sqrt{3}]x) + c_2 \exp (-[1 + i\sqrt{3}]x)$

SECTION 2.4

1. Frequency: 2 rad/s $= 1/\pi$ Hz; period: π s   **3.** Amplitude: 2 m; frequency: 5 rad/s; period: $2\pi/5$ s
7. About 10,450 ft   **11.** About 3 in.   **13.** $x = 4e^{-2t} - 2e^{-4t}$; overdamped
15. $x = 5e^{-4t} + 10te^{-4t}$; critically damped   **17.** $x = \frac{1}{3}\sqrt{313}e^{-5t/2} \cos (6t - 0.8254)$
19. (a) $k \approx 7018.39$ lb/ft (b) After about 2.49 s

SECTION 2.5

1. $y_p = \frac{1}{25}e^{3x}$ **3.** $y_p = \frac{1}{39}(\cos 3x - 5\sin 3x)$ **5.** $y_p = \frac{1}{26}(13 + 3\cos 2x - 2\sin 2x)$
7. $y_p = -\frac{1}{6}e^x + \frac{1}{6}e^{-x} = -\frac{1}{3}\sinh x$ **9.** $y_p = -\frac{1}{3} + \frac{1}{16}(2x^2 - x)e^x$ **11.** $y_p = \frac{1}{4}(3x^2 - 2x)$
13. $y_p = \frac{1}{65}e^x(7\sin x - 4\cos x)$ **15.** $y_p = -17$ **17.** $y_p = \frac{1}{10}(x^2 \sin x - 4x \cos x)$
19. $y_p = \frac{1}{8}(10x^2 - 4x^3 + x^4)$ **21.** $y_p = xe^x(A \cos x + B \sin x)$
23. $y_p = Ax \cos 2x + Bx \sin 2x + Cx^2 \cos 2x + Dx^2 \sin 2x$ **25.** $y_p = Ae^{-x} + Bx^2 e^{-x} + Cxe^{-2x} + Dx^2 e^{-2x}$
27. $y_p = Ax \cos x + Bx \sin x + Cx \cos 2x + Dx \sin 2x$ **29.** $y_p = Ax^3 e^x + Bx^4 e^x + Cxe^{2x} + Dxe^{-2x}$
31. $y_p = \cos 2x + \frac{3}{4}\sin 2x + \frac{1}{2}x$ **33.** $y = \cos 3x - \frac{2}{15}\sin 3x + \frac{1}{3}\sin 2x$
35. $y = e^x(2\cos x - \frac{5}{2}\sin x) + \frac{1}{2}x + 1$ **37.** $y = 2 - 2e^x + 2xe^x - \frac{1}{2}x^2 e^x + \frac{1}{6}x^3 e^x$
39. $y = -3 + 3x - \frac{1}{2}x^2 + \frac{1}{6}x^3 + 4e^{-x} + xe^{-x}$ **41.** (b) $y = c_1 \cos 2x + c_2 \sin 2x + \frac{1}{4}\cos x - \frac{1}{20}\cos 3x$
43. $y = c_1 \cos 3x + c_2 \sin 3x - \frac{1}{10}\cos 2x - \frac{1}{56}\cos 4x + \frac{1}{24}$ **45.** $y_p = Ax + Bx^2 + Cx^3$
47. $y_p = Axe^{5x} + Bx^2 e^{5x}$ **49.** $y_p = Ax^2 \cos 2x + Bx^2 \sin 2x$

SECTION 2.6

1. $y_2 = xe^{2x}$ **3.** $y_2 = x^{1/2} \ln x$ **5.** $y_2 = 2x^2 + 2x + 1$ **7.** $y_2 = x \sin x$ **9.** $y_2 = 2x + 1$
11. $y_2 = xe^{x/2}$ **13.** $y_2 = 1/x^3$ **15.** $y_2 = xe^x$ **17.** $y_2 = x^2 \cos x$ **21.** $y = c_1 + c_2 \ln x$
23. $y = c_1 x^3 + c_2 x^{-4}$ **25.** $y = c_1 x^{1/2} + c_2 x^{-3/2}$ **27.** $y = x[c_1 \cos (\ln x) + c_2 \sin (\ln x)]$
29. $y = c_1 \cos ([\frac{3}{2}]^{1/2} \ln x) + c_2 \sin ([\frac{3}{2}]^{1/2} \ln x)$ **31.** $y = c_1 + c_2 x^2 + c_3 x^2 \ln x$ **33.** $y = c_1 + c_2 \ln x + c_3 (\ln x)^2$
35. $y_1 = c_1 x^{-1} + x^{-1/2}\{c_1 \cos [(\sqrt{3}/2) \ln x] + c_2 \sin [(\sqrt{3}/2) \ln x]\}$

SECTION 2.7

1. $y_p = \frac{2}{3}e^x$ **3.** $y_p = x^2 e^{2x}$
5. $y_p = -\frac{1}{4}(\cos 2x \cos x - \sin 2x \sin x) + \frac{1}{20}(\cos 5x \cos 2x + \sin 5x \sin 2x) = -\frac{1}{5}\cos 3x$ (!)

7. $y_p = \frac{2}{3}x \sin 3x + \frac{2}{9}(\cos 3x) \ln |\cos 3x|$ **9.** $y_p = \frac{1}{8}(1 - x \sin 2x)$ **11.** $y_p = -e^x(1 + \ln x)$ **13.** $y_p = \frac{1}{4}x^4$

15. $y_p = \ln x$ **17.** $y_p = \frac{1}{6}x^3 e^{-x}$ **19.** $y_p = \frac{1}{2}e^x(1 - x^{-1}) - \frac{1}{2}e^{-x}\int_1^x t^{-2}e^{2t}\,dt$

21. $y_p = \frac{1}{2}\int_0^x \exp(-t^2)\sin 2(x - t)\,dt$ **23.** $y_p = x + \frac{1}{2}x^2 + (\frac{1}{2}x^2 - 2x)\ln x + e^x \int_1^x e^{-t}\ln t\,dt$ **25.** $y_p = \frac{1}{90}x^4$

27. $y_p = -x^2 + x \ln|(x + 1)/(x - 1)| + \frac{1}{2}(1 + x^2)\ln|x^2 - 1|$

29. $y = e^x\left(1 + \frac{1}{2}\int_1^x t^{-1}e^{-t}\,dt + \frac{1}{2}e^{-2x}\int_1^x t^{-1}e^t dt\right)$

SECTION 2.8

1. $x(t) = 2\cos 2t - 2\cos 3t$

3. $x(t) = \frac{1}{15}\sqrt{138{,}388}\cos(10t + \alpha) + \frac{1}{3}\cos(5t - \beta)$ where $\alpha = \tan^{-1}(\frac{1}{186}) \approx 0.0054$ and $\beta = \tan^{-1}(\frac{4}{3}) \approx 0.9273$

5. $x(t) = (x_0 - C)\cos \omega_0 t + C \cos \omega t$ where $C = F_0/(k - m\omega^2)$

7. $x_{sp} = -\frac{10}{13}\cos(3t + \alpha)$ where $\alpha = \tan^{-1}(\frac{12}{5}) \approx 1.1760$

9. $x_{sp} = -(3/\sqrt{40{,}001})\cos(10t - \alpha)$ where $\alpha = \tan^{-1}(199/20) \approx 1.4706$

11. $x_{sp} = \frac{1}{4}\sqrt{10}\cos(3t - \alpha)$ where $\tan \alpha = -3$, $-\pi/2 < \alpha < \pi$ (so $\alpha \approx 1.8925$); $x_{tr} = \frac{5}{4}\sqrt{2}\,e^{-2t}\cos(t - \beta)$ where $\tan \beta = -7$, $-\pi/2 < \beta < \pi$ (so $\beta \approx 1.7127$)

13. $x_{sp} = -(3/\sqrt{9236})\cos(10t + \alpha)$ where $\alpha = \tan^{-1}(\frac{19}{4}) \approx 0.2096$; $x_{tr} \approx (10.9761)e^{-t}\cos(t\sqrt{5} - \beta)$ where $\beta \approx 0.4181$

15. $\omega = \sqrt{384}$ rad/s (approximately 3.12 Hz) **17.** $\omega_0 = \sqrt{g/L + k/m}$

19. (a) Natural frequency: $\sqrt{10}$ rad/s ≈ 0.50 Hz (b) Amplitude: approximately $10\frac{5}{8}$ in.

SECTION 2.9

1. $I(t) = 4e^{-5t}$ **3.** $I(t) = \frac{4}{145}(\cos 60t + 12 \sin 60t - e^{-5t})$ **5.** $I(t) = \frac{5}{3}e^{-10t}\sin 60t$

7. (a) $Q(t) = E_0 C(1 - e^{-t/RC})$; $I(t) = (E_0/R)e^{-t/RC}$

9. (a) $Q(t) = \frac{1}{1480}(\cos 120t + 6 \sin 120t - e^{-20t})$; $I(t) = \frac{1}{74}(36\cos 120t - 6\sin 120t + e^{-20t})$; (b) Steady-state amplitude: $3/\sqrt{37}$

11. $I_{sp} = (1/\sqrt{37})\sin(2t - \delta)$ where $\delta = 2\pi - \tan^{-1}(\frac{1}{6}) \approx 6.1180$

13. $I_{sp} = (20/\sqrt{13})\sin(5t - \delta)$ where $\delta = 2\pi - \tan^{-1}(\frac{2}{3}) \approx 5.6952$

15. $I_0 = 33\pi[(1000 - 36\pi^2)^2 + (30\pi)^2]^{-1/2} \approx 0.1591$; $\delta = 2\pi - \tan^{-1}[(1000 - 36\pi^2)/(30\pi)] \approx 4.8576$

17. $I(t) = -25e^{-4t}\sin 3t$ **19.** $I(t) = 5e^{-20t} - 5e^{-10t} - 50te^{-10t}$

21. $I(t) = \frac{20}{13}(2\cos 5t + 3\sin 5t) - \frac{10}{39}e^{-t}(12\cos 3t + 47 \sin 3t)$ **23.** Critical frequency: $\omega_0 = 1/\sqrt{LC}$

SECTION 2.10

1. Eigenvalues: $\{(2n + 1)\pi^2/4\}$; associated eigenfunctions: $\{\cos(2n + 1)\pi x/2\}$ $(n \geq 0)$

3. Eigenvalues: $\{n^2/4\}$ for $n > 0$; associated eigenfunctions: $y_n(x) = \cos(nx/2)$ for n odd and $y_n(x) = \sin(nx/2)$ for n even

5. Eigenvalues: $\{(2n + 1)^2\pi^2/64\}$ for $n \geq 0$; associated eigenfunctions given by $y_n(x) = \cos([2n + 1]\pi x/8) + (-1)^n \sin([2n + 1]\pi x/8)$

SECTION 3.1

1. $y = c_0(1 + x + x^2/2! + x^3/3! + \cdots) = c_0 e^x$; $\rho = \infty$

3. $y = c_0(1 - 3x/2 + 3^2 x^2/2!2^2 - 3^3 x^3/3!2^3 + 3^4 x^4/4!2^4 - \cdots) = c_0 e^{-3x/2}$; $\rho = \infty$

5. $y = c_0(1 + x^3/3 + x^6/2!3^2 + x^9/3!3^3 + \cdots) = c_0 \exp(\frac{1}{3}x^3)$; $\rho = \infty$

7. $y = c_0(1 + 2x + 4x^2 + 8x^3 + \cdots) = c_0/(1 - 2x)$; $\rho = \frac{1}{2}$

9. $y = c_0(1 + 2x + 3x^2 + 4x^3 + \cdots) = c_0(1 - x)^{-2}$; $\rho = 1$

11. $y = c_0(1 + x^2/2! + x^4/4! + x^6/6! + \cdots) + c_1(x + x^3/3! + x^5/5! + x^7/7! + \cdots) = c_0 \cosh x + c_1 \sinh x$; $\rho = \infty$ for each series

13. $y = c_0(1 - 3^2 x^2/2! + 3^4 x^4/4! - 3^6 x^6/6! + \cdots) + (c_1/3)(3x - 3^3 x^3/3! + 3^5 x^5/5! - 3^7 x^7/7! + \cdots) = c_0 \cos 3x + \frac{1}{3}c_1 \sin 3x$; $\rho = \infty$ for each series

15. We find that $(n + 1)c_n = 0$ for all $n \geq 0$, so $c_n = 0$ for all $n \geq 0$.

17. We find that $c_0 = c_1 = 0$ and that $c_{n+1} = -nc_n$ for $n \geq 1$; it follows that $c_n = 0$ for all $n \geq 0$.

19. $c_n = -2^2 c_{n-2}/[n(n - 1)]$ for $n \geq 2$; $c_0 = 0$ and $c_1 = 3$; $y = \frac{3}{2}\sin 2x$

21. $c_{n+1} = (2nc_n - c_{n-1})/[n(n + 1)]$ for $n \geq 1$; $c_0 = 0$ and $c_1 = 1$; $y = xe^x$

23. $c_0 = c_1 = 0$ and the recursion relation $(n^2 - n + 1)c_n + (n - 1)c_n = 0$ for $n \geq 2$ imply that $c_n = 0$ for $n \geq 0$.

SECTION 3.2

1. $c_{n+2} = c_n$; $y = c_0 \sum\limits_{n=0}^{\infty} x^{2n} + c_1 \sum\limits_{n=0}^{\infty} x^{2n+1} = (c_0 + c_1 x)/(1 - x^2)$

2. $c_{n+2} = -\frac{1}{2}c_n$; $y = c_0 \sum\limits_{n=0}^{\infty} (-1)^n x^{2n}/2^n + c_1 \sum\limits_{n=0}^{\infty} (-1)^n x^{2n+1}/2^n$

3. $c_{n+2} = -c_n/(n + 2)$; $y = c_0 \sum\limits_{n=0}^{\infty} (-1)^n x^{2n}/(n!2^n) + c_1 \sum\limits_{n=0}^{\infty} (-1)^n x^{2n+1}/(2n + 1)!!$

4. $c_{n+2} = -[(n + 4)/(n + 2)]c_n$; $y = c_0 \sum\limits_{n=0}^{\infty} (-1)^n (n + 1)x^{2n} + \frac{1}{3}c_1 \sum\limits_{n=0}^{\infty} (-1)^n (2n + 3)x^{2n+1}$

5. $c_{n+2} = nc_n/[3(n + 2)]$; $y = c_0 + c_1 \sum\limits_{n=0}^{\infty} x^{2n+1}/[(2n + 1)3^n]$

6. $c_{n+2} = [(n - 3)(n - 4)/(n + 1)(n + 2)]c_n$; $y = c_0(1 + 6x^2 + x^4) + c_1(1 + x)$

7. $c_{n+2} = -[(n - 4)^2/3(n + 1)(n + 3)]c_n$; $y = c_0(1 - \frac{8}{3}x^2 + \frac{8}{27}x^4) + c_1(x - \frac{1}{2}x^3 + \frac{1}{120}x^5 + 9\sum\limits_{n=3}^{\infty} \{[(2n - 5)!!]^2(-1)^n/(2n + 1)!3^n\}x^{2n+1})$

8. $c_{n+2} = [(n - 4)(n + 4)/2(n + 1)(n + 2)]c_n$; $y = c_0(1 - 4x^2 + 2x^4) + c_1(x - \frac{3}{4}x^3 + \frac{7}{32}x^5 + \sum\limits_{n=3}^{\infty} [(2n - 5)!!(2n + 3)!!/(2n + 1)!2^n]x^{2n+1})$

9. $c_{n+2} = [(n + 3)(n + 4)/(n + 1)(n + 2)]c_n$; $y = c_0 \sum\limits_{n=0}^{\infty} (n + 1)(2n + 1)x^{2n} + \frac{1}{3}c_1 \sum\limits_{n=0}^{\infty} (n + 1)(2n + 3)x^{2n+1}$

10. $c_{n+2} = -[(n - 4)/3(n + 1)(n + 2)]c_n$; $y = c_0(1 + \frac{2}{3}x^2 + \frac{1}{27}x^4) + c_1(x + \frac{1}{6}x^3 + \frac{1}{360}x^5 + 3\sum\limits_{n=3}^{\infty} [(-1)^n(2n - 1)!!/(2n + 1)!3^n]x^{2n+1})$

11. $c_{n+2} = [2(n - 5)/5(n + 1)(n + 2)]c_n$; $y = c_1(x - \frac{4}{15}x^3 + \frac{4}{375}x^5) + c_0(1 - x^2 + \frac{1}{10}x^4 + \frac{1}{750}x^6 - 15\sum\limits_{n=4}^{\infty} [(2n - 7)!!2^n/(2n)!5^n]x^{2n})$

12. $c_2 = 0$; $c_{n+3} = c_n/(n + 2)$; $y = c_0(1 + \sum\limits_{n=1}^{\infty} x^{3n}/[2 \cdot 5 \cdots (3n - 1)]) + c_1 \sum\limits_{n=0}^{\infty} x^{3n+1}/(n!3^n)$

13. $c_2 = 0$; $c_{n+3} = -c_n/(n + 3)$; $y = c_0 \sum\limits_{n=0}^{\infty} (-1)^n x^{3n}/(n!3^n) + c_1 \sum\limits_{n=0}^{\infty} (-1)^n x^{3n+1}/[1 \cdot 4 \cdots (3n + 1)]$

14. $c_2 = 0$; $c_{n+3} = -c_n/[(n + 2)(n + 3)]$; $y = c_0(1 + \sum\limits_{n=1}^{\infty} (-1)^n x^{3n}/[3^n \cdot n! \cdot 2 \cdot 5 \cdots (3n - 1)]) + c_1 \sum\limits_{n=0}^{\infty} (-1)^n x^{3n+1}/[3^n \cdot n! \cdot 1 \cdot 4 \cdots (3n + 1)]$

15. $c_2 = c_3 = 0$, $c_{n+4} = -c_n/[(n + 3)(n + 4)]$; $y = c_0(1 + \sum\limits_{n=1}^{\infty} (-1)^n x^{4n}/[4^n \cdot n! \cdot 3 \cdot 7 \cdots (4n - 1)]) + c_1 \sum\limits_{n=0}^{\infty} (-1)^n x^{4n+1}/[4^n \cdot n! \cdot 5 \cdot 9 \cdots (4n + 1)]$

17. $y = \frac{1}{3} \sum\limits_{n=0}^{\infty} (2n + 3)(x - 1)^{2n+1}$; converges if $0 < x < 2$ 19. $y = 1 + 4(x + 2)^2$; converges for all x

21. $2c_2 + c_0 = 0$; $(n + 1)(n + 2)c_{n+2} + c_n + c_{n-1} = 0$ for $n \geq 1$; $y_1 = 1 - \frac{1}{2}x^2 - \frac{1}{6}x^3 + \cdots$; $y_2 = x - \frac{1}{6}x^3 - \frac{1}{12}x^4 + \cdots$

23. $c_2 = c_3 = 0$; $(n + 3)(n + 4)c_{n+4} + (n + 1)c_{n+1} + c_n = 0$ for $n \geq 0$; $y_1 = 1 - \frac{1}{12}x^4 + \frac{1}{126}x^7 + \cdots$; $y_2 = x - \frac{1}{12}x^4 - \frac{1}{20}x^5 + \cdots$

25. $y = 1 - x - \frac{1}{2}x^2 + \frac{1}{3}x^3 - \frac{1}{24}x^4 + \frac{1}{30}x^5 + \frac{29}{720}x^6 - \frac{13}{630}x^7 - \frac{143}{40,320}x^8 + \cdots$; $y(0.5) \approx 0.4156$

27. $y_1 = 1 - \frac{1}{2}x^2 + \frac{1}{720}x^6 + \cdots$; $y_2 = x - \frac{1}{6}x^3 - \frac{1}{60}x^5 + \cdots$

SECTION 3.3

1. Ordinary point   3. Irregular singular point   5. Regular singular point: $r = 0, -1$
7. Regular singular point: $r = -2, -3$ 9. Regular singular point $x = 1$
11. Regular singular points $x = +1$ and $x = -1$ 13. Regular singular points $x = +2$ and $x = -2$
15. Regular singular point $x = 2$ 17. $y_1 = \cos\sqrt{x}$, $y_2 = \sin\sqrt{x}$

18. $y_1 = \sum\limits_{n=0}^{\infty} x^n/[n!(2n + 1)!!]$, $y_2 = x^{-1/2} \sum\limits_{n=0}^{\infty} x^n/[n!(2n - 1)!!]$

19. $y_1 = x^{3/2}(1 + 3\sum\limits_{n=1}^{\infty} x^n/[n!(2n + 3)!!])$, $y_2 = 1 - x - \sum\limits_{n=2}^{\infty} x^n/[n!(2n - 3)!!]$

20. $y_1 = x^{1/3} \sum\limits_{n=0}^{\infty} (-1)^n 2^n x^n/[n! \cdot 4 \cdot 7 \cdots (3n + 1)]$, $y_2 = \sum\limits_{n=0}^{\infty} (-1)^n 2^n x^n/[n! \cdot 2 \cdot 5 \cdots (3n - 1)]$

21. $y_1 = x(1 + \sum\limits_{n=1}^{\infty} x^{2n}/[n! \cdot 7 \cdot 11 \cdots (4n+3)])$, $y_2 = x^{-1/2} \sum\limits_{n=0}^{\infty} x^{2n}/[n! \cdot 1 \cdot 5 \cdots (4n+1)]$

22. $y_1 = x^{3/2}(1 + \sum\limits_{n=1}^{\infty} (-1)^n x^{2n}/[n! \cdot 9 \cdot 13 \cdots (4n+5)])$, $y_2 = x^{-1}(1 + \sum\limits_{n=1}^{\infty} (-1)^{n-1} x^{2n}/[n! \cdot 3 \cdot 7 \cdots (4n-1)])$

23. $y_1 = x^{1/2}(1 + \sum\limits_{n=1}^{\infty} x^{2n}/[2^n \cdot n! \cdot 19 \cdot 31 \cdots (12n+7)])$, $y_2 = x^{-2/3}(1 + \sum\limits_{n=1}^{\infty} x^{2n}/[2^n \cdot n! \cdot 5 \cdot 17 \cdots (12n-7)])$

24. $y_1 = x^{1/3}(1 + \sum\limits_{n=1}^{\infty} (-1)^n x^{2n}/[2^n \cdot n! \cdot 7 \cdot 13 \cdots (6n+1)])$, $y_2 = 1 + \sum\limits_{n=1}^{\infty} (-1)^n x^{2n}/[2^n \cdot n! \cdot 5 \cdot 11 \cdots (6n-1)]$

25. $y_1 = x^{1/2} \sum\limits_{n=0}^{\infty} (-1)^n x^n/(n! 2^n) = x^{1/2} e^{-x/2}$, $y_2 = 1 + \sum\limits_{n=1}^{\infty} (-1)^n x^n/(2n-1)!!$

26. $y_1 = x^{1/2} \sum\limits_{n=0}^{\infty} x^{2n}/(n! 2^n) = x^{1/2} \exp(x^2/2)$, $y_2 = 1 + \sum\limits_{n=1}^{\infty} 2^n x^{2n}/[3 \cdot 7 \cdots (4n-1)]$

27. $y_1 = (\cos 3x)/x$, $y_2 = (\sin 3x)/x$ **28.** $y_1 = (\cosh 2x)/x$, $y_2 = (\sinh 2x)/x$

29. $y_1 = (\cos \sqrt{x})/x$, $y_2 = (\sin \sqrt{x})/x$ **30.** $y_1 = \cos x^2$, $y_2 = \sin x^2$ **31.** $y_1 = \sqrt{x} \cosh x$, $y_2 = \sqrt{x} \sinh x$

33. $y_1 = x^{-1}(1 + 10x + 5x^2 + \frac{10}{9}x^3 + \cdots)$, $y_2 = x^{1/2}(1 + \frac{11}{20}x - \frac{11}{224}x^2 + \frac{671}{24,192}x^3 + \cdots)$

SECTION 3.4

1. $y_1 = x^{-2}(1+x)$, $y_2 = 1 + 2 \sum\limits_{n=1}^{\infty} x^n/(n+2)!$

2. $y_1 = x^{-4}(1 + x + \frac{1}{2}x^2 + \frac{1}{6}x^3)$, $y_2 = 1 + 24 \sum\limits_{n=1}^{\infty} x^n/(n+4)!$

3. $y_1 = x^{-4}(1 - 3x + \frac{9}{2}x^2 - \frac{9}{2}x^3)$, $y_2 = 1 + 24 \sum\limits_{n=1}^{\infty} (-1)^n 3^n x^n/(n+4)!$

4. $y_1 = x^{-5}(1 - \frac{3}{5}x + \frac{9}{50}x^2 - \frac{9}{250}x^3 + \frac{27}{1000}x^4)$, $y_2 = 1 + 120 \sum\limits_{n=1}^{\infty} (-1)^n 3^n x^n/[(n+5)! 5^n]$

5. $y_1 = 1 + \frac{3}{4}x + \frac{1}{4}x^2 + \frac{1}{24}x^3$, $y_2 = x^5(1 + 120 \sum\limits_{n=1}^{\infty} (n+1)x^n/(n+5)!)$

6. $y_1 = x^4(1 + \frac{8}{3} \sum\limits_{n=1}^{\infty} (2n+5)!! x^n/[(n+4)! n! 2^n])$. There is no second Frobenius series solution.

7. $y_1 = x^{-2}(2 - 6x + 9x^2)$, $y_2 = \sum (-1)^{n-1} 3^n x^n/(n+2)!$

8. $y_1 = 3 + 2x + x^2$, $y_2 = x^4 \sum\limits_{n=0}^{\infty} (n+1)x^n = \dfrac{x^4}{(1-x)^2}$

9. $y_1 = 1 + x^2/2^2 + x^4/2^2 4^2 + x^6/2^2 4^2 6^2 + \cdots$, $y_2 = y_1(\ln x - x^2/4 + 5x^4/128 - 23x^6/3456 + \cdots)$

11. $y_1 = x^2(1 - 2x + \frac{3}{2}x^2 - \frac{2}{3}x^3 + \cdots)$, $y_2 = y_1(\ln x + 3x + \frac{11}{4}x^2 + \frac{49}{18}x^3 + \cdots)$

13. $y_1 = x^3(1 - 2x + 2x^2 - \frac{4}{3}x^3 + \cdots)$, $y_2 = y_1(2 \ln x - 1/2x^2 - 2/x + \frac{4}{3}x + \cdots)$

16. $y_1 = x^{3/2}(1 + \sum\limits_{n=1}^{\infty} (-1)^n x^{2n}/[2^n \cdot n! \cdot 5 \cdot 7 \cdots (2n+3)])$, $y_2 = x^{-3/2}(1 + \sum\limits_{n=1}^{\infty} (-1)^n x^{2n}/[2^n \cdot n! \cdot (-1) \cdot 1 \cdot 3 \cdots (2n-3)])$

SECTION 3.5

5. $J_4(x) = (1/x^2)(x^2 - 24)J_0(x) + (8/x^3)(6 - x^2)J_1(x)$ **11.** $x^2 J_1(x) + x J_0(x) - \int J_0(x)\, dx + C$

12. $(x^3 - 4x)J_1(x) + 2x^2 J_0(x) + C$ **13.** $(x^4 - 9x^2)J_1(x) + (3x^3 - 9x)J_0(x) + 9 \int J_0(x)\, dx + C$

14. $-x J_0(x) + \int J_0(x)\, dx + C$ **15.** $2x J_1(x) - x^2 J_0(x) + C$ **16.** $3x^2 J_1(x) + (3x - x^3)J_0(x) - 3 \int J_0(x)\, dx + C$

17. $(4x^3 - 16x)J_1(x) + (8x^2 - x^4)J_0(x) + C$ **18.** $-2J_1(x) + \int J_0(x)\, dx + C$ **19.** $J_0(x) - (4/x)J_1(x) + C$

SECTION 3.6

1. $y = x[c_1 J_0(x) + c_2 Y_0(x)]$ **2.** $y = x^{-1}[c_1 J_1(x) + c_2 Y_1(x)]$ **3.** $y = x[c_1 J_{1/2}(3x^2) + c_2 J_{-1/2}(3x^2)]$

4. $y = x^3[c_1 J_2(2x^{1/2}) + c_2 Y_2(2x^{1/2})]$ **5.** $y = x^{-1/3}[c_1 J_{1/3}(\frac{1}{3}x^{3/2}) + c_2 J_{-1/3}(\frac{1}{3}x^{3/2})]$

6. $y = x^{-1/4}[c_1 J_0(2x^{3/2}) + c_2 Y_0(2x^{3/2})]$ **7.** $y = x^{-1}[c_1 J_0(x) + c_2 Y_0(x)]$ **8.** $y = x^2[c_1 J_1(4x^{1/2}) + c_2 Y_1(4x^{1/2})]$

9. $y = x^{1/2}[c_1 J_{1/2}(2x^{3/2}) + c_2 J_{-1/2}(2x^{3/2})]$ **10.** $y = x^{-1/4}[c_1 J_{3/2}(\frac{2}{5}x^{5/2}) + c_2 J_{-3/2}(\frac{2}{5}x^{5/2})]$

11. $y = x^{1/2}[c_1 J_{1/6}(\frac{1}{3}x^3) + c_2 J_{-1/6}(\frac{1}{3}x^3)]$ **12.** $y = x^{1/2}[c_1 J_{1/5}(\frac{4}{5}x^{5/2}) + c_2 J_{-1/5}(\frac{4}{5}x^{5/2})]$

SECTION 4.1

1. $1/s^2, s > 0$ **3.** $e/(s - 3), s > 3$ **5.** $1/(s^2 - 1), s > 1$ **7.** $F(s) = \int_0^1 e^{-st}\,dt = (1 - e^{-s})/s, s > 0$

9. $(1 - e^{-s} - se^{-s})/s^2, s > 0$ **11.** $s^{-3/2}\Gamma(\frac{3}{2}) + 3s^{-2} = \frac{1}{2}\sqrt{\pi}\,s^{-3/2} + 3s^{-2}, s > 0$ **13.** $1/s^2 - 2/(s - 3), s > 3$

15. $1/s + s/(s^2 - 25), s > 5$ **17.** $\cos^2 2t = \frac{1}{2}(1 + \cos 4t)$, so $F(s) = \frac{1}{2}[1/s + s/(s^2 + 16)], s > 0$

19. $e^{-s}(3!/s^4) = 6e^{-s}/s^4, s > 0$ **21.** $-d/ds[s/(s^2 + 4)] = (s^2 - 4)/(s^2 + 4)^2, s > 0$ **23.** $f(t) = \frac{1}{2}t^3$

25. $F(s) = \Gamma(1)/s - 2\Gamma(\frac{5}{2})/\Gamma(\frac{1}{2})s^{5/2}$, so $f(t) = 1 - [2/\Gamma(\frac{1}{2})]t^{3/2} = 1 - 8t^{3/2}/3\pi^{1/2}$ **27.** $3e^{4t}$ **29.** $\frac{5}{3}\sin 3t - 3\cos 3t$

31. $\frac{2}{3}\sinh 5t - 10\cosh 5t$

SECTION 4.2

1. $x = 5\cos 2t$ **3.** $x = \frac{2}{3}e^{3t} - \frac{2}{3}e^{-t}$ **5.** $x(t) = \frac{2}{3}\sin t - \frac{1}{3}\sin 2t$ **7.** $x(t) = \frac{9}{8}\cos t - \frac{1}{8}\cos 3t$

9. $x = \frac{1}{6}e^{-3t} - \frac{1}{2}e^{-t} + \frac{1}{3}$ **11.** $f(t) = \frac{1}{3}(e^{3t} - 1)$ **13.** $f(t) = \frac{1}{4}(1 - \cos 2t)$ **15.** $f(t) = t - \sin t$

17. $f(t) = -t - \frac{1}{2}e^{-t} + \frac{1}{2}e^t = -t + \sinh t$

SECTION 4.3

1. $24/(s - \pi)^5$ **3.** $3\pi/[(s + 2)^2 + 9\pi^2]$ **5.** $\frac{3}{2}e^{2t}$ **7.** te^{-2t} **9.** $e^{3t}(3\cos 4t + \frac{7}{2}\sin 4t)$

11. $f(t) = \frac{1}{2}\sinh 2t$ **13.** $f(t) = 3e^{-2t} - 5e^{-5t}$ **15.** $f(t) = \frac{1}{25}(e^{5t} - 5t - 1)$ **17.** $f(t) = \frac{1}{16}(\sinh 2t - \sin 2t)$

19. $f(t) = \frac{1}{3}(2\cos 2t + 2\sin 2t - 2\cos t - \sin t)$ **21.** $f(t) = e^{-t}(\frac{3}{2}\sin t - t\sin t - \frac{3}{2}t\cos t)$

27. $x(t) = e^{-3t}(2\cos 4t + \frac{9}{4}\sin 4t)$ **29.** $x(t) = \frac{3}{8}\sinh 2t - \frac{3}{4}t$ **31.** $x(t) = \frac{1}{15}(6e^{2t} - e^{-3t} - 5)$

33. $x(t) = (r/2)[e^{-rt}(\cos rt + \sin rt) + e^{rt}(\cos rt - \sin rt)] = r(\cos rt \cosh rt - \sin rt \sinh rt)$ where $r = \sqrt{2}/2$

35. $x(t) = \frac{1}{16}(\sin 2t - 2t\cos 2t)$ **37.** $x(t) = \frac{1}{50}e^{-2t}\cos 3t + \frac{16}{25}e^{-2t}\sin 3t + \frac{1}{10}te^{-t} - \frac{1}{50}e^{-t}$

SECTION 4.4

1. $\frac{1}{2}t^2$ **3.** $\frac{1}{2}(\sin t - t\cos t)$ **5.** te^{at} **7.** $(e^{3t} - 1)/3$ **9.** $\frac{1}{24}(\sin 3t - 3t\cos 3t)$ **11.** $\frac{1}{4}(\sin 2t + 2t\cos 2t)$

13. $\frac{1}{10}(-3\cos t + \sin t + 3e^{3t})$ **15.** $6s/(s^2 + 9)^2, s > 0$ **17.** $(s^2 - 4s - 5)/(s^2 - 4s + 13)^2, s > 0$

19. $\pi/2 - \arctan s = \arctan(1/s), s > 0$ **21.** $\ln[s/(s - 3)], s > 3$ **23.** $f(t) = -(2/t)\sinh 2t$

25. $f(t) = (e^{-2t} - e^{3t} - 2\cos t)/t$ **27.** $f(t) = (2/t)(1 - \sin t)$

29. $(s + 1)X'(s) + 4X(s) = 0; x(t) = Ct^3e^{-t}, C \neq 0$ **31.** $(s - 2)X'(s) + 3X(s) = 0; x(t) = Ct^2e^{2t}, C \neq 0$

33. $(s^2 + 1)X'(s) + 4sX(s) = 0; x(t) = C(\sin t - t\cos t), C \neq 0$

SECTION 4.5

1. $f(t) = (t - 3)u_3(t)$; see Fig. 4.5.1.   **3.** $f(t) = e^{-2(t-1)}u_1(t)$; see Fig. 4.5.3.

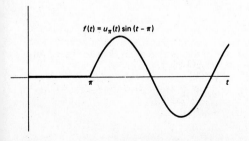

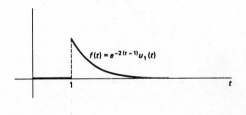

5. $f(t) = u_\pi(t)\sin(t - \pi)$; see Fig. 4.5.5.   **7.** $f(t) = [1 - u_{2\pi}(t)]\sin t$; see Fig. 4.5.7.

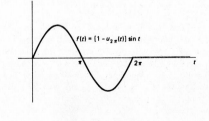

9. $f(t) = [1 + u_3(t)] \cos \pi t$ **11.** $F(s) = (2/s)(1 - e^{-3s}), s > 0$ **13.** $F(s) = (1 - e^{-2\pi s})/(s^2 + 1), s > 0$
15. $F(s) = (1 + e^{-3\pi s})/(s^2 + 1), s > 0$ **17.** $F(s) = \pi[(e^{-2s} + e^{-3s})/(s^2 + \pi^2)], s > 0$ **19.** $F(s) = e^{-s}\left(\frac{1}{s} + \frac{1}{s^2}\right), s > 0$
21. $F(s) = (1 - 2e^{-s} + e^{-2s})/s^2, s > 0$ **31.** $x(t) = \frac{1}{2}[1 - u_\pi(t)] \sin^2 t$
33. $x(t) = \frac{1}{3}[\sin t - \frac{1}{3} \sin 3t - u_{2\pi}(t) \sin t + \frac{1}{3}u_{2\pi}(t) \sin 3t]$
35. $x(t) = \frac{1}{4}\{t - 1 + (t + 1)e^{-2t} + u_2(t)[1 - t + (3t - 5)e^{-2(t-2)}]\}$ **37.** $i(t) = [1 - u_{2\pi}(t)] \sin 100t$
39. $i(t) = \frac{1}{50}[(1 - e^{-50t})^2 - u_1(t)(1 - e^{-50(t-1)})^2]$

41. $x(t) = 2 \sin^2 t + 4 \sum_{n=1}^{\infty} (-1)^n u(t - n\pi) \sin^2 (t - n\pi) = 2 |\sin t| \sin t$ for $t \geq 0$. Thus the complete solution is periodic, and so the transient solution is trivial.

SECTION 4.6

1. $x(t) = \frac{1}{2} \sin 2t$ **3.** $x(t) = \frac{1}{4}(1 - e^{-2t}) - \frac{1}{2}te^{-2t} + u_2(t)(t - 2)e^{-2(t-2)}$
5. $x(t) = 0$ if $0 \leq t < \pi$; $x(t) = -2e^{-(t-\pi)} \sin t$ if $t \geq \pi$
7. $x(t) = e^{-2t}[2 \sin t - e^{2\pi}u_\pi(t) \sin t + e^{4\pi}u_{2\pi}(t) \sin t]$ **9.** $x(t) = \int_0^t \frac{1}{2}(\sin 2\tau)f(t - \tau) \, d\tau$

11. $x(t) = \int_0^t (e^{-3\tau} \sinh \tau)f(t - \tau) \, d\tau$
17. (b) $i(t) = e^{-100(t-1)}u_1(t) - e^{-100(t-2)}u_2(t)$. If $t > 2$, then $i(t) = -(e^{100} - 1)e^{100(1-t)} < 0$.

SECTION 5.1

1. $x_1' = x_2, x_2' + 7x_1 + 3x_2 = t^2$ **3.** $x_1' = x_2, t^2x_2' + tx_1' + (t^2 - 1)x_1 = 0$
5. $x_1' = x_2, x_2' = x_3, x_3' = x_2^2 + \cos x_1$ **7.** $x_1' = x_2, x_2' = -kx_1(x_1^2 + y_1^2)^{-3/2}, y_1' = y_2, y_2' = -ky_1(x_1^2 + y_1^2)^{-3/2}$
9. $x_1' = x_2, y_1' = y_2, z_1' = z_2, x_2' = 3x_1 - y_1 + 2z_1, y_2' = x_1 + y_1 - 4z_1, z_2' = 5x_1 - y_1 - z_1$
11. $x = -\frac{1}{3} \sin 2t, y = -\frac{2}{3} \sin 2t$ **13.** $x = -\frac{2}{3} \cos 3t, y = \frac{1}{4} \cos 3t$ **15.** $x(t) = 1, y(t) = -2$
17. $x(t) = -(2/\sqrt{3}) \sinh (t/\sqrt{3}), y(t) = (1/\sqrt{3}) \sinh (t/\sqrt{3}) + \cosh (t/\sqrt{3})$
19. $x(t) = \frac{2}{3} + \frac{1}{3}[\exp(-3t/2)][\cos \frac{1}{2}t\sqrt{3} + \sqrt{3} \sin \frac{1}{2}t\sqrt{3}], y(t) = \frac{4}{3} - \frac{2}{3}e^t + \frac{2}{21}[\exp(-3t/2)][\cos \frac{1}{2}t\sqrt{3} + 4\sqrt{3} \sin \frac{1}{2}t\sqrt{3}]$
21. $x(t) = 25(1 + 1/\sqrt{33})e^{at} + 25(1 - 1/\sqrt{33})e^{bt}, y(t) = 25(1 - 15/\sqrt{33})e^{at} + 25(1 + 15/\sqrt{33})e^{bt}$ where $a = \frac{1}{40}(-9 - \sqrt{33})$, $b = \frac{1}{40}(-9 + \sqrt{33})$
22. $I_1(t) = 2 + e^{-5t}(8\sqrt{\frac{2}{3}} \sin 5t\sqrt{\frac{7}{3}} - 2 \cos 5t\sqrt{\frac{7}{3}}), I_2(t) = (20/\sqrt{6})e^{-5t} \sin 5t\sqrt{\frac{7}{3}}$
29. $2(I_1' - I_2') + 50I_1 = 100 \sin 60t, 2(I_2' - I_1') + 25I_2 = 0$

SECTION 5.2

1. $x = a_1e^{-t} + a_2e^{2t}, y = \frac{3}{2}a_2e^{2t}$ **3.** $x = \frac{4}{5}(e^{3t} - e^{-2t}), y = \frac{2}{5}(6e^{3t} - e^{-2t})$
5. $x = e^{-t}(a_1 \cos 2t + a_2 \sin 2t), y = -\frac{1}{2}e^{-t}[(a_1 + a_2) \cos 2t + (a_2 - a_1) \sin 2t]$
7. $x = a_1e^{2t} + a_2e^{3t} + \frac{1}{18} - \frac{1}{3}t, y = -2a_1e^{2t} - a_2e^{3t} - \frac{5}{3} - \frac{2}{3}t$
9. $x = a_1 \cos t + a_2 \sin t - \frac{2}{3} \cos 2t - \frac{4}{3} \sin 2t, y = \frac{1}{3}(2a_1 - a_2) \cos t + \frac{1}{3}(a_1 + 2a_2) \sin t - \frac{2}{3} \cos 2t - \frac{4}{3} \sin 2t$
11. $x = a_1 \cos 3t + a_2 \sin 3t - \frac{11}{20}e^t - \frac{1}{4}e^{-t}, y = +\frac{1}{3}(a_1 - a_2) \cos 3t + \frac{1}{3}(a_1 + a_2) \sin 3t + \frac{1}{10}e^t$
13. $x = a_1 \cos 2t + a_2 \sin 2t + b_1 \cos 3t + b_2 \sin 3t, y = \frac{1}{2}a_1 \cos 2t + \frac{1}{2}a_2 \sin 2t - 2b_1 \cos 3t - 2b_2 \sin 3t$
15. $x = a_1 \cos t + a_2 \sin t + b_1 \cos 2t + b_2 \sin 2t, y = a_2 \cos t - a_1 \sin t + 3b_2 \cos 2t - 3b_1 \sin 2t$
17. $x = a_1 \cos t + a_2 \sin t + b_1e^{2t} + b_2e^{-2t}, y = -3a_2 \cos t + 3a_1 \sin t - b_1e^{2t} + b_2e^{-2t}$
19. $x = e^{4t}(a + b \cos 4t + c \sin 4t), y = -2e^{4t}(c \cos 4t - b \sin 4t), z = -2e^{4t}(a + 3b \cos 4t + 3c \sin 4t)$
23. Infinitely many solutions   **24.** No solutions   **25.** Infinitely many solutions
27. $I_1(t) = 2 + e^{-5t}(-2 \cos \frac{5}{3}t\sqrt{6} + 4\sqrt{6} \sin \frac{5}{3}t\sqrt{6}), I_2(t) = \frac{10}{3}\sqrt{6}e^{-5t} \sin \frac{5}{3}t\sqrt{6}$
29. $I_1(t) = \frac{2}{3}(2 + e^{-60t}), I_2(t) = \frac{4}{3}(1 - e^{-60t})$
33. (a) $x = a_1 \cos 5t + a_2 \sin 5t + b_1 \cos 5t\sqrt{3} + b_2 \sin 5t\sqrt{3}, y = 2a_1 \cos 5t + 2a_2 \sin 5t - 2b_1 \cos 5t\sqrt{3} - 2b_2 \sin 5t\sqrt{3}$
(b) In the natural mode with frequency $\omega_1 = 5$ the masses move in the same direction, while in the natural mode with frequency $\omega_2 = 5\sqrt{3}$ they move in opposite directions. In each case the amplitude of the motion of m_2 is twice that of m_1.
35. $x = a_1 \cos t + a_2 \sin t + b_1 \cos 3t + b_2 \sin 3t, y = a_1 \cos t + a_2 \sin t - b_1 \cos 3t - b_2 \sin 3t$
37. $x = a_1 \cos t + a_2 \sin t + b_1 \cos t\sqrt{5} + b_2 \sin t\sqrt{5}, y = a_1 \cos t + a_2 \sin t - b_1 \cos t\sqrt{5} - b_2 \sin t\sqrt{5}$
39. $x = a_1 \cos t\sqrt{2} + a_2 \sin t\sqrt{2} + b_1 \cos 2t\sqrt{2} + b_2 \sin 2t\sqrt{2}, y = a_1 \cos t\sqrt{2} + a_2 \sin t\sqrt{2} - \frac{1}{2}b_1 \cos 2t\sqrt{2} - \frac{1}{2}b_2 \sin 2t\sqrt{2}$

SECTION 5.3

1. (a) $\begin{pmatrix} 13 & -18 \\ 23 & 17 \end{pmatrix}$ (b) $\begin{pmatrix} 0 & -1 \\ 2 & 19 \end{pmatrix}$ (c) $\begin{pmatrix} -9 & -11 \\ 47 & -9 \end{pmatrix}$ (d) $\begin{pmatrix} -10 & -37 \\ 14 & -8 \end{pmatrix}$

3. $\mathbf{AB} = \begin{pmatrix} -1 & 8 \\ 46 & -1 \end{pmatrix}$; $\mathbf{BA} = \begin{pmatrix} 11 & -12 & 14 \\ -14 & 0 & 7 \\ 0 & 8 & -13 \end{pmatrix}$

5. (a) $\begin{pmatrix} 21 & 2 & 1 \\ 4 & 44 & 9 \\ -27 & 34 & 45 \end{pmatrix}$ (b) $\begin{pmatrix} 9 & 21 & -13 \\ -5 & -8 & 24 \\ -25 & -19 & 26 \end{pmatrix}$ (c) $\begin{pmatrix} 0 & -6 & 1 \\ 10 & 31 & -15 \\ 16 & 58 & -23 \end{pmatrix}$ (d) $\begin{pmatrix} -10 & -8 & 5 \\ 18 & 12 & -10 \\ 11 & 22 & 6 \end{pmatrix}$

(e) $\begin{pmatrix} 3-t & 2 & -1 \\ 0 & 4-t & 3 \\ -5 & 2 & 7-t \end{pmatrix}$

7. $|\mathbf{A}| = |\mathbf{B}| = 0$

11. $\mathbf{x} = \begin{pmatrix} x \\ y \end{pmatrix}$, $\mathbf{P}(t) = \begin{pmatrix} 3 & -2 \\ 2 & 1 \end{pmatrix}$, $\mathbf{f}(t) = \begin{pmatrix} 0 \\ 0 \end{pmatrix}$ 13. $\mathbf{x} = \begin{pmatrix} x \\ y \end{pmatrix}$, $\mathbf{P}(t) = \begin{pmatrix} t & -e^t \\ e^{-t} & t^2 \end{pmatrix}$, $\mathbf{f}(t) = \begin{pmatrix} \cos t \\ -\sin t \end{pmatrix}$

15. $\mathbf{x} = \begin{pmatrix} x \\ y \\ z \end{pmatrix}$, $\mathbf{P}(t) = \begin{pmatrix} t & -1 & e^t \\ 2 & t^2 & -1 \\ e^{-t} & 3t & t^3 \end{pmatrix}$, $\mathbf{f}(t) = \begin{pmatrix} 0 \\ 0 \\ 0 \end{pmatrix}$ 17. $W(t) = 4$; $\mathbf{x}(t) = \begin{pmatrix} c_1 e^{2t} + c_2 e^{-2t} \\ c_1 e^{2t} + 5c_2 e^{-2t} \end{pmatrix}$

19. $W(t) = 7e^{-3t}$; $\mathbf{x}(t) = \begin{pmatrix} 3c_1 e^{2t} + c_2 e^{-5t} \\ 2c_1 e^{2t} + 3c_2 e^{-5t} \end{pmatrix}$ 21. $W(t) = 3$; $\mathbf{x}(t) = \begin{pmatrix} c_1 e^{2t} & + & c_2 e^{-t} \\ c_1 e^{2t} & + & c_3 e^{-t} \\ c_1 e^{2t} & - & (c_2 + c_3)e^{-t} \end{pmatrix}$

23. (a) $\mathbf{x}_2 = t\mathbf{x}_1$, so neither is a constant multiple of the other. (b) $W(\mathbf{x}_1, \mathbf{x}_2) \equiv 0$, whereas Theorem 2 would imply that $W \neq 0$ if $\mathbf{x}_1$ and $\mathbf{x}_2$ were independent solutions of a system of the indicated form.

29. $x_1 = a_1 \cos t + a_2 \sin t + b_1 \cos 2t + b_2 \sin 2t - \frac{7}{40} \cos 3t$, $x_2 = a_1 \cos t + a_2 \sin t - \frac{1}{2}b_1 \cos 2t - \frac{1}{2}b_2 \sin 2t + \frac{1}{40} \cos 3t$

31. $x = a_1 \cos t\sqrt{2} + a_2 \sin t\sqrt{2} + b_1 \cos 2t + b_2 \sin 2t + \frac{4}{3} \cos t$, $y = a_1 \cos t\sqrt{2} + a_2 \sin t\sqrt{2} - b_1 \cos 2t - b_2 \sin 2t + \frac{5}{3} \cos t$

SECTION 5.4

1. $x_1 = \mathcal{R}e[\frac{2}{3}e^{10it}] = \frac{2}{3} \cos 10t$, $x_2 = \mathcal{R}e[-\frac{14}{3}e^{10it}] = -\frac{14}{3} \cos 10t$

3. $x_1 = \mathcal{R}e[-\frac{1}{15}(6 - 8i)e^{10it}] = -\frac{1}{15}(6 \cos 10t + 8 \sin 10t)$, $x_2 = -x_1$

5. $x_1 = \mathcal{R}e[(0.2657 + 0.8581i)e^{10it}] = (0.2657) \cos 10t - (0.8581) \sin 10t$, $x_2 = \mathcal{R}e[(-0.7105 + 6.2980i)e^{10it}]$
$= (-0.7105) \cos 10t - (6.2980) \sin 10t$

7. Natural frequencies: $\omega_1 = 1$, $\omega_2 = 2$ (rad/s). In the natural mode with frequency $\omega_1 = 1$ the two masses move in the same direction with equal amplitudes of motion. In the natural mode with frequency $\omega_2 = 2$ the two masses move in opposite directions, with the amplitude of motion of m_1 twice that of m_2.

9. Natural frequencies: $\omega_1 = \sqrt{2}$, $\omega_2 = 2$ (rad/s). In each natural mode the two masses have equal amplitudes of oscillation. They move in the same direction at frequency $\omega_1 = \sqrt{2}$, in opposite directions at frequency $\omega_2 = 2$.

11. $x_1(t) = x_2(t) = e^{-t}(\cos 2t + \frac{1}{2} \sin 2t)$ 13. $m_2 = 0.1$ (slug)

15. Natural frequencies: $\omega_1 = \sqrt{20}$, $\omega_2 = \sqrt{30 - 10\sqrt{7}}$, and $\omega_3 = \sqrt{30 + 10\sqrt{7}}$. The ratios of amplitudes $A_i : B_i : C_i$ for the natural mode with frequency ω_i are given by $A_1 : B_1 : C_1 = 1 : 0 : -1$, $A_2 : B_2 : C_2 = 1 : (-1 + \sqrt{7}) : 2$, and $A_3 : B_3 : C_3 = 1 : (-1 - \sqrt{7}) : 2$.

19. (a) $\omega_1 \approx 1.6998$ Hz, $\omega_2 \approx 2.0990$ Hz (b) $v_1 \approx 46.36$ mi/h, $v_2 \approx 57.25$ mi/h

SECTION 5.5

1. $x_1(t) = c_1 e^{-t} + c_2 e^{3t}$, $x_2(t) = -c_1 e^{-t} + c_2 e^{3t}$ 3. $x_1(t) = -\frac{1}{2}e^{-t} + \frac{8}{3}e^{6t}$, $x_2(t) = \frac{1}{2}e^{-t} + \frac{8}{3}e^{6t}$

5. $x_1(t) = c_1 e^{-t} + 7c_2 e^{5t}$, $x_2(t) = c_1 e^{-t} + c_2 e^{5t}$ 7. $x_1(t) = c_1 e^t + 2c_2 e^{-9t}$, $x_2(t) = c_1 e^t - 3c_2 e^{-9t}$

9. $x_1(t) = 2 \cos 4t - \frac{11}{4} \sin 4t$, $x_2(t) = 3 \cos 4t + \frac{1}{2} \sin 4t$ 11. $x_1(t) = -4e^t \sin 2t$, $x_2(t) = 4e^t \cos 2t$

13. $x_1(t) = 3e^{2t}(c_1 \cos 3t - c_2 \sin 3t)$, $x_2(t) = e^{2t}[(c_1 + c_2) \cos 3t + (c_1 - c_2) \sin 3t]$

15. $x_1(t) = 5e^{5t}(c_1 \cos 4t - c_2 \sin 4t)$, $x_2(t) = e^{5t}[(2c_1 + 4c_2) \cos 4t + (4c_1 - 2c_2) \sin 4t]$

17. $x_1(t) = e^{3t}(c_1 + c_2 t)$, $x_2(t) = e^{3t}(-c_1 - \frac{1}{2}c_2 - c_2 t)$

19. $x_1(t) = e^{5t}(c_1 + c_2 t)$, $x_2(t) = e^{5t}(-2c_1 + c_2 - 2c_2 t)$

21. $x_1(t) = 6c_1 + 3c_2 e^t + 2c_3 e^{-t}$, $x_2 = 2c_1 + c_2 e^t + c_3 e^{-t}$, $x_3 = 5c_1 + 2c_2 e^t + 2c_3 e^{-t}$

23. $x_1(t) = c_1 e^{2t} + c_3 e^{3t}$, $x_2(t) = -c_1 e^{2t} + c_2 e^{-2t} - c_3 e^{3t}$, $x_3(t) = -c_2 e^{-2t} + c_3 e^{3t}$

25. $x_1(t) = c_1 + e^{2t}[(-c_2 + c_3) \cos 3t + (c_2 + c_3) \sin 3t]$, $x_2(t) = -c_1 + 2e^{2t}(c_2 \cos 3t - c_3 \sin 3t)$,
$x_3(t) = 2e^{2t}(-c_2 \cos 3t + c_3 \sin 3t)$

27. $x_1(t) = c_1 e^t + e^{2t}(c_2 + c_3 t)$, $x_2(t) = -c_1 e^t + e^{2t}(-c_2 + c_3 - c_3 t)$, $x_3(t) = e^{2t}(c_2 - c_3 + c_3 t)$

SECTION 5.6

1. $x = \frac{7}{3}, y = -\frac{8}{3}$

3. $x = \frac{216}{189}e^{-t} + \frac{1}{189}e^{6t} - \frac{2}{3}t^2 + \frac{10}{9}t - \frac{31}{27}, y = -\frac{216}{189}e^{-t} + \frac{1}{252}e^{6t} + \frac{1}{2}t^2 - \frac{7}{6}t + \frac{41}{36}$

5. $x = -4 - \frac{1}{3}(1 + 7t)e^{-t}, y = -2 - \frac{7}{3}te^{-t}$

7. $x = \frac{1}{410}(369e^t + 166e^{-9t} - 125 \cos t - 105 \sin t), y = \frac{1}{410}(369e^t - 249e^{-9t} - 120 \cos t - 150 \sin t)$

9. $x = \frac{1}{4} \sin 2t + \frac{1}{4}t \cos 2t + \frac{1}{4}t \sin 2t, y = \frac{1}{4}t \sin 2t$ **11.** $x = \frac{3}{2} + \frac{1}{4}e^{4t} - 2t - 2, y = -\frac{3}{4} + \frac{1}{4}e^{4t} + t$

13. $x = \frac{1}{2}(1 + 5t)e^t, y = -\frac{5}{2}te^t$ **15.** $\Phi(t) = \begin{pmatrix} e^t & e^{3t} \\ -e^t & e^{3t} \end{pmatrix}, \mathbf{x}(t) = \frac{1}{2}\begin{pmatrix} 5e^t + e^{3t} \\ -5e^t + e^{3t} \end{pmatrix}$

17. $\Phi(t) = \begin{pmatrix} 5 \cos 4t & -5 \sin 4t \\ 2 \cos 4t + 4 \sin 4t & 4 \cos 4t - 2 \sin 4t \end{pmatrix}, \mathbf{x}(t) = \frac{1}{4}\begin{pmatrix} -5 \sin 4t \\ 4 \cos 4t - 2 \sin 4t \end{pmatrix}$

19. $\Phi(t) = \begin{pmatrix} 2 \cos 3t & -2 \sin 3t \\ -3 \cos 3t + 3 \sin 3t & 3 \cos 3t + 3 \sin 3t \end{pmatrix}, \mathbf{x}(t) = \frac{1}{3}\begin{pmatrix} 3 \cos 3t - \sin 3t \\ -3 \cos 3t + 6 \sin 3t \end{pmatrix}$

21. $\Phi(t) = \begin{pmatrix} 6 & 3e^t & 2e^{-t} \\ 2 & e^t & e^{-t} \\ 5 & 2e^t & 2e^{-t} \end{pmatrix}, \mathbf{x}(t) = \begin{pmatrix} -12 + 12e^t + 2e^{-t} \\ -4 + 4e^t + e^{-t} \\ -10 + 8e^t + 2e^{-t} \end{pmatrix}$

23. $\mathbf{x}(t) = \frac{1}{150}\begin{pmatrix} 666 - 120t - 575e^{-t} - 91e^{5t} \\ 588 - 60t - 575e^{-t} - 13e^{5t} \end{pmatrix}$ **25.** $\mathbf{x}(t) = \frac{1}{16}\begin{pmatrix} 3e^{3t} - 3e^{-t} + 4te^{-t} \\ 3e^{3t} - 3e^{-t} + 20te^{-t} \end{pmatrix}$

27. $\mathbf{x}(t) = \frac{1}{28}\begin{pmatrix} -6 \cos 4t + 28 \sin 4t + 8 \cos 3t - 28 \sin 3t \\ -31 \cos 4t + 8 \sin 4t + 32 \cos 3t - 8 \sin 3t \end{pmatrix}$

29. $\mathbf{x}(t) = \Phi(t)\begin{pmatrix} t - \sin t + \ln|\cos t + \sin t \cos t| \\ 1 - 2t - \cos t + \ln|\cos t| \end{pmatrix}$ where $\Phi(t) = \begin{pmatrix} \cos t & -\sin t \\ 2 \cos t + \sin t & \cos t - 2 \sin t \end{pmatrix}$

31. $\mathbf{x}(t) = \frac{1}{4}e^{3t}\begin{pmatrix} -2t \sin 2t \\ \sin 2t + 2t \cos 2t \end{pmatrix}$

Note: Most of the problems in Chapter 6 call for a table of values. Space limitations prohibit reproduction of the complete tables in this book; we have chosen in most cases to give only data from the last line of the table. Your answers may differ slightly because of variations in hardware and software.

SECTION 6.1

1. $y = e^x - x - 1$; at $x = 1$, the Euler method yields 0.593742; the true value is about 0.718282.

3. $y = \exp(x^2/2)$; at $x = 2$, the Euler method yields 4.989521; the true value is about 7.389056.

5. $y = e^{-2x}$; at $x = 1$, the Euler method yields 0.107374; the true value is about 0.135335.

7. $y = -2e^x + x + 2$; at $x = 1$, the Euler method yields -2.433848; the true value is about -2.436564; the error is about -0.002716 (about 0.111%).

9. $y = \exp(x^3)$; 2.711777; 2.718282; 0.006505; 0.24%

11. $y = -\ln(2 - x)$; at $x = 2$, the Euler method yields 5.851887; the other values are not defined. At $x = 1.9$, the entries in the table should be 2.291209 (Euler), 2.302585 (true), 0.011376 (error), and 0.494068 (% error).

13. The complete table appears in Fig. 6.1.13.

X	Y (H = 0.1)	Y (H = 0.02)	Y (H = 0.004)	Y (H = 0.0008)
0.000	0.000000	0.000000	0.000000	0.000000
0.100	0.000000	0.000240	0.000314	0.000329
0.200	0.001000	0.002280	0.002587	0.002651
0.300	0.005000	0.008122	0.008824	0.008967
0.400	0.014003	0.019778	0.021039	0.021295
0.500	0.030022	0.039293	0.041286	0.041690
0.600	0.055112	0.068792	0.071709	0.072300
0.700	0.091416	0.110557	0.114629	0.115453
0.800	0.141252	0.167161	0.172681	0.173800
0.900	0.207247	0.241661	0.249035	0.250531
1.000	0.292542	0.337913	0.347733	0.349731

15. At $x = 2$, we have $y = 6.183058$ ($h = 0.1$), 6.365283 ($h = 0.02$), 6.402216 ($h = 0.004$), and 6.409622 ($h = 0.0008$).
17. As in 15, we have 2.850782, 2.868054, 2.871582, and 2.872291.
19. As in 15, we have 1.226246, 1.230001, 1.230591, and 1.230704.

SECTION 6.2

1. The solution of the initial value problem is $y = e^x - x - 1$. The complete table appears in Fig. 6.2.1.

X	Y (Euler)	Y (Imp. Euler)	Y (True)	Errors (Euler)	Errors (Imp. Euler)
0.00	0.00000	0.00000	0.00000	0.00000	0.00000
0.10	0.00000	0.00500	0.00517	0.00517	0.00017
0.20	0.01000	0.02103	0.02140	0.01140	0.00038
0.30	0.03100	0.04923	0.04986	0.01886	0.00063
0.40	0.06410	0.09090	0.09182	0.02772	0.00092
0.50	0.11051	0.14745	0.14872	0.03821	0.00127
0.60	0.17156	0.22043	0.22212	0.05056	0.00169
0.70	0.24872	0.31157	0.31375	0.06504	0.00218
0.80	0.34359	0.42279	0.42554	0.08195	0.00275
0.90	0.45795	0.55618	0.55960	0.10166	0.00342
1.00	0.59374	0.71408	0.71828	0.12454	0.00420

3. $y = 1/x$; at $x = 3$, we have $y = 0.30741$ (Euler) and $y = 0.33501$ (improved Euler).
5. $y = e^{-2x}$; at $x = 1$, we have $y = 0.10737$ (Euler) and $y = 0.13745$ (improved Euler).
7. $y = -2e^x + x + 2$; at $x = 1$, we have $y = -2.43385$ (Euler) and $y = -2.43647$.
9. $y = \exp(x^3)$; at $x = 1$, we have $y = 2.71178$ (Euler) and $y = 2.71824$ (improved Euler).
11. $y = -\ln(2 - x)$; at $x = 1.5$, we have $y = 0.69246$ (Euler) and $y = 0.69313$ (improved Euler).
13. $y = -2e^x + x + 2$; algorithm: $r = y - x - 1, s = 1 - r, y = y + h[r + (0.5)(hs)], x = x + h$. At $x = 1$, we have $y = -2.436474$ (Taylor) and $y = -2.436564$ (true value).
15. $y = \exp(x^3)$; at $x = 1$, the three-term Taylor series method gives $y = 2.717247$.
17. At $x = 1$, we find $y = 0.3518301$ ($h = 0.1$), $y = 0.3506463$ ($h = 0.05$), $y = 0.3503376$ ($h = 0.25$), $y = 0.3502586$ ($h = 0.0125$), $y = 0.3502386$ ($h = 0.00625$), and $y = 0.3502335$ ($h = 0.003125$).
19. At $x = 1$, we find $y = 0.5308372$ ($h = 0.1$), $y = 0.5290050$ ($h = 0.05$), $y = 0.5285550$ ($h = 0.25$), $y = 0.5284435$ ($h = 0.0125$), $y = 0.5284158$ ($h = 0.00625$), and $y = 0.5284088$ ($h = 0.003125$).
21. At $x = 0.5$, we find $y = 1.1296566$ ($h = 0.05$), $y = 1.1296958$ ($h = 0.025$), $y = 1.1297048$ ($h = 0.0125$), and $y = 1.1297070$ ($h = 0.00625$).

SECTION 6.3

1. $y = e^x - x - 1$. The complete results are shown in Fig. 6.3.1, and the program used to generate those results in Fig. 6.3.2.

X	Y (R-K)	Y (True)	Error	PCT Error
0.00	0.0000000	0.0000000	0.0000000	0.0000000
0.10	0.0051708	0.0051709	0.0000001	0.0016388
0.20	0.0214026	0.0214028	0.0000002	0.0008752
0.30	0.0498585	0.0498588	0.0000003	0.0006228
0.40	0.0918242	0.0918247	0.0000005	0.0004983
0.50	0.1487206	0.1487213	0.0000006	0.0004250
0.60	0.2221180	0.2221188	0.0000008	0.0003774
0.70	0.3137516	0.3137527	0.0000011	0.0003445
0.80	0.4255396	0.4255409	0.0000014	0.0003208
0.90	0.5596014	0.5596031	0.0000017	0.0003033
1.00	0.7182797	0.7182818	0.0000021	0.0002902

```
100 REM       INITIALIZE
110           X = 0.0 : Y = 0.0 : V = Y
120           E = 0.0 : P = 0.0 : H = 0.1
130 REM ––            TRUE VALUE: V
140 REM ––            ERROR: E
150 REM ––            PCT ERROR:   P
160       PRINT X, Y, V, E, P
170 REM ––            DO TEN COMPUTATIONS
180           FOR   N = 1 TO 10
190               XX = X : YY = Y
200               GOSUB 1000 : K1 = KK
210               XX = X + 0.5*H : YY = Y + 0.5*H*K1
220               GOSUB 1000 : K2 = KK
230               YY = Y + 0.5*H*K2
240               GOSUB 1000 : K3 = KK
250               X = X + H : XX = X : YY = Y + H*K3
260               GOSUB 1000 : K4 = KK
270               Y = Y + H*(K1 + K2 + K2 + K3 + K3 + K4)/6.0
280               V = EXP (X) – X – 1.0
290               E = V – Y : P = 100*E/V
300           PRINT X, Y, V, E, P
310           NEXT N
320           STOP
990 REM ––            Y' = KK = F (X, Y) IS STORED HERE
1000              KK = XX + YY
1010          RETURN
```

3. $y = 1/x$; at $x = 2$, we see $y = 0.5000003$.   **5.** $y = e^{-2x}$; at $x = 1$, we see $y = 0.1353353$.

7. $y = -2e^x + x + 2$; at $x = 1$, we see $y = -2.4365595$ with error -0.0000042 and percent error 0.00001711.

9. $y = \exp (x^3)$; at $x = 1$, we see $y = 2.7182334$.

11. $y = -\ln (2 - x)$; at $x = 1.5$, we see $y = 0.6931472$ (not distinguishable from the true value).

13. $y = x^5$; all errors are zero to 7 places, but if you carry 16 digits in your computations, you will see the percentage error increase as x increases.

15. $y = (x^2 + x + 1)^{1/3}$; at $x = 1$, we see $y = 1.9129312$ (true value about 1.9129311).

17. $y = 1/(2 - x - \ln x)$; at $x = 1.5$, we see $y = 10.5779622$ (true value about 10.5781049).

19. $y = 1/x$; the computed values agree with the true values to more than seven places.

21. Results at $x = 1$: $y = 0.35023374$ ($h = 0.1$), $y = 0.35023197$ ($h = 0.05$), $y = 0.35023185$ ($h = 0.025$), and $y = 0.35023184$ ($h = 0.0125$).

23. Results at $x = 2$: $y = 1.87660865$ ($h = 0.1$), $y = 1.87660757$ ($h = 0.05$), $y = 1.87660750$ ($h = 0.025$), and $y = 1.87660749$ ($h = 0.0125$).

25. Solution: $y = \exp (x^2)$. The complete table of values appears in Fig. 6.3.25.

X	Y (Mod.)	Y (Imp.)	Y (R–K)	Y (True)
0.00	1.000000	1.000000	1.000000	1.000000
0.20	1.040758	1.040797	1.040811	1.040811
0.40	1.173258	1.173458	1.173511	1.173511
0.60	1.432561	1.433128	1.433329	1.433329
0.80	1.894427	1.895774	1.896481	1.896481
1.00	2.712976	2.715990	2.718281	2.718282
1.20	4.206918	4.213620	4.220692	4.220696
1.40	7.062672	7.077886	7.099313	7.099327
1.60	12.834774	12.870592	12.935763	12.935817
1.80	25.242940	25.331289	25.533507	25.533722
2.00	53.719777	53.949643	54.597302	54.598150

27. $y = 3x^2 - 6x + 6 - 5e^{-x}$; at $x = 1$, we see $y = 1.160620$ (modified Euler), 1.160667 (improved Euler), 1.160603 (Runge-Kutta), and 1.160603 (true value).

29. $y = x^4$. The last three lines of the table, giving x, y (modified Euler), y (improved Euler), y (Runge-Kutta), and y (true value):

−0.20	0.001648	0.001504	0.001600	0.001600
−0.10	0.000150	0.000001	0.000100	0.000100
0.00	0.000050	−0.000100	−0.000000	0.000000

31. The Runge-Kutta method yields T to be approximately 2.0118 s.

33. The results are shown in the table in Fig. 6.3.33.

T (sec)	Velocity (mi/sec)	Altitude (mi)
0.00	6.000	0.000
10.00	5.941	59.700
20.00	5.883	118.811
30.00	5.826	177.351
40.00	5.771	235.335
50.00	5.718	292.779
60.00	5.666	349.696
70.00	5.616	406.101
80.00	5.566	462.006
90.00	5.518	517.424
100.00	5.471	572.366
110.00	5.425	626.845
120.00	5.381	680.871
130.00	5.337	734.454
140.00	5.294	787.605
150.00	5.252	840.332
160.00	5.211	892.646
170.00	5.171	944.555
180.00	5.132	996.066
190.00	5.093	1047.190
200.00	5.056	1097.932
210.00	5.019	1148.302
220.00	4.982	1198.305
230.00	4.947	1247.950
240.00	4.912	1297.243
250.00	4.878	1346.190
260.00	4.844	1394.798
270.00	4.811	1443.072
280.00	4.779	1491.019
290.00	4.747	1538.645
300.00	4.715	1585.955

The initial value problem is $y' = [(v_0)^2 + 2Rg(R/g - 1)]^{1/2}$, $y(0) = R$.

35. Time to reach maximum altitude: about 1.51 s; maximum altitude: (only!) about 296 ft.

36. At $t = 10$ (s), the motorboat has a velocity of about 17.8 ft/s and has traveled just under 224 ft.

SECTION 6.4

1. $y = e^{-x}$; algorithm:

$$y_{n+1} = \frac{3}{3+h}\left(\frac{3-h}{3}y_{n-1} - \frac{4h}{3}y_n\right).$$

At $x = 1$, we have the approximation $y = 0.3678792$.

3. $y = -1/x^{1/7}$ for $-1 \leq x < 0$; y is not determined for $x \geq 0$. Results: At $x = -0.025$, the Milne method gives $y = 1.5448155$; the true value is about 1.6938140. The Milne method generates obviously incorrect data for $x > 0$.

5. $y = 1/(x^2 + 1)$. At $x = 1$, we obtain $y = 0.4999921$.   **7.** $y = 2e^x - 1$. At $x = 2$, we obtain $y = 13.7781122$.

9. $y = \exp(x^2)$. At $x = 1$, we obtain $y = 2.7182818$.   **11.** $y = 1 + \exp(-x^2/2)$. At $x = 1$, we obtain $y = 1.6065307$.

13. $y = 1/(5 - x)$. With a Runge-Kutta predictor and Milne corrector, at $x = 3$ we obtain the poor approximation $y = 0.5023286$.

15. $y = \tan x$. With a Runge-Kutta predictor and Milne corrector, at $x = 1$ we obtain the poor approximation $y = 1.5866024$.

Comment: In Problems 13 and 15, the Runge-Kutta predictor is always a *better* estimate of the true value than the Milne corrected value. This shows that the corrector must be properly matched to the predictor.

17. $y = x + 2e^{-x}$. At $x = 2$, the Adams-Moulton method gives $y = 2.2775085$.

19. $y = 2e^x - 1$. At $x = 1$, we obtain $y = 4.3315253$.   **21.** $y = \ln(x^3 + 1)$. At $x = 1$, we obtain $y = 0.7051507$.

SECTION 6.5

1. $y = e^x \cos x$, $z = e^x(\cos x + \sin x)$. At $x = 1$, we obtain $y = 1.470978$ and $z = 3.756863$.
3. $y = (e^{-x} + e^{3x})/2$, $z = (e^{3x} - e^{-x})/2$. At $x = 1$, we obtain $y = 10.181615$, $z = 9.813920$.
5. $x = e^{5t}$, $y = 3e^{5t}$. At $x = 1$, we obtain $y = 148.413122$, $z = 445.239366$.
7. $x = [\exp(-t/r) - \exp(t/r)]/r$ where $r = \sqrt{3}$; $y = [(3 + r)\exp(t/r) + (3 - r)\exp(-t/r)]/6$. At $x = 1$, we obtain $y = -0.704326$, $z = 1.523511$.
9. $x' = \frac{3}{2}e^t - \frac{1}{10}e^{-t} - \frac{4}{3}\sin 2t - \frac{7}{3}\cos 2t$; $y = \frac{1}{2}e^t - \frac{1}{10}e^{-t} - \frac{4}{3}\sin 2t - \frac{2}{3}\cos 2t$. At $x = 1$, we obtain $y = 3.895802$ and $z = 0.761374$.
11. Your answer should have the form $ax_{n+1} + by_{n+1} = F(t_{n-1}, t_n, t_{n+1}, x_{n-1}, x_n, y_{n-1}, y_n)$, $cx_{n+1} + dy_{n+1} = G(t_{n-1}, t_n, t_{n+1}, x_{n-1}, x_n, y_{n-1}, y_n)$, where F and G depend on f and g in the equations $x' = f(t, x, y)$, $y' = g(t, x, y)$. Use Cramer's rule to solve simultaneously for x_{n+1} and y_{n+1} in your program.
13. $x(t) = (2e^{2t} - 2e^{-t} - 6te^{-t})/9$, $y(t) = (e^{2t} - e^{-t} + 6te^{-t})/9$. At $t = 1$, we obtain $x = 1.315113$, $y = 1.025449$.
15. The solution is given in the answer to Problem 9. At $t = 1$, we find $x = 3.896222$ and $y = 0.761569$.
19. From $(dy/dt)/(dx/dt) = y/(x - t)$ and $(x - t)^2 + y^2 = 100$, one can derive the initial value problem $dx/dt = (x - t)^2/[(x - t)^2 + y^2]$, $dy/dt = y(x - t)/[(x - t)^2 + y^2]$; $x(0) = 0$, $y(0) = 10$. The results, using the Runge-Kutta method with $h = 0.002$, are shown in Fig. 6.5.19.

T	X	Y
0.00	0.00000000	10.00000000
1.00	0.00332005	9.95020716
2.00	0.02624680	9.80327736
3.00	0.08687395	9.56627053
4.00	0.20051074	9.25005490
5.00	0.37882964	8.86815238
6.00	0.62950746	8.43544766
7.00	0.95632892	7.96696731
8.00	1.35964472	7.47687967
9.00	1.83704199	6.97779208
10.00	2.38409008	6.48035269

21. The long-term behavior is oscillation about $x = 0$ with period $2\pi/5$ s (the same as that of the imposed external force) and amplitude approximately 0.027. The short-term behavior is shown in Fig. 6.5.21. (*Suggestion:* Explain the reason for the unusually long persistence of the transient solution. With the Runge-Kutta method with step size $h = 0.01$, we obtained $x = 0.19721$ at $t = 10$, 0.09193035 at $t = 25$, 0.08163472 at $t = 50$, 0.07129087 at $t = 75$, and 0.05378983 at $t = 100$.)

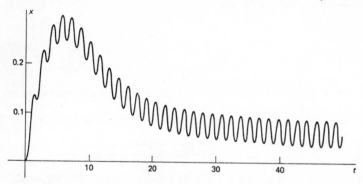

23. With $x' = y$, $y' = -2|y|y - x$, $x(0) = 1$, $y(0) = 0$, the Runge-Kutta method, and step size 0.01, we obtained the values $x = -0.10288780$, $y = -0.01780292$ at $t = 10$. The graph of $x = x(t)$ for $0 \le t \le 10$ is shown in Fig. 6.5.23.

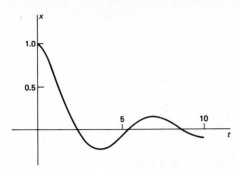

25. With the Runge-Kutta method with step size 0.01, we found the horizontal distance to be 1899.587 ft, attained at time $t = 11.632$ s. The components of velocity at impact are $v_x = 132.034$ and $v_y = -173.067$; the speed of impact is 217.681 ft/s. *Check:* With no air resistance, the results would be $t_{imp} = 12.726$, $|v| = 288$, and $x_{max} = 2592$.

SECTION 7.1

1. $\frac{2}{3}\pi$ **3.** $\frac{4}{3}\pi$ **5.** π **7.** Not periodic   **9.** π **11.** $a_0 = 2$; $a_n = b_n = 0$ for $n \geq 1$
13. $a_0 = 1$; $a_n = 0$ for $n \geq 1$; $b_n = [1 - (-1)^n]/n\pi$ **15.** $a_n = 0$ for $n \geq 0$; $b_n = (-1)^{n+1}(2/n)$
17. $b_n = 0$ for all $n \geq 1$; $a_0 = \pi$ and $a_n = [(-1)^n - 1](2/\pi n^2)$ for $n \geq 1$
19. $a_0 = \pi/2$; $a_n = [1 - (-1)^n](1/\pi n^2)$ for $n \geq 1$; $b_n = -1/n$ for $n \geq 1$
21. $a_0 = \frac{2}{3}\pi^2$; $a_n = (-1)^n(4/n^2)$ for $n \geq 1$; $b_n = 0$ for $n \geq 1$
23. $a_0 = \frac{1}{3}\pi^2$; $a_n = (-1)^n(2/n^2)$ for $n \geq 1$; $b_n = [(-1)^n - 1](2/\pi n^3) - (-1)^n(\pi/n)$ for $n \geq 1$
25. $a_0 = 1$; $a_4 = \frac{1}{2}$; $a_n = 0$ otherwise; $b_n = 0$ for $n \geq 1$

SECTION 7.2

1. $a_n = 0$ for $n \geq 0$; $b_n = [1 - (-1)^n](4/n\pi)$ for $n \geq 1$ **3.** $a_0 = 1$; $a_n = 0$ and $b_n = [1 - (-1)^n](3/n\pi)$ for $n \geq 1$
5. $a_n = 0$ for $n \geq 0$; $b_n = -(-1)^n(4/n)$ for $n \geq 1$
7. $a_0 = 1$; $a_n = [(-1)^n - 1](2/n^2\pi^2)$ for $n \geq 1$; $b_n = 0$ for all $n \geq 1$
9. $a_0 = \frac{2}{3}$; $a_n = (-1)^n(4/n^2\pi^2)$ for $n \geq 1$; $b_n = 0$ for all $n \geq 1$
11. $a_0 = 4/\pi$; $a_n = (4/\pi)[(-1)^{n+1}/(4n^2 - 1)]$; $b_n = 0$ for all $n \geq 1$
13. $a_0 = 2/\pi$; $a_n = -[1 + (-1)^n]/[\pi(n^2 - 1)]$ for $n \geq 1$; $b_1 = \frac{1}{2}$; $b_n = 0$ if $n \geq 2$

SECTION 7.3

1. Cosine series: $a_0 = 2$; $a_n = 0$ for $n \geq 1$; sine series: $b_n = [1 - (-1)^n](2/n\pi)$ for $n \geq 1$.
3. Cosine series: $a_0 = 0$; $a_n = [1 - (-1)^n](4/n^2\pi^2)$ for $n \geq 1$; sine series: $b_n = [1 + (-1)^n](2/n\pi)$.
5. Cosine series: $a_0 = \frac{2}{3}$; $a_n = 0$ if n is of any of the forms $6k + 1$, $6k + 3$, $6k + 5$, or $6k + 6$ ($k \geq 0$); $a_n = -(2/n\pi)\sqrt{3}$ if n is of the form $6k + 2$; $a_n = (2/n\pi)\sqrt{3}$ if n is of the form $6k + 4$. Thus $f(t) \sim \frac{1}{3} - (2\sqrt{3}/3)[\frac{1}{2}\cos(2\pi t/3) - \frac{1}{4}\cos(4\pi t/3) + \frac{1}{8}\cos(8\pi t/3) - \frac{1}{10}\cos(10\pi t/3) + \cdots]$. Sine series: $b_n = 0$ if n is even; $b_n = 2/n\pi$ if n is of either of the forms $6k + 1$ or $6k + 5$; if n is of the form $6k + 3$, then $b_n = -4/n\pi$
7. Cosine series: $a_0 = \pi^2/3$; $a_n = -[1 + (-1)^n](2/n^2)$ if $n \geq 1$; sine series: $b_n = [1 - (-1)^n](4/\pi n^3)$ if $n \geq 1$.
9. Cosine series: $a_0 = 4/\pi$; $a_1 = 0$; if $n \geq 2$, then $a_n = -2[1 + (-1)^n]/[(n^2 - 1)\pi]$; sine series: $b_1 = 1$, $b_n = 0$ for $n \geq 2$.
11. $x(t) = -(4/\pi) \sum_{n \text{ odd}} [1/n(n^2 - 2)] \sin nt$ **13.** $x(t) = \sum_{n=1}^{\infty} [2(-1)^n/n\pi(n^2\pi^2 - 1)] \sin n\pi t$
15. $x(t) = \pi/2 + (2/\pi) \sum_{n=1}^{\infty} \{[1 - (-1)^n]/n^2(n^2 - 2)\} \cos nt$ **17.** *Suggestion:* Substitute $u = -t$ in the left-hand integral.

SECTION 7.4

1. $x_{sp}(t) = (12/\pi) \sum\limits_{n \text{ odd}} (\sin nt)/n(5 - n^2)$ **3.** $x_{sp}(t) = 4 \sum\limits_{n=1}^{\infty} (-1)^n (\sin nt)/n(n^2 - 3)$

5. $x_{sp}(t) = (8/\pi^3) \sum\limits_{n \text{ odd}} (\sin n\pi t)/n^3(10 - n^2\pi^2)$ **7.** Resonance will occur.   **9.** Resonance will not occur.

11. Resonance will occur.

13. $x_{sp}(t) = (12/\pi) \sum\limits_{n \text{ odd}} \{\sin (nt - \delta_n)\}/\{n\sqrt{(4 - n^2)^2 + (n/10)^2}\}$ where $\delta_n = \tan^{-1}\{(n/10)/(4 - n^2)\}$, $0 \leq \delta_n \leq \pi$. The first three nonzero coefficients are (approximately) 1.27253, 0.25419, and 0.03637; the corresponding phase angles are 0.03332, 3.08166, and 3.11779.

15. The first three coefficients are (approximately) 0.08150, 0.00000, and 0.00004; the first three phase angles are 1.44692, 3.077066, and 3.10176.

17. $\alpha_n = \tan^{-1}(\{3n\pi/5\}/(36 - n^2\pi^2/4\})$; $b_n = 0$ if n is even, $b_n = (60/n\pi)[1/\sqrt{(36 - n^2\pi^2/4)^2 + 9n^2\pi^2/25}]$ if n is odd; $x(5)$ is approximately 0.248 ft—that is, about 2.98 in.

SECTION 7.5

1. $u(x, t) = 4u_2(x, t) = 4e^{-12t} \sin 2x$ **3.** $u(x, t) = 5e^{-2\pi^2 t} \sin \pi x - \frac{1}{2}e^{-18\pi^2 t} \sin 3\pi x$

5. $u(x, t) = 4e^{-8\pi^2 t/9} \cos (2\pi x/3) - 2e^{-32\pi^2 t/9} \cos (4\pi x/3)$ **7.** $u(x, t) = \frac{1}{2} + \frac{1}{2}e^{-16\pi^2 t/3} \cos 4\pi x$

9. $u(x, t) = (100/\pi) \sum\limits_{n \text{ odd}} (1/n)e^{-n^2\pi^2 t/250} \sin (n\pi x/5)$

11. $u(x, t) = 20 + (40/\pi^2) \sum\limits_{n=1}^{\infty} (1/n^2)e^{-n^2\pi^2 t/125} \cos (n\pi x/5)$

13. (a) $u(x, t) = (400/\pi) \sum\limits_{n \text{ odd}} (1/n)e^{-n^2\pi^2 kt/1600} \sin (n\pi x/40)$ (b) With $k = 1.15$, $u(20, 300) \approx 15.16°C$ (c) About 19 h 16 min

SECTION 7.6

1. $y(x, t) = \frac{1}{10} \cos 4t \sin 2x$ **3.** $y(x, t) = \frac{1}{10}(\cos t/2 + 2 \sin t/2) \sin x$

5. $y(x, t) = \frac{1}{10} \cos 5\pi t \sin \pi x + (1/\pi) \sin 10\pi t \sin 2\pi x$ **7.** $y(x, t) = (1/5\pi^2) \sum\limits_{n=1}^{\infty} [(-1)^{n+1}/n^2] \sin 10n\pi t \sin n\pi x$

9. $y(x, t) = y_B(x, t) = \sum\limits_{n \text{ odd}} (4/n^4\pi^4) \sin 2\pi nt \sin n\pi x$ **11.** Frequency 256 Hz, velocity 1024 ft/s

SECTION 7.7

1. $u(x, y) = \sum\limits_{n=1}^{\infty} c_n \sinh (n\pi x/b) \sin (n\pi y/b)$ where $c_n = [2/b \sinh (n\pi a/b)] \int_0^b g(y) \sin (n\pi y/b) \, dy$

5. $u(x, y) = (a_0/2a)(a - x) + \sum\limits_{n=1}^{\infty} a_n \{\sinh (n\pi[a - x]/b)/\sinh (n\pi a/b)\} \{\cos (n\pi y/b)\}$ where $a_n = (2/b) \int_0^b g(y) \cos (n\pi y/b) \, dy$

7. $u(x, y) = \sum\limits_{n=1}^{\infty} b_n^{-n\pi y/a} \sin (n\pi x/a)$ where $b_n = (2/a) \int_0^a f(x) \sin (n\pi x/a) \, dx$

11. $u(x, y) = \sum\limits_{n \text{ odd}} c_n \sinh (n\pi[a - x]/2b) \cos (n\pi y/2b)$ where $c_n = \{2/[b \sinh (n\pi a/2b)]\} \int_0^b g(y) \cos (n\pi y/2b) \, dy$

13. $c_n = (2/\pi a^n) \int_0^\pi f(\theta) \sin n\theta \, d\theta$ **15.** $c_n = (2/\pi a^{n/2}) \int_0^\pi f(\theta) \sin (n\theta/2) \, d\theta$

SECTION 8.1

7. $1 = 2hL \sum\limits_{n=1}^{\infty} [(1 - \cos \beta_n)/\beta_n(hL + \cos^2 \beta_n)] \sin (\beta_n x/L)$, $0 < x < L$

9. $x = 2h(1 + h) \sum\limits_{n=1}^{\infty} (\sin \beta_n \sin \beta_n x)/\beta_n^2(h + \cos^2 \beta_n)$, $0 < x < 1$

SECTION 8.2

3. $u(x, y) = \sum\limits_{n=1}^{\infty} A_n \sinh [(\beta_n/L)(L - x)] \cos (\beta_n y/L)$ where $\{\beta_n\}_1^\infty$ are the positive roots of the equation $\beta \tan \beta = hL$ and

$A_n \sinh \beta_n = [2h/(hL + \sin^2 \beta_n)] \int_0^L g(y) \cos (\beta_n y/L) \, dy$

5. $u(x, t) = \sum_{n=1}^{\infty} c_n \exp(-\beta_n^2 kt/L^2) X_n(x)$ where $\{\beta_n\}_1^{\infty}$ are the positive roots of the equation $hL \tan \beta = -\beta$, $X_n(x) = \beta_n \cos(\beta_n x/L) + hL \sin(\beta_n x/L)$, and $c_n \int_0^L [X_n(x)]^2 dx = \int_0^L f(x) X_n(x) dx$

7. $u(1, 1) \approx 31.4°$

SECTION 8.3

In the answers for Problems 1, 3, and 5, ω_n denotes the nth natural circular frequency (in radians per second).

1. $\omega_n = (n\pi/L)\sqrt{E/\delta}$ 3. $\omega_n = [(2n - 1)\pi/2L]\sqrt{E/\delta}$
5. $\omega_n = (\beta_n/L)\sqrt{E/\delta}$ where β_n is the nth positive root of the equation $AE\beta \tan \beta = kL$

SECTION 8.4

11. $u(r, t) = [F_0/\omega^2 J_0(\omega b/a)][J_0(\omega r/a) - J_0(\omega b/a)] \sin \omega t$ where b is the radius of the circular membrane

SECTION 9.1

1. $x(t) = x_0 e^{-t}$; $x = 0$ is a stable critical point. See Fig. 9.1.1.

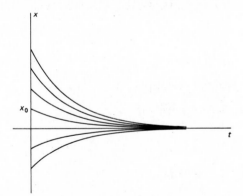

3. $x(t) = x_0[x_0 + (1 - x_0)e^{-t}]^{-1}$; $x = 0$ is an unstable critical point, while $x = 1$ is a stable critical point. See Fig. 9.1.3.

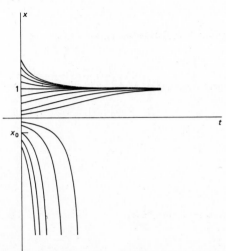

5. $x(t) = x_0[(1 + x_0)e^{-t} - x_0]^{-1}$; $x = 0$ is an unstable critical point, while $x = -1$ is a stable critical point. See Fig. 9.1.5.

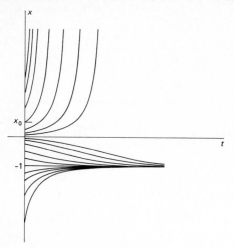

7. $x(t) = [1 + x_0 - (1 - x_0)e^{-2t}]/[1 + x_0 + (1 - x_0)e^{-2t}]$; $x = +1$ is a stable critical point, while $x = -1$ is an unstable point. See Fig. 9.1.7.

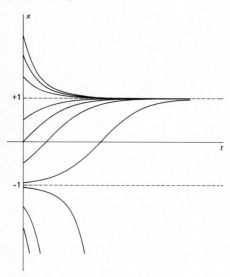

9. $x(t) = [2 + x_0 - 2(1 - x_0)e^{-3t}]/[2 + x_0 + (1 - x_0)e^{-3t}]$; $x = 1$ is a stable critical point, while $x = -2$ is an unstable critical point. See Fig. 9.1.9.

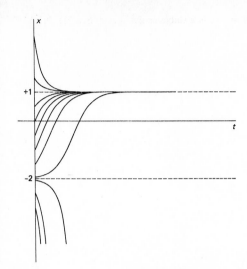

11. $(0, 0)$ **13.** $(2, 3)$ **15.** $(0, 0)$ and $(3, 2)$ **17.** $(n\pi, 0)$ for n any integer

19. $x = x_0 e^{-2t}, y = y_0 e^{-2t}$; $(0, 0)$ is an asymptotically stable proper node as illustrated in Fig. 9.3.

21. $x = x_0 e^{-2t}, y = y_0 e^{-t}$; $(0, 0)$ is an asymptotically stable improper node.

23. $x = x_0 \cos t + y_0 \sin t, y = y_0 \cos t - x_0 \sin t$; $(0, 0)$ is a stable center as illustrated in Fig. 9.6.

25. $x = x_0 \cos 2t + y_0 \sin 2t, y = y_0 \cos 2t - x_0 \sin 2t$; $(0, 0)$ is a stable center as illustrated in Fig. 9.6.

27. $x(t) = ax_0/[bx_0 + (a - bx_0)e^{abt}]$

SECTION 9.2

1. Asymptotically stable node   **3.** Unstable saddle point   **5.** Asymptotically stable node   **7.** Unstable spiral point

9. Stable center   **11.** $(2, 1)$ is an asymptotically stable node.   **13.** $(2, 2)$ is an unstable saddle point.

15. $(1, 1)$ is an asymptotically stable spiral point.   **17.** $(\frac{5}{2}, -\frac{1}{2})$ is a stable center.   **19.** Asymptotically stable node

21. Unstable saddle point   **23.** Asymptotically stable spiral point   **25.** Asymptotically stable node or spiral point

27. Either a center or a spiral point; stability is indeterminate.

29. $(0, 0)$ is an unstable saddle point; $(1, 1)$ is a center or a spiral point with indeterminate stability.

31. $(1, 1)$ is an unstable saddle point; $(-1, -1)$ is an asymptotically stable spiral point.

SECTION 9.4

1. $y_0 = 3, y_1 = 3 + 3x, y_2 = 3 + 3x + \frac{3}{2}x^2, y_3 = 3 + 3x + \frac{3}{2}x^2 + \frac{1}{2}x^3, y_4 = 3 + 3x + \frac{3}{2}x^2 + \frac{1}{2}x^3 + \frac{1}{8}x^4; y(x) = 3e^x$

3. $y_0 = 1, y_1 = 1 - x^2, y_2 = 1 - x^2 + \frac{1}{2}x^4, y_3 = 1 - x^2 + \frac{1}{2}x^4 - \frac{1}{6}x^6, y_4 = 1 - x^2 + \frac{1}{2}x^4 - \frac{1}{6}x^6 + \frac{1}{24}x^8; y = \exp(-x^2)$

5. $y_0 = 0, y_1 = 2x, y_2 = 2x + 2x^2, y_3 = 2x + 2x^2 + \frac{4}{3}x^3, y_4 = 2x + 2x^2 + \frac{4}{3}x^3 + \frac{2}{3}x^4; y(x) = e^{2x} - 1$

7. $y_0 = 0, y_1 = x^2, y_2 = x^2 + \frac{1}{2}x^4, y_3 = x^2 + \frac{1}{2}x^4 + \frac{1}{6}x^6, y_4 = x^2 + \frac{1}{2}x^4 + \frac{1}{6}x^6 + \frac{1}{24}x^8; y(x) = \exp(x^2) - 1$

9. $y_0 = 1, y_1 = (1 + x) + \frac{1}{2}x^2, y_2 = (1 + x + x^2) + \frac{1}{6}x^3, y_3 = (1 + x + x^2 + \frac{1}{3}x^3) + \frac{1}{24}x^4;$
$y(x) = 2e^x - 1 - x = 1 + x + x^2 + \frac{1}{3}x^3 + \cdots$

11. $y_0 = 1, y_1 = 1 + x, y_2 = (1 + x + x^2) + \frac{1}{3}x^3, y_3 = (1 + x + x^2 + x^3) + \frac{2}{3}x^4 + \frac{1}{3}x^5 + \frac{1}{9}x^6 + \frac{1}{63}x^7;$
$y(x) = 1/(1 - x) = 1 + x + x^2 + x^3 + \cdots$

13. $\begin{pmatrix} x_0 \\ y_0 \end{pmatrix} = \begin{pmatrix} 1 \\ -1 \end{pmatrix}, \quad \begin{pmatrix} x_1 \\ y_1 \end{pmatrix} = \begin{pmatrix} 1 + 3t \\ -1 + 5t \end{pmatrix}, \quad \begin{pmatrix} x_2 \\ y_2 \end{pmatrix} = \begin{pmatrix} 1 + 3t + \frac{1}{2}t^2 \\ -1 + 5t - \frac{1}{2}t^2 \end{pmatrix}, \quad \begin{pmatrix} x_3 \\ y_3 \end{pmatrix} = \begin{pmatrix} 1 + 3t + \frac{1}{2}t^2 + \frac{1}{3}t^3 \\ -1 + 5t - \frac{1}{2}t^2 + \frac{5}{8}t^3 \end{pmatrix}$

Index